Biology of Extracellular Matrix

Volume 11

Extracellular matrix (ECM) biology, which includes the functional complexities of ECM molecules, is an important area of cell biology. Individual ECM protein components are unique in terms of their structure, composition and function, and each class of ECM macromolecule is designed to interact with other macromolecules to produce the unique physical and signaling properties that support tissue structure and function. ECM ties everything together into a dynamic biomaterial that provides strength and elasticity, interacts with cell-surface receptors, and controls the availability of growth factors. Topics in this series include cellular differentiation, tissue development and tissue remodeling. Each volume provides an in-depth overview of a particular topic, and offers a reliable source of information for post-graduates and researchers alike.

All chapters are systematically reviewed by the series editor and respective volume editor(s).

"Biology of Extracellular Matrix" is published in collaboration with the American Society for Matrix Biology and the International Society for Matrix Biology.

Ilona Kovalszky • Marco Franchi •
Laura D. Alaniz
Editors

The Extracellular Matrix and the Tumor Microenvironment

Springer

Editors
Ilona Kovalszky
Department of Pathology & Experimental
Cancer Research
Semmelweis University
Budapest, Hungary

Marco Franchi
Department for Life Quality Study
University of Bologna
Rimini, Italy

Laura D. Alaniz
Laboratorio de Microambiente Tumoral
National University of Northwestern Buen
Junin, Argentina

ISSN 0887-3224 ISSN 2191-1959 (electronic)
Biology of Extracellular Matrix
ISBN 978-3-030-99707-6 ISBN 978-3-030-99708-3 (eBook)
https://doi.org/10.1007/978-3-030-99708-3

This Springer imprint is published by the registered company Springer Nature Switzerland AG
The registered company address is: Gewerbestrasse 11, 6330 Cham, Switzerland

Preface

The types of malignant tumors are determined by the tissue the transformed cells originate from. However, without the establishment of tumor stroma, tumor tissue could not exist. Accumulating data of cancer research shed light to the importance of the stromal components supporting the growth and survival of tumor cells. In the last 20 years, huge amounts of experimental and clinical data have been collected about the subject. Although dysregulated, cancer is an ecosystem, where cancer cells coexist with a plethora of non-tumorous cells, and blood vessels, embedded into the extracellular matrix. All of these components are in intensive communication with each other via membrane receptors and soluble factors. Furthermore, matrix proteins also have regulatory potential. Thus, to understand their interplay with the tumor cells is mandatory to step forward.

Extracellular matrix is an acellular substance synthesized by its resident cells. Its two major forms are the basement membrane and the interstitial matrix built up by structural proteins such as collagens glued together with adhesive glycoproteins, laminins, and fibronectins. Former is a sheet-like structure that provides support for various epithelial layers, or surrounds individual stromal cell like chondrocytes. Interstitial matrix is a lattice-like formation built up by collagens and fibronectins. It provides support for the resident stromal cells, blood vessels, and the epithelial cell of various organs. Proteoglycans can be found in the basement membranes, interstitial matrix, and the surface of epithelial cells. They are active members of the events taking place in the ECM. Owing to their negatively charged sugar chains, they can interact with regulatory factors, or even exist as cell surface receptors. Hyaluronic acid is responsible for the maintenance of tissue turgor. However, these molecules and several other residents of the extracellular matrix are capable to join to the wrong side and support the development and progression of cancer.

This book was created with the intention to provide examples as the results of the last 20 years altered our understanding about the role of the extracellular matrix in health and disease, including cancer. Now we know that ECM is not an inert material, but active participant of the events taking place in our body. These advancements will be presented in eighteen chapters from scientists around the world.

The **first five chapters** are dealing with proteoglycans, built up by a protein core, to which negatively charged glycosaminoglycan sugar chain(s) are covalently attached. Their effect can be beneficial, or harmful, depending on the nature of their components. Furthermore, **Chap. 5** provides an excellent example about versikine, a tryptic fragment of versican. Protease activity can liberate fragments of proteoglycans, which entering into the circulation can be utilized in diagnosis and may be in therapy as well, as it is demonstrated by **Chap. 6.** As it is presented in **Chap. 7,** hyaluronan can promote the progression of hematological and solid solid tumors. Heparanase is the sole endoglycosidase, degrading heparan sulfate chains of proteoglycans. Its activity, that can be detected in almost all types of cancer, is an indicator of poor tumor prognosis is discussed in **Chap. 8**.

The next two **chapters (Chaps. 9 and 10)** provide information as laminins, fibronectins, and collagens support tumor progression. **Chapter 11** demonstrates how matrix receptor integrins transmit information from the matrix to the tumor cells and cancer associated fibroblasts (CAFs). Unexpected negative effect of fat on the behavior of mammary cancer is reported in **Chap. 12**.

Chapter 13 sheds light on the cooperation of ECM and tumor cells in the impaired intermediary metabolism. Tumor tissue dies without blood supply. Mechanisms of cancer angiogenesis and its influencing factors are introduced in **Chaps. 14 and 15.**

Epigenetic regulation is also involved in the cooperation of extracellular matrix and tumor cells as it is demonstrated in **Chap. 16** in case of breast cancer.

Finally, the last two **chapters (Chaps. 17 and 18)** try to convince the readers, that targeting components of extracellular matrix may have supporting potential in the battle against cancer in the future.

Budapest, Hungary — Ilona Kovalszky
Rimini, Italy — Marco Franchi
Junin, Argentina — Laura D. Alaniz

Acknowledgement

The editors would like to thank to the contributors of the book, who are well-known experts of the fields. The fruits of their research, they shared with us, opens up the vivid word of extracellular matrix and stromal components indispensible for health and disease.

We are grateful for the stimulating suggestions and problem-solving interventions of our Series Editor professor Nikos Karamanos and the Associate Editor Dr Miriam Latuske, and, last but not least, for everybody from the side of Springer-Nature who helped us in the whole process to turn our chapters into a book.

Contents

Chapter 1
Small Leucine-Rich Proteoglycans Regulate Cancer Cell Growth, Apoptosis, and Associated Inflammation

Dragana Nikitovic and George Tzanakakis

Abstract The tumor microenvironment, apart from tumor cells, contains tissue nonmalignant cells, blood vessels, stromal cells, infiltrating immune cells, and the modified extracellular matrix (ECM). The altered ECM essentially regulates all tumor cells' functions and forms a permissive tumor cell environment. Proteoglycans (PGs) are an essential component of the ECM and correspond to composite molecules that comprise a protein core into which one or more glycosaminoglycans (GAG) chains are covalently bound. Small leucine-rich proteoglycans (SLRPs) are a family of 18 proteins. The SLRPs are grouped into five classes based on their homology at both the protein and genomic levels, the number of the leucine-rich repeats (LRRs), the spacing of their N-terminal cysteine residue and ear repeats, and crucially their shared functional features taking into account that some SLRPs are not classical proteoglycans. Due to their structure's specificities, the SLRPs have been indicated to participate in myriad physiological and pathological processes. This chapter provides a critical overview of biglycan and lumican roles in the reactive cancer milieu and consecutively extrapolates to potential options for developing targeted treatments.

1.1 Introduction on Cancer and Its Microenvironment

The process of carcinogenesis is highly complex with the participation of genetic, metabolic, and environmental factors (Tzanakakis et al. 2019, 2020). The malignant transformation of the cancer cells bestows them with more efficient growth, survival, motility, metastasis, and the ability to reprogram their microenvironment. Further expansion of the tumor cells is facilitated through their constant interactions with the altered tumor microenvironment (TME) (Nikitovic et al. 2015). The TME, apart from tumor cells, contains tissue nonmalignant cells, blood vessels, stromal cells,

D. Nikitovic (✉) · G. Tzanakakis
Faculty of Medicine, University of Crete, Heraklion, Greece
e-mail: nikitovic@uoc.gr; tzanakak@med.uoc.gr

I. Kovalszky et al. (eds.), *The Extracellular Matrix and the Tumor Microenvironment*, Biology of Extracellular Matrix 11,
https://doi.org/10.1007/978-3-030-99708-3_1

infiltrating immune cells, and the modified extracellular matrix (ECM) (Joshi et al. 2021; Suveges et al. 2020). The altered ECM essentially contributes to all tumor cells' functions and forms a permissive tumor cells environment (Nikitovic et al. 2018; Pickup et al. 2014). The ECM supramolecular structure is composed of fibrillar proteins such as collagen and elastin, proteoglycans (PGs), and GAGs, with the ability to specifically support the structure/function of each organ (Hynes 2009).

Thus, the altered ECM supports all the "hallmarks" of cancer as defined by Hanahan and Weinberg (2000), including sustained proliferation, evasion of growth suppression, death resistance, replicative immortality, induced angiogenesis, initiation of invasion, dysregulation of cellular energetics, avoidance of immune destruction, and chronic inflammation (Hanahan and Weinberg 2000). Indeed, the tumor ECM contributes to the tumor development and sustains the tumor immune escape forming an immunosuppressive scaffold where stromal cells intertwine with inflammatory immune cells and cells responding to the vascular system (Dunn et al. 2004). In addition, tumors commonly exhibit desmoplasia, an increased deposition and cross binding of the ECM proteins (Lu et al. 2012). Likewise, fibrosis characterized by abnormal accumulation of collagen (Sorensen et al. 1998) and increased mammographic density due to the enhanced collagen deposition (Morris et al. 2016) are correlated to cancer progression. Thus, the "stiffening" of the ECM due to the alteration of its components enhances tumor cell functions.

PGs are an essential component of the ECM. These composite molecules comprise a protein core into which one or more GAG chains are covalently bound. The attached GAGs types differ as heparan sulfate (HS), chondroitin sulfate/dermatan sulfate (CS/DS), or keratan sulfate (KS) chains can decorate the PG protein core. The localization of PGs varies as they can be associated with the plasma membranes, secreted into the ECM, or deposited inside the cells. Secreted PGs are localized to the pericellular space and the "proper" ECM where the PGs are localized away from cells. Forty-five PGs have been identified up to now, endowed with huge variability in the protein core and the glycosylation pattern (Iozzo and Schaefer 2015).

PGs expression pattern and regulatory roles are altered during tumorigenesis and have been closely correlated to cancer development and progression as previously discussed (Tzanakakis et al. 2019; Nikitovic et al. 2018). It is well established that malignant tumors have discrete PG expression profiles immediately correlated with their behavior and differentiation status. Thus, epithelial tumors exhibit a different PG phenotype compared to mesenchymal tumors (Mytilinaiou et al. 2017). The proteolysis of PGs, performed by matrix metalloproteinases (MMPs), cathepsins, and bone-morphogenetic protein-1, can liberate bioactive matrikines with different roles in tumorigenesis compared to original molecules (Pasco et al. 2004; Hope et al. 2017). Moreover, PGs are closely involved in cancer-associated inflammation (Tzanakakis et al. 2021).

1.2 SLRPS Structure and Function-Focus on Biglycan and Lumican

Small leucine-rich proteoglycans are a family of 18 proteins. Their protein core, characterized by a variable number of central leucine-rich repeat (LRR) domains, has a molecular weight between 36 kDa and 77 kDa and a horseshoe shape. LRRs exhibit different amino acid sequences across the SLRP family, their size being in the 20 to 29 residues range, whereas the N and C-terminal regions of the protein have abundant cysteine residues (Iozzo and Schaefer 2015; McEwan et al. 2006; Kobe and Kajava 2001). The penultimate of the two C-terminal LRRs is the ear-repeat characteristically being the longest LRR and protruding laterally from the core protein's main axis (Iozzo and Schaefer 2015; McEwan et al. 2006). The specific residue sequences of the LRRs, the glycosylation pattern, the number of cysteine, and the ear-repeats regulate protein structure and are involved in ligand recognition. The concave SLRPs face is formed of β-sheets, while α-helices contribute to the formation of the convex surface. Moreover, each LRR supplies one α-helix and one β-sheet that correlate to a one-turn tube-like structure, with the LRR hydrophobic residues facing inside (Scott et al. 2004). The resulting structure facilitates protein–protein interactions with the inner concave face and enables the detainment of a ligand.

Furthermore, the majority of SLRP family members undergo different post-translational glycosylation. Since SLRPs core proteins bear a multitude of substitution sites corresponding to a variable glycosylation status, this provides an exquisite specificity of SLRP interactions with diverse cell membrane receptors, cytokines, chemokines, and ECM molecules (Kram et al. 2017).

The 18 SLRP members are grouped into five classes based on their homology at both the protein and genomic levels, the number of the LRRs, the spacing of their N-terminal cysteine residue and ear repeats, and crucially their shared functional features taking into account that some SLRPs are not classical proteoglycans (Schaefer and Iozzo 2008; Dellett et al. 2012).

Indeed, the canonical class I members, decorin and biglycan, bear CS/DS side chains, whereas the asporin member does not (McEwan et al. 2006; Henry et al. 2001). In contrast, the class II members, such as lumican, are decorated by KS chains or polylactosamine bound into LRRs, while class III members can bear KS (osteoglycin), CS/DS (epiphycan), or be bereft of GAG (opticin) chains (Schaefer and Iozzo 2008; Sanders et al. 2003). Unexpectedly, the majority of the noncanonical classes IV and V members (Iozzo and Schaefer 2015) do not carry GAG chains, excluding chondroadherin which is modified with KS (Neame et al. 1994). In continuation, we will focus on the class I member biglycan and the class II member lumican structure biological roles in the tumor microenvironment (TME).

1.2.1 Biglycan and Lumican Structure

The biglycan 42-kDa protein core contains 10 to 12 leucine-rich repeats (LRRs), flanked by cysteine-rich regions. Human biglycan carries 1 to 2 GAG chains bound into sites at amino acids 5 and 10 of the N-terminus (Roughley and White 1989). The substitution is implemented by CS or DS chains in a tissue-specific manner (Roughley and White 1989). Moreover, biglycan can be expressed as a "part-time" PG considering its nonglycated form has been identified in the aging articular cartilage. The nonglycanated biglycan forms seem to result from the proteolysis of the core protein's amino-terminal region (Roughley et al. 1993). Moreover, biglycan possesses N-linked oligosaccharide chains within its central LRRs (Neame et al. 1989). The gene for biglycan is located at the Xq28 (McBride et al. 1990). The details of biglycans' structural characterizations have previously been excellently discussed (Schaefer and Iozzo 2008).

Biglycan has a significant role in the organization of ECM assembly as it interacts explicitly with various ECM components, including collagen types I, II, III, IV, and VI and elastin (Douglas et al. 2006), thereby playing a crucial structural role in the majority of tissues (Wiberg et al. 2003; Reinboth et al. 2002; Kram et al. 2020).

Lumican, a class II SLRP, has a 40-kDa core protein that bears 6 to 10 LRR. This SLRP was initially determined to be a significant KS-containing PGs of the chick cornea (Blochberger et al. 1992; Schrecengost et al. 1992). Moreover, lumican can also be substituted by polylactosamine chains of varying sizes or be bereft of glycosylation (Melching and Roughley 1999). Furthermore, it has been suggested that nonglycanated forms of lumican tend to increase with age due to the attenuation of KS synthesis (Roughley et al. 1996).

Notably, the function of the SLRPs depends on both their core protein and GAG chains (Nikitovic et al. 2008a, 2008b, 2012). Indeed, the SLRPs core proteins interact with the fibrillar collagen, mainly responsible for the tissue scaffold formation (Iozzo and Schaefer 2015). By binding, the SLRPs regulate the fibril diameter formation and their interactions with other ECM components (Svensson et al. 2000). Moreover, the targeted knock-down of the BGN gene leads to irregular morphology of the collagen fibril (Ameye et al. 2002). They also appear to control access of the collagenases to their unique cleavage site and thus, partly modulate collagen proteolysis (Monfort et al. 2006).

This chapter provides a critical overview of biglycan and lumican roles in the reactive cancer milieu and consecutively extrapolates to potential options for the development of targeted treatments.

1.3 Lumican and Biglycan Expression in Cancer and Correlation to Carcinogenesis

Biglycan and lumican exhibit altered expression in cancer tissue. Thus, biglycan is overexpressed in endometrial cancer tissue samples, both at the parenchyma and mesenchyme, compared to human normal and atypical hyperplasia endometrium (Liu et al. 2014). Furthermore, the expression of biglycan in cancerous endometrium mesenchyme was positively correlated with poor patient prognosis (Liu et al. 2014). Likewise, biglycan was found to be upregulated at both the protein and mRNA level in gastric cancer tissue compared to adjacent nontumorous tissues. Moreover, patients with biglycan-positive tumors exhibited higher disease recurrence rates and increased mortality than patients with biglycan-negative tumors following surgery (Wang et al. 2011). Indeed, it seems that biglycan mRNA is overexpressed in the majority of cancer tissues compared with normal tissues, including breast, bladder, brain and central nervous system, gastric, colorectal, esophageal, lung, head and neck, ovarian, and 28 subtypes of cancer determined by a recent study examining the Oncomine database (Zhao et al. 2020).

Notably, biglycan was a significant differentially upregulated gene correlated to gastric cancer metastases (Liu et al. 2021). Furthermore, Wei et al. (2020), utilizing the GSE75037 published data, had built the coexpression modules of genes by Weighted Gene Co-Expression Network Analysis (WGCNA) to evaluate the function and protein–protein interaction network of coexpression genes by Database for Annotation, Visualization, and Integrated Discovery (DAVID) and String database, respectively, in lung cancer. This approach determined that biglycan is one of the critical genes facilitating the occurrence of lung adenocarcinoma through putative ECM–receptor interaction (Wei et al. 2020).

In general, a good correlation was established between biglycan mRNA and protein expression in various cancer tissues, including bladder (Schulz et al. 2019; Appunni et al. 2017), esophagus (Zhu et al. 2013), or colorectal cancer (Gu et al. 2012). Thus, Zhao et al. (2020) data agree with the previous report showing that biglycan protein overexpression is correlated to poor patient prognosis in esophagic cancer (Zhu et al. 2013) and bladder cancer (Schulz et al. 2019). Biglycan was found to be overexpressed in approximately a third of prostate cancer tissues and Jacobsen et al. (2017) show that biglycan expression parallels disease progression (Jacobsen et al. 2017).

Some studies, however, had contrasting results. Indeed, in a cohort of 76 bladder cancer patients, high biglycan expression was correlated to reduced tumor cell proliferation and increased 10-year survival of patients (Niedworok et al. 2013).

Immunohistochemistry demonstrated a much higher expression of lumican in gastric cancer tissues compared to adjacent normal gastric. Moreover, lumican expression was correlated with histological type, organ metastasis, and lymphatic metastasis (Chen et al. 2017). Furthermore, upon downloading the gene expression profiles and gastric cancer patients' clinical data from The Cancer Genome Atlas (TCGA) database, a significant difference in lumican expression was identified between gastric cancer tissues and adjacent nontumor tissues by R software and quantitative real-time polymerase chain reaction (qRT-PCR) as well as subsequent comprehensive meta-analysis (Chen et al. 2020a, 2020b). Moreover, when multivariate analysis was applied, high lumican expression was identified as an independent predictor of poor overall survival (Chen et al. 2020a, 2020b).

In breast cancer, lumican mRNA was identified in the tumor stroma and correlated with higher tumor grade, lower expression of estrogen receptors, and younger age of the patients (Leygue et al. 1998). Lumican expressed in breast cancer cell lines affected their biological behavior and expression of ECM mediators (Karamanou et al. 2020). In colon cancer, lumican expression was correlated with lymph node metastasis and the extent of tumor invasion (Seya et al. 2006). Moreover, patients with a high lumican expression exhibited a lower survival rate (Seya et al. 2006).

Lumican was likewise detected in malignant melanoma's dermis and peritumoral stroma but not identified in melanoma cells (Brezillon et al. 2007). Indeed, lumican was suggested to negatively regulate the vertical progression of melanoma (Brezillon et al. 2007). Moreover, lumican was found to inhibit the metastasis of melanoma in vivo by modifying matrix effectors and invadopodia activity (Karamanou et al. 2021). On the other hand, some human melanoma cell lines were found to express lumican mRNA and secrete lumican, which could be attributed to cell culture conditions (Sifaki et al. 2006).

1.3.1 Biglycan and Lumican Affect Tumor Growth

Numerous studies have established a correlation between biglycan and lumican expression and cancer cell growth. Both positive and negative effects, putatively in a cancer-type dependent manner, have been identified. Thus, HCT116 colon cancer cells deficient in biglycan exhibit downregulated cell growth and cell cycle arrest at the G0/G1 phase. This was followed by a decrease in positive cell cycle regulators, including cyclin A and cyclin D1, and increased expression of negative cell cycle regulators such as p21 and p27 (Xing et al. 2015a, 2015b). Moreover, biglycan was shown to upregulate the growth of MG63 osteosarcoma cells through an LRP6/β-catenin/IGF-IR signaling axis with the participation of Erk1/ downstream signaling (Aggelidakis et al. 2018, 10.3389/fonc.2018.00470).

On the other hand, human urothelial carcinoma cells (J82 cells) deficient in biglycan exhibit enhanced growth, whereas biglycan overexpression and treatment with exogenous biglycan attenuated these cell proliferation. Likewise, a xenograft of biglycan-deficient J82 cells exhibited increased growth in nude mice (Niedworok et al. 2013). Moreover, Kaplan–Meier analysis indicated higher 10-year survival of patients whose biopsies showed high biglycan mRNA expression (Niedworok et al. 2013).

Notably, biglycan was one of the PG components of the desmoplastic reaction. Furthermore, it was shown that biglycan, highly expressed in pancreatic cancer, attenuates this cancer cells' growth by inducing G1-arrest (Weber et al. 2001).

The majority of studies report inhibiting effects of lumican on cancer growth. Thus, lumican was shown to prolong the doubling time and inhibit the cell growth in H460 and A549 non-small lung cancer cells. Moreover, lumican was demonstrated to inhibit the central spindle and midbody formation during cytokinesis leading to increased aneuploidy. Indeed, Yang et al. (2020) suggest that lumican controls chromosome segregation during cell division in the in vitro lung cancer model (Yang et al. 2020).

Activated pancreatic stellate cells (PSC) are the dominant stromal cell type in the microenvironment pancreatic ductal adenocarcinoma (PDA), primarily responsible for the deposition of collagen (Apte et al. 2004). Lumican secreted by PSCs cells to the PDAC ECM (Kang et al. 2016) is associated with improved patient outcome after surgical removal of localized PDAC and multimodal therapy (Li et al. 2014). Furthermore, in pancreatic ductal adenocarcinoma cells, extracellular lumican attenuated the EGFR expression and phosphorylation by inducing dimerization and internalization of EGFR, leading to the downregulation of Akt kinase activity and reduced growth. This model mimics the interaction between cancer and stromal cells resulting in the limiting of cancer growth (Li et al. 2014).

In a separate study, the prolonged exposure of pancreatic ductal adenocarcinoma cells to exogenous lumican initiated growth inhibition and G0/G1 arrest, which characterize cancer cell quiescent state. These effects were retained in xenograft and syngeneic orthotopic animal models, where lumican attenuated pancreatic tumor growth (Li et al. 2017). Moreover, a sequence of 17 amino acids within the lumican core protein, named lumcorin, strongly attenuated the monolayer and soft agar growth of mice B16F1 and human SK-MEL-28 melanoma cells (Pietraszek et al. 2013).

Interestingly, the growth of human osteosarcoma Saos 2 cells, characterized as nonaggressive, was downregulated by lumican, whereas the proliferation of the aggressive MG63 human osteosarcoma cells was not affected (Nikitovic et al. 2008a, b). The expression of lumican expression was negatively correlated with the Smad2 basal level in these cells, indicating that this SLRP may affect the bioavailability of Smad 2 activators (Nikitovic et al. 2008a, b).

On the other hand, some growth-promoting effects of lumican have been identified. An example is gastric cancer, where it was determined that lumican produced by cancer-associated fibroblasts promotes this cancer cell growth through integrin β1-FAK signaling pathways. Knock out of lumican attenuated gastric tumor growth and metastasis in vivo (Wang et al. 2017). Moreover, lumican was shown to upregulate the growth of HTB94 chondrosarcoma cells through an IGF-IR/Erk1/2 axis (Papoutsidakis et al. 2020). Some effects of biglycan and lumican on cancer cell growth are schematically depicted in Fig. 1.1.

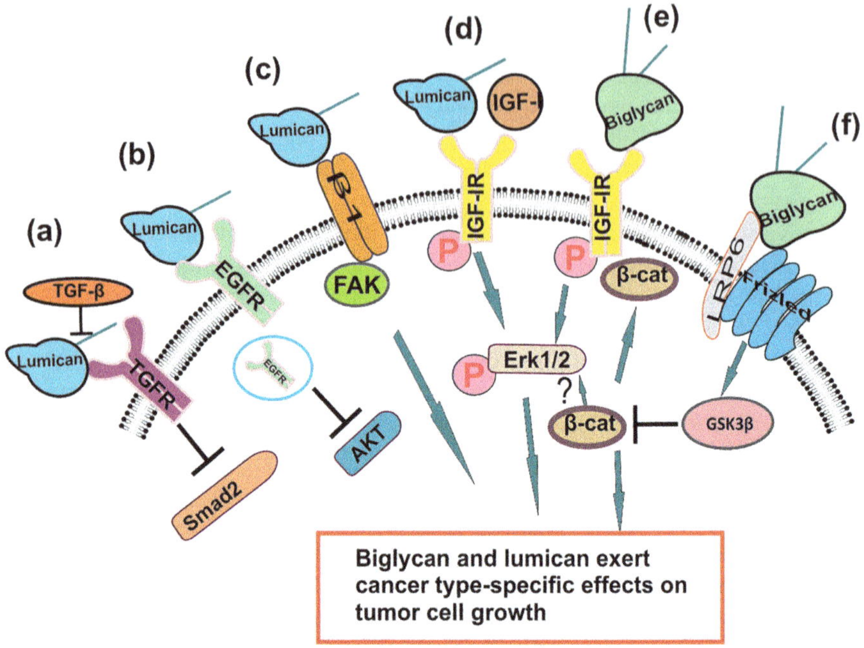

Fig. 1.1 Effects of biglycan and lumican on cancer cell growth. (**a**) Lumican inhibits osteosarcoma cell growth by affecting Smad2 signaling. (**b**) In pancreatic cancer, lumican attenuates the EGFR expression and phosphorylation by inducing dimerization and internalization of EGFR, leading to the downregulation of Akt kinase activity and reduced growth. (**c**) In gastric cancer, lumican produced by cancer-associated fibroblasts promotes this cancer cell growth through integrin β1-FAK signaling pathways. (**d**) Lumican stimulates the growth of HTB94 chondrosarcoma cells through an IGF-IR/Erk1/2 axis. (**e**) Biglycan enhances the growth of MG63 osteosarcoma cells through an LRP6/β-catenin/IGF-IR signaling axis with the participation of Erk1/ downstream signaling

1.3.2 Effects of Biglycan and Lumican on Cancer Cell Motility and Invasion

While cell migration is a normal process necessary in embryogenesis, tissue homeostasis, and regeneration, cell invasion is perpetrated during the penetration of biological barriers and metastasis (Condeelis and Segall 2003). Cells can migrate utilizing an amoeboid motion exerting weak adhesion and traction. Alternatively, cells migrate through a mesenchymal movement by forming cytoplasmic protrusions, adhering to the ECM and performing cell body translocation. The mesenchymal mode also includes extracellular proteolysis with the widening of pores and gaps (Condeelis and Segall 2003; Friedl and Wolf 2010). Cell 2D migration is facilitated by forming various locomotory cytoplasmic protrusions including filopodia, pseudopodia, and lamellipodia, promoting the "gliding" of the cells on the substrate (Taylor and Condeelis 1979). On the other hand, in 3D movement, cells form invasive plasma membrane protrusions such as invadopodia, able to adhere in an integrin-dependent manner on the ECM but also to degrade the ECM by secreting matrix metal proteinases (MMPs) (Génot and Gligorijevic 2014; Di Martino et al. 2016).

Notably, the creation of both invadopodia and pseudopodia in the tumor microenvironment is controlled by stromal cells such as macrophages and fibroblasts (Parekh et al. 2011; Gaggioli et al. 2007).

Lumican has been shown to inhibit invadopodia and lamellipodia development by prostate cells (Coulson-Thomas et al. 2013). Indeed, when prostate cancer cells were seeded on a lumican substrate, the lamellipodia decrease was detected by an attenuated rearrangement in ZO-1, keratin 8/18, integrin β1, and MT1-MMP, whereas invadopodia attenuation was identified by disruption of α-smooth muscle actin, cortactin, and N-WASP (Coulson-Thomas et al. 2013). Likewise, the inhibition of melanoma cell invadopodia formation was detected in the presence of lumican (Radwanska et al. 2008). Furthermore, lumican abrogates the Snail-dependent MMP-14 activity in B16F1 melanoma cells but not in HT-29 colon adenocarcinoma cells (Stasiak et al. 2016). Moreover, lumican inhibits melanoma cell migration by directly binding α2β1 integrin with its core protein (Zeltz et al. 2010). Furthermore, lumican affects the expression of matrix effectors, inhibits the formation of MMP-releasing invadopodia to inhibit melanoma lung metastasis in vivo, as well as cell invasion in vitro (Karamanou et al. 2021). Lumican downregulates Snail activity, considering that Snail is a significant trigger of epithelial–mesenchymal transition (Moody et al. 2005); this lumican activity could form the base of anticancer therapy.

In a breast cancer model, lumican regulated cell morphology, evoked EMT/MET reprogramming, and inhibited the expression of significant matrix effectors such as MMPs. Moreover, lumican downregulated crucial cell properties, including migration and invasion (Karamanou et al. 2017).

On the other hand, the downregulation of lumican attenuated bladder cancer cells' migration by inhibiting MAPK signaling (Mao et al. 2019). Likewise, the decrease of lumican inhibited the adhesion of lung osteotropic cancer cells to various ECM components, attenuated in vitro cell migration and invasion, and interfered in vivo lung metastasis (Hsiao et al. 2020). Furthermore, the overexpression of lumican by human colon cancer cells enhanced the formation of podosome-like structures and these cells' migration (Radwanska et al. 2012). Lumican facilitated the Saos 2 cell migration and chemotactic response to fibronectin (Nikitovic et al. 2008a, b) and promoted the TGFβ2-dependent adhesion of Saos 2 cells to the fibronectin substrate (Nikitovic et al. 2011).

Desmoplasia, a response of the cancer stroma, is characterized by the unorganized production of fibrous tissue, mainly consisting of collagen fibers and fibroblasts (Kunz-Schughart and Knuechel 2002). The fibrotic tissue is produced chiefly by the "activated" cancer-associated fibroblasts (Kunz-Schughart and Knuechel 2002; Apte et al. 2004). The secretion of the altered ECM can significantly modulate the growth, migration, and spread of cancer cells (Kunz-Schughart and Knuechel 2002). Lumican has been identified to the adjacent stromal tissues in human pancreatic cancer, and patients with lumican-positive stromal tissues had shorter survival than those with lumican-negative stromal tissues (Ishiwata et al. 2007).

Therefore, as previously summarized, lumican exhibits cancer-specific roles in carcinogenesis, and further in-depth study of its effects is necessary.

The class I SLRP biglycan was likewise shown to affect tumor cell migration and invasion abilities. Notably, biglycan was shown to affect the desmoplastic reaction in colorectal cancer, inhibiting the migration and invasion of colorectal tumor cells in 2D and 3D coculture systems (Coulson-Thomas et al. 2011).

On the other hand, biglycan facilitates endometrial cancer cell invasion and migration in vitro and a xenograft model (Sun et al. 2016). In gastric cancer in vitro and in vivo models, biglycan was demonstrated to enhance tumor invasion and metastasis by facilitating FAK phosphorylation at Tyr576/577, Tyr925, and Tyr397 as of paxillin (Hu et al. 2014). Biglycan also affects mesenchymal origin tumors motility and invasion. Indeed a cooperative mechanism of PTH (1–34) and FGF-2 action leading to an increase in biglycan expression was demonstrated to facilitate osteosarcoma cell migration (Datsis et al. 2011).

Biglycan can exhibit adverse effects on cancer cell motility. Thus, the RAC1B–SMAD3–biglycan signaling axis was found to attenuate the migration of metastatic pancreatic cancer cells (Otterbein et al. 2019). Moreover, biglycan decreases the malignant properties of HER-2/neu-transformed cells, which is inversely correlated to the PKC signaling (Recktenwald et al. 2012).

1.3.3 Biglycan and Lumican at the Crossroad between Apoptosis, Autophagy, and Cancer-Associated Inflammation

The process of tumorigenesis is intimately correlated with chronic inflammation, with a significant 20% of cancer incidences directly related to chronic infections (Aggarwal et al. 2009). All tumor types separately of etiology specifically interact with the immune system at all stages of carcinogenesis (Grivennikov et al. 2010). The ECM components play a significant role in these interactions. Indeed, the tumor microenvironment's extensive remodeling modulates the immune response (Avgustinova et al. 2016). Biglycan, either secreted or proteolytically released, is an established pro-inflammatory molecule and is characterized as a danger signal (DAMP) (Schaefer et al. 2017). Thus, soluble biglycan binds to the Toll-like receptors (TLR)-2 and − 4 on the surface of macrophages, inducing sterile inflammation or enhancing pathogen-mediated inflammation through the induced production of pro-inflammatory mediators (Schaefer et al. 2005). Utilizing TLR2/4 signaling pathways biglycan induces the adaptor molecule myeloid differentiation primary response 88 (MyD88) for the recruitment of neutrophils and macrophages or initiates Toll/interleukin (IL)-1R domain-containing adaptor inducing interferon (IFN)-β (TRIF) activities for T-lymphocyte recruitment (Zeng-Brouwers et al.

2014). Indeed, the binding of biglycan to macrophage CD14, an established GPI-anchored TLRs coreceptor, can induce a discrete response dependent on the TLR type (Roedig et al. 2019).

Notably, biglycan can regulate the generation of radical oxygen species (ROS), which are significant mediators of the tumorigenesis process (Nastase et al. 2017). Upon binding to the TLRs, biglycan was shown to stabilize NADPH oxidase (NOX) 1, 2, and 4 enzymes, modulating their ROS production and subsequently ROS-dependent IL-1β synthesis (Hsieh et al. 2016). Furthermore, the ROS production initiated by biglycan increases the secretion of chemokine (C-X-C) ligand 13 (CXCL13) which acts as an essential chemoattractant for B and B1 lymphocytes (Shaefer et al. 2005). Moreover, biglycan through TLR2/4-dependent cytokine release can induce macrophage and T cells recruitment.

These actions can be both pro- and antitumorigenic. Indeed, biglycan-dependent NOX-generated ROS can cause genomic instability and alterations of chromosomal DNA, which results in enhanced tumor cell growth and metastasis (Wu et al. 2014). On the other hand, biglycan exhibits anti-inflammatory effects promoting the resolution of acute inflammation and the initiation of acquired immunity response (Hsieh et al. 2016). Schaefer et al. (2017) summarize that biglycan actions can modulate both innate and adaptive immunity acting at different checkpoints of the tumor–immune system interactions (Schaefer et al. 2017).

The data on the effects of lumican on tumor-associated inflammation are somewhat limited (Nikitovic et al. 2014). Thus, lumican can affect peripheral monocyte extravasation and Fas–FasL signaling (Wu et al. 2007; Nikitovic et al. 2015).

1.3.3.1 Effects on Apoptosis and Autophagy

Notably, these SLRPs have been indicated in the regulation of autophagy and apoptosis. Indeed, increased expression of biglycan in gastric cancer protects against apoptosis, facilitates cell migration and invasion of tumor cells, and promotes cancer-associated angiogenesis (Pinto et al. 2021). Indeed, it was shown that lumican-deficient gastric cancer cells express increased PARP1, caspase-3 cleavage levels, and decreased mesenchymal markers (Pinto et al. 2021). Supplementation of biglycan-deficient cells with exogenous biglycan restored their survival. In another study, however, biglycan enhances the expression of pro-inflammatory mediators correlated to gastric cell necroptosis (Guo et al. 2019).

In colon cancer cells, the downregulation of biglycan was correlated to apoptosis. Interestingly, an abrogation of the p38 signaling pathway with SB203580 reversed the enhancement of apoptosis in biglycan-deficient cells (Xing et al. 2015a, b).

On the other hand, biglycan was shown to protect neuroblastoma cells from NO-induced cytotoxicity by attenuating autophagy-dependent AMPK–mTOR signaling as well as the intracellular levels of ROS (Chen et al. 2020a, b).

Lumican-overexpressing B16F1 melanoma cells exhibit attenuated anchorage-independent proliferation and the ability to invade the ECM gel. Indeed, B16F1 cells presented an induction or an increase of apoptosis in lumican presence (Vuillermoz et al. 2004).

Stromal lumican was demonstrated to decrease hypoxia-inducible factor-1α (HIF1α) expression and activity via Ak correlated to increased pancreatic cancer cell apoptosis (Li et al. 2014). Moreover, lumican was shown to initiate a quiescent pancreatic cancer state, characterized among others with apoptosis (Li et al. 2017).

The importance of cancer-stroma interactions is highlighted by a recent study showing that hypoxia induces autophagy in pancreatic cancer stellate cells, which results in AKMP/TOR/p70S6K/4EBP signaling pathway-mediated protein degradation and synthesis inhibition. Lumican secretion was strongly downregulated in these cells. Indeed, lumican mRNA expression was not altered, verifying that changes in lumican expression were perpetrated at the post-transcriptional level (Li et al. 2019). It is worth mentioning that even though hypoxia-induced autophagy was also determined in ex vivo cultures of patient-derived primary PDAC xenograft cancer cells were not affected (Li et al. 2019). The pancreatic stellate cells have been attributed significant roles in paracrine signaling, metabolism, and tumor-immunology of pancreatic ductal carcinoma (Fu et al. 2018).

1.4 Biglycan: A Regulator of Tumor-Associated Angiogenesis

Overexpression of biglycan in colon cancer cells strongly upregulated their VEGF expression and enhanced the angiogenesis and growth of the tumor xenograft. Indeed, biglycan was suggested to be a promising target for cancer antiangiogenic therapy (Xing et al. 2015a, b). Furthermore, biglycan secreted by gastric cancer cells binds to its TLR2 and TLR4 receptors on endothelial cells. This was correlated to increased endothelial cell VEGF production, migration, and tube formation (Hu et al. 2016).

A significant breakthrough was identifying bidirectional interactions between the tumor cells and the tumor endothelial cells (TECs) that exhibit discrete phenotypes compared to normal counterparts. Biglycan secreted by TECs has been characterized as an autocrine angiogenic factor of tumor endothelial cells (Yamamoto et al. 2012). Notably, tumor endothelial cells isolated from tumors with high metastatic potential were shown to promote the metastasis of the tumor cells with lower aggressivity via epigenetic dysregulation of biglycan (Maishi et al. 2016). Indeed, when biglycan was knocked out in the stroma of E0771 breast cancer-bearing mice, their metastasis to the lung was inhibited. Furthermore, the downregulation of biglycan in the tumor stroma also impaired tumor angiogenesis and normalized tumor vasculature by repressing tumor necrosis factor-α/angiopoietin 2 signaling.

On the other hand, lumican facilitates endothelial cell apoptosis through Fas-dependent signaling. Indeed, lumican-overexpressing murine fibrosarcoma (MCA102) and pancreatic adenocarcinoma (Pan02) cells exhibited attenuated in vivo growth compared to their lumican nonexpressing counterparts (Williams et al. 2010).

1.5 Effects of Biglycan and Lumican on Chemoresistance

Biglycan was shown to be overexpressed in colon cancer stem cells and induce the activation of the NF-κB pathway, contributing to colon cancer cell chemotherapy resistance. In addition, biglycan attenuated the expression of pro-apoptotic markers and facilitated the activation of the NF-kβ pathway (Liu et al. 2018). Furthermore, the overexpression of biglycan induced rapamycin resistance in WERI-Rb-1 retinoblastoma cells through activating the PI3K/Akt/NF-kappaBaxis (Fang et al. 2019). Mitz et al. suggest that tumor-driven alterations in the osteoid contribute to the development of chemoresistance in osteosarcoma and characterize biglycan as one of the differentially expressed genes (Mintz et al. 2005).

On the other hand, this SLRP was determined to be one of the 10 ECM proteins overexpressed in chemo-sensitive tissue compared to the chemo-resistant tissue (Pan et al. 2009). Biglycan was shown to revert the tumor microenvironment in breast cancer (Moreth et al. 2010).

Notably, chemotherapy agents increased the secretion of lumican in PDAC cells, which was correlated to the extent of response. In various PDAC models, including cell lines, patient-derived xenografts, and lumican knockout mice, lumican was found to enhance the anticancer chemotherapy effect (Li et al. 2016). Specifically, chemotherapeutic agents in PDAC cells facilitate autophagosome formation and enhance LC3 expression through the ROS-mediated AMP-activated kinase (AMPK) signaling pathway. Lumican attenuates AMPK activity, abrogating the protective mechanism of chemotherapy-induced autophagy in in vitro and in vivo PDAC models (Li et al. 2016).

1.6 SLRPs as Cancer Therapy Targets

Considering the many-faceted roles of biglycan and lumican in carcinogenesis, the modulation of their activities are likely targets of anticancer therapies. First steps have been taken in this direction. Thus, biglycan was shown to be the target of celastrol, a triterpene component of Chinese medicine (Guo et al. 2019). Indeed, celastrol in a biglycan-dependent manner activated receptor-interacting protein 1 and 3 (RIP1 and RIP3) and subsequently promoted the translation of mixed-lineage kinase domain-like (MLKL) from the cytoplasm to plasma membrane inducing necroptosis in HGC27 and AGS gastric cancer cells. Moreover, celastrol attenuated the secretion of pro-inflammatory cytokines TNF-α and IL-8 in HGC27 and AGS cells, which was counteracted by the overexpression biglycan. Notably, biglycan has been indicated to connect innate and adaptive immunity through TLR2/4 downstream signaling established to control neutrophil, T and B lymphocyte, macrophage, and dendritic cell activities (Babelova et al. 2009; Zeng-Brouwers 2014; Moreth et al. 2014). Indeed, an increasing amount of data suggests that the activation of TLR2/TLR4 downstream signaling stimulates the secretion of inflammation-associated cytokines by tumor and stromal cells facilitating tumor cell growth (Schaefer and Iozzo 2012; Farnebo et al. 2015; Dajon et al. 2017).

1.7 Conclusions

In summary, biglycan and lumican are significant mediators of carcinogenesis with the ability to modulate all hallmarks of cancer, including cell growth, migration, invasion, apoptosis, immunological response, and angiogenesis.

Lumican is suggested to be both promote cancer development and exhibit antitumor properties. Indeed, increasing data indicate that its role depends on the tumor origin tissue and the stage of the disease. Furthermore, it has been suggested that by regulating tumor cell behavior, lumican can predict the response to radiotherapy and modulate chemotherapy.

On the other hand, biglycan is found to be overexpressed by most solid tumors, and its expression was positively correlated to disease progression. Furthermore, biglycan seems to promote tumor-associated inflammation and facilitate tumor-related angiogenesis. In some tumor types, such as osteosarcoma, biglycan was identified as a negative prognostic biomarker for chemotherapy response. Taking into account that the TME is a composite network of different cell types, biosignaling mediators, and numerous ECM components, with the capability to regulate the biological functions of both tumor and stroma cells cell determining the role of discrete components is vital. The novel data on the multifaceted roles of these SLRPs could potentially be translated into valid therapeutic options. Therefore, further understanding of the biology of the TME is obligatory for the successful targeting of these SLRP as therapeutic strategies.

References

Aggarwal BB, Vijayalekshmi RV, Sung B (2009 Jan 15) Targeting inflammatory pathways for prevention and therapy of cancer: short-term friend, long-term foe. Clin Cancer Res 15(2): 425–430

Aggelidakis J, Berdiaki A, Nikitovic D, Papoutsidakis A, Papachristou DJ, Tsatsakis AM et al (2018) Biglycan regulates MG63 osteosarcoma cell growth through a LPR6/beta-catenin/IGFR-IR signaling Axis. Front Oncol 8:470

Ameye L, Aria D, Jepsen K, Oldberg A, Xu T, Young MF (2002 May) Abnormal collagen fibrils in tendons of biglycan/fibromodulin-deficient mice lead to gait impairment, ectopic ossification, and osteoarthritis. FASEB J 16(7):673–680

Appunni S, Anand V, Khandelwal M, Seth A, Mathur S, Sharma A (2017 May) Altered expression of small leucine-rich proteoglycans (Decorin, Biglycan and Lumican): plausible diagnostic marker in urothelial carcinoma of bladder. Tumour Biol 39(5):1010428317699112

Apte MV, Park S, Phillips PA, Santucci N, Goldstein D, Kumar RK et al (2004 Oct) Desmoplastic reaction in pancreatic cancer: role of pancreatic stellate cells. Pancreas 29(3):179–187

Avgustinova A, Iravani M, Robertson D, Fearns A, Gao Q, Klingbeil P et al (2016 Jan 18) Tumour cell-derived Wnt7a recruits and activates fibroblasts to promote tumour aggressiveness. Nat Commun 7:10305

Babelova A, Moreth K, Tsalastra-Greul W, Zeng-Brouwers J, Eickelberg O, Young MF et al (2009 Sep 4) Biglycan, a danger signal that activates the NLRP3 inflammasome via toll-like and P2X receptors. J Biol Chem 284(36):24035–24048

Blochberger TC, Cornuet PK, Hassell JR (1992 Oct 15) Isolation and partial characterization of lumican and decorin from adult chicken corneas. A keratan sulfate-containing isoform of decorin is developmentally regulated. J Biol Chem 267(29):20613–20619

Brezillon S, Venteo L, Ramont L, D'Onofrio MF, Perreau C, Pluot M et al (2007 Jul) Expression of lumican, a small leucine-rich proteoglycan with antitumour activity, in human malignant melanoma. Clin Exp Dermatol 32(4):405–416

Chen L, Zhang Y, Zuo Y, Ma F, Song H (2017 Nov) Lumican expression in gastric cancer and its association with biological behavior and prognosis. Oncol Lett 14(5):5235–5240

Chen S, Guo D, Lei B, Bi J, Yang H (2020a Apr) Biglycan protects human neuroblastoma cells from nitric oxide-induced death by inhibiting AMPK-mTOR mediated autophagy and intracellular ROS level. Biotechnol Lett 42(4):657–668

Chen X, Li X, Hu X, Jiang F, Shen Y, Xu R et al (2020b) LUM expression and its prognostic significance in gastric cancer. Front Oncol 10:605

Condeelis J, Segall JE (2003 Dec) Intravital imaging of cell movement in tumours. Nat Rev Cancer 3(12):921–930

Coulson-Thomas VJ, Coulson-Thomas YM, Gesteira TF, de Paula CA, Mader AM, Waisberg J et al (2011 Nov) Colorectal cancer desmoplastic reaction up-regulates collagen synthesis and restricts cancer cell invasion. Cell Tissue Res 346(2):223–236

Coulson-Thomas VJ, Coulson-Thomas YM, Gesteira TF, Andrade de Paula CA, Carneiro CR, Ortiz V et al (2013 Apr 15) Lumican expression, localization and antitumor activity in prostate cancer. Exp Cell Res 319(7):967–981

Dajon M, Iribarren K, Cremer I (2017 Jan) Toll-like receptor stimulation in cancer: a pro- and anti-tumor double-edged sword. Immunobiology 222(1):89–100

Datsis GA, Berdiaki A, Nikitovic D, Mytilineou M, Katonis P, Karamanos NK et al (2011 Oct) Parathyroid hormone affects the fibroblast growth factor-proteoglycan signaling axis to regulate osteosarcoma cell migration. FEBS J 278(19):3782–3792

Dellett M, Hu W, Papadaki V, Ohnuma S (2012 Apr) Small leucine rich proteoglycan family regulates multiple signalling pathways in neural development and maintenance. Develop Growth Differ 54(3):327–340

Di Martino J, Henriet E, Ezzoukhry Z, Goetz JG, Moreau V, Saltel F (2016 May 1) The microenvironment controls invadosome plasticity. J Cell Sci 129(9):1759–1768

Douglas T, Heinemann S, Bierbaum S, Scharnweber D, Worch H (2006 Aug) Fibrillogenesis of collagen types I, II, and III with small leucine-rich proteoglycans decorin and biglycan. Biomacromolecules 7(8):2388–2393

Dunn GP, Old LJ, Schreiber RD (2004 Aug) The immunobiology of cancer immunosurveillance and immunoediting. Immunity 21(2):137–148

Fang D, Lai Z, Wang Y (2019 Sep 4) Overexpression of Biglycan is associated with resistance to rapamycin in human WERI-Rb-1 retinoblastoma cells by inducing the activation of the phosphatidylinositol 3-kinases (PI3K)/Akt/nuclear factor kappa B (NF-kappaB) signaling pathway. Med Sci Monit (25):6639–6648

Farnebo L, Shahangian A, Lee Y, Shin JH, Scheeren FA, Sunwoo JB (2015 Apr 30) Targeting toll-like receptor 2 inhibits growth of head and neck squamous cell carcinoma. Oncotarget 6(12): 9897–9907

Friedl P, Wolf K (2010 Jan 11) Plasticity of cell migration: a multiscale tuning model. J Cell Biol 188(1):11–19

Fu Y, Liu S, Zeng S, Shen H (2018 Feb 19) The critical roles of activated stellate cells-mediated paracrine signaling, metabolism and onco-immunology in pancreatic ductal adenocarcinoma. Mol Cancer 17(1):62

Gaggioli C, Hooper S, Hidalgo-Carcedo C, Grosse R, Marshall JF, Harrington K et al (2007 Dec) Fibroblast-led collective invasion of carcinoma cells with differing roles for RhoGTPases in leading and following cells. Nat Cell Biol 9(12):1392–1400

Génot E, Gligorijevic B (2014 Oct) Invadosomes in their natural habitat. Eur J Cell Biol 93(10–12):367–379. https://doi.org/10.1016/j.ejcb.2014.10.002. Epub 2014 Oct 23. PMID: 25457677; PMCID: PMC4262535

Grivennikov SI, Greten FR, Karin M (2010 Mar 19) Immunity, inflammation, and cancer. Cell 140(6):883–899

Gu X, Ma Y, Xiao J, Zheng H, Song C, Gong Y et al (2012 Sep) Up-regulated biglycan expression correlates with the malignancy in human colorectal cancers. Clin Exp Med 12(3):195–199

Guo D, Zhang W, Yang H, Bi J, Xie Y, Cheng B et al (2019 Nov 14) Celastrol induces necroptosis and ameliorates inflammation via targeting Biglycan in human gastric carcinoma. Int J Mol Sci 20(22):5716

Hanahan D, Weinberg RA (2000 Jan 7) The hallmarks of cancer. Cell 100(1):57–70

Henry SP, Takanosu M, Boyd TC, Mayne PM, Eberspaecher H, Zhou W et al (2001 Apr 13) Expression pattern and gene characterization of asporin. A newly discovered member of the leucine-rich repeat protein family. J Biol Chem 276(15):12212–12221

Hope C, Emmerich PB, Papadas A, Pagenkopf A, Matkowskyj KA, Van De Hey DR et al (2017 Sep 1) Versican-derived Matrikines regulate Batf3-dendritic cell differentiation and promote T cell infiltration in colorectal cancer. J Immunol 199(5):1933–1941

Hsiao KC, Chu PY, Chang GC, Liu KJ (2020 Jan 17) Elevated expression of Lumican in lung cancer cells promotes bone metastasis through an autocrine regulatory mechanism. Cancers (Basel). 12(1):233

Hsieh LT, Frey H, Nastase MV, Tredup C, Hoffmann A, Poluzzi C et al (2016 Jan) Bimodal role of NADPH oxidases in the regulation of biglycan-triggered IL-1beta synthesis. Matrix Biol 49:61–81

Hu L, Duan YT, Li JF, Su LP, Yan M, Zhu ZG et al (2014 Apr 15) Biglycan enhances gastric cancer invasion by activating FAK signaling pathway. Oncotarget 5(7):1885–1896

Hu L, Zang MD, Wang HX, Li JF, Su LP, Yan M et al (2016 Nov) Biglycan stimulates VEGF expression in endothelial cells by activating the TLR signaling pathway. Mol Oncol 10(9): 1473–1484

Hynes RO (2009) The extracellular matrix: not just pretty fibrils. Science 326(5957):1216

Iozzo RV, Schaefer L (2015 Mar) Proteoglycan form and function: a comprehensive nomenclature of proteoglycans. Matrix Biol 42:11–55

Ishiwata T, Cho K, Kawahara K, Yamamoto T, Fujiwara Y, Uchida E et al (2007 Sep) Role of lumican in cancer cells and adjacent stromal tissues in human pancreatic cancer. Oncol Rep 18(3):537–543

Jacobsen F, Kraft J, Schroeder C, Hube-Magg C, Kluth M, Lang DS et al (2017 Sep) Up-regulation of Biglycan is associated with poor prognosis and PTEN deletion in patients with prostate cancer. Neoplasia 19(9):707–715

Joshi RS, Kanugula SS, Sudhir S, Pereira MP, Jain S, Aghi MK (2021 Mar 19) The role of cancer-associated fibroblasts in tumor progression. Cancers (Basel) 13(6):1399

Kang Y, Roife D, Lee Y, Lv H, Suzuki R, Ling J et al (2016 Oct 1) Transforming growth factor-beta limits secretion of Lumican by activated stellate cells within primary pancreatic adenocarcinoma tumors. Clin Cancer Res 22(19):4934–4946

Karamanou K, Franchi M, Piperigkou Z, Perreau C, Maquart FX, Vynios DH et al (2017 Mar 23) Lumican effectively regulates the estrogen receptors-associated functional properties of breast cancer cells, expression of matrix effectors and epithelial-to-mesenchymal transition. Sci Rep 7: 45138

Karamanou K, Franchi M, Onisto M, Passi A, Vynios DH, Brézillon S (2020 Nov) Evaluation of lumican effects on morphology of invading breast cancer cells, expression of integrins and downstream signaling. FEBS J 287(22):4862–4880. https://doi.org/10.1111/febs.15289. Epub 2020 Mar 31. PMID: 32160387

Karamanou K, Franchi M, Proult I, Rivet R, Vynios D, Brezillon S (2021 Apr 8) Lumican inhibits in vivo melanoma metastasis by altering matrix-effectors and Invadopodia markers. Cell 10(4): 841

Kobe B, Kajava AV (2001 Dec) The leucine-rich repeat as a protein recognition motif. Curr Opin Struct Biol 11(6):725–732

Kram V, Kilts TM, Bhattacharyya N, Li L, Young MF (2017 Oct 3) Small leucine rich proteoglycans, a novel link to osteoclastogenesis. Sci Rep 7(1):12627

Kram V, Shainer R, Jani P, Meester JAN, Loeys B, Young MF (2020 Nov) Biglycan in the skeleton. J Histochem Cytochem 68(11):747–762

Kunz-Schughart LA, Knuechel R (2002 Apr) Tumor-associated fibroblasts (part I): active stromal participants in tumor development and progression? Histol Histopathol 17(2):599–621

Leygue E, Snell L, Dotzlaw H, Hole K, Hiller-Hitchcock T, Roughley PJ, Watson PH, Murphy LC (1998 Apr 1) Expression of lumican in human breast carcinoma. Cancer Res 58(7):1348–1352. PMID: 9537227

Li X, Truty MA, Kang Y, Chopin-Laly X, Zhang R, Roife D et al (2014 Dec 15) Extracellular lumican inhibits pancreatic cancer cell growth and is associated with prolonged survival after surgery. Clin Cancer Res 20(24):6529–6540

Li X, Roife D, Kang Y, Dai B, Pratt M, Fleming JB (2016 Sep 15) Extracellular lumican augments cytotoxicity of chemotherapy in pancreatic ductal adenocarcinoma cells via autophagy inhibition. Oncogene 35(37):4881–4890

Li X, Kang Y, Roife D, Lee Y, Pratt M, Perez MR et al (2017 Sep 21) Prolonged exposure to extracellular lumican restrains pancreatic adenocarcinoma growth. Oncogene 36(38): 5432–5438

Li X, Lee Y, Kang Y, Dai B, Perez MR, Pratt M et al (2019 Jan) Hypoxia-induced autophagy of stellate cells inhibits expression and secretion of lumican into microenvironment of pancreatic ductal adenocarcinoma. Cell Death Differ 26(2):382–393

Liu Y, Li W, Li X, Tai Y, Lu Q, Yang N et al (2014 Mar) Expression and significance of biglycan in endometrial cancer. Arch Gynecol Obstet 289(3):649–655

Liu B, Xu T, Xu X, Cui Y, Xing X (2018 Dec) Biglycan promotes the chemotherapy resistance of colon cancer by activating NF-kappaB signal transduction. Mol Cell Biochem 449(1–2): 285–294

Liu Z, Liu S, Guo J, Sun L, Wang S, Wang Y et al (2021 Jan) Identification and analysis of key genes driving gastric cancer through bioinformatics. Genet Test Mol Biomarkers 25(1):1–11

Lu P, Weaver VM, Werb Z (2012 Feb 20) The extracellular matrix: a dynamic niche in cancer progression. J Cell Biol 196(4):395–406

Maishi N, Ohba Y, Akiyama K, Ohga N, Hamada J, Nagao-Kitamoto H et al (2016 Jun 13) Tumour endothelial cells in high metastatic tumours promote metastasis via epigenetic dysregulation of biglycan. Sci Rep 6:28039

Mao W, Luo M, Huang X, Wang Q, Fan J, Gao L et al (2019 Aug) Knockdown of Lumican inhibits proliferation and migration of bladder cancer. Transl Oncol 12(8):1072–1078

McBride OW, Fisher LW, Young MF (1990 Feb) Localization of PGI (biglycan, BGN) and PGII (decorin, DCN, PG-40) genes on human chromosomes Xq13-qter and 12q, respectively. Genomics 6(2):219–225

McEwan PA, Scott PG, Bishop PN, Bella J (2006 Aug) Structural correlations in the family of small leucine-rich repeat proteins and proteoglycans. J Struct Biol 155(2):294–305

Melching LI, Roughley PJ (1999 Aug) Modulation of keratan sulfate synthesis on lumican by the action of cytokines on human articular chondrocytes. Matrix Biol 18(4):381–390

Mintz MB, Sowers R, Brown KM, Hilmer SC, Mazza B, Huvos AG et al (2005 Mar 1) An expression signature classifies chemotherapy-resistant pediatric osteosarcoma. Cancer Res 65(5):1748–1754

Monfort J, Tardif G, Reboul P, Mineau F, Roughley P, Pelletier JP et al (2006) Degradation of small leucine-rich repeat proteoglycans by matrix metalloprotease-13: identification of a new biglycan cleavage site. Arthritis Res Ther 8(1):R26

Moody SE, Perez D, Pan TC, Sarkisian CJ, Portocarrero CP, Sterner CJ et al (2005 Sep) The transcriptional repressor Snail promotes mammary tumor recurrence. Cancer Cell 8(3):197–209

Moreth K, Brodbeck R, Babelova A, Gretz N, Spieker T, Zeng-Brouwers J et al (2010 Dec) The proteoglycan biglycan regulates expression of the B cell chemoattractant CXCL13 and aggravates murine lupus nephritis. J Clin Invest 120(12):4251–4272

Moreth K, Frey H, Hubo M, Zeng-Brouwers J, Nastase MV, Hsieh LT et al (2014 Apr) Biglycan-triggered TLR-2- and TLR-4-signaling exacerbates the pathophysiology of ischemic acute kidney injury. Matrix Biol 35:143–151

Morris BA, Burkel B, Ponik SM, Fan J, Condeelis JS, Aguirre-Ghiso JA et al (2016 Nov) Collagen matrix density drives the metabolic shift in breast cancer cells. EBioMedicine 13:146–156

Mytilinaiou M, Nikitovic D, Berdiaki A, Kostouras A, Papoutsidakis A, Tsatsakis AM, Tzanakakis GN (2017 Nov) Emerging roles of syndecan 2 in epithelial and mesenchymal cancer progression. IUBMB Life 69(11):824–833. https://doi.org/10.1002/iub.1678. Epub 2017 Sep 23. PMID: 28940845

Nastase MV, Janicova A, Wygrecka M, Schaefer L (2017 Oct 20) Signaling at the crossroads: matrix-derived proteoglycan and reactive oxygen species signaling. Antioxid Redox Signal 27(12):855–873

Neame PJ, Choi HU, Rosenberg LC (1989 May 25) The primary structure of the core protein of the small, leucine-rich proteoglycan (PG I) from bovine articular cartilage. J Biol Chem 264(15): 8653–8661

Neame PJ, Sommarin Y, Boynton RE, Heinegard D (1994 Aug 26) The structure of a 38-kDa leucine-rich protein (chondroadherin) isolated from bovine cartilage. J Biol Chem 269(34): 21547–21554

Niedworok C, Rock K, Kretschmer I, Freudenberger T, Nagy N, Szarvas T et al (2013) Inhibitory role of the small leucine-rich proteoglycan biglycan in bladder cancer. PLoS One 8(11):e80084

Nikitovic D, Berdiaki A, Zafiropoulos A, Katonis P, Tsatsakis A, Karamanos NK et al (2008a Jan) Lumican expression is positively correlated with the differentiation and negatively with the growth of human osteosarcoma cells. FEBS J 275(2):350–361

Nikitovic D, Katonis P, Tsatsakis A, Karamanos NK, Tzanakakis GN (2008b Dec) Lumican, a small leucine-rich proteoglycan. IUBMB Life 60(12):818–823

Nikitovic D, Chalkiadaki G, Berdiaki A, Aggelidakis J, Katonis P, Karamanos NK et al (2011 Jun) Lumican regulates osteosarcoma cell adhesion by modulating TGFbeta2 activity. Int J Biochem Cell Biol 43(6):928–935

Nikitovic D, Aggelidakis J, Young MF, Iozzo RV, Karamanos NK, Tzanakakis GN (2012 Oct 5) The biology of small leucine-rich proteoglycans in bone pathophysiology. J Biol Chem 287(41): 33926–33933

Nikitovic D, Papoutsidakis A, Karamanos NK, Tzanakakis GN (2014 Apr) Lumican affects tumor cell functions, tumor-ECM interactions, angiogenesis and inflammatory response. Matrix Biol 35:206–214

Nikitovic D, Tzardi M, Berdiaki A, Tsatsakis A, Tzanakakis GN (2015 Apr 14) Cancer microenvironment and inflammation: role of hyaluronan. Front Immunol 6:169. https://doi.org/10.3389/fimmu.2015.00169. PMID: 25926834; PMCID: PMC4396412

Nikitovic D, Berdiaki A, Spyridaki I, Krasanakis T, Tsatsakis A, Tzanakakis GN (2018) Proteoglycans-biomarkers and targets in cancer therapy. Front Endocrinol (Lausanne) 9:69

Otterbein H, Lehnert H, Ungefroren H (2019 Dec 6) Negative control of cell migration by Rac1b in highly metastatic pancreatic cancer cells is mediated by sequential induction of nonactivated Smad3 and Biglycan. Cancers (Basel) 11(12):1959
Pan S, Cheng L, White JT, Lu W, Utleg AG, Yan X et al (2009 Aug) Quantitative proteomics analysis integrated with microarray data reveals that extracellular matrix proteins, catenins, and p53 binding protein 1 are important for chemotherapy response in ovarian cancers. OMICS 13(4):345–354
Papoutsidakis A, Giatagana EM, Berdiaki A, Spyridaki I, Spandidos DA, Tsatsakis A et al (2020 Sep) Lumican mediates HTB94 chondrosarcoma cell growth via an IGFIR/Erk1/2 axis. Int J Oncol 57(3):791–803
Parekh A, Ruppender NS, Branch KM, Sewell-Loftin MK, Lin J, Boyer PD et al (2011 Feb 2) Sensing and modulation of invadopodia across a wide range of rigidities. Biophys J 100(3): 573–582
Pasco S, Ramont L, Maquart FX, Monboisse JC (2004 Mar) Control of melanoma progression by various matrikines from basement membrane macromolecules. Crit Rev Oncol Hematol 49(3): 221–233
Pickup MW, Mouw JK, Weaver VM (2014 Dec) The extracellular matrix modulates the hallmarks of cancer. EMBO Rep 15(12):1243–1253
Pietraszek K, Brezillon S, Perreau C, Malicka-Blaszkiewicz M, Maquart FX, Wegrowski Y (2013) Lumican - derived peptides inhibit melanoma cell growth and migration. PLoS One 8(10): e76232
Pinto F, Santos-Ferreira L, Pinto MT, Gomes C, Reis CA (2021 Mar 16) The extracellular small leucine-rich proteoglycan Biglycan is a key player in gastric cancer aggressiveness. Cancers (Basel). 13(6):1330
Radwanska A, Baczynska D, Nowak D, Brezillon S, Popow A, Maquart FX et al (2008 Nov 7) Lumican affects actin cytoskeletal organization in human melanoma A375 cells. Life Sci 83(19–20):651–660
Radwanska A, Litwin M, Nowak D, Baczynska D, Wegrowski Y, Maquart FX et al (2012 Nov 1) Overexpression of lumican affects the migration of human colon cancer cells through up-regulation of gelsolin and filamentous actin reorganization. Exp Cell Res 318(18): 2312–2323
Recktenwald CV, Leisz S, Steven A, Mimura K, Muller A, Wulfanger J et al (2012 Jul 13) HER-2/neu-mediated down-regulation of biglycan associated with altered growth properties. J Biol Chem 287(29):24320–24329
Reinboth B, Hanssen E, Cleary EG, Gibson MA (2002 Feb 8) Molecular interactions of biglycan and decorin with elastic fiber components: biglycan forms a ternary complex with tropoelastin and microfibril-associated glycoprotein 1. J Biol Chem 277(6):3950–3957
Roedig H, Nastase MV, Frey H, Moreth K, Zeng-Brouwers J, Poluzzi C et al (2019 Apr) Biglycan is a new high-affinity ligand for CD14 in macrophages. Matrix Biol 77:4–22
Roughley PJ, White RJ (1989 Sep 15) Dermatan sulphate proteoglycans of human articular cartilage. The properties of dermatan sulphate proteoglycans I and II. Biochem J 262(3): 823–827
Roughley PJ, White RJ, Magny MC, Liu J, Pearce RH, Mort JS (1993 Oct 15) Non-proteoglycan forms of biglycan increase with age in human articular cartilage. Biochem J 295(Pt 2):421–426
Roughley PJ, White RJ, Cs-Szabo G, Mort JS (1996 Sep) Changes with age in the structure of fibromodulin in human articular cartilage. Osteoarthr Cartil 4(3):153–161
Sanders EJ, Walter MA, Parker E, Aramburo C, Harvey S (2003 Dec) Opticin binds retinal growth hormone in the embryonic vitreous. Invest Ophthalmol Vis Sci 44(12):5404–5409
Schaefer L, Babelova A, Kiss E, Hausser HJ, Baliova M, Krzyzankova M, Marsche G, Young MF, Mihalik D, Götte M, Malle E, Schaefer RM, Gröne HJ (2005 Aug) The matrix component biglycan is proinflammatory and signals through toll-like receptors 4 and 2 in macrophages. J Clin Invest 115(8):2223–2233. https://doi.org/10.1172/JCI23755. Epub 2005 Jul 14. PMID: 16025156; PMCID: PMC1174916

Schaefer L, Iozzo RV (2008 Aug 1) Biological functions of the small leucine-rich proteoglycans: from genetics to signal transduction. J Biol Chem 283(31):21305–21309

Schaefer L, Iozzo RV (2012 Feb) Small leucine-rich proteoglycans, at the crossroad of cancer growth and inflammation. Curr Opin Genet Dev 22(1):56–57

Schaefer L, Tredup C, Gubbiotti MA, Iozzo RV (2017 Jan) Proteoglycan neofunctions: regulation of inflammation and autophagy in cancer biology. FEBS J 284(1):10–26

Schrecengost PK, Blochberger TC, Hassell JR (1992 Jan) Identification of chick corneal keratan sulfate proteoglycan precursor protein in whole corneas and in cultured corneal fibroblasts. Arch Biochem Biophys 292(1):54–61

Schulz GB, Grimm T, Sers C, Riemer P, Elmasry M, Kirchner T, et al. Prognostic value and association with epithelial-mesenchymal transition and molecular subtypes of the proteoglycan biglycan in advanced bladder cancer. Urol Oncol 2019 Aug;37(8):530.e9-530.e18

Scott PG, McEwan PA, Dodd CM, Bergmann EM, Bishop PN, Bella J (2004 Nov 2) Crystal structure of the dimeric protein core of decorin, the archetypal small leucine-rich repeat proteoglycan. Proc Natl Acad Sci U S A 101(44):15633–15638

Seya T, Tanaka N, Shinji S, Yokoi K, Koizumi M, Teranishi N, Yamashita K, Tajiri T, Ishiwata T, Naito Z (2006 Dec) Lumican expression in advanced colorectal cancer with nodal metastasis correlates with poor prognosis. Oncol Rep 16(6):1225–1230. PMID: 17089042

Sifaki M, Assouti M, Nikitovic D, Krasagakis K, Karamanos NK, Tzanakakis GN (2006 Oct) Lumican, a small leucine-rich proteoglycan substituted with keratan sulfate chains is expressed and secreted by human melanoma cells and not normal melanocytes. IUBMB Life 58(10): 606–610

Sorensen HT, Friis S, Olsen JH, Thulstrup AM, Mellemkjaer L, Linet M et al (1998 Oct) Risk of liver and other types of cancer in patients with cirrhosis: a nationwide cohort study in Denmark. Hepatology 28(4):921–925

Stasiak M, Boncela J, Perreau C, Karamanou K, Chatron-Colliet A, Proult I et al (2016) Lumican inhibits SNAIL-induced melanoma cell migration specifically by blocking MMP-14 activity. PLoS One 11(3):e0150226

Sun H, Wang X, Zhang Y, Che X, Liu Z, Zhang L et al (2016 Feb) Biglycan enhances the ability of migration and invasion in endometrial cancer. Arch Gynecol Obstet 293(2):429–438

Suveges S, Eftimie R, Trucu D (2020 Nov 19) Directionality of macrophages movement in tumour invasion: a multiscale moving-boundary approach. Bull Math Biol 82(12):148

Svensson L, Narlid I, Oldberg A (2000 Mar 24) Fibromodulin and lumican bind to the same region on collagen type I fibrils. FEBS Lett 470(2):178–182

Taylor DL, Condeelis JS (1979) Cytoplasmic structure and contractility in amoeboid cells. Int Rev Cytol 56:57–144

Tzanakakis G, Neagu M, Tsatsakis A, Nikitovic D (2019 Apr 24) Proteoglycans and immunobiology of cancer-therapeutic implications. Front Immunol 10:875. https://doi.org/10.3389/fimmu.2019.00875. PMID: 31068944; PMCID: PMC6491844

Tzanakakis G, Giatagana EM, Kuskov A, Berdiaki A, Tsatsakis AM, Neagu M et al (2020 Aug 24) Proteoglycans in the pathogenesis of hormone-dependent cancers: mediators and effectors. Cancers (Basel) 12(9):2401

Tzanakakis GN, Giatagana EM, Berdiaki A, Spyridaki I, Hida K, Neagu M et al (2021 May 19) The role of IGF/IGF-IR-signaling and extracellular matrix effectors in bone sarcoma pathogenesis. Cancers (Basel). 13(10):2478

Vuillermoz B, Khoruzhenko A, D'Onofrio MF, Ramont L, Venteo L, Perreau C et al (2004 Jun 10) The small leucine-rich proteoglycan lumican inhibits melanoma progression. Exp Cell Res 296(2):294–306

Wang B, Li GX, Zhang SG, Wang Q, Wen YG, Tang HM et al (2011 Nov) Biglycan expression correlates with aggressiveness and poor prognosis of gastric cancer. Exp Biol Med (Maywood) 236(11):1247–1253

Wang X, Zhou Q, Yu Z, Wu X, Chen X, Li J et al (2017 Sep 1) Cancer-associated fibroblast-derived Lumican promotes gastric cancer progression via the integrin beta1-FAK signaling pathway. Int J Cancer 141(5):998–1010
Weber CK, Sommer G, Michl P, Fensterer H, Weimer M, Gansauge F et al (2001 Sep) Biglycan is overexpressed in pancreatic cancer and induces G1-arrest in pancreatic cancer cell lines. Gastroenterology 121(3):657–667
Wei Z, Zhongqiu T, Lu S, Zhang F, Xie W, Wang Y (2020 Jan) Gene coexpression analysis offers important modules and pathway of human lung adenocarcinomas. J Cell Physiol 235(1): 454–464
Wiberg C, Klatt AR, Wagener R, Paulsson M, Bateman JF, Heinegard D et al (2003 Sep 26) Complexes of matrilin-1 and biglycan or decorin connect collagen VI microfibrils to both collagen II and aggrecan. J Biol Chem 278(39):37698–37704
Williams KE, Fulford LA, Albig AR (2010 Nov 18) Lumican reduces tumor growth via induction of fas-mediated endothelial cell apoptosis. Cancer Microenviron 4(1):115–126
Wu F, Vij N, Roberts L, Lopez-Briones S, Joyce S, Chakravarti S (2007 Sep 7) A novel role of the lumican core protein in bacterial lipopolysaccharide-induced innate immune response. J Biol Chem 282(36):26409–26417
Wu Y, Antony S, Meitzler JL, Doroshow JH (2014 Apr 10) Molecular mechanisms underlying chronic inflammation-associated cancers. Cancer Lett 345(2):164–173
Xing X, Gu X, Ma T (2015a Nov) Knockdown of biglycan expression by RNA interference inhibits the proliferation and invasion of, and induces apoptosis in, the HCT116 colon cancer cell line. Mol Med Rep 12(5):7538–7544
Xing X, Gu X, Ma T, Ye H (2015b Mar) Biglycan up-regulated vascular endothelial growth factor (VEGF) expression and promoted angiogenesis in colon cancer. Tumour Biol 36(3):1773–1780
Yamamoto K, Ohga N, Hida Y, Maishi N, Kawamoto T, Kitayama K et al (2012 Mar 13) Biglycan is a specific marker and an autocrine angiogenic factor of tumour endothelial cells. Br J Cancer 106(6):1214–1223
Yang CT, Hsu PC, Chow SE (2020 Jan) Downregulation of lumican enhanced mitotic defects and aneuploidy in lung cancer cells. Cell Cycle 19(1):97–108
Zeltz C, Brezillon S, Kapyla J, Eble JA, Bobichon H, Terryn C et al (2010 Oct 15) Lumican inhibits cell migration through alpha2beta1 integrin. Exp Cell Res 316(17):2922–2931
Zeng-Brouwers J, Beckmann J, Nastase MV, Iozzo RV, Schaefer L (2014 Apr) De novo expression of circulating biglycan evokes an innate inflammatory tissue response via MyD88/TRIF pathways. Matrix Biol 35:132–142. https://doi.org/10.1016/j.matbio.2013.12.003. Epub 2013 Dec 18. PMID: 24361484; PMCID: PMC4039567
Zhao SF, Yin XJ, Zhao WJ, Liu LC, Wang ZP (2020 Mar) Biglycan as a potential diagnostic and prognostic biomarker in multiple human cancers. Oncol Lett 19(3):1673–1682
Zhu YH, Yang F, Zhang SS, Zeng TT, Xie X, Guan XY (2013) High expression of biglycan is associated with poor prognosis in patients with esophageal squamous cell carcinoma. Int J Clin Exp Pathol 6(11):2497–2505

Chapter 2
The Role of Decorin in Cancer

Kornélia Baghy, Andrea Reszegi, Zsolt Horváth, and Ilona Kovalszky

Abstract The dynamic interactions between the tumor cells and their microenvironment consisting of stroma cells and extracellular matrix play a crucial role in cancer development. Decorin, a chondroitin sulfate/dermatan sulfate proteoglycan of the SLRP family is expressed in the stroma of cancers with various origin. Decorin was originally considered as a structural component of the ECM, but research of the last few decades discovered its essential roles in fibrotic pathologies, inflammation, and cancer. The initial observation of its being a natural inhibitor of TGFβ was followed by decorin's ability to counteract with the activity of several receptor tyrosine kinases such as EGFR, Met, IGF-IR, VEGFR, and PDGFR. Besides modulating the RTK signaling, decorin induces the degradation of these receptors by initiating their caveosomal internalization. Decorin also provokes cell cycle arrest and apoptosis, and exerts antimetastatic and antiangiogenic effects. The discovery of decorin's action in conserved catabolic processes, namely endothelial cell autophagy and tumor cell mitophagy further broadened the palette of its antitumor abilities. Therefore, in spite of the difficulties in mass producing the proteoglycan, decorin represents a promising tool against cancer, especially against those, employing altered RTK signaling.

2.1 Introduction

The decorin proteoglycan is the prototype and best-characterized member of the small leucine-rich proteoglycan (SLRP) family. Its name refers to its strong interaction with collagen fibers appearing as "decoration" of fibrils (Reese et al. 2013; Zhang et al. 2009; Chen et al. 2014; Iozzo and Schaefer 2015; Chen and Birk 2013). The gene of decorin was 1st cloned in 1986; at that time the proteoglycan was

K. Baghy (✉) · A. Reszegi · Z. Horváth · I. Kovalszky
Department of Pathology & Experimental Cancer Research, Semmelweis University, Budapest, Hungary
e-mail: kornelia.baghy@med.semmelweis-univ.hu

I. Kovalszky et al. (eds.), *The Extracellular Matrix and the Tumor Microenvironment*, Biology of Extracellular Matrix 11,
https://doi.org/10.1007/978-3-030-99708-3_2

considered only as a structural constituent of the extracellular matrix (ECM) (Krusius and Ruoslahti 1986). With time, decorin was proved to exert a plethora of cellular functions such as inhibiting proliferation, migration, differentiation, metastasis formation, modulating angiogenesis or inflammatory responses (Yamaguchi and Ruoslahti 1988; Ruoslahti and Yamaguchi 1991; Border et al. 1992; Jarvinen and Prince 2015). These first reports on the antitumor effects of decorin paved the way for further and still ongoing studies to seize the opportunity of its therapeutic utilization against cancer (Sofeu Feugaing et al. 2013).

Mammalian decorin is highly conserved across the species. This proteoglycan is mainly produced by fibroblasts, myofibroblasts, macrophages, and smooth muscle cells (Zhang et al. 2018). Decorin consists of a 42 kDa protein core containing a central domain with 12 leucine-rich repeats (LRR) flanked by cysteine-rich regions and an N-terminal attachment site for a single glycosaminoglycan (GAG) chain, that can be either chondroitin or dermatan sulfate (Krusius and Ruoslahti 1986; Mann et al. 1990). The 12 LRRs form a horseshoe or banana shape (Scott 1996; Weber et al. 1996; Brown et al. 2002; Scott et al. 2004), and this unique structure can interact with a variety of proteins and is responsible for the diverse biological functions of decorin (Neill et al. 2016) (Fig. 2.1). Monomeric decorin represents the active form of the molecule, but the proteoglycan forms homodimers in physiological solutions (Brown et al. 2002; Islam et al. 2013; Goldoni et al. 2004). Dimerization, which seems reversible, prevents the interactions with binding partners by blocking the central domain (Islam et al. 2013). Monomeric decorin acts as a soluble paracrine factor that modulates several downstream signaling pathways (Gubbiotti et al. 2016). In addition, decorin is a substrate of proteolytic enzymes called matrix metalloproteases (MMP-2, MMP-2, MMP-7, MT1-MMP) and BMP-1 peptidase that leads to the cleavage, thus inactivation of the molecule (Mimura et al. 2009; von Marschall and Fisher 2010).

The biological functions of decorin were mostly elucidated by the identification and characterization of mutations occurring in patients suffering of genetic diseases and decorin-deficient mice. Decorin-deficient mice are fertile and display skin fragility phenotype with loosely packed collagen networks that resemble the human Ehlers-Danlos syndrome (Corsi et al. 2002). Mutations in the decorin gene result in connective tissue disorders like congenital stromal corneal dystrophy (Bredrup et al. 2005; Seidler 2012), characterized by the impairment of corneal opacity and vision. The heterozygous mutation affects the ear region of the molecule causing bilateral corneal opacities in patients (Zhang et al. 2009; Pietraszek-Gremplewicz et al. 2019).

In this chapter, we will review the cellular and molecular roles of decorin in cancers including tissue distribution, signaling mechanisms, and also therapeutic applications.

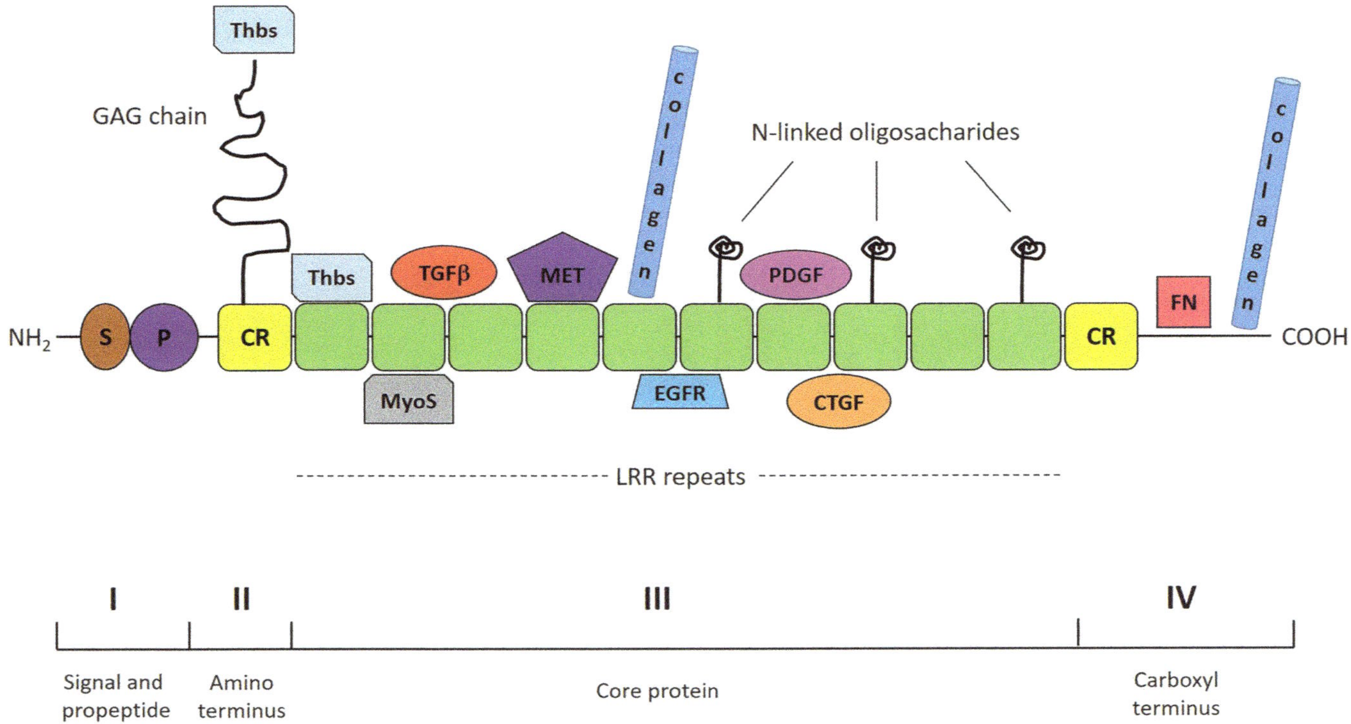

Fig. 2.1 The schematic structure and interacting partners of decorin. The proteoglycan has four domains indicated as I-IV. The core protein contains tandem leucine-rich repeats (LRR) flanked by cysteine-rich regions (CR). The molecule bears one chondroitin sulfate/dermatan sulfate glucosaminoglycan (GAG) chain and three N-linked oligosaccharides. Decorin interacts with a large number of molecules including growth factors such as TGFβ, PDGF, CTGF, receptor tyrosine kinases like EGFR and Met), myostatin (MyoS), thrombospondin, and structural components of the ECM e.g. collagens or fibronectin. As shown in the figure, decorin displays different binding sites to these signaling molecules, which suggest simultaneous modulation of different pathways for curbing tumorigenesis

2.2 The Uprise of Decorin as an Antitumor Agent

2.2.1 Genetic Evidences

The first evidences for the tumor suppressor property of decorin emerged from experimental mouse models where the decorin gene was unconditionally ablated (Danielson et al. 1997). Decorin knockout mice developed spontaneous intestinal tumors when fed with high-fat diet (Bi et al. 2008). In this model, the lack of decorin led to disturbed intestinal maturation as a result of reduced cell differentiation and increased proliferation. As underlying mechanisms, downregulation of $p21^{WAF1/CIP1}$, $p27^{KIP1}$, intestinal trefoil factor, and E-cadherin, as well as the upregulation of β-catenin signaling were identified (Bi et al. 2008). These revealed signaling events were the pioneers paving the way to further studies to uncover the antitumor mechanisms exhibited by decorin. Simultaneous disruption of both decorin and p53 genes resulted in the formation of aggressive T-cell lymphomas and premature death of these mice (Iozzo et al. 1999a). Ablation of decorin gene also resulted in enhanced tumor incidence and tumor count in livers of mice exposed to hepatocarcinogens (Horvath et al. 2014). In a model of colitis-associated cancer, the intestinal endothelium of decorin knockout mice showed an increased protein levels of Snail, Slug, Twist, and MMP2, factors associated with epithelial-mesenchymal transition. In addition to the previous, in colitis-associated tumors, ICAM-1 expression was also significantly elevated in $Dcn^{-/-}$ mice compared to wild-type animals (Mao et al. 2021).

2.2.2 Decorin Expression in Cancers

To understand the biological role of decorin in cancer, we have to discuss its expression patterns and localization within tumors. Only few studies exploring decorin expression in human tumors are available vs. in vitro investigations. However, a general phenomenon based on the available reports can be highlighted, namely that decorin expression tends to be downregulated in the parenchyma of advanced tumors of various origin, but sometimes seen with overexpression in the stroma. In human breast cancer, decorin expression is inhibited both at the mRNA and protein level when compared to either normal or non-tumorous adjacent tissues (Leygue et al. 2000). Similar observations on decorin downregulation were detected in other malignancies, such as endometrial (Smid-Koopman et al. 2000), lung (Campioni et al. 2008), and ovarian (Nash et al. 2002) cancers. In line with the previous reported data, a reduced decorin expression was associated with poor prognosis in node-negative invasive breast carcinoma and soft tissue tumors (Troup et al. 2003; Matsumine et al. 2007). Analyzing the Human Protein Atlas database, a significant reduction in decorin levels was revealed in the stroma of bladder, breast, cervical, colon, kidney, pancreas, prostate, ovary, rectal, skin,

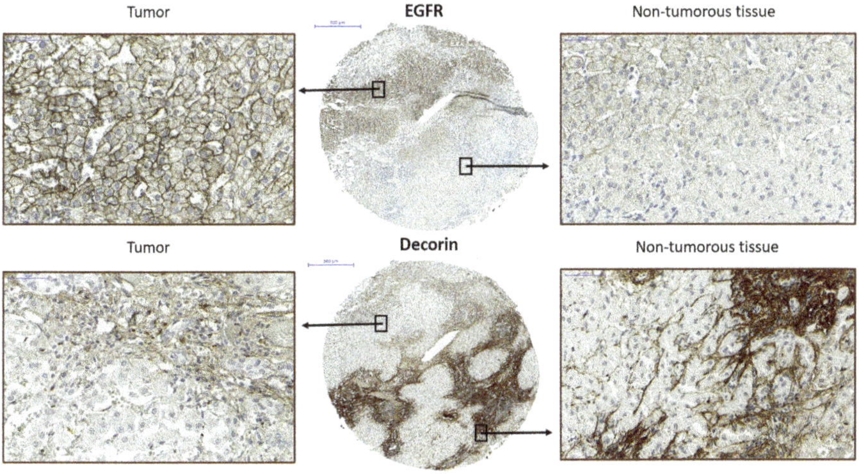

Fig. 2.2 Alternating expression of decorin and EGFR in tumorous liver tissues. In liver cancer, decorin expression is markedly reduced, even absent compared to the connective tissue of the non-tumorous liver tissue. However, EGFR, a known binding partner of decorin exerts exactly the opposite expression pattern suggesting a functional link between the two molecules. Identical tissue microarray cores of human material were simultaneously stained with decorin and EGFR antibodies

stomach, and testis tumors (Bozoky et al. 2014; Neill et al. 2015). Decorin expression was abolished in the stroma of low- and high-grade urothelial carcinoma, while elevated decorin level was detected in the submucosa and deep tumor stroma (Iozzo et al. 2011). In a colon carcinoma-liver metastasis model, normal colon exerted the highest decorin expression, but the proteoglycan level was also abundant in the stroma of the primary tumor. In contrast, liver metastasis of the same tumors generally contained reduced levels of decorin, which were associated to the aggressiveness of the disease (Reszegi et al. 2020). Decreased decorin expression was described in multiple myeloma and monoclonal gammopathy of undetermined significance (Kristensen et al. 2013), in esophageal squamous cell carcinoma (Wu et al. 2010), and in some cases of colon carcinoma (Nyman et al. 2015). Several studies reported the total absence of decorin within the tumor parenchyma in various tumors such as urothelial, prostate, myeloma, and liver cancer (Li et al. 2008; Sainio et al. 2013; Bostrom et al. 2013; Henke et al. 2012; Horvath et al. 2014; Duncan 2013) (Fig. 2.2).

In general, in tumor cells as well as in non-tumorous connective tissues decorin is missing, or its expression is very low. However, the tumorous stroma may contain high levels of decorin as reported in human colon (Iozzo et al. 1982, 1989; Iozzo and Wight 1982) and breast cancers (Santra et al. 1995; Brown et al. 1999). In addition, decorin expression is also highly enhanced in the ECM of pancreatic carcinoma, mainly detected at the border of tumorous areas. As source of the excessive decorin deposition, pancreatic stellate cells were identified (Skandalis et al. 2006; Sofeu

Feugaing et al. 2013; Koninger et al. 2004), while decorin expression was completely absent in pancreatic cancer cells.

It is a feasible hypothesis that the stromal accumulation of this proteoglycan is still able to hinder tumor growth by acting as a physical barrier similar to that seen in desmoplastic reactions. In line with that, as a paracrine inhibitor, stromal decorin can also block the action of receptor tyrosine kinases located in the membranes of tumor cells.

To add some spice to what already discussed, high decorin mRNA expression was found during the early stages of tumorigenesis as a result of transcriptional analysis of tumor progression. In line with that high decorin mRNA expression was detected in hemangiomas, while decorin production was completely abolished in malignant vascular sarcomas (Salomaki et al. 2008). The same phenomenon was reported in case of patients with B-cell chronic lymphoid leukemia (CLL), as high decorin mRNA levels were detected in early stages and with non-progressive CLL, while it was suppressed in advanced stages and in patients with the aggressive type of the disease (Campo et al. 2006). It seems conceivable that the loss of decorin expression correlates with tumor progression and aggressiveness, and could be utilized as a marker of early and late-stage cancer.

2.2.3 Signaling in Cancers

2.2.3.1 Interaction with TGF-β and Other Growth Factors

The **transforming growth factor-β (TGF-β)** (Yamaguchi et al. 1990) was the first growth factor identified as an interacting partner of decorin. The decorin protein core does not distinguish among the growth factor isotypes, as it is able to bind to TGF-β1, TGF-β2, and TGF-β3 (Hildebrand et al. 1994). The interaction of decorin with TGF-β results in attenuated proliferation and spreading of several TGF-β –dependent tumor cell lines (Jarvinen and Prince 2015; Yamaguchi et al. 1990). As a mechanism of action, it was proposed that decorin traps TGF-β before it binds to its receptors and anchors it to the collagens of the ECM (Jarvinen and Ruoslahti 2013). Several in vitro and in vivo studies confirmed that decorin inhibits TGF-β signaling pathways (Gubbiotti et al. 2016; Abdel-Wahab et al. 2002; Kolb et al. 2001; Baghy et al. 2011). The decorin-TGF-β complex is inactive, and such binding results in hindered signaling via Smad2, Smad3, and Erk1/2, and attenuates fibrogenesis and inflammation (Baghy et al. 2012). Intriguingly, TGF-β is able to inhibit decorin expression at the transcriptional level as observed in fibroblasts (Mauviel et al. 1995), suggesting a feedback mechanism responsible for the balance in matrix deposition. Another cellular process linked to TGF-β is the regulation of inflammation. Due to its ability to reduce inflammation and fibrosis, decorin was proposed as a physiological TGF-β-inhibitor that limits the duration of TGF-β responses in inflammation and tissue repair (Border et al. 1992; Jarvinen and Ruoslahti 2013; Jarvinen and Prince 2015). As the function of TGF-β in cancer is quite controversial and highly depends on the cellular context (Ikushima and Miyazono 2010), neutralization of the growth factor either by decorin or other substances as a potential antitumor strategy needs careful evaluation (Baghy et al. 2020).

In addition to TGF-β, decorin also binds to and inhibits two other members of the TGF-β superfamily, namely myostatin and activin C. In case of **myostatin**, interaction with decorin curbs myostatin-provoked inhibition of myofiber growth leading to improved muscle regeneration (Kanzleiter et al. 2014; Miura et al. 2006; Zhu et al. 2007). In addition, myostatin was reported to play a crucial role in the process of cancer cachexia, and its blocking by antagonists has a beneficial effect on survival (Argiles et al. 2017; Miyamoto et al. 2016; Sharma et al. 2015). These observations may pave the way for decorin to counteract with cancer-related wasting by antagonizing myostatin. The interaction of decorin and the oncogenic **activin-C** leads to endocytosis of the complex with subsequent degradation. The event culminates in reduced proliferative and migration capacity of colorectal carcinoma cells in vitro and in vivo (Bi et al. 2016).

Another two interacting partners of decorin belong to the connective tissue growth factor (CTGF) family. Decorin binding inhibits the action of **CTGF** (Vial et al. 2011; Gubbiotti et al. 2016) a key player in several cellular processes such as chemotaxis, proliferation and differentiation, fibrosis, regulation of ECM production, and modulation of inflammation (Lipson et al. 2012; Luft 2008; Kular et al. 2011). Interestingly, CTGF itself induces decorin expression, its own inhibitor, and at the same time is tightly regulated by TGF-β suggesting a fine-tuned autoregulation process in the system (Vial et al. 2011). The other CTGF family member, known to bind to decorin is the **Wnt-inducible signaling pathway protein-1** (**WISP-1**, alias CCN4), a known oncogene in a variety of cancers (Nivison and Meier 2018). Decorin binds to WISP via its dermatan sulfate GAG chain modulating Wnt signaling (Neill et al. 2012b; Desnoyers et al. 2001), and at the same time, it is known to inhibit **β-catenin** (Bi et al. 2008; Young et al. 1998), although a direct link between the two mechanisms has not been verified yet.

2.2.3.2 Structural Components as Binding Partners

A less explored, but not less important binding partners of decorin, are the structural components within the ECM including several types of collagens, tenascin X (Elefteriou et al. 2001) and elastin (Reinboth et al. 2002). In addition, complexes of matrilin-1 and decorin connect collagen VI microfibrils to aggrecan or collagen type II (Wiberg et al. 2003). These studies emphasize that decorin has a pivotal role in the regulation of ECM structure and may be an active participant in the desmoplastic reactions in the tumor stroma, a defensive mechanism of the host tissue against cancer (Neill et al. 2012b).

2.2.3.3 Decorin Acts as a pan-RTK Inhibitor

A whole new era of the antitumor action of decorin started with the striking observations that decorin directly interacts with a variety of receptor tyrosine kinases (RTKs). Proven with a set of studies on decorin's tumor inhibitory action via RTKs and subsequent signaling pathways led to the consideration of "the guardian from the matrix" (Neill et al. 2012b).

Epidermal growth factor receptor (EGFR) was the first RTK identified to interact with decorin (Fig. 2.2). In details, binding of the proteoglycan to EGFR at a partially overlapping site with EGF resulted in transient activation of EGFR followed by caveolin-1 mediated internalization and degradation of the receptor (Iozzo et al. 1999b; Moscatello et al. 1998; Santra et al. 2000, 2002; Zhu et al. 2005; Neill et al. 2012b). As a consequence, decorin-EGFR interaction culminated in cell cycle arrest, apoptosis, angiostasis, and prolonged suppression of oncogenes (Gubbiotti et al. 2016; Neill et al. 2012b, 2015). Downstream of EGFR, decorin enhanced the action of MAPK pathway and intracellular Ca^{2+} release (Seidler et al. 2006; Csordas et al. 2000), induced the cyclin-dependent kinase inhibitor $p21^{WAF1/CIP1}$ and the level of cleaved caspase-3 in A431 squamous carcinoma cell line (Seidler et al. 2006). The blocking action of decorin on EGFR and its downstream signaling was further corroborated by other in vitro and in vivo studies. In A549 lung carcinoma cells, inhibited EGFR action by decorin overexpression resulted in induced p53 and $p21^{WAF1/CIP1}$ expression leading to cell cycle blockade in G1 and apoptosis (Liang et al. 2013). Vice versa, genetic ablation of decorin enhances EGFR, and subsequently ERK1/2 phosphorylation in experimental hepatocarcinogenesis models (Horvath et al. 2014).

Besides EGFR, ErbB2 and ErbB4 (other members of the ErbB family) were found to interact with decorin (Goldoni et al. 2008; Gubbiotti et al. 2016; Goldoni and Iozzo 2008). Here, decorin-provoked growth arrest and stimulated differentiation occurred due to the fact that ErbB4-decorin binding hampered the dimerization of ErbB4 with ErbB2 in breast carcinoma cells (Goldoni et al. 2008).

The Met receptor tyrosine kinase (hepatocyte growth factor receptor, scatter receptor) is another important and well-described binding partner of decorin (Goldoni et al. 2009) (Fig. 2.3).

Similarly to EGFR, the interaction between decorin and Met induces autophosphorylation of the receptor leading to its proteasomal degradation (Goldoni et al. 2009; Neill et al. 2015). Inhibition of Met activity by decorin results in suppressed downstream signaling, including selective degradation of β-catenin and Myc oncoproteins, ending up with inhibited tumor growth (Buraschi et al. 2010; Goldoni et al. 2009; Neill et al. 2015). The failure of nuclear translocation of β-catenin and the subsequent transcriptional repression of its target gene MYC initiates a series of processes (Hanahan and Weinberg 2011; Buraschi et al. 2010; Horvath et al. 2019). Myc is a key transcription factor playing crucial role in the coordination of a variety of pathways that stimulates cell proliferation. Myc targets AP4, a known transcriptional repressor of the $p21^{WAF1/CIP1}$ cyclin dependent kinase

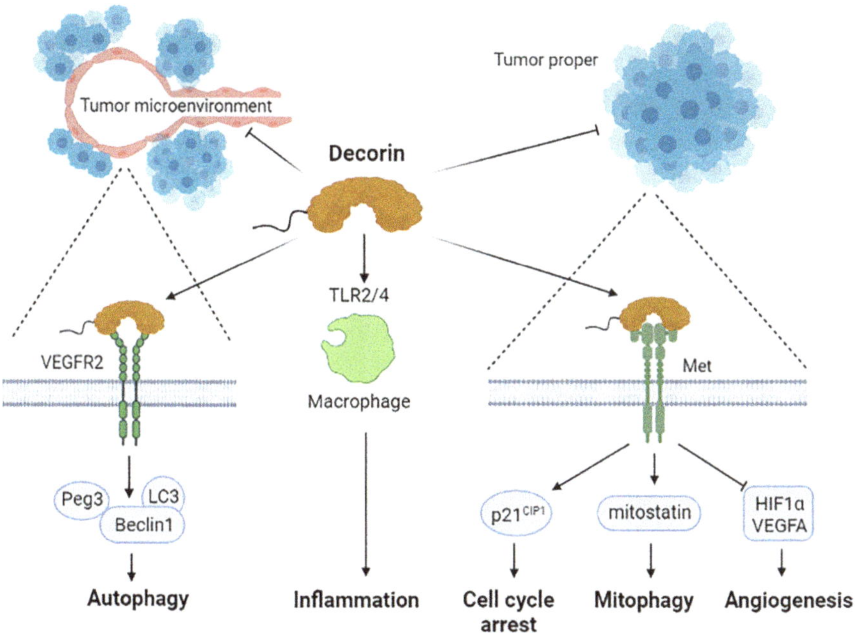

Fig. 2.3 Comprehensive illustration of interactions and signaling events provoked by decorin. The proteoglycan exerts an inhibitory effect on both the tumor microenvironment and the tumor itself. On the surface of endothelial cells, decorin blocks the action of VEGF receptor leading to autophagy via Peg3, beclin, and LC3. Within the tumor microenvironment, decorin binds to TLR2 and TLR4 on the surface of macrophages inducing inflammatory processes. Among other RTKs, decorin interacts with the Met receptor in the membrane of tumor cells. This interaction inhibits the activity of Met, which subsequently leads to cell cycle arrest via $p21^{CIP1}$ and tumor cell mitophagy via mitostatin. By hindered HIF1α and VEGF production, decorin exerts anti-angiogenic actions. (Created with BioRender.com)

inhibitor (Jung et al. 2008). Indeed, decorin was reported to block cell cycle via inducing $p21^{WAF1/CIP1}$ as a key mechanism for curbing tumorigenesis in many cancer models (Iozzo and Sanderson 2011; Neill et al. 2012b). Moreover, inhibition of Met by decorin leads to the decrease of two proangiogenic factors namely VEGFA and HIF-1α via stimulation of the angiogenesis inhibitor TIMP-3 (Neill et al. 2012a).

Since the original observation of decorin's blocking action on EGFR, several other RTKs have joined the expanding family of decorin's interacting partners. The proteoglycan can regulate the insulin signaling pathway via several mechanisms. It modulates the activity of the **insulin-like growth factor receptor-1 (IGF-1R)** and **insulin receptor (IR)** and their ligands (Morcavallo et al. 2014; Morrione et al. 2013; Schonherr et al. 2005; Baghy et al. 2013). In contrast to the caveosomal endocytosis and subsequent degradation of EGFR and Met upon decorin binding, here the proteoglycan hinders IGF-1R activity via IRS1 and IGF-1. Moreover, decorin-IGF-1 interaction prevents translocation of IGF-1R to caveosomes (Iozzo et al. 2011; Neill et al. 2012b; Schaefer et al. 2007; Schonherr et al. 2005). However, the action of decorin on IGF-1R signaling is contradictory. It seems that decorin antagonizes IGF-1R only in cancers, but enhances IGF-1R function in normal tissues (Neill et al. 2016; Morrione et al. 2013). To further nuance the mechanism, comparison of four hepatoma cell lines revealed that IGF-1R as well as IR can be either enhanced or inhibited even in tumor cell lines with the same tissue origin (Horvath et al. 2019). In addition, temporal change of receptor activation was also observed, as initial phosphorylation of IGF-1R and IR upon decorin exposure declined rapidly, and attenuated receptor activity was observed after 2 days. These reports emphasize that the role of decorin in regulating IGF-1R is complex and is dependent on unique characteristics of the tumor (Horvath et al. 2019).

Similarly to IGF-1R, the inhibitory action of decorin on **platelet-derived growth factor receptor-α (PDGFRα)** takes place via direct sequestration of its ligand PDGF (Baghy et al. 2013; Horvath et al. 2014) leading to inhibited downstream signaling in liver cancer, and hindered cellular migration and the development of intimal hyperplasia after balloon angioplasty (Nili et al. 2003).

In addition decorin is known to interact with the **vascular endothelial growth factor receptor-2** (VEGFR2, see also later) (Khan et al. 2011; Lala et al. 2012) and the macrophage stimulating protein receptor (MSPR/RON) (Horvath et al. 2014).

2.2.3.4 Intracellular Signaling Pathways

Downstream of RTK receptors discussed above, decorin effectively hinders several pathways involved in proliferation, migration, angiogenesis, or survival such as the Ras/MEK/ERK and PI3K/Akt/mTOR representing the main and widely studied pathways in a variety of cancers (Whittaker et al. 2010; Villanueva et al. 2007). In addition, the GSK-3β is a key molecule that serves as a node linking several signaling pathways (e.g. originated from Wnt and RTKs) and is known to be inactivated by Akt (Grimes and Jope 2001; Jacobs et al. 2012). The mechanisms by which decorin inhibits β-catenin and Myc, and the role of $p21^{WAF1/CIP1}$ in G1/S phase cell cycle blockade have already been discussed in this chapter. In addition to $p21^{WAF1/CIP1}$, decorin is able to induce the expression of other cyclin dependent kinase inhibitors namely $p27^{KIP1}$ (Xaus et al. 2001; Bi et al. 2008, 2012), $p15^{INK4b}$ (Liu et al. 2016) and $p57^{KIP2}$ (Hamid et al. 2013), resulting in attenuated cell cycle progression.

Besides its capability to arrest cell cycle at G1/S transition, decorin was described to block G2/M transition in HEP3B hepatoma cells (Horvath et al. 2019). This particular cell line harbors deleterious mutations of both p53 and retinoblastoma (Rb) genes (Mitry et al. 1997) that leads to continuous transition from G1 to S as the restriction point is compromised. As a mechanism of action, the researchers identified activated the ATR/Chk1/Wee1 signaling upon decorin exposure, blocking the cell cycle at the G2/M via phosphorylation of CDK1. In parallel, high phospho-AKT levels downstream of IR and IGF-1R were detected along with decreased Cdc25A phosphatase expression strengthening the novel mechanism described (Horvath et al. 2019; Benada and Macurek 2015). The genes of Rb and p53 tumor suppressor proteins are among the top most frequently mutated genes in tumors. The discovery that decorin is capable to exert its antitumor action even in cells lacking functional p53 and Rb proteins further reinforces its utilization against cancers.

In summary, the interactions of decorin with RTKs and other molecules trigger countless changes in cell signaling pathways culminated in hindered tumorigenesis (Gubbiotti et al. 2016). Thus, the introduction of decorin in tumors dependent on one or more of the proteoglycan's interacting partners may be beneficial (Goldoni et al. 2008; Bostrom et al. 2013; Horvath et al. 2014; Neill et al. 2015; Reed et al. 2005). However, it must be kept in mind that the effect of decorin is highly cell type-specific.

2.2.4 Inflammation, Immunomodulation

Several studies evidenced that decorin as well as its relative molecule biglycan is capable of regulating inflammatory and innate immune responses via TLR2 and 4 in tissue stress and injury (Merline et al. 2011; Neill et al. 2016; Schaefer 2014; Moreth et al. 2014; Zeng-Brouwers et al. 2014). Direct interaction between the proteoglycan and TLR2/4 on macrophages results in transient activation of MAPK and NFκB pathways with consequent secretion of tumor necrosis factor-α (TNFα) and IL-12p70 inflammatory molecules (Moreth et al. 2010) (Fig. 2.3). On the top of that, decorin indirectly modulates inflammation by inhibiting the action of TGF-β1, and consequently microRNA-21 (miR-21), a process that liberates macrophage-derived proinflammatory programmed cell death protein 4 (PDCD4) from translational repression (Merline et al. 2011). This mechanism leads to a proinflammatory tumor microenvironment by suppressing anti-inflammatory molecule IL-10 (Neill et al. 2012b, 2016; Gubbiotti et al. 2016; Zhang et al. 2018; Merline et al. 2011). Decorin is able to sustain the inflammatory state by enhancing CCL2 production that attracts mononuclear cells to the site of injury (Koninger et al. 2006). Decorin is also able to modulate the attachment of immune cells. By interacting with class A scavenger receptors, it enhances macrophage adhesion to the matrix (Santiago-Garcia et al. 2003). On the contrary, decorin directly inhibits the adhesion of polymorphonuclear leukocytes to the endothelium, and in parallel decreases the expression of ICAM-1 and the anti-inflammatory syndecan-1 (Seidler et al. 2011).

Generally, proinflammatory tumor microenvironment is known to retard tumor growth, thus unveiling the role of decorin in regulating inflammation is of crucial importance (Gubbiotti et al. 2016; Zhang et al. 2018; Moreth et al. 2012; Schaefer and Iozzo 2012). However, decorin was also reported to inhibit inflammatory and chemotactic genes of immune response (Buraschi et al. 2012). Decorin deficiency always results in a pro-inflammatory and pro-fibrotic phenotype as reported in several in vivo models; furthermore, treatment with exogenous decorin suppresses inflammation in experimental trials (Jarvelainen et al. 2006, 2009; Border et al. 1992; Jarvinen and Ruoslahti 2010, 2013). In addition, tumor necrosis factor-α

(TNF-α), a key proinflammatory cytokine, is known to interact with, and sequestered away by decorin, which prevents its binding to receptors (Tufvesson and Westergren-Thorsson 2002).

What determines whether decorin exerts pro or anti-inflammatory action is still far from being understood, and additional studies are to resolve the contradictions. However, it was reported that neither the protein core nor the GAG chain alone is capable to induce TNFα and IL-12p70 (Merline et al. 2011). In another study, it was the core protein that attenuated the expressions of anti-inflammatory genes (Buraschi et al. 2012). These observations give raise to the possibility that the protein core and the glycanated decorin displays different actions regarding immune responses (Zhang et al. 2018).

In addition, decorin can also be inactivated by proteases secreted by inflammatory cells. Decorin as a member of damage associated molecular patterns (DAMPs) may be recognized by pattern recognition receptors such as Toll-like receptor-2 (TLR2) and TLR4 inducing an inflammatory response (Schaefer 2014).

2.2.5 Autophagy, Mitophagy, and Angiogenesis

Decorin was discovered to indirectly induce autophagy of vascular endothelial cells, a process that attenuates tumor spread and metastasis formation via transcriptionally enhancing Peg3 (paternally expressed 3) tumor suppressor (Neill et al. 2016, 2021; Zhang et al. 2018; Buraschi et al. 2012). PEG3 is known to be hypermethylated thus inactive in a variety of tumors (Kuroiwa et al. 1996; Yamaguchi et al. 2002; Kohda et al. 2001; Dowdy et al. 2005), and can inhibit the Wnt/β-catenin signaling pathway, partly accounting for the decorin's antitumor action (Jiang et al. 2010; Neill et al. 2015). As a mechanism of action, decorin interacts with and activates VEGFR2 receptor in the membrane of endothelial cells and induces AMPK/Vps34 signaling pathway with concurrent inhibition of the anti-autophagic Akt/mTOR pathway (Schaefer and Dikic 2021; Alers et al. 2012; Goyal et al. 2014) (Fig. 2.3). In the process, decorin induces beclin-1 and the microtubule-associated protein light chain 3 (LC3) via Peg3 culminating in the formation of autophagy precursor complexes (Neill et al. 2013, 2015). In addition, decorin curbs the inhibitory Bcl-2/beclin-1 complex formation (Pattingre and Levine 2006). Importantly, beclin-1 can be inactivated by EGFR and Akt signaling leading to chemoresistance and suppression of autophagy (Wang et al. 2012; Wei et al. 2013). This process may represent a general scheme as downstream signaling of several RTKs overlap. Besides inducing autophagy on endothelial cells, decorin compromises their attachment, migration and differentiation and sprouting mainly via inhibition of VEGF expression (Grant et al. 2002; Neill et al. 2012a; Buraschi et al. 2013).

Besides endothelial autophagy, decorin is able to induce mitochondrial autophagy (mitophagy), thus directly modulate catabolic processes within the tumor (Neill et al. 2016). Decorin promotes mitophagy via induction of the Met receptor that results in enhanced expression and accumulation of the mitostatin tumor suppressor present in the mitochondrial associated membrane (Neill et al. 2014). Mitostatin is crucial for decorin to trigger mitophagy as depletion of the molecule compromises the inhibition of VEGFA by decorin. As decorin is able to raise Ca^{2+} level with subsequent depolarization of the mitochondria via interaction with EGFR, it is plausible that it is the mitostatin that mediates the process (Neill et al. 2016).

In conclusion, by inducing endothelial autophagy decorin inhibits neoangiogenesis in the tumorous stroma leading to inhibition of spreading and metastasis formation of tumors. In parallel, by interacting with RTKs, decorin enhances mitophagy modulating the catabolic processes in tumor cells (Neill et al. 2020).

2.3 Therapeutic Applications

Since the discovery of decorin's antitumor actions, enormous number of studies have been carried out to investigate the applicability of the proteoglycan as a therapeutic agent. Early studies applied in vivo virus-mediated delivery of decorin and proved its protecting effect in several different tumors. For example, decorin attenuated the growth of colon, lung, squamous cell carcinoma, inhibited invasion and metastasis formation of prostate cancer, osteosarcoma, and breast cancer as well (Reed et al. 2002; Xu et al. 2015; Shintani et al. 2008; Tralhao et al. 2003; Araki et al. 2009; Yang et al. 2015). Decorin gene therapy curbed tumor growth and had a beneficial effect on survival in an expression and time-dependent manner in a glioma model (Ma et al. 2014), and was successfully utilized in experimental models of prostate and pancreas cancer (Hu et al. 2009; Na et al. 2015).

Regarding therapeutic viral vector applications, decorin as a key player in organizing ECM, may have another profound role. As the ECM normally hampers viral spread into the tumor (Zhang et al. 2018; Parato et al. 2005), decorin is able to enhance the penetration of the virus applied (Choi et al. 2010). In parallel, by inhibiting TGF-β decorin counterweights the immunosuppressive tumor microenvironment (Zhang et al. 2018; Zwirner et al. 2010; Oh et al. 2017).

Recombinant decorin was also successfully applied in several in vitro and in vivo studies. In breast cancer both administration of decorin core protein and transfection of DCN cDNA resulted in lessened EGFR activity, enhanced apoptosis, and attenuated the growth of tumors (Csordas et al. 2000; Santra et al. 2000; Goldoni and Iozzo 2008). In line with that, systemic application of decorin led to growth retardation of breast cancer, inhibited metabolism, and decreased the number of lung metastases (Goldoni et al. 2008; Seidler et al. 2006).

Decorin is capable to modulate the effects of certain therapeutic drugs. Decorin exerts a synergistic effect with carboplatin by hindering the proliferation of ovarian cancer cells (Nash et al. 1999). On the other hand, decorin suspended the cytostatic action of carboplatin and gemcitabine in pancreatic tumor cells, but still exerted its antiproliferative action (Koninger et al. 2004).

The development of decorin into a clinical drug is yet to come, in spite of the great number of in vivo antitumor and antifibrotic investigations and preclinical studies. Its production on a large scale may be challenging as the proteoglycan is heterogeneous in size due to its GAG chains. To meet the criteria for human drugs, the recombinant decorin must contain only the core protein responsible for most of decorin's antitumor actions such as for interacting with RTKs and growth factors (Mann et al. 1990; Jarvinen and Prince 2015).

In the last years, researchers went further as to boost decorin's biological activity thus being a more attractive therapeutic candidate. As a model example, a fusion protein containing the decorin core protein and a small peptide named "CAR" was created. This CAR tag (its exact sequence is CARSKNKDC) selectively delivers large amounts of decorin to inflammatory and neoangiogenic vasculature of tumors and regenerating tissues (Jarvinen and Prince 2015; Jarvinen et al. 2015; Jarvinen and Ruoslahti 2010, 2013; Urakami et al. 2011). In addition, this CAR-DCN peptide exerts a more powerful inhibitory effect on the proliferation and spreading of tumors dependent on TGF-β (Jarvinen and Ruoslahti 2007, 2010). Because of its specificity, the CAR peptide seems to be a convenient agent for targeted delivery of therapeutic agents (Gupta et al. 2014a, b, 2015; Nahar et al. 2014; Toba et al. 2014; Urakami et al. 2011). Based on these improvements, the utilization of decorin as a therapeutic drug against cancer and diseases related to inflammation and angiogenesis is well founded (Jarvinen and Prince 2015).

2.4 Conclusions

With the emerging evidences on the important role of the tumor microenvironment in tumorigenesis, molecules of the ECM have gained more and more attention in the past decades. Decorin is the archetype, and the best-studied member of the SLRP family, originally thought to be a structural protein regulating fibrillogenesis. Since then, the proteoglycan was discovered to play essential roles as a signaling molecule in fibrotic diseases, inflammation, and cancer. Research conducted on decorin-null mouse models unveiled that the absence of decorin enhances tumorigenesis. In line with that tumors with different origin generally reduce decorin expression pointing to its tumor suppressor effect.

On the same note, application of decorin either as a recombinant protein or delivered by vectors inhibits tumor development, progression, spreading, and angiogenesis as proved in a vast number of experimental studies. The basis of decorin's antitumor action is the broad range of its interacting partners including growth factors, tyrosine kinase receptors, ECM component that results in mediating downstream signaling pathways. In addition to its negative effect on cell proliferation, migration, and angiogenesis, decorin is able to induce endothelial cell autophagy and tumor cell mitophagy. These novel regulatory mechanisms further broaden the proteoglycan's effect on tumorigenesis.

Despite of the obvious complexity of its biological actions, all of decorin's interactions culminate in inhibition of tumor progression. The antitumor action of decorin appears at multiple levels: 1) directly affecting tumor cells; 2) modifying the tumor-promoting tumor microenvironment. In addition, therapeutic utilization of decorin is on the way; also researches for improved delivery and activity of the proteoglycan, and of coping with production challenges are promising. Hence, we can say that decorin has well-earned the name "a guardian from the matrix," and will serve as a tool in the battle against cancer.

References

Abdel-Wahab N, Wicks SJ, Mason RM, Chantry A (2002) Decorin suppresses transforming growth factor-beta-induced expression of plasminogen activator inhibitor-1 in human mesangial cells through a mechanism that involves Ca2+-dependent phosphorylation of Smad2 at serine-240. Biochem J 362(Pt 3):643–649

Alers S, Loffler AS, Wesselborg S, Stork B (2012) Role of AMPK-mTOR-Ulk1/2 in the regulation of autophagy: cross talk, shortcuts, and feedbacks. Mol Cell Biol 32(1):2–11. https://doi.org/10.1128/MCB.06159-11

Araki K, Wakabayashi H, Shintani K, Morikawa J, Matsumine A, Kusuzaki K, Sudo A, Uchida A (2009) Decorin suppresses bone metastasis in a breast cancer cell line. Oncology 77(2):92–99. https://doi.org/10.1159/000228253

Argiles JM, Lopez-Soriano FJ, Stemmler B, Busquets S (2017) Novel targeted therapies for cancer cachexia. Biochem J 474(16):2663–2678. https://doi.org/10.1042/bcj20170032

Baghy K, Dezso K, Laszlo V, Fullar A, Peterfia B, Paku S, Nagy P, Schaff Z, Iozzo RV, Kovalszky I (2011) Ablation of the decorin gene enhances experimental hepatic fibrosis and impairs hepatic healing in mice. Lab Invest 91(3):439–451. https://doi.org/10.1038/labinvest.2010.172

Baghy K, Iozzo RV, Kovalszky I (2012) Decorin-TGFbeta axis in hepatic fibrosis and cirrhosis. J Histochem Cytochem 60(4):262–268. https://doi.org/10.1369/0022155412438104

Baghy K, Horvath Z, Regos E, Kiss K, Schaff Z, Iozzo RV, Kovalszky I (2013) Decorin interferes with platelet-derived growth factor receptor signaling in experimental hepatocarcinogenesis. FEBS J 280(10):2150–2164. https://doi.org/10.1111/febs.12215

Baghy K, Reszegi A, Tatrai P, Kovalszky I (2020) Decorin in the Tumor Microenvironment. Adv Exp Med Biol 1272:17–38. https://doi.org/10.1007/978-3-030-48457-6_2

Benada J, Macurek L (2015) Targeting the Checkpoint to Kill Cancer Cells. Biomolecules 5(3): 1912–1937. https://doi.org/10.3390/biom5031912

Bi X, Tong C, Dockendorff A, Bancroft L, Gallagher L, Guzman G, Iozzo RV, Augenlicht LH, Yang W (2008) Genetic deficiency of decorin causes intestinal tumor formation through disruption of intestinal cell maturation. Carcinogenesis 29(7):1435–1440

Bi X, Pohl NM, Qian Z, Yang GR, Gou Y, Guzman G, Kajdacsy-Balla A, Iozzo RV, Yang W (2012) Decorin-mediated inhibition of colorectal cancer growth and migration is associated with E-cadherin in vitro and in mice. Carcinogenesis 33(2):326–330. https://doi.org/10.1093/carcin/bgr293

Bi X, Xia X, Fan D, Mu T, Zhang Q, Iozzo RV, Yang W (2016) Oncogenic activin C interacts with decorin in colorectal cancer in vivo and in vitro. Mol Carcinog 55(11):1786–1795. https://doi.org/10.1002/mc.22427

Border WA, Noble NA, Yamamoto T, Harper JR, Yamaguchi Y, Pierschbacher MD, Ruoslahti E (1992) Natural inhibitor of transforming growth factor-beta protects against scarring in experimental kidney disease. Nature 360(6402):361–364. https://doi.org/10.1038/360361a0

Bostrom P, Sainio A, Kakko T, Savontaus M, Soderstrom M, Jarvelainen H (2013) Localization of decorin gene expression in normal human breast tissue and in benign and malignant tumors of the human breast. Histochem Cell Biol 139(1):161–171. https://doi.org/10.1007/s00418-012-1026-0

Bozoky B, Savchenko A, Guven H, Ponten F, Klein G, Szekely L (2014) Decreased decorin expression in the tumor microenvironment. Cancer Med 3(3):485–491. https://doi.org/10.1002/cam4.231

Bredrup C, Knappskog PM, Majewski J, Rodahl E, Boman H (2005) Congenital stromal dystrophy of the cornea caused by a mutation in the decorin gene. Invest Ophthalmol Vis Sci 46(2):420–426. https://doi.org/10.1167/iovs.04-0804

Brown LF, Guidi AJ, Schnitt SJ, Van De Water L, Iruela-Arispe ML, Yeo TK, Tognazzi K, Dvorak HF (1999) Vascular stroma formation in carcinoma in situ, invasive carcinoma, and metastatic carcinoma of the breast. Clin Cancer Res 5(5):1041–1056

Brown CT, Lin P, Walsh MT, Gantz D, Nugent MA, Trinkaus-Randall V (2002) Extraction and purification of decorin from corneal stroma retain structure and biological activity. Protein Expr Purif 25(3):389–399

Buraschi S, Pal N, Tyler-Rubinstein N, Owens RT, Neill T, Iozzo RV (2010) Decorin antagonizes Met receptor activity and down-regulates {beta}-catenin and Myc levels. J Biol Chem 285(53):42075–42085. https://doi.org/10.1074/jbc.M110.172841

Buraschi S, Neill T, Owens RT, Iniguez LA, Purkins G, Vadigepalli R, Evans B, Schaefer L, Peiper SC, Wang ZX, Iozzo RV (2012) Decorin protein core affects the global gene expression profile of the tumor microenvironment in a triple-negative orthotopic breast carcinoma xenograft model. PLoS One 7(9):e45559. https://doi.org/10.1371/journal.pone.0045559

Buraschi S, Neill T, Goyal A, Poluzzi C, Smythies J, Owens RT, Schaefer L, Torres A, Iozzo RV (2013) Decorin causes autophagy in endothelial cells via Peg3. Proc Natl Acad Sci U S A 110(28):E2582–E2591. https://doi.org/10.1073/pnas.1305732110

Campioni M, Ambrogi V, Pompeo E, Citro G, Castelli M, Spugnini EP, Gatti A, Cardelli P, Lorenzon L, Baldi A, Mineo TC (2008) Identification of genes down-regulated during lung cancer progression: a cDNA array study. J Exp Clin Cancer Res 27:38. https://doi.org/10.1186/1756-9966-27-38

Campo S, Campo GM, Avenoso A, D'Ascola A, Musolino C, Calabro L, Bellomo G, Quartarone E, Calatroni A (2006) Lymphocytes from patients with early stage of B-cell chronic lymphocytic leukaemia and long survival synthesize decorin. Biochimie 88(12):1933–1939. https://doi.org/10.1016/j.biochi.2006.07.010

Chen S, Birk DE (2013) The regulatory roles of small leucine-rich proteoglycans in extracellular matrix assembly. FEBS J 280(10):2120–2137. https://doi.org/10.1111/febs.12136

Chen S, Young MF, Chakravarti S, Birk DE (2014) Interclass small leucine-rich repeat proteoglycan interactions regulate collagen fibrillogenesis and corneal stromal assembly. Matrix Biol 35:103–111. https://doi.org/10.1016/j.matbio.2014.01.004

Choi IK, Lee YS, Yoo JY, Yoon AR, Kim H, Kim DS, Seidler DG, Kim JH, Yun CO (2010) Effect of decorin on overcoming the extracellular matrix barrier for oncolytic virotherapy. Gene Ther 17(2):190–201. https://doi.org/10.1038/gt.2009.142

Corsi A, Xu T, Chen XD, Boyde A, Liang J, Mankani M, Sommer B, Iozzo RV, Eichstetter I, Robey PG, Bianco P, Young MF (2002) Phenotypic effects of biglycan deficiency are linked to collagen fibril abnormalities, are synergized by decorin deficiency, and mimic Ehlers-Danlos-like changes in bone and other connective tissues. J Bone Miner Res 17(7):1180–1189. https://doi.org/10.1359/jbmr.2002.17.7.1180

Csordas G, Santra M, Reed CC, Eichstetter I, McQuillan DJ, Gross D, Nugent MA, Hajnoczky G, Iozzo RV (2000) Sustained down-regulation of the epidermal growth factor receptor by decorin. A mechanism for controlling tumor growth in vivo. J Biol Chem 275(42):32879–32887

Danielson KG, Baribault H, Holmes DF, Graham H, Kadler KE, Iozzo RV (1997) Targeted disruption of decorin leads to abnormal collagen fibril morphology and skin fragility. J Cell Biol 136(3):729–743

Desnoyers L, Arnott D, Pennica D (2001) WISP-1 binds to decorin and biglycan. J Biol Chem 276(50):47599–47607. https://doi.org/10.1074/jbc.M108339200
Dowdy SC, Gostout BS, Shridhar V, Wu X, Smith DI, Podratz KC, Jiang SW (2005) Biallelic methylation and silencing of paternally expressed gene 3 (PEG3) in gynecologic cancer cell lines. Gynecol Oncol 99(1):126–134. https://doi.org/10.1016/j.ygyno.2005.05.036
Duncan MB (2013) Extracellular matrix transcriptome dynamics in hepatocellular carcinoma. Matrix Biol 32(7–8):393–398. https://doi.org/10.1016/j.matbio.2013.05.003
Elefteriou F, Exposito JY, Garrone R, Lethias C (2001) Binding of tenascin-X to decorin. FEBS Lett 495(1-2):44–47. https://doi.org/10.1016/s0014-5793(01)02361-4
Goldoni S, Iozzo RV (2008) Tumor microenvironment: Modulation by decorin and related molecules harboring leucine-rich tandem motifs. Int J Cancer 123(11):2473–2479
Goldoni S, Owens RT, McQuillan DJ, Shriver Z, Sasisekharan R, Birk DE, Campbell S, Iozzo RV (2004) Biologically active decorin is a monomer in solution. J Biol Chem 279(8):6606–6612. https://doi.org/10.1074/jbc.M310342200
Goldoni S, Seidler DG, Heath J, Fassan M, Baffa R, Thakur ML, Owens RT, McQuillan DJ, Iozzo RV (2008) An antimetastatic role for decorin in breast cancer. Am J Pathol 173(3):844–855
Goldoni S, Humphries A, Nystrom A, Sattar S, Owens RT, McQuillan DJ, Ireton K, Iozzo RV (2009) Decorin is a novel antagonistic ligand of the Met receptor. J Cell Biol 185(4):743–754
Goyal A, Neill T, Owens RT, Schaefer L, Iozzo RV (2014) Decorin activates AMPK, an energy sensor kinase, to induce autophagy in endothelial cells. Matrix Biol 34:46–54. https://doi.org/10.1016/j.matbio.2013.12.011
Grant DS, Yenisey C, Rose RW, Tootell M, Santra M, Iozzo RV (2002) Decorin suppresses tumor cell-mediated angiogenesis. Oncogene 21(31):4765–4777
Grimes CA, Jope RS (2001) The multifaceted roles of glycogen synthase kinase 3beta in cellular signaling. Prog Neurobiol 65(4):391–426
Gubbiotti MA, Vallet SD, Ricard-Blum S, Iozzo RV (2016) Decorin interacting network: A comprehensive analysis of decorin-binding partners and their versatile functions. Matrix Biol 55:7–21. https://doi.org/10.1016/j.matbio.2016.09.009
Gupta N, Ibrahim HM, Ahsan F (2014a) Peptide-micelle hybrids containing fasudil for targeted delivery to the pulmonary arteries and arterioles to treat pulmonary arterial hypertension. J Pharm Sci 103(11):3743–3753. https://doi.org/10.1002/jps.24193
Gupta N, Patel B, Nahar K, Ahsan F (2014b) Cell permeable peptide conjugated nanoerythrosomes of fasudil prolong pulmonary arterial vasodilation in PAH rats. Eur J Pharm Biopharm 88(3): 1046–1055. https://doi.org/10.1016/j.ejpb.2014.10.012
Gupta N, Al-Saikhan FI, Patel B, Rashid J, Ahsan F (2015) Fasudil and SOD packaged in peptide-studded-liposomes: Properties, pharmacokinetics and ex-vivo targeting to isolated perfused rat lungs. Int J Pharm 488(1-2):33–43. https://doi.org/10.1016/j.ijpharm.2015.04.031
Hamid AS, Li J, Wang Y, Wu X, Ali HA, Du Z, Bo L, Zhang Y, Zhang G (2013) Recombinant human decorin upregulates p57KIP(2) expression in HepG2 hepatoma cell lines. Mol Med Rep 8(2):511–516. https://doi.org/10.3892/mmr.2013.1510
Hanahan D, Weinberg RA (2011) Hallmarks of cancer: the next generation. Cell 144(5):646–674. https://doi.org/10.1016/j.cell.2011.02.013
Henke A, Grace OC, Ashley GR, Stewart GD, Riddick AC, Yeun H, O'Donnell M, Anderson RA, Thomson AA (2012) Stromal expression of decorin, Semaphorin6D, SPARC, Sprouty1 and Tsukushi in developing prostate and decreased levels of decorin in prostate cancer. PLoS One 7(8):e42516. https://doi.org/10.1371/journal.pone.0042516
Hildebrand A, Romaris M, Rasmussen LM, Heinegard D, Twardzik DR, Border WA, Ruoslahti E (1994) Interaction of the small interstitial proteoglycans biglycan, decorin and fibromodulin with transforming growth factor beta. Biochem J 302(Pt 2):527–534. https://doi.org/10.1042/bj3020527
Horvath Z, Kovalszky I, Fullar A, Kiss K, Schaff Z, Iozzo RV, Baghy K (2014) Decorin deficiency promotes hepatic carcinogenesis. Matrix Biol 35:194–205. https://doi.org/10.1016/j.matbio.2013.11.004

Horvath Z, Reszegi A, Szilak L, Danko T, Kovalszky I, Baghy K (2019) Tumor-specific inhibitory action of decorin on different hepatoma cell lines. Cell Signal 62:109354. https://doi.org/10.1016/j.cellsig.2019.109354

Hu Y, Sun H, Owens RT, Wu J, Chen YQ, Berquin IM, Perry D, O'Flaherty JT, Edwards IJ (2009) Decorin suppresses prostate tumor growth through inhibition of epidermal growth factor and androgen receptor pathways. Neoplasia 11(10):1042–1053

Ikushima H, Miyazono K (2010) TGFbeta signalling: a complex web in cancer progression. Nat Rev Cancer 10(6):415–424. https://doi.org/10.1038/nrc2853

Iozzo RV, Sanderson RD (2011) Proteoglycans in cancer biology, tumour microenvironment and angiogenesis. J Cell Mol Med 15(5):1013–1031. https://doi.org/10.1111/j.1582-4934.2010.01236.x

Iozzo RV, Schaefer L (2015) Proteoglycan form and function: A comprehensive nomenclature of proteoglycans. Matrix Biol 42:11–55. https://doi.org/10.1016/j.matbio.2015.02.003

Iozzo RV, Wight TN (1982) Isolation and characterization of proteoglycans synthesized by human colon and colon carcinoma. J Biol Chem 257(18):11135–11144

Iozzo RV, Bolender RP, Wight TN (1982) Proteoglycan changes in the intercellular matrix of human colon carcinoma: an integrated biochemical and stereologic analysis. Lab Invest 47(2): 124–138

Iozzo RV, Sampson PM, Schmitt GK (1989) Neoplastic modulation of extracellular matrix: stimulation of chondroitin sulfate proteoglycan and hyaluronic acid synthesis in co-cultures of human colon carcinoma and smooth muscle cells. J Cell Biochem 39(4):355–378. https://doi.org/10.1002/jcb.240390403

Iozzo RV, Chakrani F, Perrotti D, McQuillan DJ, Skorski T, Calabretta B, Eichstetter I (1999a) Cooperative action of germ-line mutations in decorin and p53 accelerates lymphoma tumorigenesis. Proc Natl Acad Sci U S A 96(6):3092–3097

Iozzo RV, Moscatello DK, McQuillan DJ, Eichstetter I (1999b) Decorin is a biological ligand for the epidermal growth factor receptor. J Biol Chem 274(8):4489–4492

Iozzo RV, Buraschi S, Genua M, Xu SQ, Solomides CC, Peiper SC, Gomella LG, Owens RC, Morrione A (2011) Decorin antagonizes IGF receptor I (IGF-IR) function by interfering with IGF-IR activity and attenuating downstream signaling. J Biol Chem 286(40):34712–34721. https://doi.org/10.1074/jbc.M111.262766

Islam M, Gor J, Perkins SJ, Ishikawa Y, Bachinger HP, Hohenester E (2013) The concave face of decorin mediates reversible dimerization and collagen binding. J Biol Chem 288(49): 35526–35533. https://doi.org/10.1074/jbc.M113.504530

Jacobs KM, Bhave SR, Ferraro DJ, Jaboin JJ, Hallahan DE, Thotala D (2012) GSK-3beta: A Bifunctional Role in Cell Death Pathways. Int J Cell Biol 2012:930710. https://doi.org/10.1155/2012/930710

Jarvelainen H, Puolakkainen P, Pakkanen S, Brown EL, Hook M, Iozzo RV, Sage EH, Wight TN (2006) A role for decorin in cutaneous wound healing and angiogenesis. Wound Repair Regen 14(4):443–452

Jarvelainen H, Sainio A, Koulu M, Wight TN, Penttinen R (2009) Extracellular matrix molecules: potential targets in pharmacotherapy. Pharmacol Rev 61(2):198–223. https://doi.org/10.1124/pr.109.001289

Jarvinen TA, Prince S (2015) Decorin: A Growth Factor Antagonist for Tumor Growth Inhibition. Biomed Res Int 2015:654765. https://doi.org/10.1155/2015/654765

Jarvinen TA, Ruoslahti E (2007) Molecular changes in the vasculature of injured tissues. Am J Pathol 171(2):702–711. https://doi.org/10.2353/ajpath.2007.061251

Jarvinen TA, Ruoslahti E (2010) Target-seeking antifibrotic compound enhances wound healing and suppresses scar formation in mice. Proc Natl Acad Sci U S A 107(50):21671–21676. https://doi.org/10.1073/pnas.1016233107

Jarvinen TA, Ruoslahti E (2013) Targeted Antiscarring Therapy for Tissue Injuries. Adv Wound Care (New Rochelle) 2(2):50–54. https://doi.org/10.1089/wound.2011.0299

Jarvinen TA, May U, Prince S (2015) Systemically Administered, Target Organ-Specific Therapies for Regenerative Medicine. Int J Mol Sci 16(10):23556–23571. https://doi.org/10.3390/ijms161023556

Jiang X, Yu Y, Yang HW, Agar NY, Frado L, Johnson MD (2010) The imprinted gene PEG3 inhibits Wnt signaling and regulates glioma growth. J Biol Chem 285(11):8472–8480. https://doi.org/10.1074/jbc.M109.069450

Jung P, Menssen A, Mayr D, Hermeking H (2008) AP4 encodes a c-MYC-inducible repressor of p21. Proc Natl Acad Sci U S A 105(39):15046–15051. https://doi.org/10.1073/pnas.0801773105

Kanzleiter T, Rath M, Gorgens SW, Jensen J, Tangen DS, Kolnes AJ, Kolnes KJ, Lee S, Eckel J, Schurmann A, Eckardt K (2014) The myokine decorin is regulated by contraction and involved in muscle hypertrophy. Biochem Biophys Res Commun 450(2):1089–1094. https://doi.org/10.1016/j.bbrc.2014.06.123

Khan GA, Girish GV, Lala N, Di Guglielmo GM, Lala PK (2011) Decorin is a novel VEGFR-2-binding antagonist for the human extravillous trophoblast. Mol Endocrinol 25(8):1431–1443. https://doi.org/10.1210/me.2010-0426

Kohda T, Asai A, Kuroiwa Y, Kobayashi S, Aisaka K, Nagashima G, Yoshida MC, Kondo Y, Kagiyama N, Kirino T, Kaneko-Ishino T, Ishino F (2001) Tumour suppressor activity of human imprinted gene PEG3 in a glioma cell line. Genes Cells 6(3):237–247

Kolb M, Margetts PJ, Sime PJ, Gauldie J (2001) Proteoglycans decorin and biglycan differentially modulate TGF-beta-mediated fibrotic responses in the lung. Am J Physiol Lung Cell Mol Physiol 280(6):L1327–L1334. https://doi.org/10.1152/ajplung.2001.280.6.L1327

Koninger J, Giese NA, di Mola FF, Berberat P, Giese T, Esposito I, Bachem MG, Buchler MW, Friess H (2004) Overexpressed decorin in pancreatic cancer: potential tumor growth inhibition and attenuation of chemotherapeutic action. Clin Cancer Res 10(14):4776–4783. https://doi.org/10.1158/1078-0432.CCR-1190-03

Koninger J, Giese NA, Bartel M, di Mola FF, Berberat PO, di Sebastiano P, Giese T, Buchler MW, Friess H (2006) The ECM proteoglycan decorin links desmoplasia and inflammation in chronic pancreatitis. J Clin Pathol 59(1):21–27. https://doi.org/10.1136/jcp.2004.023135

Kristensen IB, Pedersen L, Ro TB, Christensen JH, Lyng MB, Rasmussen LM, Ditzel HJ, Borset M, Abildgaard N (2013) Decorin is down-regulated in multiple myeloma and MGUS bone marrow plasma and inhibits HGF-induced myeloma plasma cell viability and migration. Eur J Haematol 91(3):196–200. https://doi.org/10.1111/ejh.12125

Krusius T, Ruoslahti E (1986) Primary structure of an extracellular matrix proteoglycan core protein deduced from cloned cDNA. Proc Natl Acad Sci U S A 83(20):7683–7687. https://doi.org/10.1073/pnas.83.20.7683

Kular L, Pakradouni J, Kitabgi P, Laurent M, Martinerie C (2011) The CCN family: a new class of inflammation modulators? Biochimie 93(3):377–388. https://doi.org/10.1016/j.biochi.2010.11.010

Kuroiwa Y, Kaneko-Ishino T, Kagitani F, Kohda T, Li LL, Tada M, Suzuki R, Yokoyama M, Shiroishi T, Wakana S, Barton SC, Ishino F, Surani MA (1996) Peg3 imprinted gene on proximal chromosome 7 encodes for a zinc finger protein. Nat Genet 12(2):186–190. https://doi.org/10.1038/ng0296-186

Lala N, Girish GV, Cloutier-Bosworth A, Lala PK (2012) Mechanisms in decorin regulation of vascular endothelial growth factor-induced human trophoblast migration and acquisition of endothelial phenotype. Biol Reprod 87(3):59. https://doi.org/10.1095/biolreprod.111.097881

Leygue E, Snell L, Dotzlaw H, Troup S, Hiller-Hitchcock T, Murphy LC, Roughley PJ, Watson PH (2000) Lumican and decorin are differentially expressed in human breast carcinoma. J Pathol 192(3):313–320. https://doi.org/10.1002/1096-9896(200011)192:3<313::AID-PATH694>3.0.CO;2-B

Li X, Pennisi A, Yaccoby S (2008) Role of decorin in the antimyeloma effects of osteoblasts. Blood 112(1):159–168

Liang S, Xu JF, Cao WJ, Li HP, Hu CP (2013) Human decorin regulates proliferation and migration of human lung cancer A549 cells. Chin Med J (Engl) 126(24):4736–4741

Lipson KE, Wong C, Teng Y, Spong S (2012) CTGF is a central mediator of tissue remodeling and fibrosis and its inhibition can reverse the process of fibrosis. Fibrogenesis Tissue Repair 5(Suppl 1):S24. https://doi.org/10.1186/1755-1536-5-S1-S24

Liu Y, Wang X, Wang Z, Ju W, Wang D (2016) Decorin inhibits the proliferation of HepG2 cells by elevating the expression of transforming growth factor-beta receptor II. Exp Ther Med 12(4): 2191–2195. https://doi.org/10.3892/etm.2016.3572

Luft FC (2008) CCN2, the connective tissue growth factor. J Mol Med (Berl) 86(1):1–3. https://doi.org/10.1007/s00109-007-0287-x

Ma HI, Hueng DY, Shui HA, Han JM, Wang CH, Lai YH, Cheng SY, Xiao X, Chen MT, Yang YP (2014) Intratumoral decorin gene delivery by AAV vector inhibits brain glioblastomas and prolongs survival of animals by inducing cell differentiation. Int J Mol Sci 15(3):4393–4414. https://doi.org/10.3390/ijms15034393

Mann DM, Yamaguchi Y, Bourdon MA, Ruoslahti E (1990) Analysis of glycosaminoglycan substitution in decorin by site-directed mutagenesis. J Biol Chem 265(9):5317–5323

Mao L, Yang J, Yue J, Chen Y, Zhou H, Fan D, Zhang Q, Buraschi S, Iozzo RV, Bi X (2021) Decorin deficiency promotes epithelial-mesenchymal transition and colon cancer metastasis. Matrix Biol 95:1–14. https://doi.org/10.1016/j.matbio.2020.10.001

Matsumine A, Shintani K, Kusuzaki K, Matsubara T, Satonaka H, Wakabayashi T, Iino T, Uchida A (2007) Expression of decorin, a small leucine-rich proteoglycan, as a prognostic factor in soft tissue tumors. J Surg Oncol 96(5):411–418. https://doi.org/10.1002/jso.20745

Mauviel A, Santra M, Chen YQ, Uitto J, Iozzo RV (1995) Transcriptional regulation of decorin gene expression. Induction by quiescence and repression by tumor necrosis factor-alpha. J Biol Chem 270(19):11692–11700

Merline R, Moreth K, Beckmann J, Nastase MV, Zeng-Brouwers J, Tralhao JG, Lemarchand P, Pfeilschifter J, Schaefer RM, Iozzo RV, Schaefer L (2011) Signaling by the matrix proteoglycan decorin controls inflammation and cancer through PDCD4 and MicroRNA-21. Sci Signal 4(199):ra75. https://doi.org/10.1126/scisignal.2001868

Mimura T, Han KY, Onguchi T, Chang JH, Kim TI, Kojima T, Zhou Z, Azar DT (2009) MT1-MMP-mediated cleavage of decorin in corneal angiogenesis. J Vasc Res 46(6):541–550. https://doi.org/10.1159/000226222

Mitry RR, Sarraf CE, Wu CG, Pignatelli M, Habib NA (1997) Wild-type p53 induces apoptosis in Hep3B through up-regulation of bax expression. Lab Invest 77(4):369–378

Miura T, Kishioka Y, Wakamatsu J, Hattori A, Hennebry A, Berry CJ, Sharma M, Kambadur R, Nishimura T (2006) Decorin binds myostatin and modulates its activity to muscle cells. Biochem Biophys Res Commun 340(2):675–680. https://doi.org/10.1016/j.bbrc.2005.12.060

Miyamoto Y, Hanna DL, Zhang W, Baba H, Lenz HJ (2016) Molecular Pathways: Cachexia Signaling-A Targeted Approach to Cancer Treatment. Clin Cancer Res 22(16):3999–4004. https://doi.org/10.1158/1078-0432.Ccr-16-0495

Morcavallo A, Buraschi S, Xu SQ, Belfiore A, Schaefer L, Iozzo RV, Morrione A (2014) Decorin differentially modulates the activity of insulin receptor isoform A ligands. Matrix Biol 35:82–90. https://doi.org/10.1016/j.matbio.2013.12.010

Moreth K, Brodbeck R, Babelova A, Gretz N, Spieker T, Zeng-Brouwers J, Pfeilschifter J, Young MF, Schaefer RM, Schaefer L (2010) The proteoglycan biglycan regulates expression of the B cell chemoattractant CXCL13 and aggravates murine lupus nephritis. J Clin Invest 120(12): 4251–4272. https://doi.org/10.1172/JCI42213

Moreth K, Iozzo RV, Schaefer L (2012) Small leucine-rich proteoglycans orchestrate receptor crosstalk during inflammation. Cell Cycle 11(11):2084–2091. https://doi.org/10.4161/cc.20316

Moreth K, Frey H, Hubo M, Zeng-Brouwers J, Nastase MV, Hsieh LT, Haceni R, Pfeilschifter J, Iozzo RV, Schaefer L (2014) Biglycan-triggered TLR-2- and TLR-4-signaling exacerbates the pathophysiology of ischemic acute kidney injury. Matrix Biol 35:143–151. https://doi.org/10.1016/j.matbio.2014.01.010

Morrione A, Neill T, Iozzo RV (2013) Dichotomy of decorin activity on the insulin-like growth factor-I system. FEBS J 280(10):2138–2149. https://doi.org/10.1111/febs.12149

Moscatello DK, Santra M, Mann DM, McQuillan DJ, Wong AJ, Iozzo RV (1998) Decorin suppresses tumor cell growth by activating the epidermal growth factor receptor. J Clin Invest 101(2):406–412

Na Y, Choi JW, Kasala D, Hong J, Oh E, Li Y, Jung SJ, Kim SW, Yun CO (2015) Potent antitumor effect of neurotensin receptor-targeted oncolytic adenovirus co-expressing decorin and Wnt antagonist in an orthotopic pancreatic tumor model. J Control Release 220(Pt B):766–782. https://doi.org/10.1016/j.jconrel.2015.10.015

Nahar K, Absar S, Gupta N, Kotamraju VR, McMurtry IF, Oka M, Komatsu M, Nozik-Grayck E, Ahsan F (2014) Peptide-coated liposomal fasudil enhances site specific vasodilation in pulmonary arterial hypertension. Mol Pharm 11(12):4374–4384. https://doi.org/10.1021/mp500456k

Nash MA, Loercher AE, Freedman RS (1999) In vitro growth inhibition of ovarian cancer cells by decorin: synergism of action between decorin and carboplatin. Cancer Res 59(24):6192–6196

Nash MA, Deavers MT, Freedman RS (2002) The expression of decorin in human ovarian tumors. Clin Cancer Res 8(6):1754–1760

Neill T, Painter H, Buraschi S, Owens RT, Lisanti MP, Schaefer L, Iozzo RV (2012a) Decorin antagonizes the angiogenic network: concurrent inhibition of Met, hypoxia inducible factor 1alpha, vascular endothelial growth factor A, and induction of thrombospondin-1 and TIMP3. J Biol Chem 287(8):5492–5506. https://doi.org/10.1074/jbc.M111.283499

Neill T, Schaefer L, Iozzo RV (2012b) Decorin: a guardian from the matrix. Am J Pathol 181(2): 380–387. https://doi.org/10.1016/j.ajpath.2012.04.029

Neill T, Torres A, Buraschi S, Iozzo RV (2013) Decorin has an appetite for endothelial cell autophagy. Autophagy 9(10):1626–1628. https://doi.org/10.4161/auto.25881

Neill T, Torres A, Buraschi S, Owens RT, Hoek JB, Baffa R, Iozzo RV (2014) Decorin induces mitophagy in breast carcinoma cells via peroxisome proliferator-activated receptor gamma coactivator-1alpha (PGC-1alpha) and mitostatin. J Biol Chem 289(8):4952–4968. https://doi.org/10.1074/jbc.M113.512566

Neill T, Schaefer L, Iozzo RV (2015) Oncosuppressive functions of decorin. Mol Cell Oncol 2(3): e975645. https://doi.org/10.4161/23723556.2014.975645

Neill T, Schaefer L, Iozzo RV (2016) Decorin as a multivalent therapeutic agent against cancer. Adv Drug Deliv Rev 97:174–185. https://doi.org/10.1016/j.addr.2015.10.016

Neill T, Buraschi S, Kapoor A, Iozzo RV (2020) Proteoglycan-driven Autophagy: A Nutrient-independent Mechanism to Control Intracellular Catabolism. J Histochem Cytochem 68(11): 733–746. https://doi.org/10.1369/0022155420937370

Neill T, Kapoor A, Xie C, Buraschi S, Iozzo RV (2021) A functional outside-in signaling network of proteoglycans and matrix molecules regulating autophagy. Matrix Biol 100-101:118–149. https://doi.org/10.1016/j.matbio.2021.04.001

Nili N, Cheema AN, Giordano FJ, Barolet AW, Babaei S, Hickey R, Eskandarian MR, Smeets M, Butany J, Pasterkamp G, Strauss BH (2003) Decorin inhibition of PDGF-stimulated vascular smooth muscle cell function: potential mechanism for inhibition of intimal hyperplasia after balloon angioplasty. Am J Pathol 163(3):869–878. https://doi.org/10.1016/S0002-9440(10)63447-5

Nivison MP, Meier KE (2018) The role of CCN4/WISP-1 in the cancerous phenotype. Cancer Manag Res 10:2893–2903. https://doi.org/10.2147/CMAR.S133915

Nyman MC, Sainio AO, Pennanen MM, Lund RJ, Vuorikoski S, Sundstrom JT, Jarvelainen HT (2015) Decorin in Human Colon Cancer: Localization In Vivo and Effect on Cancer Cell Behavior In Vitro. J Histochem Cytochem 63(9):710–720. https://doi.org/10.1369/0022155415590830

Oh E, Choi IK, Hong J, Yun CO (2017) Oncolytic adenovirus coexpressing interleukin-12 and decorin overcomes Treg-mediated immunosuppression inducing potent antitumor effects in a weakly immunogenic tumor model. Oncotarget 8(3):4730–4746. https://doi.org/10.18632/oncotarget.13972

Parato KA, Senger D, Forsyth PA, Bell JC (2005) Recent progress in the battle between oncolytic viruses and tumours. Nat Rev Cancer 5(12):965–976. https://doi.org/10.1038/nrc1750

Pattingre S, Levine B (2006) Bcl-2 inhibition of autophagy: a new route to cancer? Cancer Res 66(6):2885–2888. https://doi.org/10.1158/0008-5472.CAN-05-4412

Pietraszek-Gremplewicz K, Karamanou K, Niang A, Dauchez M, Belloy N, Maquart FX, Baud S, Brezillon S (2019) Small leucine-rich proteoglycans and matrix metalloproteinase-14: Key partners? Matrix Biol 75-76:271–285. https://doi.org/10.1016/j.matbio.2017.12.006

Reed CC, Gauldie J, Iozzo RV (2002) Suppression of tumorigenicity by adenovirus-mediated gene transfer of decorin. Oncogene 21(23):3688–3695

Reed CC, Waterhouse A, Kirby S, Kay P, Owens RT, McQuillan DJ, Iozzo RV (2005) Decorin prevents metastatic spreading of breast cancer. Oncogene 24(6):1104–1110

Reese SP, Underwood CJ, Weiss JA (2013) Effects of decorin proteoglycan on fibrillogenesis, ultrastructure, and mechanics of type I collagen gels. Matrix Biol 32(7-8):414–423. https://doi.org/10.1016/j.matbio.2013.04.004

Reinboth B, Hanssen E, Cleary EG, Gibson MA (2002) Molecular interactions of biglycan and decorin with elastic fiber components: biglycan forms a ternary complex with tropoelastin and microfibril-associated glycoprotein 1. J Biol Chem 277(6):3950–3957. https://doi.org/10.1074/jbc.M109540200

Reszegi A, Horvath Z, Karaszi K, Regos E, Postnikova V, Tatrai P, Kiss A, Schaff Z, Kovalszky I, Baghy K (2020) The Protective Role of Decorin in Hepatic Metastasis of Colorectal Carcinoma. Biomolecules 10(8):doi:10.3390/biom10081199

Ruoslahti E, Yamaguchi Y (1991) Proteoglycans as modulators of growth factor activities. Cell 64(5):867–869

Sainio A, Nyman M, Lund R, Vuorikoski S, Bostrom P, Laato M, Bostrom PJ, Jarvelainen H (2013) Lack of decorin expression by human bladder cancer cells offers new tools in the therapy of urothelial malignancies. PLoS One 8(10):e76190. https://doi.org/10.1371/journal.pone.0076190

Salomaki HH, Sainio AO, Soderstrom M, Pakkanen S, Laine J, Jarvelainen HT (2008) Differential expression of decorin by human malignant and benign vascular tumors. J Histochem Cytochem 56(7):639–646. https://doi.org/10.1369/jhc.2008.950287

Santiago-Garcia J, Kodama T, Pitas RE (2003) The class A scavenger receptor binds to proteoglycans and mediates adhesion of macrophages to the extracellular matrix. J Biol Chem 278(9):6942–6946. https://doi.org/10.1074/jbc.M208358200

Santra M, Skorski T, Calabretta B, Lattime EC, Iozzo RV (1995) De novo decorin gene expression suppresses the malignant phenotype in human colon cancer cells. Proc Natl Acad Sci U S A 92(15):7016–7020

Santra M, Eichstetter I, Iozzo RV (2000) An anti-oncogenic role for decorin. Down-regulation of ErbB2 leads to growth suppression and cytodifferentiation of mammary carcinoma cells. J Biol Chem 275(45):35153–35161

Santra M, Reed CC, Iozzo RV (2002) Decorin binds to a narrow region of the epidermal growth factor (EGF) receptor, partially overlapping but distinct from the EGF-binding epitope. J Biol Chem 277(38):35671–35681

Schaefer L (2014) Complexity of danger: the diverse nature of damage-associated molecular patterns. J Biol Chem 289(51):35237–35245. https://doi.org/10.1074/jbc.R114.619304

Schaefer L, Dikic I (2021) Autophagy: Instructions from the extracellular matrix. Matrix Biol. https://doi.org/10.1016/j.matbio.2021.06.002

Schaefer L, Iozzo RV (2012) Small leucine-rich proteoglycans, at the crossroad of cancer growth and inflammation. Curr Opin Genet Dev 22(1):56–57. https://doi.org/10.1016/j.gde.2011.12.002

Schaefer L, Tsalastra W, Babelova A, Baliova M, Minnerup J, Sorokin L, Grone HJ, Reinhardt DP, Pfeilschifter J, Iozzo RV, Schaefer RM (2007) Decorin-mediated regulation of fibrillin-1 in the kidney involves the insulin-like growth factor-I receptor and Mammalian target of rapamycin. Am J Pathol 170(1):301–315

Schonherr E, Sunderkotter C, Iozzo RV, Schaefer L (2005) Decorin, a novel player in the insulin-like growth factor system. J Biol Chem 280(16):15767–15772

Scott JE (1996) Proteodermatan and proteokeratan sulfate (decorin, lumican/fibromodulin) proteins are horseshoe shaped. Implications for their interactions with collagen. Biochemistry 35(27): 8795–8799. https://doi.org/10.1021/bi960773t

Scott PG, McEwan PA, Dodd CM, Bergmann EM, Bishop PN, Bella J (2004) Crystal structure of the dimeric protein core of decorin, the archetypal small leucine-rich repeat proteoglycan. Proc Natl Acad Sci U S A 101(44):15633–15638. https://doi.org/10.1073/pnas.0402976101

Seidler DG (2012) The galactosaminoglycan-containing decorin and its impact on diseases. Curr Opin Struct Biol 22(5):578–582. https://doi.org/10.1016/j.sbi.2012.07.012

Seidler DG, Goldoni S, Agnew C, Cardi C, Thakur ML, Owens RT, McQuillan DJ, Iozzo RV (2006) Decorin protein core inhibits in vivo cancer growth and metabolism by hindering epidermal growth factor receptor function and triggering apoptosis via caspase-3 activation. J Biol Chem 281(36):26408–26418

Seidler DG, Mohamed NA, Bocian C, Stadtmann A, Hermann S, Schafers K, Schafers M, Iozzo RV, Zarbock A, Gotte M (2011) The role for decorin in delayed-type hypersensitivity. J Immunol 187(11):6108–6119. https://doi.org/10.4049/jimmunol.1100373

Sharma M, McFarlane C, Kambadur R, Kukreti H, Bonala S, Srinivasan S (2015) Myostatin: expanding horizons. IUBMB Life 67(8):589–600. https://doi.org/10.1002/iub.1392

Shintani K, Matsumine A, Kusuzaki K, Morikawa J, Matsubara T, Wakabayashi T, Araki K, Satonaka H, Wakabayashi H, Iino T, Uchida A (2008) Decorin suppresses lung metastases of murine osteosarcoma. Oncol Rep 19(6):1533–1539

Skandalis SS, Kletsas D, Kyriakopoulou D, Stavropoulos M, Theocharis DA (2006) The greatly increased amounts of accumulated versican and decorin with specific post-translational modifications may be closely associated with the malignant phenotype of pancreatic cancer. Biochim Biophys Acta 1760(8):1217–1225. https://doi.org/10.1016/j.bbagen.2006.03.021

Smid-Koopman E, Blok LJ, Chadha-Ajwani S, Helmerhorst TJ, Brinkmann AO, Huikeshoven FJ (2000) Gene expression profiles of human endometrial cancer samples using a cDNA-expression array technique: assessment of an analysis method. Br J Cancer 83(2):246–251. https://doi.org/10.1054/bjoc.2000.1238

Sofeu Feugaing DD, Gotte M, Viola M (2013) More than matrix: the multifaceted role of decorin in cancer. Eur J Cell Biol 92(1):1–11. https://doi.org/10.1016/j.ejcb.2012.08.004

Toba M, Alzoubi A, O'Neill K, Abe K, Urakami T, Komatsu M, Alvarez D, Jarvinen TA, Mann D, Ruoslahti E, McMurtry IF, Oka M (2014) A novel vascular homing peptide strategy to selectively enhance pulmonary drug efficacy in pulmonary arterial hypertension. Am J Pathol 184(2):369–375. https://doi.org/10.1016/j.ajpath.2013.10.008

Tralhao JG, Schaefer L, Micegova M, Evaristo C, Schonherr E, Kayal S, Veiga-Fernandes H, Danel C, Iozzo RV, Kresse H, Lemarchand P (2003) In vivo selective and distant killing of cancer cells using adenovirus-mediated decorin gene transfer. Faseb J 17(3):464–466

Troup S, Njue C, Kliewer EV, Parisien M, Roskelley C, Chakravarti S, Roughley PJ, Murphy LC, Watson PH (2003) Reduced expression of the small leucine-rich proteoglycans, lumican, and decorin is associated with poor outcome in node-negative invasive breast cancer. Clin Cancer Res 9(1):207–214

Tufvesson E, Westergren-Thorsson G (2002) Tumour necrosis factor-alpha interacts with biglycan and decorin. FEBS Lett 530(1-3):124–128. https://doi.org/10.1016/s0014-5793(02)03439-7

Urakami T, Jarvinen TA, Toba M, Sawada J, Ambalavanan N, Mann D, McMurtry I, Oka M, Ruoslahti E, Komatsu M (2011) Peptide-directed highly selective targeting of pulmonary arterial hypertension. Am J Pathol 178(6):2489–2495. https://doi.org/10.1016/j.ajpath.2011.02.032

Vial C, Gutierrez J, Santander C, Cabrera D, Brandan E (2011) Decorin interacts with connective tissue growth factor (CTGF)/CCN2 by LRR12 inhibiting its biological activity. J Biol Chem 286(27):24242–24252. https://doi.org/10.1074/jbc.M110.189365

Villanueva A, Newell P, Chiang DY, Friedman SL, Llovet JM (2007) Genomics and signaling pathways in hepatocellular carcinoma. Semin Liver Dis 27(1):55–76. https://doi.org/10.1055/s-2006-960171

von Marschall Z, Fisher LW (2010) Decorin is processed by three isoforms of bone morphogenetic protein-1 (BMP1). Biochem Biophys Res Commun 391(3):1374–1378. https://doi.org/10.1016/j.bbrc.2009.12.067

Wang RC, Wei Y, An Z, Zou Z, Xiao G, Bhagat G, White M, Reichelt J, Levine B (2012) Akt-mediated regulation of autophagy and tumorigenesis through Beclin 1 phosphorylation. Science 338(6109):956–959. https://doi.org/10.1126/science.1225967

Weber IT, Harrison RW, Iozzo RV (1996) Model structure of decorin and implications for collagen fibrillogenesis. J Biol Chem 271(50):31767–31770

Wei Y, Zou Z, Becker N, Anderson M, Sumpter R, Xiao G, Kinch L, Koduru P, Christudass CS, Veltri RW, Grishin NV, Peyton M, Minna J, Bhagat G, Levine B (2013) EGFR-mediated Beclin 1 phosphorylation in autophagy suppression, tumor progression, and tumor chemoresistance. Cell 154(6):1269–1284. https://doi.org/10.1016/j.cell.2013.08.015

Whittaker S, Marais R, Zhu AX (2010) The role of signaling pathways in the development and treatment of hepatocellular carcinoma. Oncogene 29(36):4989–5005. https://doi.org/10.1038/onc.2010.236

Wiberg C, Klatt AR, Wagener R, Paulsson M, Bateman JF, Heinegard D, Morgelin M (2003) Complexes of matrilin-1 and biglycan or decorin connect collagen VI microfibrils to both collagen II and aggrecan. J Biol Chem 278(39):37698–37704. https://doi.org/10.1074/jbc.M304638200

Wu IC, Wu DC, Huang CC, Lin HS, Chen YK, Tsai HJ, Lu CY, Chou SH, Chou YP, Li LH, Tai SY, Wu MT (2010) Plasma decorin predicts the presence of esophageal squamous cell carcinoma. Int J Cancer 127(9):2138–2146. https://doi.org/10.1002/ijc.25239

Xaus J, Comalada M, Cardo M, Valledor AF, Celada A (2001) Decorin inhibits macrophage colony-stimulating factor proliferation of macrophages and enhances cell survival through induction of p27(Kip1) and p21(Waf1). Blood 98(7):2124–2133

Xu W, Neill T, Yang Y, Hu Z, Cleveland E, Wu Y, Hutten R, Xiao X, Stock SR, Shevrin D, Kaul K, Brendler C, Iozzo RV, Seth P (2015) The systemic delivery of an oncolytic adenovirus expressing decorin inhibits bone metastasis in a mouse model of human prostate cancer. Gene Ther 22(3):247–256. https://doi.org/10.1038/gt.2014.110

Yamaguchi Y, Ruoslahti E (1988) Expression of human proteoglycan in Chinese hamster ovary cells inhibits cell proliferation. Nature 336(6196):244–246

Yamaguchi Y, Mann DM, Ruoslahti E (1990) Negative regulation of transforming growth factor-beta by the proteoglycan decorin. Nature 346(6281):281–284

Yamaguchi A, Taniguchi M, Hori O, Ogawa S, Tojo N, Matsuoka N, Miyake S, Kasai K, Sugimoto H, Tamatani M, Yamashita T, Tohyama M (2002) Peg3/Pw1 is involved in p53-mediated cell death pathway in brain ischemia/hypoxia. J Biol Chem 277(1):623–629. https://doi.org/10.1074/jbc.M107435200

Yang Y, Xu W, Neill T, Hu Z, Wang CH, Xiao X, Stock SR, Guise T, Yun CO, Brendler CB, Iozzo RV, Seth P (2015) Systemic Delivery of an Oncolytic Adenovirus Expressing Decorin for the Treatment of Breast Cancer Bone Metastases. Hum Gene Ther 26(12):813–825. https://doi.org/10.1089/hum.2015.098

Young CS, Kitamura M, Hardy S, Kitajewski J (1998) Wnt-1 induces growth, cytosolic beta-catenin, and Tcf/Lef transcriptional activation in Rat-1 fibroblasts. Mol Cell Biol 18(5):2474–2485. https://doi.org/10.1128/mcb.18.5.2474

Zeng-Brouwers J, Beckmann J, Nastase MV, Iozzo RV, Schaefer L (2014) De novo expression of circulating biglycan evokes an innate inflammatory tissue response via MyD88/TRIF pathways. Matrix Biol 35:132–142. https://doi.org/10.1016/j.matbio.2013.12.003

Zhang G, Chen S, Goldoni S, Calder BW, Simpson HC, Owens RT, McQuillan DJ, Young MF, Iozzo RV, Birk DE (2009) Genetic evidence for the coordinated regulation of collagen fibrillogenesis in the cornea by decorin and biglycan. J Biol Chem 284(13):8888–8897. https://doi.org/10.1074/jbc.M806590200

Zhang W, Ge Y, Cheng Q, Zhang Q, Fang L, Zheng J (2018) Decorin is a pivotal effector in the extracellular matrix and tumour microenvironment. Oncotarget 9(4):5480–5491. https://doi.org/10.18632/oncotarget.23869

Zhu JX, Goldoni S, Bix G, Owens RT, McQuillan DJ, Reed CC, Iozzo RV (2005) Decorin evokes protracted internalization and degradation of the epidermal growth factor receptor via caveolar endocytosis. J Biol Chem 280(37):32468–32479

Zhu J, Li Y, Shen W, Qiao C, Ambrosio F, Lavasani M, Nozaki M, Branca MF, Huard J (2007) Relationships between transforming growth factor-beta1, myostatin, and decorin: implications for skeletal muscle fibrosis. J Biol Chem 282(35):25852–25863. https://doi.org/10.1074/jbc.M704146200

Zwirner NW, Croci DO, Domaica CI, Rabinovich GA (2010) Overcoming the hurdles of tumor immunity by targeting regulatory pathways in innate and adaptive immune cells. Curr Pharm Des 16(3):255–267. https://doi.org/10.2174/138161210790170175

Chapter 3
The Mystery of Syndecan-1 in Tumor Development and Progression

Ilona Kovalszky, Kornélia Baghy, Andrea Reszegi, Péter Hollósi, Anders Hjerpe, and Katalin Dobra

Abstract Syndecan-1, a cell surface heparan sulfate proteoglycan, participates in a multitude of cellular processes, and it acts as a "fine tuner of cellular signaling." Syndecan-1 is present in most epithelial tissues and on the cellular level, it is critical for tumor cell differentiation, adhesion, growth, and invasion. Its expression carries valuable information related to the clinical outcome and prognosis of most solid tumors. Despite considerable efforts to decipher its significance in tumor development and progression, current data point toward both inhibitory and promoting effects, sometimes in a contradictory manner. A large body of evidence shows the importance of the aberrant cytoplasmic-, nuclear-, stromal localization of soluble/shed syndecan-1. In this chapter, we highlight the structure and various interactions of syndecan-1 in relation to its cellular localization and its impact on the prognosis of individual tumor types.

I. Kovalszky · K. Baghy · A. Reszegi
Department of Pathology & Experimental Cancer Research, Semmelweis University, Budapest, Hungary

P. Hollósi
Department of Pathology & Experimental Cancer Research, Semmelweis University, Budapest, Hungary

Tumor Progression Research Group of Joint Research Organization of Hungarian Academy of Sciences and Semmelweis University, Budapest, Hungary

A. Hjerpe · K. Dobra (✉)
Department of Laboratory Medicine, Division of Pathology, Karolinska University Hospital/Karolinska Institutet, Stockholm, Sweden
e-mail: katalin.dobra@ki.se

I. Kovalszky et al. (eds.), *The Extracellular Matrix and the Tumor Microenvironment*, Biology of Extracellular Matrix 11,
https://doi.org/10.1007/978-3-030-99708-3_3

3.1 Structure and Localization of Syndecans

Syndecan-1, the first member of the syndecan family, was discovered and sequenced in the late 90s (Mali et al. 1990). Its description was soon followed by that of the three other members of the family. Virtually, all epithelial tissues express syndecan-1, typically located at the basolateral surface. Syndecan-1 has been detected in cells of the skin, oral cavity, lips, tonsils, salivary glands; in the epithelium of the gastrointestinal system including the esophagus, stomach, intestine, gall bladder, pancreatic duct cells, and hepatocytes; in the urogenital system including the kidney, urinary bladder, prostate, seminal vesicle, epididymis; in the respiratory system including the lungs and bronchi; and in the reproductive organs of females including the breasts, uterus, and placenta (Kind et al. 2019). The members of the syndecan family are phylogenetically highly conserved transmembrane proteoglycans, consisting of extracellular (EC), transmembrane (TM), and intracellular cytoplasmic (CD) domains. Meanwhile, the constant C1 and C2 regions of the intracellular domains show a high degree of homology among the family members, the extracellular domains display considerable heterogeneity.

The extracellular domains of syndecan-1 and syndecan-3 carry both heparan sulfate (HS) and chondroitin sulfate (CS) glycosaminoglycan (GAG) side chains (Fig. 3.1), while syndecan-2 and -4 are exclusively glycanated by HS chains. These negatively charged GAG chains mediate most of the syndecan-1 interactions with a

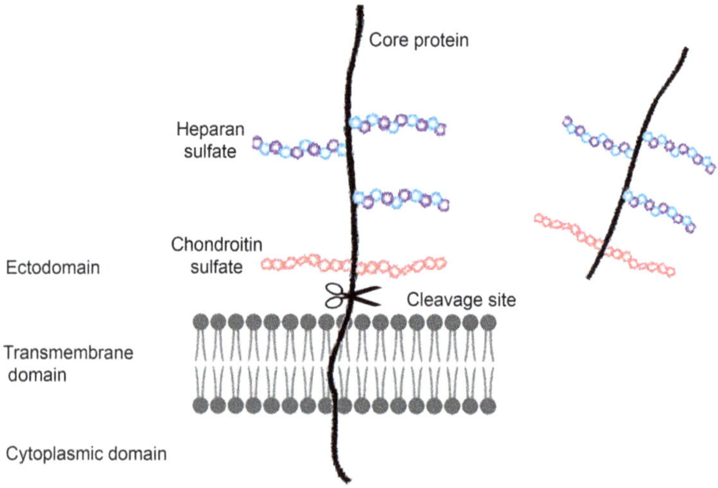

Fig. 3.1 Syndecan-1 is a transmebrane proteoglycan. According the protein is built up by intracellular, transmembrane, and extracellular domains. The latter carries two chondroitin sulfate and three heparan sulfate chains. In consequence of the negative charge of the sugar chains, syndecan-1 can establish interactions with several, basic molecules participating this way in cell regulation. The extracellular domain has more cleavage sites for proteases resulting the liberation of the EC, called shedding, that provide additional tasks for this domain. Intracellular domain, internalization of the whole molecule, or the shedded domain can exert further effect in the cytoplasm or in the cell nuclei

plethora of basic molecules. Syndecans may bind growth factors and cytokines located in the extracellular matrix (ECM), as well as cell surface molecules such as integrins. So far, around a hundred interacting partners have been identified and this number is ever-increasing (Stepp et al. 2015). In many of these interactions, syndecan-1 acts as a ternary receptor, facilitating the signaling of growth factors through their specific receptors, this way supporting the outside-in signaling from the microenvironment. To communicate with the microenvironment, besides its GAG chains, syndecan-1 also utilizes its EC domain. The cytoplasmic domain establishes interactions via its constant and variable regions, associating with cortactin, tubulin, and signaling molecules such as CASK, Fyn, Src, NF1, syntenin, synbindin, α6β4, etc. The four syndecan-family members also interact with each other, thereby creating an enormous diversity and complexity to their functions (Roper et al. 2012; Kleiser and Nyström 2020; Gondelaud and Ricard-Blum 2019; Wang et al. 2014a).

3.2 Different Locations, Different Partners, Different Functions

According to the classical description, syndecan-1 resides on the cell surface as a transmembrane proteoglycan, but it can also be found in the nuclear compartment, the cytoplasm, and the ECM, and as a soluble shed biomolecule in body fluids. Moreover, recent studies reveal that its functions are strictly related to its localization.

One of the best-known characteristics of syndecan-1 is the **shedding** of its EC domain from the cell surface. Although this, as a part of protein renewal, is a physiological process (Subramanian et al. 1997; Jalkanen et al. 1987), several pathological events go together with increased shedding. Enzymes participating in shedding include matrix metalloproteinases (MMPs), and disintegrin and metalloproteinase with thrombospondin motifs (ADAMTs), thrombin, plasmin, all of them having individual cleavage sites (Bertrand and Bollmann 2019). Besides the action of these proteases, heparanase (Rangarajan et al. 2020), the only mammalian heparin liase enzyme, also supports the shedding of syndecan-1, in part by shortening its GAG chains, and by stimulating the production of MMP9 through the ERK pathway (Purushothaman et al. 2008). Signaling molecules attached to the shed syndecan-1 are also removed from the cell surface, which suppresses their actions on the cells, while the factors dwell in the ECM. Increased amount of circulating syndecan-1 is an active participant of several pathological events, including inflammation (Gao et al. 2015), sepsis (Ostrowski et al. 2013), coagulopathies and pulmonary embolism (Lehnert et al. 2017) cardiomyopathy (Vanhoutte et al. 2007a), lung diseases (Li et al. 2019), and tumor invasion (Szarvas et al. 2018b).

The presence and the role of heparan sulfate in the **cell nucleus** has been debated when in the 80s, Ishihara and his coworkers published their results, demonstrating the presence of heparan sulfate in the nucleus of hepatocytes (Ishihara et al. 1986; Fedarko and Conrad 1986). Their results did not receive appreciation for a long time; the first supporting publications were published at the turn of the twenty-first century. It has been shown that heparan sulfate, isolated from human livers inhibits the activity of nuclear topoisomerase I, in cooperation with bFGF (Kovalszky et al. 1998). Later, with the advent of confocal microscopy, the entire syndecan-1 molecule has been shown to translocate to the nucleus in a tubulin-mediated manner (Brockstedt et al. 2002). As HS chains of syndecan-1 bind bFGF (Filla et al. 1998), the growth factor also coprecipitates with syndecan-1 in the nuclear extract. Nuclear heparan sulfate and bFGF facilitate each other's nuclear uptake (Hsia et al. 2003). The localization of syndecan-1 in the nucleus is a characteristic feature of tumor cells. This implies that it is in very low representation in the nuclei of normal cells. Nuclear HS interferes with the action of topoisomerase I and II (Kovalszky et al. 1998), as well as AP1, and Ets1 (Dudás et al. 2000; Hollósi et al. 2020). Shed syndecan-1, lacking the transmembrane and cytoplasmic domains but with preserved HS chains has also been detected in the nucleus, indicating that this proteoglycan can internalize and translocate there, promoting histone acetylation (Stewart et al. 2015). Next, it was discovered that the nuclear syndecan-1 level is regulated by heparanase, and HS chains most probably interfere with the transcriptional activity of genes in tumor cells (Purushothaman et al. 2011a). The presence of heparanase in the nucleus results in fast downregulation of nuclear syndecan-1, and consecutive enhancement of histone acetyltransferase (HAT), known to be inhibited by HS (Purushothaman et al. 2011b). This mechanism assigns an aggressive phenotype in myelomas (Purushothaman et al. 2011b). Nuclear localization was detected in malignant mesothelioma, adenocarcinoma of the breast, lung cancer, hepatocellular carcinoma, and neuroblastoma (Zong et al. 2009; Kovalszky et al. 1998; Chen and Sanderson 2009; Stewart et al. 2015).

The functions of syndecan-1 related to the nucleus were separated from the functions related to the cell surface by transfecting a fibrosarcoma cell line with various syndecan-1 constructs, one with a preserved and another with a deleted RMKKK nuclear localization signal. Nuclear translocation of syndecan-1 hampered the proliferation of fibrosarcoma cells compared to the mutant lacking the nuclear localization signal. The growth inhibitory effect of nuclear syndecan-1 was accompanied by a significant accumulation of cells in the G0/G1 phase, which indicated a possible G1/S phase arrest. By combining multiple, unsupervised global transcriptome and proteome profiling approaches with functional assays, the molecular mechanisms that governed nuclear translocation and its related functions could be followed. The transforming growth factor β (TGF-β) pathway was activated by nuclear syndecan-1, and expression levels of three genes were significantly altered with the deletion of the nuclear localization signal: early growth response 1 (EGR-1), never-in-mitosis gene a-related kinase 11 (NEK11), and dedicator of cytokinesis 8 (DOCK8). Moreover, these genes were coupled to growth and cell-cycle regulation. Nuclear translocation of syndecan-1 influenced the activity of several other transcription factors including E2F, NFκβ, and OCT-1 (Szatmári et al. 2017). Mapping the nuclear syndecan-1 interactome by proteomic analysis after coimmunoprecipitation in a malignant mesothelioma cell line identified many interacting proteins, and pathways related to cell proliferation, RNA synthesis, splicing, and transport (Kumar-Singh et al. 2020). Furthermore, loss-and gain-of-function experiments showed that syndecan-1 influences RNA levels in mesothelioma cells. The results identify a proteomic map of syndecan-1 nuclear interactors in a mesothelioma cell line and suggest a previously unknown role for syndecan-1 in RNA biogenesis.

Nuclear syndecan-1 also regulates the epithelial–mesenchymal plasticity in tumor cells. Loss of nuclear syndecan-1 associated with cellular elongation and an E-cadherin-to-N-cadherin switch during TGF-β1-induced epithelial–mesenchymal transition (EMT) in human lung adenocarcinoma cells (Kumar-Singh et al. 2021), and nuclear translocation of syndecan-1 contributed to the repression of mesenchymal and invasive properties of human B6FS fibrosarcoma cells.

Increasing number of publications call attention to the potential **of syndecan-1 in the regulation of intermediary metabolism**. In the liver, syndecan-1 functions as a cell surface receptor of lipoproteins on hepatocytes, regulating the uptake of remnant lipoproteins (Stanford et al. 2009; Wilsie et al. 2006). It also functions as a triglyceride receptor (Foley et al. 2013), and it is involved in their intracellular clearance (Reszegi et al. 2021b). Overexpression of syndecan-1 downregulates the amount of fatty acid synthase (FASN), one of the most critical player of lipid synthesis in the liver, and protects against de novo lipid production (Reszegi et al. 2021b). This raises the possibility that syndecan-1 can interfere with metabolic events taking place in the cytoplasm, such as glucose and lipid metabolism or mitochondrial functions. Experimental evidence indicates that syndecans are also implicated in the whole-body energy metabolism (De Luca et al. 2010).

The effect of syndecan-1 also depends on the cell types that carry the proteoglycan. The majority of epithelial cells express syndecan-1 on their cell surface. The functions and cellular interactions described above can be detected on all epithelial cells, but differences can also be seen. For example, syndecan-1 as a remnant lipid receptor is characteristic for hepatocytes (Stanford et al. 2009). By recruiting and binding inflammatory cytokines, shed syndecan-1 creates a cytokine gradient. As a result of its interaction with E- and P-selectin on the endothelium of blood vessels, syndecan-1 hinders the rolling of leukocytes and their subsequent extravasation (Voyvodic et al. 2014). Experiments carried out on SDC-1$^{-/-}$ mice revealed that the actions described above make syndecan-1 essential in limiting lung inflammation (Brauer et al. 2016) (Gopal 2020). Furthermore, besides attenuating inflammation, syndecan-1 protects against apoptosis of bronchial epithelial cells by facilitating Met signaling (Brauer et al. 2016). Circulating exosomes were also found to carry syndecan-1 cargo. Delivery of these exosomes into the lung attenuates the lipopolysaccharide (LPS)-induced inflammation and decreases edema via acting on pulmonary microvascular endothelial cells (Gopal 2020).

Syndecan-1 also supports the optimal resolution of myocardial infarction. It decreases the inflammatory reaction at the injured area, and attenuates the activity of MMPs, in this way protecting against type I collagen degradation, and the adverse infarct healing with concomitant cardiac dilatation (Vanhoutte et al. 2007b).

Presence of syndecan-1 regulates the motility of differentiated M2 macrophages, promotes the clearance of the inflammatory cells, and acts against the development of arteriosclerotic plaques (Voyvodic et al. 2014; Angsana et al. 2015). In the liver, shed syndecan-1 also protects against the development of fibrosis, by removing TGFβ1 and thrombospondin from the vicinity of the myofibroblasts (Regős et al. 2018).

HS chains of syndecan-1 can promote the uptake of viruses. One of the best-known examples is that of the hepatitis C virus (HCV) (Grigorov et al. 2017; Zhang et al. 2017). The virus binds to the HS chains of syndecan-1 and subsequently interacts with tetraspanin (CD81), which indicates the cooperation of the two molecules in the virus uptake. The virus in the infected hepatocytes further enhances the expression of syndecan-1, resulting in enhanced uptake of LDL and VLDL. Although HCV is capable of utilizing the HS chains of other proteoglycans as well, syndecan-1 as the major proteoglycan of hepatocytes is the most potential candidate to promote HCV entry (Shi et al. 2013). Human papilloma virus (HPV) 16 and 18 dwelling in the ECM on heparanase cleaved HS chains of syndecan-1 efficiently infect new epithelial cells of the uterine cervix (Surviladze et al. 2015). Besides CD4, syndecan-1 also acts as an attachment receptor of human immunodeficiency virus (HIV) on macrophages (Saphire et al. 2001).

3.3 Syndecan-1 and Signal Transduction

HS chains of syndecan-1 can bind growth factors such as bFGF2, hepatocyte growth factor (HGF), platelet-derived growth factor (PDGF), epidermal growth factor (EGF), as well as cytokines, interleukins, etc. These connections facilitate the interactions of growth factors with their cell surface receptors, initiating outside-in signaling. The best-known examples for these mechanisms are receptor tyrosine kinases, including EGF, cMet, FGF, PDGF VEGF receptors, utilizing the RAS-Raf-Erk, and the RAS-PIK3C-Akt-mTor pathways (Montor et al. 2018). The ligand–receptor interactions initiate signaling cascades regulating a plethora of cellular functions (Gross and Rotwein 2017). Besides the tyrosine kinase receptors, syndecan-1 is also involved in the Wnt signaling (Ren et al. 2018; Pataki et al. 2015). Furthermore, syndecan-1 and -4 with the cooperation of protein kinase Cα (PKCα) decreases calcium influx and cytosolic calcium level, by inhibition of TRCP calcium channels, a mechanism that regulates cell adhesion (Gopal et al. 2015). The effect of syndecan-1 is tissue specific and it depends on the variety of ligands and receptors present in a particular cell type. Thus, syndecan-1 orchestrates and regulates a plethora of interactions both in physiological and pathological conditions.

3.4 Syndecan-1 and Cancer; Does it Protect or Promote?

Soon after the cloning of human syndecan-1, Jalkanen and coworkers discovered that syndecan-1 inhibits the malignant transformation of human keratinocytes (Inki et al. 1994b). Since that time several experimental and clinical studies were carried out to find the mechanisms whereby syndecan-1 may inhibit or promote the development and progression of various malignant tumors. As syndecan-1 is a proteoglycan, the implication of GAG chains became one of the first potential candidates for this task. In contrast with the protein core, GAG chains and in particular HS chains, which are further modified by sulfation have a high variability (Sanderson et al. 1994; Dennissen et al. 2002; Mundhenke et al. 2002). The sulfated and acetylated regions form separated domain structures further supporting the versatility of the sugar chains (Lindahl and Li 2009). Analysis of GAG–protein interactions revealed that several proteins require well-defined HS structures for binding. Thus, different HS structures on the same core protein differently affect the actual function of the molecule (Kjellén and Lindahl 2018). This implies that syndecan-1 influences not only the behavior of the tumor cells, but also the cells in the tumor stroma, and the outcome depends on the tumor type and cellular localizations (Su et al. 2007; Ahmed Haji Omar et al. 2013; Handra-Luca 2020). Aberrant localization of syndecan-1 in the tumor cells also influences their behavior. Although the mechanism of syndecan-1 retention in the cytoplasm is not clearly understood, it is generally related to worse outcome.

Enhancement of syndecan-1 shedding is a typical feature of cancers. Sheddases, participating in this activity, are the same as those working under normal conditions; however, they act in an unregulated fashion. They remove the functional extracellular domain of syndecan-1 from the cell surface. The regulatory factors retained on the HS chains will interact with the stromal cells, interfere with their signaling, facilitate tumor angiogenesis, and proliferation. Such matrix bound syndecan-1 is capable of trapping migrating tumor cells, this way supporting the establishment of a tumor cell niche (Rangarajan et al. 2020; Manon-Jensen et al. 2010). Shed syndecan-1 in the circulation is a potential indicator of chemotherapy resistance (Szarvas et al. 2018a). Epithelial–mesenchymal transition (EMT), an indispensable process for tumor invasion, has been associated with syndecan-1. In prostate carcinoma, Zeb1 transcription factor crucial for the EMT process binds to the promoter region of syndecan-1 and silences its expression. It seems that the EMT process related to syndecan-1 does not necessarily involve TGFβ1 and vice versa, the canonical TGFβ1-initiated EMT can also be developed without the involvement of syndecan-1 (Couchman 2021; Yang et al. 2020). Syndecan-1 can be detected in the circulation as a component of the extracellular microvesicles that can discriminate between high- and low-grade gliomas (Indira Chandran et al. 2019). The concentration of circulating syndecan-1 positive myeloma microparticles correlates with the tumor burden giving a chance for individual monitoring of multiple myeloma progression (Krishnan et al. 2016; Yang et al. 2020).

3.5 miR Regulation of Syndecan-1 Expression in Tumors

An increasing number of publications report that the function of syndecan-1 is regulated by microRNAs (miR) in various cancers. Their effects can be stimulatory or inhibitory, depending on both the types of tumors and the miRNA. Downregulation of syndecan-1 by liver-derived miR 122–5p enhances mammary cancer motility (Uen et al. 2018). miR 10a via binding to the syndecan-1 3′ untranslated region supports the proliferation and migration of squamous carcinoma cells in the skin (Xiong et al. 2020). Binding of the same miR in MDA-MB-231 mammary cancer promotes cell proliferation and migration capacities by upregulating Rho-kinase activity, concluding to cytoskeleton modulation and downregulation of E-cadherin (Ibrahim et al. 2012). In case of prostate cancer, syndecan-1 can facilitate the progression of the tumor into a more aggressive phenotype, by cooperating with miR126 and miR149 (Fujii et al. 2015a). On the contrary, downregulation of syndecan-1 by the same miR results in the differentiation of urothelial cancer (Fujii et al. 2015b). Increased expression of miR302 inhibits ovarian cancer cell proliferation by targeting syndecan-1 (Guo et al. 2015). Overexpression of syndecan-1 in pancreatic cancer is related to poor prognosis and EMT. Its downregulation by miR494 inhibited tumor progression. It seems that not only miRs regulate syndecan-1 expression, but it works the other way around, as well. Syndecan-1 is able to influence the selection of exosome's miR cargo, secreted by lung cancer cells. Furthermore, depletion of syndecan-1 facilitates the production of oncogenic miR (Parimon et al. 2018). MiR-331-3p miRNA maturation, mediated by syndecan-1, promotes EMT of prostate cancer (Fujii et al. 2016).

3.6 Syndecan-1 Is a Differentiation Marker with a Prognostic Value in Solid Tumors

Syndecan-1 expression correlates with tumor cell differentiation and prognosis in solid tumors. Its cell surface reactivity serves as a differentiation marker, and it is present in all epithelial malignancies, whereas its expression is lower in mesenchymal and hematological malignancies and in less differentiated tumor components (Fig. 3.2).

The normal syndecan-1 level and its cellular localization in a particular tissue are crucial to understand the sequential changes involved in malignant transformation, tumor progression, and tumor dissemination. Experimental overexpression of full-length syndecan-1 enhances cell–ECM cohesion and restricts cell migration (Zong et al. 2011), whereas the loss of the syndecan-1 ectodomain from the cell

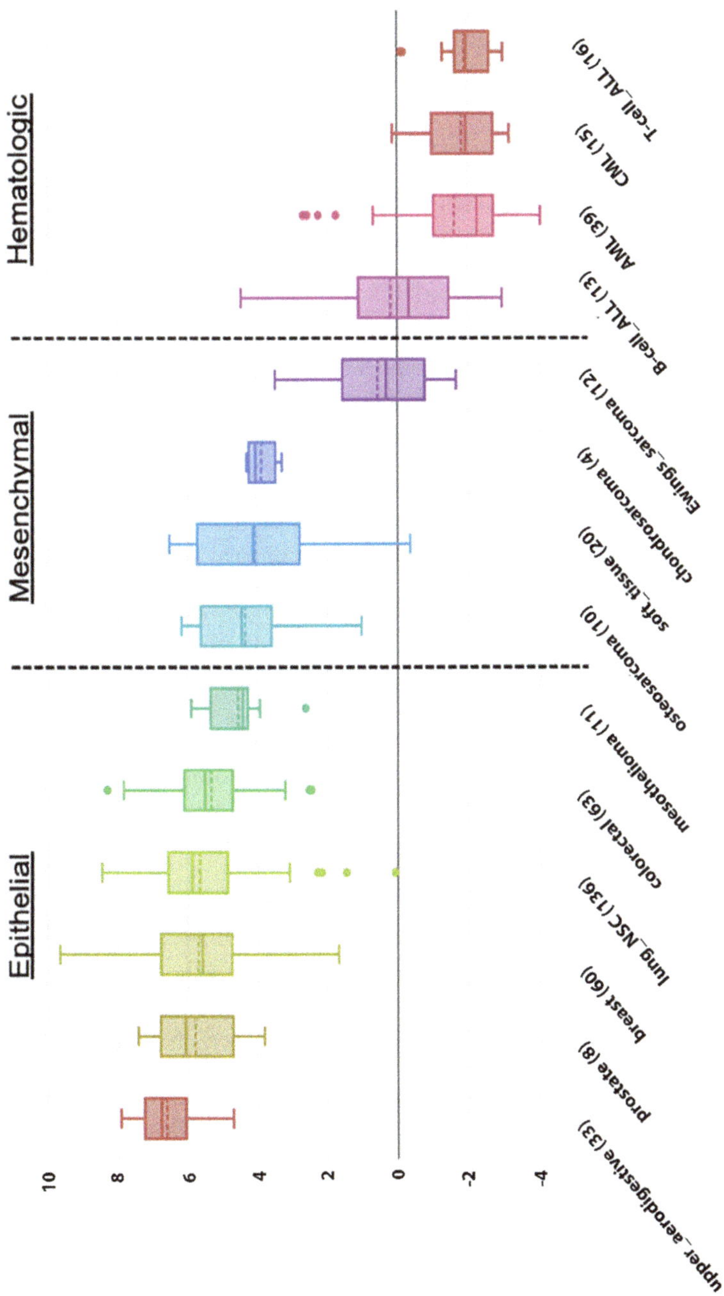

Fig. 3.2 Relative syndecan-1 mRNA expression in tumor cell lines. Data are obtained from the Broad Institute Cancer Cell Line Encyclopedia (https://portals.broadinstitute.org/ccle). Numbers within parenthesis represent the number of analyzed cell lines within the same tumor type

surface increases the migratory capacity of tumor cells (Teng et al. 2012). Thus, cell bound, stromal, and shed syndecan-1 carry tissue-specific prognostic information in individual tumor types. The tremendous body of literature related to the involvement of syndecan-1 in various cancers is seemingly contradictory but considering the localization of syndecan-1, in addition to its expression level might lead to a better understanding of the complex regulatory mechanisms and related prognostic significance in each tumor type.

The expression of **cell surface syndecan-1** in tumor tissue is context-specific. For instance, compared to normal epithelial cells, decreased syndecan-1 expression has been found during malignant transformation, and reduced cell membrane syndecan-1 immunoreactivity was observed in many epithelial malignancies connected to various stages of tumor progression.

Syndecan-1 present in **the stromal component** of different malignant tumors generally indicates poor prognosis through the promotion of tumor cell invasion, dedifferentiation, and development of metastasis (Mukunyadzi et al. 2003; Ito et al. 2003). A shift of syndecan-1 from epithelial to stromal cells was observed in various solid tumors during their progression (Mennerich et al. 2004), and a growth promoting loop could be observed between breast cancer cells and its stoma (Maeda et al. 2004), where induction of syndecan-1 expression in stromal fibroblasts stimulated the growth of breast cancer cells. High level of shed **soluble syndecan-1 is** generally associates with poor prognosis and correlates to tumor burden, cancer invasiveness, and the risk for metastasis (Yang et al. 2007; Su et al. 2007; Szarvas et al. 2018a; Kurokawa et al. 2006; Malek-Hosseini et al. 2017).

The effects of syndecan-1 in malignant tumors depend on different factors. Of major importance is the localization; whether it is present in the tumor cell membrane or in the nucleus or extracellularly in the tumor stroma. The tumor cell phenotype and tissue from which the tumor originates are other such factors.

Thus, in **squamous carcinoma** of the head neck and larynx, decreased cell surface syndecan-1 expression in epithelial cells is associated with tumor aggressiveness and poor prognosis (Kurokawa et al. 2006; Inki et al. 1994a). In nasopharyngeal carcinoma, it correlates with advanced clinical stages and poor outcomes (Pulkkinen et al. 1997). Soluble syndecan-1 levels reflect the tumor burden, and decreased serum levels of syndecan-1 is predictive for favorable outcome (Anttonen et al. 2006). In squamous oral and cutaneous carcinoma, stromal syndecan-1 was inversely correlated with tumor grade and invasiveness (Ahmed Haji Omar et al. 2013; Máthé et al. 2006).

The presence of syndecan-1 is associated with favorable outcome in both **lung cancer** (Anttonen et al. 2001) and pleural **malignant mesothelioma** (Kumar-Singh et al. 1998). Generally, adenocarcinomas show higher cell surface and soluble syndecan-1 levels than primary malignant mesotheliomas of the pleura. Considering this feature, syndecan-1 has been proposed as a diagnostic marker in distinguishing mesotheliomas from metastatic adenocarcinomas (Gulyás and Hjerpe 2003; Saqi et al. 2005; Mundt et al. 2014). In squamous cell lung carcinoma, low syndecan-1

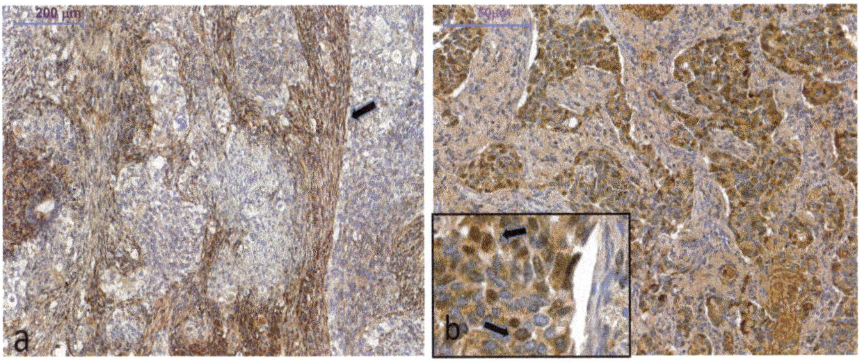

Fig. 3.3 Examples for the aberrant expression of syndecan-1 in lung adenocarcinomas. a. Vanish of syndecan-1 from the cell surface of the tumor cell parallel with its upregulation in the tumor stroma. Original magnification 10×. b. Loss of syndecan-1 from the surface of the tumor cells, and aberrant overexpression in all other compartment of the lung adenocarcinoma, showing intranuclear, cytoplasmic, and stromal positivity. Original magnification 20 ×

expression is associated with unfavorable outcome, and the majority of NSCLC express cell surface syndecan-1 reactivity as demonstrated by immunohistochemistry of tumor tissue, and elevated syndecan-1 levels in the serum associates with poor outcome in both NSCLC and SCLC (Linnerth et al. 2005; Joensuu et al. 2002; Anttonen et al. 2003) (Fig. 3.3).

The expression of syndecan-1 is induced in the stroma of **gastric cancer**, where its presence has been correlated with poor prognosis, and its immunohistochemical demonstration has been suggested as a potentially prognostic factor in patients with stage I gastric cancer (Wiksten et al. 2000). When syndecan-1 was present in the epithelial tumor cells, this correlated to a longer survival (Hu et al. 2014), whereas stromal syndecan-1 expression correlated with shorter survival (Wiksten et al. 2001). Low expression of syndecan-1 is significantly correlated with the invasion and metastasis of gastric carcinoma (Chu et al. 2008).

In the human **liver,** syndecan-1 is the major proteoglycan (Roskams et al. 1995) and its expression sometimes becomes intracellular in primary liver tumors (Roskams et al. 1998). As the majority of liver cancers develop in cirrhotic livers, characterized by elevated syndecan-1 expression, **hepatocellular carcinomas** developed on cirrhotic base express ample amounts of syndecan-1 (Regős et al. 2020). In hepatitis C-derived cancer, this increase is significant. In contrast, the loss of syndecan-1 expression is a typical feature of hepatocellular carcinomas without cirrhosis, when its expression is also reduced at the mRNA level. Both kinds of cancers have high metastatic potential and poor prognosis (Matsumoto et al. 1997). Although elevated or retained cell surface syndecan-1 indicates better prognosis, both tumors are characterized by increased proteoglycan shedding, which significantly associates with poor survival (Nault et al. 2013). Overexpression of syndecan-1 in the liver of transgenic mice exerts a protective role in a diethylnitrosamine-induced mouse hepatocellular cancer model. Compared to the wild-type mice, their tumors occurred with 6-month delay, and in spite of the enhanced syndecan-1 shedding, they retained cell surface syndecan both on hepatocytes and tumor cells (Reszegi et al. 2021a).

Another interesting question is whether the liver tissue reacts with altered syndecan-1 expression to the presence of metastatic tumors? Liver is a primary destination of colon cancers. According to the authors' personal observation, syndecan-1 expression increased in the peritumoral area, the significance of this, however, needs further evaluation (Regős et al. 2020). Syndecan-1 expression is heterogeneous in **intrahepatic cholangiocarcinoma**. High level of syndecan-1 is associated with longer survival, whereas reduced expression of syndecan-1 is correlated with lymph node metastasis and poor prognosis (Harada et al. 2003).

In **pancreatic cancer**, increased levels of membrane syndecan-1 reactivity have been found (Conejo et al. 2000). In this tumor, epithelial syndecan-1 expression predicts better prognosis in the resectable disease, whereas stromal syndecan-1 expression is an independent adverse prognostic factor (Juuti et al. 2005).

Loss of epithelial syndecan-1 is associated with advanced clinical stage and poor prognosis in **colorectal cancer** (Lundin et al. 2005). Selective expression of syndecan-1 in tumor-initiating cell lines suggests a role for syndecan-1 in cancer stem cells (Suhovskih et al. 2015). Syndecan-1 shedding is increased in colorectal cancer and it renders tumor cell's resistance to chemotherapy (Wang et al. 2014b). Moreover, different research groups found either tumor promoting or tumor inhibiting effects in colorectal cancers (Kim et al. 2015; Hashimoto et al. 2008).

The functions of syndecan-1 in **mammary cancer** depend on the location of the proteoglycan, the extent of its shedding, and the subtypes of the cancer itself (Nikolova et al. 2009). The majority of the reports emphasize that the tumor cells express increased amount of syndecan-1 on their cell surface. Its expression is also induced in the stromal cells adjacent to the cancer, particularly in tumors exhibiting an aggressive phenotype (Lendorf et al. 2011). In another publication, the loss of epithelial syndecan-1 correlated with the syndecan-1 stromal expression and was found to be significant as a poor prognostic factor (Loussouarn et al. 2008). The final conclusion of a recent meta-analysis of 1305 mammary cancer cases corroborated the general opinion that high expression of the proteoglycan indicates poor prognosis (Qiao et al. 2019), regardless if it occurs on the cell membrane, in the cytoplasm, or in the tumor stroma. Studies from an in vitro breast cancer model have also suggested that syndecan-1 directly participates in tumor cell spreading and adhesion. In a breast cancer cell line, overexpression of wild-type syndecan-1 increased cell proliferation, whereas overexpression of constitutively shed syndecan-1 influenced the migratory capacity of the cells (Nikolova et al. 2009).

Syndecan-1 expression is induced in the stroma of invasive breast carcinomas (Stanley et al. 1999), and studies link the worse prognosis of breast carcinoma patients with stromal syndecan-1 expression. Furthermore, epithelial syndecan-1 expression associates with negative ER status, whereas stromal syndecan-1 expression was associated with positive ER status (Leivonen et al. 2004). A recent study showed that the proportion of syndecan-1 positive cells correlated with tumor grade better than the amounts demonstrated by immunohistochemistry, nuclear grade, or

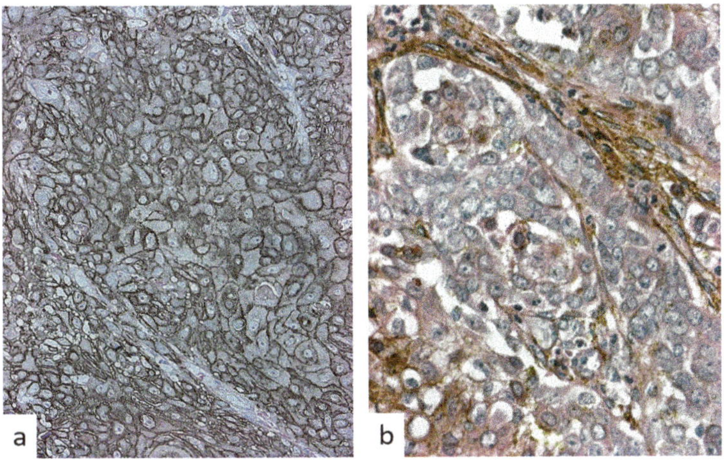

Fig. 3.4 Syndecan-1 immunostaining of squamous cancer of cervix. a. Intensive cell surface positivity of the tumor cells. Original magnification 20×. b. Loss of syndecan positivity from the tumor cell surface, together with immunopositivity of the fibroblasts in the tumor stroma. Original magnification 40 ×

localization of syndecan-1. The estrogen- and progesterone receptors both correlated negatively with syndecan-1 staining (Tiemann et al. 2014).

Syndecan-1 expression has a prognostic value also in **urothelial cancer**. Surface expression of syndecan-1 was inversely correlated with tumor stage in primary non-muscle-invasive bladder cancer (Kim and Park 2014), while high stromal syndecan-1 was associated with poor prognosis (Hatina et al. 2015; Szarvas et al. 2014).

In normal cervical squamous epithelium, cells with squamous metaplasia and koilocytotic atypia are positive for syndecan-1. During the progression of cervical intraepithelial lesions from low grade (LSIL) to high grade (HSIL), syndecan-1 expression successively diminishes. In **cervical squamous cell carcinomas (SCCs)**, syndecan-1 expression correlates with histological differentiation, being absent from most poorly differentiated tumors. Thus, loss of syndecan-1 seems to be an early event during cervical carcinogenesis and shows a clear association with histological differentiation grade (Inki et al. 1994c; Rintala et al. 1999). The progression of HSIL to early invasive squamous carcinoma and invasion into the lympho-vascular space associated with reduced syndecan-1 and HS–GAG expression along with lower overall survival rates compared to patients exhibiting high HS–GAG expression (Shinyo et al. 2005). Retained syndecan-1 expression on the tumor cell surface is a significantly positive prognostic marker, whereas its expression on tumor-associated fibroblasts indicates poor prognosis (Karászi et al. 2020) (Fig. 3.4).

Loss of epithelial syndecan-1 expression and induction of stromal syndecan-1 expression are associated with reduced survival in patients with **endometrial cancer** (Hasengaowa et al. 2005). In a xenograft model, upregulation of syndecan-1 developed proliferative and invasive/metastatic phenotypes. The growth advantage conferred by syndecan-1 overexpression was accompanied by increased tumor angiogenesis (Oh et al. 2010). Increased stromal syndecan-1 staining was similarly a poor prognostic factor in **ovarian cancer** (Davies et al. 2004).

In normal prostate tissue, syndecan-1 is expressed mainly by the epithelial cells, while in tumors, an overall increase of syndecan-1 expression is observed in the tumor stroma, along with its disappearance from tumor epithelial cells (Suhovskih et al. 2013). In **prostate cancer**, intracytoplasmic accumulation of the proteoglycan in the tumor cells indicates aggressive behavior together with the early recurrence of the tumor (Kind et al. 2020). Furthermore, syndecan-1 level inversely correlates with tumor grade (Poblete et al. 2014). It has also been suggested that syndecan-1 may promote the regression of prostate cancer by stabilizing tumor initiating cells in experimental models (Shimada et al. 2013).

In mesenchymal tumors such as **malignant mesothelioma**, the expression of syndecan-1 correlates with epithelioid morphology and favorable prognosis. The underlying molecular mechanisms are complex and involve several critical genes and pathways. By studying the effect of syndecan-1 overexpression and silencing, the responses to modulated syndecan-1 expression were monitored by microarray (Szatmári et al. 2012) and proteomic analyses. Syndecan-1 overexpression had profound effects on several genes involved in the regulation of cell growth, cell-cycle progression, adhesion, migration, and ECM organization. Particularly, the expression of growth factors, interleukins, enzymes important for establishing the heparan sulfate sulfation pattern, ECM proteins, and proteoglycans were significantly altered. Syndecan-1 silencing had less powerful effect on the transcriptome compared to overexpression, which can be explained by the already low initial syndecan-1 level of these cells. Fourteen genes showed response to both up- and downregulation of syndecan-1. The "cytokine–cytokine-receptor interaction," TGFβ, EGF, VEGF, and ERK/MAPK pathways were significantly affected in both experimental settings. Interestingly, conditioned medium from syndecan-1 overexpressing malignant mesothelioma cells also exerted a paracrine inhibitory effect on endothelial cells proliferation, migration, and tube formation (Javadi et al. 2021). Multiplex soluble biomarker analyses of pleural effusions revealed that shed syndecan-1 levels are significantly lower in malignant mesothelioma effusions compared to adenocarcinoma, but significantly higher than that of benign samples (Javadi et al. 2020). Additionally, shed syndecan-1 and VEGF have prognostic value in malignant mesothelioma patients. Overexpression of the full-length syndecan-1 enhances B6FS **fibrosarcoma** and **malignant mesothelioma** cell lines' adhesion and inhibits the proliferation and migration of these mesenchymal tumors. On the other hand, syndecan-1 enhanced the proliferation, migration, and metastatic potential of HT-1080 cells (Péterfia et al. 2012) in mouse. Expression of syndecan-1 was also found in **malignant glioma** cells (Watanabe et al. 2006), and it is highly overexpressed in **dedifferentiated liposarcoma** (Zaragosi et al. 2015).

3.7 Unpredicted Functions of Syndecan-1 in Tumors

Although the role of syndecan-1 in the development and progression of tumors is exhaustively investigated, hardly any information is available about its role in the tumor metabolism. Experimental data provide more and more evidence that intermediary metabolism supports the biological activities of cancer cells. Along with this process, tumor cells establish strict cooperation with the resident cells of the ECM (Lau and Vander Heiden 2020). The details of these interactions depend on the particular types of tumors. As the liver is one of the major organs of intermediary metabolism, hepatocarcinogenesis in syndecan-1 transgenic mice offered a good approach to study the question. Upregulation of syndecan-1 on the surface of hepatocytes inhibited insulin receptor activation, PIKC3 and mTOR signaling, remodeling of ribosome biogenesis, and activation of Foxo1 and p53 proteins already known to regulate the intermediary metabolism including glucose and lipid homeostasis (Turi et al. 2019; Pelletier et al. 2018; Iadevaia et al. 2014; Lee and Dong 2017; Tzivion et al. 2011; Liu et al. 2019). Foxo1 is regulated by 14–3-3-zeta, an adapter protein with oncogenic potential, which was also inhibited by syndecan-1 (Matta et al. 2012; Reszegi et al. 2021a). As the proteoglycan is a well-known lipoprotein receptor on the surface of hepatocytes, its overexpression most probably modulates lipoprotein uptake from the blood circulation (Stanford et al. 2009). Furthermore, overexpression of FASN, a critical component of aberrant lipid metabolism with oncogenic potential in the liver, dramatically decreased as a result of syndecan-1 upregulation (Hu et al. 2020). Syndecan-1 can succesfully inhibit FASN, a critical promoter of nonalcoholic steato-hepatitis, hindering this way the disease progression to hepatocellular cancer.

3.8 Conclusion

Taken together, syndecan-1 is present in most epithelial malignancies and it carries prognostic information. Its regulation and expression pattern in different malignancies are however complex. In some cancers, such as head and neck cancer, lung cancer, and malignant mesothelioma, it indicates favorable prognosis when anchored to the cell membrane, and in contrast, it associates with poor prognosis when recovered in shed or soluble form in body fluids. In breast carcinoma, both in the epithelium and in stroma, syndecan-1 indicates mainly poor prognosis. A comprehensive characterization of syndecan-1 in malignant tumors requires further studies that consider not only the expression level and cell surface reactivity but also its various localization as these constitute key features determining the downstream effects and clinical outcome.

References

Ahmed Haji Omar A, Haglund C, Virolainen S, Häyry V, Atula T, Kontio R, Rihtniemi J, Pihakari A, Salo T, Hagström J, Sorsa T (2013) Epithelial and stromal syndecan-1 and -2 are distinctly expressed in oral- and cutaneous squamous cell carcinomas. Journal of oral pathology & medicine : official publication of the International Association of Oral Pathologists and the American Academy of Oral Pathology 42(5):389–395. https://doi.org/10.1111/jop.12025

Angsana J, Chen J, Smith S, Xiao J, Wen J, Liu L, Haller CA, Chaikof EL (2015) Syndecan-1 modulates the motility and resolution responses of macrophages. Arterioscler Thromb Vasc Biol 35(2):332–340. https://doi.org/10.1161/ATVBAHA.114.304720

Anttonen A, Heikkilä P, Kajanti M, Jalkanen M, Joensuu H (2001) High syndecan-1 expression is associated with favourable outcome in squamous cell lung carcinoma treated with radical surgery. Lung Cancer 32(3):297–305. https://doi.org/10.1016/S0169-5002(00)00230-0

Anttonen A, Leppä S, Ruotsalainen T, Alfthan H, Mattson K, Joensuu H (2003) Pretreatment serum syndecan-1 levels and outcome in small cell lung cancer patients treated with platinum-based chemotherapy. Lung Cancer 41(2):171–177. https://doi.org/10.1016/S0169-5002(03)00196-X

Anttonen A, Leppä S, Heikkilä P, Grenman R, Joensuu H (2006) Effect of treatment of larynx and hypopharynx carcinomas on serum syndecan-1 concentrations. J Cancer Res Clin Oncol 132(7): 451–457. https://doi.org/10.1007/s00432-006-0090-z

Bertrand J, Bollmann M (2019) Soluble syndecans: biomarkers for diseases and therapeutic options. Br J Pharmacol 176(1):67–81. https://doi.org/10.1111/bph.14397

Brauer R, Ge L, Schlesinger SY, Birkland TP, Huang Y, Parimon T, Lee V, McKinney BL, McGuire JK, Parks WC, Chen P (2016) Syndecan-1 attenuates lung injury during influenza infection by potentiating c-met signaling to suppress epithelial apoptosis. Am J Respir Crit Care Med 194(3):333–344. https://doi.org/10.1164/rccm.201509-1878OC

Brockstedt U, Dobra K, Nurminen M, Hjerpe A (2002) Immunoreactivity to cell surface Syndecans in cytoplasm and nucleus: tubulin-dependent rearrangements. Exp Cell Res 274(2):235–245. https://doi.org/10.1006/excr.2002.5477

Chen L, Sanderson RD (2009) Heparanase regulates levels of syndecan-1 in the nucleus. PLoS One 4(3):e4947. https://doi.org/10.1371/journal.pone.0004947

Chu YQ, Ye ZY, Tao HQ, Wang YY, Zhao ZS (2008) Relationship between cell adhesion molecules expression and the biological behavior of gastric carcinoma. World J Gastroenterol 14(13):1990–1996. https://doi.org/10.3748/wjg.14.1990

Conejo JR, Kleeff J, Koliopanos A, Matsuda K, Zhu ZW, Goecke H, Bicheng N, Zimmermann A, Korc M, Friess H, Büchler MW (2000) Syndecan-1 expression is up-regulated in pancreatic but not in other gastrointestinal cancers. Int J Cancer 88(1):12–20. https://doi.org/10.1002/1097-0215(20001001)88:1<12::aid-ijc3>3.0.co;2-t

Couchman JR (2021) Syndecan-1 (CD138), carcinomas and EMT. Int J Mol Sci 22(8). https://doi.org/10.3390/ijms22084227

Davies EJ, Blackhall FH, Shanks JH, David G, McGown AT, Swindell R, Slade RJ, Martin-Hirsch P, Gallagher JT, Jayson GC (2004) Distribution and clinical significance of heparan sulfate proteoglycans in ovarian cancer. Clin Cancer Res 10(15):5178–5186. https://doi.org/10.1158/1078-0432.Ccr-03-0103

De Luca M, Klimentidis YC, Casazza K, Chambers MM, Cho R, Harbison ST, Jumbo-Lucioni P, Zhang S, Leips J, Fernandez JR (2010) A conserved role for syndecan family members in the regulation of whole-body energy metabolism. PLoS One 5(6):e11286. https://doi.org/10.1371/journal.pone.0011286

Dennissen MA, Jenniskens GJ, Pieffers M, Versteeg EM, Petitou M, Veerkamp JH, van Kuppevelt TH (2002) Large, tissue-regulated domain diversity of heparan sulfates demonstrated by phage display antibodies. J Biol Chem 277(13):10982–10986. https://doi.org/10.1074/jbc.M104852200

Dudás J, Ramadori G, Knittel T, Neubauer K, Raddatz D, Egedy K, Kovalszky I (2000) Effect of heparin and liver heparan sulphate on interaction of HepG2-derived transcription factors and their cis-acting elements: altered potential of hepatocellular carcinoma heparan sulphate. Biochem J 350(Pt 1):245–251

Fedarko NS, Conrad HE (1986) A unique heparan sulfate in the nuclei of hepatocytes: structural changes with the growth state of the cells. J Cell Biol 102(2):587–599. https://doi.org/10.1083/jcb.102.2.587

Filla MS, Dam P, Rapraeger AC (1998) The cell surface proteoglycan syndecan-1 mediates fibroblast growth factor-2 binding and activity. J Cell Physiol 174(3):310–321. https://doi.org/10.1002/(sici)1097-4652(199803)174:3<310::Aid-jcp5>3.0.Co;2-r

Foley EM, Gordts P, Stanford KI, Gonzales JC, Lawrence R, Stoddard N, Esko JD (2013) Hepatic remnant lipoprotein clearance by heparan sulfate proteoglycans and low-density lipoprotein receptors depend on dietary conditions in mice. Arterioscler Thromb Vasc Biol 33(9): 2065–2074. https://doi.org/10.1161/atvbaha.113.301637

Fujii T, Shimada K, Tatsumi Y, Fujimoto K, Konishi N (2015a) Syndecan-1 responsive microRNA-126 and 149 regulate cell proliferation in prostate cancer. Biochem Biophys Res Commun 456(1):183–189. https://doi.org/10.1016/j.bbrc.2014.11.056

Fujii T, Shimada K, Tatsumi Y, Hatakeyama K, Obayashi C, Fujimoto K, Konishi N (2015b) microRNA-145 promotes differentiation in human urothelial carcinoma through down-regulation of syndecan-1. BMC Cancer 15:818. https://doi.org/10.1186/s12885-015-1846-0

Fujii T, Shimada K, Tatsumi Y, Tanaka N, Fujimoto K, Konishi N (2016) Syndecan-1 up-regulates microRNA-331-3p and mediates epithelial-to-mesenchymal transition in prostate cancer. Mol Carcinog 55(9):1378–1386. https://doi.org/10.1002/mc.22381

Gao SL, Zhang Y, Zhang SY, Liang ZY, Yu WQ, Liang TB (2015) The hydrocortisone protection of glycocalyx on the intestinal capillary endothelium during severe acute pancreatitis. Shock 43(5):512–517. https://doi.org/10.1097/shk.0000000000000326

Gondelaud F, Ricard-Blum S (2019) Structures and interactions of syndecans. FEBS J 286(15): 2994–3007. https://doi.org/10.1111/febs.14828

Gopal S (2020) Syndecans in inflammation at a glance. Front Immunol 11:227–227. https://doi.org/10.3389/fimmu.2020.00227

Gopal S, Søgaard P, Multhaupt HAB, Pataki C, Okina E, Xian X, Pedersen ME, Stevens T, Griesbeck O, Park PW, Pocock R, Couchman JR (2015) Transmembrane proteoglycans control stretch-activated channels to set cytosolic calcium levels. J Cell Biol 210(7):1199–1211. https://doi.org/10.1083/jcb.201501060

Grigorov B, Reungoat E, Gentil dit Maurin A, Varbanov M, Blaising J, Michelet M, Manuel R, Parent R, Bartosch B, Zoulim F, Ruggiero F, Pécheur E-I (2017) Hepatitis C virus infection propagates through interactions between Syndecan-1 and CD81 and impacts the hepatocyte glycocalyx. Cell Microbiol 19(5):e12711. https://doi.org/10.1111/cmi.12711

Gross SM, Rotwein P (2017) Quantification of growth factor signaling and pathway cross talk by live-cell imaging. Am J Physiol Cell Physiol 312(3):C328–c340. https://doi.org/10.1152/ajpcell.00312.2016

Gulyás M, Hjerpe A (2003) Proteoglycans and WT1 as markers for distinguishing adenocarcinoma, epithelioid mesothelioma, and benign mesothelium. J Pathol 199(4):479–487. https://doi.org/10.1002/path.1312

Guo T, Yu W, Lv S, Zhang C, Tian Y (2015) MiR-302a inhibits the tumorigenicity of ovarian cancer cells by suppression of SDC1. Int J Clin Exp Pathol 8(5):4869–4880

Handra-Luca A (2020) Syndecan-1 in the tumor microenvironment. Adv Exp Med Biol 1272:39–53. https://doi.org/10.1007/978-3-030-48457-6_3

Harada K, Masuda S, Hirano M, Nakanuma Y (2003) Reduced expression of syndecan-1 correlates with histologic dedifferentiation, lymph node metastasis, and poor prognosis in intrahepatic cholangiocarcinoma. Hum Pathol 34(9):857–863. https://doi.org/10.1016/s0046-8177(03)00336-8

Hasengaowa KJ, Kusumoto T, Shinyo Y, Seki N, Hiramatsu Y (2005) Prognostic significance of syndecan-1 expression in human endometrial cancer. Ann Oncol 16(7):1109–1115. https://doi.org/10.1093/annonc/mdi224

Hashimoto Y, Skacel M, Adams JC (2008) Association of loss of epithelial syndecan-1 with stage and local metastasis of colorectal adenocarcinomas: an immunohistochemical study of clinically annotated tumors. BMC Cancer 8:185. https://doi.org/10.1186/1471-2407-8-185

Hatina J, Kripnerová M, Tuková J, Šrámek J, Dvořák P, Pešta M, Dobrá J, Babuška V, Racek J, Sobol M, Philimonenko A, Hozák P, Czuba Z, Schulz WA, Strell C, Grimm S, Jennek S, Friedrich KH (2015) Tumour-stroma interactions in urothelial cancer. Urologe A 54(4): 516–525. https://doi.org/10.1007/s00120-014-3754-3

Hollósi P, Váncza L, Karászi K, Dobos K, Péterfia B, Tátrai E, Tátrai P, Szarvas T, Paku S, Szilák L, Kovalszky I (2020) Syndecan-1 promotes hepatocyte-like differentiation of hepatoma cells targeting Ets-1 and AP-1. Biomol Ther 10(10). https://doi.org/10.3390/biom10101356

Hsia E, Richardson TP, Nugent MA (2003) Nuclear localization of basic fibroblast growth factor is mediated by heparan sulfate proteoglycans through protein kinase C signaling. J Cell Biochem 88(6):1214–1225. https://doi.org/10.1002/jcb.10470

Hu XF, Yao J, Gao SG, Yang YT, Peng XQ, Feng XS (2014) Midkine and syndecan-1 levels correlate with the progression of malignant gastric cardiac adenocarcinoma. Mol Med Rep 10(3):1409–1415. https://doi.org/10.3892/mmr.2014.2369

Hu B, Lin JZ, Yang XB, Sang XT (2020) Aberrant lipid metabolism in hepatocellular carcinoma cells as well as immune microenvironment: a review. Cell Prolif 53(3):e12772. https://doi.org/10.1111/cpr.12772

Iadevaia V, Liu R, Proud CG (2014) mTORC1 signaling controls multiple steps in ribosome biogenesis. Semin Cell Dev Biol 36:113–120. https://doi.org/10.1016/j.semcdb.2014.08.004

Ibrahim SA, Yip GW, Stock C, Pan JW, Neubauer C, Poeter M, Pupjalis D, Koo CY, Kelsch R, Schüle R, Rescher U, Kiesel L, Götte M (2012) Targeting of syndecan-1 by microRNA miR-10b promotes breast cancer cell motility and invasiveness via a rho-GTPase- and E-cadherin-dependent mechanism. Int J Cancer 131(6):E884–E896. https://doi.org/10.1002/ijc.27629

Indira Chandran V, Welinder C, Månsson AS, Offer S, Freyhult E, Pernemalm M, Lund SM, Pedersen S, Lehtiö J, Marko-Varga G, Johansson MC, Englund E, Sundgren PC, Belting M (2019) Ultrasensitive Immunoprofiling of plasma extracellular vesicles identifies Syndecan-1 as a potential tool for minimally invasive diagnosis of glioma. Clin Cancer Res 25(10):3115–3127. https://doi.org/10.1158/1078-0432.Ccr-18-2946

Inki P, Joensuu H, Grénman R, Klemi P, Jalkanen M (1994a) Association between syndecan-1 expression and clinical outcome in squamous cell carcinoma of the head and neck. Br J Cancer 70(2):319–323. https://doi.org/10.1038/bjc.1994.300

Inki P, Larjava H, Haapasalmi K, Miettinen HM, Grenman R, Jalkanen M (1994b) Expression of syndecan-1 is induced by differentiation and suppressed by malignant transformation of human keratinocytes. Eur J Cell Biol 63(1):43–51

Inki P, Stenbäck F, Grenman S, Jalkanen M (1994c) Immunohistochemical localization of syndecan-1 in normal and pathological human uterine cervix. J Pathol 172(4):349–355. https://doi.org/10.1002/path.1711720410

Ishihara M, Fedarko NS, Conrad HE (1986) Transport of heparan sulfate into the nuclei of hepatocytes. J Biol Chem 261(29):13575–13580

Ito Y, Yoshida H, Nakano K, Takamura Y, Miya A, Kobayashi K, Yokozawa T, Matsuzuka F, Matsuura N, Kuma K, Miyauchi A (2003) Syndecan-1 expression in thyroid carcinoma: stromal expression followed by epithelial expression is significantly correlated with dedifferentiation. Histopathology 43(2):157–164. https://doi.org/10.1046/j.1365-2559.2003.01656.x

Jalkanen M, Rapraeger A, Saunders S, Bernfield M (1987) Cell surface proteoglycan of mouse mammary epithelial cells is shed by cleavage of its matrix-binding ectodomain from its membrane-associated domain. J Cell Biol 105(6 Pt 2):3087–3096. https://doi.org/10.1083/jcb.105.6.3087

Javadi J, Dobra K, Hjerpe A (2020) Multiplex soluble biomarker analysis from pleural effusion. Biomol Ther 10(8). https://doi.org/10.3390/biom10081113

Javadi J, Heidari-Hamedani G, Schmalzl A, Szatmári T, Metintas M, Aspenström P, Hjerpe A, Dobra K (2021) Syndecan-1 overexpressing mesothelioma cells inhibit proliferation, wound healing, and tube formation of endothelial cells. Cancers (Basel) 13(4). https://doi.org/10.3390/cancers13040655

Joensuu H, Anttonen A, Eriksson M, Mäkitaro R, Alfthan H, Kinnula V, Leppä S (2002) Soluble syndecan-1 and serum basic fibroblast growth factor are new prognostic factors in lung cancer. Cancer Res 62(18):5210–5217

Juuti A, Nordling S, Lundin J, Louhimo J, Haglund C (2005) Syndecan-1 expression--a novel prognostic marker in pancreatic cancer. Oncology 68(2–3):97–106. https://doi.org/10.1159/000085702

Karászi K, Vigh R, Máthé M, Fullár A, Oláh L, Füle T, Papp Z, Kovalszky I (2020) Aberrant expression of Syndecan-1 in cervical cancers. Pathol Oncol Res 26(4):2255–2264. https://doi.org/10.1007/s12253-020-00816-0

Kim JH, Park J (2014) Prognostic significance of heme oxygenase-1, S100 calcium-binding protein A4, and syndecan-1 expression in primary non-muscle-invasive bladder cancer. Hum Pathol 45(9):1830–1838. https://doi.org/10.1016/j.humpath.2014.04.020

Kim SY, Choi EJ, Yun JA, Jung ES, Oh ST, Kim JG, Kang WK, Lee SH (2015) Syndecan-1 expression is associated with tumor size and EGFR expression in colorectal carcinoma: a clinicopathological study of 230 cases. Int J Med Sci 12(2):92–99. https://doi.org/10.7150/ijms.10497

Kind S, Merenkow C, Büscheck F, Möller K, Dum D, Chirico V, Luebke AM, Höflmayer D, Hinsch A, Jacobsen F, Göbel C, Weidemann S, Fraune C, Möller-Koop C, Hube-Magg C, Clauditz TS, Simon R, Sauter G, Wilczak W, Bawahab AA, Izbicki JR, Perez D, Marx A (2019) Prevalence of Syndecan-1 (CD138) expression in different kinds of human tumors and Normal tissues. Dis Markers 2019:4928315. https://doi.org/10.1155/2019/4928315

Kind S, Kluth M, Hube-Magg C, Möller K, Makrypidi-Fraune G, Lutz F, Lennartz M, Rico SD, Schlomm T, Heinzer H, Höflmayer D, Weidemann S, Uhlig R, Huland H, Graefen M, Bernreuther C, Tsourlakis MC, Minner S, Dum D, Hinsch A, Lübke AM, Simon R, Sauter G, Marx A, Polonski A (2020) Increased cytoplasmic CD138 expression is associated with aggressive characteristics in prostate cancer and is an independent predictor for biochemical recurrence. Biomed Res Int 2020:5845374. https://doi.org/10.1155/2020/5845374

Kjellén L, Lindahl U (2018) Specificity of glycosaminoglycan–protein interactions. Curr Opin Struct Biol 50:101–108. https://doi.org/10.1016/j.sbi.2017.12.011

Kleiser S, Nyström A (2020) Interplay between cell-surface receptors and extracellular matrix in skin. Biomol Ther 10(8). https://doi.org/10.3390/biom10081170

Kovalszky I, Dudás J, Oláh-Nagy J, Pogány G, Töváry J, Timár J, Kopper L, Jeney A, Iozzo RV (1998) Inhibition of DNA topoisomerase I activity by heparan sulfate and modulation by basic fibroblast growth factor. Mol Cell Biochem 183(1–2):11–23. https://doi.org/10.1023/a:1006898920637

Krishnan SR, Luk F, Brown RD, Suen H, Kwan Y, Bebawy M (2016) Isolation of human CD138 (+) microparticles from the plasma of patients with multiple myeloma. Neoplasia (New York, NY) 18(1):25–32. https://doi.org/10.1016/j.neo.2015.11.011

Kumar-Singh S, Jacobs W, Dhaene K, Weyn B, Bogers J, Weyler J, Van Marck E (1998) Syndecan-1 expression in malignant mesothelioma: correlation with cell differentiation, WT1 expression, and clinical outcome. J Pathol 186(3):300–305. https://doi.org/10.1002/(sici)1096-9896(1998110)186:3<300::Aid-path180>3.0.Co;2-q

Kumar-Singh A, Shrinet J, Parniewska MM, Fuxe J, Dobra K, Hjerpe A (2020) Mapping the Interactome of the nuclear Heparan sulfate proteoglycan Syndecan-1 in mesothelioma cells. Biomol Ther 10(7). https://doi.org/10.3390/biom10071034

Kumar-Singh A, Parniewska MM, Giotopoulou N, Javadi J, Sun W, Szatmári T, Dobra K, Hjerpe A, Fuxe J (2021) Nuclear Syndecan-1 regulates epithelial-mesenchymal plasticity in tumor cells. Biology (Basel) 10(6). https://doi.org/10.3390/biology10060521

Kurokawa H, Zhang M, Matsumoto S, Yamashita Y, Tanaka T, Takamori K, Igawa K, Yoshida M, Fukuyama H, Takahashi T, Sakoda S (2006) Reduced syndecan-1 expression is correlated with the histological grade of malignancy at the deep invasive front in oral squamous cell carcinoma. Journal of oral pathology & medicine : official publication of the International Association of Oral Pathologists and the American Academy of Oral Pathology 35(5):301–306. https://doi.org/10.1111/j.1600-0714.2006.00412.x

Lau AN, Vander Heiden MG (2020) Metabolism in the tumor microenvironment. Annual Review of Cancer Biology 4(1):17–40. https://doi.org/10.1146/annurev-cancerbio-030419-033333

Lee S, Dong HH (2017) FoxO integration of insulin signaling with glucose and lipid metabolism. J Endocrinol 233(2):R67–r79. https://doi.org/10.1530/joe-17-0002

Lehnert P, Johansson PI, Ostrowski SR, Møller CH, Bang LE, Olsen PS, Carlsen J (2017) Coagulopathy in patients with acute pulmonary embolism: a pilot study of whole blood coagulation and markers of endothelial damage. Scand J Clin Lab Invest 77(1):19–26. https://doi.org/10.1080/00365513.2016.1239130

Leivonen M, Lundin J, Nordling S, von Boguslawski K, Haglund C (2004) Prognostic value of syndecan-1 expression in breast cancer. Oncology 67(1):11–18. https://doi.org/10.1159/000080280

Lendorf ME, Manon-Jensen T, Kronqvist P, Multhaupt HA, Couchman JR (2011) Syndecan-1 and syndecan-4 are independent indicators in breast carcinoma. J Histochem Cytochem 59(6): 615–629. https://doi.org/10.1369/0022155411405057

Li D, Wu Y, Guo S, Qin J, Feng M, An Y, Zhang J, Li Y, Xiong S, Zhou H, Zeng Q, Chen L, Wen F (2019) Circulating syndecan-1 as a novel biomarker relates to lung function, systemic inflammation, and exacerbation in COPD. Int J Chron Obstruct Pulmon Dis 14:1933–1941. https://doi.org/10.2147/copd.S207855

Lindahl U, Li JP (2009) Interactions between heparan sulfate and proteins-design and functional implications. Int Rev Cell Mol Biol 276:105–159. https://doi.org/10.1016/s1937-6448(09)76003-4

Linnerth NM, Sirbovan K, Moorehead RA (2005) Use of a transgenic mouse model to identify markers of human lung tumors. Int J Cancer 114(6):977–982. https://doi.org/10.1002/ijc.20814

Liu J, Zhang C, Hu W, Feng Z (2019) Tumor suppressor p53 and metabolism. J Mol Cell Biol 11(4):284–292. https://doi.org/10.1093/jmcb/mjy070

Loussouarn D, Campion L, Sagan C, Frenel JS, Dravet F, Classe JM, Pioud-Martigny R, Berton-Rigaud D, Bourbouloux E, Mosnier JF, Bataille FR, Campone M (2008) Prognostic impact of syndecan-1 expression in invasive ductal breast carcinomas. Br J Cancer 98(12):1993–1998. https://doi.org/10.1038/sj.bjc.6604400

Lundin M, Nordling S, Lundin J, Isola J, Wiksten JP, Haglund C (2005) Epithelial syndecan-1 expression is associated with stage and grade in colorectal cancer. Oncology 68(4–6):306–313. https://doi.org/10.1159/000086969

Maeda T, Alexander CM, Friedl A (2004) Induction of syndecan-1 expression in stromal fibroblasts promotes proliferation of human breast cancer cells. Cancer Res 64(2):612–621. https://doi.org/10.1158/0008-5472.can-03-2439

Malek-Hosseini Z, Jelodar S, Talei A, Ghaderi A, Doroudchi M (2017) Elevated Syndecan-1 levels in the sera of patients with breast cancer correlate with tumor size. Breast cancer (Tokyo, Japan) 24(6):742–747. https://doi.org/10.1007/s12282-017-0773-0

Mali M, Jaakkola P, Arvilommi AM, Jalkanen M (1990) Sequence of human syndecan indicates a novel gene family of integral membrane proteoglycans. J Biol Chem 265(12):6884–6889

Manon-Jensen T, Itoh Y, Couchman JR (2010) Proteoglycans in health and disease: the multiple roles of syndecan shedding. FEBS J 277(19):3876–3889. https://doi.org/10.1111/j.1742-4658.2010.07798.x

Máthé M, Suba Z, Németh Z, Tátrai P, Füle T, Borgulya G, Barabás J, Kovalszky I (2006) Stromal syndecan-1 expression is an adverse prognostic factor in oral carcinomas. Oral Oncol 42(5): 493–500. https://doi.org/10.1016/j.oraloncology.2005.10.003

Matsumoto A, Ono M, Fujimoto Y, Gallo RL, Bernfield M, Kohgo Y (1997) Reduced expression of syndecan-1 in human hepatocellular carcinoma with high metastatic potential. Int J Cancer 74(5):482–491. https://doi.org/10.1002/(sici)1097-0215(19971021)74:5<482::aid-ijc2>3.0.co;2-#

Matta A, Siu KW, Ralhan R (2012) 14-3-3 zeta as novel molecular target for cancer therapy. Expert Opin Ther Targets 16(5):515–523. https://doi.org/10.1517/14728222.2012.668185

Mennerich D, Vogel A, Klaman I, Dahl E, Lichtner RB, Rosenthal A, Pohlenz HD, Thierauch KH, Sommer A (2004) Shift of syndecan-1 expression from epithelial to stromal cells during progression of solid tumours. Eur J Cancer 40(9):1373–1382. https://doi.org/10.1016/j.ejca.2004.01.038

Montor WR, Salas AROSE, Melo FHM (2018) Receptor tyrosine kinases and downstream pathways as druggable targets for cancer treatment: the current arsenal of inhibitors. Mol Cancer 17(1):55. https://doi.org/10.1186/s12943-018-0792-2

Mukunyadzi P, Liu K, Hanna EY, Suen JY, Fan CY (2003) Induced expression of syndecan-1 in the stroma of head and neck squamous cell carcinoma. Mod Pathol 16(8):796–801. https://doi.org/10.1097/01.Mp.0000081731.17549.53

Mundhenke C, Meyer K, Drew S, Friedl A (2002) Heparan sulfate proteoglycans as regulators of fibroblast growth factor-2 receptor binding in breast carcinomas. Am J Pathol 160(1):185–194. https://doi.org/10.1016/s0002-9440(10)64362-3

Mundt F, Heidari-Hamedani G, Nilsonne G, Metintas M, Hjerpe A, Dobra K (2014) Diagnostic and prognostic value of soluble syndecan-1 in pleural malignancies. Biomed Res Int 2014:419853. https://doi.org/10.1155/2014/419853

Nault JC, Guyot E, Laguillier C, Chevret S, Ganne-Carrie N, N'Kontchou G, Beaugrand M, Seror O, Trinchet JC, Coelho J, Lasalle P, Charnaux N, Delehedde M, Sutton A, Nahon P (2013) Serum proteoglycans as prognostic biomarkers of hepatocellular carcinoma in patients with alcoholic cirrhosis. Cancer Epidemiol Biomark Prev 22(8):1343–1352. https://doi.org/10.1158/1055-9965.Epi-13-0179

Nikolova V, Koo CY, Ibrahim SA, Wang Z, Spillmann D, Dreier R, Kelsch R, Fischgräbe J, Smollich M, Rossi LH, Sibrowski W, Wülfing P, Kiesel L, Yip GW, Götte M (2009) Differential roles for membrane-bound and soluble syndecan-1 (CD138) in breast cancer progression. Carcinogenesis 30(3):397–407. https://doi.org/10.1093/carcin/bgp001

Oh JH, Lee HS, Park SH, Ryu HS, Min CK (2010) Syndecan-1 overexpression promotes tumor growth and angiogenesis in an endometrial cancer xenograft model. Int J Gynecol Cancer 20(5):751–756. https://doi.org/10.1111/igc.0b013e3181e02faa

Ostrowski SR, Berg RMG, Windeløv NA, Meyer MAS, Plovsing RR, Møller K, Johansson PI (2013) Coagulopathy, catecholamines, and biomarkers of endothelial damage in experimental human endotoxemia and in patients with severe sepsis: a prospective study. J Crit Care 28(5):586–596. https://doi.org/10.1016/j.jcrc.2013.04.010

Parimon T, Brauer R, Schlesinger SY, Xie T, Jiang D, Ge L, Huang Y, Birkland TP, Parks WC, Habiel DM, Hogaboam CM, Gharib SA, Deng N, Liu Z, Chen P (2018) Syndecan-1 controls lung tumorigenesis by regulating miRNAs packaged in exosomes. Am J Pathol 188(4):1094–1103. https://doi.org/10.1016/j.ajpath.2017.12.009

Pataki CA, Couchman JR, Brábek J (2015) Wnt signaling cascades and the roles of Syndecan proteoglycans. J Histochem Cytochem 63(7):465–480. https://doi.org/10.1369/0022155415586961

Pelletier J, Thomas G, Volarević S (2018) Ribosome biogenesis in cancer: new players and therapeutic avenues. Nat Rev Cancer 18(1):51–63. https://doi.org/10.1038/nrc.2017.104

Péterfia B, Füle T, Baghy K, Szabadkai K, Fullár A, Dobos K, Zong F, Dobra K, Hollósi P, Jeney A, Paku S, Kovalszky I (2012) Syndecan-1 enhances proliferation, migration and metastasis of HT-1080 cells in cooperation with syndecan-2. PLoS One 7(6):e39474. https://doi.org/10.1371/journal.pone.0039474

Poblete CE, Fulla J, Gallardo M, Muñoz V, Castellón EA, Gallegos I, Contreras HR (2014) Increased SNAIL expression and low syndecan levels are associated with high Gleason grade in prostate cancer. Int J Oncol 44(3):647–654. https://doi.org/10.3892/ijo.2014.2254

Pulkkinen JO, Penttinen M, Jalkanen M, Klemi P, Grénman R (1997) Syndecan-1: a new prognostic marker in laryngeal cancer. Acta Otolaryngol 117(2):312–315. https://doi.org/10.3109/00016489709117794

Purushothaman A, Chen L, Yang Y, Sanderson RD (2008) Heparanase stimulation of protease expression implicates it as a master regulator of the aggressive tumor phenotype in myeloma. J Biol Chem 283(47):32628–32636. https://doi.org/10.1074/jbc.M806266200

Purushothaman A, Hurst DR, Pisano C, Mizumoto S, Sugahara K, Sanderson RD (2011a) Heparanase-mediated loss of nuclear syndecan-1 enhances histone acetyltransferase (HAT) activity to promote expression of genes that drive an aggressive tumor phenotype. J Biol Chem 286(35):30377–30383. https://doi.org/10.1074/jbc.M111.254789

Purushothaman A, Hurst DR, Pisano C, Mizumoto S, Sugahara K, Sanderson RD (2011b) Heparanase-mediated loss of nuclear Syndecan-1 enhances histone acetyltransferase (HAT) activity to promote expression of genes that drive an aggressive tumor phenotype*. J Biol Chem 286(35):30377–30383. https://doi.org/10.1074/jbc.M111.254789

Qiao W, Liu H, Guo W, Li P, Deng M (2019) Prognostic and clinical significance of syndecan-1 expression in breast cancer: a systematic review and meta-analysis. Eur J Surg Oncol 45(7): 1132–1137. https://doi.org/10.1016/j.ejso.2018.12.019

Rangarajan S, Richter JR, Richter RP, Bandari SK, Tripathi K, Vlodavsky I, Sanderson RD (2020) Heparanase-enhanced shedding of Syndecan-1 and its role in driving disease pathogenesis and progression. J Histochem Cytochem 68(12):823–840. https://doi.org/10.1369/0022155420937087

Regős E, Abdelfattah HH, Reszegi A, Szilák L, Werling K, Szabó G, Kiss A, Schaff Z, Kovalszky I, Baghy K (2018) Syndecan-1 inhibits early stages of liver fibrogenesis by interfering with TGFβ1 action and upregulating MMP14. Matrix Biol 68-69:474–489. https://doi.org/10.1016/j.matbio.2018.02.008

Regős E, Karászi K, Reszegi A, Kiss A, Schaff Z, Baghy K, Kovalszky I (2020) Syndecan-1 in liver diseases. Pathol Oncol Res 26(2):813–819. https://doi.org/10.1007/s12253-019-00617-0

Ren Z, van Andel H, de Lau W, Hartholt RB, Maurice MM, Clevers H, Kersten MJ, Spaargaren M, Pals ST (2018) Syndecan-1 promotes Wnt/β-catenin signaling in multiple myeloma by presenting Wnts and R-spondins. Blood 131(9):982–994. https://doi.org/10.1182/blood-2017-07-797050

Reszegi A, Karászi K, Tóth G, Rada K, Váncza L, Turiák L, Schaff Z, Kiss A, Szilák L, Szabó G, Petővári G, Sebestyén A, Dezső K, Regős E, Tátrai P, Baghy K, Kovalszky I (2021a) Overexpression of human Syndecan-1 protects against the Diethylnitrosamine-induced Hepatocarcinogenesis in mice. Cancers (Basel) 13(7):1548. https://doi.org/10.3390/cancers13071548

Reszegi A, Karászi K, Tóth G, Rada K, Váncza L, Turiák L, Schaff Z, Kiss A, Szilák L, Szabó G, Petővári G, Sebestyén A, Dezső K, Regős E, Tátrai P, Baghy K, Kovalszky I (2021b) Overexpression of human Syndecan-1 protects against the Diethylnitrosamine-induced Hepatocarcinogenesis in mice. Cancers 13(7):1548

Rintala M, Inki P, Klemi P, Jalkanen M, Grénman S (1999) Association of syndecan-1 with tumor grade and histology in primary invasive cervical carcinoma. Gynecol Oncol 75(3):372–378. https://doi.org/10.1006/gyno.1999.5595

Roper JA, Williamson RC, Bass MD (2012) Syndecan and integrin interactomes: large complexes in small spaces. Curr Opin Struct Biol 22(5):583–590. https://doi.org/10.1016/j.sbi.2012.07.003

Roskams T, Moshage H, De Vos R, Guido D, Yap P, Desmet V (1995) Heparan sulfate proteoglycan expression in normal human liver. Hepatology 21(4):950–958

Roskams T, De Vos R, David G, Van Damme B, Desmet V (1998) Heparan sulphate proteoglycan expression in human primary liver tumours. J Pathol 185(3):290–297. https://doi.org/10.1002/(sici)1096-9896(199807)185:3<290::Aid-path91>3.0.Co;2-i

Sanderson RD, Turnbull JE, Gallagher JT, Lander AD (1994) Fine structure of heparan sulfate regulates syndecan-1 function and cell behavior. J Biol Chem 269(18):13100–13106
Saphire AC, Bobardt MD, Zhang Z, David G, Gallay PA (2001) Syndecans serve as attachment receptors for human immunodeficiency virus type 1 on macrophages. J Virol 75(19): 9187–9200. https://doi.org/10.1128/JVI.75.19.9187-9200.2001
Saqi A, Yun SS, Yu GH, Alexis D, Taub RN, Powell CA, Borczuk AC (2005) Utility of CD138 (syndecan-1) in distinguishing carcinomas from mesotheliomas. Diagn Cytopathol 33(2): 65–70. https://doi.org/10.1002/dc.20297
Shi Q, Jiang J, Luo G (2013) Syndecan-1 serves as the major receptor for attachment of hepatitis C virus to the surfaces of hepatocytes. J Virol 87(12):6866–6875. https://doi.org/10.1128/JVI.03475-12
Shimada K, Anai S, Fujii T, Tanaka N, Fujimoto K, Konishi N (2013) Syndecan-1 (CD138) contributes to prostate cancer progression by stabilizing tumour-initiating cells. J Pathol 231(4):495–504. https://doi.org/10.1002/path.4271
Shinyo Y, Kodama J, Hasengaowa KT, Hiramatsu Y (2005) Loss of cell-surface heparan sulfate expression in both cervical intraepithelial neoplasm and invasive cervical cancer. Gynecol Oncol 96(3):776–783. https://doi.org/10.1016/j.ygyno.2004.11.004
Stanford KI, Bishop JR, Foley EM, Gonzales JC, Niesman IR, Witztum JL, Esko JD (2009) Syndecan-1 is the primary heparan sulfate proteoglycan mediating hepatic clearance of triglyceride-rich lipoproteins in mice. J Clin Invest 119(11):3236–3245. https://doi.org/10.1172/JCI38251
Stanley MJ, Stanley MW, Sanderson RD, Zera R (1999) Syndecan-1 expression is induced in the stroma of infiltrating breast carcinoma. Am J Clin Pathol 112(3):377–383. https://doi.org/10.1093/ajcp/112.3.377
Stepp MA, Pal-Ghosh S, Tadvalkar G, Pajoohesh-Ganji A (2015) Syndecan-1 and its expanding list of contacts. Adv Wound Care (New Rochelle) 4(4):235–249. https://doi.org/10.1089/wound.2014.0555
Stewart MD, Ramani VC, Sanderson RD (2015) Shed syndecan-1 translocates to the nucleus of cells delivering growth factors and inhibiting histone acetylation: a novel mechanism of tumor-host cross-talk. J Biol Chem 290(2):941–949. https://doi.org/10.1074/jbc.M114.608455
Su G, Blaine SA, Qiao D, Friedl A (2007) Shedding of syndecan-1 by stromal fibroblasts stimulates human breast cancer cell proliferation via FGF2 activation. J Biol Chem 282(20):14906–14915. https://doi.org/10.1074/jbc.M611739200
Subramanian SV, Fitzgerald ML, Bernfield M (1997) Regulated shedding of Syndecan-1 and -4 Ectodomains by thrombin and growth factor receptor activation*. J Biol Chem 272(23): 14713–14720. https://doi.org/10.1074/jbc.272.23.14713
Suhovskih AV, Mostovich LA, Kunin IS, Boboev MM, Nepomnyashchikh GI, Aidagulova SV, Grigorieva EV (2013) Proteoglycan expression in normal human prostate tissue and prostate cancer. ISRN Oncol 2013:680136. https://doi.org/10.1155/2013/680136
Suhovskih AV, Aidagulova SV, Kashuba VI, Grigorieva EV (2015) Proteoglycans as potential microenvironmental biomarkers for colon cancer. Cell Tissue Res 361(3):833–844. https://doi.org/10.1007/s00441-015-2141-8
Surviladze Z, Sterkand RT, Ozbun MA (2015) Interaction of human papillomavirus type 16 particles with heparan sulfate and syndecan-1 molecules in the keratinocyte extracellular matrix plays an active role in infection. J Gen Virol 96(8):2232–2241. https://doi.org/10.1099/vir.0.000147
Szarvas T, Reis H, Kramer G, Shariat SF, Vom Dorp F, Tschirdewahn S, Schmid KW, Kovalszky I, Rübben H (2014) Enhanced stromal syndecan-1 expression is an independent risk factor for poor survival in bladder cancer. Hum Pathol 45(4):674–682. https://doi.org/10.1016/j.humpath.2013.10.036
Szarvas T, Sevcenco S, Módos O, Keresztes D, Nyirády P, Kubik A, Romics M, Kovalszky I, Reis H, Hadaschik B, Shariat SF, Kramer G (2018a) Circulating syndecan-1 is associated with

chemotherapy-resistance in castration-resistant prostate cancer. Urol Oncol 36(6):312.e319–312.e315. https://doi.org/10.1016/j.urolonc.2018.03.010

Szarvas T, Sevcenco S, Módos O, Keresztes D, Nyirády P, Kubik A, Romics M, Kovalszky I, Reis H, Hadaschik B, Shariat SF, Kramer G (2018b) Circulating syndecan-1 is associated with chemotherapy-resistance in castration-resistant prostate cancer. Urologic Oncology: Seminars and Original Investigations 36(6):312.e319–312.e315. https://doi.org/10.1016/j.urolonc.2018.03.010

Szatmári T, Mundt F, Heidari-Hamedani G, Zong F, Ferolla E, Alexeyenko A, Hjerpe A, Dobra K (2012) Novel genes and pathways modulated by syndecan-1: implications for the proliferation and cell-cycle regulation of malignant mesothelioma cells. PLoS One 7(10):e48091. https://doi.org/10.1371/journal.pone.0048091

Szatmári T, Mundt F, Kumar-Singh A, Möbus L, Ötvös R, Hjerpe A, Dobra K (2017) Molecular targets and signaling pathways regulated by nuclear translocation of syndecan-1. BMC Cell Biol 18(1):34. https://doi.org/10.1186/s12860-017-0150-z

Teng YH, Aquino RS, Park PW (2012) Molecular functions of syndecan-1 in disease. Matrix Biol 31(1):3–16. https://doi.org/10.1016/j.matbio.2011.10.001

Tiemann K, Weigel MT, Alkatout I, Wenners AS, Mundhenke H, Schäfer FW, Bauer M, Schem C, Maass N, Jonat W, Mundhenke C (2014) Significance of syndecan-1 expression in ductal carcinoma in situ of the breast. Anticancer Res 34(7):3607–3616

Turi Z, Lacey M, Mistrik M, Moudry P (2019) Impaired ribosome biogenesis: mechanisms and relevance to cancer and aging. Aging (Albany NY) 11(8):2512–2540. https://doi.org/10.18632/aging.101922

Tzivion G, Dobson M, Ramakrishnan G (2011) FoxO transcription factors; regulation by AKT and 14-3-3 proteins. Biochim Biophys Acta 1813(11):1938–1945. https://doi.org/10.1016/j.bbamcr.2011.06.002

Uen Y, Wang J-W, Wang C, Jhang Y, Chung J-Y, Tseng T, Sheu M, Lee S (2018) Mining of potential microRNAs with clinical correlation - regulation of syndecan-1 expression by miR-122-5p altered mobility of breast cancer cells and possible correlation with liver injury. Oncotarget 9(46):28165–28175. https://doi.org/10.18632/oncotarget.25589

Vanhoutte D, Schellings MW, Götte M, Swinnen M, Herias V, Wild MK, Vestweber D, Chorianopoulos E, Cortés V, Rigotti A, Stepp MA, Van de Werf F, Carmeliet P, Pinto YM, Heymans S (2007a) Increased expression of syndecan-1 protects against cardiac dilatation and dysfunction after myocardial infarction. Circulation 115(4):475–482. https://doi.org/10.1161/circulationaha.106.644609

Vanhoutte D, Schellings MWM, Götte M, Swinnen M, Herias V, Wild MK, Vestweber D, Chorianopoulos E, Cortés V, Rigotti A, Stepp M-A, Werf FV, Carmeliet P, Pinto YM, Heymans S (2007b) Increased expression of Syndecan-1 protects against cardiac dilatation and dysfunction after myocardial infarction. Circulation 115(4):475–482. https://doi.org/10.1161/CIRCULATIONAHA.106.644609

Voyvodic PL, Min D, Liu R, Williams E, Chitalia V, Dunn AK, Baker AB (2014) Loss of Syndecan-1 induces a pro-inflammatory phenotype in endothelial cells with a dysregulated response to Atheroprotective flow*. J Biol Chem 289(14):9547–9559. https://doi.org/10.1074/jbc.M113.541573

Wang H, Jin H, Beauvais DM, Rapraeger AC (2014a) Cytoplasmic domain interactions of syndecan-1 and syndecan-4 with α6β4 integrin mediate human epidermal growth factor receptor (HER1 and HER2)-dependent motility and survival. J Biol Chem 289(44):30318–30332. https://doi.org/10.1074/jbc.M114.586438

Wang X, Zuo D, Chen Y, Li W, Liu R, He Y, Ren L, Zhou L, Deng T, Wang X, Ying G, Ba Y (2014b) Shed Syndecan-1 is involved in chemotherapy resistance via the EGFR pathway in colorectal cancer. Br J Cancer 111(10):1965–1976. https://doi.org/10.1038/bjc.2014.493

Watanabe A, Mabuchi T, Satoh E, Furuya K, Zhang L, Maeda S, Naganuma H (2006) Expression of syndecans, a heparan sulfate proteoglycan, in malignant gliomas: participation of nuclear factor-kappaB in upregulation of syndecan-1 expression. J Neuro-Oncol 77(1):25–32. https://doi.org/10.1007/s11060-005-9010-3

Wiksten JP, Lundin J, Nordling S, Kokkola A, Haglund C (2000) A prognostic value of syndecan-1 in gastric cancer. Anticancer Res 20(6d):4905–4907

Wiksten JP, Lundin J, Nordling S, Lundin M, Kokkola A, von Boguslawski K, Haglund C (2001) Epithelial and stromal syndecan-1 expression as predictor of outcome in patients with gastric cancer. Int J Cancer 95(1):1–6. https://doi.org/10.1002/1097-0215(20010120)95:1<1::aid-ijc1000>3.0.co;2-5

Wilsie LC, Gonzales AM, Orlando RA (2006) Syndecan-1 mediates internalization of apoE-VLDL through a low density lipoprotein receptor-related protein (LRP)-independent, non-clathrin-mediated pathway. Lipids Health Dis 5:23. https://doi.org/10.1186/1476-511x-5-23

Xiong Z, Jiang B, Li G (2020) Downregulation of miR-10a inhibits cutaneous squamous cell carcinoma cell proliferation, migration, and invasion by targeting Syndecan-1. Int J Clin Exp Pathol 13(10):2502–2512

Yang Y, Macleod V, Miao HQ, Theus A, Zhan F, Shaughnessy JD Jr, Sawyer J, Li JP, Zcharia E, Vlodavsky I, Sanderson RD (2007) Heparanase enhances syndecan-1 shedding: a novel mechanism for stimulation of tumor growth and metastasis. J Biol Chem 282(18):13326–13333. https://doi.org/10.1074/jbc.M611259200

Yang J, Antin P, Berx G, Blanpain C, Brabletz T, Bronner M, Campbell K, Cano A, Casanova J, Christofori G, Dedhar S, Derynck R, Ford HL, Fuxe J, García de Herreros A, Goodall GJ, Hadjantonakis AK, Huang RJY, Kalcheim C, Kalluri R, Kang Y, Khew-Goodall Y, Levine H, Liu J, Longmore GD, Mani SA, Massagué J, Mayor R, McClay D, Mostov KE, Newgreen DF, Nieto MA, Puisieux A, Runyan R, Savagner P, Stanger B, Stemmler MP, Takahashi Y, Takeichi M, Theveneau E, Thiery JP, Thompson EW, Weinberg RA, Williams ED, Xing J, Zhou BP, Sheng G (2020) Guidelines and definitions for research on epithelial-mesenchymal transition. Nat Rev Mol Cell Biol 21(6):341–352. https://doi.org/10.1038/s41580-020-0237-9

Zaragosi LE, Dadone B, Michiels JF, Marty M, Pedeutour F, Dani C, Bianchini L (2015) Syndecan-1 regulates adipogenesis: new insights in dedifferentiated liposarcoma tumorigenesis. Carcinogenesis 36(1):32–40. https://doi.org/10.1093/carcin/bgu222

Zhang F, Sodroski C, Cha H, Li Q, Liang TJ (2017) Infection of hepatocytes with HCV increases cell surface levels of Heparan sulfate proteoglycans, uptake of cholesterol and lipoprotein, and virus entry by up-regulating SMAD6 and SMAD7. Gastroenterology 152(1):257–270.e257. https://doi.org/10.1053/j.gastro.2016.09.033

Zong F, Fthenou E, Wolmer N, Hollósi P, Kovalszky I, Szilák L, Mogler C, Nilsonne G, Tzanakakis G, Dobra K (2009) Syndecan-1 and FGF-2, but not FGF receptor-1, share a common transport route and co-localize with heparanase in the nuclei of mesenchymal tumor cells. PLoS One 4(10):e7346. https://doi.org/10.1371/journal.pone.0007346

Zong F, Fthenou E, Mundt F, Szatmári T, Kovalszky I, Szilák L, Brodin D, Tzanakakis G, Hjerpe A, Dobra K (2011) Specific syndecan-1 domains regulate mesenchymal tumor cell adhesion, motility and migration. PLoS One 6(6):e14816. https://doi.org/10.1371/journal.pone.0014816

Chapter 4
Syndecan-2 Biology and Its Role in Colorectal Carcinoma

Eok-Soo Oh and John R. Couchman

Abstract The mammalian syndecan family of transmembrane proteoglycans comprises four members. Syndecan-2 is widely expressed, notably in mesenchymal tissues and has roles in cell adhesion, extracellular matrix assembly and its turnover, as well as concentrating ligands, such as growth factors, cytokines, and morphogens at the cell surface. Although signaling through syndecan-2 core protein is incompletely understood, research has indicated roles for this receptor in colorectal cancer. Its expression in transformed epithelia has been related to increased aggressiveness and invasive potential. At the same time, it has been shown that syndecan-2 is subject to epigenetic regulation at the level of promoter methylation. Hypermethylated SDC2 shows promise as a marker of the earliest stages of colorectal cancer and clinical trials have been established. The biology of syndecan-2 is therefore potentially relevant to the progression of this and other cancers, but further work is required to fully understand its molecular functions.

4.1 Introduction

There are four mammalian members of the syndecan family of transmembrane proteoglycans. Each can bear heparan sulfate chains on their ectodomains, while syndecans-1 and -3 can also carry chondroitin or dermatan sulfate chains (Fig. 4.1). The four syndecans are derived from a common ancestor, there being a single syndecan in invertebrates of the Bilateria (Chakravarti and Adams 2006). It appears that at the invertebrate/vertebrate boundary, two rounds of gene duplication occurred, one branch having slightly larger core proteins than the other.

E.-S. Oh
Department of Life Sciences, Ewha Womans University, Seoul, South Korea
e-mail: ohes@ewha.ac.kr

J. R. Couchman (✉)
Biotech Research & Innovation Centre, University of Copenhagen, Copenhagen, Denmark
e-mail: john.couchman@bric.ku.dk

I. Kovalszky et al. (eds.), *The Extracellular Matrix and the Tumor Microenvironment*, Biology of Extracellular Matrix 11,
https://doi.org/10.1007/978-3-030-99708-3_4

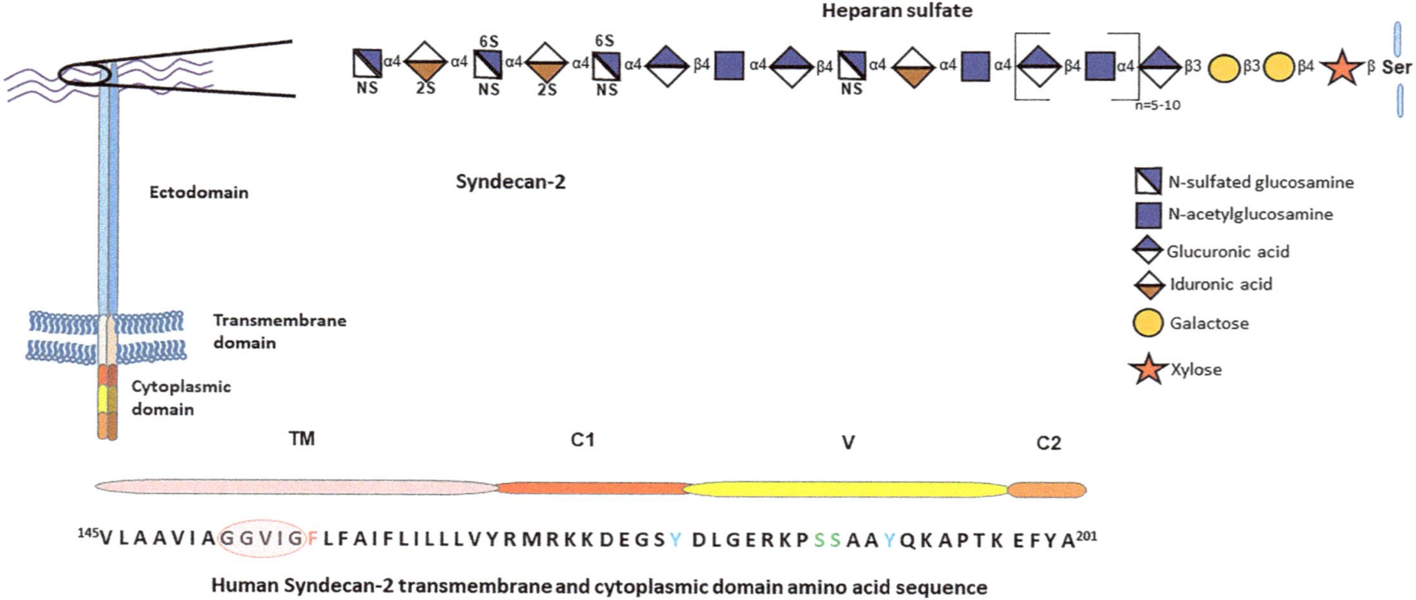

Fig. 4.1 Diagrammatic structure of syndecan 2 proteoglycan. Heparan sulfate chains are located toward the N-terminus, the composition of which is shown on the right. Below is shown the amino acid sequence of human syndecan-2 transmembrane (pink) and cytoplasmic domains. The GXXXG motif in the transmembrane domain that promotes dimerization is circled and the phenylalanine residue immediately adjacent to this motif is highlighted. This residue also contributes to dimer formation. The cytoplasmic domain has two conserved motifs, C1 and C2, either side of a variable (V) region specific to each syndecan, but nevertheless conserved across species. Tyrosine (blue) and serine (green) residues that are potentially phosphorylated are shown

Syndecan-2 is one of the smaller syndecans, along with syndecan-4 (Gopal et al. 2021). As part of having smaller core proteins, the cytoplasmic domains of these two syndecans are also smaller than that of the invertebrate ancestor as well as syndecans-1 and -3. Each of the four syndecans has a specific expression pattern, both in spatial and temporal terms (Gopal et al. 2021; Bernfield et al. 1999; Couchman 2010). Roles in neuronal guidance and stem cell regulation have been demonstrated in the invertebrate syndecans of *C. elegans* and *Drosophila* and similar trends can be seen in syndecan-3 of vertebrates (Rhiner et al. 2005; Johnson et al. 2004; Pisconti et al. 2016). However, although it is very likely that the four syndecans of mammals have embraced new roles, details are still incomplete.

Gene deletions of syndecans have been revealing. A common theme in syndecan-1 and -4 knockouts is that wound repair is delayed or impaired, though the mice are viable (Stepp et al. 2002; Echtermeyer et al. 2001). Indeed, double knockouts of syndecan-1 and -4 are also viable, but have alterations in, for example, epidermal cell morphology and junctions (Gopal et al. 2015). It has been noted several times that deletion of syndecans can lead to alterations in cadherin expression and function (Gopal et al. 2015; Kato et al. 1995; Gopal et al. 2017). Syndecan-3 knockouts have defects in hippocampus-dependent memory and enhanced long-term potentiation due to loss of interaction with heparin-binding growth-associated molecule (HB-GAM; Hienola et al. 2006). In skeletal muscle, loss of syndecan-3, which binds to Notch, leads to altered satellite cell interactions with their niche (Pisconti et al. 2010). To date, there has been a single report on syndecan-2 deletion in the mouse (Corti et al. 2019). There are defects in angiogenesis both in development and postnatal wound repair. The molecular basis appears to be that syndecan-2 forms a complex with $VEGFA_{165}$ (which is heparan sulfate-binding) and the receptor, VEGFR2. Moreover, it is proposed that the N-terminal 59 amino acids of the core protein provide information that results in enhanced 6-O-sulfation of the heparan sulfate chains, and which is required for the VEGF/VEGFR interactions. How core protein properties impact sulfation events in the Golgi apparatus is not yet understood. This region is distinct in syndecan-4 and does not enhance 6-O-sulfation, potentially explaining why it does not share the angiogenesis properties (Corti et al. 2019; Lambert et al. 2020). However, contrary data support the possibility that under some circumstances, syndecan-4 can promote angiogensis, pathogenic angiogenesis in particular (Echtermeyer et al. 2001; Pessolano et al. 2021; De Rossi et al. 2021). The syndecan-2 data are, however, consistent with angiogenesis deficits noted previously in zebrafish models using morpholinos directed to the proteoglycan core protein (Chen et al. 2004a).

4.2 Structure and Glycosylation

All four syndecans have cytoplasmic domains that can interact with the actin-based cytoskeleton and PDZ domain proteins that may regulate localization and trafficking (Gopal et al. 2021; Couchman 2010). The cytoplasmic domains of syndecans can be resolved into three regions, two highly conserved, flanking a "variable" (V) region (Fig. 4.1). Each syndecan therefore has a distinct V region, likely enabling specific interactions with cytoplasmic signaling molecules, but details still remain sparse. The exception is syndecan-4, where interactions with α-actinin and protein kinase Cα have been demonstrated that can lead to enhanced cell-extracellular matrix (ECM) adhesion and the assembly of actin-based focal adhesion/contact organelles (Gopal et al. 2021; Couchman 2010; Bass and Humphries 2002; Elfenbein and Simons 2013; Mitsou et al. 2017). Syndecan-4 cytoplasmic domain is also the only one that has been structurally resolved. It forms twisted clamp dimers that are stabilized by interactions with specific inositol phospholipids (Lee et al. 1998). It is quite possible that syndecans-1, -2, and -3 cytoplasmic domains form similar structures. A tendency to form dimers is also seen with the single pass transmembrane domains, which can form dimers resistant to dissociating detergents such as sodium dodecyl sulfate (Choi et al. 2005). An NMR study of syndecan-2 transmembrane domain monomer has been reported (Li et al. 2019), showing a conventional helical structure. The inherent ability of syndecan transmembrane domains to form dimers is derived from a GXXXG sequence, also noted in glycophorin, for example (Choi et al. 2005; Teese and Langosch 2015; Dews and Mackenzie 2007). In the case of syndecan-2, this dimer formation is reinforced by a single phenylalanine residue immediately C-terminal of the GXXXG motif (Kwon et al. 2015).

In a very recent, illuminating study, it has now been shown that the ectodomains of all four syndecans are disordered and flexible. This may permit multiple interactions with other molecules on the cell surface, and adopt different conformations that facilitate complexes with other cell surface receptors (Gondelaud et al. 2021). This work was completed on core protein ectodomains lacking glycosylation, and it will be interesting to see in the future how glycosaminoglycan chains influence conformations. Moreover, given the transmembrane domain tendency to form dimers, it may be that their presence, along with resultant dimers may also constrain ectodomain conformation. This is already hinted at by the use of a syndecan-4 ectodomain dimer possessing a single cysteine bridge between two monomers (Gondelaud et al. 2021).

By virtue of the fact that heparan sulfate chains can interact with a myriad of proteins with many different functions, syndecans can be involved with a range of cellular processes. Included are many growth factors, cytokines, chemokines, enzymes, extracellular matrix macromolecules, and mediators of lipid metabolism. The structure and synthesis of heparan sulfate has been well documented (Lindahl et al. 2017; Xu and Esko 2014) and is summarized in Fig. 4.1. In essence, the chains are a series of repeating disaccharides, consisting of N-acetyl glucosamine and glucuronic acid. However, chains are extensively modified by sulfation and epimerization of the uronic acid residues to iduronate. A number of specific transferases combine to produce heparan sulfate chains of immense complexity and potential variability. Most heparan sulfate chains have domains of high sulfation interspersed with domains of low, or no, sulfation, with intermediate sulfation at the boundaries of these two domain types (Lindahl et al. 2017; Gallagher 2015). Despite this, it is apparent that heparan sulfate from a particular source, e.g., hepatocytes, may have conserved overall characteristics (Gallagher 2015; Kjellén and Lindahl 2018). The extent of variability in chain length, and fine structure characteristics in vivo remains a challenging question.

4.3 Syndecan-2 Interactome and Signaling

As with all syndecans, syndecan-2 interactions on the cell surface or extracellular environment can involve both the glycosaminoglycan chains and regions of the core protein ectodomain (Gopal et al. 2021; Beauvais et al. 2009; Jung et al. 2019; Whiteford et al. 2011; Tsoyi et al. 2018). In the cortical cytoplasm, further interactions can involve the short cytoplasmic domain (Gopal et al. 2015, 2021; Couchman 2010; Bass and Humphries 2002; Elfenbein and Simons 2013). Since there are scores of potential ligands that can interact with heparan sulfate, derived from many families of proteins, e.g., cytokines, chemokines, enzymes, extracellular matrix macromolecules, dissecting specific signaling through syndecans has been a challenge. It is clear in some cases that syndecans can associate with other cell surface receptors, e.g., integrins, growth factor receptors, morphogen receptors, in turn regulating specific signaling pathways (Rhiner et al. 2005; Johnson et al. 2004; Bass and Humphries 2002; Jung et al. 2019; Whiteford et al. 2011; Tsoyi et al. 2018; Mundhenke et al. 2002; Bentzinger et al. 2013). Furthermore, the cytoplasmic domains of all syndecans are small and do not possess intrinsic kinase or phosphatase activities. Therefore, docking of cytoplasmic molecules to this domain is required for signaling (Gopal et al. 2021; Gopal et al. 2015). However, specific signaling has been demonstrated for syndecan-4, which involves the docking and activation of protein kinase Cα (Gopal et al. 2021; Mitsou et al. 2017; Oh et al. 1997; Keum et al. 2004). For syndecans-1 and -4, regulation of transient receptor potential canonical (TRPC) channels has also been demonstrated (Gopal et al. 2015; Mitsou et al. 2017). Alongside, all syndecans have been shown to interact with actin and tubulin cytoskeletal components, as well as junctional components and molecules involved in internalization and trafficking (Gopal et al. 2021; Gondelaud and Ricard-Blum 2019). Indeed, an important function of syndecans may be receptor/ligand internalization, with these processes being co-opted by a number of different pathogens. Examples include some viruses, bacteria, and malarial parasites (Gopal et al. 2021; Cagno et al. 2019; Teng et al. 2012; Ayres Pereira et al. 2016).

With respect to syndecan-2, there is still much to learn regarding specific signaling roles for the core protein. However, interacting proteins have been recorded and these have been curated in a comprehensive review (Gondelaud and Ricard-Blum 2019). Since then, there has been a study of the cardiac syndecan-2 interactome (Mathiesen et al. 2020). The data as a whole derive from affinity purification methods as well as interaction studies, so that in many cases the interactions could be direct or indirect.

Gondelaud and Ricard-Blum (2019) identify a core of 18 proteins that interact with all 4 mammalian syndecans. Of these, two are heparin/heparan sulfate-binding (e.g., FGF2), while four are the syndecans themselves. This has important implications, since syndecans may therefore form hetercomplexes that render specific signaling pathways difficult to ascertain. Moreover, where specific affinity probes are used, for example, the cytoplasmic domain of a syndecan, complexation with endogenous syndecans may lead to copurification of interactors with ectodomains or heparan sulfate chains. Virtually all of the remaining 12 proteins that interact with all 4 syndecans do so by binding the conserved C1 and C2 regions of their cytoplasmic domains (Fig. 4.1). This is consonant with their highly conserved sequences (Bernfield et al. 1999; Couchman 2010) and supports the idea that syndecan-specific signaling may be engendered through the V regions that are distinct to each syndecan. Among these, 12 proteins are some that form submembranous networks, e.g., syntenin, two tyrosine kinases of the Src family, the cytoskeletal proteins cortactin and tubulin, a vesicle trafficking protein (TRAPPC4/synbindin), and one integrin subunit (β4; Gondelaud and Ricard-Blum 2019; Mathiesen et al. 2020). However, in the case of syndecan-2 that is largely mesenchymal in distribution, interactions with the predominantly epithelial α6β4 integrin may be infrequent.

Of the 56 interacting partners that are listed for syndecan-2, 25% are, somewhat surprisingly, lysosomal, while many others putatively bind to the heparan sulfate chains (Gondelaud and Ricard-Blum 2019). Only three proteins are listed as interacting with syndecan-2's cytoplasmic V region. These are caveolin-2, EphB2 and Sarm-1 (sterile α and toll interleukin receptor motif containing protein-1) which hydrolyses NAD^+ (Loring and Thompson 2020). The cardiac interactome study, using three complementary affinity-based techniques revealed remarkably few heparan sulfate-interacting partners, but did include three of the four cavins (Mathiesen et al. 2020). Alongside the interactions with caveolin-2 and lysosomal proteins, this suggests an important feature of syndecan-2 function relates to trafficking. Other interactors included G protein-related, ubiquitin-related, as well as cytoskeletal proteins (e.g., cortactin, which interacts with all syndecans).

The V region interaction with EphB2 adhesion protein raises an interesting further dimension, that of phosphorylation. This interaction requires phosphorylation of two tyrosine residues ($Y^{179}$and Y^{191}—human numbering), but not that of the C2 tyrosine (Y^{200}) (Fig. 4.1; Ethell et al. 2001). This promoted syndecan clustering and dendritic spine formation. Syndecan-2 phosphorylation also featured in a study of left-right asymmetry in *Xenopus* (Kramer et al. 2002). In this case, the two adjacent serine residues were phosphorylated by protein kinase Cγ ($S^{150/151}$—*Xenopus* numbering, corresponding to $S^{187/188}$ in human) (Fig. 4.1). Phosphorylation in right, but not left animal cap cells led to noncell autonomous signaling, perhaps suggestive of syndecan clustering once again.

A specific interaction of syndecan-2 ectodomain with the phosphatase CD148 (PTPRJ) is potentially important. This direct interaction was first identified in the context of integrin-mediated cell adhesion (Whiteford et al. 2011). The mechanism in fibroblasts involved both Src tyrosine kinase and the C2β isoform of PI3-kinase (Whiteford et al. 2011). Subsequently, the syndecan-2/CD148 axis has been found to regulate fibrosis (Tsoyi et al. 2018, 2021), and the depletion of CD148 signaling exacerbates experimental lung fibrosis (Tsoyi et al. 2021). These interesting data recall previous work showing roles for syndecan-2 in extracellular matrix assembly (Klass et al. 2000; Galante and Schwarzbauer 2007; Jang et al. 2020). Moreover, syndecan-2 can mediate the internalization of TGF-βRI through a caveolin-mediated pathway (Shi et al. 2013), thereby moderating processes such as fibroblast activation and fibrosis. These properties are distinct to syndecan-2 and are not shared with its closest relative, syndecan-4. This may be advantageous in the event that syndecan-2 signaling in fibrosis becomes a potential target.

4.4 Syndecan-2 in Human Diseases

All cellular components of normal tissue are tightly regulated in response to external signals in order to perform appropriate and specific functions in tissue homeostasis. Since external stimuli vary in time and space, cells and tissues are in continual reciprocal interactions with their environment. One mechanism is to regulate the expression levels of cell surface receptors. Altered cell surface receptor expression in tissues either restores tissue homeostasis or induces changes that can become pathogenic. Syndecan-2 is a receptor now known to be aberrantly regulated in some diseases, notably cancers. When syndecan-2 was first characterized by cDNA cloning, it was named "fibroglycan," indicative of this syndecan family member having a strong bias toward mesenchymal cell expression (David et al. 1993). Indeed, it is commonly expressed in cells of mesenchymal origin, including fibroblasts, smooth muscle, endothelial, as well as neuronal cells, monocytes, and macrophages (Gopal et al. 2021; Bernfield et al. 1999; Ethell et al. 2001; Whiteford et al. 2007; Zhao et al. 2012; Noguer et al. 2009; Clasper et al. 1999; Essner et al. 2006). Therefore, acquired syndecan-2 expression in epithelial tissue that is normally low in expression creates new functions and changes tissue homeostasis, while decreased syndecan-2 expression in mesenchymal cells may also result in altered behavior.

A common trigger for altered syndecan-2 expression is inflammation, with chronic inflammation leading to diseases such as fibrosis. For instance, the inflammatory cytokines IL-1β, IL-6, and TGF-β1 upregulate syndecan-2 expression in myofibroblasts and also regulate the development of liver cirrhosis (Renga et al. 2014). However, increased expression of syndecan-2 in response to inflammation is not unique to this family member, but may be common across the family (Gopal et al. 2021; Couchman 2010; Teixeira and Götte 2020; Arokiasamy et al. 2020). For example, the expression of syndecan-2 along with other syndecan family members is also increased in myocytes after myocardial infarction (Wernly et al. 2019; Xie et al. 2012; Lunde et al. 2016), which may regulate cardiac ECM remodeling.

Syndecan-2 expression often changes in transformed cells. Overexpression of syndecan-2 is associated with the tumorigenic properties and poor prognosis of epithelial malignancies (Park et al. 2005; Mytilinaiou et al. 2013). Syndecan-2 promotes an aggressive phenotype of the MDA-MB-231 breast carcinoma cell line by combining with caveolin-2 and in part by regulating RhoGTPase (Lim and

Couchman 2014). In pancreatic cancer cells, when syndecan-2, through p120-GAP, activates Src in the absence of RACK1, mutated K-Ras activation induces cancer cell proliferation, spreading, and invasion (Betriu et al. 2021). Elevated syndecan-2 expression in skin-derived melanoma cells regulates their migratory properties (Lee et al. 2009).

Syndecan-2 expression is also increased in sarcoma progression. In the case of fibrosarcoma, a rare malignant tumor originating from fibroblasts, syndecan-2 regulates IGF-1-dependent actin cytoskeletal organization and cell motility (Mytilinaiou et al. 2017a) and TGFβ2-dependent fibrosarcoma cell functions such as cell adhesion (Mytilinaiou et al. 2013). In osteosarcoma, syndecan-2 plays a pro-apoptotic role and is therefore anti-oncogenic (Dieudonné et al. 2010). In addition, its increased expression in osteosarcoma cells was associated with reduced migration and invasion (Orosco et al. 2007). Since syndecan-2 generally has a pro-migratory function in carcinoma cells, it is possible to predict that its increased expression may have pro-oncogenic functions in these cells. Indeed, in colon cancer, one of the most studied carcinomas, the obtained syndecan-2 expression can influence a range of cancer cell behaviors. Syndecan-2 expression on the cell surface causes carcinoma cells to reorganize the cytoskeleton (Kusano et al. 2004), remodel the ECM (Jang et al. 2020), promote a more migratory and invasive cell phenotype or epithelial-to-mesenchymal transition (EMT) (Hua et al. 2020; Contreras et al. 2010), and angiogenesis (Noguer et al. 2009), all of which are all critical for carcinoma cells to establish primary tumors and secondary metastases.

4.5 Syndecan-2 in Colorectal Cancer

One of the best studied carcinomas related to roles of syndecan-2 is colorectal cancer. Several studies have shown that increased syndecan-2 is involved in the regulation of adhesion, proliferation, migration and invasion, cytoskeletal organization, and ECM assembly to regulate the tumorigenic activity of colon carcinoma (Klass et al. 2000; Han et al. 2004; Park et al. 2002).

As a cell surface receptor, syndecan-2 regulates the behavior of colon cancer cells in three different ways. First, as an adhesion receptor, syndecan-2 mediates intracellular signaling events in response to the external environment. Upon interaction with external ligands through the glycosaminoglycan (GAG) chains attached to the extracellular domain and/or the extracellular core protein itself, syndecan-2 triggers the transmission of signals from the extracellular environment into the cytosol to regulate the intracellular environment (Chen et al. 2004b; Mytilinaiou et al. 2017b), probably together with integrin-dependent signaling (Choi et al. 2009). The C-terminal EFYA sequence of syndecan-2 cytoplasmic domain interacts with syntenin-1, a scaffold protein containing two PDZ domains, which controls membrane localization and activation of Tiam-1, a guanine nucleotide exchange factor (GEF) of Rac (Grootjans et al. 1997; Lee et al. 2011; Choi et al. 2010). Syndecan-2 thus induces syntenin-1-Tiam1-mediated Rac activation, which leads to actin filament assembly, for example, the production of filopodia, morphological changes, and increased cell motility (Choi et al. 2010). Syndecan-2 also regulates the expression and secretion of ECM components that have various functions in modulating cancer progression. For example, the engagement of the extracellular domain of syndecan-2 and the ECM transduces signals to activate PKCγ/FAK/ERK signaling pathway to induce the expression of MMP-7, which degrading a wide variety of ECM molecules, contributes to the remodeling of the matrix and favors colon cancer progression (Jang et al. 2016, 2017). Therefore, syndecan-2 expression regulates both intracellular and extracellular signaling events (Fig. 4.2).

Second, as a docking receptor, syndecan-2 regulates multiple extracellular events including receptor activation and the turnover of ECM. For instance, in the ECM, the binding of FGF-2 to syndecan-2 initiates a high local concentration of FGF-2 on the cell surface. This interaction facilitates efficient FGF-2 binding to the FGFRs with higher affinity than FGF-2 alone and induces FGF receptor activation which signals for cancer cell elongation (polarity) and motility (Clasper et al. 1999; Essner et al. 2006). Similarly, syndecan-2 increases the binding of $VEGF_{165}$ to VEGF receptor-2 (VEFGR2) and activation of VEGFR2 (Corti et al. 2019), which is necessary for angiogenesis in zebrafish development (Chen et al. 2004a). On the other hand, the binding of syndecan-2 to the pro-form of MMP-7 triggers the processing of pro-MMP-7 into active MMP-7 (Ryu et al. 2009), which results in activation of extracellular proteases. Activated MMP-7 may then enhance degradation of proteins of both ECM and cell surface (Gopal et al. 2015; Choi et al. 2012). As an ECM-degrading enzyme, activated MMP-7 degrades protein components such as collagen, fibronectin, and proteoglycans (Nagase and Woessner 1999; Miyazaki et al. 1990), as part of cell-induced ECM remodeling. Activated MMP-7 also cleaves cell adhesion receptors such as E-cadherin (Jang et al. 2016; McGuire et al. 2003), the major cell–cell adhesion receptor of epithelial adherens junctions (McCawley and Matrisian 2001). The resulting functional loss of E-cadherin in early-stage E-cadherin-expressing colon cancer cells may induce EMT with subsequently enhanced cell migration (Jang et al. 2016; Bates and Mercurio 2005). Similarly, MMP-7 mediates the cleavage of the extracellular domain of syndecan-2 which results in loss of syndecan-2 function at the cell surface, under conditions whereby cancer cell migration and invasion are increased (Choi et al. 2012, 2015).

Third, remodeling of the ECM by MMPs releases various bioactive fragments that further regulate numerous biological processes, including tumor growth and metastasis. Since syndecan-2 is itself cleaved by MMP-2, -7, and MT1-MMP, the shed proteoglycan is also involved in the colon cancer progression. First, shed syndecan-2 may not only release signal proteins engaged at the cell surface, but also increases their activity. FGF-2 bound to shed syndecan-2 interacts more efficiently with the FGF receptor than FGF alone. As a result, syndecan-2 can enhance angiogenesis via the FGFR signaling pathway (Clasper et al. 1999; Villena et al. 2003). In addition, shed syndecan-2 can bind to (unknown) receptors to regulate cell functions as ligands to impact gene expression. In fact, treatment with shed syndecan-2 directly increases the migration of HCT116 cells (Choi et al. 2015) and promotes capillary tube formation in microvascular endothelial cells of the brain

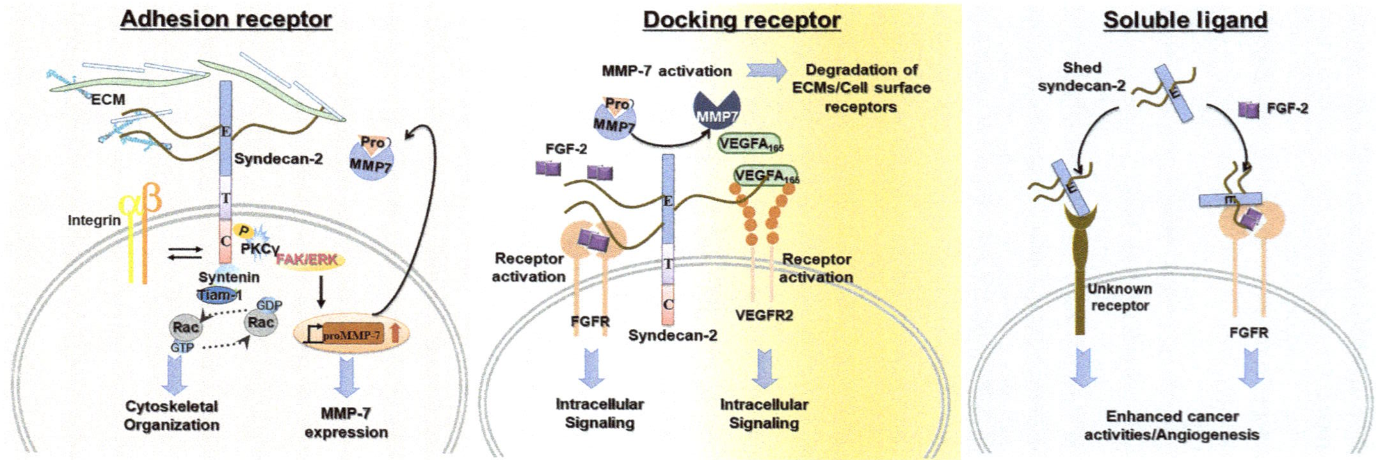

Fig. 4.2 The regulatory mechanisms of syndecan-2 in cancer. (**a**) In the context of an adhesion receptor, the extracellular domain of a syndecan-2 transduces signals from the extracellular environment to the cytosol, thereby regulating intracellular signaling events such as (i) cytoskeletal organization, and (ii) expression of ECM molecules including MMP-7. (**b**) As a docking receptor, a syndecan-2 can regulate extracellular events. When extracellular ligands bind to syndecan-2, (i) extracellular ligands such as growth factors bind with high affinities to their receptors and activate their receptor signaling and (ii) extracellular enzymes such as MMP-7 become activated. Activated MMP-7 cleaves ECM molecules and cell surface receptor such as syndecan-2, which produces shed syndecan-2 fragment in the ECM. (**c**) As a soluble ligand, (i) shed syndecan-2 bound FGF-2 activates FGF receptor signaling and (ii) shed syndecan-2 directly regulates various cancer activities including angiogenesis by unknown mechanisms

(Fears et al. 2006). Therefore, shed syndecan-2 can intensify angiogenic processes, thereby facilitating the growth and metastasis of cancer cells and promoting tumorigenic activities as a soluble ligand.

4.6 SDC2 Gene Methylation and Colon Cancer

In 2013, Oh et al. (Oh et al. 2013) published a study showing that SDC2 promoter hypermethylation could be detected in serum samples from colon cancer patients (i.e., cell free) with high specificity and sensitivity. Subsequently, several studies and literature surveys have supported this notion (Rasmussen et al. 2016; Rasmussen et al. 2017; Kim and Park 2018; Han et al. 2019; Su et al. 2021), with most research coming from South Korea and Denmark. There have been suggestions that SDC2 gene analysis can be used alongside other genes (e.g., TFPI2 or SEPT9; Zhang et al. 2021; Zhao et al. 2019). Promoter methylation can be common in early carcinogenesis, and since cancer cell exfoliation can occur before blood leakage into the colon, the hope is that potential markers such as this may detect the earliest manifestations of malignancy. Consistent with this, analysis has moved from serum to colon lavage fluid and stool DNA samples.

Two clinical trials are in progress (NCT03146520, NCT04304131) to assess the utility of stool-derived SDC2 methylation status as a basis for colorectal carcinoma detection (Kim et al. 2021). A test kit based on this premise has now been developed by Genomictree Inc., Daejon, S. Korea. It will be interesting to see whether such epigenetic analysis will supplant or accompany the current immunological tests for fecal blood. Hypermethylation of CpG islands in gene promoters is an epigenetic mechanism to repress transcription (Weber et al. 2007). It would therefore appear that syndecan-2 expression is downregulated in colon carcinoma, at least as far as serum and lumenal colon samples are concerned. However, this is most likely not as simple as it appears at first sight. As noted above, syndecan-2 may function to enhance the invasive potential of colon carcinoma cells, which suggests that perhaps the proteoglycan is expressed in the metastatic or premetastatic niche. There is clearly much more to learn regarding the regulation of syndecan-2 expression in this disease.

4.7 Conclusions

Colorectal cancer is the third most common malignancy and fourth leading cause of cancer-related death worldwide. As with many cancers, early detection can be an essential component of successful treatment. While syndecans have not been an intensively studied family of transmembrane receptors so far, it is fascinating to consider that syndecan-2 may not only be highly relevant to tumor progression, but epigenetic alterations in its methylation status may be a very useful tool for early diagnosis. Syndecan-1 is now known to be a key receptor in the progression of myeloma and is a target for therapy (Yang et al. 2007; Jagannath et al. 2019; Schönfeld et al. 2017; Yu et al. 2020). Therefore, two of the four syndecans are connected strongly to cancers. Given that other ongoing work connects syndecans of invertebrates and vertebrates with stem cell regulation (Nakato and Li 2016; Gopal et al. 2016; Kumar Katakam et al. 2021), for example, through Notch signaling (Pisconti et al. 2010; Zhao et al. 2012), it is likely that increased relevance of syndecans to human health and disease will emerge. The linkage between syndecan function and their regulation of TRPC calcium channels (Gopal et al. 2015, 2016; Mitsou et al. 2017) is also very likely to impact cell behaviors. Therefore, even though there are little data to link syndecan core protein mutations with disease, the relevance of this family, with its long evolutionary history, both for normal and pathogenic tissues has justified their study.

References

Arokiasamy S, Balderstone MJM, De Rossi G, Whiteford JR (2020) Syndecan-3 in inflammation and angiogenesis. Front Immunol 10:3031. https://doi.org/10.3389/fimmu.2019.03031

Ayres Pereira M, Mandel Clausen T, Pehrson C et al (2016) Placental sequestration of Plasmodium falciparum malaria parasites is mediated by the interaction between VAR2CSA and chondroitin sulfate A on syndecan-1. PLoS Pathog 12:e1005831. https://doi.org/10.1371/journal.ppat.1005831

Bass MD, Humphries MJ (2002) Cytoplasmic interactions of syndecan-4 orchestrate adhesion receptor and growth factor receptor signalling. Biochem J 368:1–15. https://doi.org/10.1042/BJ20021228

Bates RC, Mercurio AM (2005) The epithelial-mesenchymal transition (EMT) and colorectal cancer progression. Cancer Biol Ther 4:365–370. https://doi.org/10.4161/cbt.4.4.1655

Beauvais DM, Ell BJ, McWhorter AR, Rapraeger AC (2009) Syndecan-1 regulates alphavbeta3 and alphavbeta5 integrin activation during angiogenesis and is blocked by synstatin, a novel peptide inhibitor. J Exp Med 206:691–705. https://doi.org/10.1084/jem.20081278

Bentzinger CF, Wang YX, von Maltzahn J et al (2013) Fibronectin regulates Wnt7a signaling and satellite cell expansion. Cell Stem Cell 12:75–87. https://doi.org/10.1016/j.stem.2012.09.015

Bernfield M, Götte M, Park PW et al (1999) Functions of cell surface heparan sulfate proteoglycans. Annu Rev Biochem 68:729–777. https://doi.org/10.1146/annurev.biochem.68.1.729

Betriu N, Bertran-Mas J, Andreeva A, Semino CE (2021) Syndecans and pancreatic ductal adenocarcinoma. Biomolecules 11:349. https://doi.org/10.3390/biom11030349

Cagno V, Tseligka ED, Jones ST, Tapparel C (2019) Heparan sulfate proteoglycans and viral attachment: true receptors or adaptation bias? Viruses 11:596. https://doi.org/10.3390/v11070596

Chakravarti R, Adams JC (2006) Comparative genomics of the syndecans defines an ancestral genomic context associated with matrilins in vertebrates. BMC Genomics 7:83. https://doi.org/10.1186/1471-2164-7-83

Chen E, Hermanson S, Ekker SC (2004a) Syndecan-2 is essential for angiogenic sprouting during zebrafish development. Blood 103:1710–1719. https://doi.org/10.1182/blood-2003-06-1783

Chen L, Klass C, Woods A (2004b) Syndecan-2 regulates transforming growth factor-beta signaling. J Biol Chem 279:15715–15718. https://doi.org/10.1074/jbc.C300430200

Choi S, Lee E, Kwon S et al (2005) Transmembrane domain-induced oligomerization is crucial for the functions of syndecan-2 and syndecan-4. J Biol Chem 280:42573–42579. https://doi.org/10.1074/jbc.280.52:42573

Choi S, Kim Y, Park H et al (2009) Syndecan-2 overexpression regulates adhesion and migration through cooperation with integrin alpha2. Biochem Biophys Res Commun 26(384):231–235. https://doi.org/10.1016/j.bbrc.2009.04.093

Choi Y, Kim H, Chung H et al (2010) Syndecan-2 regulates cell migration in colon cancer cells through Tiam1-mediated Rac activation. Biochem Biophys Res Commun 391:921–925. https://doi.org/10.1016/j.bbrc.2009.11.165

Choi S, Kim JY, Park JH et al (2012) The matrix metalloproteinase-7 regulates the extracellular shedding of syndecan-2 from colon cancer cells. Biochem Biophys Res Commun 417:1260–1264. https://doi.org/10.1016/j.bbrc.2011.12.120

Choi S, Choi Y, Jun E et al (2015) Shed syndecan-2 enhances tumorigenic activities of colon cancer cells. Oncotarget 6:3874–3886. https://doi.org/10.18632/oncotarget.2885

Clasper S, Vekemans S, Fiore M et al (1999) Inducible expression of the cell surface heparan sulfate proteoglycan syndecan-2 (fibroglycan) on human activated macrophages can regulate fibroblast growth factor action. J Biol Chem 274:24113–24123. https://doi.org/10.1074/jbc.274.34.24113

Contreras HR, Ledezma RA, Vergara J et al (2010) The expression of syndecan-1 and -2 is associated with Gleason score and epithelial-mesenchymal transition markers, E-cadherin and beta-catenin, in prostate cancer. Urol Oncol 28:534–540. https://doi.org/10.1016/j.urolonc.2009.03.018

Corti F, Wang Y, Rhodes JM et al (2019) N-terminal syndecan-2 domain selectively enhances 6-O heparan sulfate chains sulfation and promotes VEGFA165-dependent neovascularization. Nat Commun 10:1562. https://doi.org/10.1038/s41467-019-09605-z. Erratum in: Nat Commun (2019) 10:2124.

Couchman JR (2010) Transmembrane signaling proteoglycans. Annu Rev Cell Dev Biol 26:89–114. https://doi.org/10.1146/annurev-cellbio-100109-104126

David G, Bai XM, Van der Schueren B et al (1993) Spatial and temporal changes in the expression of fibroglycan (syndecan-2) during mouse embryonic development. Development 119:841–854

De Rossi G, Vähätupa M, Cristante E et al (2021) Pathological angiogenesis requires Ssyndecan-4 for efficient VEGFA-induced VE-cadherin internalization. Arterioscler Thromb Vasc Biol 41: 1374–1389. https://doi.org/10.1161/ATVBAHA.121.315941

Dews IC, Mackenzie KR (2007) Transmembrane domains of the syndecan family of growth factor coreceptors display a hierarchy of homotypic and heterotypic interactions. Proc Natl Acad Sci U S A 104:20782–20787. https://doi.org/10.1073/pnas.0708909105

Dieudonné FX, Marion A, Haÿ E et al (2010) High Wnt signaling represses the proapoptotic proteoglycan syndecan-2 in osteosarcoma cells. Cancer Res 70:5399–5408. https://doi.org/10.1158/0008-5472.CAN-10-0090

Echtermeyer F, Streit M, Wilcox-Adelman S et al (2001) Delayed wound repair and impaired angiogenesis in mice lacking syndecan-4. J Clin Invest 107:R9–R14. https://doi.org/10.1172/JCI10559

Elfenbein A, Simons M (2013) Syndecan-4 signaling at a glance. J Cell Sci 126:3799–3804. https://doi.org/10.1242/jcs.124636

Essner JJ, Chen E, Ekker SC (2006) Syndecan-2. Int J Biochem Cell Biol 38:152–156. https://doi.org/10.1016/j.biocel.2005.08.012

Ethell IM, Irie F, Kalo MS et al (2001) EphB/syndecan-2 signaling in dendritic spine morphogenesis. Neuron 31:1001–1013. https://doi.org/10.1016/s0896-6273(01)00440-8

Fears CY, Gladson CL, Woods A (2006) Syndecan-2 is expressed in the microvasculature of gliomas and regulates angiogenic processes in microvascular endothelial cells. J Biol Chem 281: 14533–14536. https://doi.org/10.1074/jbc.C600075200

Galante LL, Schwarzbauer JE (2007) Requirements for sulfate transport and the diastrophic dysplasia sulfate transporter in fibronectin matrix assembly. J Cell Biol 179:999–1009. https://doi.org/10.1083/jcb.200707150

Gallagher J (2015) Fell-Muir lecture: Heparan sulphate and the art of cell regulation: a polymer chain conducts the protein orchestra. Int J Exp Pathol 96:203–231. https://doi.org/10.1111/iep.12135

Gondelaud F, Ricard-Blum S (2019) Structures and interactions of syndecans. FEBS J 286:2994–3007. https://doi.org/10.1111/febs.14828

Gondelaud F, Bouakil M, Le Fèvre A, Miele A, Chirot, F, Duclos B, Ricard-Blum S (2021) Extended disorder at the cell surface: the conformational landscape of the ectodomains of syndecans. Matrix Biol Plus In Press.

Gopal S, Søgaard P, Multhaupt HA et al (2015) Transmembrane proteoglycans control stretch-activated channels to set cytosolic calcium levels. J Cell Biol 210:1199–1211. https://doi.org/10.1083/jcb.201501060

Gopal S, Couchman J, Pocock R (2016) Redefining the role of syndecans in C. elegans biology. Worm 5:e1142042. https://doi.org/10.1080/21624054.2016.1142042

Gopal S, Multhaupt HAB, Pocock R, Couchman JR (2017) Cell-extracellular matrix and cell-cell adhesion are linked by syndecan-4. Matrix Biol 60-61:57. https://doi.org/10.1016/j.matbio.2016.10.006

Gopal S, Arokiasamy S, Pataki C, Whiteford JR, Couchman JR (2021) Syndecan receptors: pericellular regulators in development and inflammatory disease. Open Biol 11:200377. https://doi.org/10.1098/rsob.200377

Grootjans JJ, Zimmermann P, Reekmans G et al (1997) Syntenin, a PDZ protein that binds syndecan cytoplasmic domains. Proc Natl Acad Sci U S A 94:13683–13688. https://doi.org/10.1073/pnas.94.25.13683

Han I, Park H, Oh ES (2004) New insights into syndecan-2 expression and tumourigenic activity in colon carcinoma cells. J Mol Histol 35:319–326. https://doi.org/10.1023/b:hijo.0000032363.78829.4e

Han YD, Oh TJ, Chung TH et al (2019) Early detection of colorectal cancer based on presence of methylated syndecan-2 (SDC2) in stool DNA. Clin Epigenetics 15(11):51. https://doi.org/10.1186/s13148-019-0642-0

Hienola A, Tumova S, Kulesskiy E, Rauvala H (2006) N-syndecan deficiency impairs neural migration in brain. J Cell Biol 174:569–580. https://doi.org/10.1083/jcb.200602043

Hua R, Yu J, Yan X, Ni Q et al (2020) Syndecan-2 in colorectal cancer plays oncogenic role via epithelial-mesenchymal transition and MAPK pathway. Biomed Pharmacother 121:109630. https://doi.org/10.1016/j.biopha.2019.109630

Jagannath S, Heffner LT Jr, Ailawadhi S et al (2019) Indatuximab Ravtansine [BT062] Monotherapy in Patients With Relapsed and/or Refractory Multiple Myeloma. Clin Lymphoma Myeloma Leuk 19:372–380. https://doi.org/10.1016/j.clml.2019.02.006

Jang B, Jung H, Chung H et al (2016) Syndecan-2 enhances E-cadherin shedding and fibroblast-like morphological changes by inducing MMP-7 expression in colon cancer cells. Biochem Biophys Res Commun 477:47–53. https://doi.org/10.1016/j.bbrc.2016.06.019

Jang B, Jung H, Choi S et al (2017) Syndecan-2 cytoplasmic domain up-regulates matrix metalloproteinase-7 expression via the protein kinase Cγ-mediated FAK/ERK signaling pathway in colon cancer. J Biol Chem 292:16321–16332. https://doi.org/10.1074/jbc.M117.793752

Jang B, Kim A, Hwang J, Song HK, Kim Y, Oh ES (2020) Emerging role of syndecans in extracellular matrix remodeling in cancer. J Histochem Cytochem 68:863–870. https://doi.org/10.1369/0022155420930112

Johnson KG, Ghose A, Epstein E et al (2004) Axonal heparan sulfate proteoglycans regulate the distribution and efficiency of the repellent slit during midline axon guidance. Curr Biol 14:499–504. https://doi.org/10.1016/j.cub.2004.02.005

Jung O, Beauvais DM, Adams KM, Rapraeger AC (2019) VLA-4 phosphorylation during tumor and immune cell migration relies on its coupling to VEGFR2 and CXCR4 by syndecan-1. J Cell Sci 132:jcs232645. https://doi.org/10.1242/jcs.232645

Kato M, Saunders S, Nguyen H, Bernfield M (1995) Loss of cell surface syndecan-1 causes epithelia to transform into anchorage-independent mesenchyme-like cells. Mol Biol Cell 6: 559–576. https://doi.org/10.1091/mbc.6.5.559

Keum E, Kim Y, Kim J et al (2004) Syndecan-4 regulates localization, activity and stability of protein kinase C-alpha. Biochem J 378:1007–1014. https://doi.org/10.1042/BJ20031734

Kim JH, Park SC (2018) Syndecan-2 methylation as a new biomarker for early detection of colorectal neoplasm. Gut Liver 12:479–480. https://doi.org/10.5009/gnl18286

Kim CW, Kim H, Kim HR et al (2021) Colorectal cancer screening using a stool DNA-based SDC2 methylation test: a multicenter, prospective trial. BMC Gastroenterol 21:173. https://doi.org/10.1186/s12876-021-01759-9

Kjellén L, Lindahl U (2018) Specificity of glycosaminoglycan-protein interactions. Curr Opin Struct Biol 50:101–108. https://doi.org/10.1016/j.sbi.2017.12.011

Klass CM, Couchman JR, Woods A (2000) Control of extracellular matrix assembly by syndecan-2 proteoglycan. J Cell Sci 113:493–506

Kramer KL, Barnette JE, Yost HJ (2002) PKCgamma regulates syndecan-2 inside-out signaling during xenopus left-right development. Cell 111:981–990. https://doi.org/10.1016/s0092-8674(02)01200-x

Kumar Katakam S, Tria V, Sim WC et al (2021) The heparan sulfate proteoglycan syndecan-1 regulates colon cancer stem cell function via a focal adhesion kinase-Wnt signaling axis. FEBS J 288:486–506. https://doi.org/10.1111/febs.15356

Kusano Y, Yoshitomi Y, Munesue S et al (2004) Cooperation of syndecan-2 and syndecan-4 among cell surface heparan sulfate proteoglycans in the actin cytoskeletal organization of Lewis lung carcinoma cells. J Biochem 35:129–137. https://doi.org/10.1093/jb/mvh015

Kwon MJ, Choi Y, Yun JH et al (2015) A unique phenylalanine in the transmembrane domain strengthens homodimerization of the syndecan-2 transmembrane domain and functionally regulates syndecan-2. J Biol Chem 290:5772–5782. https://doi.org/10.1074/jbc.M114.5

Lambert J, Makin K, Akbareian S et al (2020) ADAMTS-1 and syndecan-4 intersect in the regulation of cell migration and angiogenesis. J Cell Sci 133:jcs235762. https://doi.org/10.1242/jcs.235762

Lee D, Oh ES, Woods A, Couchman JR, Lee W (1998) Solution structure of a syndecan-4 cytoplasmic domain and its interaction with phosphatidylinositol 4,5-bisphosphate. J Biol Chem 273:13022–13029. https://doi.org/10.1074/jbc.273.21.13022

Lee JH, Park H, Chung H et al (2009) Syndecan-2 regulates the migratory potential of melanoma cells. J Biol Chem 284:27167–27175. https://doi.org/10.1074/jbc.M109.034678

Lee H, Kim Y, Choi Y et al (2011) Syndecan-2 cytoplasmic domain regulates colon cancer cell migration via interaction with syntenin-1. Biochem Biophys Res Commun 409:148–153. https://doi.org/10.1016/j.bbrc.2011.04.135

Li Q, Ng HQ, Kang C (2019) Secondary structure and topology of the transmembrane domain of syndecan-2 in detergent micelles. FEBS Lett 593:554–561. https://doi.org/10.1002/1873-3468.13335

Lim HC, Couchman JR (2014) Syndecan-2 regulation of morphology in breast carcinoma cells is dependent on RhoGTPases. Biochim Biophys Acta 1840:2482–2490. https://doi.org/10.1016/j.bbagen.2014.01.018

Lindahl U, Couchman J, Kimata K, Esko JD (2017) Proteoglycans and sulfated glycosaminoglycans. In: Essentials of glycobiology, 3rd edn. Cold Spring Harbor Laboratory Press, Cold Spring Harbor, NY, pp 2015–2017. Chapter 17

Loring HS, Thompson PR (2020) Emergence of SARM1 as a Potential Therapeutic Target for Wallerian-type Diseases. Cell Chem Biol 27:1–13. https://doi.org/10.1016/j.chembiol.2019.11.002

Lunde IG, Herum KM, Carlson CC, Christensen G (2016) Syndecans in heart fibrosis. Cell Tissue Res 365:539–552. https://doi.org/10.1007/s00441-016-2454-2

Mathiesen SB, Lunde M, Stensland M et al (2020) The cardiac syndecan-2 interactome. Front Cell Dev Biol 8:792. https://doi.org/10.3389/fcell.2020.00792

McCawley LJ, Matrisian LM (2001) Matrix metalloproteinases: they're not just for matrix anymore! Curr Opin Cell Biol 13:534–540. https://doi.org/10.1016/s0955-0674(00)00248-9

McGuire JK, Li Q, Parks WC (2003) Matrilysin (matrix metalloproteinase-7) mediates E- cadherin ectodomain shedding in injured lung epithelium. Am J Pathol 162:1831–1843. https://doi.org/10.1016/S0002-9440(10)64318-0

Mitsou I, Multhaupt HAB, Couchman JR (2017) Proteoglycans, ion channels and cell-matrix adhesion. Biochem J 474:1965–1979. https://doi.org/10.1042/BCJ20160747

Miyazaki K, Hattori Y, Umenishi F et al (1990) Purification and characterization of extracellular matrix-degrading metalloproteinase, matrin (pump-1), secreted from human rectal carcinoma cell line. Cancer Res 50:7758–7764

Mundhenke C, Meyer K, Drew S, Friedl A (2002) Heparan sulfate proteoglycans as regulators of fibroblast growth factor-2 receptor binding in breast carcinomas. Am J Pathol 160:185–194. https://doi.org/10.1016/S0002-9440(10)64362-3

Mytilinaiou M, Bano A, Nikitovic D et al (2013) Syndecan-2 is a key regulator of transforming growth factor beta 2/Smad2-mediated adhesion in fibrosarcoma cells. IUBMB Life 65:134–143. https://doi.org/10.1002/iub.1112

Mytilinaiou M, Nikitovic D, Berdiaki A et al (2017a) IGF-I regulates HT1080 fibrosarcoma cell migration through a syndecan-2/Erk/ezrin signaling axis. Exp Cell Res 361:9–18. https://doi.org/10.1016/j.yexcr.2017.09.035

Mytilinaiou M, Nikitovic D, Berdiaki A et al (2017b) Emerging roles of syndecan 2 in epithelial and mesenchymal cancer progression. IUBMB Life 69:824–833. https://doi.org/10.1002/iub.1678

Nagase H, Woessner JF Jr (1999) Matrix metalloproteinases. J Biol Chem 274:21491–21494. https://doi.org/10.1074/jbc.274.31.21491

Nakato H, Li JP (2016) Functions of heparan sulfate proteoglycans in development: insights from drosophila models. Int Rev Cell Mol Biol 325:275–293. https://doi.org/10.1016/bs.ircmb.2016.02.008

Noguer O, Villena J, Lorita J et al (2009) Syndecan-2 downregulation impairs angiogenesis in human microvascular endothelial cells. Exp Cell Res 315:795–808. https://doi.org/10.1016/j.yexcr.2008.11.016

Oh ES, Woods A, Couchman JR (1997) Syndecan-4 proteoglycan regulates the distribution and activity of protein kinase C. J Biol Chem 272:8133–8136. https://doi.org/10.1074/jbc.272.13.8133

Oh T, Kim N, Moon Y et al (2013) Genome-wide identification and validation of a novel methylation biomarker, SDC2, for blood-based detection of colorectal cancer. J Mol Diagn 15:498–507. https://doi.org/10.1016/j.jmoldx.2013.03.004

Orosco A, Fromigué O, Bazille C et al (2007) Syndecan-2 affects the basal and chemotherapy-induced apoptosis in osteosarcoma. Cancer Res 67:3708–3715. https://doi.org/10.1158/0008-5472.CAN-06-4164

Park H, Kim Y, Lim Y, Han I, Oh ES (2002) Syndecan-2 mediates adhesion and proliferation of colon carcinoma cells. J Biol Chem 277:29730–29736. https://doi.org/10.1074/jbc.M202435200

Park H, Han I, Kwon HJ, Oh ES (2005) Focal adhesion kinase regulates syndecan-2-mediated tumorigenic activity of HT1080 fibrosarcoma cells. Cancer Res 65:9899–9905. https://doi.org/10.1158/0008-5472.CAN-05-1386

Pessolano E, Belvedere R, Novizio N et al (2021) Mesoglycan connects Syndecan-4 and VEGFR2 through Annexin A1 and formyl peptide receptors to promote angiogenesis in vitro. FEBS J. In Press. https://doi.org/10.1111/febs.16043

Pisconti A, Cornelison DD, Olguín HC, Antwine TL, Olwin BB (2010) Syndecan-3 and Notch cooperate in regulating adult myogenesis. J Cell Biol 190:427–441. https://doi.org/10.1083/jcb.201003081

Pisconti A, Banks GB, Babaeijandaghi F et al (2016) Loss of niche-satellite cell interactions in syndecan-3 null mice alters muscle progenitor cell homeostasis improving muscle regeneration. Skelet Muscle 6:34. https://doi.org/10.1186/s13395-016-0104-8

Rasmussen SL, Krarup HB, Sunesen KG et al (2016) Hypermethylated DNA as a biomarker for colorectal cancer: a systematic review. Color Dis 18:549–561. https://doi.org/10.1111/codi.13336
Rasmussen SL, Krarup HB, Sunesen KG et al (2017) Hypermethylated DNA, a circulating biomarker for colorectal cancer detection. PLoS One 12(7):e0180809. https://doi.org/10.1371/journal.pone.0180809
Renga B, Francisci D, Schiaroli E et al (2014) The HIV matrix protein p17 promotes the activation of human hepatic stellate cells through interactions with CXCR2 and Syndecan-2. PLoS One 9:e94798. https://doi.org/10.1371/journal.pone.0094798
Rhiner C, Gysi S, Frohli E, Hengartner MO, Hajna A (2005) Syndecan regulates cell migration and axon guidance in C. elegans. Development 132:4621–4633. https://doi.org/10.1242/dev.02042
Ryu HY, Lee J, Yang S et al (2009) Syndecan-2 functions as a docking receptor for pro-matrix metalloproteinase-7 in human colon cancer cells. J Biol Chem 284:35692–35701. https://doi.org/10.1074/jbc.M109.054254
Schönfeld K, Zuber C, Pinkas J et al (2017) Indatuximab ravtansine [BT062] combination treatment in multiple myeloma: pre-clinical studies. J Hematol Oncol 10:13. https://doi.org/10.1186/s13045-016-0380-0
Shi Y, Gochuico BR, Yu G et al (2013) Syndecan-2 exerts antifibrotic effects by promoting caveolin-1-mediated transforming growth factor-β receptor I internalization and inhibiting transforming growth factor-β1 signaling. Am J Respir Crit Care Med 188:831–841. https://doi.org/10.1164/rccm.201303-0434OC
Stepp MA, Gibson HE, Gala PH et al (2002) Defects in keratinocyte activation during wound healing in the syndecan-1-deficient mouse. J Cell Sci 115:4517–4531. https://doi.org/10.1242/jcs.00128
Su WC, Kao WY, Chang TK et al (2021) Stool DNA test targeting methylated syndecan-2 (SDC2) as a noninvasive screening method for colorectal cancer. Biosci Rep 41:BSR20201930. https://doi.org/10.1042/BSR20201930
Teese MG, Langosch D (2015) Role of GxxxG motifs in transmembrane domain interactions. Biochemistry 54:5125–5135. https://doi.org/10.1021/acs.biochem.5b00495
Teixeira FCOB, Götte M (2020) Involvement of Syndecan-1 and Heparanase in Cancer and Inflammation. Adv Exp Med Biol 1221:97–135. https://doi.org/10.1007/978-3-030-34521-1_4
Teng YH, Aquino RS, Park PW (2012) Molecular functions of syndecan-1 in disease. Matrix Biol 31:3–16. https://doi.org/10.1016/j.matbio.2011.10.001
Tsoyi K, Chu SG, Patino-Jaramillo NG et al (2018) Syndecan-2 attenuates radiation-induced pulmonary fibrosis and inhibits fibroblast activation by regulating PI3K/Akt/ROCK pathway via CD148. Am J Respir Cell Mol Biol 58:208–215. https://doi.org/10.1165/rcmb.2017-0088OC
Tsoyi K, Liang X, De Rossi G et al (2021) CD148 deficiency in fibroblasts promotes the development of pulmonary fibrosis. Am J Respir Crit Care Med Epub ahead of print. https://doi.org/10.1164/rccm.202008-3100OC
Villena J, Berndt C, Granés F, Reina M, Vilaró S (2003) Syndecan-2 expression enhances adhesion and proliferation of stably transfected Swiss 3T3 cells. Cell Biol Int 27:1005–1010. https://doi.org/10.1016/j.cellbi.2003.09.004
Weber M, Hellmann I, Stadler MB et al (2007) Distribution, silencing potential and evolutionary impact of promoter DNA methylation in the human genome. Nat Genet 39:457–466. https://doi.org/10.1038/ng1990
Wernly B, Fuernau G, Masyuk M et al (2019) Syndecan-1 predicts outcome in patients with ST-segment elevation infarction independent from infarct-related myocardial injury. Sci Rep 9:18367. https://doi.org/10.1038/s41598-019-54937-x
Whiteford JR, Behrends V, Kirby H et al (2007) Syndecans promote integrin-mediated adhesion of mesenchymal cells in two distinct pathways. Exp Cell Res 313(18):3902–3913. https://doi.org/10.1016/j.yexcr.2007.08.002

Whiteford JR, Xian X, Chaussade C et al (2011) Syndecan-2 is a novel ligand for the protein tyrosine phosphatase receptor CD148. Mol Biol Cell 22:3609–3624. https://doi.org/10.1091/mbc.E11-02-0099

Xie J, Wang J, Li R et al (2012) Syndecan-4 over-expression preserves cardiac function in a rat model of myocardial infarction. J Mol Cell Cardiol 53:250–258. https://doi.org/10.1016/j.yjmcc.2012.04.014

Xu D, Esko JD (2014) Demystifying heparan sulfate-protein interactions. Annu Rev Biochem 83:129–157. https://doi.org/10.1146/annurev-biochem-060713-035314

Yang Y, MacLeod V, Dai Y et al (2007) The syndecan-1 heparan sulfate proteoglycan is a viable target for myeloma therapy. Blood 110:2041–2048. https://doi.org/10.1182/blood-2007-04-082495

Yu T, Chaganty B, Lin L et al (2020) VIS832, a novel CD138-targeting monoclonal antibody, potently induces killing of human multiple myeloma and further synergizes with IMiDs or bortezomib in vitro and in vivo. Blood Cancer J 10:110. https://doi.org/10.1038/s41408-020-00378-z

Zhang W, Yang C, Wang S et al (2021) SDC2 and TFPI2 methylation in stool samples as an integrated biomarker for early detection of colorectal cancer. Cancer Manag Res 13:3601–3617. https://doi.org/10.2147/CMAR.S300861

Zhao N, Liu H, Lilly B (2012) Reciprocal regulation of syndecan-2 and notch signaling in vascular smooth muscle cells. J Biol Chem 287:16111–16120. https://doi.org/10.1074/jbc.M111.322107

Zhao G, Li H, Yang Z et al (2019) Multiplex methylated DNA testing in plasma with high sensitivity and specificity for colorectal cancer screening. Cancer Med 8:5619–5628. https://doi.org/10.1002/cam4.2475

Chapter 5
Versican in Tumor Progression, Tumor–Host Interactions, and Cancer Immunotherapy

Athanasios Papadas, Alexander Cicala, Sean G. Kraus, Garrett Arauz, Alexander Tong, Dustin Deming, and Fotis Asimakopoulos

Abstract Versican (VCAN) is a large matrix proteoglycan whose very name acts as a testament to its extensive versatility in tissue sculpting, inflammation, and cell fate specification. The broad spectrum of activity of the parent molecule is further diversified through the actions of VCAN's bioactive proteolytic products (VCAN–matrikines). This chapter, while focusing on the functions of VCAN in tumor progression and tumor immunity, contrasts VCAN's roles in adult tissue remodeling with VCAN's nonredundant functions in embryonic morphogenesis. We propose that the interplay between VCAN and VCAN–matrikines provides crucial survival, differentiation, and activation cues wherever tissue architecture is forged or altered, whether in the embryo or in the adult organism. Better understanding of VCAN's mechanisms in the last decade has led to the conceptualization and testing of novel

A. Papadas
Division of Blood and Marrow Transplantation, Department of Medicine and Moores Cancer Center, University of California San Diego (UCSD), La Jolla, CA, USA

Graduate School, University of Wisconsin-Madison, Madison, WI, USA

A. Cicala · G. Arauz · A. Tong · F. Asimakopoulos (✉)
Division of Blood and Marrow Transplantation, Department of Medicine and Moores Cancer Center, University of California San Diego (UCSD), La Jolla, CA, USA
e-mail: fotis@health.ucsd.edu

S. G. Kraus
Graduate School, University of Wisconsin-Madison, Madison, WI, USA

McArdle Laboratory for Cancer Research, University of Wisconsin-Madison, Madison, WI, USA

D. Deming
McArdle Laboratory for Cancer Research, University of Wisconsin-Madison, Madison, WI, USA

Division of Hematology/Oncology, Department of Medicine, UW Comprehensive Cancer Center, University of Wisconsin-Madison, Madison, WI, USA

I. Kovalszky et al. (eds.), *The Extracellular Matrix and the Tumor Microenvironment*, Biology of Extracellular Matrix 11,
https://doi.org/10.1007/978-3-030-99708-3_5

biomarkers and therapies, particularly in the fast-paced field of cancer immunotherapy.

5.1 VCAN: Structure and Variant Forms

Versican (VCAN) is a large aggregating chondroitin sulfate (CS) proteoglycan and a member of the hyalectan family of matrix proteoglycans, which derive their name from their ability to bind hyaluronan (HA). Unlike many of its family members, such as aggrecan, neurocan, and brevican, VCAN is expressed in a variety of tissues throughout the body (Wu et al. 2005b). Its uniquely wide-spread expression predicts its vital, nonredundant roles in organ development and disease (Wight 2002). VCAN's importance is also exemplified by its highly conserved nature, with 89% of its amino acid sequence being identical between the human and mouse protein (Naso et al. 1995). The human gene for VCAN (*VCAN, CSPG2*) is encoded from a single locus on chromosome 5, comprises of 15 exons spanning over 90 kb of contiguous DNA, and is regulated by a typical TATA box, a promoter sequence between −209 and − 445 base pairs, and a negative regulatory sequence between −445 and − 632 base pairs (Iozzo et al. 1992; Wu et al. 2005b).

Characteristically for a hyalectan, VCAN displays globular motifs at its N-terminus (G1 domain) and C-terminus (G3 domain), bridged by a CS chain-binding region. The G1 domain is made up of an immunoglobulin (Ig)-like subunit followed by two HA-binding domains (link modules) (Fig. 5.1). VCAN's G3 domain is comprised of two epidermal growth factor (EGF)-like repeats, followed by carbohydrate recognition domain (lectin-like, CRD) and complement binding

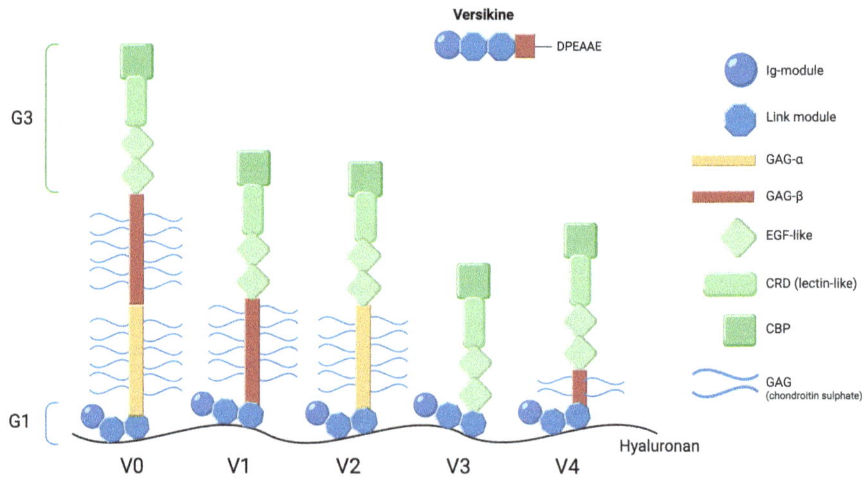

Fig. 5.1 Versican (VCAN) functional domains and splice isoforms. *CRD* carbohydrate recognition domain, *CBP* complement binding protein, *EGF* epidermal growth factor, *GAG* glycosaminoglycan, *Ig* immunoglobulin

protein (CBP)-like subunits. The CS-binding domain is further subdivided into glycosaminoglycan (GAG)-α and GAG-β regions (Wu et al. 2005b). Four common splice variants of VCAN have been identified: V0, V1, V2, and V3. These are distinct in the number and species of GAG regions present. The GAG-bearing domain of the VCAN core protein can be alternately spliced at exon 7 and exon 8, which code for the GAG-α and GAG-β regions, respectively. The V0 isoform is produced when neither exon 7 nor 8 are spliced out. When only exon 7 is spliced out, the V1 variant is generated, while V2 is formed when only exon 8 is spliced out. Translation of mRNA with both exons 7 and 8 spliced out results in VCAN V3. Since V3 is devoid of GAG regions, and thus also CS chains, it cannot technically be considered a proteoglycan. Despite this, it is frequently grouped with proteoglycans and characterized as such (Wight 2002, 2017). An additional isoform, V4, consisting of the G1 domain, the first 398 amino acids of the GAG-β-region and the G3 domain, has been detected in breast cancers (Kischel et al. 2010).

5.2 VCAN Binding Partners in the Tumor Microenvironment

VCAN engages binding partners both in the extracellular matrix and on the surface of cells in the tumor microenvironment. This section will cover key players that VCAN binds to, in both of these compartments, grouped by the domain of VCAN that they interact with.

The G1 region of VCAN is involved with binding hyaluronan and link protein (Matsumoto et al. 2003). Link protein has also been found to bind to other hyalectans like aggrecan in cartilage matrices. It plays an important role in connecting and building the scaffolding of extracellular matrix. Hyaluronan is a polysaccharide made up of 200 to 10,000 repeating units of the disaccharide β1,3 N-acetylglucosaminyl-β1,4 glucuronide (Toole 1990). While hyaluronan plays a role in the architecture of the extracellular matrix, it is also thought to be involved with cell signaling. A signaling role is suggested by its relatively high turnover rate for a matrix molecule—its half-life ranges from half a day to a few days—and its increased expression at sites of high cell activity such as angiogenesis, hematopoiesis, limb-bud formation, inflammation, and tissue remodeling (Laurent and Fraser 1992; Nandadasa et al. 2021a). Of the domains that make up the G1 region, the HA-binding regions are the most crucial, as they are necessary and sufficient for binding both HA and link protein. However, VCAN's binding affinity for both HA and link protein was shown to increase when the Ig-like region is present (Matsumoto et al. 2003).

Whereas the G1 region of VCAN is known for binding extracellular matrix molecules, the CS-binding domain is known for binding cell surface molecules. VCAN's GAG regions do not form extrinsic interactions through the protein backbone itself but through the branching chondroitin sulfate chains bound to the core protein. For example, VCAN's ability to bind L- and P-selectins (cell adhesion molecules) is lost when VCAN undergoes chondroitinase digestion or is treated with blocking antibody (Kawashima et al. 2000). VCAN also binds chemokines via its GAG chains. Secondary lymphoid tissue chemokine (SLC) was shown to have a specific affinity for GAG-β and was significantly less efficient at inducing $\alpha_4\beta_7$ integrin activation the presence of VCAN (Hirose et al. 2001). CD44 has a few similarities to VCAN: both are chondroitin sulfate proteoglycans that bind to HA. CD44, in contrast to the secreted VCAN, is a transmembrane molecule, and much of its functionality is derived from its ability to endocytose and degrade HA (Hua et al. 1993). CD44 functions to mediate, among other things, angiogenesis, and cell migration out of the bloodstream into inflamed tissues (Lokeshwar et al. 1995; Savani et al. 2001). These combine to contribute to tumor progression and metastasis. A common theme of the aforementioned CS-binding cell surface molecules appears to be some involvement with leukocyte trafficking.

The G3 domain of VCAN is the most versatile region of the molecule. It has binding partners in both the extracellular matrix and on the cell surface. VCAN's EGF-like tandem repeats in its G3 region bind to the extracellular molecules fibulin-1 and -2, fibrillin, as well as the cell surface molecule P-selectin glycoprotein ligand-1 (PSGL-1). Fibrillin-1 is a predominant extracellular matrix molecule, and its epidermal growth factor-like domains 11 and 21 are likely where VCAN binds to microfibrils. Disruptions of these domains are associated with Marfan disease and can cause cardiovascular disease (Isogai et al. 2002). Outside of cardiovascular tissues, VCAN and fibrillin are also colocalized in blood vessel and dermal tissues (Wu et al. 2005b). These highly ordered and elastic structures also heavily feature fibulin, and may function as a bridge between VCAN and fibrillin (Wight 2002). Fibulin-1 and fibulin-2's calcium-binding EGF-like domains have been shown to be the ligand for VCAN's EGF-like tandem repeats (Aspberg et al. 1999).

P-selectin glycoprotein ligand-1 (PSGL-1) is a homodimeric cell surface molecule that is nearly ubiquitously expressed in the blood leukocyte compartment. The process of inflammation and tissue infiltration by immune cells begins with circulating cells sticking to the endothelium of blood vessels, followed by leukocyte rolling. It is this action of leukocyte rolling that PSGL-1 is involved with. When various version G3 constructs were exposed to PSGL-1, it was found that versions EGF-like tandem repeats and CRD motifs bound PSGL-1. The presence of the CBP motif suppressed their ability to bind PSGL-1 individually, but when both the tandem repeats and CRD were present, they resisted the CBP negative binding effect (Zheng et al. 2004a).

As previously mentioned, the CRD motif of VCAN's G3 binds PSGL-1. Its other primary binding partner is tenascin-R. This is a commonality in the lectican family, as brevican, neurocan, syndecan, and others bind to tenascin (Aspberg et al. 1997; Grumet et al. 1994; Salmivirta et al. 1991). Tenascin-R is expressed in the central nervous system, and it has been proposed to play a role in the macromolecular organization of the adult brain extracellular matrix based on its binding properties (Pesheva et al. 1991). The ultimate goal of the resulting matrix architecture, including HA and lecticans such as VCAN, is to create a physical barrier around neural tissue, nonpermeable to other cells (Yamaguchi 2000).

The most C-terminal motif of VCAN's G3, the CBP domain, does not appear to have as many binding partners as the tandem repeats or CRD domains. One such partner, however, is fibronectin (Wu et al. 2005b). Fibronectin is a component of the extracellular matrix that is also an integrin ligand and has been shown to increase cell adhesion. VCAN reduces fibronectin's ability to adhere to pancreatic cancer cell lines, with this effect being diminished when VCAN is treated with chondroitinase, implying that VCAN's GAG chains play a part in this function (Wu et al. 2005b). Somewhat paradoxically, despite acting to limit fibronectin's ability to bind tumor cells in vitro, it has been suggested that VCAN's G3 domain may simultaneously upregulate expression of fibronectin in vivo to stimulate tumor propagation and angiogenesis (Zheng et al. 2004a). Finally, VCAN and fibronectin can complex along with VEGF to increase endothelial cell adhesion. Preventing the formation of the complex with an antibody against fibronectin abrogated G3's ability to induce this behavior (Wu et al. 2005b).

VCAN can bind to integrin β1, although it is not clear specifically which motif of the VCAN's G3 domain is involved (Wu et al. 2002). Integrin is a cell surface receptor and anchoring protein comprised of noncovalently linked heterodimer α and β subunits, each of which has several isoforms. The various αβ dimers have different binding specificities and cell signaling characteristics, including but not limited to, tumor initiation, proliferation, survival, and therapeutic resistance (Cooper and Giancotti 2019). Integrin β1 is also able to colocalize with EGFR to alter its downstream signaling pathways (Yamada and Even-Ram 2002). When the pheochromocytoma PC12 cell line was transfected to produce VCAN V1, it led to increased EGFR and integrin expression as well as PC12 neuronal differentiation and neurite outgrowth. This effect was significantly abrogated by inhibition of EGFR, integrin, or Src, a downstream cell signal. Thus, VCAN appears to alter cell differentiation via signaling through both integrin and EGFR pathways (Lee et al. 2015).

5.3 Post-Translational Modifications, Proteolysis, and VCAN–Matrikines

VCAN is cleaved by a disintegrin-like and metalloproteinase with thrombospondin-1 motifs (ADAMTS) and matrix metalloproteinase (MMP) enzymes to generate bioactive fragments, *VCAN–matrikines* (Sandy et al. 2001; Westling et al. 2004; Jonsson-Rylander et al. 2005). Matrix metalloproteinase MMP-1, MMP-2, MMP-3, MMP-7, and MMP-9 from the proteoglycan-cleaving superfamily degrade VCAN in vitro (Perides et al. 1995; Passi et al. 1999; Halpert et al. 1996). Plasmin degrades VCAN core protein in vitro (Kenagy et al. 2002). Six ADAMTS family proteinases (ADAMTS-1, −4, −5, −9, −15, and − 20) (Apte 2009) (Apte 2020; Mead and Apte

2018) are known to cleave VCAN, hence they are named "versicanases" (Nandadasa et al. 2014). Interestingly, ADAMTS-8 has also potential of cleaving VCAN since it shares significant protein homology with a subgroup of ADAMTSs that includes ADAMTS-1, −4, −5, and −15 and has been demonstrated to cleave "aggrecanase-susceptible" sites within the extracellular matrix (ECM) proteoglycan aggrecan (Collins-Racie et al. 2004). While VCAN core protein undergoes cleavage by different proteinases, regulated ADAMTS cleavage at specific sites is important for development and disease (see below).

Most of the characterized VCAN–matrikines contain the HA-binding G1 domain. The best-studied is generated by cleavage of a Glu^{441}-Ala^{442} bond (V1 enumeration) in the GAG-β domain of V0/V1, generating a neoepitope, DPEAAE (Sandy et al. 2001) and can be detected with an anti-DPEAAE antibody (Fig. 5.1). This proteolytic event generates a G1-$DPEAAE^{441}$ fragment of ~80 kDa from the V1 isoform and a G1-$DPEAAE^{1428}$ fragment of ~220 kDa (Sandy et al. 2001) from the V0 isoform. The G1-$DPEAAE^{441}$ fragment is named "versikine" and has roles in various cellular processes and in the regulation of immune phenotypes in the inflammatory milieu of tumors (Papadas et al. 2020; Hope et al. 2016; Nandadasa et al. 2014; Kern et al. 2006; Kern et al. 2007). The mode of action of versikine has not been fully determined yet. Since it derives from the V1 variant, versikine consists of the G1 fragment and a short stretch (~94 amino acid residue) of GAG-β. Despite the known function of the G1 domain in binding HA and expanding HA-rich cellular glycocalyx, whether versikine solely acts through binding to HA has not yet been definitively determined (Islam and Watanabe 2020). MMP-8 and -12 cleave N-terminally at position 3306 to generate the neoepitope KTFGKMKPRY found at atherosclerotic plaques in individuals with varying degrees of atherosclerosis (Barascuk et al. 2013). High performance liquid chromatography and sequence analysis demonstrated that ADAMTS-1 cleaves V0/V2 at Glu^{950}-Gly^{951} and V0/V1 at Tyr^{1410}-Ile^{1411} and Tyr^{423}-Ile^{424} in vitro, but the in vivo physiological consequences of these events are unclear (Jonsson-Rylander et al. 2005). Glial hyaluronate-binding protein (GHAP), found mainly in the brain ECM, is a 64 kDa HA-binding fragment, which is generated through cleavage of Glu^{405}-Gln^{406} bond on V0/V2 by ADAMTS-4 (Westling et al. 2004). Immunological analysis of purified human GHAP showed that it contained the C-terminal sequence, $NIVSFE^{405}$, and arose as a product of cleavage and digestion of human cerebellum VCAN with ADAMTS-4. Besides the Glu^{405} fragment, which was the major immunoreactive species, at least six other C-terminally truncated V2 fragments of varying GAG-α domain length were identified in brain tissue (Westling et al. 2004).

Multiple more cleavage sites have been identified or predicted but whether they generate activities that impact on their tissue and organ microenvironment is unknown (Hope et al. 2016). There is undeniable evidence that certain VCAN–matrikines can act as key drivers of cell-fate determination, immunity, and inflammation in a variety of physiological and disease processes (Mushtaq et al. 2018). However, their overall effect seems to be context-dependent.

5.4 VCAN and VCAN–Matrikines in Development

VCAN and its ADAMTS-generated fragments play a crucial role in cardiovascular, neural crest cell migration, and skeletal development (Papadas and Asimakopoulos 2020). VCAN's proteolytic processing is essential for morphogenesis at sites such as interdigital webs, secondary palate shelves prior to the midline fusion, myocardial contraction, and the fashioning of mature heart leaflets (Nandadasa et al. 2014). VCAN is much more prevalent in embryonic than adult ECM, except for brain tissue. VCAN positively regulates the formation of a mesenchymal matrix, ideal for cell invasion, migration, and morphogenesis in the embryo (Kamiya et al. 2006). As also discussed later in this chapter with regard to adult tissues, the spectrum of activities of parental VCAN and its proteolytic products in embryonic tissues often does not overlap and in some cases, could be even characterized as antagonistic (Nandadasa et al. 2014).

VCAN is essential for cardiac development, contributing to the formation of endocardial cushion mesenchyme by epithelial–mesenchymal transformation (EMT) (Kern et al. 2006). Altered ADAMTS expression plays a crucial role in the stoichiometric balance of intact versus cleaved VCAN in the stroma and overall ECM structure and organization of the developing heart. VCAN proteolytic fragments generated through the actions of ADAMTS proteases can be detected in the cardiac cushions. Endocardial cushions are then promptly remodeled to reach structural maturity and VCAN fragments are distributed around cushion mesenchymal cells. Kern and colleagues reported that *Adamts9*-haploinsufficient and *Adamts5*-null mice had intact VCAN accumulation in the mitral valves, which led to congenital valve anomalies due to defects and alterations in extracellular matrix remodeling (Kern et al. 2007). Intact VCAN accumulated also in the aortic and pulmonary artery walls and was associated with several pathologies including increased adventitial thickness, interruption of the aortic wall and reduction of aortic interlamellar elasticity (Kern et al. 2010). Myocardial trabeculation is another crucial context of VCAN–matrikine activities (Stankunas et al. 2008). During trabeculation, both myocardial and endothelial cells must undergo extensive cellular movements to form the long thin projections protruding into the ventricular cavity (Stankunas et al. 2008). Thus, establishment of a favorable extracellular milieu that promotes changes in migration, cell shape, and adhesion may be critical for trabeculation. In particular, the upregulation of ADAMTS-1 caused VCAN reduction and premature breakdown of the cardiac jelly as well as termination of trabeculation. Contrarily, trabeculation was rescued in *Adamts1*-null mice, as the ventricles of the latter exhibited increased trabecular muscle compared to wild type (Stankunas et al. 2008). Loss of fibulin-1, an ADAMTS-1 cofactor, also led to diminished VCAN cleavage and increased trabecular growth (Cooley et al. 2012). The equilibrium between intact and cleaved VCAN is a central regulator of trabeculation, and ADAMTS-1 activity is suppressed by endocardial Bgr1, a chromatin-based transcriptional regulator, until trabeculation is completed (Stankunas et al. 2008). Finally, immunohistochemical studies demonstrated colocalization of VCAN accumulation and degradation in developing mouse lungs as detected by antibodies against intact VCAN and DPEAAE, respectively, suggesting functional roles of VCAN deposition and degradation in lung morphogenesis (Snyder et al. 2015).

VCAN proteolysis by ADAMTS-9 in vascular endothelium and by ADAMTS-20 in palate mesenchyme results in palatal shelf sculpting and extension. Co-ordination of ADAMTS-9 and ADAMTS-20 contributes to secondary palate closure (Enomoto et al. 2010). Mesenchyme of *Adamts9*$^{+/-}$ and *Adamts20*$^{bt/bt}$ mutant murine palatal shelves had reduced cell proliferation, lower cell density, and decreased processing of VCAN. *Vcan* haploinsufficiency in the *Adamts20*$^{bt/bt}$ background produced cleft palate, providing proof that VCAN is a ADAMTS partner during palate closure, possibly operating through versikine (Enomoto et al. 2010). VCAN and hyaluronan are also associated with emerging Flk1^{+} hematoendothelial progenitors at gastrulation. Apte and colleagues recently demonstrated that the mouse VCAN mutant *Vcan*^{hdf} lacked yolk sac vasculature, with attenuated yolk sac hematopoiesis. In addition, hyaluronan was severely depleted in *Vcan*^{hdf} embryos. Conversely, hyaluronan-deficient mouse embryos also had vasculogenic suppression but with increased VCAN proteolysis. $VEGF_{165}$ and Indian hedgehog, crucial vasculogenic factors, utilized the VCAN–hyaluronan matrix, specifically VCAN chondroitin sulfate chains, for binding. VCAN–hyaluronan ECM is thus essential for vasculogenesis and primitive hematopoiesis (Nandadasa et al. 2021b).

VCAN cleavage may involve sequential events (Montgomery et al. 1993; Arner 2002), first generating G1 and GAG-β-G3 fragments, followed by the loss of GAG-β and CS chains, freeing globular domains that may persist indefinitely (Kenagy et al. 2005, 2006). The initial cleavage site may be the site generating versikine from the V1 isoform (corresponding to Glu^{1428}-Ala^{1429} in V0, and Glu^{441}-Ala^{442} in V1 and V4 variants in humans, respectively). The Watanabe group evaluated the relevance of the initial cleavage site in VCAN turnover by generating knock-in mice, V1R, whose VCAN is mutated at the initial cleavage site in V0 and V1 (Islam et al. 2020). V1R homozygote (R/R) mice showed fewer littermates. When studied, some embryos were small, and the others exhibited organ hemorrhage and syndactyly (Islam et al. 2020). The presentation of hindlimbs and fused digits was consistent with the earlier Apte data implicating VCAN proteolysis in interdigital web regression (McCulloch et al. 2009). More recently, the Apte group reported a new mouse transgene, *Vcan*^{AA}, with validated mutations in the GAG-β domain that specifically abolishes the generation of versikine (Nandadasa et al. 2021a). As a result, *Vcan*$^{AA/AA}$ mice presented with partially penetrant hindlimb soft tissue syndactyly. However, *Adamts20* inactivation in *Vcan*$^{AA/AA}$ mice led to fully penetrant, more severe phenotype affecting all limbs, suggesting that ADAMTS-20 cleavage of VCAN at other sites or of other substrates is an additional requirement for web regression. Interestingly, mice with deletion of *Vcan* exon 8, encoding the GAG-β domain, consistently developed soft tissue syndactyly, while mice lacking exon 7, encoding the GAG-α domain in Vcan transcripts, consistently had fully separated digits. These findings suggest that VCAN is cleaved within each GAG-bearing domain during web regression but generation of versikine through proteolysis in the GAG-β domain retains essential role in interdigital web regression (Nandadasa et al. 2021a).

5.5 VCAN in Tumor Progression and Metastasis

VCAN supports tumor growth and facilitates metastatic dissemination. Increased VCAN expression has been described in a wide variety of malignant tumors, and it correlates with both cancer relapse and poor patient outcomes (Papadas and Asimakopoulos 2020). There are diverse sources of VCAN production in the tumor microenvironment: the tumor cells, the stromal cells, the tumor-associated myeloid cells, and possibly lymphoid cells in some contexts. In lung cancer, VCAN's principal source of secretion appears to be the tumor cell (Kim et al. 2009). Tumor cells also show an elevated expression of VCAN in ovarian cancer (Li et al. 2013), leiomyosarcoma (Keire et al. 2014), hepatocellular carcinoma (Xia et al. 2014), colon carcinoma (Bögels et al. 2012), glioma (Hu et al. 2015), and bladder cancer (Said and Theodorescu 2012). In other contexts, stromal cells constitute the main source of VCAN production, such as in prostate cancer (Ricciardelli et al. 1998; Sakko et al. 2003; Sakko et al. 2001), ovarian cancer (Yeung et al. 2013), pharyngeal cancer (Pukkila et al. 2004), breast cancer (de Lima et al. 2012; Kischel et al. 2010; Ricciardelli et al. 2002; Takahashi et al. 2012), and colon cancer (Iozzo 1995). In some tumors, the source of VCAN could be both the stromal and tumor cells. The combination of tumor and stromal expression of VCAN correlates with shortened disease-free survival and overall survival (Kodama et al. 2007). Moreover, in spontaneous breast cancer murine models, $CD11b^{+}Ly6C^{high}$ monocytic cells are the main producers of VCAN (Gao et al. 2012), whereas in Sezary syndrome, a leukemic variant of cutaneous T cell lymphoma, VCAN was identified as one of the most upregulated genes in lymphocytes isolated from patients (Fujii et al. 2015).

VCAN plays a central role in almost each and every hallmark of cancer (Hanahan and Weinberg 2011). Each of the four functional domains/motifs has been implicated in diverse cancer-promoting processes. The precise VCAN domain/motif that influences tumor cell behavior may be dependent on the specific context and microenvironment (Islam and Watanabe 2020; Wight et al. 2020). VCAN is an important player in tumor cell proliferation and self-renewal: the G1 domain of VCAN has been demonstrated to promote proliferation by interrupting cell adhesion, whereas the G3 domain promotes proliferation through two EGF-like motifs, which are implicated in triggering cell growth. VCAN V0/V1 interacts with TGF-β2 and stimulates proliferation and migration of high-grade gliomas. Silencing of VCAN-isoform V1 by siRNA reduced tumor cell proliferation in human glioma cells (Onken et al. 2014). The G3 domain of VCAN interacted with beta-1 (β1) integrin, activated focal adhesion kinase, enhanced integrin expression, and promoted cell adhesion in glioma cells rendering them resistant to free radical-induced apoptosis (Wu et al. 2002, 2005a). Finally, coexpression of G1 and G3 domains of VCAN protects cells from apoptosis induced by death receptor ligands or cytotoxic drugs (Cattaruzza et al. 2004).

VCAN has been implicated in the formation of new blood vessels in nascent tumors. Tumor neo-angiogenesis provides developing tumors with adequate oxygen supply and nutrients. In Lewis Lung Carcinoma (LLC) tumors, stromal cell-derived VCAN and versikine are associated with increased angiogenesis (Asano et al. 2017). A significant regression of tumor volume as well as capillary formation in the *Vcan*$^{hdf/+}$ mice at 10 days and 13 days post-tumor inoculation compared to wild-type mice was observed (Asano et al. 2017). Accumulation of VCAN in tumors is also positively correlated with the number of microvessels within tumor stroma (Labropoulou et al. 2006; Ghosh et al. 2010). For instance, human stromal stem cells that produce elevated levels of VCAN were found to form an extensive vascular network enriched in hyaluronan and VCAN when cultured with vascular endothelial cells (Kreutziger et al. 2011). In the context of a highly vascularized tumor, glioblastoma, VCAN also appears to promote angiogenesis. The VCAN G3 domain augmented angiogenesis both in vitro and in vivo. G3-expressing cells and tumors formed by these cells expressed very high levels of fibronectin and VEGF. Moreover, the G3 domain after interacting with fibronectin carved a complex together with VEGF. This complex promoted vascular formation in endothelial cells, which lined the vessel walls in human astrocytoma tumors (Zheng et al. 2004b). Consistent with the finding that G3 domain binds fibronectin, the V2 VCAN isoform enhanced substantial blood vessel formation by upregulating and binding to fibronectin (Yang and Yee 2013). Silencing fibronectin expression by siRNA abrogated V2 VCAN's effect in promoting tube-like vascularization (Yang and Yee 2013).

VCAN promotes local tumor cell motility and invasion (Ricciardelli et al. 2009). The G1 and G3 domains seem to exert different effects in cancer cell motility. VCAN enhances cancer cell migration and reduces cell adhesion through its G1 domain. VCAN stimulates locomotion and reduces cell adhesion of astrocytoma cells through binding of its G1 domain to HA and link protein (HAPLN1) (Ang et al. 1999; Yang et al. 1999). On the other hand, VCAN G3 domain appears to be more important in local and systemic tumor invasiveness of breast cancer (Yee et al. 2007). Moreover, VCAN proteolysis plays an important role in cancer cell motility and invasion via signaling pathways in the tumor microenvironment. Arslan and his colleagues demonstrated in immunochemical analyses of glioma tumors that exogenous TGF-β2 not only induced V1 expression but also increased MMP-2 activity and the migration rate of glioma cells (Arslan et al. 2007). However, treatment of glioma tumor spheroids with a blocking anti-DPEAAE antibody reversed the enhancement of glioma migration induced by TGF-β2 in a dose-dependent manner. Thus, VCAN proteolysis facilitates TGF-β2's effect on tumor cells (Arslan et al. 2007). In addition, hyaluronectin, a glycoprotein with an amino acid sequence corresponding to nucleotides of 1 to 1372 of the human VCAN N-terminus, was shown to exert opposing effects on lung cancer cell metastatic spread depending on its primary tumor concentration: elevated levels reduced migration in vitro and metastasis in vivo, while low levels boosted migration and metastasis (Paris et al. 2006). The formation of an HA–VCAN pericellular matrix augmented prostate cancer motility in Boyden chamber motility assays using fibronectin as a chemoattractant. Thus, prostate cancer cells in vitro have the ability to recruit ECM components, such as VCAN, secreted by prostatic stromal cells to promote their motility (Ricciardelli et al. 2007). This observation is consistent with a scenario whereby the formation of a pericellular sheath in vivo by prostate cancer cells utilizing VCAN laid down by prostate stromal cells may contribute to the development of metastatic disease. In glioma, VCAN V0 and V1 expression in tumor vessels was increased and was described to promote local invasion (Paulus et al. 1996). In chondrosarcoma, V1 isoform transfection significantly enhanced tumor cell motility and migration (Wasa et al. 2012). Elevated VCAN expression in tumor-associated stroma also promoted local invasion in cervical cancer and decreased the numbers of intraepithelial CD8+ T cells (Gorter et al. 2010), whereas increased expression of CD44 and VCAN correlated with reduced expression of E-cadherin in endometrial cancer (Hanekamp et al. 2003). Finally, HA oligomers were found to impede the formation of pericellular matrix by ovarian cells and the increased motility and invasion triggered by recombinant VCAN. Therefore, HA oligomers could be a potential adjuvant therapeutic modality, administered intraperitoneally together with chemotherapy drugs, after surgical ovarian cancer debulking (Ween et al. 2011).

VCAN accumulation is associated with tumor metastasis to distant organs. Michael Karin and colleagues have showed that by activating TLR2/TLR6 complexes and inducing TNF-α secretion by myeloid cells, VCAN strongly enhanced metastatic growth in the murine lung cancer model, Lewis Lung Carcinoma (LLC). Metastatic enhancement was abrogated in $Tlr2^{-/-}$ mice (Kim et al. 2009). TNF-α is one of the main pro-metastatic factors secreted by host myeloid cells. TNF-α can repress the apoptosis of cancer cells and promote their proliferation through NF-κB activation (Luo et al. 2004). Moreover, by increasing vascular permeability, TNF-α can augment recruitment of leukocytes as well as intravasation and extravasation of cancer cells. In mouse mammary tumor virus-polyoma middle T-antigen (MMTV-PyMT)-induced cancers, VCAN produced from $CD11b^+$ $Ly6C^{high}$ monocytes was essential in advancing metastasis to the lung in a TGF-β-dependent manner (Gao et al. 2012). Consistently, VCAN knockdown in the bone marrow significantly impaired lung metastases in vivo, without impacting their recruitment to the lungs or altering the immune microenvironment (Gao et al. 2012). In the 4 T1 murine mammary carcinoma model, VCAN expression in the lung correlated with tumor-associated macrophage (TAM) abundance and higher numbers of pulmonary metastatic nodules (Dos Reis et al. 2019). Finally, Mitsui and colleagues highlighted the oncogenic role of VCAN in clear cell carcinoma by demonstrating that VCAN expression was upregulated in patients with clear cell renal carcinoma (ccRCC), and this upregulation was associated with adverse patient outcomes and high rate of metastasis. In a study of 84 matched sporadic ccRCC and normal renal tissues, patients with high VCAN expression had a significantly worse 5-year OS (overall survival) (p-value $= 0.007$) and a higher rate of systemic metastasis than those with low VCAN expression (p-value $= 0.0139$). VCAN significantly promoted ccRCC cell migration and invasion through MMP7 and CXCR4 induction (Mitsui et al. 2017).

5.6 VCAN in Tumor Inflammation and Antitumor Immunity

VCAN is produced by, and conversely regulates, crucial components of the myeloid regulatory cell network in the tumor microenvironment (Wang et al. 2009; Tzanakakis et al. 2019). In particular, several key studies in the last decade demonstrated VCAN's role in controlling the function of intratumoral dendritic cells and macrophages. Dendritic cells (DCs) are professional antigen-presenting cells that fine-tune immune responses and help determine the balance between immunity versus tolerance. With regard to tumor contexts in particular, the important cross-presenting Batf3-DC (also known as conventional dendritic cell type 1, cDC1) subset has been shown to be essential for intratumoral effector T cell response and immunotherapy efficacy (Spranger et al. 2017). Batf3-DC are indispensable for "innate sensing" of tumors and priming of the adaptive immune response (Merad et al. 2013). However, Batf3-DC can be mediators of tolerogenic adaptive immune responses in different contexts. It is interesting that VCAN, through its multifaceted activities, can alternatively induce tolerogenic vs. immunogenic polarization of tumor DC.

Tumor-derived VCAN leads to DC dysfunction (tolerogenic polarization) through TLR2 activation. TLR2 ligation has dual consequences: first, secretion of autocrine IL-10 and IL-6, and second, enhanced expression of their cytokine receptors, which decreases the threshold required to activate STAT3. This positive feedback loop renders DCs tolerogenic which results in blunted antitumor responses through priming of regulatory or ineffective T cells. Thus, VCAN–TLR2 signaling impedes immunogenic DC maturation and limits Th1 and cytotoxic lymphocyte (CTL) responses (Tang et al. 2015, 2017).

By stark contrast, immunogenic DC polarization can be induced through VCAN's proteolytic fragments, VCAN–matrikines (the best studied among them being versikine) (Papadas et al. 2021). Matrikines have been defined as extracellular matrix-derived fragments that regulate cell activity, often in a manner distinct from that of their parent macromolecule (Maquart et al. 2004). VCAN proteolysis at Glu^{441}-Ala^{442} is associated with robust CD8+ infiltration in multiple myeloma bone marrow (Dhakal et al. 2019; Hope et al. 2016) as well as solid tumors (Hope et al. 2017). This specific proteolytic event can produce CD8+ infiltration through either local VCAN depletion or the generation of versikine (or both). Indeed, a role for VCAN depletion in independently regulating immune cell influx into tumors would seem to be supported by earlier data showing that reduced stromal VCAN correlated with immune effector infiltration in cervical cancer biopsies (Gorter et al. 2010). However, there is strong emerging evidence to suggest that versikine, far from being an inert remnant of proteolysis, may constitute the more crucial entity responsible for CD8+ T cell mobilization. Consistent with this scenario, versikine triggers type I interferon-dependent transcription in myeloid cells (Hope et al. 2016). In addition, versikine promotes IRF8-dependent Batf3-DC (Grajales-Reyes et al. 2015; Murphy 2013) generation from Flt3L-mobilized BM in vitro (Hope et al.

2017) and Batf3-DC density in vivo (Papadas et al. 2019; Papadas et al. 2021). Whereas the precise mechanisms of versikine's actions in vivo remain to be elucidated, the available data suggest that versikine may regulate intratumoral CD8+ influx through regulation of abundance, and activation of Batf3-DC (Papadas et al. 2021). It is interesting that versikine appears to act a tug-of-war with its parent macromolecule, nonproteolyzed VCAN, with regard to CD8+ infiltration. Whereas most matrikines possess activities that merely differ from those of their parent macromolecules, versikine appears to actively antagonize its parent.

The role of VCAN and VCAN–matrikines in macrophage regulation is also well established, both in the context of tumor and non-neoplastic inflammatory conditions. VCAN acts as a danger-associated molecular pattern (DAMP) molecule that interacts with Toll-like receptors (TLRs), such as TLR2 on alveolar macrophages, to promote production of inflammatory cytokines, including tumor necrosis factor-α (TNFα), IL-6, and other pro-inflammatory cytokines (Gill et al. 2010; Wight et al. 2014; Wang et al. 2009; Kim et al. 2009). LPS and poly I:C, two TLR agonists, stimulate VCAN expression in both murine bone marrow-derived macrophages in vitro and in murine alveolar macrophages (Chang et al. 2014, 2017) as well as in stromal cells in vivo (Kang et al. 2017). In mesothelioma, VCAN enhances tumor progression through pro-tumoral macrophage polarization. Mice harboring VCAN-deficient tumors presented fewer tumor/pleural macrophages and neutrophils, and fewer pleural T-regulatory cells, compared to control animals (Pappas et al. 2019). In multiple myeloma (MM), VCAN is robustly expressed and processed in the bone marrow (Hope et al. 2014). We previously proposed a model in which VCAN activates myeloma-associated monocytes/macrophages through TLR2/6 signaling, thus triggering trophic IL-1β and IL-6 upregulation (Hope et al. 2014). A role for the VCAN pathway in human MM has been highlighted by two recent reports: Bailur et al. produced a high-resolution analysis of the human immune microenvironment in MM showing that myeloid-derived VCAN transcription was correlated with MM progression, loss of protective T cell stem-like (Tcf1+) memory and expansion of dysfunctional/exhausted T effectors (Bailur et al. 2019). Moreover, the Paiva group showed that immunosuppressive macrophages (expressing VCAN, ENTPD1, and STAB1) were associated with persistence of minimal residual disease (MRD) postautologous stem cell transplant for MM, and higher probability of relapse (Arana et al. 2016). Additional levels of complexity of VCAN's actions have been suggested through elegant studies by the Wight group, using VCAN conditional-null mice: VCAN produced by myeloid cells can have a distinct spectrum of activity compared to stromal-derived VCAN in a model of lung inflammation (Wight et al. 2020). Similar studies in tumor models are not yet available but should yield exciting insights about the diversity of VCAN's actions on the immune microenvironment.

It is important to note that versikine is likely only one of multiple bioactive VCAN–matrikines, as noted earlier in this chapter within the context of embryonic development. VCAN–matrikines other than versikine have been reported, but are much less-well studied. A proteolytic fragment of VCAN detected in neural tissue, glial hyaluronan-binding protein (GHAP) may possess immunomodulatory activities (Bignami et al. 1993). GHAP is a matrikine generated through ADAMTS proteolysis of VCAN-V0 and -V2 at a consensus cleavage site (Glu^{405}-Gln^{406}) distinct from the site cleaved to release versikine (Westling et al. 2004). In preliminary unpublished work from the Asimakopoulos laboratory, GHAP could stimulate type I interferon-dependent transcripts in human macrophages. Further investigation is needed to unravel the full spectrum of VCAN activities through the compendium of VCAN–matrikines.

5.7 VCAN Biomarkers in Tumor Immunology and Immunotherapy

While VCAN appears to have a broad range of immunoregulatory effects in the tumor microenvironment, several publications have focused more specifically on the association between VCAN and tumor-infiltrating lymphocytes (TILs) in patient tumors. TIL abundance is an important characteristic within patient tumors, demonstrating the ability in different disease settings to predict disease progression, immune therapy response, and chemotherapy response (Shaban et al. 2019; Bai et al. 2020; Kong et al. 2018). However, TIL infiltration is not a direct measure of these cells' functionality, only their ability to migrate into tumors. Accordingly, patient tumors with high TIL infiltration do not respond to immune therapy equally and can have substantial heterogeneity in the functionality of effector populations (Paijens et al. 2021). This indicates the need for a biomarker that gives a more complete assessment of both immune infiltration and function. With regard to immune function, this chapter has already outlined several of VCAN's immunosuppressive functions; however, VCAN has also been shown to predict TIL infiltration in patient tumors. Together this indicates VCAN can be both a predictor of TIL functionality and infiltration, and possibly a better predictor of immune therapy response in patients.

The regulation of the abundance of TILs in tumors is multifaceted with the tumor microenvironment playing a very important role in this process. Recently, VCAN accumulation has been shown to inversely correlate with TIL abundance in a large cohort of colorectal cancer patient samples (Hope et al. 2017). In this study, it was shown that patient tumors with high levels of VCAN accumulation have significantly fewer TILs. In addition to the overall abundance of VCAN, the proteolysis of VCAN into versikine was shown to be a strong predictor of TIL abundance. Patient tumors with low levels of intact VCAN and high levels of versikine showed the highest average number of TILs. Similar analyses were also extended into breast, pancreatic, esophageal, and neuroendocrine tumors ($N = 1780$ in total) showing the same inverse trend between VCAN accumulation and the abundance of TILs within patient tumors (Emmerich et al. 2020). Two additional studies also identified the same inverse trend between VCAN accumulation and CD8+ T cell infiltration in both cervical cancer patient samples and post-ACST multiple myeloma core biopsies (Gorter et al. 2010; Dhakal et al. 2019). Altogether, these findings across several cancer settings, including both solid and hematological, show that the accumulation of VCAN may be altering the tumor microenvironment to one that may be hindering either CD8+ T cell infiltration or function. These studies also demonstrate that VCAN proteolysis may have the opposite effect, correlating with higher levels of antitumor immune activation. At this point, it is unclear if the lack of TILs found in tumors with high levels of VCAN is the result of a direct or indirect mechanism. Future research is needed to evaluate this mechanism and whether or not it can be disrupted to improve treatment outcomes.

The lack of immune infiltration seen in many cancers is widely believed to be large contributor to immune checkpoint blockade therapy (ICB) resistance, and with VCAN demonstrating many immunosuppressive functions, it is possible VCAN accumulation or expression within tumors may be a marker for ICB resistance in patients. A recent study correlated the fraction of VCAN positive tumor-associated macrophages with patient's response to immune checkpoint blockade in melanoma, finding higher levels of VCAN+ macrophages in nonresponding patients (Cheng et al. 2021). An additional piece of evidence supporting VCAN as a biomarker for immune therapy response was found in a small phase II study comparing progression-free survival at 12 months between patients with high and low levels of VCAN accumulation or VCAN proteolysis (Deming et al. 2020). Prior to surgery, patients were treated with stereotactic body radiation therapy (SBRT) and anti-PD-1 therapy, followed postoperatively with several additional rounds of anti-PD-1 treatment. Patients with low-to-moderate levels of VCAN had a significantly lower rate of disease recurrence, with 67% of patients ($N = 12$) remaining disease-free at 12 months postoperatively. Conversely, both of the patients in this study with high levels of VCAN had disease recurrence by the same time-point. Additionally, an even stronger difference in recurrence-free survival at 1 year was seen between patients with high and low levels of VCAN proteolysis. Patients with high levels of VCAN proteolysis had significantly lower rates of disease recurrence at 12 months, with 80% of patients remaining disease-free, whereas only 44% of VCAN-proteolysis-weak (VPW) patients made it to 1 year without recurrence. Together these data indicate that VCAN's immunoregulatory properties may make it an excellent marker for ICB response or disease recurrence following SBRT and anti-PD-1 treatment. It is important to note, however, that the cohort for the latter study was small and did not directly correlate VCAN accumulation or proteolysis with response to ICB therapy. Future studies are currently underway to address these pitfalls in order to evaluate more directly if VCAN can predict patient response to ICB therapy.

5.8 Prospects for VCAN Therapeutics in the Immuno-Oncology Era

A decade following the initial reports of checkpoint inhibitor efficacy in common cancer types, most patients are still refractory (Sharma and Allison 2020; Waldman et al. 2020; Havel et al. 2019). Early attempts at elucidating the molecular and physiologic underpinnings of response pointed to the importance of an "inflamed' immune microenvironment, i.e., evidence of pre-existing adaptive antitumor immunity that could be potentiated through immunotherapy (Ochoa de Olza et al. 2020).

The role of tumor matrix in regulating key players in adaptive immunity has not received adequate scrutiny. By contrast, the notion that matrix components and their bioactive fragments can act as endogenous "alarmins" has been firmly established (Schaefer 2014). Small leucine-rich proteoglycans were among the prototypes of matrix components acting as danger-associated molecular patterns (DAMPs) to activate Toll-like receptors in settings of inflammatory tissue damage (Schaefer and Iozzo 2012). Of the large matrix proteoglycans, VCAN is perhaps best studied as a regulator and amplifier of inflammation in the adult.

The interplay between VCAN and VCAN–matrikines provides an exciting model to understand the impact of matrix remodeling in regulating the tumor "immune thermostat." VCAN and its proteolytic "daughters" actively antagonize each other in the framework of the tumor immune milieu. Whereas stromal nonproteolyzed VCAN promotes T cell exclusion, the bioactive fragment versikine promotes Batf3-DC recruitment, survival, and activation. Therefore, versikine helps sculpt an immune microenvironment where Batf3-DC are encouraged to home in, and whose homeostasis is locally supported. Versikine may be useful in updialing the immunogenicity of a given tumor, toward a more inflamed "hot" setpoint, thus lowering threshold for immunotherapy efficacy. Thus, versikine could be a useful agonist in in situ vaccination approaches either alone or in combination with locally administered immunotherapy.

The yin vs. yang of VCAN vs. versikine appears critical in the regulation of adaptive immune responses through the crucial Batf3-DC subset (Schmitt 2016). Boosting and/or activating Batf3-DC at the tumor site is a crucial requisite for efficacy of several modern immunotherapy modalities. Versikine could synergize with diverse contemporary immunotherapy approaches to boost endogenous or engineered responses, in both hematopoietic and solid cancers.

References

Ang LC, Zhang Y, Cao L, Yang BL, Young B, Kiani C, Lee V, Allan K, Yang BB (1999) Versican enhances locomotion of astrocytoma cells and reduces cell adhesion through its G1 domain. J Neuropathol Exp Neurol 58(6):597–605. https://doi.org/10.1097/00005072-199906000-00004

Apte SS (2009) A disintegrin-like and metalloprotease (reprolysin-type) with thrombospondin type 1 motif (ADAMTS) superfamily: functions and mechanisms. J Biol Chem 284(46): 31493–31497. https://doi.org/10.1074/jbc.R109.052340

Apte SS (2020) ADAMTS proteins: concepts, challenges, and prospects. Methods Mol Biol 2043: 1–12. https://doi.org/10.1007/978-1-4939-9698-8_1

Arana P, Zabaleta A, Lasa M, Maiso P, Alignani D, Jelinek T, Segura V, van den Bossche W, Teodosio C, Damasceno D, Rodriguez-Otero P, Prosper F, Mateos M-V, Lahuerta JJ, Bladé J, van Dongen JJM, San Miguel J, Orfao A, Paiva B (2016) High-throughput characterization and new insight into the role of tumor associated macrophages (TAMs) in multiple myeloma (MM). Blood 128(22):482–482

Arner EC (2002) Aggrecanase-mediated cartilage degradation. Curr Opin Pharmacol 2(3):322–329. https://doi.org/10.1016/s1471-4892(02)00148-0

Arslan F, Bosserhoff AK, Nickl-Jockschat T, Doerfelt A, Bogdahn U, Hau P (2007) The role of versican isoforms V0/V1 in glioma migration mediated by transforming growth factor-beta2. Br J Cancer 96(10):1560–1568. https://doi.org/10.1038/sj.bjc.6603766

Asano K, Nelson CM, Nandadasa S, Aramaki-Hattori N, Lindner DJ, Alban T, Inagaki J, Ohtsuki T, Oohashi T, Apte SS, Hirohata S (2017) Stromal Versican regulates tumor growth by promoting angiogenesis. Sci Rep 7(1):17225. https://doi.org/10.1038/s41598-017-17613-6

Aspberg A, Miura R, Bourdoulous S, Shimonaka M, Heinegârd D, Schachner M, Ruoslahti E, Yamaguchi Y (1997) The C-type lectin domains of lecticans, a family of aggregating chondroitin sulfate proteoglycans, bind tenascin-R by protein-protein interactions independent of carbohydrate moiety. Proc Natl Acad Sci U S A 94(19):10116–10121. https://doi.org/10.1073/pnas.94.19.10116

Aspberg A, Adam S, Kostka G, Timpl R, Heinegard D (1999) Fibulin-1 is a ligand for the C-type lectin domains of aggrecan and versican. J Biol Chem 274(29):20444–20449. https://doi.org/10.1074/jbc.274.29.20444

Bai R, Lv Z, Xu D, Cui J (2020) Predictive biomarkers for cancer immunotherapy with immune checkpoint inhibitors. Biomark Res 8:34. https://doi.org/10.1186/s40364-020-00209-0

Bailur JK, McCachren SS, Doxie DB, Shrestha M, Pendleton K, Nooka AK, Neparidze N, Parker TL, Bar N, Kaufman JL, Hofmeister CC, Boise LH, Lonial S, Kemp ML, Dhodapkar KM, Dhodapkar MV (2019) Early alterations in stem-like/resident T cells, innate and myeloid cells in the bone marrow in preneoplastic gammopathy. JCI Insight 5(11):e127807. https://doi.org/10.1172/jci.insight.127807

Barascuk N, Genovese F, Larsen L, Byrjalsen I, Zheng Q, Sun S, Hosbond S, Poulsen TS, Diederichsen A, Jensen JM, Mickley H, Register TC, Rasmussen LM, Leeming DJ, Christiansen C, Karsdal MA (2013) A MMP derived versican neo-epitope is elevated in plasma from patients with atherosclerotic heart disease. Int J Clin Exp Med 6(3):174–184

Bignami A, Hosley M, Dahl D (1993) Hyaluronic acid and hyaluronic acid-binding proteins in brain extracellular matrix. Anat Embryol (Berl) 188(5):419–433. https://doi.org/10.1007/BF00190136

Bögels M, Braster R, Nijland PG, Gül N, van de Luijtgaarden W, Fijneman RJ, Meijer GA, Jimenez CR, Beelen RH, van Egmond M (2012) Carcinoma origin dictates differential skewing of monocyte function. Onco Targets Ther 1(6):798–809. https://doi.org/10.4161/onci.20427

Cattaruzza S, Schiappacassi M, Kimata K, Colombatti A, Perris R (2004) The globular domains of PG-M/versican modulate the proliferation-apoptosis equilibrium and invasive capabilities of tumor cells. FASEB J 18(6):779–781. https://doi.org/10.1096/fj.03-0660fje

Chang MY, Tanino Y, Vidova V, Kinsella MG, Chan CK, Johnson PY, Wight TN, Frevert CW (2014) Reprint of: a rapid increase in macrophage-derived versican and hyaluronan in infectious lung disease. Matrix Biol 35:162–173. https://doi.org/10.1016/j.matbio.2014.04.003

Chang MY, Kang I, Gale M, Manicone AM, Kinsella MG, Braun KR, Wigmosta T, Parks WC, Altemeier WA, Wight TN, Frevert CW (2017) Versican is produced by Trif- and type I interferon-dependent signaling in macrophages and contributes to fine control of innate immunity in lungs. Am J Physiol Lung Cell Mol Physiol 313(6):L1069–L1086. https://doi.org/10.1152/ajplung.00353.2017

Cheng S, Li Z, Gao R, Xing B, Gao Y, Yang Y, Qin S, Zhang L, Ouyang H, Du P, Jiang L, Zhang B, Yang Y, Wang X, Ren X, Bei JX, Hu X, Bu Z, Ji J, Zhang Z (2021) A pan-cancer single-cell transcriptional atlas of tumor infiltrating myeloid cells. Cell 184 (3):792-809:e723. https://doi.org/10.1016/j.cell.2021.01.010

Collins-Racie LA, Flannery CR, Zeng W, Corcoran C, Annis-Freeman B, Agostino MJ, Arai M, DiBlasio-Smith E, Dorner AJ, Georgiadis KE, Jin M, Tan XY, Morris EA, LaVallie ER (2004) ADAMTS-8 exhibits aggrecanase activity and is expressed in human articular cartilage. Matrix Biol 23(4):219–230. https://doi.org/10.1016/j.matbio.2004.05.004

Cooley MA, Fresco VM, Dorlon ME, Twal WO, Lee NV, Barth JL, Kern CB, Iruela-Arispe ML, Argraves WS (2012) Fibulin-1 is required during cardiac ventricular morphogenesis for versican cleavage, suppression of ErbB2 and Erk1/2 activation, and to attenuate trabecular cardiomyocyte proliferation. Dev Dyn 241(2):303–314. https://doi.org/10.1002/dvdy.23716

Cooper J, Giancotti FG (2019) Integrin signaling in cancer: Mechanotransduction, Stemness, epithelial plasticity, and therapeutic resistance. Cancer Cell 35(3):347–367. https://doi.org/10.1016/j.ccell.2019.01.007

de Lima CR, de Arimatéa dos Santos Junior J, Nazário AC, Michelacci YM (2012) Changes in glycosaminoglycans and proteoglycans of normal breast and fibroadenoma during the menstrual cycle. Biochim Biophys Acta 1820(7):1009–1019. https://doi.org/10.1016/j.bbagen.2012.04.010

Deming DA, Emmerich P, Turk AA, Lubner SJ, Uboha NV, LoConte NK, Mulkerin D, Kim DH, Matkowskyj KA, Weber SM, Abbott D, Eickhoff JC, Bassetti MF (2020) Pembrolizumab (Pem) in combination with stereotactic body radiotherapy (SBRT) for resectable liver oligometastatic MSS/MMR proficient colorectal cancer (CRC). J Clin Oncol 38(15_Suppl):4046. https://doi.org/10.1200/JCO.2020.38.15_suppl.4046

Dhakal B, Pagenkopf A, Umair Mushtaq M, Cunningham AM, Flietner E, Morrow Z, Papadas A, Hope C, Leith C, Hematti P, Hari P, Callander NS, Asimakopoulos F (2019) Versican proteolysis predicts immune effector infiltration and post-transplant survival in myeloma. Leuk Lymphoma 60(10):2558–2562. https://doi.org/10.1080/10428194.2019.1585836

Dos Reis DC, Damasceno KA, de Campos CB, Veloso ES, Pêgas GRA, Kraemer LR, Rodrigues MA, Mattos MS, Gomes DA, Campos PP, Ferreira E, Russo RC, Cassali GD (2019) Versican and tumor-associated macrophages promotes tumor progression and metastasis in canine and murine models of breast carcinoma. Front Oncol 9:577. https://doi.org/10.3389/fonc.2019.00577

Emmerich P, Matkowskyj KA, McGregor S, Kraus S, Bischel K, Qyli T, Buehler D, Pasch C, Babiarz C, Depke M, Clipson L, Wisinski KB, Burkard ME, Baschnagel A, Uboha NV, Asimakopoulos F, Deming DA (2020) VCAN accumulation and proteolysis as predictors of T lymphocyte-excluded and permissive tumor microenvironments. J Clin Oncol 38(15_Suppl):3127. https://doi.org/10.1200/JCO.2020.38.15_suppl.3127

Enomoto H, Nelson CM, Somerville RP, Mielke K, Dixon LJ, Powell K, Apte SS (2010) Cooperation of two ADAMTS metalloproteases in closure of the mouse palate identifies a requirement for versican proteolysis in regulating palatal mesenchyme proliferation. Development 137(23):4029–4038. https://doi.org/10.1242/dev.050591

Fujii K, Karpova MB, Asagoe K, Georgiev O, Dummer R, Urosevic-Maiwald M (2015) Versican upregulation in Sézary cells alters growth, motility and resistance to chemotherapy. Leukemia 29(10):2024–2032. https://doi.org/10.1038/leu.2015.103

Gao D, Joshi N, Choi H, Ryu S, Hahn M, Catena R, Sadik H, Argani P, Wagner P, Vahdat LT, Port JL, Stiles B, Sukumar S, Altorki NK, Rafii S, Mittal V (2012) Myeloid progenitor cells in the premetastatic lung promote metastases by inducing mesenchymal to epithelial transition. Cancer Res 72(6):1384–1394. https://doi.org/10.1158/0008-5472.CAN-11-2905

Ghosh S, Albitar L, LeBaron R, Welch WR, Samimi G, Birrer MJ, Berkowitz RS, Mok SC (2010) Up-regulation of stromal versican expression in advanced stage serous ovarian cancer. Gynecol Oncol 119(1):114–120. https://doi.org/10.1016/j.ygyno.2010.05.029

Gill S, Wight TN, Frevert CW (2010) Proteoglycans: key regulators of pulmonary inflammation and the innate immune response to lung infection. Anat Rec (Hoboken) 293(6):968–981. https://doi.org/10.1002/ar.21094

Gorter A, Zijlmans HJ, van Gent H, Trimbos JB, Fleuren GJ, Jordanova ES (2010) Versican expression is associated with tumor-infiltrating CD8-positive T cells and infiltration depth in cervical cancer. Mod Pathol 23(12):1605–1615. https://doi.org/10.1038/modpathol.2010.154. Epub 2010 Aug 20

Grajales-Reyes GE, Iwata A, Albring J, Wu X, Tussiwand R, Kc W, Kretzer NM, Briseno CG, Durai V, Bagadia P, Haldar M, Schonheit J, Rosenbauer F, Murphy TL, Murphy KM (2015) Batf3 maintains autoactivation of Irf8 for commitment of a CD8alpha(+) conventional DC clonogenic progenitor. Nat Immunol 16(7):708–717. https://doi.org/10.1038/ni.3197

Grumet M, Milev P, Sakurai T, Karthikeyan L, Bourdon M, Margolis RK, Margolis RU (1994) Interactions with tenascin and differential effects on cell adhesion of neurocan and phosphacan, two major chondroitin sulfate proteoglycans of nervous tissue. J Biol Chem 269(16): 12142–12146

Halpert I, Sires UI, Roby JD, Potter-Perigo S, Wight TN, Shapiro SD, Welgus HG, Wickline SA, Parks WC (1996) Matrilysin is expressed by lipid-laden macrophages at sites of potential rupture in atherosclerotic lesions and localizes to areas of versican deposition, a proteoglycan substrate for the enzyme. Proc Natl Acad Sci U S A 93(18):9748–9753. https://doi.org/10.1073/pnas.93.18.9748

Hanahan D, Weinberg RA (2011) Hallmarks of cancer: the next generation. Cell 144(5):646–674. https://doi.org/10.1016/j.cell.2011.02.013

Hanekamp EE, Gielen SC, Smid-Koopman E, Kühne LC, de Ruiter PE, Chadha-Ajwani S, Brinkmann AO, Grootegoed JA, Burger CW, Huikeshoven FJ, Blok LJ (2003) Consequences of loss of progesterone receptor expression in development of invasive endometrial cancer. Clin Cancer Res 9(11):4190–4199

Havel JJ, Chowell D, Chan TA (2019) The evolving landscape of biomarkers for checkpoint inhibitor immunotherapy. Nat Rev Cancer 19(3):133–150. https://doi.org/10.1038/s41568-019-0116-x

Hirose J, Kawashima H, Yoshie O, Tashiro K, Miyasaka M (2001) Versican interacts with chemokines and modulates cellular responses. J Biol Chem 276(7):5228–5234. https://doi.org/10.1074/jbc.M007542200

Hope C, Ollar SJ, Heninger E, Hebron E, Jensen JL, Kim J, Maroulakou I, Miyamoto S, Leith C, Yang DT, Callander N, Hematti P, Chesi M, Bergsagel PL, Asimakopoulos F (2014) TPL2 kinase regulates the inflammatory milieu of the myeloma niche. Blood 123(21):3305–3315. https://doi.org/10.1182/blood-2014-02-554071

Hope C, Foulcer S, Jagodinsky J, Chen SX, Jensen JL, Patel S, Leith C, Maroulakou I, Callander N, Miyamoto S, Hematti P, Apte SS, Asimakopoulos F (2016) Immunoregulatory roles of versican proteolysis in the myeloma microenvironment. Blood 128(5):680–685. https://doi.org/10.1182/blood-2016-03-705780

Hope C, Emmerich PB, Papadas A, Pagenkopf A, Matkowskyj KA, Van De Hey DR, Payne SN, Clipson L, Callander NS, Hematti P, Miyamoto S, Johnson MG, Deming DA, Asimakopoulos F (2017) Versican-derived Matrikines regulate Batf3-dendritic cell differentiation and promote T cell infiltration in colorectal cancer. J Immunol 199(5):1933–1941. https://doi.org/10.4049/jimmunol.1700529

Hu F, Dzaye O, Hahn A, Yu Y, Scavetta RJ, Dittmar G, Kaczmarek AK, Dunning KR, Ricciardelli C, Rinnenthal JL, Heppner FL, Lehnardt S, Synowitz M, Wolf SA, Kettenmann H (2015) Glioma-derived versican promotes tumor expansion via glioma-associated microglial/macrophages toll-like receptor 2 signaling. Neuro-Oncology 17(2):200–210. https://doi.org/10.1093/neuonc/nou324

Hua Q, Knudson CB, Knudson W (1993) Internalization of hyaluronan by chondrocytes occurs via receptor-mediated endocytosis. J Cell Sci 106(Pt 1):365–375
Iozzo RV (1995) Tumor stroma as a regulator of neoplastic behavior. Agonistic and antagonistic elements embedded in the same connective tissue. Lab Investig 73(2):157–160
Iozzo RV, Naso MF, Cannizzaro LA, Wasmuth JJ, McPherson JD (1992) Mapping of the versican proteoglycan gene (CSPG2) to the long arm of human chromosome 5 (5q12-5q14). Genomics 14(4):845–851. https://doi.org/10.1016/s0888-7543(05)80103-x
Islam S, Watanabe H (2020) Versican: a dynamic regulator of the extracellular matrix. J Histochem Cytochem 68(11):763–775. https://doi.org/10.1369/0022155420953922
Islam S, Chuensirikulchai K, Khummuang S, Keratibumrungpong T, Kongtawelert P, Kasinrerk W, Hatano S, Nagamachi A, Honda H, Watanabe H (2020) Accumulation of versican facilitates wound healing: implication of its initial ADAMTS-cleavage site. Matrix Biol 87:77–93. https://doi.org/10.1016/j.matbio.2019.10.006
Isogai Z, Aspberg A, Keene DR, Ono RN, Reinhardt DP, Sakai LY (2002) Versican interacts with fibrillin-1 and links extracellular microfibrils to other connective tissue networks. J Biol Chem 277(6):4565–4572. https://doi.org/10.1074/jbc.M110583200
Jonsson-Rylander AC, Nilsson T, Fritsche-Danielson R, Hammarstrom A, Behrendt M, Andersson JO, Lindgren K, Andersson AK, Wallbrandt P, Rosengren B, Brodin P, Thelin A, Westin A, Hurt-Camejo E, Lee-Sogaard CH (2005) Role of ADAMTS-1 in atherosclerosis: remodeling of carotid artery, immunohistochemistry, and proteolysis of versican. Arterioscler Thromb Vasc Biol 25(1):180–185. https://doi.org/10.1161/01.ATV.0000150045.27127.37
Kamiya N, Watanabe H, Habuchi H, Takagi H, Shinomura T, Shimizu K, Kimata K (2006) Versican/PG-M regulates chondrogenesis as an extracellular matrix molecule crucial for mesenchymal condensation. J Biol Chem 281(4):2390–2400. https://doi.org/10.1074/jbc.M509341200
Kang I, Harten IA, Chang MY, Braun KR, Sheih A, Nivison MP, Johnson PY, Workman G, Kaber G, Evanko SP, Chan CK, Merrilees MJ, Ziegler SF, Kinsella MG, Frevert CW, Wight TN (2017) Versican deficiency significantly reduces lung inflammatory response induced by Polyinosine-Polycytidylic acid stimulation. J Biol Chem 292(1):51–63. https://doi.org/10.1074/jbc.M116.753186
Kawashima H, Hirose M, Hirose J, Nagakubo D, Plaas AH, Miyasaka M (2000) Binding of a large chondroitin sulfate/dermatan sulfate proteoglycan, versican, to L-selectin, P-selectin, and CD44. J Biol Chem 275(45):35448–35456. https://doi.org/10.1074/jbc.M003387200
Keire PA, Bressler SL, Lemire JM, Edris B, Rubin BP, Rahmani M, McManus BM, van de Rijn M, Wight TN (2014) A role for versican in the development of leiomyosarcoma. J Biol Chem 289(49):34089–34103. https://doi.org/10.1074/jbc.M114.607168
Kenagy RD, Fischer JW, Davies MG, Berceli SA, Hawkins SM, Wight TN, Clowes AW (2002) Increased plasmin and serine proteinase activity during flow-induced intimal atrophy in baboon PTFE grafts. Arterioscler Thromb Vasc Biol 22(3):400–404. https://doi.org/10.1161/hq0302.105376
Kenagy RD, Fischer JW, Lara S, Sandy JD, Clowes AW, Wight TN (2005) Accumulation and loss of extracellular matrix during shear stress-mediated intimal growth and regression in baboon vascular grafts. J Histochem Cytochem 53(1):131–140. https://doi.org/10.1177/002215540505300115
Kenagy RD, Plaas AH, Wight TN (2006) Versican degradation and vascular disease. Trends Cardiovasc Med 16(6):209–215. https://doi.org/10.1016/j.tcm.2006.03.011
Kern CB, Twal WO, Mjaatvedt CH, Fairey SE, Toole BP, Iruela-Arispe ML, Argraves WS (2006) Proteolytic cleavage of versican during cardiac cushion morphogenesis. Dev Dyn 235(8):2238–2247. https://doi.org/10.1002/dvdy.20838
Kern CB, Norris RA, Thompson RP, Argraves WS, Fairey SE, Reyes L, Hoffman S, Markwald RR, Mjaatvedt CH (2007) Versican proteolysis mediates myocardial regression during outflow tract development. Dev Dyn 236(3):671–683. https://doi.org/10.1002/dvdy.21059

Kern CB, Wessels A, McGarity J, Dixon LJ, Alston E, Argraves WS, Geeting D, Nelson CM, Menick DR, Apte SS (2010) Reduced versican cleavage due to Adamts9 haploinsufficiency is associated with cardiac and aortic anomalies. Matrix Biol 29(4):304–316. https://doi.org/10.1016/j.matbio.2010.01.005

Kim S, Takahashi H, Lin WW, Descargues P, Grivennikov S, Kim Y, Luo JL, Karin M (2009) Carcinoma-produced factors activate myeloid cells through TLR2 to stimulate metastasis. Nature 457(7225):102–106. https://doi.org/10.1038/nature07623

Kischel P, Waltregny D, Dumont B, Turtoi A, Greffe Y, Kirsch S, De Pauw E, Castronovo V (2010) Versican overexpression in human breast cancer lesions: known and new isoforms for stromal tumor targeting. Int J Cancer 126(3):640–650. https://doi.org/10.1002/ijc.24812

Kodama J, Hasengaowa KT, Seki N, Matsuo T, Ojima Y, Nakamura K, Hongo A, Hiramatsu Y (2007) Prognostic significance of stromal versican expression in human endometrial cancer. Ann Oncol 18(2):269–274. https://doi.org/10.1093/annonc/mdl370

Kong JCH, Guerra GR, Millen RM, Roth S, Xu H, Neeson PJ, Darcy PK, Kershaw MH, Sampurno S, Malaterre J, Liu DSH, Pham TD, Narasimhan V, Wang M, Huang Y-K, Visvanathan K, McCormick J, Lynch AC, Warrier S, Michael M, Desai J, Murray W, Mitchell C, Ngan S, Phillips WA, Heriot AG, Ramsay RG (2018) Tumor-infiltrating lymphocyte function predicts response to neoadjuvant Chemoradiotherapy in locally advanced rectal cancer. JCO Precis Oncol 2:1–15. https://doi.org/10.1200/PO.18.00075

Kreutziger KL, Muskheli V, Johnson P, Braun K, Wight TN, Murry CE (2011) Developing vasculature and stroma in engineered human myocardium. Tissue Eng Part A 17(9–10): 1219–1228. https://doi.org/10.1089/ten.TEA.2010.0557

Labropoulou VT, Theocharis AD, Ravazoula P, Perimenis P, Hjerpe A, Karamanos NK, Kalofonos HP (2006) Versican but not decorin accumulation is related to metastatic potential and neovascularization in testicular germ cell tumours. Histopathology 49(6):582–593. https://doi.org/10.1111/j.1365-2559.2006.02558.x

Laurent TC, Fraser JR (1992) Hyaluronan. FASEB J 6(7):2397–2404

Lee HC, Lo HC, Lo DM, Su MY, Hu JR, Wu CC, Chang SN, Dai MS, Tsai CT, Tsai HJ (2015) Amiodarone induces overexpression of similar to Versican b to repress the EGFR/Gsk3b/snail signaling Axis during cardiac valve formation of zebrafish embryos. PLoS One 10(12): e0144751. https://doi.org/10.1371/journal.pone.0144751

Li D, Wang X, Wu JL, Quan WQ, Ma L, Yang F, Wu KY, Wan HY (2013) Tumor-produced versican V1 enhances hCAP18/LL-37 expression in macrophages through activation of TLR2 and vitamin D3 signaling to promote ovarian cancer progression in vitro. PLoS One 8(2): e56616. https://doi.org/10.1371/journal.pone.0056616

Lokeshwar BL, Lokeshwar VB, Block NL (1995) Expression of CD44 in prostate cancer cells: association with cell proliferation and invasive potential. Anticancer Res 15(4):1191–1198

Luo JL, Maeda S, Hsu LC, Yagita H, Karin M (2004) Inhibition of NF-kappaB in cancer cells converts inflammation- induced tumor growth mediated by TNFalpha to TRAIL-mediated tumor regression. Cancer Cell 6(3):297–305. https://doi.org/10.1016/j.ccr.2004.08.012

Maquart FX, Pasco S, Ramont L, Hornebeck W, Monboisse JC (2004) An introduction to matrikines: extracellular matrix-derived peptides which regulate cell activity. Implication in tumor invasion. Crit Rev Oncol Hematol 49(3):199–202. https://doi.org/10.1016/j.critrevonc.2003.06.007

Matsumoto K, Shionyu M, Go M, Shimizu K, Shinomura T, Kimata K, Watanabe H (2003) Distinct interaction of versican/PG-M with hyaluronan and link protein. J Biol Chem 278(42): 41205–41212. https://doi.org/10.1074/jbc.M305060200

McCulloch DR, Nelson CM, Dixon LJ, Silver DL, Wylie JD, Lindner V, Sasaki T, Cooley MA, Argraves WS, Apte SS (2009) ADAMTS metalloproteases generate active versican fragments that regulate interdigital web regression. Dev Cell 17(5):687–698. https://doi.org/10.1016/j.devcel.2009.09.008

Mead TJ, Apte SS (2018) ADAMTS proteins in human disorders. Matrix Biol 71-72:225–239. https://doi.org/10.1016/j.matbio.2018.06.002

Merad M, Sathe P, Helft J, Miller J, Mortha A (2013) The dendritic cell lineage: ontogeny and function of dendritic cells and their subsets in the steady state and the inflamed setting. Annu Rev Immunol 31:563–604. https://doi.org/10.1146/annurev-immunol-020711-074950

Mitsui Y, Shiina H, Kato T, Maekawa S, Hashimoto Y, Shiina M, Imai-Sumida M, Kulkarni P, Dasgupta P, Wong RK, Hiraki M, Arichi N, Fukuhara S, Yamamura S, Majid S, Saini S, Deng G, Dahiya R, Nakajima K, Tanaka Y (2017) Versican promotes tumor progression, metastasis and predicts poor prognosis in renal carcinoma. Mol Cancer Res 15(7):884–895. https://doi.org/10.1158/1541-7786.MCR-16-0444

Montgomery AM, De Clerck YA, Langley KE, Reisfeld RA, Mueller BM (1993) Melanoma-mediated dissolution of extracellular matrix: contribution of urokinase-dependent and metalloproteinase-dependent proteolytic pathways. Cancer Res 53(3):693–700

Murphy KM (2013) Transcriptional control of dendritic cell development. Adv Immunol 120:239–267. https://doi.org/10.1016/B978-0-12-417028-5.00009-0

Mushtaq MU, Papadas A, Pagenkopf A, Flietner E, Morrow Z, Chaudhary SG, Asimakopoulos F (2018) Tumor matrix remodeling and novel immunotherapies: the promise of matrix-derived immune biomarkers. J Immunother Cancer 6(1):65. https://doi.org/10.1186/s40425-018-0376-0

Nandadasa S, Foulcer S, Apte SS (2014) The multiple, complex roles of versican and its proteolytic turnover by ADAMTS proteases during embryogenesis. Matrix Biol 35:34–41. https://doi.org/10.1016/j.matbio.2014.01.005

Nandadasa S, Burin des Roziers C, Koch C, Tran-Lundmark K, Dours-Zimmermann MT, Zimmermann DR, Valleix S, Apte SS (2021a) A new mouse mutant with cleavage-resistant versican and isoform-specific versican mutants demonstrate that proteolysis at the Glu(441)-ala(442) peptide bond in the V1 isoform is essential for interdigital web regression. Matrix Biol Plus 10:100064. https://doi.org/10.1016/j.mbplus.2021.100064

Nandadasa S, O'Donnell A, Murao A, Yamaguchi Y, Midura RJ, Olson L, Apte SS (2021b) The versican-hyaluronan complex provides an essential extracellular matrix niche for Flk1(+) hematoendothelial progenitors. Matrix Biol 97:40–57. https://doi.org/10.1016/j.matbio.2021.01.002

Naso MF, Morgan JL, Buchberg AM, Siracusa LD, Iozzo RV (1995) Expression pattern and mapping of the murine versican gene (Cspg2) to chromosome 13. Genomics 29(1):297–300. https://doi.org/10.1006/geno.1995.1251

Ochoa de Olza M, Navarro Rodrigo B, Zimmermann S, Coukos G (2020) Turning up the heat on non-immunoreactive tumours: opportunities for clinical development. Lancet Oncol 21(9):e419–e430. https://doi.org/10.1016/S1470-2045(20)30234-5

Onken J, Moeckel S, Leukel P, Leidgens V, Baumann F, Bogdahn U, Vollmann-Zwerenz A, Hau P (2014) Versican isoform V1 regulates proliferation and migration in high-grade gliomas. J Neuro-Oncol 120(1):73–83. https://doi.org/10.1007/s11060-014-1545-8

Paijens ST, Vledder A, de Bruyn M, Nijman HW (2021) Tumor-infiltrating lymphocytes in the immunotherapy era. Cell Mol Immunol 18(4):842–859. https://doi.org/10.1038/s41423-020-00565-9

Papadas A, Asimakopoulos F (2020) Versican in the tumor microenvironment. Adv Exp Med Biol 1272:55–72. https://doi.org/10.1007/978-3-030-48457-6_4

Papadas A, Flietner E, Morrow Z, Wiesner J, Cicala A, Pagenkopf A, Hope CL, Rajagopalan A, Wen Z, Emmerich P, Deming D, Miyamoto S, Hematti P, Callander NS, Rakhmilevich A, Otto M, Capitini CM, Zhang J, Asimakopoulos F (2019) Versican proteolytic fragments (Matrikines) regulate the Intratumoral dendritic cell milieu in vivo: implications for in situ tumor vaccination. Blood 134(Supplement_1):1210. https://doi.org/10.1182/blood-2019-131582

Papadas A, Arauz G, Cicala A, Wiesner J, Asimakopoulos F (2020) Versican and Versican-matrikines in cancer progression, inflammation, and immunity. J Histochem Cytochem 68(12):871–885. https://doi.org/10.1369/0022155420937098

Papadas A, Deb G, Officer A, Cicala A, Hope C, Emmerich P, Wiesner J, Pagenkopf A, Arauz G, Bansal V, Matkowskyj KA, Deming D, Politi K, Abrams SI, Harismendy O, Asimakopoulos F

(2021) Stromal remodeling regulates dendritic cell abundance and activity in the tumor microenvironment. bioRxiv:2021.2011.2010.467836. https://doi.org/10.1101/2021.11.10.467836
Pappas AG, Magkouta S, Pateras IS, Skianis I, Moschos C, Vazakidou ME, Psarra K, Gorgoulis VG, Kalomenidis I (2019) Versican modulates tumor-associated macrophage properties to stimulate mesothelioma growth. Onco Targets Ther 8(2):e1537427. https://doi.org/10.1080/2162402X.2018.1537427
Paris S, Sesboue R, Chauzy C, Maingonnat C, Delpech B (2006) Hyaluronectin modulation of lung metastasis in nude mice. Eur J Cancer 42(18):3253–3259. https://doi.org/10.1016/j.ejca.2006.06.012
Passi A, Negrini D, Albertini R, Miserocchi G, De Luca G (1999) The sensitivity of versican from rabbit lung to gelatinase a (MMP-2) and B (MMP-9) and its involvement in the development of hydraulic lung edema. FEBS Lett 456(1):93–96. https://doi.org/10.1016/s0014-5793(99)00929-1
Paulus W, Baur I, Dours-Zimmermann MT, Zimmermann DR (1996) Differential expression of versican isoforms in brain tumors. J Neuropathol Exp Neurol 55(5):528–533. https://doi.org/10.1097/00005072-199605000-00005
Perides G, Asher RA, Lark MW, Lane WS, Robinson RA, Bignami A (1995) Glial hyaluronate-binding protein: a product of metalloproteinase digestion of versican? Biochem J 312(Pt 2):377–384. https://doi.org/10.1042/bj3120377
Pesheva P, Probstmeier R, Spiess E, Schachner M (1991) Divalent cations modulate the inhibitory substrate properties of murine glia-derived J1-160 and J1-180 extracellular matrix glycoproteins for neuronal adhesion. Eur J Neurosci 3(4):356–365. https://doi.org/10.1111/j.1460-9568.1991.tb00823.x
Pukkila MJ, Kosunen AS, Virtaniemi JA, Kumpulainen EJ, Johansson RT, Kellokoski JK, Nuutinen J, Kosma VM (2004) Versican expression in pharyngeal squamous cell carcinoma: an immunohistochemical study. J Clin Pathol 57(7):735–739. https://doi.org/10.1136/jcp.2003.014589
Ricciardelli C, Mayne K, Sykes PJ, Raymond WA, McCaul K, Marshall VR, Horsfall DJ (1998) Elevated levels of versican but not decorin predict disease progression in early-stage prostate cancer. Clin Cancer Res 4(4):963–971
Ricciardelli C, Brooks JH, Suwiwat S, Sakko AJ, Mayne K, Raymond WA, Seshadri R, LeBaron RG, Horsfall DJ (2002) Regulation of stromal versican expression by breast cancer cells and importance to relapse-free survival in patients with node-negative primary breast cancer. Clin Cancer Res 8(4):1054–1060
Ricciardelli C, Russell DL, Ween MP, Mayne K, Suwiwat S, Byers S, Marshall VR, Tilley WD, Horsfall DJ (2007) Formation of hyaluronan- and versican-rich pericellular matrix by prostate cancer cells promotes cell motility. J Biol Chem 282(14):10814–10825. https://doi.org/10.1074/jbc.M606991200
Ricciardelli C, Sakko AJ, Ween MP, Russell DL, Horsfall DJ (2009) The biological role and regulation of versican levels in cancer. Cancer Metastasis Rev 28(1–2):233–245. https://doi.org/10.1007/s10555-009-9182-y
Said N, Theodorescu D (2012) RhoGDI2 suppresses bladder cancer metastasis via reduction of inflammation in the tumor microenvironment. Onco Targets Ther 1(7):1175–1177. https://doi.org/10.4161/onci.20594
Sakko AJ, Ricciardelli C, Mayne K, Tilley WD, Lebaron RG, Horsfall DJ (2001) Versican accumulation in human prostatic fibroblast cultures is enhanced by prostate cancer cell-derived transforming growth factor beta1. Cancer Res 61(3):926–930
Sakko AJ, Ricciardelli C, Mayne K, Suwiwat S, LeBaron RG, Marshall VR, Tilley WD, Horsfall DJ (2003) Modulation of prostate cancer cell attachment to matrix by versican. Cancer Res 63(16):4786–4791
Salmivirta M, Elenius K, Vainio S, Hofer U, Chiquet-Ehrismann R, Thesleff I, Jalkanen M (1991) Syndecan from embryonic tooth mesenchyme binds tenascin. J Biol Chem 266(12):7733–7739

Sandy JD, Westling J, Kenagy RD, Iruela-Arispe ML, Verscharen C, Rodriguez-Mazaneque JC, Zimmermann DR, Lemire JM, Fischer JW, Wight TN, Clowes AW (2001) Versican V1 proteolysis in human aorta in vivo occurs at the Glu441-Ala442 bond, a site that is cleaved by recombinant ADAMTS-1 and ADAMTS-4. J Biol Chem 276(16):13372–13378. https://doi.org/10.1074/jbc.M009737200

Savani RC, Cao G, Pooler PM, Zaman A, Zhou Z, DeLisser HM (2001) Differential involvement of the hyaluronan (HA) receptors CD44 and receptor for HA-mediated motility in endothelial cell function and angiogenesis. J Biol Chem 276(39):36770–36778. https://doi.org/10.1074/jbc.M102273200

Schaefer L (2014) Complexity of danger: the diverse nature of damage-associated molecular patterns. J Biol Chem 289(51):35237–35245. https://doi.org/10.1074/jbc.R114.619304. Epub 2014 Nov 12

Schaefer L, Iozzo RV (2012) Small leucine-rich proteoglycans, at the crossroad of cancer growth and inflammation. Curr Opin Genet Dev 22 (1):56–57. https://doi.org/10.1016/j.gde.2011.12.002. Epub 2012 Feb 9

Schmitt M (2016) Versican vs versikine: tolerance vs attack. Blood 128(5):612–613. https://doi.org/10.1182/blood-2016-06-721092

Shaban M, Khurram SA, Fraz MM, Alsubaie N, Masood I, Mushtaq S, Hassan M, Loya A, Rajpoot NM (2019) A novel digital score for abundance of tumour infiltrating lymphocytes predicts disease free survival in Oral squamous cell carcinoma. Sci Rep 9(1):13341. https://doi.org/10.1038/s41598-019-49710-z

Sharma P, Allison JP (2020) Dissecting the mechanisms of immune checkpoint therapy. Nat Rev Immunol 20(2):75–76. https://doi.org/10.1038/s41577-020-0275-8

Snyder JM, Washington IM, Birkland T, Chang MY, Frevert CW (2015) Correlation of Versican expression, accumulation, and degradation during embryonic development by quantitative immunohistochemistry. J Histochem Cytochem 63(12):952–967. https://doi.org/10.1369/0022155415610383

Spranger S, Dai D, Horton B, Gajewski TF (2017) Tumor-residing Batf3 dendritic cells are required for effector T cell trafficking and adoptive T cell therapy. Cancer Cell 31(5):711–723.e714. https://doi.org/10.1016/j.ccell.2017.04.003

Stankunas K, Hang CT, Tsun ZY, Chen H, Lee NV, Wu JI, Shang C, Bayle JH, Shou W, Iruela-Arispe ML, Chang CP (2008) Endocardial Brg1 represses ADAMTS1 to maintain the microenvironment for myocardial morphogenesis. Dev Cell 14(2):298–311. https://doi.org/10.1016/j.devcel.2007.11.018

Takahashi Y, Kuwabara H, Yoneda M, Isogai Z, Tanigawa N, Shibayama Y (2012) Versican G1 and G3 domains are upregulated and latent transforming growth factor-β binding protein-4 is downregulated in breast cancer stroma. Breast Cancer 19(1):46–53. https://doi.org/10.1007/s12282-011-0264-7

Tang M, Diao J, Gu H, Khatri I, Zhao J, Cattral MS (2015) Toll-like receptor 2 activation promotes tumor dendritic cell dysfunction by regulating IL-6 and IL-10 receptor signaling. Cell Rep 13(12):2851–2864. https://doi.org/10.1016/j.celrep.2015.11.053

Tang M, Diao J, Cattral MS (2017) Molecular mechanisms involved in dendritic cell dysfunction in cancer. Cell Mol Life Sci 74(5):761–776. https://doi.org/10.1007/s00018-016-2317-8

Toole BP (1990) Hyaluronan and its binding proteins, the hyaladherins. Curr Opin Cell Biol 2(5):839–844. https://doi.org/10.1016/0955-0674(90)90081-o

Tzanakakis G, Neagu M, Tsatsakis A, Nikitovic D (2019) Proteoglycans and Immunobiology of cancer-therapeutic implications. Front Immunol 10:875. https://doi.org/10.3389/fimmu.2019.00875

Waldman AD, Fritz JM, Lenardo MJ (2020) A guide to cancer immunotherapy: from T cell basic science to clinical practice. Nat Rev Immunol 20(11):651–668. https://doi.org/10.1038/s41577-020-0306-5

Wang W, Xu GL, Jia WD, Ma JL, Li JS, Ge YS, Ren WH, Yu JH, Liu WB (2009) Ligation of TLR2 by versican: a link between inflammation and metastasis. Arch Med Res 40(4):321–323. https://doi.org/10.1016/j.arcmed.2009.04.005

Wasa J, Nishida Y, Shinomura T, Isogai Z, Futamura N, Urakawa H, Arai E, Kozawa E, Tsukushi S, Ishiguro N (2012) Versican V1 isoform regulates cell-associated matrix formation and cell behavior differentially from aggrecan in swarm rat chondrosarcoma cells. Int J Cancer 130(10):2271–2281. https://doi.org/10.1002/ijc.26230

Ween MP, Oehler MK, Ricciardelli C (2011) Role of versican, hyaluronan and CD44 in ovarian cancer metastasis. Int J Mol Sci 12(2):1009–1029. https://doi.org/10.3390/ijms12021009

Westling J, Gottschall PE, Thompson VP, Cockburn A, Perides G, Zimmermann DR, Sandy JD (2004) ADAMTS4 (aggrecanase-1) cleaves human brain versican V2 at Glu405-Gln406 to generate glial hyaluronate binding protein. Biochem J 377(Pt 3):787–795. https://doi.org/10.1042/BJ20030896

Wight TN (2002) Versican: a versatile extracellular matrix proteoglycan in cell biology. Curr Opin Cell Biol 14(5):617–623. https://doi.org/10.1016/s0955-0674(02)00375-7

Wight TN (2017) Provisional matrix: a role for versican and hyaluronan. Matrix Biol 60-61:38–56. https://doi.org/10.1016/j.matbio.2016.12.001

Wight TN, Kinsella MG, Evanko SP, Potter-Perigo S, Merrilees MJ (2014) Versican and the regulation of cell phenotype in disease. Biochim Biophys Acta 1840(8):2441–2451. https://doi.org/10.1016/j.bbagen.2013.12.028

Wight TN, Kang I, Evanko SP, Harten IA, Chang MY, Pearce OMT, Allen CE, Frevert CW (2020) Versican-a critical extracellular matrix regulator of immunity and inflammation. Front Immunol 11:512. https://doi.org/10.3389/fimmu.2020.00512

Wu Y, Chen L, Zheng PS, Yang BB (2002) Beta 1-integrin-mediated glioma cell adhesion and free radical-induced apoptosis are regulated by binding to a C-terminal domain of PG-M/versican. J Biol Chem 277(14):12294–12301. https://doi.org/10.1074/jbc.M110748200

Wu Y, Wu J, Lee DY, Yee A, Cao L, Zhang Y, Kiani C, Yang BB (2005a) Versican protects cells from oxidative stress-induced apoptosis. Matrix Biol 24(1):3–13. https://doi.org/10.1016/j.matbio.2004.11.007

Wu YJ, La Pierre DP, Wu J, Yee AJ, Yang BB (2005b) The interaction of versican with its binding partners. Cell Res 15(7):483–494. https://doi.org/10.1038/sj.cr.7290318

Xia L, Huang W, Tian D, Zhang L, Qi X, Chen Z, Shang X, Nie Y, Wu K (2014) Forkhead box Q1 promotes hepatocellular carcinoma metastasis by transactivating ZEB2 and VersicanV1 expression. Hepatology 59(3):958–973. https://doi.org/10.1002/hep.26735

Yamada KM, Even-Ram S (2002) Integrin regulation of growth factor receptors. Nat Cell Biol 4(4): E75–E76. https://doi.org/10.1038/ncb0402-e75

Yamaguchi Y (2000) Lecticans: organizers of the brain extracellular matrix. Cell Mol Life Sci 57(2):276–289. https://doi.org/10.1007/PL00000690

Yang W, Yee AJ (2013) Versican V2 isoform enhances angiogenesis by regulating endothelial cell activities and fibronectin expression. FEBS Lett 587(2):185–192. https://doi.org/10.1016/j.febslet.2012.11.023

Yang BL, Zhang Y, Cao L, Yang BB (1999) Cell adhesion and proliferation mediated through the G1 domain of versican. J Cell Biochem 72(2):210–220

Yee AJ, Akens M, Yang BL, Finkelstein J, Zheng PS, Deng Z, Yang B (2007) The effect of versican G3 domain on local breast cancer invasiveness and bony metastasis. Breast Cancer Res 9(4):R47. https://doi.org/10.1186/bcr1751

Yeung TL, Leung CS, Wong KK, Samimi G, Thompson MS, Liu J, Zaid TM, Ghosh S, Birrer MJ, Mok SC (2013) TGF-β modulates ovarian cancer invasion by upregulating CAF-derived versican in the tumor microenvironment. Cancer Res 73(16):5016–5028. https://doi.org/10.1158/0008-5472.CAN-13-0023

Zheng PS, Vais D, Lapierre D, Liang YY, Lee V, Yang BL, Yang BB (2004a) PG-M/versican binds to P-selectin glycoprotein ligand-1 and mediates leukocyte aggregation. J Cell Sci 117(Pt 24): 5887–5895. https://doi.org/10.1242/jcs.01516

Zheng PS, Wen J, Ang LC, Sheng W, Viloria-Petit A, Wang Y, Wu Y, Kerbel RS, Yang BB (2004b) Versican/PG-M G3 domain promotes tumor growth and angiogenesis. FASEB J 18(6): 754–756. https://doi.org/10.1096/fj.03-0545fje

Chapter 6
Circulating Proteoglycans/Glycosaminoglycans as Cancer Biomarkers

Antonio Junior Lepedda, Gabriele Nieddu, Nikos Karamanos, and Marilena Formato

Abstract Cancer propagation and progression are associated with remarkable remodeling of the extracellular microenvironment with the formation of a permissive matrix for tumor growth, enriched in inflammatory mediators and matrix-degrading enzymes. Proteoglycans, which represent major structural and functional extracellular matrix components, play pivotal roles by affecting tissue organization, cell–matrix interactions, cell signaling, and, ultimately, cell behavior in physiological conditions as well as during cancer development and progression. In this respect, the proteolytic activity in tumor stroma liberates bioactive fragments, which can be detected in serum and may be useful as diagnostic and prognostic markers. Furthermore, since cell surface proteoglycans are specifically expressed by cancer cells, as in the case of glypican-3 in hepatocellular carcinoma, they can represent potential promising targets for immunotherapy.

This chapter reviews and critically addresses the major clinical literature and provides an overview of the significance of circulating proteoglycans (PGs) and glycosaminoglycans (GAGs), especially their soluble forms originating from the cell surface heparan sulfate proteoglycans syndecans and glypicans, in both prognosis and diagnosis of various malignancies responsible for high mortality and disability, including hepatocellular carcinoma, multiple myeloma, breast cancer, and renal cell carcinoma.

A. J. Lepedda · G. Nieddu · M. Formato (✉)
Department of Biomedical Sciences, University of Sassari, Sassari, Italy
e-mail: ajlepedda@uniss.it; ganieddu@uniss.it; formato@uniss.it

N. Karamanos
Biochemistry, Biochemical Analysis & Matrix Pathobiology Research Group, Laboratory of Biochemistry, Department of Chemistry, University of Patras, Patras, Greece
e-mail: n.k.karamanos@upatras.gr

I. Kovalszky et al. (eds.), *The Extracellular Matrix and the Tumor Microenvironment*, Biology of Extracellular Matrix 11,
https://doi.org/10.1007/978-3-030-99708-3_6

6.1 Introduction

Cancer is an ever-growing healthcare problem, being the second leading cause of death globally, with an estimated 9.6 million deaths in 2018 (https://www.who.int/health-topics/cancer). It is a group of related diseases caused by genetic mutations leading to an abnormal cell proliferation, which can potentially invade other parts of the body.

Tumor cells, together with inflammatory cells and activated resident fibroblasts or cancer-associated fibroblasts (CAFs), create a microenvironment enriched in growth factors, cytokines, and extracellular matrix (ECM)-degrading enzymes. ECM, which physiologically provides cells with the chemical and mechanical signals essential for maintaining tissue homeostasis, is a key player undergoing deep remodeling that, in turn, supports tumor growth and spreading (Yuzhalin et al. 2018). In this respect, proteoglycans (PGs), as major functional ECM components, are crucial for tumor cells–matrix interactions and modulation of the activity of soluble factors (Theocharis and Karamanos 2019).

PGs are a family of ubiquitously distributed macromolecules, composed of specific core protein moieties bearing one or more O- or N-linked sulfated glycosaminoglycan (GAG) chains, primarily responsible for their biological properties (Raman et al. 2005), which can also be found in body fluids, i.e., plasma and urine.

GAGs are a group of complex anionic and unbranched heteropolysaccharides composed of repeating disaccharide units containing an *N*-acetylated hexosamine, either an *N*-AcetylGalactosamine (GalNAc) or an *N*-AcetylGlucosamine (GlcNAc), and an hexuronic acid (D-Glucuronic acid (GlcA) or its carbon-5 epimer L-Iduronic acid (IdoA)) or, exclusively in keratan sulfate, Galactose. GAGs are key structural and functional components of the ECM of all connective tissues, being involved in numerous events including embryonic development, ECM assembly, and regulation of cell signaling in various pathophysiological conditions. Their structural heterogeneity, in terms of repeating disaccharide unit, chain length, degree and pattern of sulfation, as well as degree of epimerization, is such huge that, virtually, there are not two identical glycosaminoglycans in the body, and is responsible for most of the numerous biological functions of PGs (Karamanos et al. 2018). Indeed, it has been recently shown that GAGs may interact with more than 800 different proteins (Vallet et al. 2021), including both cytokines and chemokines, as well as enzymes and enzyme inhibitors, ECM proteins, and membrane receptors (Kjellen and Lindahl 2018). Because of their pleiotropic functions and physicochemical properties, GAGs have also been employed as therapeutic agents as well as in many tissue engineering applications (Lepedda et al. 2021a).

According to the type of repeating disaccharide unit, six main classes of GAGs have been described so far: hyaluronan or hyaluronic acid (HA), chondroitin sulfate (CS), dermatan sulfate (DS), keratan sulfate (KS), heparan sulfate (HS), and heparin (Hep) (Fig. 6.1).

PGs that can be found in the ECM include versican (belonging to the hyalectans subfamily), decorin, and biglycan (belonging to the small leucine-rich PGs (SLRPs)

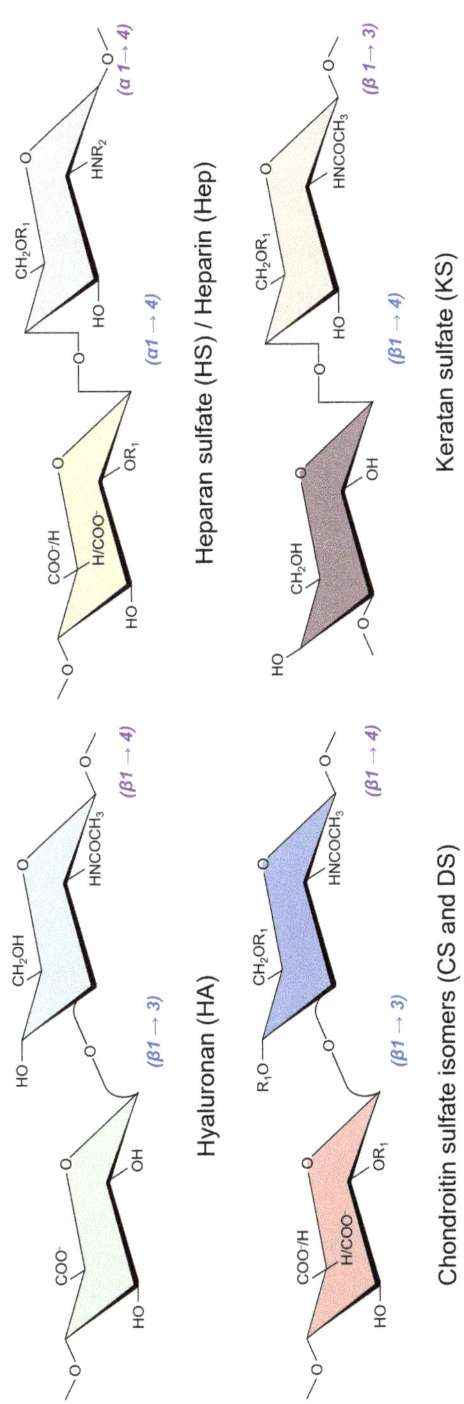

Fig. 6.1 Disaccharides structures representative of the six GAGs classes (modified from Vynios et al. 2002). Both DS and HS/Hep are hybrid structures containing GlcA as well as its C-5 epimer IdoA. HS and Hep differ in terms of sulfation and epimerization, being both higher in the latter. Except for HA, GAGs are variously sulfated, showing higher negative charge densities. Saccharides are reported as chair conformations. Configurations of the O-glycosidic bonds are reported either in blue (within disaccharide units) or in purple (between adjacent disaccharide units). $R_1 = SO_3^-$; $R_2 = COCH_3/SO_3^-$

subfamily); the pericellular PGs are perlecan, agrin, collagens XV and XVIII; those connected to the cell surface comprise glypicans (GPCs) (belonging to the GPI-anchored PGs subfamily), syndecans (SDCs) CSPG4, betaglycan, phosphacan, and CD44 (belonging to the transmembrane PGs subfamily), whereas the only known intracellular PG is serglycin (Fig. 6.2) (Theocharis et al. 2010; Karamanos et al. 2021).

Regarding their function, SDCs and GPCs play multiple crucial roles in both physiological and pathological conditions, ranging from regulation of cell–matrix interactions to cell signaling, when anchored to the cell as well as when in the free form as a result of shedding.

The whole ECM undergoes substantial remodeling during tumor formation and growth by the action of both proteolytic and glycosidic enzymes secreted in the tumor stroma by inflammatory cells. As a consequence of these events, either ECM PGs, cell surface PGs, or their degradation products, may be released into circulation, representing potential markers with diagnostic, prognostic, or therapeutic significance. Their high blood levels may be either the result of an intense proteolytic activity on constitutively expressed PGs (e.g., those physiologically present in ECM or glycocalyx) or to a de novo expression in tumor stroma (e.g., the expression of the oncofetal GPC-3 by hepatocellular carcinoma).

Recently, we performed a literature investigation to evaluate the clinical utility of measuring circulating HS/HSPGs in pathophysiological conditions showing interesting results on several inflammation-based pathologies such as cardiovascular disease, diabetes, obesity and kidney disease, trauma, sepsis, but also multiple sclerosis, preeclampsia, pathologies requiring surgery, pulmonary disease, and others (Lepedda et al. 2021b).

Several evidences show that the soluble forms, originating from shedding, are detectable in blood and other body fluids and represent useful diagnostic and/or prognostic tools of various malignancies responsible for high mortality and disability including, among others, hepatocellular carcinoma (HCC) and multiple myeloma (MM).

This chapter reviews the major clinical literature on this topic and provides an overview of the main findings. The items for PubMed search included: ECM, circulating PGs/HSPGs, HA, HS, GAGs, SDCs, GPCs, glycocalyx/shedding, circulating cancer biomarkers. More than 300 articles were reviewed. Among them, about 150 clinical studies dealt with the association between circulating GAGs/PGs levels, mainly originated from glycocalyx degradation, and several cancer types. Noteworthy, in almost all of them, the HSPGs SDCs, GPCs or their ectodomains from the cellular glycocalyx degradation, were chosen as biomarkers.

For more detailed discussion, readers are referred to excellent reviews on topics such as ECM structure (Theocharis et al. 2019; Karamanos et al. 2021), PGs structure and function (Theocharis et al. 2010; Theocharis and Karamanos 2019; Karamanos et al. 2021; Zolota et al. 2021), as well as endothelial glycocalyx structure and function (Weinbaum et al. 2007; Zhang et al. 2018b), in both physiological and pathological conditions.

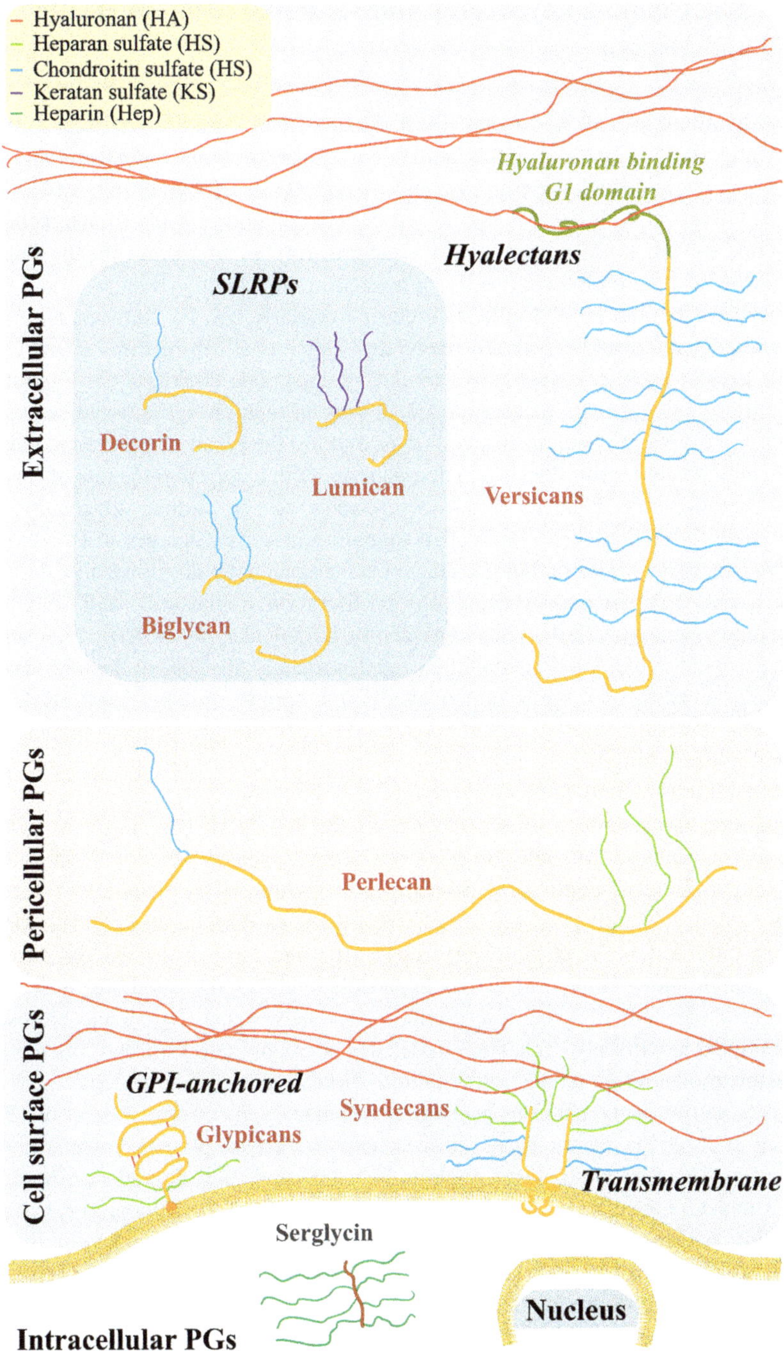

Fig. 6.2 Schematic representation of the four classes of PGs according to their localization: intracellular, cell surface, pericellular, and extracellular. Some representative PGs and the corresponding subfamilies are reported. Protein moieties are shown as yellow backbones (versican G1 domain is highlighted in green)

6.2 The Cellular Glycocalyx and Its Remodeling in Health and Disease

Every mammalian cell is covered by a dynamic and complex gel-like network, consisting of PGs (mainly SDCs and GPCs), glycoproteins (selectins, integrins, and immunoglobulins), glycolipids, and GAGs (HA, HS, and CS), collectively termed "glycocalyx," which is implicated in cell–cell and cell–matrix interactions.

Bennett et al. (1959) showed this structure, for the first time, on the luminal surface of vascular endothelial cells using electron microscopy (Bennett et al. 1959). Due to the high content of polysaccharides, particularly HS, CS isomers, and HA, it was called "glycocalyx" (literally meaning "sugar coat" from the ancient Greek words glykys/γλυκύς = sweet, kylix/κύλιξ = cup). The thickness (approximately 0.5–5.0μm) and structure of the glycocalyx vary in relation to the tissue, as well as with several pathophysiological conditions, and are the result of a dynamic balance between persistent biosynthesis and degradation, with a considerable rate of renewal of its components. PGs are major components of the cellular glycocalyx and participate to biochemical and mechanosignaling in cells. Changes in PGs expression in cancer and other pathologies have been associated with glycocalyx derangement and with intense effects on proliferation, adhesion, and migration (Barkovskaya et al. 2020). HS is covalently bound to core proteins to form the main glycocalyx PGs, SDCs, and GPCs (approximately 50–90% of the total amount of PGs), whereas HA binds noncovalently membrane proteins, such as CD44, providing structural support to the network. The cellular glycocalyx interacts with several matrix components, enzymes, proteins of plasma origin, pro-inflammatory cytokines, and growth factors, regulating cell–cell and cell–matrix crosstalk for maintaining tissue homeostasis (Zhang et al. 2018b). Derangement of the cellular glycocalyx structure, by either physical or chemical insults, plays essential roles in several pathological conditions, as reviewed with regard to pathologies such as cardiovascular disease (Machin et al. 2019), diabetes (Dogne et al. 2018), kidney disease (Jourde-Chiche et al. 2019), sepsis (Iba and Levy 2019; Uchimido et al. 2019), and trauma (Tuma et al. 2016). In particular, glycocalyx shedding, a regulated proteolytic cleavage of some of its components (Becker et al. 2015), may expose, in pathological conditions, adhesion molecules such as selectins and intercellular adhesion molecule 1, on the endothelium surface, thus inducing leukocytes and platelets recruitment, therefore leading to vascular dysfunction (Celie et al. 2009; Kumar et al. 2015). Furthermore, all these events increase plasma concentration of several glycocalyx components, such as HA, HS, and their degradation products, SDCs and GPCs ectodomains, which may in turn affect both local and systemic signaling pathways (Yang et al. 2017), in some cases lasting even for days as reported for highly sulfated HS fragments released in circulation in patients with respiratory failure (Schmidt et al. 2014).

Under various acute and chronic clinical conditions, the glycocalyx shedding primarily involves various MMPs and A disintegrin metalloproteinases with thrombospondin motifs (ADAMTS) -1 and -4, as well as heparanase and hyaluronidase, released by inflammatory cells and activated endothelial cells, reactive oxygen and nitrogen species (ROS and RNS). Also serine proteases such as thrombin, elastase, proteinase 3, plasminogen, and cathepsin B are involved in glycocalyx shedding (Becker et al. 2015).

MMPs (MT1-MMP, MMP-2, -7, and -9) and heparanase are secreted following inflammatory stimuli and contribute to the cleavage of matrix components as well as of cell surface PGs. Heparanase cleaves HS chains, mainly at low sulfation sites, therefore releasing fragments of 4–7 kDa (Meirovitz et al. 2013). Besides the action of heparanase and other specific endoglycosidases, also ROS and RNS can lead to GAGs fragmentation (Moseley et al. 1997; Soltes et al. 2006) with fragments which can be released in circulation, representing potential useful markers of disease.

MMPs activity is regulated by tissue inhibitors of metalloproteinases (TIMPs) which can bind HS and CS on the glycocalyx, whereas Hep is an inhibitor of heparanase (Oduah et al. 2016). In this respect, pharmaceutical mixtures of low-molecular weight Hep and CS isomers, such as sulodexide, can be used as antithrombotic and pro-fibrinolytic drugs with anti-inflammatory, antioxidant, and vasculoprotective properties, probably preserving endothelial glycocalyx from shedding by heparanase and MMP-9 activities (Weiss et al. 2007).

HA is a ubiquitous ECM and glycocalyx component that plays key roles during embryogenesis, inflammation, wound healing, and, particularly, in neoplasia where it modulates the crosstalk between cancer cells and the host peritumor stroma (Stern 2008; Passi et al. 2019; Caon et al. 2020). It is a linear, high-molecular weight polymer (up to 3–4 MDa) composed of long unbranched and nonsulfated free chains (1–2μm) of repeating GlcA and GlcNAc residues connected by β-1,3- and β-1,4-glycosidic bonds (Passi et al. 2019). Unlike the other GAGs, which are synthesized in the Golgi apparatus bound to a protein moiety, HA is synthesized by membrane-bound synthases at the inner side of the plasma membrane and then it is extruded through pore-like structures to the cell surface without undergoing any further postsynthetic modifications (i.e., sulfation or epimerization) (Vigetti et al. 2014). HA interacts with cell surface by specific receptors, such as the transmembrane CD44 and the receptor for HA-mediated motility (RHAMM), modulating cell behavior differently, according to its molecular weight (Tavianatou et al. 2019a; Caon et al. 2020). In particular, it is widely acknowledged that high-molecular weight HA has antiangiogenic, antiproliferative, and immunosuppressive properties, whereas low-molecular weight HA fragments are pro-inflammatory, pro-angiogenic, and could favor cancer cell aggressive phenotype (Tavianatou et al. 2019b, 2021). HA is synthesized at high levels in several tumors by the action of HA synthase-2, the most catalytically active isoform, strongly related to tumor aggressiveness and unfavorable outcome (Caon et al. 2020). Moreover, its degradation by the action of hyaluronidase 1 (the most active somatic hyaluronidase), and hyaluronidase 2 (a glycosylphosphatidylinositol (GPI)-anchored enzyme) during extracellular remodeling in tumorigenesis is increased and causes the release of bioactive fragments. Accordingly, both high- and low-molecular weight HA have been found at high levels in plasma, particularly in breast cancer patients (Velesiotis et al. 2019), where they have been associated with metastasis onset and poor prognosis, and used as useful markers for monitoring the response to therapy (Wu et al. 2015; Peng et al. 2016).

Besides changes in HA metabolism, also the biosynthetic pathways of the other GAGs may undergo consistent alterations in cancer, as elegantly shown in clear cell renal cell carcinoma (ccRCC), resulting in altered metabolite levels in plasma. Indeed, Gatto et al. (2016b), using genome-scale metabolic modeling, evidenced deep changes about GAGs biosynthesis in ccRCC and identified distinct GAGs profiles in plasma and urine of patients. They calculated scores based on 18 different GAGs properties, including CS and HS levels, disaccharides composition, and charge density, with diagnostic accuracy ranging from 93.1 to 100% (Gatto et al. 2016b) and high prognostic value (Gatto et al. 2016a), also in surgically treated patients, independently from stage, grade, or histology (Gatto et al. 2018).

6.2.1 Cell Surface PGs

Type I transmembrane SDCs, together with glycosyl-phosphatidylinositol-anchored GPCs, represent the two major families of cell surface PGs. They are structural glycocalyx components that play major roles in regulating cell behavior, cell signaling, and cell–matrix interactions.

SDCs are a well-conserved family of single-span transmembrane HSPGs (both SDC-1 and -3 bear also CS chains) with pleiotropic roles in development, inflammation, and tumor progression by modulating cell proliferation, differentiation, adhesion, and migration (Chung et al. 2016; Afratis et al. 2017). Their core proteins, ranging from 20 to 40 kDa (33, 23, 41, and 22 kDa for SDC-1, -2, -3, and -4, respectively), are coded by four genes in vertebrates with distinct spatio-temporal pattern of expression (Gondelaud and Ricard-Blum 2019). Accordingly, SDC-1 is widely expressed in both epithelial and plasma cells, while SDC-2 is mainly expressed in mesenchymal cells, such as fibroblasts and smooth muscle cells, SDC-3 is expressed in neural tissues and developing musculoskeletal tissues, whereas SDC-4 is abundant in most tissues (Afratis et al. 2017). Their expression is finely regulated, since SDCs are involved in several signaling events, including tumorigenesis. Each isoform consists of a short C-terminal cytoplasmic domain, a transmembrane domain and an ectodomain carrying three to five HS or CS chains (Fig. 6.2). A model of full-length SDC-3 has been generated though scanty structural data are currently available, probably because of the high percentages of disordered amino acid residues (Peysselon et al. 2011). Following alternative splicing removal of the transmembrane domain, SDC-4 may occur also as a soluble protein isoform (Xing et al. 2003). The short cytoplasmic domain, consisting of the two conserved C1 and C2 regions, located respectively near the transmembrane domain and at the C-terminal, and a variable V region in between specific to each SDC and highly conserved across different species (Fig. 6.2), interacts with several intracellular kinases, as well as with actin cytoskeleton, thus modulating signal transduction (Couchman et al. 2015). Both C1 and C2 domains are highly conserved throughout the animal kingdom (Chakravarti and Adams 2006), being C1 identical in all four mammalian SDCs. This latter participates in SDC di- or oligomerization, as well as in binding of several intracellular proteins. In SDC-4, this region includes a phosphatidylinositol-4,5-bisphosphate (PIP_2) binding site implicated in dimerization and required for protein kinase C activation and intracellular signaling (Horowitz et al. 1999, 2002; Murakami et al. 2002). C2 domain has a well-conserved carboxyl-terminal EFYA tetrapeptide sequence able to bind some PDZ domain-containing proteins that may have a role as scaffold proteins for the recruitment of signaling and cytoskeletal proteins to the plasma membrane (Cheng et al. 2016). Besides, it is also implicated in intracellular transduction by organizing signaling complexes at cell membrane. The transmembrane domain is highly conserved and contains a GXXXG motif that promotes self-association, thus leading to the formation of dimers/oligomers essential for protein kinase C activation (Oh et al. 1997). The ectodomain, interacts with the extracellular milieu primarily by its GAGs chains and is fundamental for the mechanisms of signal transduction (Chu et al. 2004). Its shedding is mainly carried out by MMPs and occurs constitutively at specific sites (Becker et al. 2015), but it is fairly accelerated consequently to both inflammatory stimuli and pathological conditions (Schmidt et al. 2005; Piperigkou et al. 2016; Zhang et al. 2018b). This leads to a rapid modulation of intracellular response, by reducing surface receptors as well as generating soluble bioactive ectodomains. HS chains are covalently O-linked to serine residues in a serine-glycine motif surrounded by acidic residues near the N-terminal of ectodomain. Furthermore, SDC-1 and -3 carry CS/DS chains at sites closer to the transmembrane domain. Polymerization of HS and CS/DS chains occurs in the Golgi apparatus, starting from a xylose–galactose–galactose–uronic acid tetrasaccharide linker, and it is carried out by the sequential and repetitive addition of GlcA and either GlcNAc (HS chain) or GalNAc (CS chain) residues, to reach 50–200 disaccharides in length. During polymerization, both iduronation and sulfation at specific positions occur, mediated by glucuronyl C-5 epimerase and by sulfotransferases, respectively, hence leading to extensive chain modification. With regard to HS, sulfation may occur at the 2-O position of the hexuronic acid residues, whereas N-glucosamine residues may be N-sulfated, sulfated at 6-O position or, less frequently, at the 3-O position. The different disaccharides species are not randomly distributed but produce alternating subdomains of low (N-acetylated domains) and high sulfation (N-sulfated domains) with intermediate sulfation sequences in between (N-sulfated/N-acetylated domains) (Fig. 6.3).

After being exposed at the cell surface, HS chains may be further modified by specific sulfatases, which selectively remove 6-O sulfates (Morimoto-Tomita et al. 2002; Bishop et al. 2007; Nadanaka and Kitagawa 2008). Due to their high structural heterogeneity, HS can interact with a wide range of ligands such as ECM proteins (e.g., collagens, fibronectin), plasma proteins (e.g., albumin, antithrombin), cytokines, chemokines, growth factors (e.g., FGF, VEGF, TGF-β, and platelet-derived growth factor (PDGF)), and enzymes, which are primarily involved in signal transduction by forming ternary complexes with both growth factors and their receptors, e.g., with FGF2-FGFR1 (Qiao et al. 2003; Tkachenko et al. 2005). During glycocalyx shedding, the endo-β-D-glucuronidase heparanase may cleave HS chains

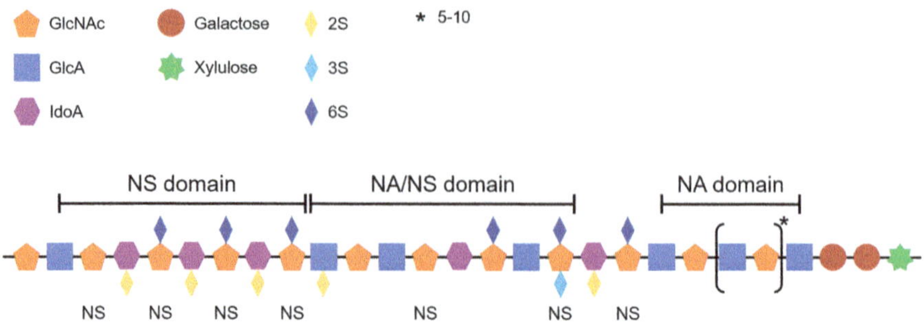

Fig. 6.3 Repetitive domain organization of HS. Highly sulfated/iduronated regions (N-sulfated domains or NS domains) are flanked by moderately modified regions (N-acetylated/N-sulfated domains or NA/NS domains) with interspersed unmodified sequences (N-acetylated domains or NA domains) (Esko and Selleck 2002)

at sites of low-sulfation, with the release of 10–20 units-long oligosaccharides, thereby converting HS from an inhibitor to a potent activator of FGF-2 (Kato et al. 1998; Barash et al. 2010). So far, little is known about the functions of SDCs CS chains, for which has been suggested a role in modulating cell adhesion to the ECM, together with HS chains (Okamoto et al. 2003). Recently, Gondelaud and Ricard-Blum have provided a comprehensive SDC interactome consisting of 351 partners, identified also by high-throughput affinity purification-mass spectrometry (Gondelaud and Ricard-Blum 2019). Very interestingly, it has been shown that most of these interactions are mediated by the GAGs moiety (71% for SDC-1, 67% for SDC-2, 91% for SDC-3, and 68% for SDC-4).

GPCs are a family of six (in mammals) GPI-anchored HSPGs (GPC-1 to -6), organized into two subfamilies (GPCs -1, -2, -4, -6 and GPCs -3 and -5). They have core proteins ranging from 60 to 70 kDa with about 25% amino-acid identity with each other, and very similar three-dimensional structures stabilized by 7 disulfide bridges, originated by 14 well-conserved cysteine residues distributed in a large globular cysteine-rich domain (Filmus and Selleck 2001; Filmus et al. 2008). An hydrophobic C-terminal domain, required for the binding of the GPI anchor, links GPCs to specific plasma membrane lipid rafts enriched in cholesterol and sphingolipids, and is involved in vesicular transport and cell signaling (Belting 2003). The last 50 amino acids sequence in the C-terminus is close to the membrane and contains attachment sites for 3–4 HS chains. During maturation in the secretory pathway, the endoprotease Furin may cleave GPCs (between Arg^{358} and Ser^{359} in GPC-3) in 2 fragments of 40 and 30 kDa (the N-terminal and the C-terminal subunits, respectively) connected by disulfide bridges (De Cat et al. 2003; Hippo et al. 2004). GPCs can also be found in circulation either as full-length glycated form following cleavage by Notum, a kind of lipase that cleaves GPI-anchored proteins (Traister et al. 2008), or as N-terminal fragments lacking HS side chains (Hippo et al. 2004). GPCs are usually expressed at high levels during embryonic development with tissue-specific and stage-specific expression, suggesting a fundamental role in various biological processes, such as interaction between cell and ECM, as well as the control of cell division, differentiation, and morphogenesis (Filmus and Selleck 2001). GPC-1 is expressed in bone, bone marrow, muscle, epithelium, and kidney; GPC-2 is specifically present in nervous system; GPC-3 and GPC-6 are produced in most tissues; GPC-4 is produced in brain, kidney, and lung, whereas GPC-5 is expressed in brain, lung, liver, kidney, and limbs. In adults, significant alterations in the pattern of GPCs expression have been reported. In particular, GPC-1, GPC-4, and GPC-6 are widely expressed in various tissues, while GPC-2 is no longer expressed; GPC-3 is reported only in the ovary, mammary gland, mesothelium, lung, and kidney, whereas GPC-5 is specifically found in the brain (Filmus 2001). However, among all GPCs, only GPC-1 is expressed by endothelial cells (Rosenberg et al. 1997). GPCs are involved in the regulation of pathways including Wnt, FGF, Hedgehog, bone morphogenic protein, Slit, and insulin-like growth factor, with either stimulatory or inhibitory activity depending on the biological context (Filmus et al. 2008; Theocharis et al. 2010).

6.3 Circulating PGs/GAGs as Cancer Biomarkers

As mentioned above, ECM is tightly involved in tissue homeostasis as well as in the pathogenesis of several diseases such as cancer and undergoes deep remodeling by the activities of both proteolytic and glycosidic enzymes secreted by inflammatory cells. Besides, also the synthesis rate and the expression pattern of some of its components may vary greatly.

PGs play key roles in tumorigenesis, from regulation of cell–matrix interactions to cell signaling. Their soluble forms, detectable in plasma or other fluids, may be useful diagnostic and/or prognostic tools for several malignancies, overall responsible for about 14% of global deaths and 7.5% of total disability-adjusted life year (DALY). This parameter shows severity of a given disease by combining years of life lost due to premature mortality and disability (https://www.who.int/data/gho/indicator-metadata-registry/imr-details/158). A picture of the burden of these diseases is shown in Table 6.1 where percentages of global deaths and total DALY for each cancer type are reported.

The following discussion is focused on HCC and MM, two cancers responsible for high mortality and disability. In these cancers, the cell surface HSPGs GPC-3 and SDC-1 are well-established markers, whereas the main findings on the significance of circulating PGs/GAGs in prognosis and diagnosis of other tumor types are summarized in Table 6.2.

6.3.1 *Hepatocellular Carcinoma*

According to the International Agency for Research on Cancer, liver cancer is the third most common cause of cancer death leading to 781,631 deaths worldwide

Table 6.1 List of cancer types in which a significant relationship with plasma GAGs/PGs levels has been reported, in the last 20 years (https://pubmed.ncbi.nlm.nih.gov). Percentages of global deaths and of total disability-adjusted life years (DALYs) for 2019 are shown

Cancer type	% of global deaths	% of total DALY
Liver cancer	0.86	0.49
Multiple myeloma	0.20	0.099
Prostate cancer	0.86	0.34
Pancreatic cancer	0.94	0.46
Larynx cancer	0.22	0.13
Esophageal cancer	0.88	0.46
Stomach cancer	1.69	0.88
Breast cancer	1.24	0.81
Kidney cancer	0.29	0.16
Colon and rectum cancer	1.92	0.96
Tracheal, bronchus, and lung cancer	3.61	1.81
Leukemia	0.59	0.46
Ovarian cancer	0.35	0.21
Hodgkin's lymphoma	0.05	0.045
Malignant skin melanoma	0.11	0.067
Bladder cancer	0.40	0.17
Total	14.21	7.55

Source: https://vizhub.healthdata.org/gbd-compare/

(8.2%) (Source: Bray et al. 2018), being HCC the major contributor. Chronic persistent infections of hepatitis B or C virus are primary risk factors for HCC, but an association with environmental factors, such as alcohol consumption or exposure to some chemicals and aflatoxin B1 has been reported as well. HCC is a very aggressive cancer characterized by a considerably low 5-year survival rates (El-Serag 2011).

As demonstrated in 1997 by Hsu et al., GPC-3 mRNA is highly expressed in embryonal tissues such as placenta, liver, lung, and kidney, whereas it is undetectable (e.g., liver) or barely detectable (e.g., lung and kidney) in almost all adult tissues. Furthermore, Hsu et al. (1997) showed that it is highly and specifically expressed in HCC (Hsu et al. 1997). These findings were followed by many studies indicating GPC-3 as a reliable immunohistochemical diagnostic marker of prognostic significance in predicting overall survival and disease-free survival in HCC patients (Table 6.3). Indeed, a GPC-3 expression has been associated with tumor stage, differentiation, and presence of vascular invasion and metastasis (Li et al. 2014; Xiao et al. 2014; Liu et al. 2018; Zhang et al. 2018a; Moudi et al. 2019). Interestingly, as it is highly and specifically expressed on HCC cells' surface, it could also represent a promising target for immunotherapies, treatment with peptide/DNA vaccines, immunotoxin, or genetic therapies (Zhou et al. 2018b; Guo et al. 2020).

As mentioned above, during maturation, the endoprotease Furin can cleave GPC-3 between Arg^{358} and Ser^{359} residues, thus generating 2 fragments of 40 and

Table 6.2 Main findings of studies correlating circulating PGs/GAGs levels with tumors other than HCC or MM

References	Tumor typology	PGs/GAGs	Casuistry	Main findings
Grindel et al. (2016)	Prostate Cancer (PC)	Perlecan	288 PC vs. 12 healthy controls	Perlecan fragments associated with MMP-7 in prostate cancer tissues; domain IV perlecan in stage IV, but absent in normal sera. Perlecan fragments in sera and MMP-7 in tissues are measures of invasive prostate cancer
Szarvas et al. (2016)		SDC-1	99 PC	SDC-1 levels as promising tool for pre-operative risk-stratification and/or therapy monitoring
Arslan et al. (2017)		Endocan	86 PC vs. 80 controls	High serum endocan levels ($\geq$1.8 ng/ml) is a significant predictor of biochemical progression-free survival (hazard ratio 2.44; 95% confidence interval 1.78–3.23; $p = 0.001$)
da Silva et al. (2018)		HA	44 PC vs. 14 controls	39.68 ± 30.00 ng/ml vs. 15.04 ± 7.11 ng/ml ($p < 0.004$); potentially useful biomarker for the diagnosis and prognosis of prostate cancer
Levin et al. (2018)		GPC-1	15 PC vs. 15 benign prostatic hyperplasia vs. controls	GPC-1 reduction in prostate cancer patients ($p < 0.05$)
Szarvas et al. (2018)		SDC-1	75 PC who received docetaxel therapy until the appearance of therapy resistance	Serum SDC-1 may help to facilitate clinical decision-making regarding the type and timing of therapy for patients with castration-resistant prostate cancer (CRPC); positive correlation between SDC-1 and MMP7

(continued)

Table 6.2 (continued)

References	Tumor typology	PGs/GAGs	Casuistry	Main findings
Melo et al. (2015)	Pancreatic Ductal Adenocarcinoma (PDAC)	GPC-1$^+$ circulating exosomes (crExos)	190 PDAC vs. 32 breast cancer vs. 100 healthy donors	GPC-1$^+$ crExos potential marker for detection of early stages of pancreatic cancer; correlation with tumor burden and the survival of pre- and post-surgical patients
Lai et al. (2017)		Exosomal GPC-1	3 PDAC vs. 3 chronic pancreatitis vs. 6 healthy controls	GPC-1 is not diagnostic for Pancreatic ductal adenocarcinoma
Frampton et al. (2018)		GPC-1$^+$ circulating exosomes (crExos)	27 PDAC vs. 16 benign pancreatic disease	GPC-1 levels enriched in PDAC crExos; GPC-1 association with PDAC tumor size and disease burden
Zhou et al. (2018a)		GPC-1	156 PDAC vs. 199 non-cancer controls vs. 240 patients with other cancers	High GPC-1 levels association with poor prognosis in PDAC
Chen et al. (2020)		HA	809 PDAC vs. 44 ampullary carcinoma vs. 31 distal biliary tract cancer vs. 15 chronic pancreatitis vs. 41 intraductal papillarymucinous neoplasm vs. 7 duodenal adenoma vs. 25 no cancer	Higher baseline serum HA in pancreatic cancer than healthy subjects and patients with benign conditions; association with overall survival
Xiao et al. (2020)		Exosomal GPC-1 (panel of exosomal proteins)	24 pancreatic cancer vs. 6 chronic pancreatitis vs. 23 acute pancreatitis vs. 9 benign breast tumor vs. 11 breast cancer vs. 26 Healthy people	Specific, sensitive, and reproducible detection panel for the diagnosis of pancreatic cancer

Anttonen et al. (2006)	Upper gastrointestinal cancers	SDC-1	44 larynx and hypopharynx carcinomas	Higher SDC-1 levels correlate with cancer recurrence
Wu et al. (2010)		Decorin	275 esophageal squamous cell carcinoma vs. 295 controls	Decorin levels significantly lower in patients than in controls ($p < 0.0001$)
Ruan et al. (2011)		HA	50 gastric cancer vs. 41 precancerous lesion vs. 30 control subjects	Higher HA levels in patients with gastric cancer than in patients with precancerous lesion and control group ($p < 0.05$, $p < 0.0001$)
Aghcheli et al. (2012)		HA	20 gastric cardia cancer vs. 23 gastric noncardia cancer vs. 20 esophageal squamous cell carcinoma incident cases vs. 25 controls	Higher HA levels in cancer ($p < 0.01$); useful to identify potentially high-risk groups of upper gastrointestinal cancers
Shegefti et al. (2016)		SDC-1	43 Tongue squamous cell carcinoma (SCC) vs. 46 controls	Lower levels of SDC-1 in patients (91.17 ± 88.60 vs. 158.17 ± 103.47 ng/ml, $p = 0.002$); no correlation with the tumor progression in the tongue SCC
Li et al. (2017)	Colorectal Cancer (CRC)	Exosomal GPC-1	102 CRC vs. 80 healthy controls	Plasma GPC-1+ exosomes, miR-96-5p and miR-149 are specific markers for the diagnosis of CRC and targets for the therapy
Wang et al. (2014a)		SDC-1	52 resectable primary CRC vs. 11 controls	High SDC-1 levels correlated with chemotherapy resistance; potential new prognostic marker in colorectal cancer
Jary et al. (2016)		SDC-1	Development cohort: 126 metastatic CRC, validation cohort: 51 patients	SDC-1 is an independent prognostic factor in metastatic CRC patients
Wu et al. (2019)		~6 kDa HA	184 CRC vs. 75 benign diseases vs. 63 controls	Potential biomarker for diagnosing CRC and predicting early relapse, cancer progression, and lymph node metastasis; prompt decrease after tumor removal

(continued)

Table 6.2 (continued)

References	Tumor typology	PGs/GAGs	Casuistry	Main findings
Joensuu et al. (2002)	Lung cancer (LC)	SDC-1	138 nonsmall cell LC vs. 46 small cell LC	Survival: SDC-1 > 59 ng/ml vs. SDC-1 < 59 ng/ml, 4 months vs. 11 months ($p < 0.0001$); high serum SDC-1 levels at diagnosis are associated with poor outcome, independently from the histological subtype
Anttonen et al. (2003)		SDC-1	88 small cell LC	High pre-treatment serum SDC-1 level association with poor prognosis in patients treated with platinum-based chemotherapy
Nishio et al. (2011)		HS	83 stage IV nonsmall cell LC	Higher HS levels are significantly associated with shorter progression-free survival, and overall survival ($p = 0.0012$ and $p = 0.0003$)
Jilani et al. (2009)	Chronic lymphocytic leukemia (CLL)	SDC-1	104 CLL vs. 32 controls	52.8 ng/ml (13.4–252.7 ng/ml) vs. 19.86 ng/ml (14.49–33.14 ng/ml) ($p < 0.01$)
Wolowiec et al. (2006)		SDC-1	52 CLL vs. 12 healthy controls	Higher SDC-1 levels in early-stage B-CLL patients as well as in patients with more indolent disease course than in healthy controls
Molica et al. (2006)		SDC-1	67 CLL (Binet stage A, 46; stage B, 7; stage C, 14) vs. 15 healthy controls	Lower levels in patients than in controls ($p = 0.02$); no correlation with clinical stages
Wolowiec et al. (2006)		SDC-1	52 CLL vs. 12 healthy controls	149.0 ng/ml (13.2–257.1 ng/ml) vs. 36.7 ng/ml (17.4–135.8 ng/ml) ($p = 0.0002$)

Larsen et al. (2013)	Acute myeloid leukemia (AML)	SDC-1	49 AML	High levels of SDC-1 associated with bleeding, impaired platelet function, higher age, endothelial cell activation and damage, and leukocytosis
Anagnostopoulou et al. (2017)		HA	51 AML, B-acute lymphoblastic leukemia (BALL), and mantle-cell lymphoma (MCL) vs. 52 controls	High HA levels in patients with AML, B-ALL, and MCL; inverse correlation between overall survival and HA levels
Alghandour et al. (2020)		SDC-1	24 AML vs. 16 acute lymphoblastic leukemia vs. 15 healthy subjects	Useful diagnostic and prognostic marker for AML but not for acute lymphoblastic leukemia
Thylen et al. (1999)	Malignant pleural mesothelioma	HA	19 Malignant pleural mesotheliomas	Positive correlation between plasma and pleural fluid levels of HA. Positive correlation between HA levels and tumor volume in HA-producing group of mesotheliomas ($p = 0.01$)
Burchardt et al. (2003)	Melanoma	HA	6 malignant melanomas (stage I/II and 6 stage IV)	Significant increase in all stages of disease
Ikuta et al. (2005)		GPC-3	109 melanoma vs. 5 large congenital melanocytic nevus vs. 61 age-matched healthy donors vs. 13 disease-free patients after undergoing a surgical removal	Potential marker of melanoma, especially at an early stage
Vassilakopoulos et al. (2005)	Hodgkin's lymphoma	SDC-1	66 Hodgkin's lymphoma vs. 14 healthy controls	100.2 ± 35.9 ng/ml vs. 67.9 ± 24.5 ng/ml ($p < 0.001$)
Pothacharoen et al. (2006)	Ovarian cancer	CS and HA	91 ovarian epithelial cancer vs. 39 noncancer gynecological disorders vs. 30 healthy women	CS levels highly increased in epithelial types of ovarian cancer and at all stages of development ($p < 0.005$); HA levels significantly higher in ovarian cancer patients than in normal controls ($p < 0.05$)
Obayashi et al. (2008)		HA, SHAP-HA complex, UTI	45 ovarian cancer vs. 22 benign ovarian tumors vs. 50 healthy women	Levels of HA and SHAP-HA complex higher in ovarian cancer; elevated SHAP-HA complex associated with shorter disease-free survival

(continued)

Table 6.2 (continued)

References	Tumor typology	PGs/GAGs	Casuistry	Main findings
Zhou et al. (2017)	Hepatoblastoma (HB)	GPC-3	134 HB vs. 30 benign hepatobiliary disorders vs. 20 controls	1.93 ng/ml (0–31.19) vs. 1.74 ng/ml (0–25.95) ($p = 0.6$) vs. 0.59 ng/ml (0–6.20) ($p = 0.003$); GPC-3 is inferior to AFP as a serum marker for HB
Appunni et al. (2017)	Bladder cancer	Decorin, Biglycan and Lumican	30 bladder cancer vs. 30 healthy controls	Higher plasma levels of Biglycan ($p = 0.0038$) and Lumican ($p < 0.0001$) and reduced levels of Decorin ($p < 0.0001$) in urothelial carcinoma of bladder
Indira Chandran et al. (2019)	Glioblastoma	SDC-1 extracellular vesicles	136 high-grade glioblastoma multiforme and low-grade glioma	Potential tool for minimally invasive diagnosis of glioma

Table 6.3 Meta-analysis studies on the prognostic significance of high hepatic tissue GPC-3 expression in predicting overall survival (OS) and disease-free survival (DFS) of HCC patients

References	Casuistry	Assay	Overall survival[a]	Disease-free survival[b]
Li et al. (2014)	493 patients/5 studies	IHC	HR = 2.18, 95% CI: 1.47–3.24, $p = 0.0001$	HR = 2.05, 95% CI: 1.43–2.93, $p < 0.0001$
Xiao et al. (2014)	1070 patients/ 8 studies	IHC	HR = 1.96, 95% CI: 1.51–2.55, $p = 0.000$	HR = 1.99, 95% CI: 1.57–2.51, $p = 0.000$
Liu et al. (2018)	2336 patients/ 15 studies	IHC	HR = 1.38 95% CI: 1.05–1.80, $p = 0.02$	HR = 1.98, 95% CI: 1.08–3.62, $p = 0.027$
Zhang et al. (2018a)	2364 patients/ 14 studies	IHC	HR = 1.40, 95% CI: 1.07–1.85, $p = 0.02$	HR = 1.61, 95% CI: 1.13–2.30, $p = 0.008$
Moudi et al. (2019)	2618 patients/ 17 studies	IHC	HR = 1.57, 95% CI: 1.18–2.10, $p = 0.002$	HR = 1.93, 95% CI: 1.09–3.43, $p = 0.02$

IHC immunohistochemistry, *HR* hazard ratio, *CI* confidence interval
[a]Association between overexpression of GPC-3 and decreased overall survival
[b]Association between overexpression of GPC-3 and decreased disease-free survival

a 30 kDa, respectively, the NH_2-terminal and the COOH-terminal, which carries the sites for HS chains attachment, linked together by disulfide bonds (De Cat et al. 2003; Hippo et al. 2004). The former can be released in serum of HCC patients, representing an effective circulating marker for early-stage HCC (Hippo et al. 2004).

Taking its aggressiveness and the associated poor prognosis into account, identification of reliable early markers is mandatory. With this aim, several studies have focused on validating serum GPC-3 as a noninvasive diagnostic marker of HCC (Table 6.4), most of which have been included in the six meta-analysis studies reported in Table 6.5 (Xu et al. 2013, 2019; Jia et al. 2014; Liu et al. 2014, 2015; Yang et al. 2014).

These studies calculated pooled accuracy parameters showing a wide range of sensitivity (from 53 to 69%), specificity (from 58 to 93%), and diagnostic odds ratio (from 3.6 to 31), thus suggesting serum GPC-3 as a promising diagnostic marker for HCC. However, the great heterogeneity among studies (I^2 values were considerably higher than the cut-off value of 50%, as reported in Table 6.5) and major design deficiencies, including lack of clear exclusion and inclusion criteria, have therefore led to nondefinitive conclusions.

To improve the diagnostic power of GPC-3 for HCC, Attallah et al. (2016) developed a GPC–HCC model based on the combination of GPC-3 levels and routine laboratory tests, including the α-fetoprotein actually considered the gold standard tumor marker for HCC, alanine transaminase, aspartate transaminase, alkaline phosphatase, total bilirubin, albumin, and platelet count (Attallah et al. 2016). A cohort of 138 HCC patients, 56 with cirrhosis *versus* 62 with liver fibrosis were assessed, showing high diagnostic power with 93% sensitivity, 93% specificity, 89% positive predictive value, 95% negative predictive value, and 93% efficiency.

Despite being conducted on a smaller cohort of patients, namely 39 HCC patients *versus* 62 affected by liver fibrosis, also Jing et al. (2017) obtained high diagnostic

Table 6.4 Main findings of studies on the association of serum PGs/GAGs levels with HCC, carried out in the last 20 years

References	PGs/GAGs	Casuistry	Main findings
Nakatsura et al. (2003)	GPC-3	40 HCC vs. 13 liver cirrhosis vs. 34 chronic hepatitis vs. 60 healthy donors	GPC-3 positivity in serum: 40.0% in HCC vs. 0% in the other groups
Xu et al. (2013)	GPC-3	34 HCC vs. 20 hepatitis plus liver cirrhosis vs. 18 hepatitis vs. 53 controls	GPC-3 undetectable in both healthy individuals and patients with hepatitis; elevated GPC-3 in 53% of patients with HCC; elevated GPC-3 in 5% of patients with hepatitis plus cirrhosis
Nakatsura et al. (2003)	HA	28 chronic hepatitis vs. 43 liver cirrhosis vs. 57 HCC vs. 60 healthy controls	HA levels correlate with the extent of liver fibrosis and severity of cirrhosis
Hippo et al. (2004)	GPC-3	69 HCC vs. 38 liver cirrhosis vs. 96 healthy controls	4.84 ± 8.91 ng/ml vs. 1.09 ± 0.74 ng/ml ($p < 0.01$) vs. 0.65 ± 0.32 ng/ml ($p < 0.001$)
Zolota et al. (2021)	GPC-3	50 HCC with cirrhosis vs. 41 cirrhosis	161.41 ± 422.33 ng/ml vs. 125.41 ± 281.05 ng/ml (not significant)
Yasuda et al. (2010)	GPC-3	200 HCC vs. 200 chronic liver disease	924.8 pg/ml (495.2–1335.6 pg/ml) vs. 1161.6 pg/ml (762.0–1784.0 pg/ml) ($p < 0.0001$); lower expression in HCC
Tangkijvanich et al. (2010)	GPC-3	100 HCC vs. 50 intrahepatic cholangiocarcinoma vs. 50 metastatic carcinoma vs. 50 liver cirrhosis vs. 50 chronic hepatitis vs. 40 healthy controls	Elevated serum GPC-3 in 53% of HCC (ranging 35.5–7826.6 ng/ml) but undetectable in the other groups; Serum GPC-3 is highly specific for detecting HCC
Suzuki et al. (2010)	GPC-3	HCC vs. Chronic liver disease	800 ± 200 ng/ml vs. 350 ± 50 ng/ml ($p < 0.05$)
Liu et al. (2010)	GPC-3	37 HCC vs. 32 liver cirrhosis	GPC-3 is a sensitive, specific serum and tissue marker for the diagnosis of early HCC
Zhang et al. (2010)	GPC-3	36 HCC vs. 20 secondary liver cancer vs. 25 hepatitis B vs. 20 hepatitis C vs. 28 cirrhosis vs. 56 controls	116.8 ± 98.6 ng/ml vs. 24.60 ± 24.01 ng/ml vs. 13.67 ± 15.68 ng/ml vs. 6.73 ± 1.22 ng/ml vs. 0.86 ± 1.12 ng/ml, $p < 0.05$
Qiao et al. (2011)	GPC-3	101 HC vs. 40 cirrhosis vs. 18 hepatitis vs. 30 controls	GPC-3 is a useful tumor marker complementary to AFP for clinical diagnosis of HCC

Ozkan et al. (2011)	GPC-3	75 HCC vs. 55 cirrhosis vs. 28 healthy controls	5.13 pg/ml (53.9–93.2 pg/ml) vs. 5.51 pg/ml (53.9–236.2 pg/ml) vs. 3.9 pg/ml (53.9–7.7 pg/ml); Diagnostic values: Sensitivity 61.33%, Specificity 41.82%, positive predictive value 58.97%, negative predictive values 44.43%; GPC-3 not useful as diagnostic and prognostic marker for HCC
Xia et al. (2012)	HA	152 small HCC underwent complete radiofrequency ablation	High HA levels are main prognostic factors of local recurrence after complete radiofrequency ablation
Metwaly et al. (2012)	SDC-1	40 HCC vs. 31 liver cirrhosis vs. 15 controls	128.64 ± 16.45 ng/ml vs. 83.23 ± 10.13 ng/ml vs. 31.52 ± 15.3 ng/ml ($p < 0.05$)
Sadik et al. (2012)	HA	19 noncirrhotic HCC vs. 50 cirrhotic HCC vs. 36 cirrhosis vs. 21 controls	HA levels significantly increased in all three patients' groups
Abdelgawad et al. (2013)	GPC-3	40 HCC vs. 10 cirrhosis vs. 10 controls	Diagnostic values: Sensitivity 95%, Specificity 95%, Diagnostic accuracy 95%, Positive predictive value 97.5%, Negative predictive value 90.5%; promising diagnostic marker with high sensitivity and specificity for HCC
Nault et al. (2013)	Endocan, SDC-1, and GPC-3	58 early HCC vs. 67 advanced HCC vs. 170 alcoholic cirrhosis	Serum endocan and SDC-1 are prognostic serum biomarkers of overall survival in alcoholic cirrhosis with and without HCC
Li et al. (2013)	GPC-3	605 HCC vs. 25 controls	20.20 ± 5.41μg/l vs. 1.92 ± 0.95μg/l, $p < 0.01$
Yao et al. (2013)	GPC-3	123 Hepatocellular Carcinoma vs. 70 liver cirrhosis vs. 70 chronic hepatitis vs. 56 acute hepatitis vs. 50 nonliver tumor vs. 30 healthy controls	Diagnostic values: Sensitivity 52.8%, Specificity 98.8%, Diagnostic accuracy 83.5%, Positive predictive value 95.6%, Negative predictive value 80.7%
Chen et al. (2013)	GPC-3	155 HCC vs. 180 chronic hepatitis vs. 124 liver cirrhosis vs. 442 non-HCC cancer vs. 136 healthy controls	99.94 ± 267.2 ng/ml vs. 10.45 ± 46.02 ng/ml ($p < 0.0001$) vs. 19.44 ± 50.88 ng/ml ($p = 0.0013$) vs. 20.50 ± 98.33 ng/ml ($p < 0.0001$) vs. 4.14 ± 31.65 ng/ml ($p < 0.0001$); Application as a single marker in the diagnosis of HCC limited by sGPC-3 presence in lung cancer and thyroid cancer

(continued)

Table 6.4 (continued)

References	PGs/GAGs	Casuistry	Main findings
Lee et al. (2014)	GPC-3	120 HCC vs. 40 chronic liver disease	75.8 ng/ml (21.7–482.5 ng/ml) vs. 66.4 ng/ml (2.33–66.4 ng/ml) ($p < 0.020$)
Abd El Gawad et al. (2014)	GPC-3	40 HCC vs. 10 cirrhosis vs. 10 healthy controls	7.7 ng/ml (4.9–11 ng/ml) vs. 2.74 ng/ml (1.99–5.93 ng/ml) vs. 0.99 ng/ml (0.86–1.67 ng/ml) ($p < 0{,}0001$); Diagnostic values: Sensitivity 95%, Specificity 95%, Diagnostic accuracy 95%, Positive predictive value 97%, Negative predictive value 90.5%
Wang et al. (2014b)	GPC-3	84 HCC vs. 80 cirrhosis vs. 32 hepatitis B vs. 61 healthy controls	GPC-3 tests negative in all 84 HCC patients
Badr et al. (2014)	GPC-3	30 HCC vs. 30 hepatitis C virus (HCV) cirrhosis vs. 20 healthy controls	551.47 ± 185.25 ng/ml vs. 98.23 ± 73.54 ng/ml ($p < 0.01$); Diagnostic values: Sensitivity 100%, Specificity 93.3%, Diagnostic accuracy 96.7%, Positive predictive value 93.8%, Negative predictive value 100%
Mima et al. (2014)	HA	506 patients	High preoperative serum HA levels predict poor prognosis in patients with HCC after hepatic resection
Yu et al. (2015)	GPC-3	192 HCC vs. 54 gastric cancer vs. 92 colorectal cancer vs. 57 hepatitis vs. 31 esophageal cancer vs. 44 cirrhosis vs. 48 controls	GPC-3 levels in HCC significantly higher than that in other liver diseases as well as healthy controls
Haruyama et al. (2015)	GPC-3 N-terminal subunit antigen (sGPC-3N)	115 HCC vs. 25 controls	sGPC-3N levels significantly higher in HCC with respect to controls ($p < 0.0001$); high pre-operative sGPC-3N associated with shorter OS and DFS after hepatectomy ($p < 0.01$)
Jia et al. (2016)	GPC-3	283 HCC vs. 445 chronic hepatic diseases vs. 162 controls	Higher levels in both HCC and liver cirrhosis than controls ($p < 0.05$); low diagnostic potential alone but useful in combination with alpha-fetoprotein
Shimizu et al. (2016)	HA	400 HCC underwent liver resection	High HA levels (>200 ng/ml) as useful predictor of postoperative large amount of ascites, which are associated with poor prognosis

Jeon et al. (2016)	GPC-3	157 newly diagnosed HCC vs. 156 liver cirrhosis	0.80 ng/ml (0–3.09 ng/ml) vs. 0.60 ng/ml (0.07–7.40 ng/ml) ($p = 0.255$)
Attallah et al. (2016)	GPC-3	138 HCC vs. 56 liver cirrhosis vs. 62 liver fibrosis	Diagnostic power: sensitivity 93%, specificity 93%, positive predictive value 89%, negative predictive value 95%, efficiency 93%
Jing et al. (2017)	GPC-3	39 HCC vs. 31 liver cirrhosis	Higher GPC-3 levels in HCC ($p < 0.0001$); Diagnostic values: Sensitivity 89.74%, Specificity 96.77%, Positive predictive value 97.2%, Negative predictive value 88.2%; effective noninvasive diagnostic indicators of HCC
Sun et al. (2017)	GPC-3	76 HCC vs. 30 Cirrhosis vs. 40 Chronic liver disease vs. 30 controls	272.5 ± 13.3 ng/ml vs. 661.9 ± 6.8 ng/ml vs. 57.5 ± 8.3 ng/ml vs. 56.2 ± 6.1 ng/ml ($p < 0.01$)
Ofuji et al. (2017)	GPC-3	25 stage I HCC who underwent surgical resection	33.7 ng/ml (7.5–1729.3 ng/ml) (postoperation) vs. 60.1 ng/ml (6.4–3466.8 ng/ml) ($p < 0.001$); Postoperative recurrence associated with high preoperative plasma GPC-3
Farag et al. (2018)	GPC-3	145 HCC vs. 105 chronic liver cirrhosis vs. 50 healthy controls	Diagnostic values: Sensitivity 95%, Specificity 100%, Diagnostic accuracy 95%, Positive predictive value 97.5%, Negative predictive value 90.5%
Wu et al. (2019)	Fucosylated GPC-3 (fuc-GPC-3)	51 HCC vs. 47 benign liver disease	fuc-GPC-3 effective and useful tumor biomarker
Tahon et al. (2019)	GPC-3	40 cirrhotic patients with primary HCC vs. 30 cirrhotic patients without HCC vs. 15 healthy controls	12.08 ± 14.11 ng/ml vs. 2.07 ± 1.44 ng/ml vs. 1.25 ± 0.32 ng/ml ($p < 0.001$)
Hagag et al. (2020)	GPC-3	25 HCC vs. 75 liver cirrhosis vs. 50 controls	Higher serum levels of GPC-3 in HCC ($p < 0.001$); positive correlation between GPC-3 and miR-1291
Liu et al. (2020)	GPC-3	210 HCC vs. 36 intrahepatic cholangiocarcinoma vs. 9 combined Hepatocellular cholangiocarcinoma vs. 10 metastatic liver cancer vs. 134 normal controls	Serum GPC-3 superior to AFP for the diagnosis of early-stage HCC

(continued)

Table 6.4 (continued)

References	PGs/GAGs	Casuistry	Main findings
Tang et al. (2020)	GPC-3	166 HCC vs. 94 health controls vs. 50 benign controls	0.210 (0.048, 0.801) mg/l vs. 0.029(0.019, 0.052) mg/l vs. 0.033(0.021, 0.043) mg/l ($p < 0.001$)
Shimizu et al. (2020)	GPC-3	25 hepatitis C virus-related HCC vs. 15 hepatitis B virus-related HCC vs. 16 HCC	9.9 pg/ml (2.8–273 pg/ml) vs. 2.6 pg/ml (0.5–384 pg/ml) vs. 3.0 pg/ml (0.5–22.1 pg/ml); association between GPC-3 expression and secretion and the virus type

Table 6.5 Meta-analysis studies on the diagnostic significance of high serum GPC-3 levels in HCC patients

References	Casuistry	Assay	Sensitivity	Specificity	Positive likelihood ratio (PLR)	Negative likelihood ratio (NLR)	Diagnostic odds ratio (DOR)
Xu et al. (2013)	590 HCC patients/10 studies	ELISA/ radioimmunology	0.59 (95% CI: 0.55–0.63)	0.85 (95% CI: 0.82–0.87)	6.290 (95% CI: 2.649–14.933)	0.488 (95% CI: 0.386–0.617)	17.988 (95% CI: 5.361–60.353)
Liu et al. (2014)	898 HCC patients/12 studies	ELISA/ radioimmunology	0.53 (95% CI: 0.49–0.57; I^2: 79.9%)	0.77 (95% CI: 0.74–0.81; I^2: 95.9%)	NR (I^2: 91.5%)	NR (I^2: 86.1%)	10 (95% CI: 2–38; I^2: 90.6%)
Yang et al. (2014)	NR/18 studies	ELISA/ Chemiluminescent immunoassay	0.69 (95% CI: 0.55–0.80; I^2: 90.71%)	0.93 (95% CI: 0.85–0.97; I^2: 97.30%)	10.50	0.34	31 (95% CI: 11–92)
Jia et al. (2014)	1935 HCC patients/19 studies	ELISA/ radioimmunology	0.55 (95% CI: 0.53–0.57)	0.84 (CI: 0.82–0.86)	5.22 (CI: 3.09–8.84)	0.54 (CI: 0.46–0.63)	13.8 (95% CI: 6.6–28.8)
Liu et al. (2015)	1201 HCC patients/17 studies	ELISA/ Chemiluminescent immunoassay	0.56 (95% CI: 0.53–0.59; I^2: 90.7%)	0.89 (95% CI: 0.87–0.90; I^2:94.3%)	7.82 (CI: 3.86–15.85; I^2: 93.9%)	0.48 (CI: 0.39–0.59; I^2: 85.9%)	26.73 (95% CI: 10.31–69.26; I^2: 90.0%)
Xu et al. (2019)	11 studies	ELISA	0.55 (95% CI: 0.52–0.58; I^2: 93.3%)	0.58 (95% CI: 0.54–0.61; I^2: 90.2%)	1.69 (95% CI: 1.20–2.39)	0.67 (95% CI: 0.50–0.90)	3.64 (95% CI: 1.74–7.60; I^2: 85.3%)

Positive likelihood ratio (PLR): true positive/false positive
Negative likelihood ratio (NLR): false negative/true negative
Diagnostic odds ratio (DOR): the odds of positive test results in patients with HCC/ the odds of positive test results in patients without HCC
I^2: a quantitative measurement of inconsistency across different studies. I^2 value typically ranges from 0 (no observed heterogeneity) to 100% (maximal heterogeneity), and an I^2 value $\geq$50% is considered to represent substantial heterogeneity
NR Not reported

values, being sensitivity, specificity, positive predictive value, and negative predictive value 89.74%, 96.77%, 97.2%, and 88.2%, respectively. Similar results have been obtained by Farag et al. (2018) by analyzing GPC-3 levels in 145 HCC patients, 105 with chronic liver cirrhosis vs. 50 healthy controls (Farag et al. 2018).

Interestingly, Ofuji et al. (2017) reported an association between perioperative serum GPC-3 levels and tumor recurrence rate in 25 patients with stage I HCC who underwent surgical resection (Ofuji et al. 2017). However, Shimizu et al. (2020) who recently evidenced a positive correlation among serum GPC-3 levels, its expression in HCC, and the comorbidity with hepatitis C virus infection in 56 patients affected by HCC, suggested the utility of serum GPC-3 (as predictive of tissue GPC-3) to select the most appropriate therapy (e.g., immunotherapies targeting GPC-3) (Shimizu et al. 2020). Overall, all the mentioned studies have highlighted the usefulness of serum GPC-3, in combination with α-fetoprotein, as an effective noninvasive diagnostic and prognostic marker of HCC.

6.3.2 Multiple Myeloma

MM is a hematological cancer characterized by uncontrolled proliferation of malignant plasma cells that spread throughout the skeletal system where they crowd out normal blood cells disrupting hematopoiesis and bone physiology and creating lytic lesions. Although the etiology of MM is unknown in most patients, it may evolve from benign monoclonal gammopathy and some risk factors including obesity, alcohol consumption, radiation exposure, and family history have been reported (International Agency for Research on Cancer (IARC) 2020).

Despite the many progresses that have been done in treatment because of its capacity to develop drug resistance, MM is still incurable, as it ultimately leads almost all patients to death, with a median survival of 4–6 years. MM is characterized by a strong interaction between tumor cells and bone marrow microenvironment that supports their survival and growth through signals mediated by adhesion molecules, cytokines, and growth factors (Hideshima et al. 2007).

It is well known that SDC-1 is involved in multiple cell processes, including proliferation, migration, adhesion, and angiogenesis. In this respect, it may act both as coreceptor for various Hep-binding growth factors, such as bFGF/FGF2, VEGF, TGF-β, as well as PDGF, and as receptor primarily through its HS chains. Dysregulation of its expression, as well as its shedding, may lead to cancer progression in different ways, as recently reported by Teixeira and Götte (2020). Generally, if downregulated, it may enhance cell motility and invasion by reducing cell–matrix interactions, whereas in some cases, its upregulation may contribute to angiogenesis (Akl et al. 2015). Furthermore, the release of its ectodomain by the activity of sheddases, such as MMPs and heparanase, promotes stemness and tumorigenesis (Ramani et al. 2013; Tripathi et al. 2020; Teixeira and Götte 2020). Indeed, the free extracellular moiety of SDC-1 is still bioactive, modulating the binding between growth factors and their receptors in tumor stroma, which acts as a storage of growth factors and chemokines promoting cell proliferation and tissue invasion, as well as sequestering inhibitory molecules (Akl et al. 2015).

Since SDC-1 plays key roles in maturation of normal B cells, as well as in their differentiation and function, its expression is tightly regulated (Reijmers et al. 2013). Besides, SDC-1 is essential for the survival of long-lived plasma cells and MM plasma cells (Mahtouk et al. 2006) which express it at high levels. The shedding of SDC-1, carried out by matrix proteases and/or endoglycosidases, releases in plasma its ectodomain and bioactive HS fragments of 5–10 disaccharides. In this respect, the high expression of heparanase, compared to normal tissues, is strongly correlated with a poor prognosis (Mahtouk et al. 2007; Purushothaman et al. 2010; Tripathi et al. 2020; Purushothaman and Sanderson 2020), representing a potential therapeutic target. Actually, new heparanase inhibitors, such as SST0001 developed to prevent glycocalyx degradation, are promising candidates for inhibiting myeloma growth (Ritchie et al. 2011).

Numerous studies have shown that serum SDC-1 levels, which are physiologically low in healthy people, following glycocalyx shedding are elevated in plasma of patients with MM (Table 6.6). Overall, these studies show that soluble SDC-1 represents a strong independent prognostic factor (Seidel et al. 2000; Bjoro et al. 2018) associated with both poor survival and resistance to chemotherapies (Aref et al. 2003; Janosi et al. 2004; Lovell et al. 2005). Nonetheless, its involvement in the crosstalk between MM cells and bone marrow microenvironment makes it also a promising molecular target for therapeutic strategies (Yang et al. 2007; Katz 2010; Ritchie et al. 2011; Reijmers et al. 2013).

6.4 Conclusions

It is currently and widely accepted that ECM, besides being a key modulator of cell–cell communication and cell function, greatly contributes to tissue homeostasis as well as to pathogenesis of several diseases, including cancer. HSPGs in particular, as major functional ECM components, may alter many critical cell processes including proliferation, transformation, and tissue invasion during tumorigenesis. Accordingly, they can be highly and specifically expressed in some tumors, representing useful diagnostic markers and potential therapeutic targets. During inflammation, the shedding by matrix enzymes releases in circulation bioactive fragments that, in turn, can be potentially valuable as noninvasive early markers. However, among the large number of published studies, data are sometimes conflicting or of low clinical relevance. In this respect, the absence of both common experimental design including standard protocols for enrollment and accurate diagnostics kits represent confounding issues that have to be solved.

Table 6.6 Main findings of studies on the association of serum PGs/GAGs levels with MM, carried out in the last 20 years

References	PGs/ GAGs	Casuistry	Main findings
Dahl et al. (1999)	HA	386 MM (58% in stage III, 33% in stage II, and 9% in stage I)	Patients with abnormally low or high HA levels had more advanced disease and a shorter median survival
Seidel et al. (2000)	SDC-1	174 MM vs. 40 healthy controls	643 units/ml (401–2022 units/ml) vs. 128 units/ml (76–208 units/ml), $p < 0.0001$; survival: 20 months vs. 44 months ($p < 0.0001$)
Aref et al. (2003)	SDC-1	25 newly diagnosed MM	Soluble SDC-1 higher in nonresponders to chemotherapy vs. responders ($p < 0.01$), and in nonsurvivors vs. survivors ($p < 0.001$); Plasma cells SDC-1 expression lower in nonresponders to chemotherapy vs. responders ($p < 0.01$), in nonsurvivors vs. survivors ($p < 0.05$)
Kyrtsonis et al. (2004)	SDC-1	27 MM vs. 11 healthy controls	177.5 ng/ml (34–3500 ng/ml) vs. 40 ng/ml (28–75 ng/ml), $p = 0.001$; correlation with stage and shorter survival
Kumar et al. (2004)	SDC-1	501 MM	Survival: SDC-1 > 158 ng/ml vs. SDC-1 < 158 ng/ml, 36.3 months vs. 49.3 months ($p < 0.0001$)
Janosi et al. (2004)	SDC-1	13 monoclonal gammopathy of undetermined significance vs. 4 solitary plasmocytoma vs. 50 MM; 6 months follow-up of MM patients receiving chemotherapy	77.9 ng/ml (33–122 ng/ml) vs. 65.6 ng/ml (33.8–94.5 ng/ml) vs. 223.8 ng/ml (36–508 ng/ml); Responders to chemotherapy ($n = 11$): 106 ng/ml (57.3–440 ng/ml) (after 6 months) vs. 258 ng/ml (97.3–460 ng/ml) (baseline), nonresponders ($n = 4$): 361.7 ng/ml (251–486 ng/ml) (after 6 months) vs. 327 ng/ml (245–466 ng/ml) (baseline)
Andersen et al. (2005)	SDC-1	67 newly diagnosed MM vs. 18 controls	1053.92 ± 292.72 ng/ml vs. 81.28 ± 8.83 ng/ml, $p < 0.0001$; SDC-1 is a strong independent prognostic factor of shorter survival

(continued)

Table 6.6 (continued)

References	PGs/ GAGs	Casuistry	Main findings
Lovell et al. (2005)	SDC-1	324 MM at presentation vs. 154 MM at plateau phase	336 ng/ml (143–1635 ng/ml) vs. 192 ng/ml (111–422 ng/ml) ($p < 0.0006$)
Schaar et al. (2005)	SDC-1	66 MM vs. 54 monoclonal gammopathies of undetermined significance (MGUS) vs. 69 provisional MGUS (no bone marrow examination performed) vs. 36 controls	SDC-1 levels widely variable and of limited discriminatory value in patients with newly diagnosed monoclonal proteinemia
Janosi et al. (2005)	SDC-1	MM vs. plasmocytoma vs. MGUS	Strong independent indicator of poor prognosis
Maisnar et al. (2006)	SDC-1	17 MM vs. 14 MGUS	1542 ng/ml (10–17,300 ng/ml) vs. 32 ng/ml (5–128 ng/ml), $p < 0.001$; high levels associated with worse prognosis in MM
Kim et al. (2010)	SDC-1	28 MM vs. 50 controls; 6-month follow-up of MM patients receiving chemotherapy	265 ng/ml (98–1049 ng/ml) vs. 81 ng/ml (27–192 ng/ml) ($p < 0.0001$); responders to chemotherapy ($n = 20$): 97 ng/ml (60–470 ng/ml) (after 6 months) vs. 220 ng/ml (98–1,049 ng/ml) (baseline), nonresponders ($n = 8$): 266 ng/ml (172–629 ng/ml) (after 6 months) vs. 337 ng/ml (265–785 ng/ml) (baseline)
Scudla et al. (2010)	SDC-1	179 MM vs. 89 MGUS	189.1 (2.5–256.0) ng/ml vs. 42.2 (2.5–256.0) ng/ml ($p < 0.00001$)
Skliris et al. (2011)	Serglycin	37 MM vs. 21 healthy controls	Higher levels in MM patients ($p = 0.0001$)
Minarik et al. (2012)	SDC-1	156 MM vs. 88 MGUS	Serum SDC-1 correlates with the activity of MM and might become useful in differentiation of MGUS, asymptomatic MM, and overt/symptomatic form of MM
Gupta et al. (2015)	Versican (VCAN)	25 MM vs. 10 idiopathic thrombocytopenic perpurae vs. 25 healthy controls	Higher levels of VCAN in MM also in relation with disease severity; potential marker of active disease
Cigliana et al. (2015)	SDC-1	50 Intact Immunoglobulin MM vs. 34 Light Chain MM vs. 40 healthy controls	90.73 ± 103.80 ng/ml vs. 40.21 ± 55.22 ng/ml vs. 15 ± 9.02 ng/ml ($p = 0.012$; $p < 0.0001$; $p = 0.006$)

Acknowledgments Lepedda A.J. thanks Regione Autonoma della Sardegna for its financial support (POR-FSE 2014-2020-Asse Prioritario 3 "Istruzione e Formazione" Obiettivo Tematico: 10, Priorità d'investimento: 10ii, Obiettivo Specifico: 10.5, Azione dell'Accordo di Partenariato 10.5.12-C.U.P. J86C18000270002).

Nieddu G. thanks "Programma Operativo Nazionale (PON) Ricerca e Innovazione 2014–2020—Asse I "Capitale Umano", Azione I.2—A.I.M. Attraction and International Mobility, Linea 1 Mobilità dei ricercatori, AIM1874325-3, CUP: J54I18000110001" for its financial support.

Lepedda A.J., Nieddu G. and Formato M. thank the University of Sassari (Fondo di Ateneo per la Ricerca 2020) and the Fondazione di Sardegna (BANDO FONDAZIONE DI SARDEGNA 2022 E 2023 – PROGETTI DI RICERCA DI BASE DIPARTIMENTALI (D.R.16/2022)) for their financial support.

References

Abdelgawad IA, Mossallam GI, Radwan NH, Elzawahry HM, Elhifnawy NM (2013) Can Glypican3 be diagnostic for early hepatocellular carcinoma among Egyptian patients? Asian Pac J Cancer Prev 14(12):7345–7349

Abd El Gawad IA, Mossallam GI, Radwan NH, Elzawahry HM, Elhifnawy NM (2014) Comparing prothrombin induced by vitamin K absence-II (PIVKA-II) with the oncofetal proteins glypican-3, Alpha feto protein and carcinoembryonic antigen in diagnosing hepatocellular carcinoma among Egyptian patients. J Egypt Natl Canc Inst 26(2):79–85

Afratis NA, Nikitovic D, Multhaupt HA, Theocharis AD, Couchman JR, Karamanos NK (2017) Syndecans—key regulators of cell signaling and biological functions. FEBS J 284(1):27–41

Aghcheli K, Parsian H, Qujeq D, Talebi M, Mosapour A, Khalilipour E, Islami F, Semnani S, Malekzadeh R (2012) Serum hyaluronic acid and laminin as potential tumor markers for upper gastrointestinal cancers. Eur J Intern Med 23(1):58–64

Akl MR, Nagpal P, Ayoub NM, Prabhu SA, Gliksman M, Tai B, Hatipoglu A, Goy A, Suh KS (2015) Molecular and clinical profiles of syndecan-1 in solid and hematological cancer for prognosis and precision medicine. Oncotarget 6(30):28693–28715

Alghandour R, Ebrahim MA, Ghazy H, Shamaa S, Emarah Z, Al-Gayyar MM (2020) Evaluation of the diagnostic and prognostic value of syndecan-1 in acute leukemia patients. Cureus 12(9): e10594

Anagnostopoulou E, Papanastasopoulou C, Papastamataki M, Kotsiou A, Topouzoglou Z, Anagnostopoulos N, Sitaras N (2017) Serum hyaluronic acid levels are altered in acute leukemia patients: potential prognostic implications. Acta Haematol 138(1):44–51

Andersen NF, Standal T, Nielsen JL, Heickendorff L, Borset M, Sorensen FB, Abildgaard N (2005) Syndecan-1 and angiogenic cytokines in multiple myeloma: correlation with bone marrow angiogenesis and survival. Br J Haematol 128(2):210–217

Anttonen A, Leppa S, Ruotsalainen T, Alfthan H, Mattson K, Joensuu H (2003) Pretreatment serum syndecan-1 levels and outcome in small cell lung cancer patients treated with platinum-based chemotherapy. Lung Cancer 41(2):171–177

Anttonen A, Leppa S, Heikkila P, Grenman R, Joensuu H (2006) Effect of treatment of larynx and hypopharynx carcinomas on serum syndecan-1 concentrations. J Cancer Res Clin Oncol 132(7): 451–457

Appunni S, Anand V, Khandelwal M, Seth A, Mathur S, Sharma A (2017) Altered expression of small leucine-rich proteoglycans (Decorin, Biglycan and Lumican): Plausible diagnostic marker in urothelial carcinoma of bladder. Tumour Biol 39(5):1010428317699112

Aref S, Goda T, El-Sherbiny M (2003) Syndecan-1 in multiple myeloma: relationship to conventional prognostic factors. Hematology 8(4):221–228

Arslan B, Onuk O, Hazar I, Aydin M, Cilesiz NC, Eroglu A, Nuhoglu B (2017) Prognostic value of endocan in prostate cancer: clinicopathologic association between serum endocan levels and biochemical recurrence after radical prostatectomy. Tumori 103(2):204–208

Attallah AM, El-Far M, Omran MM, Abdelrazek MA, Attallah AA, Saeed AM, Farid K (2016) GPC-HCC model: a combination of glybican-3 with other routine parameters improves the diagnostic efficacy in hepatocellular carcinoma. Tumour Biol 37(9):12571–12577

Badr EAE, Korah TE, Ghani AA, El-Sayed S, Badr S (2014) Role of serum glypican-3 in the diagnosis and differentiation of small hepatocellular carcinoma from hepatitis-C virus cirrhosis. Alexandria J Med 50(3):221–226

Barash U, Cohen-Kaplan V, Dowek I, Sanderson RD, Ilan N, Vlodavsky I (2010) Proteoglycans in health and disease: new concepts for heparanase function in tumor progression and metastasis. FEBS J 277(19):3890–3903

Barkovskaya A, Buffone A Jr, Zidek M, Weaver VM (2020) Proteoglycans as mediators of cancer tissue mechanics. Front Cell Dev Biol 8:569377

Becker BF, Jacob M, Leipert S, Salmon AH, Chappell D (2015) Degradation of the endothelial glycocalyx in clinical settings: searching for the sheddases. Br J Clin Pharmacol 80(3):389–402

Belting M (2003) Heparan sulfate proteoglycan as a plasma membrane carrier. Trends Biochem Sci 28(3):145–151

Bennett HS, Luft JH, Hampton JC (1959) Morphological classifications of vertebrate blood capillaries. Am J Physiol 196(2):381–390

Bishop JR, Schuksz M, Esko JD (2007) Heparan sulphate proteoglycans fine-tune mammalian physiology. Nature 446(7139):1030–1037

Bjoro B, Dalgard O, Midgard H, Verbaan H, Smastuen MC, Rustoen T (2018) Increased hope following successful treatment for hepatitis C infection. J Adv Nurs 74(3):724–733

Bray F, Ferlay J, Soerjomataram I, Siegel RL, Torre LA, Jemal A (2018) Global cancer statistics 2018: GLOBOCAN estimates of incidence and mortality worldwide for 36 cancers in 185 countries. CA Cancer J Clin 68(6):394–424. https://doi.org/10.3322/caac.21492. Epub 2018 Sep 12

Burchardt ER, Hein R, Bosserhoff AK (2003) Laminin, hyaluronan, tenascin-C and type VI collagen levels in sera from patients with malignant melanoma. Clin Exp Dermatol 28(5): 515–520

Caon I, Bartolini B, Parnigoni A, Carava E, Moretto P, Viola M, Karousou E, Vigetti D, Passi A (2020) Revisiting the hallmarks of cancer: The role of hyaluronan. Semin Cancer Biol 62:9–19

Celie JW, Beelen RH, van den Born J (2009) Heparan sulfate proteoglycans in extravasation: assisting leukocyte guidance. Front Biosci (Landmark Ed) 14:4932–4949

Chakravarti R, Adams JC (2006) Comparative genomics of the syndecans defines an ancestral genomic context associated with matrilins in vertebrates. BMC Genomics 7:83

Chen M, Li G, Yan J, Lu X, Cui J, Ni Z, Cheng W, Qian G, Zhang J, Tu H (2013) Reevaluation of glypican-3 as a serological marker for hepatocellular carcinoma. Clin Chim Acta 423:105–111

Chen IM, Willumsen N, Dehlendorff C, Johansen AZ, Jensen BV, Hansen CP, Hasselby JP, Bojesen SE, Pfeiffer P, Nielsen SE, Hollander NH, Yilmaz MK, Karsdal M, Johansen JS (2020) Clinical value of serum hyaluronan and propeptide of type III collagen in patients with pancreatic cancer. Int J Cancer 146(10):2913–2922

Cheng B, Montmasson M, Terradot L, Rousselle P (2016) Syndecans as cell surface receptors in cancer biology. A focus on their interaction with PDZ domain proteins. Front Pharmacol 7:10

Chu CL, Buczek-Thomas JA, Nugent MA (2004) Heparan sulphate proteoglycans modulate fibroblast growth factor-2 binding through a lipid raft-mediated mechanism. Biochem J 379 (Pt 2):331–341

Chung H, Multhaupt HA, Oh ES, Couchman JR (2016) Minireview: Syndecans and their crucial roles during tissue regeneration. FEBS Lett 590(15):2408–2417

Cigliana G, Torti E, Gulli F, De Santis E, Dell'Abate MT, Colacicco L, Pisani F, Conti L, Basile U (2015) Relationship between circulating syndecan-1 levels (CD138s) and serum free light chains in monoclonal gammopathies. J Exp Clin Cancer Res 34:37

Couchman JR, Gopal S, Lim HC, Norgaard S, Multhaupt HA (2015) Fell-Muir Lecture: Syndecans: from peripheral coreceptors to mainstream regulators of cell behaviour. Int J Exp Pathol 96(1): 1–10

Dahl IM, Turesson I, Holmberg E, Lilja K (1999) Serum hyaluronan in patients with multiple myeloma: correlation with survival and Ig concentration. Blood 93(12):4144–4148
da Silva MNR, Mendes A, Martins JRM, Tobias-Machado M, Pinhal M (2018) Prospective evaluation of chondroitin sulfate, heparan sulfate and hyaluronic acid in prostate cancer. Int Braz J Urol 44(6):1139–1146
De Cat B, Muyldermans SY, Coomans C, Degeest G, Vanderschueren B, Creemers J, Biemar F, Peers B, David G (2003) Processing by proprotein convertases is required for glypican-3 modulation of cell survival, Wnt signaling, and gastrulation movements. J Cell Biol 163(3): 625–635
Dogne S, Flamion B, Caron N (2018) Endothelial glycocalyx as a shield against diabetic vascular complications: involvement of hyaluronan and hyaluronidases. Arterioscler Thromb Vasc Biol 38(7):1427–1439
El-Serag HB (2011) Hepatocellular carcinoma. N Engl J Med 365(12):1118–1127
Esko JD, Selleck SB (2002) Order out of chaos: assembly of ligand binding sites in heparan sulfate. Annu Rev Biochem 71:435–471
Farag RMMA, Al-Ayobi D, Alsaleh KA, Kwon HJ, El-Ansary A, Dawoud EA (2018) Influence of glypican-3 as a newly diagnostic biomarker in early detection of hepatocellular carcinoma among Saudi patients. Biomed Pharmacol J 11(4)
Filmus J (2001) Glypicans in growth control and cancer. Glycobiology 11(3):19R–23R
Filmus J, Selleck SB (2001) Glypicans: proteoglycans with a surprise. J Clin Invest 108(4):497–501
Filmus J, Capurro M, Rast J (2008) Glypicans. Genome Biol 9(5):224
Frampton AE, Prado MM, Lopez-Jimenez E, Fajardo-Puerta AB, Jawad ZAR, Lawton P, Giovannetti E, Habib NA, Castellano L, Stebbing J, Krell J, Jiao LR (2018) Glypican-1 is enriched in circulating-exosomes in pancreatic cancer and correlates with tumor burden. Oncotarget 9(27):19006–19013
Gatto F, Maruzzo M, Magro C, Basso U, Nielsen J (2016a) Prognostic value of plasma and urine glycosaminoglycan scores in clear cell renal cell carcinoma. Front Oncol 6:253
Gatto F, Volpi N, Nilsson H, Nookaew I, Maruzzo M, Roma A, Johansson ME, Stierner U, Lundstam S, Basso U, Nielsen J (2016b) Glycosaminoglycan profiling in patients' plasma and urine predicts the occurrence of metastatic clear cell renal cell carcinoma. Cell Rep 15(8): 1822–1836
Gatto F, Blum KA, Hosseini SS, Ghanaat M, Kashan M, Maccari F, Galeotti F, Hsieh JJ, Volpi N, Hakimi AA, Nielsen J (2018) Plasma glycosaminoglycans as diagnostic and prognostic biomarkers in surgically treated renal cell carcinoma. Eur Urol Oncol 1(5):364–377
Gondelaud F, Ricard-Blum S (2019) Structures and interactions of syndecans. FEBS J 286(15): 2994–3007
Grindel B, Li Q, Arnold R, Petros J, Zayzafoon M, Muldoon M, Stave J, Chung LW, Farach-Carson MC (2016) Perlecan/HSPG2 and matrilysin/MMP-7 as indices of tissue invasion: tissue localization and circulating perlecan fragments in a cohort of 288 radical prostatectomy patients. Oncotarget 7(9):10433–10447
Guo M, Zhang H, Zheng J, Liu Y (2020) Glypican-3: a new target for diagnosis and treatment of hepatocellular carcinoma. J Cancer 11(8):2008–2021
Gupta N, Khan R, Kumar R, Kumar L, Sharma A (2015) Versican and its associated molecules: potential diagnostic markers for multiple myeloma. Clin Chim Acta 442:119–124
Hagag NA, Ali YBM, Elsharawy AA, Talaat RM (2020) Clinical impact of circulated miR-1291 in plasma of patients with liver cirrhosis (LC) and hepatocellular carcinoma (HCC): implication on glypican-3 expression. J Gastrointest Cancer 51(1):234–241
Haruyama Y, Yorita K, Yamaguchi T, Kitajima S, Amano J, Ohtomo T, Ohno A, Kondo K, Kataoka H (2015) High preoperative levels of serum glypican-3 containing N-terminal subunit are associated with poor prognosis in patients with hepatocellular carcinoma after partial hepatectomy. Int J Cancer 137(7):1643–1651

Hideshima T, Mitsiades C, Tonon G, Richardson PG, Anderson KC (2007) Understanding multiple myeloma pathogenesis in the bone marrow to identify new therapeutic targets. Nat Rev Cancer 7(8):585–598

Hippo Y, Watanabe K, Watanabe A, Midorikawa Y, Yamamoto S, Ihara S, Tokita S, Iwanari H, Ito Y, Nakano K, Nezu J, Tsunoda H, Yoshino T, Ohizumi I, Tsuchiya M, Ohnishi S, Makuuchi M, Hamakubo T, Kodama T, Aburatani H (2004) Identification of soluble NH2-terminal fragment of glypican-3 as a serological marker for early-stage hepatocellular carcinoma. Cancer Res 64(7):2418–2423

Horowitz A, Murakami M, Gao Y, Simons M (1999) Phosphatidylinositol-4,5-bisphosphate mediates the interaction of syndecan-4 with protein kinase C. Biochemistry 38(48): 15871–15877

Horowitz A, Tkachenko E, Simons M (2002) Fibroblast growth factor-specific modulation of cellular response by syndecan-4. J Cell Biol 157(4):715–725

Hsu HC, Cheng W, Lai PL (1997) Cloning and expression of a developmentally regulated transcript MXR7 in hepatocellular carcinoma: biological significance and temporospatial distribution. Cancer Res 57(22):5179–5184

Iba T, Levy JH (2019) Derangement of the endothelial glycocalyx in sepsis. J Thromb Haemost 17(2):283–294

Ikuta Y, Nakatsura T, Kageshita T, Fukushima S, Ito S, Wakamatsu K, Baba H, Nishimura Y (2005) Highly sensitive detection of melanoma at an early stage based on the increased serum secreted protein acidic and rich in cysteine and glypican-3 levels. Clin Cancer Res 11(22): 8079–8088

Indira Chandran V, Welinder C, Mansson AS, Offer S, Freyhult E, Pernemalm M, Lund SM, Pedersen S, Lehtio J, Marko-Varga G, Johansson MC, Englund E, Sundgren PC, Belting M (2019) Ultrasensitive immunoprofiling of plasma extracellular vesicles identifies syndecan-1 as a potential tool for minimally invasive diagnosis of glioma. Clin Cancer Res 25(10):3115–3127

International Agency for Research on Cancer (IARC) (2020) World cancer report 2020

Janosi J, Sebestyen A, Mikala G, Nemeth J, Kiss Z, Valyi-Nagy I (2004) Soluble syndecan-1 levels in different plasma cell dyscrasias and in different stages of multiple myeloma. Haematologica 89(3):370–371

Janosi J, Sebestyen A, Mikala G, Peto M, Jako J, Domjan G, Nemeth J, Kis Z, Kopper L, Valyi-Nagy I (2005) [Soluble syndecan-1 levels in different plasma cell dyscrasias]. Orv Hetil 146(4): 165–168

Jary M, Lecomte T, Bouche O, Kim S, Dobi E, Queiroz L, Ghiringhelli F, Etienne H, Leger J, Godet Y, Balland J, Lakkis Z, Adotevi O, Bonnetain F, Borg C, Vernerey D (2016) Prognostic value of baseline seric Syndecan-1 in initially unresectable metastatic colorectal cancer patients: a simple biological score. Int J Cancer 139(10):2325–2335

Jeon Y, Jang ES, Choi YS, Kim JW, Jeong SH (2016) Glypican-3 level assessed by the enzyme-linked immunosorbent assay is inferior to alpha-fetoprotein level for hepatocellular carcinoma diagnosis. Clin Mol Hepatol 22(3):359–365

Jia X, Liu J, Gao Y, Huang Y, Du Z (2014) Diagnosis accuracy of serum glypican-3 in patients with hepatocellular carcinoma: a systematic review with meta-analysis. Arch Med Res 45(7): 580–588

Jia X, Gao Y, Zhai D, Liu J, Cai J, Wang Y, Jing L, Du Z (2016) Assessment of the clinical utility of glypican 3 as a serum marker for the diagnosis of hepatocellular carcinoma. Technol Cancer Res Treat 15(6):780–786

Jilani I, Wei C, Bekele BN, Zhang ZJ, Keating M, Wierda W, Ferrajoli A, Estrov Z, Kantarjian H, O'Brien SM, Giles FJ, Albitar M (2009) Soluble syndecan-1 (sCD138) as a prognostic factor independent of mutation status in patients with chronic lymphocytic leukemia. Int J Lab Hematol 31(1):97–105

Jing JS, Ye W, Jiang YK, Ma J, Zhu MQ, Ma JM, Zhou H, Yu LQ, Yang YF, Wang SC (2017) The value of GPC3 and GP73 in clinical diagnosis of hepatocellular carcinoma. Clin Lab 63(11): 1903–1909

Joensuu H, Anttonen A, Eriksson M, Mäkitaro R, Alfthan H, Kinnula V, Leppä S (2002) Soluble syndecan-1 and serum basic fibroblast growth factor are new prognostic factors in lung cancer. Cancer Research 62(18):5210–5217

Jourde-Chiche N, Fakhouri F, Dou L, Bellien J, Burtey S, Frimat M, Jarrot PA, Kaplanski G, Le Quintrec M, Pernin V, Rigothier C, Sallee M, Fremeaux-Bacchi V, Guerrot D, Roumenina LT (2019) Endothelium structure and function in kidney health and disease. Nat Rev Nephrol 15(2): 87–108

Karamanos NK, Piperigkou Z, Theocharis AD, Watanabe H, Franchi M, Baud S, Brezillon S, Gotte M, Passi A, Vigetti D, Ricard-Blum S, Sanderson RD, Neill T, Iozzo RV (2018) Proteoglycan chemical diversity drives multifunctional cell regulation and therapeutics. Chem Rev 118(18):9152–9232

Karamanos NK, Theocharis AD, Piperigkou Z, Manou D, Passi A, Skandalis SS, Vynios DH, Orian-Rousseau V, Ricard-Blum S, Schmelzer CEH, Duca L, Durbee M, Afratis NA, Troeberg L, Franchi M, Masola V, Onisto M (2021) A guide to the composition and functions of the extracellular matrix. FEBS J 288(24):6850–6912

Kato M, Wang H, Kainulainen V, Fitzgerald ML, Ledbetter S, Ornitz DM, Bernfield M (1998) Physiological degradation converts the soluble syndecan-1 ectodomain from an inhibitor to a potent activator of FGF-2. Nat Med 4(6):691–697

Katz BZ (2010) Adhesion molecules—The lifelines of multiple myeloma cells. Semin Cancer Biol 20(3):186–195

Kim JM, Lee JA, Cho IS, Ihm CH (2010) Soluble syndecan-1 at diagnosis and during follow up of multiple myeloma: a single institution study. Korean J Hematol 45(2):115–119

Kjellen L, Lindahl U (2018) Specificity of glycosaminoglycan-protein interactions. Curr Opin Struct Biol 50:101–108

Kumar S, Blood E, Oken MM, Greipp PR (2004) Prognostic value of syndecan-1 in multiple myeloma and its relationship with other prognostic factors. Blood 104(11):2402–2402

Kumar AV, Katakam SK, Urbanowitz AK, Gotte M (2015) Heparan sulphate as a regulator of leukocyte recruitment in inflammation. Curr Protein Pept Sci 16(1):77–86

Kyrtsonis MC, Vassilakopoulos TP, Siakantaris MP, Kokoris SI, Gribabis DA, Dimopoulou MN, Angelopoulou MK, Pangalis GA (2004) Serum syndecan-1, basic fibroblast growth factor and osteoprotegerin in myeloma patients at diagnosis and during the course of the disease. Eur J Haematol 72(4):252–258

Lai X, Wang M, McElyea SD, Sherman S, House M, Korc M (2017) A microRNA signature in circulating exosomes is superior to exosomal glypican-1 levels for diagnosing pancreatic cancer. Cancer Lett 393:86–93

Larsen AM, Leinoe EB, Johansson PI, Birgens H, Ostrowski SR (2013) High syndecan-1 levels in acute myeloid leukemia are associated with bleeding, thrombocytopathy, endothelial cell damage, and leukocytosis. Leuk Res 37(7):777–783

Lee HJ, Yeon JE, Suh SJ, Lee SJ, Yoon EL, Kang K, Yoo YJ, Kim JH, Seo YS, Yim HJ, Byun KS (2014) Clinical utility of plasma glypican-3 and osteopontin as biomarkers of hepatocellular carcinoma. Gut Liver 8(2):177–185

Lepedda AJ, Nieddu G, Formato M, Baker MB, Fernandez-Perez J, Moroni L (2021a) Glycosaminoglycans: from vascular physiology to tissue engineering applications. Front Chem 9:680836

Lepedda AJ, Nieddu G, Piperigkou Z, Kyriakopoulou K, Karamanos N, Formato M (2021b) Circulating heparan sulfate proteoglycans as biomarkers in health and disease. Semin Thromb Hemost 47(3):295–307

Levin RA, Lund ME, Truong Q, Wu A, Shore ND, Saltzstein DR, Concepcion RS, Paivanas TA, van Breda A, Beebe-Dimmer J, Ruterbusch JJ, Wissmueller S, Campbell DH, Walsh BJ (2018) Development of a reliable assay to measure glypican-1 in plasma and serum reveals circulating glypican-1 as a novel prostate cancer biomarker. Oncotarget 9(32):22359–22367

Li B, Liu H, Shang HW, Li P, Li N, Ding HG (2013) Diagnostic value of glypican-3 in alpha fetoprotein negative hepatocellular carcinoma patients. Afr Health Sci 13(3):703–709

Li J, Gao JZ, Du JL, Wei LX (2014) Prognostic and clinicopathological significance of glypican-3 overexpression in hepatocellular carcinoma: a meta-analysis. World J Gastroenterol 20(20): 6336–6344

Li J, Chen Y, Guo X, Zhou L, Jia Z, Peng Z, Tang Y, Liu W, Zhu B, Wang L, Ren C (2017) GPC1 exosome and its regulatory miRNAs are specific markers for the detection and target therapy of colorectal cancer. J Cell Mol Med 21(5):838–847

Liu H, Li P, Zhai Y, Qu CF, Zhang LJ, Tan YF, Li N, Ding HG (2010) Diagnostic value of glypican-3 in serum and liver for primary hepatocellular carcinoma. World J Gastroenterol 16(35):4410–4415

Liu XF, Hu ZD, Liu XC, Cao Y, Ding CM, Hu CJ (2014) Diagnostic accuracy of serum glypican-3 for hepatocellular carcinoma: a systematic review and meta-analysis. Clin Biochem 47(3): 196–200

Liu JW, Zuo XL, Wang S (2015) Diagnosis accuracy of serum Glypican-3 level in patients with hepatocellular carcinoma and liver cirrhosis: a meta-analysis. Eur Rev Med Pharmacol Sci 19(19):3655–3673

Liu H, Yang C, Lu W, Zeng Y (2018) Prognostic significance of glypican-3 expression in hepatocellular carcinoma: A meta-analysis. Medicine (Baltimore) 97(4):e9702

Liu S, Wang M, Zheng C, Zhong Q, Shi Y, Han X (2020) Diagnostic value of serum glypican-3 alone and in combination with AFP as an aid in the diagnosis of liver cancer. Clin Biochem 79: 54–60

Lovell R, Dunn JA, Begum G, Barth NJ, Plant T, Moss PA, Drayson MT, Pratt G, Working Party on Leukaemia in Adults of the National Cancer Research Institute Haematological Oncology Clinical Studies (2005) Soluble syndecan-1 level at diagnosis is an independent prognostic factor in multiple myeloma and the extent of fall from diagnosis to plateau predicts for overall survival. Br J Haematol 130(4):542–548

Machin DR, Phuong TT, Donato AJ (2019) The role of the endothelial glycocalyx in advanced age and cardiovascular disease. Curr Opin Pharmacol 45:66–71

Mahtouk K, Cremer FW, Reme T, Jourdan M, Baudard M, Moreaux J, Requirand G, Fiol G, De Vos J, Moos M, Quittet P, Goldschmidt H, Rossi JF, Hose D, Klein B (2006) Heparan sulphate proteoglycans are essential for the myeloma cell growth activity of EGF-family ligands in multiple myeloma. Oncogene 25(54):7180–7191

Mahtouk K, Hose D, Raynaud P, Hundemer M, Jourdan M, Jourdan E, Pantesco V, Baudard M, De Vos J, Larroque M, Moehler T, Rossi JF, Reme T, Goldschmidt H, Klein B (2007) Heparanase influences expression and shedding of syndecan-1, and its expression by the bone marrow environment is a bad prognostic factor in multiple myeloma. Blood 109(11):4914–4923

Maisnar V, Touskova M, Tichy M, Krejsek J, Chrobak L, Voglova J, Maly J (2006) The significance of soluble CD138 in diagnosis of monoclonal gammopathies. Neoplasma 53(1): 26–29

Meirovitz A, Goldberg R, Binder A, Rubinstein AM, Hermano E, Elkin M (2013) Heparanase in inflammation and inflammation-associated cancer. FEBS J 280(10):2307–2319

Melo SA, Luecke LB, Kahlert C, Fernandez AF, Gammon ST, Kaye J, LeBleu VS, Mittendorf EA, Weitz J, Rahbari N, Reissfelder C, Pilarsky C, Fraga MF, Piwnica-Worms D, Kalluri R (2015) Glypican-1 identifies cancer exosomes and detects early pancreatic cancer. Nature 523(7559): 177–182

Metwaly HA, Al-Gayyar MM, Eletreby S, Ebrahim MA, El-Shishtawy MM (2012) Relevance of serum levels of interleukin-6 and syndecan-1 in patients with hepatocellular carcinoma. Sci Pharm 80(1):179–188

Mima K, Beppu T, Ishiko T, Chikamoto A, Nakagawa S, Hayashi H, Watanabe M, Sakamaki K, Baba H (2014) Preoperative serum hyaluronic acid level as a prognostic factor in patients undergoing hepatic resection for hepatocellular carcinoma. Br J Surg 101(3):269–276

Minarik J, Pika T, Bacovsky J, Petrova P, Langova K, Scudla V (2012) Prognostic value of hepatocyte growth factor, syndecan-1, and osteopontin in multiple myeloma and monoclonal gammopathy of undetermined significance. ScientificWorldJournal 2012:356128

Molica S, Vitelli G, Mirabelli R, Digiesu G, Giannarelli D, Cuneo A, Ribatti D, Vacca A (2006) Serum levels of syndecan-1 in B-cell chronic lymphocytic leukemia: correlation with the extent of angiogenesis and disease-progression risk in early disease. Leuk Lymphoma 47(6): 1034–1040

Morimoto-Tomita M, Uchimura K, Werb Z, Hemmerich S, Rosen SD (2002) Cloning and characterization of two extracellular heparin-degrading endosulfatases in mice and humans. J Biol Chem 277(51):49175–49185

Moseley R, Waddington RJ, Embery G (1997) Degradation of glycosaminoglycans by reactive oxygen species derived from stimulated polymorphonuclear leukocytes. Biochim Biophys Acta 1362(2-3):221–231

Moudi B, Heidari Z, Mahmoudzadeh-Sagheb H (2019) Meta-analysis and systematic review of prognostic significance of Glypican-3 in patients with hepatitis B-related hepatocellular carcinoma. Virusdisease 30(2):193–200

Murakami M, Horowitz A, Tang S, Ware JA, Simons M (2002) Protein kinase C (PKC) delta regulates PKCalpha activity in a Syndecan-4-dependent manner. J Biol Chem 277(23): 20367–20371

Nadanaka S, Kitagawa H (2008) Heparan sulphate biosynthesis and disease. J Biochem 144(1): 7–14

Nakatsura T, Yoshitake Y, Senju S, Monji M, Komori H, Motomura Y, Hosaka S, Beppu T, Ishiko T, Kamohara H, Ashihara H, Katagiri T, Furukawa Y, Fujiyama S, Ogawa M, Nakamura Y, Nishimura Y (2003) Glypican-3, overexpressed specifically in human hepatocellular carcinoma, is a novel tumor marker. Biochem Biophys Res Commun 306(1):16–25

Nault JC, Guyot E, Laguillier C, Chevret S, Ganne-Carrie N, N'Kontchou G, Beaugrand M, Seror O, Trinchet JC, Coelho J, Lasalle P, Charnaux N, Delehedde M, Sutton A, Nahon P (2013) Serum proteoglycans as prognostic biomarkers of hepatocellular carcinoma in patients with alcoholic cirrhosis. Cancer Epidemiol Biomarkers Prev 22(8):1343–1352

Nishio M, Yamanaka T, Matsumoto K, Kimura H, Sakai K, Sakai A, Sone T, Horiike A, Koizumi F, Kasahara K, Ohira T, Ikeda N, Saijo N, Arao T, Nishio K (2011) Serum heparan sulfate concentration is correlated with the failure of epidermal growth factor receptor tyrosine kinase inhibitor treatment in patients with lung adenocarcinoma. J Thorac Oncol 6(11): 1889–1894

Obayashi Y, Yabushita H, Kanyama K, Noguchi M, Zhuo L, Kimata K, Wakatsuki A (2008) Role of serum-derived hyaluronan-associated protein-hyaluronan complex in ovarian cancer. Oncol Rep 19(5):1245–1251

Oduah EI, Linhardt RJ, Sharfstein ST (2016) Heparin: past, present, and future. Pharmaceuticals (Basel) 9(3)

Ofuji K, Saito K, Suzuki S, Shimomura M, Shirakawa H, Nobuoka D, Sawada Y, Yoshimura M, Tsuchiya N, Takahashi M, Yoshikawa T, Tada Y, Konishi M, Takahashi S, Gotohda N, Nakamoto Y, Nakatsura T (2017) Perioperative plasma glypican-3 level may enable prediction of the risk of recurrence after surgery in patients with stage I hepatocellular carcinoma. Oncotarget 8(23):37835–37844

Oh ES, Woods A, Couchman JR (1997) Multimerization of the cytoplasmic domain of syndecan-4 is required for its ability to activate protein kinase C. J Biol Chem 272(18):11805–11811

Okamoto O, Bachy S, Odenthal U, Bernaud J, Rigal D, Lortat-Jacob H, Smyth N, Rousselle P (2003) Normal human keratinocytes bind to the alpha3LG4/5 domain of unprocessed laminin-5 through the receptor syndecan-1. J Biol Chem 278(45):44168–44177

Ozkan H, Erdal H, Kocak E, Tutkak H, Karaeren Z, Yakut M, Koklu S (2011) Diagnostic and prognostic role of serum glypican 3 in patients with hepatocellular carcinoma. J Clin Lab Anal 25(5):350–353

Passi A, Vigetti D, Buraschi S, Iozzo RV (2019) Dissecting the role of hyaluronan synthases in the tumor microenvironment. FEBS J 286(15):2937–2949

Peng C, Wallwiener M, Rudolph A, Cuk K, Eilber U, Celik M, Modugno C, Trumpp A, Heil J, Marme F, Madhavan D, Nees J, Riethdorf S, Schott S, Sohn C, Pantel K, Schneeweiss A,

Chang-Claude J, Yang R, Burwinkel B (2016) Plasma hyaluronic acid level as a prognostic and monitoring marker of metastatic breast cancer. Int J Cancer 138(10):2499–2509

Peysselon F, Xue B, Uversky VN, Ricard-Blum S (2011) Intrinsic disorder of the extracellular matrix. Mol Biosyst 7(12):3353–3365

Piperigkou Z, Mohr B, Karamanos N, Gotte M (2016) Shed proteoglycans in tumor stroma. Cell Tissue Res 365(3):643–655

Pothacharoen P, Siriaunkgul S, Ong-Chai S, Supabandhu J, Kumja P, Wanaphirak C, Sugahara K, Hardingham T, Kongtawelert P (2006) Raised serum chondroitin sulfate epitope level in ovarian epithelial cancer. J Biochem 140(4):517–524

Purushothaman A, Sanderson RD (2020) Heparanase: a dynamic promoter of myeloma progression. Adv Exp Med Biol 1221:331–349

Purushothaman A, Uyama T, Kobayashi F, Yamada S, Sugahara K, Rapraeger AC, Sanderson RD (2010) Heparanase-enhanced shedding of syndecan-1 by myeloma cells promotes endothelial invasion and angiogenesis. Blood 115(12):2449–2457

Qiao D, Meyer K, Mundhenke C, Drew SA, Friedl A (2003) Heparan sulfate proteoglycans as regulators of fibroblast growth factor-2 signaling in brain endothelial cells. Specific role for glypican-1 in glioma angiogenesis. J Biol Chem 278(18):16045–16053

Qiao SS, Cui ZQ, Gong L, Han H, Chen PC, Guo LM, Yu X, Wei YH, Ha SA, Kim JW, Jin ZT, Li S, Peng JR, Leng XS (2011) Simultaneous measurements of serum AFP, GPC-3 and HCCR for diagnosing hepatocellular carcinoma. Hepatogastroenterology 58(110-111):1718–1724

Raman R, Sasisekharan V, Sasisekharan R (2005) Structural insights into biological roles of protein-glycosaminoglycan interactions. Chem Biol 12(3):267–277

Ramani VC, Purushothaman A, Stewart MD, Thompson CA, Vlodavsky I, Au JL, Sanderson RD (2013) The heparanase/syndecan-1 axis in cancer: mechanisms and therapies. FEBS J 280(10):2294–2306

Reijmers RM, Spaargaren M, Pals ST (2013) Heparan sulfate proteoglycans in the control of B cell development and the pathogenesis of multiple myeloma. FEBS J 280(10):2180–2193

Ritchie JP, Ramani VC, Ren Y, Naggi A, Torri G, Casu B, Penco S, Pisano C, Carminati P, Tortoreto M, Zunino F, Vlodavsky I, Sanderson RD, Yang Y (2011) SST0001, a chemically modified heparin, inhibits myeloma growth and angiogenesis via disruption of the heparanase/syndecan-1 axis. Clin Cancer Res 17(6):1382–1393

Rosenberg RD, Shworak NW, Liu J, Schwartz JJ, Zhang L (1997) Heparan sulfate proteoglycans of the cardiovascular system. Specific structures emerge but how is synthesis regulated? J Clin Invest 100(11 Suppl):S67–S75

Ruan HL, Hong RT, Xie HJ, Hu NZ, Xu JM, Zhang W (2011) Significance of elevated levels of collagen type IV and hyaluronic acid in gastric juice and serum in gastric cancer and precancerous lesion. Dig Dis Sci 56(7):2001–2008

Sadik NA, Ahmed A, Ahmed S (2012) The significance of serum levels of adiponectin, leptin, and hyaluronic acid in hepatocellular carcinoma of cirrhotic and noncirrhotic patients. Hum Exp Toxicol 31(4):311–321

Schaar CG, Vermeer HJ, Wijermans PW, Huisman W, le Cessie S, Kluin-Nelemans HC (2005) Serum syndecan-1 in patients with newly diagnosed monoclonal proteinemia. Haematologica 90(10):1437–1438

Schmidt A, Echtermeyer F, Alozie A, Brands K, Buddecke E (2005) Plasmin- and thrombin-accelerated shedding of syndecan-4 ectodomain generates cleavage sites at Lys(114)-Arg(115) and Lys(129)-Val(130) bonds. J Biol Chem 280(41):34441–34446

Schmidt EP, Li G, Li L, Fu L, Yang Y, Overdier KH, Douglas IS, Linhardt RJ (2014) The circulating glycosaminoglycan signature of respiratory failure in critically ill adults. J Biol Chem 289(12):8194–8202

Scudla V, Pika T, Budikova M, Petrova J, Minarik J, Bacovsky J, Langova K, Zivna J, Czech Myeloma G (2010) The importance of serum levels of selected biological parameters in the diagnosis, staging and prognosis of multiple myeloma. Neoplasma 57(2):102–110

Seidel C, Sundan A, Hjorth M, Turesson I, Dahl IM, Abildgaard N, Waage A, Borset M (2000) Serum syndecan-1: a new independent prognostic marker in multiple myeloma. Blood 95(2): 388–392

Shegefti MS, Malekzadeh M, Malek-Hosseini Z, Khademi B, Ghaderi A, Doroudchi M (2016) Reduced serum levels of syndecan-1 in patients with tongue squamous cell carcinoma. Laryngoscope 126(5):E191–E195

Shimizu T, Ishizuka M, Kato M, Aoki T, Kubota K (2016) Hyaluronic acid predicts postoperative large amount of ascites after liver resection for patients with hepatocellular carcinoma. Dig Surg 33(6):520–528

Shimizu Y, Mizuno S, Fujinami N, Suzuki T, Saito K, Konishi M, Takahashi S, Gotohda N, Tada T, Toyoda H, Kumada T, Miura M, Suto K, Yamaji T, Matsuda T, Endo I, Nakatsura T (2020) Plasma and tumoral glypican-3 levels are correlated in patients with hepatitis C virus-related hepatocellular carcinoma. Cancer Sci 111(2):334–342

Skliris A, Happonen KE, Terpos E, Labropoulou V, Borset M, Heinegard D, Blom AM, Theocharis AD (2011) Serglycin inhibits the classical and lectin pathways of complement via its glycosaminoglycan chains: implications for multiple myeloma. Eur J Immunol 41(2):437–449

Soltes L, Mendichi R, Kogan G, Schiller J, Stankovska M, Arnhold J (2006) Degradative action of reactive oxygen species on hyaluronan. Biomacromolecules 7(3):659–668

Stern R (2008) Hyaluronan in cancer biology. Semin Cancer Biol 18(4):237

Sun B, Huang Z, Wang B, Yu Y, Lin S, Luo L, Wang Y, Huang Z (2017) Significance of glypican-3 (GPC3) expression in hepatocellular cancer diagnosis. Med Sci Monit 23:850–855

Suzuki M, Sugimoto K, Tanaka J, Tameda M, Inagaki Y, Kusagawa S, Nojiri K, Beppu T, Yoneda K, Yamamoto N, Ito M, Yoneda M, Uchida K, Takase K, Shiraki K (2010) Up-regulation of glypican-3 in human hepatocellular carcinoma. Anticancer Res 30(12): 5055–5061

Szarvas T, Reis H, Vom Dorp F, Tschirdewahn S, Niedworok C, Nyirady P, Schmid KW, Rubben H, Kovalszky I (2016) Soluble syndecan-1 (SDC1) serum level as an independent pre-operative predictor of cancer-specific survival in prostate cancer. Prostate 76(11):977–985

Szarvas T, Sevcenco S, Modos O, Keresztes D, Nyirady P, Kubik A, Romics M, Kovalszky I, Reis H, Hadaschik B, Shariat SF, Kramer G (2018) Circulating syndecan-1 is associated with chemotherapy-resistance in castration-resistant prostate cancer. Urol Oncol 36(6):312 e319–312 e315

Tahon AM, El-Ghanam MZ, Zaky S, Emran TM, Bersy AM, El-Raey F, Elsayed AZ, El Kharsawy AM, Johar D (2019) Significance of glypican-3 in early detection of hepatocellular carcinoma in cirrhotic patients. J Gastrointest Cancer 50(3):434–441

Tang XY, Wang YC, Lu RQ, Guo L (2020) [The value of serum glypican-3 level in aided diagnosis of patients with primary hepatocellular carcinoma]. Zhonghua Yu Fang Yi Xue Za Zhi 54(9): 998–1002

Tangkijvanich P, Chanmee T, Komtong S, Mahachai V, Wisedopas N, Pothacharoen P, Kongtawelert P (2010) Diagnostic role of serum glypican-3 in differentiating hepatocellular carcinoma from non-malignant chronic liver disease and other liver cancers. J Gastroenterol Hepatol 25(1):129–137

Tavianatou AG, Caon I, Franchi M, Piperigkou Z, Galesso D, Karamanos NK (2019a) Hyaluronan: molecular size-dependent signaling and biological functions in inflammation and cancer. FEBS J 286(15):2883–2908

Tavianatou AG, Piperigkou Z, Barbera C, Beninatto R, Masola V, Caon I, Onisto M, Franchi M, Galesso D, Karamanos NK (2019b) Molecular size-dependent specificity of hyaluronan on functional properties, morphology and matrix composition of mammary cancer cells. Matrix Biol Plus 3:100008

Tavianatou AG, Piperigkou Z, Koutsakis C, Barbera C, Beninatto R, Franchi M, Karamanos NK (2021) The action of hyaluronan in functional properties, morphology and expression of matrix effectors in mammary cancer cells depends on its molecular size. FEBS J 288(14):4291–4310

Teixeira F, Götte M (2020) Involvement of syndecan-1 and heparanase in cancer and inflammation. Adv Exp Med Biol 1221:97–135
Theocharis AD, Karamanos NK (2019) Proteoglycans remodeling in cancer: Underlying molecular mechanisms. Matrix Biol 75–76:220–259
Theocharis AD, Skandalis SS, Tzanakakis GN, Karamanos NK (2010) Proteoglycans in health and disease: novel roles for proteoglycans in malignancy and their pharmacological targeting. FEBS J 277(19):3904–3923
Theocharis AD, Manou D, Karamanos NK (2019) The extracellular matrix as a multitasking player in disease. FEBS J 286(15):2830–2869
Thylen A, Wallin J, Martensson G (1999) Hyaluronan in serum as an indicator of progressive disease in hyaluronan-producing malignant mesothelioma. Cancer 86(10):2000–2005
Tkachenko E, Rhodes JM, Simons M (2005) Syndecans: new kids on the signaling block. Circ Res 96(5):488–500
Traister A, Shi W, Filmus J (2008) Mammalian Notum induces the release of glypicans and other GPI-anchored proteins from the cell surface. Biochem J 410(3):503–511
Tripathi K, Ramani VC, Bandari SK, Amin R, Brown EE, Ritchie JP, Stewart MD, Sanderson RD (2020) Heparanase promotes myeloma stemness and in vivo tumorigenesis. Matrix Biol 88:53–68
Tuma M, Canestrini S, Alwahab Z, Marshall J (2016) Trauma and endothelial glycocalyx: the microcirculation helmet? Shock 46(4):352–357
Uchimido R, Schmidt EP, Shapiro NI (2019) The glycocalyx: a novel diagnostic and therapeutic target in sepsis. Crit Care 23(1):16
Vallet SD, Clerc O, Ricard-Blum S (2021) Glycosaminoglycan-protein interactions: the first draft of the glycosaminoglycan interactome. J Histochem Cytochem 69(2):93–104
Vassilakopoulos TP, Kyrtsonis M-C, Papadogiannis A, Nadali G, Angelopoulou MK, Tzenou T, Dimopoulou MN, Siakantaris MP, Kontopidou FN, Kalpadakis C, Kokoris SI, Dimitriadou EM, Tsaftaridis P, Pizzolo G, Pangalis GA (2005) Serum levels of soluble syndecan-1 in Hodgkin's lymphoma. Anticancer Res 25(6C):4743–4746
Velesiotis C, Vasileiou S, Vynios DH (2019) A guide to hyaluronan and related enzymes in breast cancer: biological significance and diagnostic value. FEBS J 286(15):3057–3074
Vigetti D, Karousou E, Viola M, Deleonibus S, De Luca G, Passi A (2014) Hyaluronan: biosynthesis and signaling. Biochim Biophys Acta 1840(8):2452–2459
Vynios DH, Karamanos NK, Tsiganos CP (2002) Advances in analysis of glycosaminoglycans: its application for the assessment of physiological and pathological states of connective tissues. J Chromatogr B Analyt Technol Biomed Life Sci 781(1–2):21–38
Wang X, Zuo D, Chen Y, Li W, Liu R, He Y, Ren L, Zhou L, Deng T, Wang X, Ying G, Ba Y (2014a) Shed Syndecan-1 is involved in chemotherapy resistance via the EGFR pathway in colorectal cancer. Br J Cancer 111(10):1965–1976
Wang Y, Yang H, Xu H, Lu X, Sang X, Zhong S, Huang J, Mao Y (2014b) Golgi protein 73, not Glypican-3, may be a tumor marker complementary to alpha-Fetoprotein for hepatocellular carcinoma diagnosis. J Gastroenterol Hepatol 29(3):597–602
Weinbaum S, Tarbell JM, Damiano ER (2007) The structure and function of the endothelial glycocalyx layer. Annu Rev Biomed Eng 9:121–167
Weiss R, Niecestro R, Raz I (2007) The role of sulodexide in the treatment of diabetic nephropathy. Drugs 67(18):2681–2696
Wolowiec D, Dybko J, Wrobel T, Urbaniak-Kujda D, Jazwiec B, Tomaszewska-Toporska B, Kapelko-Slowik K, Potoczek S, Kuliczkowski K (2006) Circulating sCD138 and some angiogenesis-involved cytokines help to anticipate the disease progression of early-stage B-cell chronic lymphocytic leukemia. Mediators Inflamm 2006(3):42394
Wu IC, Wu DC, Huang CC, Lin HS, Chen YK, Tsai HJ, Lu CY, Chou SH, Chou YP, Li LH, Tai SY, Wu MT (2010) Plasma decorin predicts the presence of esophageal squamous cell carcinoma. Int J Cancer 127(9):2138–2146

Wu M, Cao M, He Y, Liu Y, Yang C, Du Y, Wang W, Gao F (2015) A novel role of low molecular weight hyaluronan in breast cancer metastasis. FASEB J 29(4):1290–1298

Wu M, Liu Z, Zhang A, Li N (2019) Associated measurement of fucosylated levels of AFP, DCP, and GPC3 for early diagnosis in hepatocellular carcinoma. Int J Biol Markers 34(1):20–26

Xia F, Lai EC, Lau WY, Ma K, Li X, Bie P, Qian C (2012) High serum hyaluronic acid and HBV viral load are main prognostic factors of local recurrence after complete radiofrequency ablation of hepatitis B-related small hepatocellular carcinoma. Ann Surg Oncol 19(4):1284–1291

Xiao WK, Qi CY, Chen D, Li SQ, Fu SJ, Peng BG, Liang LJ (2014) Prognostic significance of glypican-3 in hepatocellular carcinoma: a meta-analysis. BMC Cancer 14:104

Xiao D, Dong Z, Zhen L, Xia G, Huang X, Wang T, Guo H, Yang B, Xu C, Wu W, Zhao X, Xu H (2020) Combined exosomal GPC1, CD82, and serum CA19-9 as multiplex targets: a specific, sensitive, and reproducible detection panel for the diagnosis of pancreatic cancer. Mol Cancer Res 18(2):300–310

Xing Y, Xu Q, Lee C (2003) Widespread production of novel soluble protein isoforms by alternative splicing removal of transmembrane anchoring domains. FEBS Lett 555(3):572–578

Xu C, Yan Z, Zhou L, Wang Y (2013) A comparison of glypican-3 with alpha-fetoprotein as a serum marker for hepatocellular carcinoma: a meta-analysis. J Cancer Res Clin Oncol 139(8): 1417–1424

Xu D, Su C, Sun L, Gao Y, Li Y (2019) Performance of serum glypican 3 in diagnosis of hepatocellular carcinoma: a meta-analysis. Ann Hepatol 18(1):58–67

Yang Y, MacLeod V, Dai Y, Khotskaya-Sample Y, Shriver Z, Venkataraman G, Sasisekharan R, Naggi A, Torri G, Casu B, Vlodavsky I, Suva LJ, Epstein J, Yaccoby S, Shaughnessy JD Jr, Barlogie B, Sanderson RD (2007) The syndecan-1 heparan sulfate proteoglycan is a viable target for myeloma therapy. Blood 110(6):2041–2048

Yang SL, Fang X, Huang ZZ, Liu XJ, Xiong ZF, Liu P, Yao HY, Li CH (2014) Can serum glypican-3 be a biomarker for effective diagnosis of hepatocellular carcinoma? A meta-analysis of the literature. Dis Markers 2014:127831

Yang Y, Haeger SM, Suflita MA, Zhang F, Dailey KL, Colbert JF, Ford JA, Picon MA, Stearman RS, Lin L, Liu X, Han X, Linhardt RJ, Schmidt EP (2017) Fibroblast growth factor signaling mediates pulmonary endothelial glycocalyx reconstitution. Am J Respir Cell Mol Biol 56(6): 727–737

Yao M, Yao DF, Bian YZ, Wu W, Yan XD, Yu DD, Qiu LW, Yang JL, Zhang HJ, Sai WL, Chen J (2013) Values of circulating GPC-3 mRNA and alpha-fetoprotein in detecting patients with hepatocellular carcinoma. Hepatobiliary Pancreat Dis Int 12(2):171–179

Yasuda E, Kumada T, Toyoda H, Kaneoka Y, Maeda A, Okuda S, Yoshimi N, Kozawa O (2010) Evaluation for clinical utility of GPC3, measured by a commercially available ELISA kit with Glypican-3 (GPC3) antibody, as a serological and histological marker for hepatocellular carcinoma. Hepatol Res 40(5):477–485

Yu JP, Xu XG, Ma RJ, Qin SN, Wang CR, Wang XB, Li M, Li MS, Ma Q, Xu WW (2015) Development of a clinical chemiluminescent immunoassay for serum GPC3 and simultaneous measurements alone with AFP and CK19 in diagnosis of hepatocellular carcinoma. J Clin Lab Anal 29(2):85–93

Yuzhalin AE, Lim SY, Kutikhin AG, Gordon-Weeks AN (2018) Dynamic matrisome: ECM remodeling factors licensing cancer progression and metastasis. Biochim Biophys Acta Rev Cancer 1870(2):207–228

Zhang Q, Xiao Q, Lin Z, Ying X, Li Z, Lin JM (2010) Development of a competitive radioimmunoassay for glypican-3 and the clinical application in diagnosis of hepatocellular carcinoma. Clin Biochem 43(12):1003–1008

Zhang J, Zhang M, Ma H, Song X, He L, Ye X, Li X (2018a) Overexpression of glypican-3 is a predictor of poor prognosis in hepatocellular carcinoma: An updated meta-analysis. Medicine (Baltimore) 97(24):e11130

Zhang X, Sun D, Song JW, Zullo J, Lipphardt M, Coneh-Gould L, Goligorsky MS (2018b) Endothelial cell dysfunction and glycocalyx—a vicious circle. Matrix Biol 71–72:421–431

Zhou S, O'Gorman MR, Yang F, Andresen K, Wang L (2017) Glypican 3 as a serum marker for hepatoblastoma. Sci Rep 7:45932
Zhou CY, Dong YP, Sun X, Sui X, Zhu H, Zhao YQ, Zhang YY, Mason C, Zhu Q, Han SX (2018a) High levels of serum glypican-1 indicate poor prognosis in pancreatic ductal adenocarcinoma. Cancer Med 7(11):5525–5533
Zhou F, Shang W, Yu X, Tian J (2018b) Glypican-3: A promising biomarker for hepatocellular carcinoma diagnosis and treatment. Med Res Rev 38(2):741–767
Zolota V, Tzelepi V, Piperigkou Z, Kourea H, Papakonstantinou E, Argentou Mu I, Karamanos NK (2021) Epigenetic alterations in triple-negative breast cancer-the critical role of extracellular matrix. Cancers (Basel) 13(4)

Chapter 7
Hyaluronan in the Extracellular Matrix of Hematological and Solid Tumors. Its Biological Effects

Antonella Icardi, Silvina L. Lompardia, Daniela L. Papademetrio, Paolo Rosales, Mariángeles Díaz, Matías A. Pibuel, Laura Alaniz, and Elida Alvarez

Abstract Tumors are most frequently induced by environmental factors with oncogenic potential. Beside tumor cells, cancer tissue contains several nonmalignant cells with well-defined function, embedded into the acellular extracellular matrix playing an important role in tumor progression. The extracellular matrix (ECM) is an indispensable element of all tissues which can modulate cell growth, survival, migration, immune response, and drug resistance. Among its main molecular components are glycosaminoglycans (GAGs) and proteoglycans (PGs), which have been shown to play fundamental roles either in physiological or malignant tissue. An important GAG of the ECM in the tumor microenvironment (TME) is hyaluronic acid, whose synthesis and degradation are altered in a variety of tumors. The biological effects of the abnormal accumulation of HA and the subsequent remodeling of the associated ECM differ according to the type of tumor. In this sense, the ECM-derived HA biological effects in different types of tumors are discussed below in this chapter. Additionally, we make it clear that approaches that interfere with HA metabolism may be a potential therapeutic approach for the treatment of cancer.

A. Icardi · P. Rosales · L. Alaniz (✉)
Laboratorio de Microambiente Tumoral, Centro de Investigaciones Básicas y Aplicadas (CIBA), Universidad Nacional del Noroeste de la Provincia de Buenos Aires (UNNOBA), Centro de Investigaciones y Transferencia del Noroeste de la Provincia de Buenos Aires (CITNOBA), UNNOBA-UNSAdA-CONICET, Junin, Buenos Aires, Argentina
e-mail: ldalaniz@comunidad.unnoba.edu.ar

S. L. Lompardia · D. L. Papademetrio · M. Díaz · M. A. Pibuel · E. Alvarez (✉)
Departamento de Microbiología, Inmunología, Biotecnología y Genética, Cátedra de Inmunología, Universidad de Buenos Aires Facultad de Farmacia y Bioquímica, Consejo Nacional de Investigaciones Científicas y Técnicas (CONICET), Instituto de Estudios de la Inmunidad Humoral (IDEHU), Junín, CABA, Argentina
e-mail: elialv@ffyb.uba.ar

I. Kovalszky et al. (eds.), *The Extracellular Matrix and the Tumor Microenvironment*, Biology of Extracellular Matrix 11,
https://doi.org/10.1007/978-3-030-99708-3_7

7.1 Introduction

The tumor microenvironment (TME) is essential part of tumors, being either solid, as carcinomas, sarcomas, lymphomas, or liquid as leukemia. Within this, TME are composed by an extracellular matrix (ECM) which altered composition directly influences in malignant transformation and/or progression. The components from ECM as proteins, glycosaminoglycans (GAGs) and proteoglycans (PGs), have been shown to play fundamental roles in different physiological processes and malignancies. Among the GAGs, HA, which is our molecule of discussion during this chapter, is a key factor that organized the integrity in the ECM. HA is a member of GAGs family and is synthesized by all vertebrate organisms. It is present in almost all mature tissues, within the interstitial matrix but also at intracellular level. Besides, HA has multiple interactions with other proteins and PGs that also impact the function of ECM (Wei et al. 2020). Stavros Garantziotis and Rashmin C. Savani identified at least four distinctive characteristics of HA that defined its mechanism of action, its structure (weight and size), its binding patterns and chemical modifications, its metabolism, and the induction of intracellular signal (Garantziotis and Savani 2019).

The HA's chemical structure is well defined, it is a linear polysaccharide composed of 2000–25,000 repeating disaccharide units [D-glucuronic acid β (1→3) and *N*-acetyl-D-glucosamine β (1→4)] (Fig. 7.1). This large glycosaminoglycan (GAG) exhibits a molecular weight ranging from 10^5 to 107 Da (Laurent and Fraser 1992).

The large quantity of HA in the mammalian ECM impacts in hydration, lubrication, and physical properties of tissues (Toole 2004). In addition, it mediates cell proliferation and differentiation, regulates cell adhesion and motility, making it not only a structural component, but also an effective signaling biomolecule. In many types of tumors, with enhanced HA production, HA surface binding generate a physical barrier, which covers tumor cells and may contribute to the failure of treatments.

HA has a high rate of turnover in vertebrates, for example, in human, 5 g of the 15 g of total body HA renew daily (Stern 2008). It is synthesized by one of the three hyaluronan synthases (Has1/Has2/Has3). These are multipass transmembrane

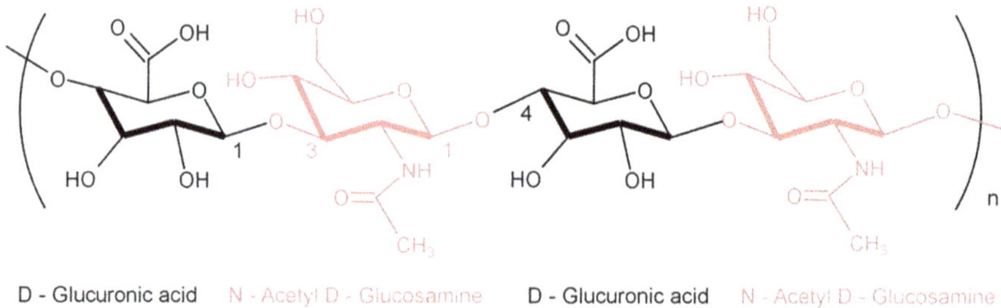

Fig. 7.1 Hyaluronic acid chemical structure. Disaccharide unit [D-glucuronic acid β (1→3) and *N*-acetyl-D-glucosamine β (1→4)] that forms the linear nonsulfated polysaccharide

enzymes that extrude HA while it is being synthesized either onto the cell surface or into the ECM (Monslow et al. 2015). Elevated levels of HA are associated with hyperproliferative and malignant phenotypes in melanomas and various carcinomas (Chanmee et al. 2016).

The breakdown of HA (HMW) in tissues can be accomplished by other mechanisms, either enzymatically by hyaluronidases, or by cleavage that are nonenzymatic oxidation reactions. In cancer cells, the type and levels of hyaluronidases are variable; in some cases, it is elevated, in others suppressed, compared to normal tissues (Toole 2004). A result of hyaluronidase activation, low molecular HA (LMW) is present in the ECM, which promotes inflammation, immune cell recruitment, and the epithelial cell migration. On contrary, elevated HMW-HA production, in the absence of fragmentation, is linked to cancer resistance (Toole 2009). Interestingly, Xiao Tian et al. have been observed that naked mole-rat fibroblasts secrete extremely high-molecular-mass HA, which is over five times larger than human or mouse HA. This sort of high-molecular-mass HA accumulates, abundantly in naked mole-rat tissues, providing cancer resistance and longevity to this species (Tian et al. 2013). Additionally, the authors observed that inhibition of HA synthesis enhances in these cells the susceptibility to malignant transformation.

HA interacts with cell surfaces in at least two ways. First, it can bind to specific cell surface receptors, to induce transduction of several intracellular pathway directly or by interacting with other receptors. HA interacts with many proteins, so termed as hyaladherins, several of which are known as surface receptors. The interaction depends on a binding domain presents every hyaladherins that contains seven nonacidic amino acids. This domain is common to all CD44 isoforms (CD44s and CD44v), the lymphatic vessel endothelial HA receptor-1 (LYVE-1), RHAMM (Receptor for hyaluronate-mediated motility), HARE (Hyaluronan receptor for endocytosis), and TOLL4, Stabilin (Orian-Rousseau 2010). Interestingly, layilin is a transmembrane protein widely expressed in different cell types and tissue extracts. It is found in peripheral ruffles of spreading. Layilin colocalizes with talin in ruffles and binds to talin's ~50-kDa head domain (amino acids 280–435), thus represents a membrane-binding site for talin in cell ruffles (Bono et al. 2001). However, this protein has not sequence homology to the other HA receptors and represents a novel member of the HA-binding protein family (Bono et al. 2001).

Among them, CD44 and RHAMM are the best characterized receptors to date as modulators in cancer. CD44 is a single-pass, glycosylated class-I transmembrane protein involved in multiple cellular functions, including interaction with the matrix microenvironment and intracellular signaling. The extracellular portion of CD44 primarily binds to glycosaminoglycan HA, thereby contributing to cell adhesion, migration, angiogenesis, inflammation, wound healing, and downstream signaling that promotes cell growth and survival (Ponta et al. 2003; Nikitovic et al. 2015; Karamanos et al. 2021).

Many studies have found a correlation between HA expression and the malignant properties of diverse tumors, and disruption of HA production or receptor interaction decreased tumor growth (Hosono et al. 2007; Stern 2008). In fact, the interaction of HA with its receptors drives to the activation of oncogenic pathways as the MAP kinases and PI3K kinase/Akt inducing survival and cell proliferation, as well as various transport that participate in drug resistance and malignant cell properties (Toole and Slomiany 2008; Bouguignon 2009). Various adaptor proteins such as Vav2, Grb2, and Gab-1 mediate interaction of CD44 with upstream effectors as RhoA, Rac1, and Ras, which drive these pathways (Bourguignon 2008).

7.1.1 HA and Tumor Progression

HA participates in cancer initiation and tumor progression (Chanmee et al. 2016). The increased deposition of HA in tumor tissues is not an inactive process, rather, it triggers signaling events and promotes the association between CD44 and other cell surface receptors, driving to the activation of antiapoptotic pathways mediated by receptor tyrosine kinases that finally promote survival of tumor cells (Bourguignon 2008). Therefore, increase of HA relates to poor patient prognosis and facilitates tumor proliferation, invasion, and angiogenesis among others (Toole and Slomiany 2008). However, contradictory data also exist. In some contexts, accumulation of HA decreases tumorigenicity (Chanmee et al. 2016), while the expression of hyaluronidases, enzymes that degrade HA, can stimulate dissimilar tumor behavior (Stern and Jedrzejas 2006).

HA Oligosaccharides (oHA) were reported to have suppressive effects on various malignant tumors via disruption of the interactions between HA and receptors. oHA has been shown to suppress PI3K/Akt signaling pathway and module cell behavior as growth, cell survival, and expression of multidrug-resistance genes. In some lymphoma model as well as in breast carcinoma, oHA were able to induce apoptosis and reduce tumor growth in vivo (Russo et al. 2008). Besides, it was reported that oHA disaccharides suppressed progression of bone metastasis in breast cancer (Urakawa et al. 2012).

Contrary, HMW-HA or native HA contributes to flexibility of extracellular matrices (Solis et al. 2012) which is important for regulating cell trafficking. Moreover, HA contributes to tumor resistance in normal tissues by anti-inflammatory and antiproliferative effects (Cowman et al. 2015). HMW-HA present in either the peri-tumoral stroma or tumor parenchyma possess several functions that favor tumor growth by protection from apoptosis or inhibition of autophagy, not only by stopping the synthesis of autophagosomes but also inducing their degradation (Kuang et al. 2007). Moreover, LMW-HA can activate signaling cascades that promote proliferation, cell migration, neo-angiogenesis, immune cell influx, and mesenchymal cell trafficking (Cyphert et al. 2015). LMW-HA also attracts macrophages, which in the microenvironment, polarize into subpopulation M2 that protect tumor cells from adaptive immune cell killing (Kuang et al. 2007). However, the HA MW-dependent actions can be controversial and are associated with the tumor context.

In the next sections, we will discuss into the different studies that demonstrate the participation of HA in tumor progression, making a distinction between solid tumors and those of hematological origin, considering as liquids leukemias and myelomas, which do not produce solid masses, and lymphoma.

7.2 HA in Hematological Malignancy

Hematologic malignancies comprise a complex and diverse group of pathologies, including leukemia, myelomas, lymphomas that affect blood-forming cells. In this section, the features of each pathology will be summarized first, and then, the impact of HA and its main receptors CD44 and RHAMM on the progression of each kind of hematological cancer will be described.

7.2.1 HA and Leukemia

Leukemia is an uncommon and heterogeneous group of diseases characterized by infiltration of the bone marrow, blood, and visceral organs by neoplastic cells of the hematopoietic system (Menghrajani and Tallman 2018). Leukemia is among the ten types of cancer with high number of death in both sexes and all ages, according to The Globocan Cancer Observatory (https://gco.iarc.fr/today/data/factsheets/cancers/36-Leukaemia-fact-sheet.pdf). In general, leukemia is classified into lymphoid or myeloid (depending on the cell line of origin) and acute or chronic (according to the type of evolution), thus forming four main groups: Acute Lymphoblastic Leukemia (ALL), Acute Myeloid Leukemia (AML), Chronic Lymphocytic Leukemia (CLL), and Chronic Myeloid Leukemia (CML).

The ECM of the bone marrow (BM) is the largest anatomical component which contains factors that contribute to adhesion, functionality and even regulation of the hematopoietic stem cells (HSC) (Karantanou et al. 2018). In this way, HA, as part of this ECM, actively contributes to these effects. A decrease in the levels of HA in the BM impairs the ability of microenvironment to maintain normal hematopoiesis, and recruit mesenchymal hematopoietic cells into the medulla in alive (Goncharova et al. 2012; Khaldoyanidi et al. 2014). Similarly, the HA would appear to be critical in the niches of these cells as it would protect them from DNA damage by activating efflux pumps capable of extruding genotoxic compounds (Csoka and Stern 2013). In addition, the viscoelastic properties of HA contribute to the porosity and malleability of the ECM of stem cell niches which is important to resist somatic mutations and mechanical damage (Simpson et al. 2016; Gesteira et al. 2017). However, these properties are also used by tumor cells to establish multidrug resistance (García et al. 2009; Lompardía et al. 2013).

ALL results from the acquisition of mutations in hematopoietic progenitors that confer a proliferative and/or survival advantage and impair hematopoietic differentiation (Ley et al. 2013). Genetics is considered one of the main players in the etiology of acute leukemia (Tebbi 2021). However, other factors, such as age, environment and occupation, infections, radiation, among others, have been implicated in leukemogenesis (Tebbi 2021). Interestingly, recent studies suggested that leukemia results from a type of gene–environmental interaction that can cooperate with a genetic predisposition, not by inducing mutations, but by reprogramming the epigenome to modulate gene expression (González-Herrero et al. 2018). ALL is diagnosed by the detection of equal or higher than 20% of blasts in the peripheral blood or bone marrow. However, this determination may not accurately define the biology of the disease (Menghrajani and Tallman 2018).

ALL is a hematological malignancy originating from B- or T-lymphoid progenitor cells and represents the most common type of cancer in children and adolescents (80% of cases). The 5-year survival rates of these groups of diseases have increased from 10 to 85% in the last 50 years. However, there are a significant number of relapses and patients who do not respond to therapy. In this sense, 20% of pediatric patients die from the disease and this number rises to 50% in adults (Terwilliger and Abdul-Hay 2017). The B-cell subtype (B-ALL) accounts for about 75–80% of ALL cases and it mainly develops in children with a peak incidence of around 2–5 years (Simioni et al. 2021). T-ALL represents 15–25% of all cases of acute leukemia of children and adults, respectively. Despite being the least frequent subtype, it results to be more aggressive than B-ALL and is considered a risk factor for poor prognosis for patients (Vadillo et al. 2018; Follini et al. 2019).

AML constitutes 15–20% of leukemia in children and adolescents and 80% of acute leukemia in adults (Fiegl 2016). There is a complex classification of AML that created by the combination of clinical, morphologic, immunophenotypic, and genetic features (according with WHO).

Both in bone marrow samples and in the serum of patients diagnosed with acute leukemia, HA levels are increased compared to samples of healthy patients. For this reason, HA has been proposed as a prognostic and follow-up marker for early detection of relapse (Sundström et al. 2005, 2010; Anagnostopoulou et al. 2017). CD44 and RHAMM are the HA receptors most studied in the context of ALL. CD44-HA interactions play a key role in the adhesion, homing, and migration of leukemia initiating cells in bone marrow niches (Zöller 2015; Schepers et al. 2015). Moreover, the ability of CD44 to interact with different components of the ECM is used by leukemic cells to maintain their niches favoring disease relapse and chemoresistance (Hanke et al. 2014; Zöller 2015; Izzi et al. 2017). Indeed, clinical experience indicates that high levels CD44 are associated with a poor prognosis and high relapse rates in AML patients (Hanke et al. 2014). In line with this, it was suggested that CD44 binds to HA promoting an inside-out activation of VLA-4 in AML cells, causing clustering and stabilization of the integrin, leading to the interaction with VCAM-1 in stromal cells. This AML cell–stromal cell interaction triggers survival signaling involving Akt, MAPK, and NF-kB pathway activation (Gutjahr et al. 2020). NOTCH1 plays a crucial role in T-ALL pathogenesis, as illustrated by the fact that over 60% of human T-ALLs contain gain-of-function NOTCH1 mutations that lead to ligand-independent NOTCH1 signaling (García-Peydró et al. 2018). In human T-ALL xenografts, it was demonstrated that CD44 is a direct NOTCH1 transcriptional target that mediates crucial cell interactions with the

BM microenvironment that result in preleukemic engraftment and further support T-ALL LIC (Leukemic Incited Cell) activity and disease progression (García-Peydró et al. 2018). For this reason, targeted therapy against CD44 aiming to block its interaction with ECM or decrease its expression in tumor cells has been studied (Gul-Uludağ et al. 2014; Vey et al. 2016; Amanzadeh et al. 2017). However, it is well known that different sizes of HA display different biological effects, it has been observed in B-precursor leukemia cells with high surface CD44 expression, that high levels of ultra LMW-HA (<10 kDa) trigger necrosis (Kasai et al. 2017).

Besides, RHAMM is considered a leukemia-associated antigen (Greiner et al. 2002) and it was proposed as an important molecular target for design of therapeutic vaccines in patients with AML (Schmitt et al. 2008; Willemen et al. 2016). A recent study showed that acute leukemia patients with high percentages of RHAMM-positive blasts had more postinduction blasts, blasts in minimal residual disease, and poorer prognosis (Shalini et al. 2018). However, little is known about the implication of this receptor in the biological mechanisms that contribute to the progression of the disease. Similarly, CD38 has become an interesting target for the treatment of ALL and HA its ligand (Lato et al. 2021). Daratumumab, a human monoclonal antibody that binds specifically to CD38, in addition to standard chemotherapy, is under investigation in a phase II trial for pediatric and young adult participants with relapsed and/or refractory T- or B-cell ALL (Clinical Trials. gov identifier: NCT03384654) (Lato et al. 2021). However, the implication of its interaction with HA in the progression of acute leukemia is an empty field for future research.

CLL is the most common leukemia in older patients. TP53 aberrations were associated with aggressiveness and resistance to therapy. For that reason, different first-line therapeutic strategies are used depending on the presence of mutation in TP53, as well as physical fit of patients (Hallek et al. 2018). However, the microenvironment also enhances CCL progression.

Certainly, *ex vivo* assays have established that CLL cells require immune or stromal cells in order to not die by apoptosis, showing the importance of CCL cells and their milieu interaction for leukemic cell survival (Collins et al. 1989). Moreover, HA and its receptors, CD44 and RHAMM, participate in CLL progression. Both CD44 and RHAMM were found to be increased in patients with CLL and have been associated with a worse prognosis, which is why they have been proposed as therapeutic targets (Giannopoulos et al. 2009; Tabarkiewicz and Giannopoulos 2010; Gutjahr et al. 2015). It was described that HA-RHAMM interaction, in presence of IL-8, is able to induce CCL cells migration (Till et al. 1999). Likewise, activation of CD40 enhances CD44 adhesion to HA promoting leukemic cells retention in lymph nodes and enhancing their proliferation and survival (Girbl et al. 2013). Moreover, CD44 is able to complex with key prognostic factors of CLL (CD38 and CD49d) providing a possible nexus between prognosis and leukemic biology (Gutjahr et al. 2015).

On the other hand, CML is a myeloproliferative neoplasm. It is characterized by the presence of Philadelphia chromosome, a product of the reciprocal translocation between chromosomes 9 and 22 (Apperley 2015). This aberration gives rise to a fusion gene called BCR-ABL which encodes a constitutively activated kinase capable of activating multiple signaling pathways initiating the leukemogenic process due to the clonal expansion of malignant cells. Although tyrosine kinase inhibitor (TKIs) are very effective drugs and significantly increase the mean survival of patients, in many cases, the prolonged use of them leads to the selection of resistant leukemic cells by BCR-ABL-dependent mechanisms (gene amplification, protein overexpression, mutations in the active site of the enzyme) as well as by independent mechanisms (such as Pgp, PI3K, and ERK activation) (Bavaro et al. 2019). However, not only these stimuli favor drug resistance, but also the interaction with the bone marrow milieu enhances CML progression. In this sense, Graham et al. 2002 described that the leukemic stem cells of CML patients were resistant to imatinib and this resistance would be based on the niche in which they are found, which prevents the entry of drug, as well as provides survival signaling (Graham et al. 2002; Schepers et al. 2013).

Regarding HA, it also plays an important role in the progression of CML. In patients with therapeutic failure, after prolonged treatment with imatinib, a gelatinous transformation of the bone marrow characterized by high levels of HA has been described (Hong et al. 2010). In agreement, in vitro studies demonstrated that CML cell lines synthesize HA which interacts with CD44 and RHAMM, triggering PI3K/Akt and MEK/ERK signaling pathways avoiding the induction of senescence and favoring resistance to vincristine and imatinib (mediated by PI3K and Pgp activation) (Lompardía et al. 2013, 2019). It is worth to note that such proteins are involved in BCR-ABL-independent resistance to imatinib. Furthermore, the use of HA oligosaccharides (which are capable of binding to receptors without crosslinking them) allowed the sensitization of CML cell lines to imatinib by inducing apoptosis, as well as senescence through PI3K inhibition (Lompardía et al. 2016). In the same way, the inhibitor of HA synthesis, 4-methylumbelliferone, synergizes with imatinib inhibiting CML cell lines proliferation and enhances the induction of senescence (Lompardía et al. 2017, 2019). Considering the gelatinous transformation of bone marrow observed in patients with therapeutic failure and the in vitro findings described, it could be suggested that HA participates in resistance to the first-line drug. This appreciation is in concordance with in vitro studies which show that HA treatment was able to counteract the antiproliferative and pro-senescent effect of imatinib (Lompardía et al. 2019). In the same way, CD44 and RHAMM expression have been associated with worse prognostic and TKI resistance in CML patients (Greiner et al. 2002; Hu and Li 2016; Zhou et al. 2017). Therefore, reducing HA levels or mitigating its effects blocking CD44 and RHAMM would be a promising therapeutic approach to improve current therapy.

Considering that the bone marrow niche provides a sanctuary for leukemic stem cells since it favors immune and chemotherapeutic resistance, the looking for improvement in the leukemic therapy could result from understanding this complex microenvironment in which HA is a key player. In this way, the high levels of HA in bone marrow and the background described, let us to hypothesize that the leukemic cells are in a microenvironment that feedback their survival, favoring disease progression.

7.2.2 HA and Multiple Myeloma

Multiple myeloma (MM), a plasma cell neoplasia characterized by clonal proliferation of malignant plasma cells mainly in the bone marrow (BM). MM is an aggressive malignancy characterized by the clonal expansion of terminally differentiated B cells in the BM. It is the second most common hematological cancer. MM is clinically defined by increased BM plasmacytosis, serum and/or urine monoclonal immunoglobulin, secretion of free light chains, hypercalcemia, renal insufficiency, anemia, and bone pain due to osteolytic disease (Palumbo and Anderson 2011; Rajkumar et al. 2014).

Plasma cell's (PC) survival depends on different signals from neighboring cells within the BM (Slifka and Ahmed 1998). It was suggested that both the myeloma cells and the microenvironment have undergone alterations as early as during precursor stages of the disease (García-Ortiz et al. 2021). In addition, MM is associated with immune deficiencies, suggesting that the evolution of the disease from a precursor state is related with an immunosuppressive environment that allows tumor growth.

The interaction among myeloma cells and the components of the microenvironment is considered crucial in multiple myeloma pathogenesis. Myeloma cells make use of the supportive surrounding stromal cells, osteocytes, and endothelial cells for their own growth. Myeloma precursor cells have been shown to mediate progressive growth using in vivo experimental models that establish the relationship between the microenvironment and tumor signals in regulating tumor growth (Das et al. 2016). Adhesion molecules, cytokines, and the ECM play a critical role in the interplay among genetically transformed clonal plasma cells and stromal cells, leading to the proliferation, progression, and survival of myeloma cells.

The bone marrow microenvironment offers a structure, the ECM, which acts as a feeder and support for myeloma progenitor cells. ECM is constituted by fibronectin, collagen, osteopontin, HA, and laminin. Myeloma cells binding into ECM components proved to be important for survival and drug resistance (Katz 2010).

The glycosaminoglycan HA is a critical component of the hematopoietic microenvironment as mentioned above. It is possible that HA binds either growth factors or other ECM components. HA-expressing stromal cells create hematopoietic foci and act as feeder of hematopoietic cells in vitro (Goncharova et al. 2012). Moreover, B cells from MM patients showed HA matrix around the cells.

CD44/HA interactions are implicated in the regulation of homing of normal HSC and malignant cells into BM (Avigdor 2004; Krause et al. 2006). While myeloma cells migrate to the BM (Cook et al. 1997), their interactions with the microenvironment may significantly influence disease progression. CD44 and RHAMM are the major HA receptors of multiple myeloma cells. CD38, another hyaluronic acid interacting protein, is also expressed in high levels in myeloma cells (Costa et al. 2019). On the contrary, its expression is low in other lymphoid and myeloid cells making it an effective target for immunotherapies.

Vincent and Mechti demonstrate that IL-6, a growth factor for myeloma cells, greatly increases CD44 gene expression. The authors shown that IL-6 modulates CD44 RNA alternative splicing and induces the overexpression of all CD44 variant exons. As IL-6 secretion, induced from bone marrow stromal cells by myeloma cells, is mediated through direct cell-to-cell communication involving CD44 adhesion molecules. As consequence, it was suggested that a CD44/IL-6 amplification loop plays a crucial role in myeloma cell survival (Vincent and Mechti 2004).

In addition, adhesion of myeloma cells to the ECM increases angiogenesis, and the expression of adhesion molecules such as VLA4, LFA1, and CD44 are implicated in this event (Vacca et al. 1995). In some cases, the myeloma cells are able to degrade basement membrane and ECM. They must also reduce their affinity to ECM and cells, which are specific to the bone marrow, soon after upregulate migratory proteins. Subsequently, they alter their adhesion properties to extravasate and migrate through the ECM. Besides, in myeloma, plasma cells decrease expression of CD56 while increasing expression of certain CD44 isoforms that are important for proliferation and motility (Dahl et al. 2002). Finally, myeloma cells must degrade the basement membrane to allow passage through gaps of endothelial cells, and when arriving into the circulation, they may penetrate through the vasculature and form tumors in organs. Extramedullary growth is a feature of advanced MM but circulating tumor cells can also be detected even in early stages of MM.

The role for CD44 as a promoter of cell adhesion-mediated drug resistance has been described in several cancers, and it may in part mediate dexamethasone resistance in myeloma (Ohwada et al. 2008).

Lenalidomide therapy is used in frontline, relapsed/refractory, and maintenance settings for multiple myeloma. Lenalidomide-resistant models were found to overexpress the HA. In addition, resistant cells were more adhesive to bone marrow stroma and HA-coated plates. Blockade of CD44 with monoclonal antibodies, free HA, or CD44 knockdown reduce cellular adhesion and sensitized to lenalidomide by affecting Wnt/β-catenin signal. Bjorklund et al. showed a strong association between CD44 expression and clinical resistance to lenalidomide, suggesting that CD44 levels could serve as biomarkers of lenalidomide resistance (Bjorklund et al. 2014).

Inhibitions of the interactions of MM cells with the BM represent an interesting therapeutic strategy in MM (Bjorklund et al. 2019). The development of new tumor therapies uses HYAL as an adjuvant, agent to depolymerize HA from the ECM. As consequence of the degradation of HA, the absorption of the chemotherapeutic agent of choice is favored. To that end, the administration of Darzalex uses daratumumab plus human recombinant HYALs to treat multiple myeloma (Kaul et al. 2021).

7.2.3 HA and Lymphoma

Lymphoma is a group of hematological malignancies that begins in cells of the lymphoid system. The two main types are Hodgkin lymphoma (HL) and non-Hodgkin lymphoma (NHL). The majority of lymphoma cases derive from the

B cell lineage and 90% of them originate from different stages of maturation and differentiation during their transit through the germinal center (GC) (Hamel et al. 2012). HL and different subtypes of NHL, such as Diffuse Large B Cell (DLBCL), Burkitt's Lymphoma (BL), and Follicular Lymphoma (FL), derive from this compartment (Klein and Dalla-Favera 2008).

It has been observed that an increase in the HA content in patient biopsies would correlate with high-grade lymphomas (Bertrand et al. 2005), indicating that different types of lymphoma cells produce dissimilar levels of HA and this fact would be related to their malignancy. Additionally, it has been demonstrated that lymphoma cells synthesize this GAG (Funamoto et al. 2002; Kuwabara et al. 2003). Among HA receptors, CD44 has been proposed as a diagnostic marker, as well as a therapeutic target in certain types of lymphomas (Ma et al. 2008; Eberth et al. 2010). In murine NHL lymphoma cell lines, the interaction of HA with CD44 would activate the NF-kB pathway and this would correlate with an increase in MMP-9 activity, which would favor cell invasion (Alaniz et al. 2004) Moreover, the inhibition of HA synthesis by 4-methylumbelliferone (4-MU) suppresses MMP-9 activity on lymphoma cells (Nakamura et al. 2007). On the other hand, blocking CD44-HA interaction with HA oligosaccharides inhibits the activation of the PI3K/Akt pathway and modulates efflux pumps sensitizing lymphoma resistant cells to chemotherapy treatment (Russo et al. 2008). In contrast, exogenous LMW-HA might increase intracellular doxorubicin accumulation in CD44 positive lymphoma cells by the modulation of ABC drug transporters expression involved in drug efflux in hematopoietic malignancies (Vitale et al. 2018). These data suggest that HA–CD44 interactions also play a key role in invasion and chemoresistance events in lymphoma and that different sizes of HA may exert dissimilar biological effects on lymphoma progression.

In cHL, HRS cells actively modulate their microenvironment to support their survival and proliferation, and to create an immunosuppressive environment (Weniger and Küppers 2021). In this context, it was suggested that HRS cells activates endothelial cells to increase HA levels in the perivascular zone which facilitate the recruitment of naïve T cells into lymph nodes through CD44-HA adhesive pathways (Fhu et al. 2014). Moreover, it was demonstrated that a downregulation of CD44 after inhibition of NF-kB signaling may decrease cHL cells adhesion to HL-associated fibroblast, highlighting the role of this receptor in with chemotherapy resistance (Celegato et al. 2014).

The literature regarding the role of HA in lymphoma progression is scarce. However, the interplay between HA and CD44 represent an interesting starting point to delve further in the studies of the biology of this GAG in the context of lymphoma and to investigate its potential as a new therapeutic target for this disease.

7.3 Biological Effects of HA from ECM in Solid Tumors: Carcinomas, Sarcomas, and Gliomas

Malignant transformation of epithelial cells results in the development of carcinomas, being the most frequent solid tumors. Sarcomas arise from bone and connective tissues, being the osteosarcoma the most common malignancy of bone cells. Gliomas, the primary brain tumors, are caused by decontrolled growth of glial cells, which are the supporting cells of neurons. In this section, we describe and discus the role of HA in the most frequent solids tumor mentioned.

7.3.1 Carcinomas

7.3.1.1 HA and Breast Cancer

After lung cancer, breast carcinomas are the second most leading cause of mortality in females and correspond to the most frequently diagnosed malignancy, affecting 2.3 million new cases each year. Hyaluronic acid (HA) is known to be involved in breast cancer progression to metastasis (Toole 2004; Eberth et al. 2010). In breast cancer, a dramatic HA increase is observed, and HA levels have been suggested to predict clinical outcome. Even more, it has been observed that high levels of HA, and in particular small HA oligosaccharides, are associated with poor prognosis (Auvinen et al. 2000) and have been detected that invasive breast cancer cells synthesize and accumulate larger amounts of HA compared to normal breast epithelial tissues (Li et al. 2007). In line with these antecedents, Corte et al. have compared the HA accumulation in early and later stage breast tumors, specifically in ductal carcinoma in situ (DCIS), DCIS with microinvasion and invasive carcinoma, for determining if altered HA production is linked to invasion events in breast cancer. They demonstrated that HA levels are significantly increased in late-stage DCIS compared to early-stage, suggesting a key role for this glycosaminoglycan in the early invasive stage of breast carcinomas (Corte et al. 2010). These antecedents added to the key role of HA in modulating the inflammation, angiogenesis and fibrosis in breast cancer (Heldin et al. 2013), processes involved in its progression to metastasis (Toole 2004). Thus the regulation of HA metabolism will be a potential therapeutic target for the treatment of breast cancer. Numerous studies have demonstrated that inhibition of HA synthesis reduces breast cancer tumor cell proliferation and migration (Kultti et al. 2009; Urakawa et al. 2012; Brett et al. 2018). Brett et al. showed that the invasive and metastatic potential of breast carcinoma cells correlates to the presence of a HA-rich pericellular matrix, and that inhibition of HA synthesis significantly inhibits carcinoma cell extravasation and invasion (Brett et al. 2018). Kultti et al. showed that the growth of MCF-7 cells is sensitive to inhibition of HA synthesis, and that inhibiting HA production inhibits migration of the noninvasive MCF-7 breast cancer cells (Kultti et al. 2009). Even more, Urakawa showed that suppression of HA synthesis and accumulation by 4-MU inhibited cell proliferation, motility and invasiveness, and induced apoptosis in the highly aggressive MDA-MB-231 breast cancer cell line (Urakawa et al. 2012). Furthermore, inhibition of HA production or its degradation was shown to improve drug delivery in breast cancer (Shpilberg and Jackisch 2013; Kohli et al. 2014; Clift et al. 2019). In this sense, has been shown that the use of hyaluronidases to degrade HA in the TME improves drug delivery of antibody-based trastuzumab-targeted therapy in HER2+ breast cancer (Shpilberg and Jackisch 2013). Similarly, Kohli et al. showed a more heterogeneous distribution of Doxil and reduced tumor growth by inhibiting of HA synthesis (Kohli et al. 2014). Even more, a recent study has shown that HA degradation by pegrhyaluronidase allows to remodel the TME in a murine model of breast cancer increasing the uptake of anti-Programmed Death-Ligand 1 (PD-L1) therapeutic antibody and reducing the tumor growth (Clift et al. 2019). It has also been shown that HA allows to overcome the pro-apoptotic effects of SP-D surfactant protein D and can be used as an escape mechanism in breast cancer. In this sense, when a possible interaction between SP-D, with immune surveillance function against tumor cells, and hyaluronic acid, was studied, it was shown that, in the presence of HA, rfhSP-D is incapable of inducing apoptosis in triple positive breast cancer cell lines that overexpress HER2. In addition, HA-bound rfhSP-D was able to restore cell growth, suggesting that these breast cancer cells may use HA as an escape mechanism to overcome pro-apoptotic effects of surfactant protein D rfhSP-D (Murugaiah et al. 2020). Even more, Witschen et al. found that production of the chemokine CCL2 by breast cancer cells was significantly decreased after depletion of either CD44 or HA. In vivo, they found that CD44 deletion in breast cancer cells resulted in a delay in tumor formation and localized progression. This finding was accompanied by a decrease in infiltrating CD206+ macrophages, which are typically associated with tumor promoting functions (Witschen et al. 2020). Furthermore, a recent study has shown the importance of altering HA metabolism to malignant progression of breast carcinoma. Arnold et al. proved that increased HA synthesis requires reprogramming of the glucose metabolism and that disrupting of this metabolic reprogramming through the genetic and pharmacological depletion of the HA precursor UDP-glucuronic acid significantly reduces cellular invasion and colony formation in vitro in MDA-231 cells. Even more, results of in vivo mesenchymal-like breast cancer models revealed a reduction in tumor growth and metastasis (Arnold et al. 2020).

These data suggest that HA plays a key role in proliferation, invasion, and chemoresistance events in breast cancer and that it does so through different mechanisms.

7.3.1.2 HA and Ovarian Cancer

Epithelial ovarian cancer (EOC) is the commonest cause of gynecological cancer-associated death and the fourth cancer-related deaths in women in the developed world (Jayson et al. 2014). Approximately 90% of all ovarian cancers are epithelial and of these ~80% represent the serous histological subtype (Hiltunen et al. 2002). Conventional ovarian cancer treatment comprises first-line chemotherapy with a carboplatin-paclitaxel regimen and cytoreductive surgery (Marchetti et al. 2010; Vaughan et al. 2011). However, although this regimen is initially effective, there is a high recurrence rate and most ovarian cancer patients suffer at least one relapse within 12–18 months (Colombo et al. 2006, 2017; Kartal-Yandim et al. 2016; Yousefi et al. 2016) owing to chemoresistance and metastasis (Lengyel 2010; Amoroso et al. 2017). Increased expression of HA has been shown to be closely correlated with the epithelial ovarian cancer (EOC) degree and metastatic potential (Anttila et al. 2000; Ween et al. 2011), with a 100-fold increase in HA expression in grade three EOC (Hiltunen et al. 2002). Different antecedents suggest that targeting the synthesis or accumulation of HA represents a promising therapeutic approach for the treatment of ovarian cancer. Lokman et al. showed that inhibition of HA production by 4-MU in combination with chemotherapy with carboplatin (CBP) significantly reduced chemo-resistant serous ovarian cancer cells survival and spheroid formation and increased apoptosis of this cells compared to CBP alone. Even more, showed that 4-MU alone or combined with CBP significantly decreased the invasion ability of these cells in vivo compared to control treatment. The fact that the inhibitor of the synthesis of HA 4-MU allows to overcome the chemoresistance to the chemotherapeutic drug CBP and that they determined that the production of HA increases in ovarian cancers resistant to chemotherapy indicates that the inhibition of HA is, therefore, a promising new strategy to overcome chemoresistance and improve ovarian cancer survival (Lokman et al. 2019). Even more, the reduction of HA-pericellular coat was related with the inhibition of cell migration, proliferation, and invasion (Anttila et al. 2000). On the other hand, clear-cell ovarian cancers, an epithelial ovarian cancer subtype, show a spherule-like mucoid stroma with a hollow acellular space. Despite the absence of stromal cells, both the mucoid stroma and hollow spheroids contain abundant ECM, mainly composed of HA that plays a crucial role in the formation of that structures and tumor progression. In this sense, Kato et al. showed that when HA synthesis was inhibited by 4-MU in HAC-2, a clear cell carcinoma cell line, the spherule-like accumulation of HA, or hollow spheroids were not observed. Moreover, the inhibition of HA synthesis was associated with the reduction of cell growth (Kato et al. 2016). Even more, Lin et al. found that upregulation of UDP-glucose dehydrogenase (UGDH) is related to ovarian cancer metastasis and that knockdown of UGDH significantly decreased wound healing and migration and proliferation in three ovarian cancer cell models, TOV21G, A2780, and HeyA8 cells. Furthermore, in an in vivo xenograft model from TOV21GHI cells, they demonstrated that UGDH removal significantly decreased ovarian cancer tumor growth (Lin et al. 2020). Given that the UDP-glucose dehydrogenase converts UDP-glucose to UDP-glucuronic acid, a precursor of several GAGs and PGs presents in the ECM, as well as cell proliferation and migration, it is likely that the elimination of UGDH decreased HA levels in ovarian cancer and thus affected the metastatic ability of ovarian cancer. However, further studies are needed to demonstrate the biological effects of HA from ECM in ovarian cancer.

7.3.1.3 HA and Prostate Cancer

Patients with clinically localized prostate cancer (CaP) are often treated with radical prostatectomy or radiation with curative intent. However, the disease recurs in a substantial number of the patients and becomes hormone refractory.

Lipponen et al. found high levels of HA in the stroma of prostate cancer and showed that stromal HA accumulation is related to poor prognosis in prostate cancer (Lipponen et al. 2001). Posey et al. showed that in radical prostatectomy specimens, HA expression is elevated in CaP tissues, but it is not an independent predictor of biochemical recurrence (Posey et al. 2003), besides has been observed that HA and HYAL-1 expression in prostate biopsy specimens and correlated in with disease recurrence (Gomez et al. 2009). Lokeshwar et al. studied the effects of inhibition of HA synthesis by 4-MU on five prostate cancer cell lines, LNCaP, DU145, PC3-ML, LAPC-4, and C4-2B. They demonstrated that 4-MU inhibited the proliferation of all cell lines by inducing apoptosis, in a dose-dependent manner, and that this effect of 4-MU on cell growth and apoptosis was due to the inhibition of HA synthesis. Even more, they demonstrated that 4-MU inhibited invasive activity and chemotactic motility of PC3-ML and DU145 cells by inhibiting HA production. In PC3-ML xenograft mice model, oral administration of 4-MU significantly inhibited tumor growth and reduced microvessel density in tumors from 4-MU-treated animals (Lokeshwar et al. 2010). In a transgenic adenocarcinoma of the prostate model, oral administration of 4-MU also significantly decreased microvessel density and prostate cancer cells proliferative index. Furthermore, daily gavage of 4-MU inhibited tumor growth in a DU145 subcutaneous xenograft model and inhibited skeletal metastasis in the jaw, pelvis, femur, and spinal cord in a PC3-ML intracardiac bone metastasis model. Even more, they showed that all these effects were the result of inhibition HA synthesis and signaling (Yates et al. 2015). Different studies showed that, in addition to the accumulation of HA in the tumor stroma, the alteration of hyaluronic acid synthase and hyaluronidase in tumor epithelial cells are associated with increased cell proliferation, invasion, metastasis, and poor outcome in men who have undergone radical prostatectomy (Simpson and Lokeshwar 2008; Bharadwaj et al. 2009).

7.3.1.4 HA and Pancreatic Cancer

Pancreatic cancer is one of the most difficult conditions to treat, although it only accounts for 4% of all cancers; 5-year survival is less than 10% in patients with the disease and has remained relatively unchanged over the past 25 years (Siegel et al. 2016). Most patients present with locally advanced or metastatic disease, and such individuals have a grim median survival of 6–10 months, and 3–6 months, respectively. Although 10–15% of patients have potentially rejectable tumors, the recurrence rate of the disease following surgery is high (Wong and Lemoine 2009). The vast majority (90%) of pancreatic cancers (PC) are tumors originating from pancreatic ductal cells (Sener et al. 1999; Sharma et al. 2011).

The most common subtype of human pancreatic malignancies is Pancreatic Adenocarcinoma (PDAC). This type of cancer originates in the ductal epithelium and evolves from premalignant lesions to fully invasive cancer. It has been established that PDAC is preceded by the evolution of precursor lesions called pancreatic intraepithelial neoplasia (PanIN 1A/B, 2, and 3) (Hruban et al. 2001), and, under certain conditions, Acinar Ductal Metaplasia (ADM) might be critical for the development of PanIN lesions (Means et al. 2005). The progression from histologically normal ductal epithelium to low-grade PanIN to high-grade PanIN is associated with the accumulation of specific genetic changes (Guerra et al. 2007; Hidalgo 2010). It has also been reported that elevated expression of autophagy in cancer cells has been implicated in the development of PDAC (Rui Kang 2012).

PDAC is characterized by a dense stromal matrix also referred to as TME. This stomal matrix starts to evolve early around PanIN lesions. Moreover, in established carcinomas, the TME constitutes up to 90% of the tumor mass. PDAC TME is composed by various cellular and acellular components. Between the acellular components, large amounts of collagen, fibronectin, proteoglycans, and HA are present (Provenzano et al. 2012; Jacobetz et al. 2013). Particularly, HA is consistently increased in pancreatic cancer stroma, where exerts a pro-tumoral action (Sato et al. 2016), including tumor growth (Toole and Hascall 2002), cell proliferation, invasion (Cheng et al. 2016), and metastasis (Zhang et al. 1995; Toole 2002). These actions, associated with tumor progression, require that HA binds to its specific receptors. The main binding receptor, CD44 is located on the cell surface, and the intracellular RHAMM (receptor for hyaluronan-mediated motility) in the cytoplasm, where it activates the PI3K/Akt and ERK1/2 signaling pathways (Zhu et al. 2013; Lokeshwar et al. 2014). Moreover, CD44 has been recognized as a cancer stem cell marker of PDAC, involved in both epithelial mesenchymal transition and multidrug resistance; mechanisms involved in cancer cells protection from chemotherapeutic agents (Zhang et al. 2012; Wei et al. 2013). The secretion of HA by tumor cells facilitates the action of cancer-associated fibroblast. These cells produce growth factors, hormones, and cytokines to stimulate HA production by PDAC cells, which stimulates and promote their malignant behavior. As a result, the interaction between HA and its receptors provides tumor cells the suitable microenvironment to survive, proliferate, and invade. HA not only works as a tumor promoting factor, but also provides a barrier to the arrival of chemotherapeutic agents. In this sense, the accumulation of HA in TME increases the interstitial fluid pressure (IFP), which makes drug perfusion very difficult (Dufort et al. 2016a, b). It was described that hyaluronidase treatment can rapidly reduce the IFP in a mouse model of PDAC and significantly improve the appropriated response to gemcitabine (Provenzano et al. 2012). It is correct to think that reducing the level of HA in the ECM may offer an effective approach to treating PDAC.

It is attractive to presume that the inhibition of HAS individually with specific antibodies or the use of promoters of each hyaluronidase would be an adequate strategy to inhibit tumor progression and induce remission. However, the expression of these enzymes in pancreatic cancer cells is heterogeneous (Cheng et al. 2016), suggesting that this strategy could not result in a successful response in most patients with PDAC. Rather, the use of compounds that reduce the synthesis of HA could be suitable as a strategy of HA reduction. Among them, 4-methylumbelliferone (4-MU), which stably suppresses the synthesis of HA in vitro and in vivo, has been widely accepted because of its mechanism of action and because it is innocuous to organisms (Kudo et al. 2017). Two therapeutic strategies were proposed: the combination of 4-MU with gemcitabine, the first-line chemotherapeutic agent used for decades, and decombination of 4-MU with immunotherapy. For the first option, it was described that 4-MU inhibited the formation enveloped HA-rich matrix, promoting the perfusion of gemcitabine (Nakazawa et al. 2006). In another study, in a xenograft mouse model, electron microscopic observation revealed that 4-MU reduced the amount of HA in tumors surrounding, on the backs of animals and altered the intercellular space (Yoshida et al. 2016). Moreover, 4-MU orally administrated, enhanced the survival rate of mice that were intraperitoneally inoculated with human PDAC cells. In these mice, intratumoral HA level was reduced in a 30% compared to control mice (Yoshida et al. 2016). This result is really encouraging, because it is supposed that orally administered 4-MU is distributed throughout the whole body, assuming that the beneficial effect would be not only on the primary tumor but also on distant metastases, causing suppression of proliferation, migration, and invasion activities.

The other therapeutic strategy involves the combination of 4-MU and immunotherapy. Within the last, immune checkpoint inhibitor therapy (targeting CTLA4, PD1, PDL1) or the perfusion of immune cells have not been remarkably successful against PDAC probably because of the strong and extensive ECM rich in HA. In this context, HA plays an integral role in this immune evasion. As explained above, the reduction of HA in the tumor ECM reduces the IFP, promoting drug perfusion as well as the arrival of immune cells. It was described that the administration in vivo of 4-MU reduced intratumoral HA level and promoted infiltration of inoculated γδ T-cells into tumor tissue, with the consequent suppression of tumor growth (Suto et al. 2019). However, further research is needed to achieve an adequate potential combined therapy. But it is promising that the control of HA will become an adjuvant treatment that can be used in combination with different immunotherapies, enhancing their efficacy through the remodeling of the ECM around cancer cells.

7.3.1.5 HA and Colorectal Cancer

Colorectal cancer (CRC) is the third most common cancer and the fourth most common cancer cause of death globally (Brenner et al. 2014). The global burden of CRC is projected to increase by 60% within 2030, especially in countries with medium and high human development index (HDI) scores (Wong et al. 2021). CRC originate primarily from genetic and epigenetic alterations in epithelial cells, leading to the development of benign polyposis, develops into adenoma and then carcinoma, and finally the invasion of the interstitial matrix. Those processes require the degradation and modification of the existing ECM (Onfroy-Roy et al. 2020).

Several studies have shown the importance of HA in this type of cancer. HA levels are higher in cancer tissues than in noncancerous tissues. The stromal tissue stains highly intensely and the tumor epithelium is mostly HA negative, and this feature is associated with tumor grade, thus high-grade tumors have strong HA intensity. Furthermore, patients with strong HA intensity in tumor epithelium have a low survival rate (Wang et al. 1996; Ropponen et al. 1998). Also, cytosolic HA levels are correlated with clinical tumor stage, thus, patients with more advanced tumors have higher HA levels. However, HA content is not correlated with age, sex, tumor location, or histological grade (Llaneza et al. 2000). Zhang et al. evaluated the molecular weight (MW) of abnormally elevated HA in cancer tissues. HA MW was approximately 6 kDa and it was positively correlated with metastasis and invasion of CRC. Besides, 6-kDa HA was found elevated in sera from CRC patients and they observed a significantly positive correlation between 6-kDa HA expression in primary CRC lesions and matched serum samples from these patients (Zhang et al. 2019).

HA binds to CD44 that is expressed on the plasma membrane of many cells but also binds to other receptors including toll like receptor 4 (TLR4) that is widely distributed in the gastrointestinal tract. HA binding to TLR4 promotes epithelial repair in the DSS-colitis and radiation injury models (Riehl et al. 2012) and mediates the normal growth of the intestine and colon including the proliferation of must repair. b or p? epithelial stem cells (Riehl et al. 2015). However, HA–CD44–TLR4 interactions are involved in pathogenesis of CRC (Kim et al. 2004). Makkar et al. demonstrated that HA binding to TLR4 and CD44 promotes cell proliferation and blocks spontaneous apoptosis. In addition, TLR4 and CD44 knockdown or HA blocking by PEP1 slows tumor growth and reduces tumor volume (Makkar et al. 2019). Even more, the decrease in GAGs synthesis due to UDP-glucose dehydrogenase (UGDH) downregulation and 4-MU treatment reduce CRC cells growth and motility, indicating that invasive ability of CRC cells is HA dependent (Wang et al. 2010).

HA synthases (HAS) are also involved in the development and progression of CRC. The expression of HAS3 was known to produce smaller HA fragments with molecular size between 50 and 1000 kDa, as well as total HA production were found to be increased in metastatic CRC cells when compared with cells isolated from a primary tumor. Furthermore, downregulation of HAS3 in a highly tumorigenic CRC cell line decreased subcutaneous colon cancer growth in a mouse model by increasing apoptosis rate of these cells (Bullard et al. 2003). Even more, Kim observed a correlation between HAS2 levels and malignant phenotypes of CRC cells. HAS2 expression was significantly induced in CRC tissue samples and CRC cell lines. Furthermore, HAS2 depletion increased apoptosis, therapeutic sensitivity, and decreased epithelial to mesenchymal transition-related migration and invasive ability of CRC cells (Kim et al. 2019).

HA may influence metastasis of colon carcinoma cells. Laurich et al. demonstrated that metastatic colon carcinoma cells that express high levels of HAS and synthesize large pericellular HA matrices adhere more avidly to laminin than cells isolated from a primary tumor that secrete and retain less HA (Laurich et al. 2004).

7.3.1.6 HA and Lung Cancer

Lung cancer is the leading cause of cancer death around the world. It arises from the cells of the respiratory epithelium and can be classified into two categories. Small cell lung cancer (SCLC) is a highly malignant tumor derived from cells exhibiting neuroendocrine characteristics and accounts for 15% of lung cancer cases and non-small cell lung cancer (NSCLC), which accounts for the remaining 85% of cases (dela Cruz et al. 2011). The pulmonary ECM provides mechanical stability and elasticity, which are essential for physiological lung function during the breathing, inhalation, and exhalation. Lung-ECM is mainly composed of collagens, proteoglycans, and glycoproteins and serves as a reservoir for several growth factors and cytokines, which are crucial for cell differentiation and proliferation. Lung tumors are surrounded by an extensive stroma and interactions between cancer cells-ECM protects cells from chemotherapy-induced apoptosis (Burgstaller et al. 2017).

The content of GAGs in cancer tissues is higher compared to normal tissues and the amount of these polysaccharides is different in every histologic type (Horai et al. 1981). HA concentration is increased mainly in squamous cell carcinoma, and it is related to the low survival of patients (Pirinen et al. 2001; Rangel et al. 2015). Moreover, sputum from lung cancer patients has higher HA concentration levels compared to cancer-free and healthy people (Rangel et al. 2015). Also, HA receptors are related to the development of lung cancer. RHAMM is upregulated in non-small cell lung carcinomas, especially in metastatic tumors (Wang et al. 2016) and there is an increase in LYVE-1 levels in tumor tissue and periphery related to lymph nodes metastasis (Renyi-Vamos et al. 2005).

HA is produced by tumor cells and stroma cells and several studies have shown the importance of HA in the progression of lung cancer. Song et al. demonstrated that conditioned medium of NSCLC cells increases HA production by lung cancer-associated fibroblasts (LCAFs) and the inhibition of HA production by these cells reduces tumor cells growth and proliferation. Even more, simultaneous silencing of HAS2 and HAS3 and CD44/RHAMM silencing reduced HA production and in vitro proliferation of NSCLC cells (Song et al. 2019).

On the other hand, HA production is regulated by cytokines and growth factors like tumor necrosis factor alpha (TNF-α), tumor growth factor beta (TGF-β), and endothelial growth factor (EGF), produced by tumor cells, fibroblasts, and immune cells (Tammi et al. 2011). Mulshine et al. demonstrated that NSCLC cells treatment with TGF/IL enhance HA production by malignant cells and this increase is related to epithelial to mesenchymal-like transition in NSCLC cells (Mulshine et al. 2010).

7.3.2 HA and Sarcomas: Osteosarcoma, Chondrosarcoma, and Fibrosarcoma

Osteosarcoma (OS) is the most common primary bone tumor and responsible for considerable morbidity and mortality due to its high rates of pulmonary metastasis. The prognosis of OS patients has improved dramatically with the introduction of chemotherapy; however, cases with metastases or unresectable tumors still have a poor prognosis (Arai et al. 2011). Several studies have implicated HA-rich ECM on osteosarcoma progression (Nishida et al. 2005; Tofuku et al. 2006; Hosono et al. 2007). Tofuku et al. found that HA synthesized by HAS3 promoted proliferation, invasion and degradation of ECM, biological functions crucial for metastasis and showed that the inhibition of HA synthesis by 4-MU suppressed proliferation and invasion of LM8 cells in vitro (Tofuku et al. 2006). Nishida et al. showed that the selective inhibition of HAS-2 mRNA in MG-63 osteosarcoma cell line by antisense phosphorothioate oligonucleotides reduced HA production by these cells what disrupted the assembly of cell-associated matrices and, consequently, hindered cell proliferation, motility, and invasiveness (Nishida et al. 2005). In a later study, these same authors examined the effects of exogenously added HA oligosaccharides on tumorigenicity of murine osteosarcoma cells LM-8, highly metastatic in mice, and human osteoblastic osteosarcoma cells, MG-63 and proposed that these small oligosaccharides compete with the binding of high-molecular-mass HA for cell surface receptors such as CD44 as a mechanism to deplete HA-rich matrices from cells. They determined that the treatment of MG-63, which have abundant HA-rich cell-associated matrix, as well as LM-8, with HA oligos inhibited the formation of HA-rich cell-associated matrix, resulting in the inhibition of growth, motility, and invasiveness and induction of apoptotic activity in vitro. Even more, daily application of HA oligosaccharides showed a trend toward inhibition of LM-8 tumor growth in vivo. These data suggest that the abrogation of hyaluronan-rich cell-associated matrices have potent antitumor effects (Hosono et al. 2007). Similarly, Arai et al. showed that the inhibition of HA retention by 4-MU reduced the formation of functional cell-associated matrices in OS cells, and inhibited cell proliferation, migration, and invasion, resulting in the reduction of tumorigenicity and lung metastasis. Even more, although 4-MU showed only a mild inhibitory effect on the growth of the primary tumor in vivo, it markedly inhibited the development of lung metastasis by inhibiting HA retention (Arai et al. 2011). Even more, Gvozdenovic et al. showed that overexpression of CD44 in SaOS-2 cells enhances intratibial primary tumor growth and formation of pulmonary metastases using an intratibial xenograft OS mouse model. These malignancy-enhancing effect of CD44s was HA-dependent and was reflected in the significant increase of primary tumor volume and the numbers of pulmonary micro- and macrometastases when compared to that of control SaOS-2 cells. These results suggest that CD44s/HA interactions play a key role in OS (Gvozdenovic et al. 2013). Even more, it was shown that hyaluronidase increased the transcapillary pressure gradient in OS xenografts by degrading extracellular HA and remodeling ECM, thus improving the uptake and distribution of liposomal doxorubicin (Eikenes et al. 2005). Taken together, these reports highlight the involvement of HA in the progression and metastasis of OS.

Chondrosarcoma, the second most common primary malignant bone tumor, particularly when low-grade, is also characterized by the formation of a HA-rich ECM which has been considered as one explanation of drug resistance (van Oosterwijk et al. 2012). In this sense, Hamada et al. showed that the inhibition of HA synthesis by 4-MU suppressed cell proliferation, migration, and invasiveness of chondrosarcoma cells. Even more, in an in vivo model of chondrosarcoma, the inhibition of HA production markedly inhibited grafted tumor growth (Hamada et al. 2018). Recently, Koike et al. evaluated the effects of a novel hyaluronidase, KIAA1199, on ECM formation as well as antitumor effects on chondrosarcoma. They demonstrated that the KIAA1199 forced expression did not affect proliferation or apoptosis but inhibited migration and invasion of RCS cells in vitro. In contrast, the expression of KIAA1199 significantly inhibited the growth of grafted tumors. Although there was no direct inhibitory effect on proliferation in vitro, induction of KIAA1199 showed the antitumor effects in grafted tumor growth in vivo possibly due to changes in the TME such as inhibition of ECM formation. Together, KIAA1199 could be a novel promising therapeutic tool for low-grade chondrosarcoma, mediated by the degradation of HA (Koike et al. 2020).

Other antecedents have also provided direct evidence for the role of HA in tumorigenicity in fibrosarcoma, another of the most common sarcomas. In this case, Kosaki et al. showed that overproduction of HA by expression of the HAS2 enhances growth of tumors in xenograft models (Kosaki et al. 1999).

7.3.3 HA and Glioblastoma

Glioblastoma (GBM), also known as grade IV astrocytoma by the World Health Organization classification (Louis et al. 2016), is the most frequent primary tumor of the central nervous system (CNS) in adults. It is characterized by fast growth, invasiveness, and high mortality, with a median survival of less than 15 months after diagnosis (Anjum et al. 2017). Currently, there are few therapeutic options (le Rhun et al. 2019; Perus et al. 2019; Strobel et al. 2019), with the first-line therapy being surgical resection and radiotherapy combined with cycles of temozolomide (TMZ). Unfortunately, TMZ therapy causes severe adverse effects, and almost 50% of patients exhibit resistance to the treatment (Anjum et al. 2017; Philteos et al. 2019; Strobel et al. 2019). Furthermore, high inter- and intratumor heterogeneity, individual variability and different stages of disease at diagnosis time complicate GBM treatment (Rajaratnam et al. 2020).

For several years there is a growing interest in studying the CNS's ECM and the mechanisms through which it impacts in the development and progression of brain tumors, with obvious implications for the development of new therapeutic alternatives. In the CNS, HA is the main component of ECM, beside with proteoglycans (such as aggrecan, versican, neurocan, and brevican without collagen), the tenascins and link proteins, and other GAGs, including chondroitin sulfate, heparan sulfate, and keratan sulfate (Eikenes et al. 2005). In the CNS, HA participates in the correct generation, proliferation, and maturation of neural stem cell progenitors (NSCP) during brain development and repair (Su et al. 2019). Reports published in the 70s and 80s established that HA has another, undesirable role: these studies demonstrated that production of GAGs—and particularly HA—by malignant glioma cells was higher than those in normal glial cell lines (Pibuel et al. 2021b), and which was associated with a higher rate of cell proliferation and several publications did confirm a correlation between the addition of HA and an enhanced rate of invasion across multiple glioma cell lines (Chintala et al. 1996; Nakagawa et al. 1996; Radotra and McCormick 1997; Pibuel et al. 2021a).

After that, several approaches have been used in the study of the relationship between HA and GBM including in vitro 2D and 3D assays, in vivo models and patient samples (Pibuel et al. 2021b). In GBM, CD44 is strongly involved in cell invasion (Mooney et al. 2016), and it was demonstrated that optimal levels of this receptor were necessary for GBM cells to generate highly infiltrative tumors in a mouse model (Klank et al. 2017), while the treatment with an anti-CD44 monoclonal antibody inhibited tumor growth of local glioma (Breyer et al. 2000). Regarding RHAMM, it has been associated with an increase in migration and proliferation of glioma cells, and its levels have been correlated with tumor grade (Akiyama et al. 2001; Virga et al. 2017; Lim et al. 2017). Although the importance of CD44 and RHAMM in mediating HA's functions is clear, other receptors such as EGFR in association with the HA receptors could explain some of the effects of HA on GBM progression. It was demonstrated that CD44 through the bind of EGFR form complex, that might provide a mechanism for HA-mediated cell invasion and proliferation (Tsatas et al. 2002).

After these interactions, HA activates several signaling pathways being the most relevant MEK/ERK, c-MYC, PI3K/Akt, FAK, RhoA/ROK, and the transcription factor NF-κB (Pibuel et al. 2021b), enhancing all the malignant features of GBM, mostly proliferation, migration, and invasion but also chemoresistance (Kim et al. 2005). The activation of these signaling pathways increases the activity of MMPs, the CD44 cleavage and in turn an increase of migration and invasion of this tumor type (Chetty et al. 2012).

Regarding the tissue samples of patients, glioma lesion showed HA content much higher than in adult normal brains (Delpech et al. 1993); however, the proportion of HA was inversely associated with glioma grade (Sadeghi et al. 2003). Interestingly, although HA failed as a molecular marker, HAS2 and HYAL-2 were associated with tumor grade and even correlate with patient outcomes (Valkonen et al. 2018; Pibuel et al. 2021b).

Considering the poor outcomes of the patients with GBM, several studies focused its efforts on the development of new therapeutic alternatives and molecular targets. The increases of HA in tumor tissues compared to the healthy brain, discussed above, may not have obvious clear cut diagnostic potential, but they do implicate HA in GBM progression, the results suggest its potential as a therapeutic target. HA degradation has been explored as a therapeutic target in several studies, and some clinical trials have shown that treatment with HYAL as adjuvant chemotherapy contributes to clinically relevant remissions in high-grade astrocytoma and to the efficacy of chemotherapy in pediatric brain tumors (Baumgartner et al. 1998; Martinez-Quintanilla et al. 2015). In addition, it was recently reported that treatment with HYAL, alone or in combination with TMZ, showed a cytotoxic effect on the GSC population (Hartheimer et al. 2019). Interesting results were also obtained with oHA. The oHA treatment improved the effects of both radiation and methotrexate and even TMZ on GBM cells (Maria et al. 2008; Karbownik et al. 2014). Finally, as a new alternative, the use of 4MU showed a marked effect on GBM cells decreasing metabolic activity, cell proliferation, migration, and MMP-2 activity while causing high levels of apoptosis in murine GL26 GBM cells. Moreover, our group showed that 4MU partially diminishes HA synthesis, although several of its effects would be independent of this inhibition (Pibuel et al. 2021a).

Due to this, it is one of the main targets for the investigation is to find alternative therapies that seek to control tumor spread and reduce mortality, one way is understanding the interaction of HA with GBM cells.

In this sense, the mechanical interplay between the recruited cell stroma, the ECM, and the tumor cells has therefore attracted significant attention as a new avenue of therapy, termed stromal therapy.

7.4 Conclusion and Perspectives

To sum up, we could conclude that the ECM component GAG molecule, hyaluronan has tumor specific characteristics, and is an important factor in the development and progression in different hematological and solid tumors. It is involved either in direct or indirect way in all process defined as the hallmarks of cancer (Fig. 7.2).

It is possible to conclude that, molecules involved in HA metabolism and signals could be considered as clinical targets for tumors beyond consideration of the etiology. However, for the clinical application, it is important to consider that the impact of its modulation effect depends on the type of tumor. Thus, its application could be associated with induction of tumor cell death, avoidance of drug resistance to the enhancement of immunotherapy. With this perspective, we have described and discussed the different aspects of the role of HA in the biology and therapy of various tumors in this chapter. Finally, we aimed to support help the strategies for pharmacological application able modulate HA and consequently different processes in cancer progression.

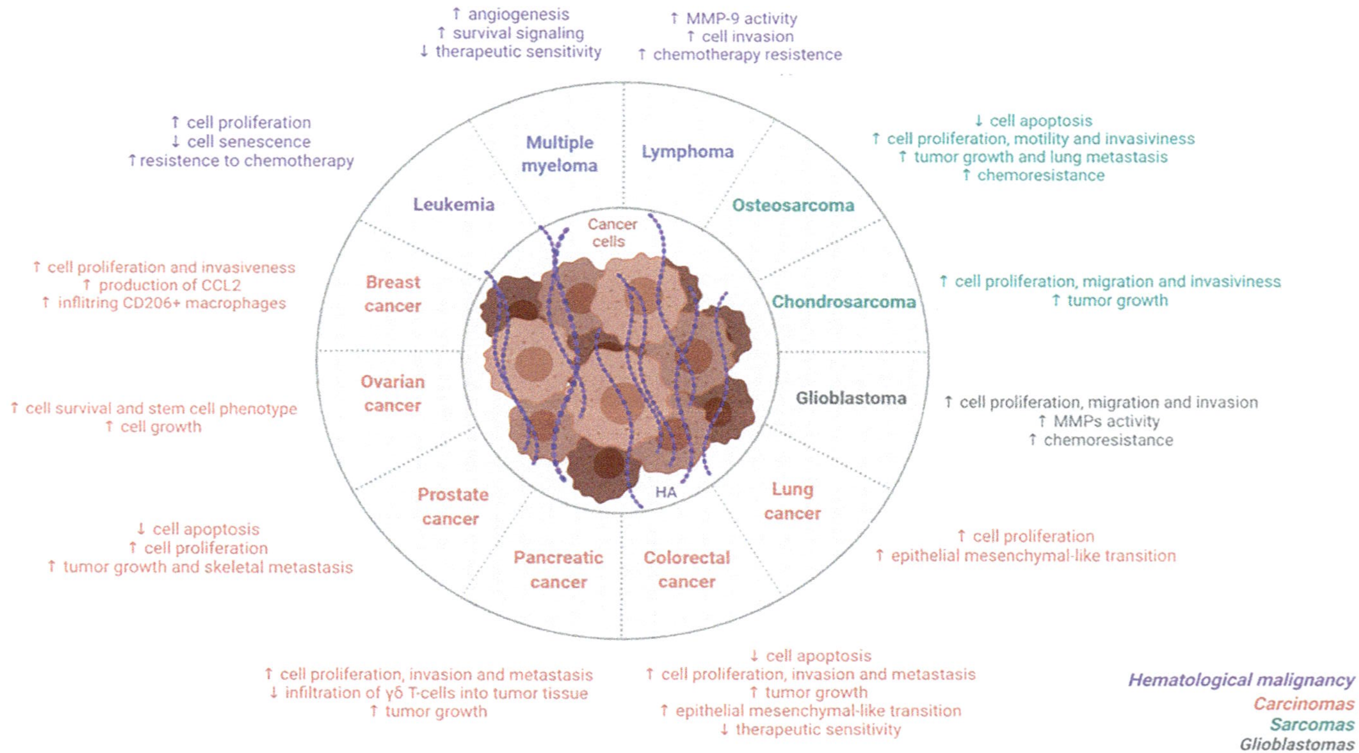

Fig. 7.2 Biological effects of ECM Hyaluronan (HA) in hematological malignancy and in carcinomas, sarcomas and gliomas

References

Akiyama Y, Jung S, Salhia B et al (2001) Hyaluronate receptors mediating glioma cell migration and proliferation. J Neurooncol 53:115–127. https://doi.org/10.1023/A:1012297132047

Alaniz L, García M, Cabrera P et al (2004) Modulation of matrix metalloproteinase-9 activity by hyaluronan is dependent on NF-κB activity in lymphoma cell lines with dissimilar invasive behavior. Biochem Biophys Res Commun 324:736–743. https://doi.org/10.1016/j.bbrc.2004.09.120

Amanzadeh A, Heidarnejad F, Abdollahpour-Alitappeh M et al (2017) Development of high-affinity monoclonal antibody using CD44 overexpressed cells as a candidate for targeted immunotherapy and diagnosis of acute myeloid leukemia. Hum Antibodies 26:7–15. https://doi.org/10.3233/HAB-170315

Amoroso MR, Matassa DS, Agliarulo I et al (2017) Stress-adaptive response in ovarian cancer drug resistance. Adv Protein Chem Struct Biol 108:163–198

Anagnostopoulou E, Papanastasopoulou C, Papastamataki M et al (2017) Serum hyaluronic acid levels are altered in acute leukemia patients: potential prognostic implications. Acta Haematol 138:44–51. https://doi.org/10.1159/000477574

Anjum K, Shagufta BI, Abbas SQ et al (2017) Current status and future therapeutic perspectives of glioblastoma multiforme (GBM) therapy: a review. Biomed Pharmacother 92:681–689. https://doi.org/10.1016/j.biopha.2017.05.125

Anttila MA, Tammi RH, Tammi MI et al (2000) High levels of stromal hyaluronan predict poor disease outcome in epithelial ovarian cancer. Cancer Res 60

Apperley JF (2015) Chronic myeloid leukaemia. Lancet 385:1447–1459. https://doi.org/10.1016/S0140-6736(13)62120-0

Arai E, Nishida Y, Wasa J et al (2011) Inhibition of hyaluronan retention by 4-methylumbelliferone suppresses osteosarcoma cells in vitro and lung metastasis in vivo. Br J Cancer 105:1839–1849. https://doi.org/10.1038/bjc.2011.459

Arnold JM, Gu F, Ambati CR et al (2020) UDP-glucose 6-dehydrogenase regulates hyaluronic acid production and promotes breast cancer progression. Oncogene 39:3089–3101. https://doi.org/10.1038/s41388-019-0885-4

Auvinen P, Tammi R, Parkkinen J et al (2000) Hyaluronan in peritumoral stroma and malignant cells associates with breast cancer spreading and predicts survival. Am J Pathol 156:529–536. https://doi.org/10.1016/S0002-9440(10)64757-8

Avigdor A (2004) CD44 and hyaluronic acid cooperate with SDF-1 in the trafficking of human CD34+ stem/progenitor cells to bone marrow. Blood 103:2981–2989. https://doi.org/10.1182/blood-2003-10-3611

Baumgartner G, Gomar-Höss C, Sakr L et al (1998) The impact of extracellular matrix on the chemoresistance of solid tumors—Experimental and clinical results of hyaluronidase as additive to cytostatic chemotherapy. Cancer Lett 131(1):85–99

Bavaro L, Martelli M, Cavo M, Soverini S (2019) Mechanisms of disease progression and resistance to tyrosine kinase inhibitor therapy in chronic myeloid leukemia: an update. Int J Mol Sci 20:6141. https://doi.org/10.3390/ijms20246141

Bertrand P, Courel MN, Maingonnat C et al (2005) Expression of HYAL2 mRNA, hyaluronan and hyaluronidase in B-cell non-Hodgkin lymphoma: relationship with tumor aggressiveness. Int J Cancer 113:207–212. https://doi.org/10.1002/ijc.20562

Bharadwaj AG, Kovar JL, Loughman E et al (2009) Spontaneous metastasis of prostate cancer is promoted by excess hyaluronan synthesis and processing. Am J Pathol 174:1027–1036. https://doi.org/10.2353/ajpath.2009.080501

Bjorklund CC, Baladandayuthapani V, Lin HY et al (2014) Evidence of a role for CD44 and cell adhesion in mediating resistance to lenalidomide in multiple myeloma: therapeutic implications. Leukemia 28(2):373–383. https://doi.org/10.1038/leu.2013.174

Bjorklund CC, Kang J, Amatangelo M et al (2019) Iberdomide (CC-220) is a potent cereblon E3 ligase modulator with antitumor and immunostimulatory activities in lenalidomide- and

pomalidomide-resistant multiple myeloma cells with dysregulated CRBN. Leukemia 34(4): 1197–1201. https://doi.org/10.1038/s41375-019-0620-8
Bono P, Rubin K, Higgins JMG, Hynes RO (2001) Layilin, a novel integral membrane protein, is a hyaluronan receptor. Mol Biol Cell 12:891–900. https://doi.org/10.1091/mbc.12.4.891
Bouguignon LYW (2009) Hyaluronan-mediated CD44 interaction with receptor and non-receptor kinases promotes oncogenic signaling, cytoskeleton activation and tumor progression. Hyaluronan Cancer Biol 89–107. https://doi.org/10.1016/B978-012374178-3.10006-7
Bourguignon LYW (2008) Hyaluronan-mediated CD44 activation of RhoGTPase signaling and cytoskeleton function promotes tumor progression. Semin Cancer Biol 18:251–259. https://doi.org/10.1016/j.semcancer.2008.03.007
Brenner H, Kloor M, Pox CP (2014) Colorectal cancer. Lancet 383:1490–1502. https://doi.org/10.1016/S0140-6736(13)61649-9
Brett M-E, Bomberger HE, Doak GR et al (2018) In vitro elucidation of the role of pericellular matrix in metastatic extravasation and invasion of breast carcinoma cells. Integr Biol 10:242–252. https://doi.org/10.1039/c7ib00173h
Breyer R, Hussein S, Radu DL et al (2000) Disruption of intracerebral progression of C6 rat glioblastoma by in vivo treatment with anti-CD44 monoclonal antibody. J Neurosurg 92:140–149. https://doi.org/10.3171/jns.2000.92.1.0140
Bullard KM, Kim HR, Wheeler MA et al (2003) Hyaluronan synthase-3 is upregulated in metastatic colon carcinoma cells and manipulation of expression alters matrix retention and cellular growth. Int J Cancer 107:739–746. https://doi.org/10.1002/ijc.11475
Burgstaller G, Oehrle B, Gerckens M et al (2017) The instructive extracellular matrix of the lung: basic composition and alterations in chronic lung disease. Eur Respir J 50. https://doi.org/10.1183/13993003.01805-2016
Celegato M, Borghese C, Casagrande N et al (2014) Bortezomib down-modulates the survival factor interferon regulatory factor 4 in Hodgkin lymphoma cell lines and decreases the protective activity of Hodgkin lymphoma-associated fibroblasts. Leuk Lymphoma 55:149–159. https://doi.org/10.3109/10428194.2013.800196
Chanmee T, Ontong P, Itano N (2016) Hyaluronan: a modulator of the tumor microenvironment. Cancer Lett 375:20–30. https://doi.org/10.1016/j.canlet.2016.02.031
Cheng X, Kohi S, Koga A et al (2016) Hyaluronan stimulates pancreatic cancer cell motility. Oncotarget 7:4829–4840. https://doi.org/10.18632/oncotarget.6617
Chetty C, Vanamala SK, Gondi CS et al (2012) MMP-9 induces CD44 cleavage and CD44 mediated cell migration in glioblastoma xenograft cells. Cell Signal 24:549–559. https://doi.org/10.1016/j.cellsig.2011.10.008
Chintala SK, Gokaslan ZL, Go Y et al (1996) Role of extracellular matrix proteins in regulation of human glioma cell invasion in vitro. Clin Exp Metastasis 14:358–366. https://doi.org/10.1007/BF00123395
Clift R, Souratha J, Garrovillo SA et al (2019) Remodeling the tumor microenvironment sensitizes breast tumors to anti-programmed death-ligand 1 immunotherapy. Cancer Res 79:4149–4159. https://doi.org/10.1158/0008-5472.CAN-18-3060
Collins RJ, Verschuer LA, Harmon B et al (1989) Spontaneous programmed death (apoptosis) of B-chronic lymphocytic leukaemia cells following their culture in vitro. Br J Haematol 71:343–350. https://doi.org/10.1111/j.1365-2141.1989.tb04290.x
Colombo N, van Gorp T, Parma G et al (2006) Ovarian cancer. Crit Rev Oncol Hematol 60:159–179. https://doi.org/10.1016/j.critrevonc.2006.03.004
Colombo N, Lorusso D, Scollo P (2017) Impact of recurrence of ovarian cancer on quality of life and outlook for the future. Int J Gynecol Cancer 27:1134–1140. https://doi.org/10.1097/IGC.0000000000001023
Cook G, Dumbar M, Franklin IM (1997) The role of adhesion molecules in multiple myeloma. Acta Haematol 97:81–89. https://doi.org/10.1159/000203663

Corte MD, González LO, Junquera S et al (2010) Analysis of the expression of hyaluronan in intraductal and invasive carcinomas of the breast. J Cancer Res Clin Oncol 136:745–750. https://doi.org/10.1007/s00432-009-0713-2
Costa F, Dalla Palma B, Giuliani N (2019) CD38 expression by myeloma cells and its role in the context of bone marrow microenvironment: modulation by therapeutic agents. Cell 8. https://doi.org/10.3390/cells8121632
Cowman MK, Lee H-G, Schwertfeger KL et al (2015) The content and size of hyaluronan in biological fluids and tissues. Front Immunol 6. https://doi.org/10.3389/fimmu.2015.00261
Csoka AB, Stern R (2013) Hypotheses on the evolution of hyaluronan: a highly ironic acid. Glycobiology 23:398–411. https://doi.org/10.1093/glycob/cws218
Cyphert JM, Trempus CS, Garantziotis S (2015) Size matters: molecular weight specificity of hyaluronan effects in cell biology. Int J Cell Biol 2015:1–8. https://doi.org/10.1155/2015/563818
Dahl IMS, Rasmussen T, Kauric G, Husebekk A (2002) Differential expression of CD56 and CD44 in the evolution of extramedullary myeloma. Br J Haematol 116:273–277. https://doi.org/10.1046/j.1365-2141.2002.03258.x
Das R, Strowig T, Verma R, Koduru S, Hafemann A, Hopf S et al (2016) Microenvironment dependent growth of preneoplastic and malignant plasma cells in humanized mice. Nat Med 22(11):1351–1357
dela Cruz CS, Tanoue LT, Matthay RA (2011) Lung cancer: epidemiology, etiology, and prevention. Clin Chest Med 32:605–644. https://doi.org/10.1016/j.ccm.2011.09.001
Delpech B, Maingonnat C, Girard N et al (1993) Hyaluronan and hyaluronectin in the extracellular matrix of human brain tumour stroma. Eur J Cancer 29:1012–1017. https://doi.org/10.1016/S0959-8049(05)80214-X
Dufort CC, DelGiorno KE, Carlson MA et al (2016a) Interstitial pressure in pancreatic ductal adenocarcinoma is dominated by a gel-fluid phase. Biophys J 110:2106–2119. https://doi.org/10.1016/j.bpj.2016.03.040
Dufort CC, Delgiorno KE, Hingorani SR (2016b) Mounting pressure in the microenvironment: fluids, solids, and cells in pancreatic ductal adenocarcinoma. Gastroenterology 150:1545–1557.e2
Eberth S, Schneider B, Rosenwald A et al (2010) Epigenetic regulation of CD44 in Hodgkin and non-Hodgkin lymphoma. BMC Cancer 10:1–14. https://doi.org/10.1186/1471-2407-10-517
Eikenes L, Tari M, Tufto I et al (2005) Hyaluronidase induces a transcapillary pressure gradient and improves the distribution and uptake of liposomal doxorubicin (Caelyx™) in human osteosarcoma xenografts. Br J Cancer 93(1):81–88. https://doi.org/10.1038/sj.bjc.6602626
Fhu CW, Graham AM, Yap CT et al (2014) Reed-Sternberg cell-derived lymphotoxin-a activates endothelial cells to enhance T-cell recruitment in classical Hodgkin lymphoma. Blood 124: 2973–2982. https://doi.org/10.1182/blood-2014-05-576140
Fiegl M (2016) Epidemiology, pathogenesis, and etiology of acute leukemia. Handb Acute Leukemia 3–13. https://doi.org/10.1007/978-3-319-26772-2_2
Follini E, Marchesini M, Roti G (2019) Strategies to overcome resistance mechanisms in T-cell acute lymphoblastic leukemia. Int J Mol Sci 20:3021. https://doi.org/10.3390/ijms20123021
Funamoto Y, Nagai M, Haba R et al (2002) Hyaluronan synthesis by anaplastic large cell lymphoma with massive lymphomatous effusion: a case report. Acta Cytol 46:864–868. https://doi.org/10.1159/000327060
Garantziotis S, Savani RC (2019) Hyaluronan biology: a complex balancing act of structure, function, location and context. Matrix Biol 78–79:1–10. https://doi.org/10.1016/j.matbio.2019.02.002
García MG, Alaniz LD, Cordo Russo RI et al (2009) PI3K/Akt inhibition modulates multidrug resistance and activates NF-κB in murine lymphoma cell lines. Leuk Res 33. https://doi.org/10.1016/j.leukres.2008.06.010

García-Ortiz A, Rodríguez-García Y, Encinas J et al (2021) The role of tumor microenvironment in multiple myeloma development and progression. Cancer 13:217. https://doi.org/10.3390/cancers13020217

García-Peydró M, Fuentes P, Mosquera M et al (2018) The NOTCH1/CD44 axis drives pathogenesis in a T cell acute lymphoblastic leukemia model. J Clin Investig 128:2802–2818. https://doi.org/10.1172/JCI92981

Gesteira TF, Sun M, Coulson-Thomas YM et al (2017) Hyaluronan rich microenvironment in the limbal stem cell niche regulates limbal stem cell differentiation. Invest Ophthalmol Vis Sci 58: 4407. https://doi.org/10.1167/IOVS.17-22326

Giannopoulos K, Mertens D, Bühler A et al (2009) The candidate immunotherapeutical target, the receptor for hyaluronic acid-mediated motility, is associated with proliferation and shows prognostic value in B-cell chronic lymphocytic leukemia. Leukemia 23(3):519–527. https://doi.org/10.1038/leu.2008.338

Girbl T, Hinterseer E, Grössinger EM et al (2013) CD40-mediated activation of chronic lymphocytic leukemia cells promotes their CD44-dependent adhesion to hyaluronan and restricts CCL21-induced motility. Cancer Res 73:561–570. https://doi.org/10.1158/0008-5472.CAN-12-2749

Gomez CS, Gomez P, Knapp J et al (2009) Hyaluronic acid and HYAL-1 in prostate biopsy specimens: predictors of biochemical recurrence. J Urol 182:1350–1356. https://doi.org/10.1016/J.JURO.2009.06.070

Goncharova V, Serobyan N, Iizuka S et al (2012) Hyaluronan expressed by the hematopoietic microenvironment is required for bone marrow hematopoiesis. J Biol Chem 287:25419–25433. https://doi.org/10.1074/jbc.M112.376699

González-Herrero I, Rodríguez-Hernández G, Luengas-Martínez A et al (2018) The making of leukemia. Int J Mol Sci 19. https://doi.org/10.3390/ijms19051494

Graham SM, Jørgensen HG, Allan E et al (2002) Primitive, quiescent, Philadelphia-positive stem cells from patients with chronic myeloid leukemia are insensitive to STI571 in vitro. Blood 99: 319–325. https://doi.org/10.1182/blood.V99.1.319

Greiner J, Ringhoffer M, Taniguchi M et al (2002) Receptor for hyaluronan acid-mediated motility (RHAMM) is a new immunogenic leukemia-associated antigen in acute and chronic myeloid leukemia. Exp Hematol 30:1029–1035. https://doi.org/10.1016/S0301-472X(02)00874-3

Guerra C, Schuhmacher AJ, Cañamero M et al (2007) Chronic pancreatitis is essential for induction of pancreatic ductal adenocarcinoma by K-Ras oncogenes in adult mice. Cancer Cell 11:291–302. https://doi.org/10.1016/j.ccr.2007.01.012

Gul-Uludağ H, Valencia-Serna J, Kucharski C et al (2014) Polymeric nanoparticle-mediated silencing of CD44 receptor in CD34+ acute myeloid leukemia cells. Leuk Res 38:1299–1308. https://doi.org/10.1016/j.leukres.2014.08.008

Gutjahr JC, Greil R, Hartmann TN (2015) The role of CD44 in the pathophysiology of chronic lymphocytic leukemia. Front Immunol 6. https://doi.org/10.3389/FIMMU.2015.00177

Gutjahr JC, Bayer E, Yu X et al (2020) CD44 engagement enhances acute myeloid leukemia cell adhesion to the bone marrow microenvironment by increasing VLA-4 avidity. Haematologica 106. https://doi.org/10.3324/haematol.2019.231944

Gvozdenovic A, Arlt MJ, Campanile C et al (2013) CD44 enhances tumor formation and lung metastasis in experimental osteosarcoma and is an additional predictor for poor patient outcome. J Bone Miner Res 28:838–847. https://doi.org/10.1002/jbmr.1817

Hallek M, Shanafelt TD, Eichhorst B (2018) Chronic lymphocytic leukaemia. Lancet 391:1524–1537. https://doi.org/10.1016/S0140-6736(18)30422-7

Hamada S, Nishida Y, Zhuo L et al (2018) Suppression of hyaluronan synthesis attenuates the tumorigenicity of low-grade chondrosarcoma. J Orthop Res 36:1573–1580. https://doi.org/10.1002/jor.23794

Hamel KM, Liarski VM, Clark MR (2012) Germinal center B-cells. Autoimmunity 45:333–347

Hanke M, Hoffmann I, Christophis C et al (2014) Differences between healthy hematopoietic progenitors and leukemia cells with respect to CD44 mediated rolling versus adherence behavior on hyaluronic acid coated surfaces. Biomaterials 35:1411–1419. https://doi.org/10.1016/j.biomaterials.2013.11.011

Hartheimer JS, Park S, Rao SS, Kim Y (2019) Targeting hyaluronan interactions for glioblastoma stem cell therapy. Cancer Microenviron 12:47–56. https://doi.org/10.1007/s12307-019-00224-2

Heldin P, Basu K, Olofsson B et al (2013) Deregulation of hyaluronan synthesis, degradation and binding promotes breast cancer. J Biochem 154:395–408. https://doi.org/10.1093/jb/mvt085

Hidalgo M (2010) Pancreatic cancer. N Engl J Med 21:1605–1618. https://doi.org/10.1016/j.ccr.2012.04.025.Tumor-derived

Hiltunen ELJ, Anttila M, Kultti A et al (2002) Elevated hyaluronan concentration without hyaluronidase activation in malignant epithelial ovarian tumors. Cancer Res 62:6410–6413

Hong FS, Mitchell CA, Zantomio D (2010) Gelatinous transformation of the bone marrow as a late morphological change in imatinib mesylate treated chronic myeloid leukaemia. Pathology 42:84–85. https://doi.org/10.3109/00313020903434686

Horai T, Nakamura N, Tateishi R, Hattori S (1981) Glycosaminoglycans in human lung cancer. Cancer 48:2016–2021. https://doi.org/10.1002/1097-0142(19811101)48:9<2016::AID-CNCR2820480918>3.0.CO;2-A

Hosono K, Nishida Y, Knudson W et al (2007) Hyaluronan oligosaccharides inhibit tumorigenicity of osteosarcoma cell lines MG-63 and LM-8 in vitro and in vivo via perturbation of hyaluronan-rich pericellular matrix of the cells. Am J Pathol 171:274–286. https://doi.org/10.2353/ajpath.2007.060828

Hruban RH, Adsay NV, Albores-Saavedra J et al (2001) Pancreatic intraepithelial neoplasia: a new nomenclature and classification system for pancreatic duct lesions. Am J Surg Pathol 25:579–586. https://doi.org/10.1097/00000478-200105000-00003

Hu Y, Li S (2016) Survival regulation of leukemia stem cells. Cell Mol Life Sci 73:1039–1050. https://doi.org/10.1007/s00018-015-2108-7

Izzi V, Lakkala J, Devarajan R et al (2017) An extracellular matrix signature in leukemia precursor cells and acute myeloid leukemia. Haematologica 102:e245–e248. https://doi.org/10.3324/haematol.2017.167304

Jacobetz MA, Chan DS, Neesse A et al (2013) Hyaluronan impairs vascular function and drug delivery in a mouse model of pancreatic cancer. Gut 62:112–120. https://doi.org/10.1136/gutjnl-2012-302529

Jayson GC, Kohn EC, Kitchener HC, Ledermann JA (2014) Ovarian cancer. Lancet (London, England) 384:1376–1388. https://doi.org/10.1016/S0140-6736(13)62146-7

Karamanos NK, Theocharis AD, Piperigkou Z et al (2021) A guide to the composition and functions of the extracellular matrix. FEBS J. https://doi.org/10.1111/febs.15776

Karantanou C, Godavarthy PS, Krause DS (2018) Targeting the bone marrow microenvironment in acute leukemia. Leuk Lymphoma 59:2535–2545. https://doi.org/10.1080/10428194.2018.1434886

Karbownik MS, Pietras T, Szemraj J et al (2014) The ability of hyaluronan fragments to reverse the resistance of C6 rat glioma cell line to temozolomide and carmustine. Wspolczesna Onkologia 18:323–328. https://doi.org/10.5114/wo.2014.43493

Kartal-Yandim M, Adan-Gokbulut A, Baran Y (2016) Molecular mechanisms of drug resistance and its reversal in cancer. Crit Rev Biotechnol 36:716–726. https://doi.org/10.3109/07388551.2015.1015957

Kasai S, Furuichi Y, Ando N et al (2017) Inflammatory mediator ultra-low-molecular-weight hyaluronan triggers necrosis of B-precursor leukemia cells with high surface CD44 expression. Cell Death Dis 8:e2857–e2857. https://doi.org/10.1038/cddis.2017.249

Kato N, Shibata K, Uchigasaki S, Kurose A (2016) Relation between hyaluronan synthesis and cell morphology in ovarian clear cell carcinomas. Pathol Int 66:218–223. https://doi.org/10.1111/pin.12405

Katz B-Z (2010) Adhesion molecules—The lifelines of multiple myeloma cells. Semin Cancer Biol 20:186–195. https://doi.org/10.1016/j.semcancer.2010.04.003

Kaul A, Short WD, Wang X, Keswani SG (2021) Hyaluronidases in human diseases. Int J Mol Sci 22:1–11. https://doi.org/10.3390/IJMS22063204

Khaldoyanidi SK, Goncharova V, Mueller B, Schraufstatter IU (2014) Hyaluronan in the healthy and malignant hematopoietic microenvironment. Adv Cancer Res 123:149–189. https://doi.org/10.1016/B978-0-12-800092-2.00006-X

Kim HR, Wheeler MA, Wilson CM et al (2004) Hyaluronan facilitates invasion of colon carcinoma cells in vitro via interaction with CD44. Cancer Res 64:4569–4576. https://doi.org/10.1158/0008-5472.CAN-04-0202

Kim MS, Park MJ, Kim SJ et al (2005) Emodin suppresses hyaluronic acid-induced MMP-9 secretion and invasion of glioma cells. Int J Oncol 27:839–846. https://doi.org/10.3892/ijo.27.3.839

Kim YH, Lee SB, Shim S et al (2019) Hyaluronic acid synthase 2 promotes malignant phenotypes of colorectal cancer cells through transforming growth factor beta signaling. Cancer Sci 110: 2226–2236. https://doi.org/10.1111/cas.14070

Klank RL, Decker Grunke SA, Bangasser BL et al (2017) Biphasic dependence of glioma survival and cell migration on CD44 expression level. Cell Rep 18:23–31. https://doi.org/10.1016/j.celrep.2016.12.024

Klein U, Dalla-Favera R (2008) Germinal centres: role in B-cell physiology and malignancy. Nat Rev Immunol 8:22–33

Kohli AG, Kivimäe S, Tiffany MR, Szoka FC (2014) Improving the distribution of Doxil® in the tumormatrix by depletion of tumor hyaluronan. J Control Release 191:105–114. https://doi.org/10.1016/j.jconrel.2014.05.019

Koike H, Nishida Y, Shinomura T et al (2020) Forced expression of KIAA1199, a novel hyaluronidase, inhibits tumorigenicity of low-grade chondrosarcoma. J Orthop Res 38:1942–1951. https://doi.org/10.1002/jor.24629

Kosaki R, Watanabe K, Yamaguchi Y (1999) Overproduction of hyaluronan by expression of the hyaluronan synthase Has2 enhances anchorage-independent growth and tumorigenicity. Cancer Res 59:1141

Krause DS, Lazarides K, von Andrian UH, van Etten RA (2006) Requirement for CD44 in homing and engraftment of BCR-ABL-expressing leukemic stem cells. Nat Med 12:1175–1180. https://doi.org/10.1038/nm1489

Kuang D-M, Wu Y, Chen N et al (2007) Tumor-derived hyaluronan induces formation of immunosuppressive macrophages through transient early activation of monocytes. Blood 110: 587–595. https://doi.org/10.1182/blood-2007-01-068031

Kudo D, Suto A, Hakamada K (2017) The development of a novel therapeutic strategy to target hyaluronan in the extracellular matrix of pancreatic ductal adenocarcinoma. Int J Mol Sci 18

Kultti A, Pasonen-Seppänen S, Jauhiainen M et al (2009) 4-Methylumbelliferone inhibits hyaluronan synthesis by depletion of cellular UDP-glucuronic acid and downregulation of hyaluronan synthase 2 and 3. Exp Cell Res 315:1914–1923. https://doi.org/10.1016/j.yexcr.2009.03.002

Kuwabara H, Yoneda M, Nagai M et al (2003) High levels of hyaluronan production by a malignant lymphoma cell line with primary effusion lymphoma immunophenotype OHK. Br J Haematol 120:1055–1057. https://doi.org/10.1046/j.1365-2141.2003.04206.x

Lato MW, Przysucha A, Grosman S et al (2021) The new therapeutic strategies in pediatric T-cell acute lymphoblastic leukemia. Int J Mol Sci 22. https://doi.org/10.3390/ijms22094502

Laurent TC, Fraser JRE (1992) Hyaluronan 1. FASEB J 6:2397–2404. https://doi.org/10.1096/FASEBJ.6.7.1563592

Laurich C, Wheeler MA, Iida J et al (2004) Hyaluronan mediates adhesion of metastatic colon carcinoma cells. J Surg Res 122:70–74. https://doi.org/10.1016/j.jss.2004.05.018

Lengyel E (2010) Ovarian cancer development and metastasis. Am J Pathol 177:1053–1064. https://doi.org/10.2353/ajpath.2010.100105

le Rhun E, Preusser M, Roth P et al (2019) Molecular targeted therapy of glioblastoma. Cancer Treat Rev 80:101896. https://doi.org/10.1016/j.ctrv.2019.101896

Ley TJ, Miller C, Ding L et al (2013) Genomic and epigenomic landscapes of adult de novo acute myeloid leukemia. N Engl J Med 368:2059–2074. https://doi.org/10.1056/NEJMoa1301689

Li Y, Li L, Brown TJ, Heldin P (2007) Silencing of hyaluronan synthase 2 suppresses the malignant phenotype of invasive breast cancer cells. Int J Cancer 120:2557–2567. https://doi.org/10.1002/ijc.22550

Lim EJ, Suh Y, Yoo KC et al (2017) Tumor-associated mesenchymal stem-like cells provide extracellular signaling cue for invasiveness of glioblastoma cells. Oncotarget 8:1438–1448. https://doi.org/10.18632/oncotarget.13638

Lin L-H, Chou H-C, Chang S-J et al (2020) Targeting UDP-glucose dehydrogenase inhibits ovarian cancer growth and metastasis. J Cell Mol Med 24:11883–11902. https://doi.org/10.1111/JCMM.15808

Lipponen P, Aaltomaa S, Tammi R, Tammi M, Ågren U, Kosma VM (2001) High stromal hyaluronan level is associated with poor differentiation and metastasis in prostate cancer. Eur J Cancer 37(7)

Llaneza A, Vizoso F, Rodríguez JC, Raigoso P, García-Muñiz JL, Allende MT, García Morán M (2000) Hyaluronic acid as prognostic marker in resectable colorectal cancer. Br J Surg 87(12): 1690–1696

Lokeshwar VB, Lopez LE, Munoz D et al (2010) Antitumor activity of hyaluronic acid synthesis inhibitor 4-methylumbelliferone in prostate cancer cells. Cancer Res 70:2613–2623. https://doi.org/10.1158/0008-5472.CAN-09-3185

Lokeshwar VB, Mirza S, Jordan A (2014) Targeting hyaluronic acid family for cancer chemoprevention and therapy, 1st edn. Elsevier

Lokman NA, Price ZK, Hawkins EK et al (2019) 4-methylumbelliferone inhibits cancer stem cell activation and overcomes chemoresistance in ovarian cancer. Cancer 11. https://doi.org/10.3390/cancers11081187

Lompardía SL, Papademetrio DL, Mascaró M et al (2013) Human leukemic cell lines synthesize hyaluronan to avoid senescence and resist chemotherapy. Glycobiology 23:1463–1476. https://doi.org/10.1093/glycob/cwt074

Lompardía SL, Díaz M, Papademetrio DL et al (2016) Hyaluronan oligomers sensitize chronic myeloid leukemia cell lines to the effect of Imatinib. Glycobiology 26:343–352. https://doi.org/10.1093/glycob/cwv107

Lompardía SL, Díaz M, Papademetrio DL et al (2017) 4-methylumbelliferone and imatinib combination enhances senescence induction in chronic myeloid leukemia cell lines. Invest New Drugs 35:1–10. https://doi.org/10.1007/s10637-016-0397-9

Lompardía S, Díaz M, Pibuel M et al (2019) Hyaluronan abrogates imatinib-induced senescence in chronic myeloid leukemia cell lines. Sci Rep 9:10930. https://doi.org/10.1038/s41598-019-47248-8

Louis DN, Perry A, Reifenberger G et al (2016) The 2016 World Health Organization classification of tumors of the central nervous system: a summary. Acta Neuropathol 131:803–820. https://doi.org/10.1007/s00401-016-1545-1

Ma Y, Visser L, Roelofsen H et al (2008) Proteomics analysis of Hodgkin lymphoma: identification of new players involved in the cross-talk between HRS cells and infiltrating lymphocytes. Blood 111:2339–2346. https://doi.org/10.1182/blood-2007-09-112128

Makkar S, Riehl TE, Chen B et al (2019) Hyaluronic acid binding to TLR4 promotes proliferation and blocks apoptosis in colon cancer. Mol Cancer Ther 18:2446–2456. https://doi.org/10.1158/1535-7163.MCT-18-1225

Marchetti C, Pisano C, Facchini G et al (2010) First-line treatment of advanced ovarian cancer: current research and perspectives. Expert Rev Anticancer Ther 10:47–60. https://doi.org/10.1586/era.09.167

Maria BL, Gupta N, Gilg AG et al (2008) Targeting hyaluronan interactions in spinal cord astrocytomas and diffuse pontine gliomas. J Child Neurol 1214–1220

Martinez-Quintanilla J, He D, Wakimoto H et al (2015) Encapsulated stem cells loaded with hyaluronidase-expressing oncolytic virus for brain tumor therapy. Mol Ther 23:108–118. https://doi.org/10.1038/mt.2014.204

Means AL, Meszoely IM, Suzuki K et al (2005) Pancreatic epithelial plasticity mediated by acinar cell transdifferentiation and generation of nestin-positive intermediates. Development 132: 3767–3776. https://doi.org/10.1242/dev.01925

Menghrajani K, Tallman MS (2018) New therapeutic strategies for high-risk acute myeloid leukemia. Curr Opin Hematol 25:90–94. https://doi.org/10.1097/MOH.0000000000000409

Monslow J, Govindaraju P, Puré E (2015) Hyaluronan—a functional and structural sweet spot in the tissue microenvironment. Front Immunol 6. https://doi.org/10.3389/FIMMU.2015.00231

Mooney KL, Choy W, Sidhu S et al (2016) The role of CD44 in glioblastoma multiforme. J Clin Neurosci 34:1–5. https://doi.org/10.1016/j.jocn.2016.05.012

Mulshine JL, Chow G, Tauler J (2010) Cytokines and growth factors stimulate hyaluronan production: role of hyaluronan in epithelial to mesenchymal-like transition in non-small cell lung cancer. J Biomed Biotechnol. https://doi.org/10.1155/2010/485468

Murugaiah V, Agostinis C, Varghese PM et al (2020) Hyaluronic acid present in the tumor microenvironment can negate the pro-apototic effect of a recombinant fragment of human surfactant protein D on breast cancer cells. Front Immunol 11:1171. https://doi.org/10.3389/fimmu.2020.01171

Nakagawa T, Kubota T, Kabuto M, Kodera T (1996) Hyaluronic acid facilitates glioma cell invasion in vitro. Anticancer Res 16:2917–2922

Nakamura R, Kuwabara H, Yoneda M et al (2007) Suppression of matrix metalloproteinase-9 by 4-methylumbelliferone. Cell Biol Int 31:1022–1026. https://doi.org/10.1016/j.cellbi.2007.03.016

Nakazawa H, Yoshihara S, Kudo D et al (2006) 4-methylumbelliferone, a hyaluronan synthase suppressor, enhances the anticancer activity of gemcitabine in human pancreatic cancer cells. Cancer Chemother Pharmacol 57:165–170. https://doi.org/10.1007/s00280-005-0016-5

Nikitovic D, Tzardi M, Berdiaki A et al (2015) Cancer microenvironment and inflammation: role of hyaluronan. Front Immunol 0:169. https://doi.org/10.3389/FIMMU.2015.00169

Nishida Y, Knudson W, Knudson CB, Ishiguro N (2005) Antisense inhibition of hyaluronan synthase-2 in human osteosarcoma cells inhibits hyaluronan retention and tumorigenicity. Exp Cell Res 307:194–203. https://doi.org/10.1016/j.yexcr.2005.03.026

Ohwada C, Nakaseko C, Koizumi M et al (2008) CD44 and hyaluronan engagement promotes dexamethasone resistance in human myeloma cells. Eur J Haematol 80:245–250. https://doi.org/10.1111/j.1600-0609.2007.01014.x

Onfroy-Roy L, Hamel D, Foncy J et al (2020) Extracellular matrix mechanical properties and regulation of the intestinal stem cells: when mechanics control fate. Cell 9:1–23. https://doi.org/10.3390/cells9122629

Orian-Rousseau V (2010) CD44, a therapeutic target for metastasising tumours. Eur J Cancer 46: 1271–1277. https://doi.org/10.1016/j.ejca.2010.02.024

Palumbo A, Anderson K (2011) Multiple myeloma. N Engl J Med 364:1046–1060. https://doi.org/10.1056/NEJMra1011442

Perus LJM, Walsh LA, Walsh LA (2019) Microenvironmental heterogeneity in brain malignancies. Front Immunol 10:2294. https://doi.org/10.3389/fimmu.2019.02294

Philteos J, Karmur BS, Mansouri A (2019) MGMT testing in glioblastomas pitfalls and opportunities. Am J Clin Oncol 42:117–122. https://doi.org/10.1097/COC.0000000000000490

Pibuel MA, Díaz M, Molinari Y et al (2021a) 4-Methylumbelliferone as a potent and selective antitumor drug on a glioblastoma model. Glycobiology 31:29–43. https://doi.org/10.1093/glycob/cwaa046

Pibuel MA, Poodts D, Díaz M et al (2021b) The scrambled story between hyaluronan and glioblastoma. J Biol Chem 296

Pirinen R, Tammi R, Tammi M et al (2001) Prognostic value of hyaluronan expression in non-small-cell lung cancer: increased stromal expression indicates unfavorable outcome in patients with adenocarcinoma. Int J Cancer 95. https://doi.org/10.1002/1097-0215(20010120)95:1<12::AID-IJC1002>3.0.CO;2-E

Ponta H, Sherman L, Herrlich PA (2003) CD44: from adhesion molecules to signalling regulators. Nat Rev Mol Cell Biol 4(4):33–45. https://doi.org/10.1038/nrm1004

Posey JT, Soloway MS, Ekici S, Sofer M, Civantos F, Duncan RC, Lokeshwar VB (2003) Evaluation of the prognostic potential of hyaluronic acid and hyaluronidase (HYAL1) for prostate cancer. Cancer Res 63(10):2638–2644

Provenzano PP, Cuevas C, Chang AE et al (2012) Enzymatic targeting of the stroma ablates physical barriers to treatment of pancreatic ductal adenocarcinoma. Cancer Cell 21:418–429. https://doi.org/10.1016/j.ccr.2012.01.007

Radotra B, McCormick D (1997) Glioma invasion in vitro is mediated by CD44–hyaluronan interactions. J Pathol 181:434–438. https://doi.org/10.1002/(SICI)1096-9896(199704)181:4<434::AID-PATH797>3.0.CO;2-S

Rajaratnam V, Islam MM, Yang M et al (2020) Glioblastoma: pathogenesis and current status of chemotherapy and other novel treatments. Cancer 12:937. https://doi.org/10.3390/cancers12040937

Rajkumar SV, Dimopoulos MA, Palumbo A et al (2014) International Myeloma Working Group updated criteria for the diagnosis of multiple myeloma. Lancet Oncol 15:e538–e548. https://doi.org/10.1016/S1470-2045(14)70442-5

Rangel MP, de Sá VK, Martins V et al (2015) Tissue hyaluronan expression, as reflected in the sputum of lung cancer patients, is an indicator of malignancy. Braz J Med Biol Res 48:557–567. https://doi.org/10.1590/1414-431X20144300

Renyi-Vamos F, Tovari J, Fillinger J et al (2005) Lymphangiogenesis correlates with lymph node metastasis, prognosis, and angiogenic phenotype in human non-small cell lung cancer. Clin Cancer Res 11:7344–7353. https://doi.org/10.1158/1078-0432.CCR-05-1077

Riehl TE, Ee X, Stenson WF (2012) Hyaluronic acid regulates normal intestinal and colonic growth in mice. Am J Physiol Gastrointest Liver Physiol 303:377–388. https://doi.org/10.1152/ajpgi.00034.2012

Riehl TE, Santhanam S, Foster L et al (2015) CD44 and TLR4 mediate hyaluronic acid regulation of Lgr5+ stem cell proliferation, crypt fission, and intestinal growth in postnatal and adult mice. Am J Physiol Gastrointest Liver Physiol 309:G874–G887. https://doi.org/10.1152/ajpgi.00123.2015

Ropponen K, Tammi M, Parkkinen J, Eskelinen M, Tammi R, Lipponen P et al (1998) Tumor cell-associated hyaluronan as an unfavorable prognostic factor in colorectal cancer. Cancer Res 58(2):342–347

Rui Kang DT (2012) Autophagy in pancreatic cancer pathogenesis and treatment—PubMed. Am J Cancer Res 2:383–396

Russo RIC, García MG, Alaniz L et al (2008) Hyaluronan oligosaccharides sensitize lymphoma resistant cell lines to vincristine by modulating P-glycoprotein activity and PI3K/Akt pathway. Int J Cancer 122:1012–1018. https://doi.org/10.1002/ijc.23122

Sadeghi N, Camby I, Goldman S et al (2003) Effect of hydrophilic components of the extracellular matrix on quantifiable diffusion-weighted imaging of human gliomas: preliminary results of correlating apparent diffusion coefficient values and hyaluronan expression level. Am J Roentgenol 181:235–241. https://doi.org/10.2214/ajr.181.1.1810235

Sato N, Kohi S, Hirata K, Goggins M (2016) Role of hyaluronan in pancreatic cancer biology and therapy: once again in the spotlight. Cancer Sci 107:569–575

Schepers K, Pietras EM, Reynaud D et al (2013) Myeloproliferative neoplasia remodels the endosteal bone marrow niche into a self-reinforcing leukemic niche. Cell Stem Cell 13:285–299. https://doi.org/10.1016/j.stem.2013.06.009

Schepers K, Campbell TB, Passegué E (2015) Normal and leukemic stem cell niches: insights and therapeutic opportunities. Cell Stem Cell 16:254–267. https://doi.org/10.1016/J.STEM.2015.02.014

Schmitt M, Schmitt A, Rojewski MT et al (2008) RHAMM-R3 peptide vaccination in patients with acute myeloid leukemia, myelodysplastic syndrome, and multiple myeloma elicits immunologic and clinical responses. Blood 111:1357–1365. https://doi.org/10.1182/blood-2007-07-099366

Sener SF, Fremgen A, Menck HR, Winchester DP (1999) Pancreatic cancer: a report of treatment and survival trends for 100,313 patients diagnosed from 1985-1995, using the National Cancer Database. J Am Coll Surg 189:1–7. https://doi.org/10.1016/S1072-7515(99)00075-7

Shalini CNS, Suman FR, Jacob JS et al (2018) Prognostic significance of receptor for hyaluronan acid-mediated motility (CD168) in acute pediatric leukemias—assessment of clinical outcome, post induction, end of treatment and minimal residual disease. Hematol Transfus Cell Ther 40: 310–316. https://doi.org/10.1016/j.htct.2018.01.008

Sharma C, Eltawil KM, Renfrew PD et al (2011) Advances in diagnosis, treatment and palliation of pancreatic carcinoma: 1990-2010. World J Gastroenterol 17:867–897. https://doi.org/10.3748/wjg.v17.i7.867

Shpilberg O, Jackisch C (2013) Subcutaneous administration of rituximab (MabThera) and trastuzumab (Herceptin) using hyaluronidase. Br J Cancer 109:1556–1561. https://doi.org/10.1038/bjc.2013.371

Siegel RL, Miller KD, Jemal A (2016) Cancer statistics, 2016. CA Cancer J Clin 66:7–30. https://doi.org/10.3322/caac.21332

Simioni C, Conti I, Varano G et al (2021) The complexity of the tumor microenvironment and its role in acute lymphoblastic leukemia: implications for therapies. Front Oncol 11:673506. https://doi.org/10.3389/fonc.2021.673506

Simpson MA, Lokeshwar VB (2008) Hyaluronan and hyaluronidase in genitourinary tumors. Front Biosci 13:5664

Simpson RML, Hong X, Wong MM et al (2016) Hyaluronan is crucial for stem cell differentiation into smooth muscle lineage. Stem Cells 34:1225–1238. https://doi.org/10.1002/STEM.2328

Slifka MK, Ahmed R (1998) Long-lived plasma cells: a mechanism for maintaining persistent antibody production. Curr Opin Immunol 10:252–258. https://doi.org/10.1016/s0952-7915(98)80162-3

Solis MA, Chen YH, Wong TY et al (2012) Hyaluronan regulates cell behavior: a potential niche matrix for stem cells. Biochem Res Int. https://doi.org/10.1155/2012/346972

Song JM, Im J, Nho RS et al (2019) Hyaluronan-CD44/RHAMM interaction-dependent cell proliferation and survival in lung cancer cells. Mol Carcinog 58:321–333. https://doi.org/10.1002/mc.22930

Stern R (2008) Hyaluronan in cancer biology. Semin Cancer Biol 18:237. https://doi.org/10.1016/j.semcancer.2008.04.001

Stern R, Jedrzejas MJ (2006) Hyaluronidases: their genomics, structures, and mechanisms of action. Chem Rev 106:818–839. https://doi.org/10.1021/cr050247k

Strobel H, Baisch T, Fitzel R et al (2019) Temozolomide and other alkylating agents in glioblastoma therapy. Biomedicine 69:1–17

Su W, Matsumoto S, Sorg B, Sherman LS (2019) Distinct roles for hyaluronan in neural stem cell niches and perineuronal nets. Matrix Biol 78–79:272–283. https://doi.org/10.1016/j.matbio.2018.01.022

Sundström G, Dahl IMS, Hultdin M et al (2005) Bone marrow hyaluronan distribution in patients with acute myeloid leukemia. Med Oncol 22:071–078. https://doi.org/10.1385/MO:22:1:071

Sundström G, Hultdin M, Engström-Laurent A, Dahl IMS (2010) Bone marrow hyaluronan and reticulin in patients with malignant disorders. Med Oncol 27:618–623. https://doi.org/10.1007/s12032-009-9257-1

Suto A, Kudo D, Yoshida E et al (2019) Increase of tumor infiltrating γδ T-cells in pancreatic ductal adenocarcinoma through remodeling of the extracellular matrix by a hyaluronan synthesis suppressor, 4-methylumbelliferone. Pancreas 48. https://doi.org/10.1097/MPA.0000000000001211

Tabarkiewicz J, Giannopoulos K (2010) Definition of a target for immunotherapy and results of the first peptide vaccination study in chronic lymphocytic leukemia. Transplant Proc 42:3293–3296. https://doi.org/10.1016/j.transproceed.2010.07.022

Tammi RH, Passi AG, Rilla K et al (2011) Transcriptional and post-translational regulation of hyaluronan synthesis. FEBS J 278:1419–1428. https://doi.org/10.1111/j.1742-4658.2011.08070.x
Tebbi CK (2021) Etiology of acute leukemia: a review. Cancers 13:2256. https://doi.org/10.3390/CANCERS13092256
Terwilliger T, Abdul-Hay M (2017) Acute lymphoblastic leukemia: a comprehensive review and 2017 update. Blood Cancer J 7:e577–e577. https://doi.org/10.1038/bcj.2017.53
Tian X, Azpurua J, Hine C et al (2013) High-molecular-mass hyaluronan mediates the cancer resistance of the naked mole rat. Nature 499(7458):346–349. https://doi.org/10.1038/nature12234
Till KJ, Zuzel M, Cawley JC (1999) The role of hyaluronan and interleukin 8 in the migration of chronic lymphocytic leukemia cells within lymphoreticular tissues. Cancer Res 59:4419
Tofuku K, Yokouchi M, Murayama T et al (2006) HAS3-related hyaluronan enhances biological activities necessary for metastasis of osteosarcoma cells. Int J Oncol 29:175–183. https://doi.org/10.3892/IJO.29.1.175
Toole BP (2002) Hyaluronan promotes the malignant phenotype. Glycobiology 12:37R–42R
Toole BP (2004) Hyaluronan: from extracellular glue to pericellular cue. Nat Rev Cancer 4(7):528–539. https://doi.org/10.1038/nrc1391
Toole BP (2009) Hyaluronan-CD44 interactions in cancer: paradoxes and possibilities. Clin Cancer Res 15
Toole BP, Hascall VC (2002) Hyaluronan and tumor growth. Am J Pathol 161:745–747
Toole BP, Slomiany MG (2008) Hyaluronan, CD44 and emmprin: partners in cancer cell chemoresistance. Drug Resist Updat 11:110–121. https://doi.org/10.1016/j.drup.2008.04.002
Tsatas D, Kanagasundaram V, Kaye A, Novak U (2002) EGF receptor modifies cellular responses to hyaluronan in glioblastoma cell lines. J Clin Neurosci 9:282–288. https://doi.org/10.1054/jocn.2001.1063
Urakawa H, Nishida Y, Wasa J et al (2012) Inhibition of hyaluronan synthesis in breast cancer cells by 4-methylumbelliferone suppresses tumorigenicity in vitro and metastatic lesions of bone in vivo. Int J Cancer 130:454–466. https://doi.org/10.1002/ijc.26014
Vacca A, di Loreto M, Ribatti D et al (1995) Bone marrow of patients with active multiple myeloma: angiogenesis and plasma cell adhesion molecules LFA-1, VLA-4, LAM-1, and CD44. Am J Hematol 50:9–14. https://doi.org/10.1002/ajh.2830500103
Vadillo E, Dorantes-Acosta E, Pelayo R, Schnoor M (2018) T cell acute lymphoblastic leukemia (T-ALL): new insights into the cellular origins and infiltration mechanisms common and unique among hematologic malignancies. Blood Rev 32:36–51. https://doi.org/10.1016/j.blre.2017.08.006
Valkonen M, Haapasalo H, Rilla K et al (2018) Elevated expression of hyaluronan synthase 2 associates with decreased survival in diffusely infiltrating astrocytomas. BMC Cancer 18:664. https://doi.org/10.1186/s12885-018-4569-1
van Oosterwijk JG, Herpers B, Meijer D et al (2012) Restoration of chemosensitivity for doxorubicin and cisplatin in chondrosarcoma in vitro: BCL-2 family members cause chemoresistance. Ann Oncol 23:1617–1626. https://doi.org/10.1093/annonc/mdr512
Vaughan S, Coward JI, Bast RC et al (2011) Rethinking ovarian cancer: recommendations for improving outcomes. Nat Rev Cancer 11:719–725. https://doi.org/10.1038/nrc3144
Vey N, Delaunay J, Martinelli G et al (2016) Phase I clinical study of RG7356, an anti-CD44 humanized antibody, in patients with acute myeloid leukemia. Oncotarget 7:32532–32542. https://doi.org/10.18632/oncotarget.8687
Vincent T, Mechti N (2004) IL-6 regulates CD44 cell surface expression on human myeloma cells. Leukemia 18:967–975. https://doi.org/10.1038/sj.leu.2403333
Virga J, Bognár L, Hortobágyi T et al (2017) Tumor grade versus expression of invasion-related molecules in astrocytoma. Pathol Oncol Res. https://doi.org/10.1007/s12253-017-0194-6
Vitale DL, Spinelli FM, del Dago D et al (2018) Co-treatment of tumor cells with hyaluronan plus doxorubicin affects endothelial cell behavior independently of VEGF expression. Oncotarget 9:36585–36602. https://doi.org/10.18632/oncotarget.26379

Wang C, Tammi M, Guo H, Tammi R (1996) Hyaluronan distribution in the normal epithelium of esophagus, stomach, and colon and their cancers. Am J Pathol 148:1861–1869
Wang TP, Pan YR, Fu CY, Chang HY (2010) Down-regulation of UDP-glucose dehydrogenase affects glycosaminoglycans synthesis and motility in HCT-8 colorectal carcinoma cells. Exp Cell Res 316:2893–2902. https://doi.org/10.1016/j.yexcr.2010.07.017
Wang D, Narula N, Azzopardi S et al (2016) Expression of the receptor for hyaluronic acid mediated motility (RHAMM) is associated with poor prognosis and metastasis in non-small cell lung carcinoma. Oncotarget 7:39957–39969. https://doi.org/10.18632/oncotarget.9554
Ween MP, Oehler MK, Ricciardelli C (2011) Role of versican, hyaluronan and CD44 in ovarian cancer metastasis. Int J Mol Sci 12:1009–1029
Wei X, Senanayake TH, Warren G, Vinogradov S, v. (2013) Hyaluronic acid-based nanogel-drug conjugates with enhanced anticancer activity designed for the targeting of cd44-positive and drug-resistant tumors. Bioconjug Chem 24:658–668. https://doi.org/10.1021/bc300632w
Wei J, Hu M, Huang K et al (2020) Roles of proteoglycans and glycosaminoglycans in cancer development and progression. Int J Mol Sci 21. https://doi.org/10.3390/ijms21175983
Weniger MA, Küppers R (2021) Molecular biology of Hodgkin lymphoma. Leukemia 35:968. https://doi.org/10.1038/S41375-021-01204-6
Willemen Y, van den Bergh JMJ, Bonte SM et al (2016) The tumor-associated antigen RHAMM (HMMR/CD168) is expressed by monocyte-derived dendritic cells and presented to T cells. Oncotarget 7:73960–73970. https://doi.org/10.18632/ONCOTARGET.12170
Witschen PM, Chaffee TS, Brady NJ et al (2020) Tumor cell associated hyaluronan-CD44 signaling promotes pro-tumor inflammation in breast cancer. Cancer 12. https://doi.org/10.3390/cancers12051325
Wong HH, Lemoine NR (2009) Pancreatic cancer: molecular pathogenesis and new therapeutic targets. Nat Rev Gastroenterol Hepatol 6:412–422. https://doi.org/10.1038/nrgastro.2009.89
Wong MCS, Huang J, Lok V et al (2021) Differences in incidence and mortality trends of colorectal cancer worldwide based on sex, age, and anatomic location. Clin Gastroenterol Hepatol 19:955–966.e61. https://doi.org/10.1016/j.cgh.2020.02.026
Yates TJ, Lopez LE, Lokeshwar SD et al (2015) Dietary supplement 4-methylumbelliferone: an effective chemopreventive and therapeutic agent for prostate cancer. J Natl Cancer Inst 107:1–10. https://doi.org/10.1093/jnci/djv085
Yoshida E, Kudo D, Nagase H et al (2016) Antitumor effects of the hyaluronan inhibitor 4-methylumbelliferone on pancreatic cancer. Oncol Lett 12:2337–2344. https://doi.org/10.3892/ol.2016.4930
Yousefi B, Zarghami N, Samadi N, Majidinia M (2016) Peroxisome proliferator-activated receptors and their ligands in cancer drug-resistance: opportunity or challenge. Anticancer Agents Med Chem 16:1541–1548. https://doi.org/10.2174/1871520616666160204112941
Zhang L, Underhill CB, Chen L (1995) Hyaluronan on the surface of tumor cells is correlated with metastatic behavior. Cancer Res 55:428–433
Zhang YE, Wei J, Wang H et al (2012) Epithelial mesenchymal transition correlates with CD24+CD44+ and CD133+ cells in pancreatic cancer. Oncol Rep 27:1599–1605. https://doi.org/10.3892/or.2012.1681
Zhang G, Lu R, Wu M et al (2019) Colorectal cancer-associated ~ 6 kDa hyaluronan serves as a novel biomarker for cancer progression and metastasis. FEBS J 286:3148–3163. https://doi.org/10.1111/febs.14859
Zhou H, Mak PY, Mu H et al (2017) Combined inhibition of β-catenin and Bcr–Abl synergistically targets tyrosine kinase inhibitor-resistant blast crisis chronic myeloid leukemia blasts and progenitors in vitro and in vivo. Leukemia 31:2065–2074. https://doi.org/10.1038/leu.2017.87
Zhu R, Wang SC, Sun C et al (2013) Hyaluronan-CD44 interaction promotes growth of decidual stromal cells in human first-trimester pregnancy. PLoS One 8. https://doi.org/10.1371/journal.pone.0074812
Zöller M (2015) CD44, hyaluronan, the hematopoietic stem cell, and leukemia-initiating cells. Front Immunol 6. https://doi.org/10.3389/FIMMU.2015.00235/FULL

Chapter 8
Heparanase: A Paramount Enzyme for Cancer Initiation, Progression, and Metastasis

Valentina Masola, Nicola Greco, Giovanni Gambaro, Marco Franchi, and Maurizio Onisto

Abstract Consolidated data indicate that tumor bulk is not only made up of a heterogeneous set of neoplastic cells but also of a variety of resident and infiltrating host cells, soluble factors, and components of the extracellular matrix which as a whole is defined as the tumor microenvironment. In this context, the extracellular matrix plays a fundamental role in tumor progression as it acts as a repository for various biomolecules such as growth factors, cytokines, enzymes, and inhibitors which are mainly linked to heparan sulfate proteoglycans (HSPG) and whose release can regulate the response or not of cancer cells. Among the various enzymes involved in the degradation of the ECM, heparanase (HPSE) has been shown to be particularly involved in tumor progression and metastatic invasion. This enzyme, capable of cutting heparan sulfate (HS) chains, is overexpressed in practically all solid tumors, clearly demonstrating that it has pro-invasive and pro-angiogenic characteristics for neoplastic cells.

Furthermore, considering that heparanase is released not only by tumor cells but also by platelets, endothelial cells, and immune cells, we can admit that its enzymatic activity has a strong impact on the tumor microenvironment. Here, we discuss the recent development in the study of heparanase in cancer progression as well as on novel mechanisms by which heparanase regulates the nature of the tumor microenvironment.

V. Masola
Renal Unit, Department of Medicine, University Hospital of Verona, Verona, Italy

Department of Biomedical Sciences, University of Padova, Padua, Italy

N. Greco · M. Onisto (✉)
Department of Biomedical Sciences, University of Padova, Padua, Italy
e-mail: maurizio.onisto@unipd.it

G. Gambaro
Renal Unit, Department of Medicine, University Hospital of Verona, Verona, Italy

M. Franchi
Department for Life Quality Studies, University of Bologna, Rimini, Italy

I. Kovalszky et al. (eds.), *The Extracellular Matrix and the Tumor Microenvironment*, Biology of Extracellular Matrix 11,
https://doi.org/10.1007/978-3-030-99708-3_8

8.1 Introduction

The life of almost all cells that make up the tissues of multicellular organisms is influenced by the surrounding microenvironment. The signals arriving at individual cells can be both physical and chemical in nature and can depend on contacts with adjacent cells and structural molecules or on soluble factors released by cells or tissues located nearby. This bidirectional communication between the microenvironment and the cells that are immersed in it therefore plays a fundamental role in tissue homeostasis both in normal and pathological conditions such as in the case of tumors (Quail and Joyce 2013). Tumor microenvironment (TME) has been defined as the environment containing the cancer cells and extracellular matrix (ECM). Beside the growing tumor includes blood vessels, fibroblasts, immune and inflammatory cells, signal molecules and the various structural molecules that make up the interstitial stroma (Baghban et al. 2020). The importance of the tumor microenvironment is demonstrated by the fact that evolving environmental conditions and signals from growing tumors continually induce changes in the composition of TME, increasingly underlining the need to develop new therapeutic strategies to target the microenvironment rather than the tumor itself (Labani-Motlagh et al. 2020).

For this reason, it is important to stress the concept that, in addition to the structural function, the tumor ECM can also acts as a storage reserve for numerous bioactive molecules, such as growth factors, cytokines, chemokines, and enzymes which bind mainly to the HS proteoglycans (HSPG) owing to the negative charges present on the GAGs (De Pasquale and Pavone 2020; Bartolini et al. 2020). The heparan sulfate proteoglycans (HSPG) constitute a heterogeneous group of molecules consisting of various types of proteins to which heparan sulfate (HS) linear chains of variable length are covalently linked. HS consists of the repetition of a disaccharide unit formed by a glucuronic acid (GlcA) or its iduronic acid epimer (IdoA) and by an *N*-acetyl glucosamine (GlcNAc) or *N*-sulfo glucosamine (GlcNS). The bond that joins the disaccharides is of type α $(1 \rightarrow 4)$ or β $(1 \rightarrow 4)$, while the interdisaccharide bonds are always α $(1 \rightarrow 4)$. HSPGs can be mainly present in the intracellular environment or on the cell membranes and in the extracellular space. HSPGs associated with cell surface are classified into two groups: syndecans (with transmembrane domain) and glypicans (which bind through glycosylfosfatidyl-inositol (GPI) anchors) (Karamanos et al. 2018, 2021). Membrane HSPGs can undergo a proteolytic cut of the protein core with consequent release from the cell surface (shedded ectodomain) or they can act as scaffold protein, coreceptor, endocytic, and adhesion receptors (Manon-Jensen et al. 2010). Pericellular HSPGs, on the other hand, refer those components of the extracellular matrix that are associated with the membrane of different cell types through integrins and/or cell surface receptors. This group of HSPGs includes perlecan, agrin, collagen XV and XVIII, the components of basement membranes. Perlecan is an HSPG that contains three HS chains at its N-terminus and which, in addition to its structural role, can be a storage tank for various growth factors including vascular endothelial growth factor

(VEGF), TGF-β, and FGF family members (Iozzo and Schaefer 2015). Agrin has an organization similar to perlecan with three side chains of HS. Together with laminin, collagen IV and nidogens, agrin, and perlecan are the fundamental constituents of the basement membrane (BM), which is the specialized laminar structure of the extracellular matrix that separates a connective tissue from a nonconnective tissue such as the epithelial layer (Iozzo 2005). Collagen XV and XVIII, having the structural properties of both a collagen and a proteoglycan, are other components of the basement membranes (Karamanos et al. 2021). In this scenario, the only enzyme capable of degrading the HS chains of HSPGs is heparanase, described and characterized since the 1980s by the group of I. Vlodavsky (Vlodavsky et al. 1999).

Heparanase-1 (HPSE) is an endo-β-D-glucuronidase, cutting heparan sulfate chain at the level of the β-1,4-glycosidic bond between glucuronic acid and glycosamine residue, generating 5–10 kDa fragments of HS. The human heparanase-1 gene is located on chromosome 4q21.3 and by alternative splicing express two mRNAs containing the same open reading frame (ORF) (Vlodavsky et al. 1999). In addition to HPSE-1, there is HPSE-2, a protein 40% similar to heparanase-1, which has no glycosidase activity but seems to act as inhibitor for HPSE-1 (McKenzie et al. 2000).

Heparanase is synthesized in the endoplasmic reticulum as a pre-HPSE precursor and has a very complex pathway of maturation.

The inactive pre-HPSE form is converted into pro-HPSE of approximately 68 kDa, through the proteolytic cleavage of the signal peptide located in the N-terminal region and subsequently processed in the Golgi apparatus. The precursor is then internalized in vesicles to be secreted into the extracellular space where it interacts with many membrane-bound molecules including mannose 6-phosphate and low-density lipoprotein receptors as well as syndecan-1 (Levy-Adam et al. 2003; Simizu et al. 2004; Ben-Zaken et al. 2008). From the extracellular space, it is carried inside the cell by endocytosis. In the lysosome, the pro-HPSE form undergoes a proteolytic cut by the enzyme cathepsin-L which leads to the formation of two active peptides (Abboud-Jarrous et al. 2008).

Once the crystalline structure of human HPSE-1 has been determined, it has been shown that this enzyme includes a TIM-barrel domain where the catalytic site is located, and a C-terminal domain essential for secretion and for the regulation of its enzymatic and nonenzymatic activity (Wu et al. 2015).

As far as protein expression is concerned, in healthy tissues and in physiological conditions, HPSE shows very low levels of expression and is limited to a few cell types such as keratinocytes, trophoblast, platelets, mast cells, and leukocytes. Conversely, in pathological conditions such as tumor progression and metastasis, inflammation, EMT, and fibrosis, HPSE is overexpressed. By means of immune-histochemical analysis, in situ hybridization, and real-time PCR, it has been demonstrated that heparanase is upregulated essentially in almost all human tumors examined (Masola et al. 2020; Vlodavsky et al. 2012).

Since more recent studies have shown that heparanase is not only produced by tumor cells but also by endothelial cells and by activated immune cells, there is more and more evidence that its activity has a strong impact on the tumor microenvironment by promoting growth, neo-angiogenesis, and formation of a metastatic niche. In confirmation of this hypothesis, it has been shown that the overexpression of heparanase in transgenic mice (Hpa-Tg) makes the tumor microenvironment more favorable to the development of tumors in various experimental animal models (Boyango et al. 2014). Collectively, these results led to the hypothesis that heparanase plays a fundamental role in promoting the early stages of tumorigenesis.

The present chapter intends to summarize all the concepts and experimental evidences so far described concerning the involvement of HPSE in the initiation, progression, and metastatic process of cancer cells in relation to the tumor microenvironment.

8.2 Heparanase, Tumor Invasiveness, and Metastasis

Tumor invasion and metastasis can be defined as the result of a series of steps of interaction between the neoplastic cells and the microenvironment that surrounds them, and which implies their ability to modify the scaffolding of the tissues through the degradation or remodeling of the molecular structures that form and give rise to the extracellular matrix (Hanahan and Coussens 2012; Hinshaw and Shevde 2019). These invasive and metastatic capacities of tumor cells constitute one if not the main hallmark of malignant tumors compared to "in situ" tumors and are the cause of the majority of cancer-related deaths. It is now a consolidated fact that for this function, cancer cells have at their disposal a battery of degradative enzymes capable of "cutting" all the various macromolecules that make up the ECM surrounding the neoplastic mass, and whose overexpression correlates directly with the tumor progression and the formation of metastases. Among these degradative enzymes, the various matrix metalloproteases (MMPs), aspartic and cysteine proteases, together with the urokinase-type Plasminogen Activator (uPA) should be mentioned (Gialeli et al. 2011; Stetler-Stevenson and Yu 2001; McMahon and Kwaan 2015). In cooperation with these proteolytic enzymes, heparanase acts and contributes to tumor progression thanks to its ability to cut HS chains, thus altering the structure and function of HSPG and consequently contributing in a fundamental way both to the remodeling of the extracellular matrix and to the cleavage of syndecans and glypicans bound to the cell surface. From an experimental point of view, several studies have confirmed the role of HPSE in promoting tumor invasiveness by demonstrating how the overexpression of this enzyme increased the metastatic potential of different tumor cell lines, and correlated with reduced survival in various tumor xenograft models (Vlodavsky et al. 1999; Ilan et al. 2006). As counter-proof of this role, by inhibiting the activity of HPSE by means of gene silencing or specific inhibitors, a reduction in the invasive capacity and metastasis of various tumor cell lines was obtained both in vivo and in vitro (Edovitsky et al. 2004; Masola et al. 2009). A further confirmation of HPSE as a promoter of tumor invasiveness was obtained by the immunohistochemical analysis of tumor samples which then highlighted how the higher HPSE positivity was attributable to the areas of tumor invasiveness, while, on the contrary, the adjacent healthy tissue showed no detectable signal level (Beckhove et al. 2005; Tang et al. 2002).

It should also be borne in mind that, in particular for tumors of epithelial origin, the possibility of invading adjacent tissues requires the tumor cell to lose its adhesion with other tumor cells and to change shape by acquiring mesenchymal cell traits. This epithelial to mesenchymal transition (EMT) process, originally described during gastrulation in embryonic development, was subsequently observed and described in various pathological events including tissue fibrosis, tumor invasiveness, and metastasis (Di Gregorio et al. 2020; Dongre and Weinberg 2019). During EMT, epithelial cells tend to lose cell–cell adhesion junctions and apical–basal polarity and acquire a mesenchymal phenotype with a marked propensity for motility (Kalluri 2009). Various experimental studies on different animal models of tumors strongly support the concept that EMT is a key mechanism in tumor progression aimed at invasion and metastasis (Dongre and Weinberg 2019). In this context, the role played by heparanase in promoting EMT has been well documented in both renal fibrosis and tumor pathology (Masola et al. 2016; Li et al. 2016). Bearing in mind that among the major inducers of EMT there are both FGF and TGF-β and that these factors are present in the extracellular space mainly linked to the HS chains of proteoglycans, the action of heparanase allows the cutting of the HS chains and consequently the release of these factors which can then act on the tumor cells inducing the EMT (Masola et al. 2012, 2014; Pang et al. 2016).

Finally, it should be mentioned that a fundamental step of the metastatic cascade is the process of intravasation and extravasation of invasive tumor cells that enter the circulatory system directly or through the lymphatic network (Wong and Hynes 2006). These steps therefore imply the ability of tumor cells to degrade the ECM and more particularly the subendothelial basement membrane (BM). BM is a particular form of ECM in which, in addition to laminin, type IV collagen, and nidogen, there are also three typical HSPGs such as agrin, perlecan, and collagen XVIII (Iozzo and Schaefer 2015). In concert with the role played by MMPs (in particular, MMP-2 and MMP-9 specifically for collagen IV), the activity of HPSE toward the HS chains contributes decisively to regulating the passage of tumor cells through the BM in the metastatic process (Shuman Moss et al. 2012; Reiland et al. 2004). This hypothesis has been increasingly confirmed over the years by several follow-up clinical studies in which patients with high HPSE expression showed increased tumor angiogenesis, a higher incidence of metastases, and a worse prognosis than patients with low or no HPSE expression. Furthermore, in experimental xenograft models, it has been shown that nude mice inoculated with HPSE-silenced tumor cells show a longer survival, a lower rate of liver infiltration/metastasis, and a lower rate of neo-angiogenesis with respect to controls (Vlodavsky et al. 2012). Overall, all of this experimental and clinical evidence allows us to affirm that HPSE is, for all intents and purposes, a key player in the process of invasiveness and metastasis of tumors.

8.3 Heparanase and Tumor Angiogenesis

The process of neo-angiogenesis, which is the formation of a new network of blood vessels, is fundamental to supply the tumor cells with nutrients and oxygen, to allow the growth of the neoplasm and to initiate metastatic dissemination. The enzymatic activity of heparanase toward HS chains contributes significantly to neo-angiogenesis because, as previously described, agrin, perlecan, and collagen XVIII HSPGs are structural components of the subendothelial basement membranes of blood capillaries and, therefore, their degradation allows endothelial cells to proliferate and migrate in response to angiogenic stimuli. In all this, the action that heparanase can carry out is twofold and synergistic: on the one hand, the cutting of the HS chains involves the remodeling of the ECM but, on the other hand, it determines the release and diffusion of growth factors and molecules linked to the ECM, such as VEGF-A, FGF-2 which are potent pro-angiogenic factors (Ilan et al. 2006; Vlodavsky et al. 1996). One of the first confirmations of the pro-angiogenic role of HPSE was obtained from the analysis of transgenic mice in which the relative overexpression of HPSE was accompanied by increased vascularization. In particular, in the mammary glands of hpse-tg mice, there was an excessive branching and grouping of alveoli connected to an interruption of the basement membrane and to a greater vascularization. The hpa-tg mice also showed an accelerated rate of hair growth, correlated with high expression of heparanase in hair follicle keratinocytes and increased vascularity (Zcharia et al. 2004). To confirm these effects on the neo-angiogenesis process, it was then demonstrated with immunohistochemical analysis that a preferential staining of the HPSE protein was evident in the endothelium of the capillaries and small germinative vessels, while the mature quiescent blood vessels did not show any positivity (Elkin et al. 2001).

In 2016, a new HPSE-dependent mechanism for inducing neo-angiogenesis was unveiled in myeloma cells. In these tumor cells, most of the pro-tumorigenic effects of HPSE reside in the ability to increase expression and shedding of syndecan-1, a proteoglycan which is critical for the growth of myeloma cells (Ramani et al. 2013). Specifically, shortening the HS chains on syndecan-1 HPSE provides greater access to the core protein of syndecan-1 for metalloproteinase 9 (MMP9), thus facilitating cleavage of syndecan-1. The extracellular cleavage of syndecan-1 allows its release from the cell surface thus uncovering a cryptic domain which is an unveiled site for both the VEGF-2 receptor and the $\alpha4\beta1$ integrin present on the cell membrane of both myeloma cells and endothelial cells. This coupling of the receptor with the integrin leads to the activation of VEGF-2 receptor signaling which promotes augmented invasion and angiogenesis of myeloma. Roneparstat, an HPSE inhibitor, reduces the spread of syndean-1 and consequently inhibits tumor invasion and neo-angiogenesis (Jung et al. 2016).

Also on another experimental myeloma model, it has recently been shown that heparanase overexpression is associated with reduced levels of CXCL10, a member of the CXC chemokine family that binds to the CXCR3 receptor and is involved in the induction of apoptosis, chemotaxis and which suppresses the growth of myeloma by attenuating the proliferation of endothelial cells and, therefore, neo-angiogenesis (Strieter et al. 2006). Since the expression of heparanase is inversely proportional to the expression of CXCL10, this experimental evidence highlights a new mechanism that allows this enzyme to control tumor angiogenesis in myeloma (Barash et al. 2014).

8.4 Heparanase and Tumor Microenvironment

ECM represents a reservoir for bioactive molecules and it provides a strong support to tumor growth and metastasis (Folkman et al. 1988). Heparan sulfate is an essential element in cell signaling and communication. HS has the ability to bind growth factors and cytokines and thereby create a reservoir of signaling molecules in the ECM and on the surface of cells (Bartolini et al. 2020; Iozzo and Sanderson 2011). HS controls growth-factor activity in the microenvironment by regulating the half-life of the ligand, its diffusion, and its interaction with tyrosine kinase receptors (Casu et al. 2010). ECM changes modulate the activation of intracellular signaling pathways (Kim et al. 2011), causing downstream variations in gene expression (Fattet et al. 2020). Consequently, it regulates cellular adhesion, cytoskeletal dynamics, cell invasion and migration, cell proliferation and survival, differentiation, stem cell-like phenotype, as well as epithelial–mesenchymal transition (EMT), cell metabolism, and chemosensitivity (Cox 2021). Cell surface HS proteoglycans such as syndecans, especially syndecan-1, are actively involved in tumor onset and cancer progression. In particular, syndecans are shedded by proteases (Fears and Woods 2006) and the subsequent action of heparanase changes their physiological roles (Piperigkou et al. 2016). For instance, it has been shown that HPSE-mediated shedding of syndecan-1 liberates peptide fragments with VEGF receptor activity, and this sustains angiogenesis (Jung et al. 2016).

Therefore, HS remodeling by HPSE has important implications for remodeling of the cellular microenvironment during inflammation and cancer progression (Iozzo and Sanderson 2011).

Several works have proved that HPSE is expressed not only by tumor cells but also by cells of the tumor microenvironment (stroma, fibroblasts, and inflammatory cells) (Vlodavsky et al. 2012; Weissmann et al. 2016; Hammond et al. 2014). Thus, targeting the tumor microenvironment by heparanase inhibitors enhances the antitumor activity of approved therapies (Ritchie et al. 2011; Noseda and Barbieri 2020). For instance, in the bone marrow microenvironment, heparanase regulates the retention and proliferation of hematopoietic progenitor cells (Spiegel et al. 2008), modulates clonogenicity, proliferative potential, and migration of mesenchymal stem cells (Cheng et al. 2014), as well as shifting the differentiation potential of osteoblast progenitors within the myeloma bone microenvironment from osteoblastogenesis to adipogenesis (Ruan et al. 2013). Inflammatory cells are a key component of the tumor microenvironment, and they exert many activities promoting metastatic dissemination and metastatic niche formation. Since heparanase is involved in both leukocyte and macrophage activation/migration, it is therefore no surprise that there is a tight relation among heparanase, inflammation, and cancer microenvironment (Higashi et al. 2020; Elkin 2020).

Heparanase also contributes to the formation of metastases-regulating exosomes in the tumor microenvironment (Thompson et al. 2013). Exosomes mediate tumor–host cell interactions both locally and at metastatic sites (Peinado et al. 2017). It has been proved that the increased amount of HPSE in the tumor microenvironment greatly increases exosome secretion through modulation of syndecan-1 signaling (Thompson et al. 2013; Baietti et al. 2012). In addition, heparanase is present on exosome surfaces (Sanderson et al. 2019), and this is one of the mechanisms by means of which heparanase induces chemoresistance in myeloma (Purushothaman and Sanderson 2020).

In summary, heparanase expressed by tumor cells and by the cells of the tumor microenvironment regulates the aggressiveness of cancer, and represents an important contributor to bad outcomes among cancer patients.

8.5 Heparanase, Autophagy, and Evading Cell Death

Autophagy is an evolutionarily conserved catabolic pathway required to remove unfolded proteins and damaged organelles, and thus maintain cellular homeostasis. In cancer cells, however, autophagy is further induced by starvation and stress, sustaining their high metabolic rate and promoting cell survival (Levy et al. 2017). Recent data suggest that heparanase enhances tumor growth and chemoresistance by promoting autophagy (Shteingauz et al. 2015). The mechanism by means of which heparanase regulates autophagy has not been completely described but it involves mTOR1 (Shteingauz et al. 2015).

Another mechanism by means of which heparanase may contribute to cancer cells evading cell death is the inhibition of apoptosis, indeed it has been documented that HPSE inhibition induces massive expression of cleaved caspase-3 in xenografts of medulloblastoma (Spyrou et al. 2017).

8.6 Heparanase in Inflammation

The tumor microenvironment is characterized by chronic inflammation and tumors have been described as wounds that never heal (Singel and Segal 2016). Numerous immune cells have been shown to be intimately involved with the tumor microenvironment (TME). While it is true that some infiltrating immune cells function in eliminating tumors, on the other hand, there exists a pool of immune cells that promote tumor growth and progression. Indeed, inflammatory cells persisting in the tumor establish a cross-talk with tumor cells that may result in a phenotype switch into tumor-supporting cells (Dehne et al. 2017; Marzagalli et al. 2019; Gajewski et al. 2013).

The heparan sulfate proteoglycans (HSPG) plays an important role in cancer inflammation. Heparan sulfate (HS) is known to control inflammatory responses at multiple levels, including sequestration of cytokines/chemokines in the extracellular matrix (Jayatilleke and Hulett 2020; Xie and Li 2019), modulation of leukocyte interactions with endothelium and ECM (Higashi et al. 2020), and initiation of innate immune response through interactions with toll-like receptors 4 (TLR4) (Elkin 2020; Goodall et al. 2014). As a consequence, heparanase through the cleavage of HS of ECM may affect several aspects of inflammatory reactions, such as the regulation of pro-inflammatory immune-cell migration and activation, establishing acute and chronic inflammation, regulating secretion of cytokines and chemokines anchored within the ECM, and promoting lymphoangiogenesis (Higashi et al. 2020; Giese et al. 2019; Vlodavsky et al. 2016; Jayatilleke and Hulett 2020).

Leukocyte migration is characterized by adhesive interactions with endothelial cells leading to arrest, adhesion strengthening, crawling, and the migration of cells through the vessel wall and into inflammation sites (Higashi et al. 2020). The adhesion of leukocytes to the endothelial wall is facilitated by HS and other cell surface molecules (i.e., selectin and integrin), leading to cell arrest and the initiation of infiltration (Higashi et al. 2020; Collins and Troeberg 2019). A number of pro-inflammatory chemokines bind to HS, and this chemokine presentation plays a critical role in leukocyte recruitment. The enzymatic activity of HPSE produces a gradient of HS-bound pro-inflammatory cytokines such as interleukin (IL)-1β, IL-6, IL-8, IL-10, and TNF-α that stimulates the recruitment of leukocytes (Collins and Troeberg 2019; Xie and Li 2019). Studies on HS degradation have shown that HPSE is overexpressed in different types of innate immune cells such as neutrophils (Higashi et al. 2020), macrophages (Elkin 2020), DCs (Poon et al. 2014), and mast cells that mediate both acute and chronic inflammatory responses.

Neutrophils are the prevalent type of innate immune cells that are involved in acute inflammation. In model organisms like zebrafish, it has been demonstrated that neutrophils are often the first cells to arrive at sites of developing inflammation both at a wound, and during the early initiation phases of carcinogenesis. Neutrophils are the major mediators of acute inflammation and related tissue injury (Dehne et al. 2017). For example, HPSE expression has been shown to increase the expression of pancreatic cytokines (TNF-α, IL-6, etc.) and signaling molecule (phospho-STAT3) activity, along with enhanced edema and inflammation marked by neutrophil infiltration, which ultimately leads to acute pancreatitis (Khamaysi et al. 2017). In addition, studies on the mouse model of sepsis-associated inflammatory lung disease and heparanase null-mice suggest that HPSE activity induces neutrophil infiltration into pulmonary microvascular endothelial cells (Schmidt et al. 2012). In an interesting contrast, the enzymatic activity of HPSE may produce an anti-inflammatory effect on neutrophils, probably by disrupting the chemokine gradients with consequent impairment of recruitment (Massena et al. 2010). When in acute inflammation, the action of neutrophiles is not sufficient to completely resolve the inflammatory state, the infiltrating leukocytes changes from neutrophils to macrophages, which are the major players in chronic inflammation. A large proportion of tumor-infiltrating immune cells are tumor-associated macrophages (TAMs), which are key promoters of inflammation and contribute strongly to cancer progression (Szebeni et al. 2017; Nathan and Ding 2010).

HPSE participates in macrophage activation resulting in increased production of pro-inflammatory cytokines (Elkin 2020). Macrophage activation is mediated by toll-like receptor (TLRs) activation through the soluble HS fragments produced by HPSE (Goodall et al. 2014). HPSE has been implicated in a number of macrophages-induced HPSE expressions in the colonic epithelial cells, promoting inflammation that drives the progression of colitis to colon cancer enhancing cancer progression (Putz et al. 2017). Macrophages interact with endothelial cells in an HPSE-mediated manner to maintain a chronic inflammatory condition, which aids the formation of a tumor-promoting microenvironment with NF-κB signaling and induction of STAT3 expression (Waterman et al. 2007). HPSE has been shown to generate a vicious cycle by stimulating macrophages, which induce the production and activation of epithelial–HPSE via TNF-α and cathepsin-L (Lerner et al. 2011; Menzel et al. 2006). HPSE expression has also been shown to promote macrophage activation, leading to TNF-α production in macrophages as well as in renal tissue, and to enhancement of chronic inflammation associated with diabetic nephropathy (Goldberg et al. 2014). Using a genetic approach, mice lacking HPSE have been shown to possess macrophages that express lower levels of cytokines such as TNFα, IL-1β, IL-6, and IL-10. Macrophages lacking HPSE show impaired phagocytic activity and reduced infiltrative capacity (Waterman et al. 2007).

Natural killer (NK) lymphocytes are lymphoid cells characterized by potent cytolytic activity against bacterial infection and tumors. In vivo experiment on a HPSE-deficient mouse has proved that the lack of HPSE in NK reduced their capacity to infiltration (Putz et al. 2017). On the other hand, it seems that the HS fragment, produced by HPSE activity, can interfere with NK-cell toxicity receptors (NCR) and thus with NK activation (Mayes et al. 2017).

Interestingly, HPSE plays a relevant role in CAR therapy. In fact, the data suggest that expression of HPSE in long-term ex vivo-expanded T cells coexpressing a tumor-specific CAR (chimeric antigen receptor) improves their capacity to degrade the ECM. Thus, HPSE can improve CAR-T antitumor activity without compromising their viability, expansion, or effector function (Caruana et al. 2015).

8.7 Heparanase, Exosome, and Chemoresistance

High expression of HPSE is strongly associated with tumor aggressiveness, greater tumor growth, angiogenesis, and metastasis with poor prognosis. It was demonstrated that exogenous HPSE induces tumor cells to increase exosome secretion, regulates their composition, and promotes tumor progression (Salem et al. 2016; Thompson et al. 2013).

Exosoms are membrane-derived particles belonging to the extracellular vesicles (EVs) and involved in signaling processes and intercellular cross-talk. They are released into the extracellular space from different cells in normal conditions but their secretion is highly upregulated in diseases (Mustonen et al. 2021) or tumors (Vlodavsky et al. 2020; Zahavi et al. 2021). EVs have been generally classified in three groups based on their mechanism of biogenesis and mainly considering their size: (a) exosomes, ranging from 30 to 100 nm in diameter, (b) microvesicles or microparticles, measuring from 100 to 1000 nm, and (c) larger apoptotic bodies (1–5 μm) (Rilla et al. 2014). EVs and also exosomes have been found in extracellular fluid filtrates like plasma or bile (Witek et al. 2009), breast milk (Admyre et al. 2007), saliva (Ogawa et al. 2008), urine (Pisitkun et al. 2004), semen (Brody et al. 1983), blood (Johnstone et al. 1987), ascites (Andre et al. 2002), cerebrospinal fluid (Vella et al. 2008), and synovial fluid (Mustonen et al. 2016). Recently, EVs and in particular exosomes were found in extracellular matrix (ECM) of most of connective tissues (Arasu et al. 2017).

Exosomes which derive from the luminal membranes of the endosome play crucial roles in mediating local and systemic cell communication (Salem et al. 2016). Exosomes are able to transfer their cargo including nucleic acids as microRNA and mRNA, lipids, proteins and signaling molecules to target cells and ECM in distant sites within the body. They may act as intercellular messengers or carriers of regulatory molecules into target cells, thus providing cell-to-cell distant communications, but they are also able to transfer important information from cells to the surrounding microenvironment (Salem et al. 2016; Rilla et al. 2014; Arasu et al. 2017; Madhusoodanan 2019).

Secretion and shedding of EVs into the surrounding ECM originates from both the apical and basal surfaces of the epithelial cells. EVs secretion represents a way to reverse intracellular content or molecules into the ECM. For instance, cells release Hyaluronan (HA) in extracellular spaces by shedding HA-coated EVs from the tip of short cytoplasmic filopodia or microvilli (Mustonen et al. 2016; Rilla et al. 2013). It is of interest that an increase of HA expression is related to an increased aggressiveness of cancers, so that EVs are evaluated as clinical biomarkers (Rilla et al. 2013, 2014; Properzi et al. 2013) or future therapeutics (Alvarez-Erviti et al. 2011; Kordelas et al. 2014).

Even though the modality and regulation of exosome uptake by target cells is still under discussion, it was fascinatingly suggested that filopodia could be involved in an exosome uptake which resembles the virus enveloping: exosomes were observed to surf along the filopodia surface to the body cell surface (Heusermann et al. 2016). Here, exosomes seem to entry into recipient cells by two main mechanisms: endocytosis and membrane fusion (Raposo and Stoorvogel 2013; Turturici et al. 2014; James 2016). It was suggested that exosome uptake is HSPGs-dependent, following the same mode previously described for several viruses. These proteoglycans are sorted to and associate with exosomes and act as internalizing receptors of cancer cell-derived exosomes or receptors of exosome cargo (Christianson et al. 2013).

When cancer cells develop a more aggressive phenotype, the secretion of exosomes radically increases. To confirm this data, it was reported that also high amounts of enzymatically inactive forms of HPSE do not evidently increase exosome secretion (Thompson et al. 2013). Exosomes might be able to modify the composition of tumor microenvironment, directly digesting ECM components or indirectly controlling target cells in synthesizing or even degrading ECM molecules. In fact, exosomes are so small that they can easily penetrate into the depth of collagen meshwork of ECM, invade endothelial cells, and travel to lymph nodes and distant tissues to promote cancer metastasis (Conigliaro and Cicchini 2019). Exosomes interact with tumor microenvironment. Degradation of peritumoral ECM is supported by some proteinases which have been described in exosomes: matrix metalloproteinases (MMPs) are able to digest both basement membranes and other ECM molecules, but also aggrecanase which specifically degrade the aggrecan-rich ECM in some tumors and diseases (Arasu et al. 2017; Dolo et al. 1999). In exosomes is present also another enzyme: heparanase the sole heparan sulfate degrading endoglycosidase (Thompson et al. 2013; Vlodavsky et al. 2016). Both exogenous heparanase, which cleaves HS chains on proteoglycans in ECM, or also its increased expression develop exosomes production and secretion (Thompson et al. 2013). An increased expression of heparanase observed in many cancers including ovarian, pancreatic, myeloma, colon, bladder, brain, prostate, breast, liver tumors, and rhabdomyosarcoma represents a clinical marker associated to a high risk of metastasis and poor prognosis.

Heparanase can also affect the exosome protein cargo, which shows higher content of syndecan-1, VEGF, and HGF in exosomes secreted by heparanase-high expressing cells (Thompson et al. 2013; Baietti et al. 2012). In particular, it has been reported that HPSE trims long heparan sulfate chains into shorter ones and favors syndecans complexes that, by recruiting syntenin-1 (a cytoplasmic adaptor of HSPGs) and ALIX–ESCRT (endosomal-sorting complex required for transport machinery), improve endosomal intraluminal budding and production of exosomes (Baietti et al. 2012; Friand et al. 2015).

Intercellular communications between cancer cells and circulating lymphocytes is also suggested, as circulating lymphocytes in breast cancer patients can secrete heparan sulfate syndecan-1 which upregulating heparanase expression, and exosomes secretion could be a fundamental regulator of molecular mechanisms in carcinogenesis (Roucourt et al. 2015; Theodoro et al. 2019), together with syndecan-2 which can affect breast carcinoma progression through a regulation of cytoskeleton, cell adhesion, and invasion (Lim et al. 2015).

It was demonstrated that exosomes released by cells expressing high levels of HSPE favor spreading of cancer cells on fibronectin and promote intravasation much more than exosomes secreted by heparanase-low cells (Thompson et al. 2013). Data reporting that the effect of heparanase/exosomes is related to fibronectin, which in its fibrillar form seems to act as a scaffold favoring cancers cell invasion (Erdogan et al. 2017), and the fact that a dramatic syndecan-1 expression in stromal fibroblasts of invasive breast carcinoma is associated with an aligned ECM fiber array to favor cancer cell invasion (Yang and Friedl 2016) suggest that any HSPE/exosomes effect could be mediated by the other ECM components. Moreover, monoclonal antibodies which are able to neutralize HSPE can inhibit myeloma and lymphoma tumor growth, and dissemination is the result of a combined effect on tumor cells and cells of the tumor microenvironment (Vlodavsky et al. 2016).

Anticancer drugs can induce therapy resistance by promoting the development of autophagy (Larrue et al. 2016; Sanderson et al. 2017) and although intracellular HSPE is primarily located in lysosomes, late endosomes, it is also found in autophagosomes (Coombe and Gandhi 2019). Heparanase enhances tumor growth and chemoresistance by favoring both exosomes secretion and inducing autophagy (Shteingauz et al. 2015). Interesting results suggest that exosomes interplay with tumor cells, and acting as barriers to anticancer therapy, play a dramatic role in supporting tumor chemoresistance (Zhang and Grizzle 2014; Au Yeung et al. 2016). Consequently, heparanase favoring the secretion and uptaking of exosomes and inducing autophagy promotes tumor progression and chemoresistance by affecting both tumor and host cells behavior (Shteingauz et al. 2015).

When human myeloma cells are exposed to chemotherapy, the increase of HSPE expression promotes exosomes secretion and enhances their association together in cells which survive chemotherapy. Exosomes induced by anticancer drugs, also called chemoexosomes, allow an easy degradation of heparan sulfate within ECM as they exhibit a different protease profile compared the untreated cells and an increase of HSPE on their surface. Metastatic breast cancer cells showing a relation between exosomes secretion and invadopodia development suggest that both ECM degradation and invasion finally depend on exosome MT1-MMP and other proteases (Hoshino et al. 2013).

Chemoexosomes trapped within the tumor microenvironment can immediately induce changing in neighboring tumor cells by increasing signaling pathways (e.g., ERK, P38) that promote chemoresistance. For instance, chemoexosome HSPG can be readily transferred to macrophages which improve their migration but also secretion of TNF-α, an important myeloma growth factor (Bandari et al. 2018). Chemoexosomes can also be stored within tumor ECM to exert their action on local ECM or migrate from the tumor microenvironment to distant tissues to develop premetastatic niches only later (Bandari et al. 2018; Peinado et al. 2011; Becker et al. 2016).

To confirm the crucial role of HPSE in regulating exosomes secretion and chemoresistance, some authors recently reported that in multiple myeloma cells overexpressing EVs carrying miR-1252-5p, a consistent reduction in the levels of HSPE occurred. This data suggest that this microRNA may contribute to negative regulation of HSPE expression and can represent a novel therapeutic approach to sensitize multiple myeloma cells to bortezomib treatment (Rodrigues-Junior et al. 2021).

8.8 Concluding Remarks

Along with cardiovascular diseases, cancer constitutes a major cause of death among people in high-income countries. Despite the numerous therapeutic approaches developed so far, the complete recovery of the patient is often hindered and impeded by the complexity and variability of the disease considering not only the molecular changes within the tumor cell itself but the alterations that occur within the tumor microenvironment.

In particular, what makes malignant tumors so difficult to eradicate both surgically and with radio-chemotherapy is their ability to invade surrounding tissues and metastasize, to promote angiogenesis, to regulate inflammation, and to evade the immune response and apoptosis. In all these aspects of tumor pathology, heparanase plays a fundamental role both at the level of neoplastic cells and also at the level of the tumor microenvironment where its neutralization seems to be sufficient to slow down tumor growth as demonstrated in some lymphoma models. Consequently, a possible pharmacological strategy should be aimed at the inactivation of HSPE both within the tumor mass and in the tumor microenvironment where the use of the heparanase inhibitors has proved particularly decisive in the control of minimal residual disease.

Several HPSE inhibitors are already in clinical trials as anticancer drugs and, for some, a certain therapeutic efficacy and good tolerability for the patient have already been proven. For the near future, we therefore hope that the development of new drugs aimed at inhibiting the activity of HPSE may have a better therapeutic value not only in the oncology field, but also for all the other pathologies for which its involvement has been amply demonstrated.

References

Abboud-Jarrous G, Atzmon R, Peretz T, Palermo C, Gadea BB, Joyce JA, Vlodavsky I (2008) Cathepsin L is responsible for processing and activation of proheparanase through multiple cleavages of a linker segment. J Biol Chem 283(26):18167–18176

Admyre C, Johansson SM, Qazi KR, Filén J-J, Lahesmaa R, Norman M, Neve EP, Scheynius A, Gabrielsson S (2007) Exosomes with immune modulatory features are present in human breast milk. J Immunol 179:1969–1978

Alvarez-Erviti L, Seow Y, Yin H, Betts C, Lakhal S, Wood MJA (2011) Delivery of siRNA to the mouse brain by systemic injection of targeted exosomes. Nat Biotechnol 29:341–345. https://doi.org/10.1038/nbt.1807

Andre F, Schartz NE, Movassagh M, Flament C, Pautier P, Morice P, Pomel C, Lhomme C, Escudier B, Le Chevalier T, Tursz T, Amigorena S, Raposo G, Angevin E, Zitvogel L (2002) Malignant effusions and immunogenic tumour-derived exosomes. Lancet 360:295–305

Arasu UT, Kärnä R, Härkönen K, Oikari S, Koistinen A, Kröger H, Qu C, Lammi MJ, Rilla K (2017) Human mesenchymal stem cells secrete hyaluronan-coated extracellularvesicles. Matrix Biol 64:54–68. https://doi.org/10.1016/j.matbio.2017.05.001. Epub 2017 May 5

Au Yeung CL et al (2016) Exosomal transfer of stroma-derived miR21 confers paclitaxel resistance in ovarian cancer cells through targeting APAF1. Nat Commun 7:11150

Baghban R, Roshangar L, Jahanban-Esfahlan R, Seidi K, Ebrahimi-Kalan A, Jaymand M, Kolahian S, Javaheri T, Zare P (2020) Tumor microenvironment complexity and therapeutic implications at a glance. Cell Commun Signal 18(1):59

Baietti MF, Zhang Z, Mortier E, Melchior A, Degeest G, Geeraerts A, Ivarsson Y, Depoortere F, Coomans C, Vermeiren E et al (2012) Syndecan-syntenin-ALIX regulates the biogenesis of exosomes. Nat Cell Biol 14:677–685

Bandari SK, Purushothaman A, Ramani VC, Brinkley GJ, Chandrashekar DS, Varambally S, Mobley JA, Zhang Y, Brown EE, Vlodavsky I, Sanderson RD (2018) Chemotherapy induces secretion of exosomes loaded with heparanase that degrades extracellular matrix and impacts tumor and host cell behavior. Matrix Biol 65:104–118. https://doi.org/10.1016/j.matbio.2017.09.001

Barash U, Zohar Y, Wildbaum G, Beider K, Nagler A, Karin N, Ilan N, Vlodavsky I (2014) Heparanase enhances myeloma progression via CXCL10 downregulation. Leukemia 28(11): 2178–2187

Bartolini B, Caravà E, Caon I, Parnigoni A, Moretto P, Passi A, Vigetti D, Viola M, Karousou E (2020) Heparan sulfate in the tumor microenvironment. Adv Exp Med Biol 1245:147–161

Becker A, Thakur BK, Weiss JM, Kim HS, Peinado H, Lyden D (2016) Extracellular vesicles in cancer: cell-to-cell mediators of metastasis. Cancer Cell 30:836–848. PubMed: 27960084

Beckhove P, Helmke BM, Ziouta Y, Bucur M, Dorner W, Mogler C et al (2005) Heparanase expression at the invasion front of human head and neck cancers and correlation with poor prognosis. Clin Cancer Res 11:2899–2906

Ben-Zaken O, Shafat I, Gingis-Velitski S, Bangio H, Kelson IK, Alergand T, Amor Y, Maya RB, Vlodavsky I, Ilan N (2008) Low and high affinity receptors mediate cellular uptake of heparanase. Int J Biochem Cell Biol 40(3):530–542

Boyango I, Barash U, Naroditsky I, Li JP, Hammond E, Ilan N, Vlodavsky I (2014) Heparanase cooperates with Ras to drive breast and skin tumorigenesis. Cancer Res 74(16):4504–4514

Brody I, Ronquist G, Gottfries A (1983) Ultrastructural localization of the prostasome—an organelle in human seminal plasma. Ups J Med Sci 88:63–80

Caruana I, Savoldo B, Hoyos V et al (2015) Heparanase promotes tumor infiltration and antitumor activity of CAR-redirected T lymphocytes. Nat Med 21:524–529. https://doi.org/10.1038/nm.3833

Casu B, Naggi A, Torri G (2010) Heparin-derived heparan sulfate mimics to modulate heparan sulfate-protein interaction in inflammation and cancer. Matrix Biol 29(6):442–452

Cheng CC, Lee YH, Lin SP, Huangfu WC, Liu IH (2014) Cell-autonomous heparanase modulates self-renewal and migration in bone marrow-derived mesenchymal stem cells. J Biomed Sci 21: 21

Christianson HC, Svensson KJ, van Kuppevelt TH, Li JP, Belting M (2013) Cancer cell exosomes depend on cell-surface heparin sulfate proteoglycans for their internalization and functional activity. PNAS 110(43):17380–17385. ww.pnas.org/cgi/doi/10.1073/pnas.1304266110

Collins LE, Troeberg L (2019) Heparan sulfate as a regulator of inflammation and immunity. J Leukoc Biol 105(1):81–92

Conigliaro A, Cicchini C (2019) Exosome-mediated signaling in epithelial to mesenchymal transition and tumor progression. J Clin Med 8:26. https://doi.org/10.3390/jcm8010026

Coombe DR, Gandhi NS (2019) Heparanase: a challenging cancer drug target. Front Oncol 9:article 1316. www.frontiersin.org. https://doi.org/10.3389/fonc.2019.01316

Cox TR (2021) The matrix in cancer. Nat Rev Cancer 21(4):217–238. https://doi.org/10.1038/s41568-020-00329-7

Dehne N, Mora J, Namgaladze D, Weigert A, Brüne B (2017) Cancer cell and macrophage cross-talk in the tumor microenvironment. Curr Opin Pharmacol 35:12–19. https://doi.org/10.1016/j.coph.2017.04.007

De Pasquale V, Pavone LM (2020) Heparan sulfate proteoglycan signaling in tumor microenvironment. Int J Mol Sci 21(18):6588

Di Gregorio J, Robuffo I, Spalletta S, Giambuzzi G, De Iuliis V, Toniato E, Martinotti S, Conti P, Flati V (2020) The epithelial-to-mesenchymal transition as a possible therapeutic target in fibrotic disorders. Front Cell Dev Biol 8:607483

Dolo V, D'Ascenzo S, Violini S, Pompucci L, Festuccia C, Ginestra A, Vittorelli ML, Canevari S, Pavan A (1999) Matrix degrading proteinases are shed in membrane vesicles by ovarian cancer cells in vivo and in vitro. Clin Exp Metastasis 17:131–140

Dongre A, Weinberg RA (2019) New insights into the mechanisms of epithelial-mesenchymal transition and implication for cancer. Nat Rev Mol Cell Biol 20(2):69–84

Edovitsky E, Elkin M, Zcharia E, Peretz T, Vlodavsky I (2004) Heparanase gene silencing, tumor invasiveness, angiogenesis, and metastasis. J Natl Cancer Inst 96(16):1219–1230

Elkin M (2020) Role of heparanase in macrophage activation. Adv Exp Med Biol 1221:445–460

Elkin M, Ilan N, Ishai-Michaeli R, Friedmann Y, Papo O, Pecker I, Vlodavsky I (2001) Heparanase as mediator of angiogenesis: mode of action. FASEB J 15(9):1661–1663

Erdogan B, Ao M, White LM, Means AL, Brewer BM, Yang L, Washington MK, Shi C, Franco OE, Weaver AM, Hayward SW, Li D, Webb J (2017) Cancer-associated fibroblasts promote directional cancer cell migration by aligning fibronectin. Cell Biol 216(11):3799–3816. https://doi.org/10.1083/jcb.201704053. Epub 2017 Oct 11

Fattet L et al (2020) Matrix rigidity controls epithelial-mesenchymal plasticity and tumor metastasis via a mechanoresponsive EPHA2/LYN complex. Dev Cell 54:302–316.e7

Fears CY, Woods A (2006) The role of syndecans in disease and wound healing. Matrix Biol 25(7): 443–456

Folkman J, Klagsbrun M, Sasse J, Wadzinski M, Ingber D, Vlodavsky I (1988) A heparin-binding angiogenic protein—basic fibroblast growth factor—is stored within basement membrane. Am J Pathol 130:393–400

Friand V, David G, Zimmermann P (2015) Syntenin and syndecan in the biogenesis of exosomes. Biol Cell 107:331–341. [PubMed: 26032692]

Gajewski TF, Schreiber H, Fu YX (2013) Innate and adaptive immune cells in the tumor microenvironment. Nat Immunol 14(10):1014–1022 5

Gialeli C, Theocharis AD, Karamanos NK (2011) Roles of matrix metalloproteinases in cancer progression and their pharmacological targeting. FEBS J 278(1):16–27

Giese MA, Hind LE, Huttenlocher A (2019) Neutrophil plasticity in the tumor microenvironment. Blood 133(20):2159–2167

Goldberg R, Rubinstein AM, Gil N, Hermano E, Li JP, van der Vlag J, Atzmon R, Meirovitz A, Elkin M (2014) Role of heparanase-driven inflammatory cascade in pathogenesis of diabetic nephropathy. Diabetes 63(12):4302–4313

Goodall KJ, Poon IK, Phipps S, Hulett MD (2014) Soluble heparan sulfate fragments generated by heparanase trigger the release of pro-inflammatory cytokines through TLR-4. PLoS One 9(10): e109596

Hammond E, Khurana A, Shridhar V, Dredge K (2014) The role of heparanase and sulfatases in the modification of heparan sulfate proteoglycans within the tumor microenvironment and opportunities for novel cancer therapeutics. Front Oncol 4:195

Hanahan D, Coussens LM (2012) Accessories to the crime: functions of cells recruited to the tumor microenvironment. Cancer Cell 21(3):309–322

Heusermann W, Hean J, Trojer D, Steib E, von Bueren S, Graff-Meyer A, Genoud C, Martin K, Pizzato N, Voshol J, Morrissey DV, Andaloussi SE, Wood MJ, Meisner-Kober NC (2016) Exosomes surf on filopodia to enter cells at endocytic hot spots, traffic within endosomes, and are targeted to the ER. J Cell Biol 213:173–184

Higashi N, Irimura T, Nakajima M (2020) Heparanase is involved in leukocyte migration. Adv Exp Med Biol 1221:435–444

Hinshaw DC, Shevde LA (2019) The tumor microenvironment innately modulates cancer progression. Cancer Res 79(18):4557–4566

Hoshino D, Kirkbride KC, Costello K, Clark ES, Sinha S, Grega-Larson N, Tyska MJ, Weaver AM (2013) Exosome secretion is enhanced by invadopodia and drives invasive behavior. Cell Rep 5: 1159–1168. [PubMed: 24290760]

Ilan N, Elkin M, Vlodavsky I (2006) Regulation, function and clinical significance of heparanase in cancer metastasis and angiogenesis. Int J Biochem Cell Biol 38(12):2018–2039

Iozzo RV (2005) Basement membrane proteoglycans: from cellar to ceiling. Nat Rev Mol Cell Biol 6(8):646–656

Iozzo RV, Sanderson RD (2011) Proteoglycans in cancer biology, tumour microenvironment and angiogenesis. J Cell Mol Med 15:1013–1031

Iozzo RV, Schaefer L (2015) Proteoglycan form and function: a comprehensive nomenclature of proteoglycans. Matrix Biol 42:11–55

James R (2016) Edgar Q&A: what are exosomes, exactly? BMC Biol 14:46

Jayatilleke KM, Hulett MD (2020) Heparanase and the hallmarks of cancer. J Transl Med 18(1):453

Johnstone RM, Adam M, Hammond JR, Orr L, Turbide C (1987) Vesicle formation during reticulocyte maturation. Association of plasma membrane activities with released vesicles (exosomes). J Biol Chem 262:9412–9420

Jung O, Trapp-Stamborski V, Purushothaman A, Jin H, Wang H, Sanderson RD, Rapraeger AC (2016) Heparanase-induced shedding of syndecan-1/CD138 in myeloma and endothelial cells activates VEGFR2 and an invasive phenotype: prevention by novel synstatins. Oncogenesis 5: e202. https://doi.org/10.1038/oncsis.2016.5

Kalluri R (2009) EMT: when epithelial cells decide to become mesenchymal-like cells. J Clin Invest 119:1417–1419

Karamanos NK, Piperigkou Z, Theocharis AD, Watanabe H, Franchi M, Baud S, Brézillon S, Götte M, Passi A, Vigetti D, Ricard-Blum S, Sanderson RD, Neill T, Iozzo RV (2018) Proteoglycan chemical diversity drives multifunctional cell regulation and therapeutics. Chem Rev 118(18):9152–9232

Karamanos NK, Theocharis AD, Piperigkou Z, Manou D, Passi A, Skandalis SS, Vynios DH, Orian-Rousseau V, Ricard-Blum S, Schmelzer CEH, Duca L, Durbeej M, Afratis NA, Troeberg L, Franchi M, Masola V, Onisto M (2021) A guide to the composition and functions of the extracellular matrix. FEBS J. https://doi.org/10.1111/febs.15776. Epub ahead of print. PMID: 33605520

Khamaysi I, Singh P, Nasser S, Awad H, Chowers Y, Sabo E, Hammond E, Gralnek I, Minkov I, Noseda A, Ilan N, Vlodavsky I, Abassi Z (2017) The role of heparanase in the pathogenesis of acute pancreatitis: a potential therapeutic target. Sci Rep 7(1):715

Kim S-H, Turnbull J, Guimond S (2011) Extracellular matrix and cell signalling: the dynamic cooperation of integrin, proteoglycan and growth factor receptor. J Endocrinol 209:139–151

Kordelas L, Rebmann V, Ludwig A-K, Radtke S, Ruesing J, Doeppner TR, Epple M, Horn PA, Beelen DW, Giebel B (2014) MSC-derived exosomes: a novel tool to treat therapy-refractory graft-versus-host disease. Leukemia 28:970–973

Labani-Motlagh A, Ashja-Mahdavi M, Loskog A (2020) The tumor microenvironment: a milieu hindering and obstructing antitumor immune responses. Front Immunol 11:940

Larrue C, Saland E, Boutzen H, Vergez F, David M, Joffre C, Hospital MA, Tamburini J, Delabesse E, Manenti S, Sarry JE, Recher C (2016) Proteasome inhibitors induce FLT3-ITD degradation through autophagy in AML cells. Blood 127:882–892. [PubMed: 26286850]

Lerner I, Hermano E, Zcharia E, Rodkin D, Bulvik R, Doviner V, Rubinstein AM, Ishai-Michaeli R, Atzmon R, Sherman Y, Meirovitz A, Peretz T, Vlodavsky I, Elkin M (2011) Heparanase powers a chronic inflammatory circuit that promotes colitis-associated tumorigenesis in mice. J Clin Invest 121(5):1709–1721

Levy JMM, Towers CG, Thorburn A (2017) Targeting autophagy in cancer. Nat Rev Cancer 17(9): 528–542

Levy-Adam F, Miao HQ, Heinrikson RL, Vlodavsky I, Ilan N (2003) Heterodimer formation is essential for heparanase enzymatic activity. Biochem Biophys Res Commun 308(4):885–891

Li J, Pan Q, Rowan PD, Trotter TN, Peker D, Regal KM, Javed A, Suva LJ, Yang Y (2016) Heparanase promotes myeloma progression by inducing mesenchymal features and motility of myeloma cells. Oncotarget 7(10):11299–11309

Lim HC, Multhaupt HA, Couchman JR (2015) Cell surface proteoglycans control adhesion and invasion of breast carcinoma cells. Mol Cancer 14:15

Madhusoodanan J (2019) Matrix mimics cell shape studies. Nature 566:563–565

Manon-Jensen T, Itoh Y, Couchman JR (2010) Proteoglycans in health and disease: the multiple roles of syndecan shedding. FEBS J 277(19):3876–3889

Marzagalli M, Ebelt ND, Manuel ER (2019) Unraveling the crosstalk between melanoma and immune cells in the tumor microenvironment. Semin Cancer Biol 59:236–250. https://doi.org/10.1016/j.semcancer.2019.08.002

Masola V, Maran C, Tassone E, Zin A, Rosolen A, Onisto M (2009) Heparanase activity in alveolar and embryonal rhabdomyosarcoma: implications for tumor invasion. BMC Cancer 9:304

Masola V, Gambaro G, Tibaldi E, Brunati AM, Gastaldello A, D'Angelo A, Onisto M, Lupo A (2012) Heparanase and syndecan-1 interplay orchestrates fibroblast growth factor-2-induced epithelial-mesenchymal transition in renal tubular cells. J Biol Chem 287(2):1478–1488

Masola V, Zaza G, Secchi MF, Gambaro G, Lupo A, Onisto M (2014) Heparanase is a key player in renal fibrosis by regulating TGF-β expression and activity. Biochim Biophys Acta 1843(9): 2122–2128

Masola V, Zaza G, Gambaro G, Onisto M, Bellin G, Vischini G, Khamaysi I, Hassan A, Hamoud S, Nativ O, Heyman SN, Lupo A, Vlodavsky I, Abassi Z (2016) Heparanase: a potential new factor involved in the renal epithelial mesenchymal transition (EMT) induced by ischemia/reperfusion (I/R) injury. PLoS One. 11(7):e016074

Masola V, Zaza G, Gambaro G, Franchi M, Onisto M (2020) Role of heparanase in tumor progression: molecular aspects and therapeutic options. Semin Cancer Biol 62:86–98

Massena S, Christoffersson G, Hjertström E, Zcharia E, Vlodavsky I, Ausmees N, Rolny C, Li JP, Phillipson M (2010) A chemotactic gradient sequestered on endothelial heparan sulfate induces directional intraluminal crawling of neutrophils. Blood 116(11):1924–1931

Mayes K, Elsayed Z, Alhazmi A, Waters M, Alkhatib SG, Roberts M, Song C, Peterson K, Chan V, Ailaney N, Malapati P, Blevins T, Lisnić B, Dumur CI, Landry JW (2017) BPTF inhibits NK cell activity and the abundance of natural cytotoxicity receptor co-ligands. Oncotarget 8(38): 64344–64357

McKenzie E et al (2000) Cloning and expression profiling of Hpa2, a novel mammalian heparanase family member. Biochem Biophys Res Commun 276(3):1170–1177

McMahon BJ, Kwaan HC (2015) Components of the plasminogen-plasmin system as biologic markers for cancer. Adv Exp Med Biol 867:145–156

Menzel K, Hausmann M, Obermeier F, Schreiter K, Dunger N, Bataille F et al (2006) Cathepsins B, L and D in inflammatory bowel disease macrophages and potential therapeutic effects of cathepsin inhibition in vivo. Clin Exp Immunol 146(1):169–180

Mustonen AM, Nieminen P, Joukainen A, Jaroma A, Kääriäinen T, Kröger H, Lázaro-Ibáñez E, Siljander PR, Kärjä V, Härkönen K, Koistinen A, Rilla K (2016) First in vivo detection and characterization of hyaluronan-coated extracellular vesicles in human synovial fluid. J Orthop Res 34:1960–1968

Mustonen AM, Capra J, Rilla K, Lehenkari P, Oikari S, Kääriäinen T, Joukainen A, Kröger H, Paakkonen T, Matilainen J, Nieminen P (2021) Characterization of hyaluronan-coated extracellular vesicles in synovial fluid of patients with osteoarthritis and rheumatoid arthritis. BMC Musculoskelet Disord 22:Article number 247

Nathan C, Ding A (2010) Nonresolving inflammation. Cell 140(6):871–882

Noseda A, Barbieri P (2020) Roneparstat: development, preclinical and clinical studies. Adv Exp Med Biol 1221:523–538

Ogawa Y, Kanai-Azuma M, Akimoto Y, Kawakami H, Yanoshita R (2008) Exosome-like vesicles with dipeptidyl peptidase IV in human saliva. Biol Pharm Bull 31:1059–1062

Pang MF, Georgoudaki AM, Lambut L, Johansson J, Tabor V, Hagikura K, Jin Y, Jansson M, Alexander JS, Nelson CM, Jakobsson L, Betsholtz C, Sund M, Karlsson MC, Fuxe J (2016) TGF-β1-induced EMT promotes targeted migration of breast cancer cells through the lymphatic system by the activation of CCR7/CCL21-mediated chemotaxis. Oncogene 35(6):748–760

Peinado H, Lavotshkin S, Lyden D (2011) The secreted factors responsible for pre-metastatic niche formation: old sayings and new thoughts. Semin Cancer Biol 21:139–146. [PubMed: 21251983]

Peinado H, Zhang H, Matei IR, Costa-Silva B, Hoshino A, Rodrigues G, Psaila B, Kaplan RN, Bromberg JF, Kang Y et al (2017) Pre-metastatic niches: organ-specific homes for metastases. Nat Rev Cancer 17:302–317

Piperigkou Z, Mohr B, Karamanos N, Götte M (2016) Shed proteoglycans in tumor stroma. Cell Tissue Res 365(3):643–655

Pisitkun T, Shen RF, Knepper MA (2004) Identification and proteomic profiling of exosomes in human urine. Proc Natl Acad Sci U S A 101:13368–13373

Poon IK, Goodall KJ, Phipps S, Chow JD, Pagler EB, Andrews DM, Conlan CL, Ryan GF, White JA, Wong MK, Horan C, Matthaei KI, Smyth MJ, Hulett MD (2014) Mice deficient in heparanase exhibit impaired dendritic cell migration and reduced airway inflammation. Eur J Immunol 44(4):1016–1030

Properzi F, Logozzi M, Fais S (2013) Exosomes: the future of biomarkers in medicine. Biomark Med 7:769–778. https://doi.org/10.2217/bmm.13.63

Purushothaman A, Sanderson RD (2020) Heparanase: a dynamic promoter of myeloma progression. Adv Exp Med Biol 1221:331–349

Putz EM, Mayfosh AJ, Kos K, Barkauskas DS, Nakamura K, Town L, Goodall KJ, Yee DY, Poon IK, Baschuk N, Souza-Fonseca-Guimaraes F, Hulett MD, Smyth MJ (2017) NK cell heparanase controls tumor invasion and immune surveillance. J Clin Invest 127(7):2777–2788

Quail D, Joyce J (2013) Microenvironmental regulation of tumor progression and metastasis. Nat Med 19:1423–1437

Ramani VC, Purushothaman A, Stewart MD, Thompson CA, Vlodavsky I, Au JL, Sanderson RD (2013) The heparanase/syndecan-1 axis in cancer: mechanisms and therapies. FEBS J 280(10): 2294–2306

Raposo G, Stoorvogel W (2013) Extracellular vesicles: exosomes, microvesicles, and friends. J Cell Biol 200:373–383

Reiland J, Sanderson RD, Waguespack M, Barker SA, Long R, Carson DD, Marchetti D (2004) Heparanase degrades syndecan-1 and perlecan heparan sulfate: functional implications for tumor cell invasion. J Biol Chem 279(9):8047–8055

Rilla K, Pasonen-Seppanen S, Deen AJ, Koistinen VV, Wojciechowski S, Oikari S et al (2013) HA production enhances shedding of plasma membrane-derived microvesicles. Exp Cell Res 319: 2006–2018

Rilla K, Siiskonen H, Tammi M, Tammi R (2014) Hyaluronan-coated extracellular vesicles—a novel link between hyaluronan and cancer. Adv Cancer Res 123:121–148. https://doi.org/10.1016/B978-0-12-800092-2.00005-8

Ritchie JP, Ramani VC, Ren Y, Naggi A, Torri G, Casu B et al (2011) SST0001, a chemically modified heparin, inhibits myeloma growth and angiogenesis via disruption of the heparanase/syndecan-1 axis. Clin Cancer Res 17:1382–1393

Rodrigues-Junior DM, Pelarin MFA, Nader HB, Vettore AL, Pinhal MAS (2021) MicroRNA-1252-5p associated with extracellular vesicles enhances bortezomib sensitivity in multiple myeloma cells by targeting heparanase. OncoTargets Ther 14

Roucourt B, Meeussen S, Bao J, Zimmermann P, David G (2015) Heparanase activates the syndecan-syntenin-ALIX exosome pathway. Cell Res 25:412–428

Ruan J, Trotter TN, Nan L, Luo R, Javed A, Sanderson RD, Suva LJ, Yang Y (2013) Heparanase inhibits osteoblastogenesis and shifts bone marrow progenitor cell fate in myeloma bone disease. Bone 57(1):10–17

Salem KZ, Moschetta M, Sacco A, Imberti L, Rossi G, Ghobrial IM, Manier S, Roccaro AM (2016) Exosomes in tumor angiogenesis. Methods Mol Biol 1464:25–34

Sanderson RD, Elkin M, Rapraeger AC, Ilan N, Vlodavsky I (2017) Heparanase regulation of cancer, autophagy and inflammation: new mechanisms and targets for therapy. FEBS J 284(1): 42–55. https://doi.org/10.1111/febs.13932
Sanderson RD, Bandari SK, Vlodavsky I (2019) Proteases and glycosidases on the surface of exosomes: newly discovered mechanisms for extracellular remodeling. Matrix Biol 75–76:160–169
Schmidt EP, Yang Y, Janssen WJ, Gandjeva A, Perez MJ, Barthel L, Zemans RL, Bowman JC, Koyanagi DE, Yunt ZX, Smith LP, Cheng SS, Overdier KH, Thompson KR, Geraci MW, Douglas IS, Pearse DB, Tuder RM (2012) The pulmonary endothelial glycocalyx regulates neutrophil adhesion and lung injury during experimental sepsis. Nat Med 18:1217–1223
Shteingauz A, Boyango I, Naroditsky I, Hammond E, Gruber M, Doweck I, Ilan N, Vlodavsky I (2015) Heparanase enhances tumor growth and chemoresistance by promoting autophagy. Cancer Res 75:3946–3957. [PubMed: 26249176]
Shuman Moss LA, Jensen-Taubman S, Stetler-Stevenson WG (2012) Matrix metalloproteinases: changing roles in tumor progression and metastasis. Am J Pathol 181(6):1895–1899
Simizu S, Ishida K, Wierzba MK, Osada H (2004) Secretion of heparanase protein is regulated by glycosylation in human tumor cell lines. J Biol Chem 279(4):2697–2703
Singel KL, Segal BH (2016) Neutrophils in the tumor microenvironment: trying to heal the wound that cannot heal. Immunol Rev 273(1):329–343
Spiegel A, Zcharia E, Vagima Y, Itkin T, Kalinkovich A, Dar A, Kollet O, Netzer N, Golan K, Shafat I, Ilan N, Nagler A, Vlodavsky I, Lapidot T (2008) Heparanase regulates retention and proliferation of primitive Sca-1+/c-kit+/Lin- cells via modulation of the bone marrow microenvironment. Blood 111(10):4934–4943
Spyrou A, Kundu S, Haseeb L, Yu D, Olofsson T, Dredge K, Hammond E, Barash U, Vlodavsky I, Forsberg-Nilsson K (2017) Inhibition of heparanase in pediatric brain tumor cells attenuates their proliferation, invasive capacity, and in vivo tumor growth. Mol Cancer Ther 16:1705–1716
Stetler-Stevenson WG, Yu AE (2001) Proteases in invasion: matrix metalloproteinases. Semin Cancer Biol 11(2):143–152
Strieter RM, Burdick MD, Mestas J, Gomperts B, Keane MP, Belperio JA (2006) Cancer CXC chemokine networks and tumour angiogenesis. Eur J Cancer 42(6):768–778
Szebeni GJ, Vizler C, Kitajka K, Puskas LG (2017) Inflammation and cancer: extra- and intracellular determinants of tumor-associated macrophages as tumor promoters. Mediators Inflamm 2017:9294018
Tang W, Nakamura Y, Tsujimoto M, Sato M, Wang X, Kurozumi K et al (2002) Heparanase: a key enzyme in invasion and metastasis of gastric carcinoma. Mod Pathol 15:593–598
Theodoro TR, Matos LL, Cavalheiro RP, Justo GZ, Nader HB, Pinhal MAS (2019) Crosstalk between tumor cells and lymphocytes modulates heparanase expression. J Transl Med 17:103. https://doi.org/10.1186/s12967-019-1853-z
Thompson CA, Purushothaman A, Ramani VC, Vlodavsky I, Sanderson RD (2013) Heparanase regulates secretion, composition, and function of tumor cell-derived exosomes. J Biol Chem 288:10093–10099
Turturici G, Tinnirello R, Sconzo G, Geraci F (2014) Extracellular membrane vesicles as a mechanism of cell-to-cell communication: advantages and disadvantages. Am J Physiol Cell Physiol 306:C621–C633
Vella LJ, Greenwood DL, Cappai R, Scheerlinck JP, Hill AF (2008) Enrichment of prion protein in exosomes derived from ovine cerebral spinal fluid. Vet Immunol Immunopathol 124:385–393
Vlodavsky I, Miao HQ, Medalion B, Danagher P, Ron D (1996) Involvement of heparan sulfate and related molecules in sequestration and growth promoting activity of fibroblast growth factor. Cancer Metastasis Rev 15:177–186
Vlodavsky I, Friedmann Y, Elkin M, Aingorn H, Atzmon R, Ishai-Michaeli R, Bitan M, Pappo O, Peretz T, Michal I, Spector L, Pecker I (1999) Mammalian heparanase: gene cloning, expression and function in tumor progression and metastasis. Nat Med 5(7):793–802

Vlodavsky I, Beckhove P, Lerner I, Pisano C, Meirovitz A, Ilan N, Elkin M (2012) Significance of heparanase in cancer and inflammation. Cancer Microenviron 5(2):115–132

Vlodavsky I, Singh P, Boyango I, Gutter-Kapon L, Elkin M, Sanderson RD, Ilan N (2016) Heparanase: from basic research to therapeutic applications in cancer and inflammation. Drug Resist Updat 29:54–75. https://doi.org/10.1016/j.drup.2016.10.001

Vlodavsky I, Ilan N, Sanderson RD (2020) Forty years of basic and translational heparanase research. In: Heparanase. Chapter first online: 10 April 2020 part of the Advances in experimental medicine and biology book series (AEMB, vol 1221). Springer, Cham, pp 3–59

Waterman M, Ben-Izhak O, Eliakim R, Groisman G, Vlodavsky I, Ilan N (2007) Heparanase upregulation by colonic epithelium in inflammatory bowel disease. Mod Pathol 20(1):8–14

Weissmann M, Arvatz G, Horowitz N, Feld S, Naroditsky I, Zhang Y et al (2016) Heparanase-neutralizing antibodies attenuate lymphoma tumor growth and metastasis. Proc Natl Acad Sci U S A 113:704–709

Witek RP, Yang L, Liu R, Jung Y, Omenetti A, Syn WK, Choi SS, Cheong Y, Fearing CM, Agboola KM, Chen W, Diehl AM (2009) Liver cell-derived microparticles activate hedgehog signaling and alter gene expression in hepatic endothelial cells. Gastroenterology 136:320–330

Wong SY, Hynes RO (2006) Lymphatic or hematogenous dissemination: how does a metastatic tumor cell decide? Cell Cycle 5(8):812–817. https://doi.org/10.4161/cc.5.8.2646. Epub 2006 Apr 17

Wu L, Viola CM, Brzozowski AM, Davies GJ (2015) Structural characterization of human heparanase reveals insights into substrate recognition. Nat Struct Mol Biol 22(12):1016–1022

Xie M, Li J-P (2019) Heparan sulfate proteoglycan—a common receptor for diverse cytokines. Cell Signal 54:115–121

Yang N, Friedl A (2016) Syndecan-1-Induced ECM Fiber alignment requires integrin αvβ3 and Syndecan-1 ectodomain and heparan sulfate chains. PLoS One 11(2):e0150132

Zahavi T, Salmon-Divon M, Salgado R, Elkin M, Hermano E, Rubinstein AM, Francis PA, Di Leo A, Viale G, de Azambuja E, Ameye L, Sotiriou C, Salmon A, Kravchenko-Balasha N, Sonnenblick A (2021) Heparanase: a potential marker of worse prognosis in estrogen receptor-positive breast cancer. NPJ Breast Cancer 7(1):67. https://doi.org/10.1038/s41523-021-00277-x

Zcharia E, Metzger S, Chajek-Shaul T, Aingorn H, Elkin M, Friedmann Y, Weinstein T, Li JP, Lindahl U, Vlodavsky I (2004) Transgenic expression of mammalian heparanase uncovers physiological functions of heparan sulfate in tissue morphogenesis, vascularization, and feeding behavior. FASEB J 18(2):252–263

Zhang HG, Grizzle WE (2014) Exosomes: a novel pathway of local and distant intercellular communication that facilitates the growth and metastasis of neoplastic lesions. Am J Pathol 184:28–41. [PubMed: 24269592]

Chapter 9
Laminins and Matrix Metalloproteinases Connection: A Subtle Relationship That Can Go Wrong in a Tumor Context, Particularly If CD44 Gets Involved

Patricia Rousselle and Konrad Beck

Abstract Laminins (LM) are large extracellular glycoproteins involved in several biological processes, including cellular interactions, self-polymerization, and binding with other extracellular matrix proteins. LMs influence cell function by inducing various signaling pathways via cell membrane receptors and have multiple, often cell type-specific, functions in, for example, adhesion, differentiation, migration, and phenotype maintenance, and they also provide resistance to apoptosis. They are also important components of basement membranes. The basement membrane is partly degraded in the course of tumor growth, facilitating the invasion of budding cells and their migration to lymphatic or blood vessels. In this context, LMs undergo proteolytic cleavage, which disrupts their involvement in maintaining the structural and biological properties of the basement membrane. LMs are also involved via their participation in cancer cell adhesion and migration processes. These events are either supported by their major cell binding domains or triggered by cryptic interaction sites revealed by matrix metalloproteinase (MMP)-induced proteolytic cleavage. While being ideal targets for MMPs, LM can enhance their expression and activity. They appear to be key matrix elements in the regulation of MMP activity via the recruitment of the CD44 receptor, a multiple MMP-interacting and activating platform playing an important role in cancer progression.

P. Rousselle (✉)
Laboratoire de Biologie Tissulaire et Ingénierie Thérapeutique, Institut de Biologie et Chimie des Protéines, UMR 5305, CNRS, Univ. Lyon 1, SFR BioSciences Gerland-Lyon Sud, Lyon, France
e-mail: patricia.rousselle@ibcp.fr

K. Beck
School of Dentistry, Cardiff University, Cardiff, UK

I. Kovalszky et al. (eds.), *The Extracellular Matrix and the Tumor Microenvironment*, Biology of Extracellular Matrix 11,
https://doi.org/10.1007/978-3-030-99708-3_9

Abbreviations

ECM	Extracellular matrix
EMT	Epithelial–mesenchymal transition
FAK	Focal adhesion kinase
GAG	Glycosaminoglycan
HA	Hyaluronic acid (hyaluronan)
HBS	Heparin binding site
HS	Heparan sulfate
LG	Laminin (C-terminal α-chain) globular domain
LM	Laminin
MAPK	Mitogen-activated protein kinase
MMP	Matrix metalloproteinase
PI3K	Phosphatidylinositol-3-kinase

9.1 The Laminin Protein Family

Laminins (LMs) are a family of glycoproteins found in basement membranes with biological and structural properties that attract a growing interest. As essential structural constituents of the extracellular matrix (ECM), LMs also maintain close relationships with cells and transmit important morphogenetic information (Miner and Yurchenco 2004) (Domogatskaya et al. 2012). All LMs are composed of three different gene products, termed α, β, and γ chains, which are assembled into a cross-shaped heterotrimer αβγ. The three chains assemble within the endoplasmic reticulum via their C-terminal regions to form a triple stranded α-helical coiled coil rod and are covalently linked by disulfide bonds at both extremities of the coiled coil (Beck et al. 1993) (Matsui et al. 1995). The primary structures of five α chains (α1, α2, α3, α4, and α5), four β chains (β1, β2, β3, and β4), and three γ chains (γ1, γ2, and γ3) were identified by cDNA sequencing, leading to at least 18 heterotrimeric isoforms (Table 9.1, Fig. 9.1). Some genes encoding LMs may produce more than one subunit isoform due to alternative splicing; short-chain and long-chain splice variants are annotated with A and B, respectively (Aumailley et al. 2005). All basement membranes contain at least one LM isoform, and some may have two or three different isoforms. The expression and assembly of the various subunits vary in a spatio-temporal manner, which suggests biological and structural roles specific to each isoform (Yap et al. 2019). Some isoforms appear very early in embryogenesis and are markers for a given ECM, whereas others appear later and in a tissue-specific manner. In general, expression of the α5 subunit is the most widespread, whereas that of the α1 subunit is the most restricted (Miner et al. 1997). The α2 subunit is particularly abundant in tissues of mesodermal origin (cardiac and skeletal muscle), and the α4 subunit is abundant in endothelial and mesenchymal cells.

Table 9.1 Laminin isoforms

Name	Chain assembly	Previous	Original name
Laminin-111	α1β1γ1	1	EHS laminin
Laminin-211	α2β1γ1	2	Merosin
Laminin-121	α1β2γ1	3	S-laminin
Laminin-221	α2β2γ1	4	S-merosin
Laminin-3A32	α3Aβ3γ2	5 or 5A	Kalinin, epilegrin, nicein, ladsin
Laminin-3B32	α3Bβ3γ2	5B	
Laminin-3A11	α3Aβ1γ1	6 or 6A	K-laminin
Laminin-3A21	α3Aβ2γ1	7 or 7A	KS-laminin
Laminin-411	α4β1γ1	8	
Laminin-421	α4β2γ1	9	
Laminin-511	α5β1γ1	10	*Drosophila*-like laminin
Laminin-521	α5β2γ1	11	
Laminin-213	α2β1γ3	12	
Laminin-423	α4β2γ3	14	
Laminin-523	α5β2γ3	15	
Laminin-522[a]	α5β2γ2		
Laminin-212[b]	α2β1γ2		
Laminin-222[b]	α2β2γ2		
Laminin-333[c]	α3β3γ3		

[a]Laminin-522 reported so far solely in bone marrow

[b]The existence of either LM-212 or LM-222 is proposed based on studies of peripheral nerves in wild-type and LM α2 chain-deficient mice

[c]Laminin-333 is expressed at the apical specialization of adult rat testes, rather than in the testicular basement membrane

The α3 subunit, similar to β3 and γ2, is concentrated in epithelial basement membranes. The β1 and β2 subunits often have mutually exclusive expression, whereas the γ1 chain is present in all basement membranes (Durbeej 2010). LMs are essential at various stages of development. Thus, most LM mutants are fatal to the embryo and have made it difficult to study their functions at later stages (Yao 2017).

Although LMs are differentially recognized by cellular receptors, they all share a molecular organization based on the repetition of structural modules, giving them a common molecular organizational pattern. LM-111, which was initially extracted from the Engelbreth–Holm–Swarm tumor, was the first LM described as an asymmetric cross when visualized by electron microscopy comprising one long arm and three short arms carrying globular domains linked by linear segments (Fig. 9.2a). Other LMs do not exhibit all of the short arms, as they include "truncated" subunits at the N-terminal ends (α3, α4, β3, and γ2) (Figs. 9.1 and 9.2b). The N-terminus of each full-length LM chain consists of cysteine-poor sequences and are named LN domain. The short-arm rod-like portions consist of tandem arrays of LE modules, which sequences are related to the epidermal growth factor but contain one additional C-terminal disulfide bond. Inner globular structures within the short arms are

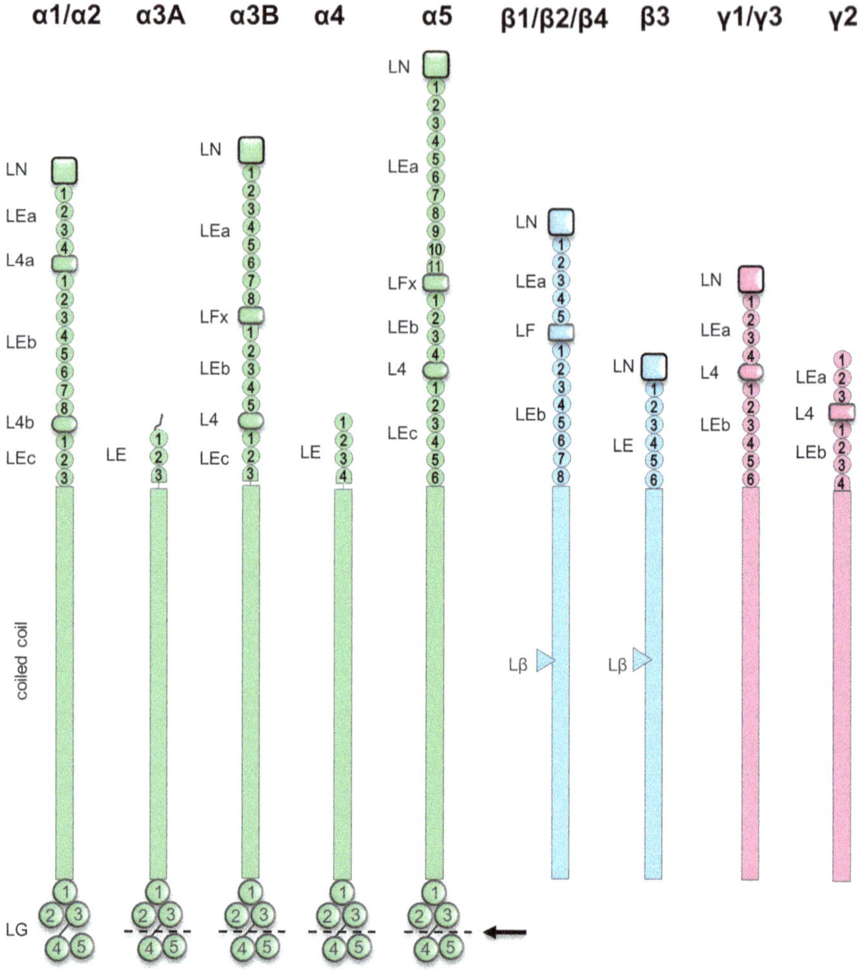

Fig. 9.1 Schematical models of the different laminin chains. LM chains contain tandem arrays of globular and rod-like regions. Chains with identical order of domains are combined (α1/2, β1/2/4, γ1/3). The N-terminal ends of all chains contain tandem arrays of a variable number of epidermal growth factor-like repeats LE with eight cysteine residues in short rod-like domains (LEa–LEc), as well as various globular domains (LN, L4, L4a, L4b, LFa, LFb, LFx), some of which can interact with integrins. Truncated LE domains containing less than 8 cysteines are depicted by half-circles. LE numbering follows the laminin nomenclature (Aumailley et al. 2005) though frequently, including in data base entries, they are counted consecutively starting at the N-terminus. The α chains have five globular domains (LG1–LG5) at the C-terminus. LG1–LG3 bind mostly to integrins, with some requiring the presence of the γ chain C-terminal region. LG4 and LG5 contain binding sites for dystroglycan (αDG), syndecans, and sulfated glycolipids. The LG4–LG5 tandem is thought to be cleaved off extracellularly (arrow) in α3, α4, and α5, though this is not the case for the α1 and α2 chains; LG domains of α2 stay attached to the molecule after cleavage. A small six cysteine residues containing domain Lβ of yet unknown biological function is present in the coiled coil regions of the β chains; originally the regions N- and C-terminal of Lβ were denoted by roman numerals II and I, respectively, but as this is unique to the β chains, this has been abolished. The LAMB4 gene coding for the β4 chain might be a pseudogene as no transcripts has yet been found

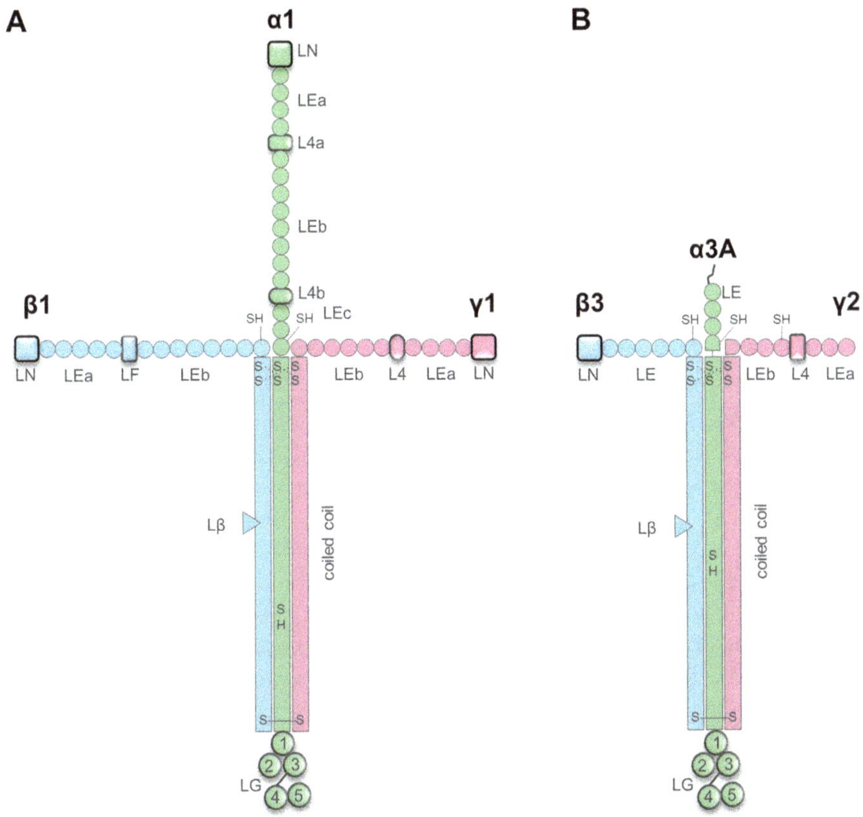

Fig. 9.2 Prototype assembly model of laminins. The α, β, and γ chains assemble to form a coiled coil in at least 18 combinations (Table 9.1). **(a)** LM-111 is a prototype heterotrimer able to self-assemble and form a network due to the presence of LN domains at the N-terminal ends of its three subunits. **(b)** LM-3A32 is a prototype heterotrimer unable to self-assemble as its α and β subunits are truncated. S-S at the termini of the coiled coil regions denote the presence of disulfide bridges. SH indicates the location of free cysteine residues

the L4 domains resulting from a long insert between cysteines 3 and 4 of the canonical 8-cysteine pattern of an LE domain. The LF domains have a distant structural similarity to L4 but are unique with respect to their cysteine pattern (Pulido et al. 2017). The LFx domains share similar sequences at the N- and C-termini with LF, but have further insertions of two internally similar sequence regions. C-terminal regions of the α, β, and γ chains assemble into a three-strand α-helical coiled coil forming the long arm (Macdonald et al. 2010).

The amino acids involved in this structure are organized in a heptad repeat in which the first and fourth residue are of hydrophobic, and the fifth and seventh frequently of charged character (Paulsson et al. 1985). Within the β chains, the coiled coil region is interrupted by domain Lβ consisting of 30–40 residues including 6 cysteines, which biological function is not yet known. All LM α chains possess a large globule at the C-terminal end that consists of five similar domains (LG1–LG5), each containing approximately 200 residues (Timpl et al. 2000; Hohenester 2019). LG3 and LG4 are separated by a flexible hinge providing access for proteolytical attack. All LMs are glycosylated, which regulates their biological functions in addition to stabilizing them (Morita et al. 1985; Inamori et al. 2016).

9.2 Laminins as Multifunctional Elements Within Basement Membranes

Early molecular and in vivo studies in *Drosophila*, *C. elegans*, and mice support the idea that LMs are the basic building block for initiating basement membrane formation (McKee et al. 2007; Jayadev and Sherwood 2017). Most LMs can self-associate into polymeric sheet-like networks that are tightly associated with the cell surface through their LG domains (Yurchenco et al. 1992; Li et al. 2003). This polymerization process is reversible and depends on both the concentration and presence of divalent cations. This assembly model involves and absolutely requires the N-terminal domain LN of the α, β, and γ subunits (Cheng et al. 1997; Hohenester and Yurchenco 2013) (Fig. 9.2a). LM networks are noncovalent in nature and appear more dynamic than the cross-linked collagen IV networks described as being responsible for the mechanical strength of basement membranes. Thus, collagen IV recruitment and further basement membrane assembly appear to depend on LM self-assembly and its LG-mediated cell surface anchorage. LMs containing at least one truncated subunit, and therefore lacking one LN domain (i.e., the α3A, α4, and γ2 chains), are nonpolymerizing LMs and form other types of supramolecular arrays to integrate the basement membrane (Fig. 9.2b). The epithelial LM heterotrimers α3Aβ3γ2, α3Aβ1γ1, and α3Aβ2γ1 establish specific molecular interactions leading to the formation of anchoring structures characteristic of epithelial basement membranes (Champliaud et al. 1996; Rousselle and Beck 2013). The matrix integration of LMs comprising a truncated α chain assembled with the β1 and γ1 subunits, such as LM-3A11 and LM-411, may involve known interactions between the β1 and γ1 chains with agrin, nidogen (entactin), and perlecan.

LMs provide interaction sites for many other constituents including cell surface receptors. Some LM isoforms are modified extracellularly by proteolytic processing at the N- or C-terminal ends prior to binding cellular receptors or other matrix molecules (Tzu and Marinkovich 2008). Cellular receptors, such as integrins, syndecans, α-dystroglycan, Lutheran glycoprotein, or sulfated glycolipids, predominantly bind to the five LG domains at the C-terminal end of the α chains. The LG1–LG3 region contains the integrin binding domains, which often require the context of the LM heterotrimer and, most importantly, the C-terminal end of the γ chain to be fully effective (Deutzmann et al. 1990; Rousselle et al. 1995; Ido et al. 2004, 2007, 2008; Navdaev et al. 2008). The major LM-binding integrins are α3β1, α6β1, α7β1, and α6β4 (Nishiuchi et al. 2006). The LG1–LG3 region of the LM α5 chain also binds to the Lutheran blood group antigen/basal cell adhesion molecule, a cell surface protein consisting of five immunoglobulin-like domains (Udani et al. 1998; Vainionpää et al. 2006; Kikkawa et al. 2007). The LG45 region contains binding sites for α-dystroglycan, syndecans, and sulfated glycolipids (Gee et al. 1993; Talts et al. 1999; Smirnov et al. 2002; Wizemann et al. 2003; Yamashita et al. 2004; Utani et al. 2003; Hozumi et al. 2009; Carulli et al. 2012; Suzuki et al. 2005). The N-terminal globular LN domains of the α1 (Colognato-Pyke et al. 1995) and α2 chains (Colognato et al. 1997) as well as LN (Nielsen and Yamada 2001) and L4 (Sasaki and Timpl 2001) of the α5 chain can bind to α1β1, α2β1, α3β1, and αVβ3 integrins.

Deregulated expression and involvement of LMs and their integrin receptors in tumor development and aggressiveness have been demonstrated in recent years. They will not be the subjects of this review as they have been exhaustively covered recently (Maltseva and Rodin 2018; Qin et al. 2017; Ramovs et al. 2017; Rousselle and Scoazec 2020). Instead, here we will focus on the link between LMs and matrix metalloproteinases (MMPs) and demonstrate how it can become toxic in cancer by amplifying the tumor process. We will also examine the current state of understanding the status of LMs as ligands for the CD44 receptor and seek for a potential link with MMP activity.

9.3 Laminins Are Involved in the Regulation of MMP Expression and Activity

MMPs are a family of secreted and membrane-bound zinc-dependent endopeptidases that have the capacity to degrade all components of the ECM. MMPs play central roles in morphogenesis, wound healing, tissue repair and remodeling in response to injury, and in the progression of diseases, such as inflammatory and neoplastic diseases (Nagase et al. 2006). The MMP family comprises 25 related but distinct vertebrate gene products, 24 of which are found in mammals and can be divided into six groups: collagenases (MMPs 1, 8, 13, 18), gelatinases (MMPs 2, 9), stromelysins (MMPs 3, 10, 11, 17), matrilysins (MMPs 7, 26), membrane-type MMPs (MT-MMPs; MMPs 14, 15, 16, 17, 24, 25), and other MMPs (MMPs 12, 19, 20). The expression of MMP genes is transcriptionally regulated by a variety of extracellular factors including cytokines, growth factors, and ECM proteins (Gaffney et al. 2015). A number of studies have revealed that LMs or their proteolytic fragments can be involved in the regulation of MMP expression and/or activity. These subtle regulations that occur in a physiological context can be amplified in pathological situations such as cancer, contributing to tumorigenesis and invasion (Karamanos et al. 2021).

Early studies revealed that an interaction between LM-111 and the 67-kDa LM receptor promotes MDA-MB231 breast carcinoma cell aggressiveness through the upregulation of MMP-14 and enhanced gelatinolytic activity of MMP-2 (Berno et al.

2005). Subsequent studies have shown that LM-111 can increase MMP activity in human neural stem cells (Sypecka et al. 2009). Later, activation of MMP-9 expression was associated with an α2β1 integrin-dependent migration process when the human cervical cancer SiHa cell line was cultured on an LM-111-coated surface (Maity et al. 2011). A mechanism involving focal adhesion kinase (FAK), integrin linked kinase, phosphatidylinositol-3-kinase (PI3K), and extracellular signal regulated kinase was followed by increased DNA-binding activity of NF-κB and Ap1 and subsequent stimulation of MMP-9 gene expression (Maity et al. 2011). A similar signaling pathway was revealed when human breast cancer MCF-7 cells were plated on an LM-111-coated surface (Pal et al. 2014). In this study, the concomitant downregulation of TIMP1, a negative regulator of MMP-9 activity, led to enhanced MMP-9 proteolytic activity. Another context in which a link between LM and MMPs was established is the kidney. The LM α2 chain was shown to contribute to pathogenesis in the Alport glomerular syndrome via activation of FAK on glomerular podocytes leading to the downstream activation of MMP-9, MMP-10, and MMP-12 gene expression (Delimont et al. 2014). A β1 chain fragment released through MMP-2 processing, LN–LE1-LE4, was shown to modulate the behavior of pluripotent stem cells and promote the epithelial–mesenchymal transition (EMT) in an α3β1-integrin-dependent fashion, whatever the fragment was delivered to the cells as a soluble or immobilized factor (Horejs et al. 2014). The β1 chain fragment triggered the downregulation of MMP-2 in human and mouse cells, as well as the upregulation of E-cadherin and MMP-9 in mouse cells during spontaneous stem cell differentiation.

The cryptic domains of LMs (i.e., hidden in full-length LMs and exposed upon proteolysis) can elicit biological responses and be involved in the regulation of MMPs. For example, peptides containing the sequence S_{2099}IKVAV, which is adjacent to the N-terminus the C-terminal LG1 domain of the LM α1 chain, have been shown to induce the expression of MMP-9 by monocytes/macrophages (Corcoran et al. 1995; Khan and Falcone 1997), whereas intact LM-111 does not (Khan and Falcone 2000). A mouse α5 LG1 domain-derived peptide was shown to induce macrophage and neutrophil chemotaxis both in vitro and in vivo and to enhance MMP-9 and MMP-14 activities (Adair-Kirk et al. 2003). Furthermore, human γ2 chain fragments and LG4-derived peptides from the α3, α4, and α5 chains exhibit strong chemotactic activity toward leukocytes, neutrophils, and monocytes (Mydel et al. 2008; Kenne et al. 2010; Senyürek et al. 2014). Whether these activities are associated with activation of MMP expression or activity is an interesting question that remains to be answered. Another study revealed that LM-332 potentiates human monocyte differentiation to tumor-associated macrophage-like cells secreting a high level of MMP-9 (Kamoshida et al. 2014). Upregulation of MMP-1 and MMP-9 expression was also reported at both the transcriptional and protein levels in epidermal keratinocytes upon treatment with recombinant C-terminal LG4 or LG45 domains of the α3 chain, an event that relies on an IL-1β autocrine loop through the mitogen-activated protein kinase (MAPK) pathway (Momota et al. 2005; Utani et al. 2003; Michopoulou et al. 2020). Several peptide sequences involved in this mechanism were identified as heparin binding sites (HBSs) and characterized as ligands for the proteoglycan-type cell surface receptors, the syndecans (Utani et al. 2003; Carulli et al. 2012; Rousselle and Beck 2013). However, the role of LM was not limited to this transcriptional activation as the α3 LG45 domain participated in establishing the gelatinolytic activity of epithelial podosomes in migrating keratinocytes during wound healing (Michopoulou et al. 2020). We showed that the α3 LG45 domain triggers the proteolytic activity of MMP-9 and MMP-14 in epithelial podosomes via a mechanism involving the cellular receptor syndecan-1 (Michopoulou et al. 2020).

9.4 Laminins Are Cell Adhesion Ligands for CD44

The cell surface glycoprotein CD44, a hyaluronic acid (HA) receptor, plays a role in the regulation of cell–cell interactions, cell adhesion, cell growth, migration, and invasion (Knudson et al. 1996; Ponta et al. 2003). CD44 is an 85- to 200-kDa transmembrane glycoprotein ubiquitously expressed throughout the body and is detected in both normal and tumor cells. CD44 is encoded by a single gene with 20 exons but has many different isoforms (Fig. 9.3a). It is composed of a distal extracellular N-terminal domain, a stem region, a transmembrane domain, and an intracellular cytoplasmic C-terminal domain (Fig. 9.3b). The standard form CD44s is the shortest and most commonly expressed isoform. It results from expression of the constant exons 1–10, which are found in all isoforms. A number of splicing variants (CD44v) result from the insertion of variable exons in the extracellular region near the membrane, called the stem region.

There are ten variant exons expressed in mice (designated v1–v10), whereas humans express only variant exons v2–v10 due to the presence of an in-frame stop codon in exon v1 (Screaton et al. 1993). The N-terminal portion of the ectodomain is of a compact, ordered structure, stabilized by three disulfide bridges, that contains a HA binding region and glycosaminoglycan (GAG) chains (Fig. 9.4a). The stem region varies according to the expression of the splicing domains in the CD44v isoforms, and the transmembrane and intracellular domains both play important roles in CD44 functionality and connection to the actin cytoskeleton through ezrin/radixin/moesin (ERM) proteins (Tsukita et al. 1994; Gal et al. 2003; Medrano-González et al. 2021). Importantly, upregulation of CD44 is often closely associated with abnormal tumor cell behavior (e.g., proliferation, survival, migration/invasion, and chemoresistance) (Bourguignon 2019; Medrano-González et al. 2021).

Understanding the various interactions of CD44 on a molecular structural level that would allow for reliable predictions of binding sites is difficult. The complex post-translational modifications by N- and O-linked glycosylation greatly depend on cell types, developmental, and environmental conditions. Large chondroitin and dermatan sulfate GAG chains attached to the stem region in CD44s, and heparan sulfate chains found within the variable regions are highly flexible making high-resolution structural analysis challenging. Except for the most N-terminal *ca.*

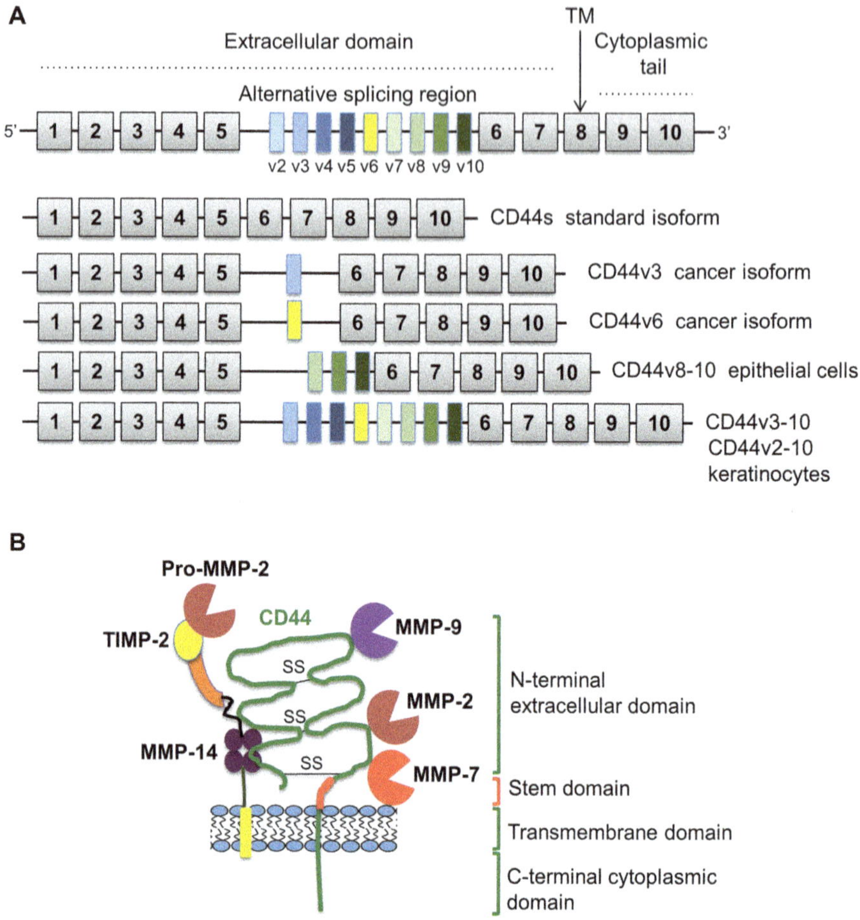

Fig. 9.3 CD44 as a platform for MMP binding. (**a**) Structure of the human CD44 gene, which consists of ten constant exons expressed in all CD44 mRNAs and proteins (grey boxes) and nine variant exons v (colored boxes) expressed in CD44 splice variants (CD44v). Other vertebrate genes contain an additional exon v1. The examples shown are the smallest isoform (standard CD44s, expressed ubiquitously in vertebrates) and the larger variant isoforms expressed in some epithelial cells and cancers. (**b**) CD44 protein structural domains. The CD44 protein is composed of an extracellular link domain (green), a stem domain (red), where the variant exon products are inserted, the transmembrane region, and the intracellular cytoplasmic domain. The CD44 extracellular domain is a platform for MMP-2, MMP-7, MMP-9, and MMP-14 binding and can indirectly activate MMP-2 through MMP-14 cleavage. Figure adapted from references Ponta et al. (2003); Cauwe et al. (2007); Medrano-González et al. (2021)

160 amino acids, even the unmodified ectodomain evades exploration by showing a high degree of intrinsic disorder with nearly 30% of residues being serine and threonine (Fig. 9.4b). In contrast to the classical view of structure–function relation as a lock-and-key system, the resulting flexibility on the structural level allows the

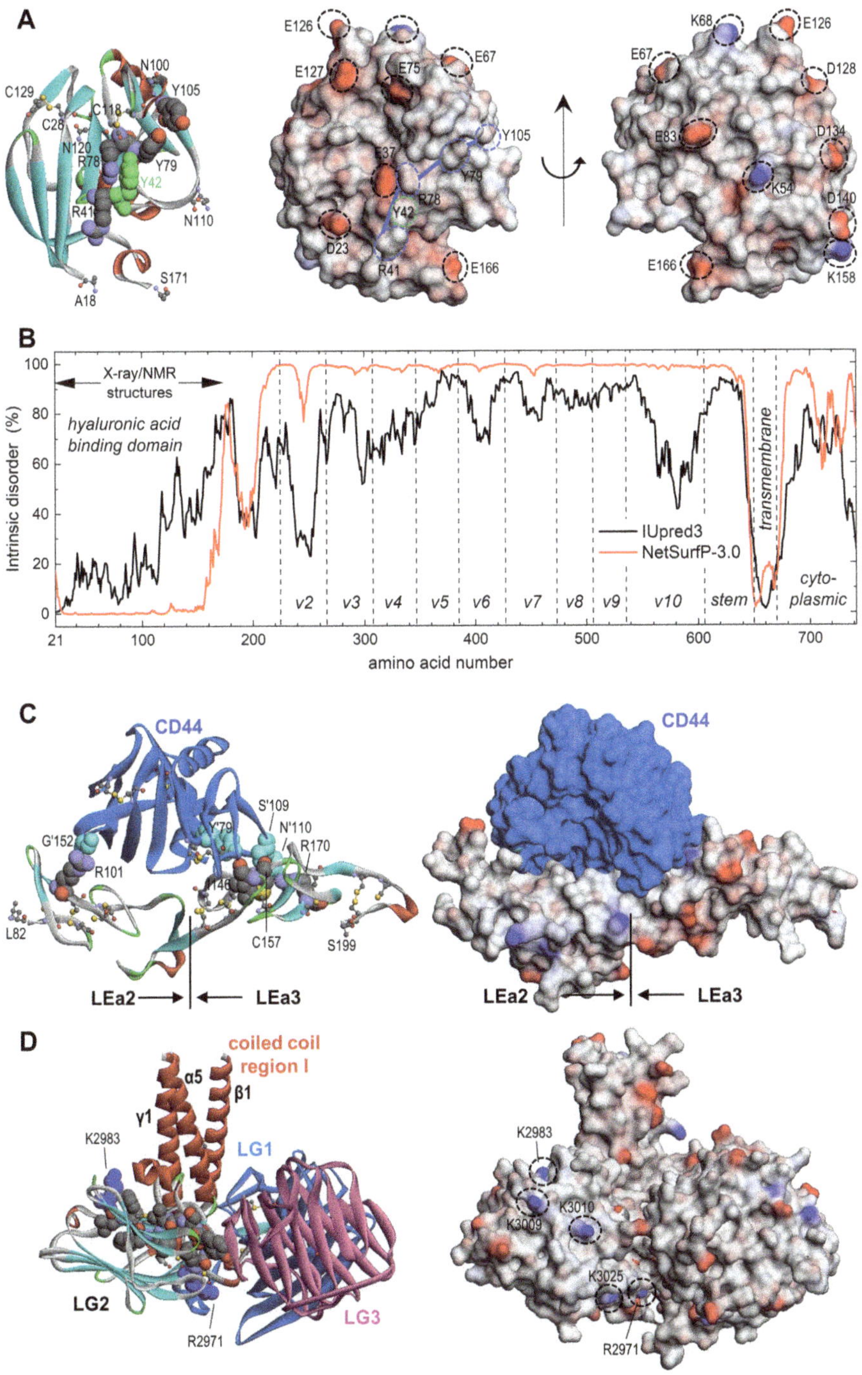

Fig. 9.4 Molecular constraints of CD44 and laminin interactions. (**a**) The protein backbone of the HA binding domain of human CD44 is shown on the left with α-helices in red and β-strands in cyan (PDB code: 4PZ4, ref. Liu and Finzel 2014). Terminal residues, disulfide bonds and putative

protein to adapt various conformations for optimal ligand binding (Habchi et al. 2014).

Even the various structures solved at high resolution by x-ray crystallography may not be regarded as static. Molecular dynamics simulations suggest that Y42 can act as a molecular switch that upon binding to HA changes the overall HA binding domain structure from a closed to open conformation that further increases affinity for HA (Guvench 2015).

CD44 is upregulated in a variety of cancers and can be expressed as CD44s or as a number of alternatively spliced CD44v isoforms. CD44 mediates its effects on the cancer cell by activating signaling pathways, such as protein kinases and transcription factors. The functional role of CD44 is pleiotropic, including induction of EMT and antiapoptosis, alterations in the cellular cytoskeleton, and promotion of drug resistance. CD44 is a compelling marker of cancer stem cells from many solid malignancies (Takaishi et al. 2009; Du et al. 2008; Zöller 2011). In addition, interactions between HA and CD44 promote epidermal growth factor receptor-mediated pathways, consequently leading to tumor cell growth, tumor cell migration, and chemotherapy resistance in solid cancers (Thapa and Wilson 2016; Yaghobi et al. 2021).

Fig. 9.4 (continued) acceptor sites for N-linked glycosylation are depicted as ball-and-sticks. Disulfide bonds follow a C1–C6, C2–C5, C3–C4 arrangement. Residues involved in HA binding are presented in CPK style as based on the similarity with the mouse sequence (R41, R78, Y79, Y105) (Banerji et al. 2007) Y42 (green) has been implicated in acting as a molecular switch inducing a conformational change upon interaction of R41 with HA to a higher affinity state (Guvench 2015). A solvent-accessible surface view of the same orientation is shown in the middle, and one obtained by rotation of 180° around the vertical axis on the right. The electrostatic potential is represented by red and blue for negative and positive charges, respectively. The position of various residues is indicated by dashed lines with those involved in HA binding highlighted and connected by blue lines. (**b**) Prediction of the intrinsic disorder of full-length human CD44 was performed using the IUpred3 (black line) (Dosztányi 2018) (https://iupred.elte.hu/) and NetSurfP-3.0 (red line) (Klausen et al. 2019) (https://biolib.com/dtu/nsp3) algorithms. (**c**) Putative interactions between the CD44 HA binding domain (blue) and LM γ2 LEa2–LEa3 domains are shown for the protein backbones on the left and the corresponding solvent-accessible surfaces on the right. Residues within hydrogen bonding distance between proteins are represented in CPK style, where the apostrophe denotes CD44 residues (R101-G′152, I146-Y′79, C157-S′109, R170-N′110). Simulations were performed using PatchDock (Schneidman-Duhovny et al. 2005) (https://bioinfo3d.cs.tau.ac.il/PatchDock/) and refined with FireDock (Andrusier et al. 2007) (https://bioinfo3d.cs.tau.ac.il/FireDock/). LM γ2 LEa2–LEa3 was modeled with SWISS-MODEL (Waterhouse et al. 2018) (https://swissmodel.expasy.org/) using PDB structure 4AQS of the mouse LM β1 LEa1–LEa4 as a template (Carafoli et al. 2012). (**d**) The protein backbone (left) and the corresponding solvent-accessible surface (right) of the C-terminal region of human LM 511 is shown (PDB code 5XAU, chains A, B, C) (Takizawa et al. 2017). The LM α5 sequence R_{2971}LVSYSGVLFFLK within the LG2 domain corresponding to the mouse peptide A5G27 (R_{2892}LVSYNGIIFFLK) shown to be involved in CD44 binding (Hibino et al. 2004) extends over one β-strand and is depicted in CPK style with the basic terminal residues highlighted in blue. The surface view indicates that further lysine residues (K3009, K3010, K3025) are arranged along a track that could facilitate interactions with GAG chains. Sequence numbering follows UniProtKB/Swiss-Prot data base entries P16070.3 (CD44), Q13753.2 (LM γ2) and O15230.8 (LM α5)

In addition to its direct signaling function, CD44 forms a platform at the cell membrane for the assembly of various MMPs with their substrates, ultimately to modulate cell migration (Cauwe et al. 2007) (Fig. 9.3b). CD44 is an important partner of MMP-9, playing a role in its anchorage to the cell membrane and preservation of its proteolytic activity (Yu and Stamenkovic 1999, 2000). Evidence suggests that the association of MMP-9 and CD44 in mouse and human tumor cells promotes invasion (Yu and Stamenkovic 1999; Gupta et al. 2013). Clustering of CD44 mediates MMP-9 relocation in breast tumor cells promoting its proteolytic activity on the membrane (Peng et al. 2007). The interaction of CD44 and the proteolytic form of MMP-9 is involved in the invasion activity of prostate cancer PC3 cells (Desai et al. 2007).

CD44 is also a molecular partner of MMP-14, playing a role in its translocation to the front of migrating cells (Mori et al. 2002). MMP-14 has been clearly characterized as a CD44 shedding enzyme with migration-promoting activity (Kajita et al. 2001). MMP-14 and CD44 are coexpressed at the lamellopodia of cells undergoing EMT (Cho et al. 2012). In normal mouse breast and uterine epithelium, the heparan sulfate (HS) chains of CD44v containing the exon v3 derived sequence bind active MMP-7 (Yu et al. 2002; Seiki 2002). CD44 also binds MMPs 2, 15, 16, 24, and 25 through their hemopexin domains in lamellipodia edges (Mori et al. 2002; Suenaga et al. 2005; Samanna et al. 2006) and can act indirectly by activating MMP-2 through MMP-14 cleavage (Zöller 2015) (Fig. 9.3b).

CD44 has been identified as a component of podosomes, which are protrusive adhesion structures involved in ECM degradation and mechanosensing (Murphy and Courtneidge 2011; Linder and Wiesner 2016). It was shown to colocalize with the F-actin core of podosomes in osteoclasts (Chabadel et al. 2007), human primary epidermal keratinocytes (Michopoulou et al. 2020), and in 3D-cultured macrophages (Van Goethem et al. 2011). By demonstrating a direct interaction between the LM α3 LG45 domain pair and the CD44 receptor expressed by human primary keratinocytes, we have revealed a determining role for this LM isoform in regulation of the proteolytic activity of MMP-14 and -9 in epithelial podosomes (Michopoulou et al. 2020). The 200-kDa CD44 isoform identified in our study is the long form v3–10 expressed by keratinocytes (Bourguignon and Bikle 2015). It is conceivable that the LM-332 LG45 domain concentrates active MMP-9 and -14 within epithelial podosomes through recruitment of CD44. In addition, the combination of LG45 and CD44 could facilitate binding of adaptor molecules to the cytoplasmic region in CD44 and activate multiple pathways involved in cell adhesion, migration, and proliferation including Ras, MAPK, and PI3K (Xu et al. 2020). Although this mechanism has been demonstrated in the physiological context of skin repair, it could undoubtedly be implemented in cancer cells in which the expression of LM-332 is frequently upregulated (Rousselle and Scoazec 2020). In cancer cells, the α3LG45 domain could regulate the expression and concentrate active MMP-9 and MMP-14 in invadopodia, the adhesive actin-rich membrane protrusions with high proteolytic activity found in cancer cells (Augoff et al. 2020). An argument in favor of this hypothesis is based on our study showing that Ras/IκBα-transformed keratinocytes lacking the LG45 domain exhibit a deficiency in MMP-9 and MMP-1 expression and a decreased invasive capacity (Tran et al. 2008). This deficiency was reversed by replacing LG45 through retroviral transduction. Whether the CD44 isoforms expressed in cancer cells can bind to α3LG45 is an important question (Chen et al. 2018). We already know that, in addition to the CD44v3-10 isoform, the 100-kDa standard CD44s expressed on melanoma A375 cells also binds the α3LG45 domain (Michopoulou et al. 2020; Takahashi et al. 1999), and we may wonder whether the various CD44 isoforms will bind to α3LG45 with different affinities. For example, the presence of variable exon 3, which encodes an HS chain carrying region, may impact the interaction (Bennett et al. 1995).

A link between LM-332 and the MMP-14/CD44 tandem was already revealed a few years earlier in a study showing that interactions between mature human thymocytes and LM-332 induces strong upregulation of active MMP-14 leading to CD44 cleavage and thymocyte migration (Vivinus-Nebot et al. 2004). As the LG45 domain was cleaved off and removed from the LM-332 used in this study, we can evaluate the possible participation of the N-terminal end of the γ2 chain, which also comprises an HBS, in this mechanism (Sasaki et al. 2001). One study revealed that the LN domain of the LM γ2 chain induces breast cancer MDA-MB-231 cell migration as a result of its binding to CD44 and phosphorylation of its cytoplasmic tail (Sato et al. 2015). Surprisingly, interaction studies uncovered a binding site for CD44 within the LEa2–LEa3 region, away from the HBS of the γ2 chain identified within the L4 and LEa domains (Sasaki et al. 2001; Ogawa et al. 2007), suggesting the involvement of a LM binding domain distinct from the GAGs in the CD44 ectodomain (Sato et al. 2015). However, the study did not mention any MMP-9 or MMP-14 recruitment, though it clearly reported a γ2 chain-dependent pro-migratory activation of CD44 reinforcing the hypothesis that LM-332 is a multisite ligand for CD44. Based on these data, we performed rigid body molecular docking simulations using a model of the LM γ2 LEa2–LEa3 pair and the coordinates of the HBS domain of CD44.The best result with a good global minimal energy conformation shows a good steric fit of CD44 within the groove of the LEa2-LEa3 connection and suggests four pairs of residues that come into hydrogen bonding distance (Fig. 9.4c). Though this model is in agreement with the experimental data, as any glycosylation and the stem portion of CD44s were not considered, it should be viewed with caution.

LMs were identified as ligands for CD44 already nearly 30 years ago. The binding of CD44 to a commercial LM preparation was detected in an ELISA-based assay using CD44 purified from lymphocytes (Jalkanen and Jalkanen 1992). The epithelial isoform CD44v8-10 (Brown et al. 1991), which is expressed in carcinoma cells such as human colorectal carcinoma KM-12, CCL 188, and MIP-101 cells, was identified as an LM-111 receptor (Ishii et al. 1993). These studies already reported that the interaction is mediated by the proteoglycan-rich region of the extracellular domain of CD44, an interaction site distinct from that of HA, which is located in the disulfide-rich N-terminus of the molecule (Ponta et al. 2003). In a study screening the adhesive properties of overlapping short peptides covering the entire LG1–LG5 domains of the α5 chain toward B16–F10 mouse melanoma cells, peptide A5G27 R_{2892}LVSYNGIIFFLK corresponding to one β strand of the LG2 module with the terminal basic residues in the surface exposed loop regions was identified as a ligand for CD44 via its GAG moieties (Hibino et al. 2004). In the human sequence, this peptide relates to R_{2971}LVSYSGVLFFLK. When visualized within the solved terminal structure of LM 511 (Takizawa et al. 2017), it is evident that the basic side chains are surface exposed at opposite ends of the LG2 domain (Fig. 9.4d). Further positively charged surface areas are generated by K3009, K3010, and K3025. These five residues form a track that could further the alignment of a GAG chain. A similar track of basic residues crucial for heparin binding has been previously reported for the LM LG4 domain (Yamashita et al. 2004). The LM α5 chain interaction with CD44 stimulates melanoma cell growth, angiogenesis, migration, and metastasis (Hibino et al. 2005). CD44 overexpression in neural precursor cells significantly improved their spreading over LMs and increased the formation and length of filopodia, suggesting a role in migration (Deboux et al. 2013). Using a protein database approach, an important role for CD44 interactions with LM-111 was predicted in the axon growth of retinal ganglion cells during nerve regeneration (Ries et al. 2007). As CD44 is overexpressed on a wide variety of neuronal, epithelial, and hematopoietic malignancies, CD44 may be involved in either the malignant transformation of cells or the migration of malignant cells from the primary site to sites of metastasis. The mode of interaction of CD44 with LMs is very complex and can take place in a number of ways. Further work is now required to better understand these interactions and apprehend their biological consequences.

9.5 Laminins Are Targets of MMPs Activity

Like any other ECM protein, LMs are subject to total or partial proteolysis by MMPs, leading to the release of protein fragments, which may play a role in tumor progression. Over the years, MMPs have been found to play a remarkable number of regulatory roles at the cellular level, including pathways involved in apoptosis, immunity, cellular migration, and angiogenesis (Fingleton 2006; Itoh and Nagase 2002; Shay et al. 2015). Thus, MMPs play a major role in carcinogenesis. MMP functionality often complements classical tumor properties, leading to invasion, immune system avoidance, and metastasis (Winkler et al. 2020). The specific role of each MMP in cancer and their dysregulation in many cancer types have been accurately documented recently (Gobin et al. 2019; Piperigkou et al. 2021; Karamanos et al. 2021). An exhaustive gene expression study of 15 different types of cancers compared to normal tissue confirmed that MMP expression has a large degree of heterogeneity across cancers and revealed that the most prevalent gene expression changes were upregulation as opposed to downregulation in tumor tissue versus control tissue (Gobin et al. 2019). Facilitation of cell migration and invasion is thought to be the principal effect of the MMP-mediated breakdown of basement membranes surrounding cancer cells (Jayadev and Sherwood 2017; Hanahan and Weinberg 2011). All molecular networks, including LM networks, as well as ECM/cell adhesion complexes are targets of MMP-mediated proteolysis. Notably, even if cells can physically breach basement membrane barriers in the absence of MMPs (Kelley et al. 2019), the massive production of MMPs by cancer cells greatly enhances their invasive capacity. Another barrier to tumor cell extravasation and invasion is the basement membrane underlying the endothelium of the vessel wall, where LM-411 and LM-511 are expressed (Spessotto et al. 2001). LMs can be cleaved by MMPs 2, 3, 7, 9, 10–16, 19, and 20 (Okada et al. 1986; Bejarano et al. 1988; Wilson and Matrisian 1996; Chandler et al. 1996; Ohuchi et al. 1997; Pirilä et al. 2003; Sadowski et al. 2005; Itoh and Seiki 2006; Laronha and Caldeira 2020), but invasive tumor cells primarily use MMP-2, MMP-9, and MMP-14 within invadopodia to cleave or degrade ECM proteins including LMs (Jacob and Prekeris 2015). Expression of MMP-14 in cancer cells correlates with their metastatic potential (Poincloux et al. 2009) and their ability to degrade fibrillar collagen and LMs. MMP-14 has the ability to activate MMP-2, MMP-9, and MMP-13 (Nishida et al. 2008; Knäuper et al. 2002; Li et al. 2017). The destruction of LM-111 by MMP-9 is thought to play a detrimental role in the early stage of breast cancer as the basement membrane no longer provides the appropriate signals to restrain epithelial cell proliferation (Beliveau et al. 2010).

The effects of LM degradation on cancer cell behavior, however, are more complex in the end because proteolytic cleavage leads to the release or exposure of either ECM-sequestered cytokines or a functional ECM fragment. Some LMs can express cryptic biological functions after proteolysis thereby altering cellular behavior and phenotypes (Niland and Eble 2021). The combined elastase/MMP-9-mediated proteolytic remodeling of LM-111 has been shown to unmask an epitope that triggers proliferation of dormant breast cancer cells through an α3β1 integrin signaling mechanism leading to aggressive lung metastases (Albrengues et al. 2018). This fascinating study demonstrated that the proteases present in neutrophil extracellular traps, together with DNA, are delivered to LM-111 through a DNA/LM interaction. LM-211, LM-411, and LM-511 could be degraded in the same way (Albrengues et al. 2018).

LM α1, β1, and γ1 chains are subject to MMP-2 and MMP-9 cleavage, releasing proteolytic fragments endowed with integrin- or syndecan-mediated cellular functions (Kikkawa et al. 2013). MMP-2 cleavage of the short arm of the β1 chain is predicted to disassemble the LM network, releasing a β1 LN–LE1-4 fragment that mediates α3β1-integrin-dependent embryonic stem cell adhesion (Horejs et al. 2014). Structural changes occur after LM-511 cleavage by MMP-9 (Gu et al. 2005), and fragments with activity toward inflammatory or cancer cells were identified within the LM α5 chain (Adair-Kirk et al. 2003; Hibino et al. 2004; Kusuma et al. 2011). An 80-kDa LM α2 C-terminal LG3–LG5 fragment possibly generated by MMP-9 is thought to play a role in reshaping the blood–testis barrier and spermatogenic function in a rat model (Gao et al. 2017). Analysis of the secretome from MDCK cells undergoing Ras-induced EMT resulted in the identification of two LM α5 internal and C-terminal fragments generated from MMP-1 cleavage with pro-migratory and pro-angiogenic activities (Gopal et al. 2016). LM α5 has also been reported to be cleaved by MMP-14, leading to prostate cancer cell migration (Bair et al. 2005).

LM-332 has been the focus of increasing interest over the last few years in the cancer research field, particularly the various proteolytic cleavages of its three subunits (Rousselle and Beck 2013). Many proteases degrade LM-332 and have been exhaustively documented (Rousselle and Scoazec 2020). To fulfill its essential function of anchoring proteins in epithelial basement membranes (Rousselle et al. 1991, 1997; Champliaud et al. 1996), LM-332 undergoes several physiological post-translational modification events. This maturation process allows its supramolecular integration and promotion of cell adhesion (Rousselle and Beck 2013). In cancers, additional cleavage events occur, compromising both these adhesive and connecting functions and additionally delivering pro-migratory signals to cells (Rousselle and Scoazec 2020). Cleavage of the β3 chain by MMP-7 or MMP-14 has been reported to enhance colon or prostate carcinoma cell migration and invasion, respectively (Remy et al. 2006; Udayakumar et al. 2003). MMP-2 or MMP-14 cleavage of the γ2 chain is reported to be a pro-tumorigenic signaling event (Giannelli et al. 1997; Koshikawa et al. 2000; Gilles et al. 2001). The released N-terminal LE repeats have been shown to induce EGF signaling and downstream MAPK activity in cancer cells (Koshikawa et al. 2004). The involvement of MMPs 3, 12, 13, 19, and 20 in the cleavage of the γ2 chain reinforces the status of LM-332 and its 2 subunit as an ideal target for MMPs in tumorigenesis (Pirilä et al. 2003; Sadowski et al. 2005; Väänänen et al. 2001).

Studies of the testis have reported that MMP-2 cleavage of the LM γ3 chain, part of the α3β3γ3 LM heterotrimer localized at the apical ectoplasmic specialization of Sertoli cells (Yan and Cheng 2006), induces dysfunction in the Sertoli cell tight junction permeability barrier through a 50-amino-acid residue peptide sequence in domain L4 (Siu and Cheng 2004; Yan et al. 2008; Su et al. 2012; Li et al. 2020). This process is thought to provide an efficient mechanism for modulating different aspects of spermatogenic function in response to changes in the epithelial cycle of spermatogenesis (Li et al. 2020). During zebrafish skeletal muscle development, LM was shown to act as an MMP-11 mediator and regulator of fibronectin levels at the myotendinous junction (Jenkins et al. 2016).

9.6 Conclusion

ECM remodeling is an important contributor to tumor progression and metastasis by contributing to the dissemination of cancer cells to distant tissues as well as angiogenesis. Among the many extracellular proteases expressed by human cells, MMPs are the most prominent enzymes supporting ECM proteolytic degradation and remodeling, as well as the modulation of cellular and ECM interactions. Whether MMPs are soluble/secreted in the tumor microenvironment or anchored to the cell membrane, their dysregulated expression in tumor cells, cancer-associated fibroblasts, and infiltrating immune cells is a hallmark of cancer. Being ideal targets of MMPs, the cleavage or degradation of LMs has structural and biological impacts favoring tumor development and metastasis. The fragments or peptides that are generated can induce the production of more MMPs in various cell types, setting up a vicious cascade. The proteolytic activity of MMPs is often localized in adhesion structures, such as invadopodia, or occurs in proximity to the cell membrane where LMs are expressed. Cellular receptors, such as CD44, play an important role in addressing and regulating MMP activity. Their connection to LMs could promote the establishment of an active MMP-rich microenvironment prone to efficient degradation and remodeling of the ECM. Thus, LMs can be targets of MMPs, bioactive factors stimulating their expression, or even interacting platforms capable of recruiting and concentrating them.

Acknowledgments Original work leading to this review was generously supported by research grants from the French Agence Nationale de la Recherche (grant number ANR-18-CE92-0035-01, awarded to PR) and Rhône-Auvergne Pack Ambition Recherche 2020.

References

Adair-Kirk TL, Atkinson JJ, Broekelmann TJ, Doi M, Tryggvason K, Miner JH, Mecham RP, Senior RM (2003) A site on laminin alpha 5, AQARSAASKVKVSMKF, induces inflammatory cell production of matrix metalloproteinase-9 and chemotaxis. J Immunol 171(1):398–406. https://doi.org/10.4049/jimmunol.171.1.398

Albrengues J, Shields MA, Ng D, Park CG, Ambrico A, Poindexter ME, Upadhyay P, Uyeminami DL, Pommier A, Küttner V, Bružas E, Maiorino L, Bautista C, Carmona EM, Gimotty PA, Fearon DT, Chang K, Lyons SK, Pinkerton KE, Trotman LC, Goldberg MS, Yeh JTH, Egeblad M (2018) Neutrophil extracellular traps produced during inflammation awaken dormant cancer cells in mice. Science 361(6409):eaao4227. https://doi.org/10.1126/science.aao4227

Andrusier N, Nussinov R, Wolfson HJ (2007) FireDock: fast interaction refinement in molecular docking. Proteins 69(1):139–159. https://doi.org/10.1002/prot.21495

Augoff K, Hryniewicz-Jankowska A, Tabola R (2020) Invadopodia: clearing the way for cancer cell invasion. Ann Transl Med 8(14):902. https://doi.org/10.21037/atm.2020.02.157

Aumailley M, Bruckner-Tuderman L, Carter WG, Deutzmann R, Edgar D, Ekblom P, Engel J, Engvall E, Hohenester E, Jones JC, Kleinman HK, Marinkovich MP, Martin GR, Mayer U, Meneguzzi G, Miner JH, Miyazaki K, Patarroyo M, Paulsson M, Quaranta V, Sanes JR, Sasaki T, Sekiguchi K, Sorokin LM, Talts JF, Tryggvason K, Uitto J, Virtanen I, von der Mark K, Wewer UM, Yamada Y, Yurchenco PD (2005) A simplified laminin nomenclature. Matrix Biol 24(5):326–332. https://doi.org/10.1016/j.matbio.2005.05.006

Bair EL, Chen ML, McDaniel K, Sekiguchi K, Cress AE, Nagle RB, Bowden GT (2005) Membrane type 1 matrix metalloprotease cleaves laminin-10 and promotes prostate cancer cell migration. Neoplasia 7(4):380–389. https://doi.org/10.1593/neo.04619

Banerji S, Wright AJ, Noble M, Mahoney DJ, Campbell ID, Day AJ, Jackson DG (2007) Structures of the Cd44–hyaluronan complex provide insight into a fundamental carbohydrate-protein interaction. Nat Struct Mol Biol 14(3):234–239. https://doi.org/10.1038/nsmb1201

Beck K, Dixon TW, Engel J, Parry DA (1993) Ionic interactions in the coiled-coil domain of laminin determine the specificity of chain assembly. J Mol Biol 231(2):311–323. https://doi.org/10.1006/jmbi.1993.1284

Bejarano PA, Noelken ME, Suzuki K, Hudson BG, Nagase H (1988) Degradation of basement membranes by human matrix metalloproteinase 3 (stromelysin). Biochem J 256(2):413–419. https://doi.org/10.1042/bj2560413

Beliveau A, Mott JD, Lo A, Chen EI, Koller AA, Yaswen P, Muschler J, Bissell MJ (2010) Raf-induced MMP9 disrupts tissue architecture of human breast cells in three-dimensional culture and is necessary for tumor growth in vivo. Genes Dev 24(24):2800–2811. https://doi.org/10.1101/gad.1990410

Bennett KL, Jackson DG, Simon JC, Tanczos E, Peach R, Modrell B, Stamenkovic I, Plowman G, Aruffo A (1995) CD44 isoforms containing exon V3 are responsible for the presentation of heparin-binding growth factor. J Cell Biol 128(4):687–698. https://doi.org/10.1083/jcb.128.4.687

Berno V, Porrini D, Castiglioni F, Campiglio M, Casalini P, Pupa SM, Balsari A, Ménard S, Tagliabue E (2005) The 67 kDa laminin receptor increases tumor aggressiveness by remodeling laminin-1. Endocr Relat Cancer 12(2):393–406. https://doi.org/10.1677/erc.1.00870

Bourguignon LYW (2019) Matrix hyaluronan-CD44 interaction activates microRNA and LncRNA Signaling associated with chemoresistance, invasion, and tumor progression. Front Oncol 9:492. https://doi.org/10.3389/fonc.2019.00492

Bourguignon LY, Bikle D (2015) Selective hyaluronan-CD44 signaling promotes miRNA-21 expression and interacts with vitamin D function during cutaneous squamous cell carcinomas progression following UV irradiation. Front Immunol 6:224. https://doi.org/10.3389/fimmu.2015.00224

Brown TA, Bouchard T, St John T, Wayner E, Carter WG (1991) Human keratinocytes express a new CD44 core protein (CD44E) as a heparan-sulfate intrinsic membrane proteoglycan with additional exons. J Cell Biol 113(1):207–221. https://doi.org/10.1083/jcb.113.1.207

Carafoli F, Hussain SA, Hohenester E (2012) Crystal structures of the network-forming short-arm tips of the laminin β1 and γ1 chains. PLoS One 7(7):e42473. https://doi.org/10.1371/journal.pone.0042473

Carulli S, Beck K, Dayan G, Boulesteix S, Lortat-Jacob H, Rousselle P (2012) Cell surface proteoglycans syndecan-1 and -4 bind overlapping but distinct sites in laminin α3 LG45 protein domain. J Biol Chem 287(15):12204–12216. https://doi.org/10.1074/jbc.M111.300061

Cauwe B, Van den Steen PE, Opdenakker G (2007) The biochemical, biological, and pathological kaleidoscope of cell surface substrates processed by matrix metalloproteinases. Crit Rev Biochem Mol Biol 42(3):113–185. https://doi.org/10.1080/10409230701340019

Chabadel A, Bañon-Rodríguez I, Cluet D, Rudkin BB, Wehrle-Haller B, Genot E, Jurdic P, Anton IM, Saltel F (2007) CD44 and beta3 integrin organize two functionally distinct actin-based domains in osteoclasts. Mol Biol Cell 18(12):4899–4910. https://doi.org/10.1091/mbc.e07-04-0378

Champliaud MF, Lunstrum GP, Rousselle P, Nishiyama T, Keene DR, Burgeson RE (1996) Human amnion contains a novel laminin variant, laminin 7, which like laminin 6, covalently associates with laminin 5 to promote stable epithelial-stromal attachment. J Cell Biol 132(6):1189–1198. https://doi.org/10.1083/jcb.132.6.1189

Chandler S, Cossins J, Lury J, Wells G (1996) Macrophage metalloelastase degrades matrix and myelin proteins and processes a tumour necrosis factor-alpha fusion protein. Biochem Biophys Res Commun 228(2):421–429. https://doi.org/10.1006/bbrc.1996.1677

Chen C, Zhao S, Karnad A, Freeman JW (2018) The biology and role of CD44 in cancer progression: therapeutic implications. J Hematol Oncol 11(1):64. https://doi.org/10.1186/s13045-018-0605-5

Cheng YS, Champliaud MF, Burgeson RE, Marinkovich MP, Yurchenco PD (1997) Self-assembly of laminin isoforms. J Biol Chem 272(50):31525–31532. https://doi.org/10.1074/jbc.272.50.31525

Cho SH, Park YS, Kim HJ, Kim CH, Lim SW, Huh JW, Lee JH, Kim HR (2012) CD44 enhances the epithelial-mesenchymal transition in association with colon cancer invasion. Int J Oncol 41(1):211–218. https://doi.org/10.3892/ijo.2012.1453

Colognato H, MacCarrick M, O'Rear JJ, Yurchenco PD (1997) The laminin α2-chain short arm mediates cell adhesion through both the α1β1 and α2β1 integrins. J Biol Chem 272(46):29330–29336. https://doi.org/10.1074/jbc.272.46.29330

Colognato-Pyke H, O'Rear JJ, Yamada Y, Carbonetto S, Cheng YS, Yurchenco PD (1995) Mapping of network-forming, heparin-binding, and alpha 1 beta 1 integrin-recognition sites within the alpha-chain short arm of laminin-1. J Biol Chem 270(16):9398–9406. https://doi.org/10.1074/jbc.270.16.9398

Corcoran ML, Kibbey MC, Kleinman HK, Wahl LM (1995) Laminin SIKVAV peptide induction of monocyte/macrophage prostaglandin E2 and matrix metalloproteinases. J Biol Chem 270(18):10365–10368. https://doi.org/10.1074/jbc.270.18.10365

Deboux C, Ladraa S, Cazaubon S, Ghribi-Mallah S, Weiss N, Chaverot N, Couraud PO, Baron-Van Evercooren A (2013) Overexpression of CD44 in neural precursor cells improves trans-endothelial migration and facilitates their invasion of perivascular tissues in vivo. PLoS One 8(2):e57430. https://doi.org/10.1371/journal.pone.0057430

Delimont D, Dufek BM, Meehan DT, Zallocchi M, Gratton MA, Phillips G, Cosgrove D (2014) Laminin α2-mediated focal adhesion kinase activation triggers Alport glomerular pathogenesis. PLoS One 9(6):e99083. https://doi.org/10.1371/journal.pone.0099083

Desai B, Rogers MJ, Chellaiah MA (2007) Mechanisms of osteopontin and CD44 as metastatic principles in prostate cancer cells. Mol Cancer 6:18. https://doi.org/10.1186/1476-4598-6-18

Deutzmann R, Aumailley M, Wiedemann H, Pysny W, Timpl R, Edgar D (1990) Cell adhesion, spreading and neurite stimulation by laminin fragment E8 depends on maintenance of secondary and tertiary structure in its rod and globular domain. Eur J Biochem 191(2):513–522. https://doi.org/10.1111/j.1432-1033.1990.tb19151.x

Domogatskaya A, Rodin S, Tryggvason K (2012) Functional diversity of laminins. Annu Rev Cell Dev Biol 28:523–553. https://doi.org/10.1146/annurev-cellbio-101011-155750

Dosztányi Z (2018) Prediction of protein disorder based on IUPred. Protein Sci 27(1):331–340. https://doi.org/10.1002/pro.3334

Du L, Wang H, He L, Zhang J, Ni B, Wang X, Jin H, Cahuzac N, Mehrpour M, Lu Y, Chen Q (2008) CD44 is of functional importance for colorectal cancer stem cells. Clin Cancer Res 14(21):6751–6760. https://doi.org/10.1158/1078-0432.Ccr-08-1034

Durbeej M (2010) Laminins. Cell Tissue Res 339(1):259. https://doi.org/10.1007/s00441-009-0838-2

Fingleton B (2006) Matrix metalloproteinases: roles in cancer and metastasis. Front Biosci Landmark 11(1):479–491. https://doi.org/10.2741/1811

Gaffney J, Solomonov I, Zehorai E, Sagi I (2015) Multilevel regulation of matrix metalloproteinases in tissue homeostasis indicates their molecular specificity in vivo. Matrix Biol 44–46:191–199. https://doi.org/10.1016/j.matbio.2015.01.012

Gal I, Lesley J, Ko W, Gonda A, Stoop R, Hyman R, Mikecz K (2003) Role of the extracellular and cytoplasmic domains of CD44 in the rolling interaction of lymphoid cells with hyaluronan under physiologic flow. J Biol Chem 278(13):11150–11158. https://doi.org/10.1074/jbc.M210661200

Gao Y, Mruk D, Chen H, Lui WY, Lee WM, Cheng CY (2017) Regulation of the blood-testis barrier by a local axis in the testis: role of laminin α2 in the basement membrane. FASEB J 31(2):584–597. https://doi.org/10.1096/fj.201600870R

Gee SH, Blacher RW, Douville PJ, Provost PR, Yurchenco PD, Carbonetto S (1993) Laminin-binding protein 120 from brain is closely related to the dystrophin-associated glycoprotein, dystroglycan, and binds with high affinity to the major heparin binding domain of laminin. J Biol Chem 268(20):14972–14980. https://doi.org/10.1016/S0021-9258(18)82427-9

Giannelli G, Falk-Marzillier J, Schiraldi O, Stetler-Stevenson WG, Quaranta V (1997) Induction of cell migration by matrix metalloprotease-2 cleavage of laminin-5. Science 277(5323):225. https://doi.org/10.1126/science.277.5323.225

Gilles C, Polette M, Coraux C, Tournier JM, Meneguzzi G, Munaut C, Volders L, Rousselle P, Birembaut P, Foidart JM (2001) Contribution of MT1-MMP and of human laminin-5 gamma2 chain degradation to mammary epithelial cell migration. J Cell Sci 114(16):2967–2976. https://doi.org/10.1242/jcs.114.16.2967

Gobin E, Bagwell K, Wagner J, Mysona D, Sandirasegarane S, Smith N, Bai S, Sharma A, Schleifer R, She J-X (2019) A pan-cancer perspective of matrix metalloproteases (MMP)

gene expression profile and their diagnostic/prognostic potential. BMC Cancer 19(1):581. https://doi.org/10.1186/s12885-019-5768-0

Gopal SK, Greening DW, Zhu H-J, Simpson RJ, Mathias RA (2016) Transformed MDCK cells secrete elevated MMP1 that generates LAMA5 fragments promoting endothelial cell angiogenesis. Sci Rep 6(1):28321. https://doi.org/10.1038/srep28321

Gu Z, Cui J, Brown S, Fridman R, Mobashery S, Strongin AY, Lipton SA (2005) A highly specific inhibitor of matrix metalloproteinase-9 rescues laminin from proteolysis and neurons from apoptosis in transient focal cerebral ischemia. J Neurosci 25(27):6401–6408. https://doi.org/10.1523/jneurosci.1563-05.2005

Gupta A, Cao W, Sadashivaiah K, Chen W, Schneider A, Chellaiah MA (2013) Promising noninvasive cellular phenotype in prostate cancer cells knockdown of matrix metalloproteinase 9. ScientificWorldJournal 2013:493689. https://doi.org/10.1155/2013/493689

Guvench O (2015) Revealing the mechanisms of protein disorder and N-glycosylation in CD44-hyaluronan binding using molecular simulation. Front Immunol 6:305. https://doi.org/10.3389/fimmu.2015.00305

Habchi J, Tompa P, Longhi S, Uversky VN (2014) Introducing protein intrinsic disorder. Chem Rev 114(13):6561–6588. https://doi.org/10.1021/cr400514h

Hanahan D, Weinberg RA (2011) Hallmarks of cancer: the next generation. Cell 144(5):646–674. https://doi.org/10.1016/j.cell.2011.02.013

Hibino S, Shibuya M, Engbring JA, Mochizuki M, Nomizu M, Kleinman HK (2004) Identification of an active site on the laminin alpha5 chain globular domain that binds to CD44 and inhibits malignancy. Cancer Res 64(14):4810–4816. https://doi.org/10.1158/0008-5472.Can-04-0129

Hibino S, Shibuya M, Hoffman MP, Engbring JA, Hossain R, Mochizuki M, Kudoh S, Nomizu M, Kleinman HK (2005) Laminin alpha5 chain metastasis- and angiogenesis-inhibiting peptide blocks fibroblast growth factor 2 activity by binding to the heparan sulfate chains of CD44. Cancer Res 65(22):10494–10501. https://doi.org/10.1158/0008-5472.Can-05-0314

Hohenester E (2019) Structural biology of laminins. Essays Biochem 63(3):285–295. https://doi.org/10.1042/ebc20180075

Hohenester E, Yurchenco PD (2013) Laminins in basement membrane assembly. Cell Adh Migr 7(1):56–63. https://doi.org/10.4161/cam.21831

Horejs C-M, Serio A, Purvis A, Gormley AJ, Bertazzo S, Poliniewicz A, Wang AJ, DiMaggio P, Hohenester E, Stevens MM (2014) Biologically-active laminin-111 fragment that modulates the epithelial-to-mesenchymal transition in embryonic stem cells. Proc Natl Acad Sci 111(16):5908–5913. https://doi.org/10.1073/pnas.1403139111

Hozumi K, Suzuki N, Uchiyama Y, Katagiri F, Kikkawa Y, Nomizu M (2009) Chain-specific heparin-binding sequences in the laminin alpha chain LG45 modules. Biochemistry 48(23):5375–5381. https://doi.org/10.1021/bi900542u

Ido H, Harada K, Futaki S, Hayashi Y, Nishiuchi R, Natsuka Y, Li S, Wada Y, Combs AC, Ervasti JM, Sekiguchi K (2004) Molecular dissection of the alpha-dystroglycan- and integrin-binding sites within the globular domain of human laminin-10. J Biol Chem 279(12):10946–10954. https://doi.org/10.1074/jbc.M313626200

Ido H, Nakamura A, Kobayashi R, Ito S, Li S, Futaki S, Sekiguchi K (2007) The requirement of the glutamic acid residue at the third position from the carboxyl termini of the laminin gamma chains in integrin binding by laminins. J Biol Chem 282(15):11144–11154. https://doi.org/10.1074/jbc.M609402200

Ido H, Ito S, Taniguchi Y, Hayashi M, Sato-Nishiuchi R, Sanzen N, Hayashi Y, Futaki S, Sekiguchi K (2008) Laminin isoforms containing the gamma3 chain are unable to bind to integrins due to the absence of the glutamic acid residue conserved in the C-terminal regions of the gamma1 and gamma2 chains. J Biol Chem 283(42):28149–28157. https://doi.org/10.1074/jbc.M803553200

Inamori KI, Beedle AM, de Bernabé DB, Wright ME, Campbell KP (2016) LARGE2-dependent glycosylation confers laminin-binding ability on proteoglycans. Glycobiology 26(12):1284–1296. https://doi.org/10.1093/glycob/cww075

Ishii S, Ford R, Thomas P, Nachman A, Steele G Jr, Jessup JM (1993) CD44 participates in the adhesion of human colorectal carcinoma cells to laminin and type IV collagen. Surg Oncol 2(4): 255–264. https://doi.org/10.1016/0960-7404(93)90015-q

Itoh Y, Nagase H (2002) Matrix metalloproteinases in cancer. Essays Biochem 38:21–36. https://doi.org/10.1042/bse0380021

Itoh Y, Seiki M (2006) MT1-MMP: a potent modifier of pericellular microenvironment. J Cell Physiol 206(1):1–8. https://doi.org/10.1002/jcp.20431

Jacob A, Prekeris R (2015) The regulation of MMP targeting to invadopodia during cancer metastasis. Front Cell Dev Biol 3:4–4. https://doi.org/10.3389/fcell.2015.00004

Jalkanen S, Jalkanen M (1992) Lymphocyte CD44 binds the COOH-terminal heparin-binding domain of fibronectin. J Cell Biol 116(3):817–825. https://doi.org/10.1083/jcb.116.3.817

Jayadev R, Sherwood DR (2017) Basement membranes. Curr Biol 27(6):R207–R211. https://doi.org/10.1016/j.cub.2017.02.006

Jenkins MH, Alrowaished SS, Goody MF, Crawford BD, Henry CA (2016) Laminin and Matrix metalloproteinase 11 regulate Fibronectin levels in the zebrafish myotendinous junction. Skelet Muscle 6(1):18. https://doi.org/10.1186/s13395-016-0089-3

Kajita M, Itoh Y, Chiba T, Mori H, Okada A, Kinoh H, Seiki M (2001) Membrane-type 1 matrix metalloproteinase cleaves CD44 and promotes cell migration. J Cell Biol 153(5):893–904. https://doi.org/10.1083/jcb.153.5.893

Kamoshida G, Ogawa T, Oyanagi J, Sato H, Komiya E, Higashi S, Miyazaki K, Tsuji T (2014) Modulation of matrix metalloproteinase-9 secretion from tumor-associated macrophage-like cells by proteolytically processed laminin-332 (laminin-5). Clin Exp Metastasis 31(3):285–291. https://doi.org/10.1007/s10585-013-9627-0

Karamanos NK, Piperigkou Z, Passi A, Götte M, Rousselle P, Vlodavsky I (2021) Extracellular matrix-based cancer targeting. Trends Mol Med. https://doi.org/10.1016/j.molmed.2021.07.009

Kelley LC, Chi Q, Cáceres R, Hastie E, Schindler AJ, Jiang Y, Matus DQ, Plastino J, Sherwood DR (2019) Adaptive F-actin polymerization and localized ATP production drive basement membrane invasion in the absence of MMPs. Dev Cell 48(3):313–328.e318. https://doi.org/10.1016/j.devcel.2018.12.018

Kenne E, Soehnlein O, Genové G, Rotzius P, Eriksson EE, Lindbom L (2010) Immune cell recruitment to inflammatory loci is impaired in mice deficient in basement membrane protein laminin alpha4. J Leukoc Biol 88(3):523–528. https://doi.org/10.1189/jlb.0110043

Khan KM, Falcone DJ (1997) Role of laminin in matrix induction of macrophage urokinase-type plasminogen activator and 92-kDa metalloproteinase expression. J Biol Chem 272(13): 8270–8275. https://doi.org/10.1074/jbc.272.13.8270

Khan KM, Falcone DJ (2000) Selective activation of MAPK(erk1/2) by laminin-1 peptide alpha1: Ser(2091)-Arg(2108) regulates macrophage degradative phenotype. J Biol Chem 275(6): 4492–4498. https://doi.org/10.1074/jbc.275.6.4492

Kikkawa Y, Sasaki T, Nguyen MT, Nomizu M, Mitaka T, Miner JH (2007) The LG1-3 tandem of laminin alpha5 harbors the binding sites of Lutheran/basal cell adhesion molecule and alpha3beta1/alpha6beta1 integrins. J Biol Chem 282(20):14853–14860. https://doi.org/10.1074/jbc.M611706200

Kikkawa Y, Hozumi K, Katagiri F, Nomizu M, Kleinman HK, Koblinski JE (2013) Laminin-111-derived peptides and cancer. Cell Adh Migr 7(1):150–256. https://doi.org/10.4161/cam.22827

Klausen MS, Jespersen MC, Nielsen H, Jensen KK, Jurtz VI, Sønderby CK, Sommer MOA, Winther O, Nielsen M, Petersen B, Marcatili P (2019) NetSurfP-2.0: improved prediction of protein structural features by integrated deep learning. Proteins 87(6):520–527. https://doi.org/10.1002/prot.25674

Knäuper V, Bailey L, Worley JR, Soloway P, Patterson ML, Murphy G (2002) Cellular activation of proMMP-13 by MT1-MMP depends on the C-terminal domain of MMP-13. FEBS Lett 532(1-2):127–130. https://doi.org/10.1016/s0014-5793(02)03654-2

Knudson W, Aguiar DJ, Hua Q, Knudson CB (1996) CD44-anchored hyaluronan-rich pericellular matrices: an ultrastructural and biochemical analysis. Exp Cell Res 228(2):216–228. https://doi.org/10.1006/excr.1996.0320

Koshikawa N, Giannelli G, Cirulli V, Miyazaki K, Quaranta V (2000) Role of cell surface metalloprotease MT1-MMP in epithelial cell migration over laminin-5. J Cell Biol 148(3): 615–624. https://doi.org/10.1083/jcb.148.3.615

Koshikawa N, Schenk S, Moeckel G, Sharabi A, Miyazaki K, Gardner H, Zent R, Quaranta V (2004) Proteolytic processing of laminin-5 by MT1-MMP in tissues and its effects on epithelial cell morphology. FASEB J 18(2):364–366. https://doi.org/10.1096/fj.03-0584fje

Kusuma N, Anderson RL, Pouliot N (2011) Laminin α5-derived peptides modulate the properties of metastatic breast tumour cells. Clin Exp Metastasis 28(8):909–921. https://doi.org/10.1007/s10585-011-9422-8

Laronha H, Caldeira J (2020) Structure and function of human matrix metalloproteinases. Cell 9(5): 1076. https://doi.org/10.3390/cells9051076

Li S, Edgar D, Fässler R, Wadsworth W, Yurchenco PD (2003) The role of laminin in embryonic cell polarization and tissue organization. Dev Cell 4(5):613–624. https://doi.org/10.1016/s1534-5807(03)00128-x

Li Z, Takino T, Endo Y, Sato H (2017) Activation of MMP-9 by membrane type-1 MMP/MMP-2 axis stimulates tumor metastasis. Cancer Sci 108(3):347–353. https://doi.org/10.1111/cas.13134

Li H, Liu S, Wu S, Li L, Ge R, Cheng CY (2020) Bioactive fragments of laminin and collagen chains: lesson from the testis. Reproduction 159(3):R111–r123. https://doi.org/10.1530/rep-19-0288

Linder S, Wiesner C (2016) Feel the force: podosomes in mechanosensing. Exp Cell Res 343(1): 67–72. https://doi.org/10.1016/j.yexcr.2015.11.026

Liu L-K, Finzel B (2014) High-resolution crystal structures of alternate forms of the human CD44 hyaluronan-binding domain reveal a site for protein interaction. Acta Crystallogr F Struct Biol Commun 70(9):1155–1161. https://doi.org/10.1107/S2053230X14015532

Macdonald PR, Lustig A, Steinmetz MO, Kammerer RA (2010) Laminin chain assembly is regulated by specific coiled-coil interactions. J Struct Biol 170(2):398–405. https://doi.org/10.1016/j.jsb.2010.02.004

Maity G, Sen T, Chatterjee A (2011) Laminin induces matrix metalloproteinase-9 expression and activation in human cervical cancer cell line (SiHa). J Cancer Res Clin Oncol 137(2):347–357. https://doi.org/10.1007/s00432-010-0892-x

Maltseva DV, Rodin SA (2018) Laminins in metastatic cancer. Mol Biol (Mosk) 52(3):350–371. https://doi.org/10.1134/S0026893318030093

Matsui C, Wang CK, Nelson CF, Bauer EA, Hoeffler WK (1995) The assembly of laminin-5 subunits. J Biol Chem 270(40):23496–23503. https://doi.org/10.1074/jbc.270.40.23496

McKee KK, Harrison D, Capizzi S, Yurchenco PD (2007) Role of laminin terminal globular domains in basement membrane assembly. J Biol Chem 282(29):21437–21447. https://doi.org/10.1074/jbc.M702963200

Medrano-González PA, Rivera-Ramírez O, Montaño LF, Rendón-Huerta EP (2021) Proteolytic processing of CD44 and its implications in cancer. Stem Cells Int 2021:6667735. https://doi.org/10.1155/2021/6667735

Michopoulou A, Montmasson M, Garnier C, Lambert E, Dayan G, Rousselle P (2020) A novel mechanism in wound healing: laminin 332 drives MMP9/14 activity by recruiting syndecan-1 and CD44. Matrix Biol 94:1–17. https://doi.org/10.1016/j.matbio.2020.06.004

Miner JH, Yurchenco PD (2004) Laminin functions in tissue morphogenesis. Annu Rev Cell Dev Biol 20:255–284. https://doi.org/10.1146/annurev.cellbio.20.010403.094555

Miner JH, Patton BL, Lentz SI, Gilbert DJ, Snider WD, Jenkins NA, Copeland NG, Sanes JR (1997) The laminin alpha chains: expression, developmental transitions, and chromosomal locations of alpha1-5, identification of heterotrimeric laminins 8-11, and cloning of a novel alpha3 isoform. J Cell Biol 137(3):685–701. https://doi.org/10.1083/jcb.137.3.685

Momota Y, Suzuki N, Kasuya Y, Kobayashi T, Mizoguchi M, Yokoyama F, Nomizu M, Shinkai H, Iwasaki T, Utani A (2005) Laminin alpha3 LG4 module induces keratinocyte migration: involvement of matrix metalloproteinase-9. J Recept Signal Transduct Res 25(1):1–17. https://doi.org/10.1081/rrs-200047870

Mori H, Tomari T, Koshikawa N, Kajita M, Itoh Y, Sato H, Tojo H, Yana I, Seiki M (2002) CD44 directs membrane-type 1 matrix metalloproteinase to lamellipodia by associating with its hemopexin-like domain. EMBO J 21(15):3949–3959. https://doi.org/10.1093/emboj/cdf411

Morita A, Sugimoto E, Kitagawa Y (1985) Post-translational assembly and glycosylation of laminin subunits in parietal endoderm-like F9 cells. Biochem J 229(1):259–264. https://doi.org/10.1042/bj2290259

Murphy DA, Courtneidge SA (2011) The 'ins' and 'outs' of podosomes and invadopodia: characteristics, formation and function. Nat Rev Mol Cell Biol 12(7):413–426. https://doi.org/10.1038/nrm3141

Mydel P, Shipley JM, Adair-Kirk TL, Kelley DG, Broekelmann TJ, Mecham RP, Senior RM (2008) Neutrophil elastase cleaves laminin-332 (laminin-5) generating peptides that are chemotactic for neutrophils. J Biol Chem 283(15):9513–9522. https://doi.org/10.1074/jbc.M706239200

Nagase H, Visse R, Murphy G (2006) Structure and function of matrix metalloproteinases and TIMPs. Cardiovasc Res 69(3):562–573. https://doi.org/10.1016/j.cardiores.2005.12.002

Navdaev A, Heitmann V, Desantana Evangelista K, Mörgelin M, Wegener J, Eble JA (2008) The C-terminus of the gamma 2 chain but not of the beta 3 chain of laminin-332 is indirectly but indispensably necessary for integrin-mediated cell reactions. Exp Cell Res 314(3):489–497. https://doi.org/10.1016/j.yexcr.2007.10.027

Nielsen PK, Yamada Y (2001) Identification of cell-binding sites on the laminin α5 N-terminal domain by site-directed mutagenesis. J Biol Chem 276(14):10906–10912. https://doi.org/10.1074/jbc.M008743200

Niland S, Eble JA (2021) Hold on or cut? Integrin- and MMP-mediated cell-matrix interactions in the tumor microenvironment. Int J Mol Sci 22(1). https://doi.org/10.3390/ijms22010238

Nishida Y, Miyamori H, Thompson EW, Takino T, Endo Y, Sato H (2008) Activation of matrix metalloproteinase-2 (MMP-2) by membrane type 1 matrix metalloproteinase through an artificial receptor for proMMP-2 generates active MMP-2. Cancer Res 68(21):9096–9104. https://doi.org/10.1158/0008-5472.Can-08-2522

Nishiuchi R, Takagi J, Hayashi M, Ido H, Yagi Y, Sanzen N, Tsuji T, Yamada M, Sekiguchi K (2006) Ligand-binding specificities of laminin-binding integrins: a comprehensive survey of laminin-integrin interactions using recombinant alpha3beta1, alpha6beta1, alpha7beta1 and alpha6beta4 integrins. Matrix Biol 25(3):189–197. https://doi.org/10.1016/j.matbio.2005.12.001

Ogawa T, Tsubota Y, Hashimoto J, Kariya Y, Miyazaki K (2007) The Short arm of laminin γ2 chain of laminin-5 (laminin-332) binds syndecan-1 and regulates cellular adhesion and migration by suppressing phosphorylation of integrin β4 chain. Mol Biol Cell 18(5):1621–1633. https://doi.org/10.1091/mbc.e06-09-0806

Ohuchi E, Imai K, Fujii Y, Sato H, Seiki M, Okada Y (1997) Membrane type 1 matrix metalloproteinase digests interstitial collagens and other extracellular matrix macromolecules. J Biol Chem 272(4):2446–2451. https://doi.org/10.1074/jbc.272.4.2446

Okada Y, Nagase H, Harris ED (1986) A metalloproteinase from human rheumatoid synovial fibroblasts that digests connective tissue matrix components. Purification and characterization. J Biol Chem 261(30):14245–14255. https://doi.org/10.1016/S0021-9258(18)67011-5

Pal S, Moulik S, Dutta A, Chatterjee A (2014) Extracellular matrix protein laminin induces matrix metalloproteinase-9 in human breast cancer cell line mcf-7. Cancer Microenviron 7(1-2):71–78. https://doi.org/10.1007/s12307-014-0146-6

Paulsson M, Deutzmann R, Timpl R, Dalzoppo D, Odermatt E, Engel J (1985) Evidence for coiled-coil alpha-helical regions in the long arm of laminin. EMBO J 4(2):309–316. https://doi.org/10.1002/j.1460-2075.1985.tb03630.x

Peng ST, Su CH, Kuo CC, Shaw CF, Wang HS (2007) CD44 crosslinking-mediated matrix metalloproteinase-9 relocation in breast tumor cells leads to enhanced metastasis. Int J Oncol 31(5):1119–1126. https://doi.org/10.3892/ijo.31.5.1119

Piperigkou Z, Kyriakopoulou K, Koutsakis C, Mastronikolis S, Karamanos NK (2021) Key matrix remodeling enzymes: functions and targeting in cancer. Cancers (Basel) 13(6). https://doi.org/10.3390/cancers13061441

Pirilä E, Sharabi A, Salo T, Quaranta V, Tu H, Heljasvaara R, Koshikawa N, Sorsa T, Maisi P (2003) Matrix metalloproteinases process the laminin-5 gamma 2-chain and regulate epithelial cell migration. Biochem Biophys Res Commun 303(4):1012–1017. https://doi.org/10.1016/s0006-291x(03)00452-2

Poincloux R, Lizárraga F, Chavrier P (2009) Matrix invasion by tumour cells: a focus on MT1-MMP trafficking to invadopodia. J Cell Sci 122(17):3015–3024. https://doi.org/10.1242/jcs.034561

Ponta H, Sherman L, Herrlich PA (2003) CD44: from adhesion molecules to signalling regulators. Nat Rev Mol Cell Biol 4(1):33–45. https://doi.org/10.1038/nrm1004

Pulido D, Briggs DC, Hua J, Hohenester E (2017) Crystallographic analysis of the laminin β2 short arm reveals how the LF domain is inserted into a regular array of LE domains. Matrix Biol 57–58:204–212. https://doi.org/10.1016/j.matbio.2016.06.006

Qin Y, Rodin S, Simonson OE, Hollande F (2017) Laminins and cancer stem cells: partners in crime? Semin Cancer Biol 45:3–12. https://doi.org/10.1016/j.semcancer.2016.07.004

Ramovs V, Te Molder L, Sonnenberg A (2017) The opposing roles of laminin-binding integrins in cancer. Matrix Biol 57-58:213–243. https://doi.org/10.1016/j.matbio.2016.08.007

Remy L, Trespeuch C, Bachy S, Scoazec JY, Rousselle P (2006) Matrilysin 1 influences colon carcinoma cell migration by cleavage of the laminin-5 beta3 chain. Cancer Res 66(23):11228–11237. https://doi.org/10.1158/0008-5472.Can-06-1187

Ries A, Goldberg JL, Grimpe B (2007) A novel biological function for CD44 in axon growth of retinal ganglion cells identified by a bioinformatics approach. J Neurochem 103(4):1491–1505. https://doi.org/10.1111/j.1471-4159.2007.04858.x

Rousselle P, Beck K (2013) Laminin 332 processing impacts cellular behavior. Cell Adh Migr 7(1):122–134. https://doi.org/10.4161/cam.23132

Rousselle P, Scoazec JY (2020) Laminin 332 in cancer: when the extracellular matrix turns signals from cell anchorage to cell movement. Semin Cancer Biol 62:149–165. https://doi.org/10.1016/j.semcancer.2019.09.026

Rousselle P, Lunstrum GP, Keene DR, Burgeson RE (1991) Kalinin: an epithelium-specific basement membrane adhesion molecule that is a component of anchoring filaments. J Cell Biol 114(3):567–576. https://doi.org/10.1083/jcb.114.3.567

Rousselle P, Golbik R, van der Rest M, Aumailley M (1995) Structural requirement for cell adhesion to kalinin (laminin-5). J Biol Chem 270(23):13766–13770. https://doi.org/10.1074/jbc.270.23.13766

Rousselle P, Keene DR, Ruggiero F, Champliaud MF, Rest M, Burgeson RE (1997) Laminin 5 binds the NC-1 domain of type VII collagen. J Cell Biol 138(3):719–728. https://doi.org/10.1083/jcb.138.3.719

Sadowski T, Dietrich S, Koschinsky F, Ludwig A, Proksch E, Titz B, Sedlacek R (2005) Matrix metalloproteinase 19 processes the laminin 5 gamma 2 chain and induces epithelial cell migration. Cell Mol Life Sci CMLS 62(7):870–880. https://doi.org/10.1007/s00018-005-4478-8

Samanna V, Wei H, Ego-Osuala D, Chellaiah MA (2006) Alpha-V-dependent outside-in signaling is required for the regulation of CD44 surface expression, MMP-2 secretion, and cell migration by osteopontin in human melanoma cells. Exp Cell Res 312(12):2214–2230. https://doi.org/10.1016/j.yexcr.2006.03.022

Sasaki T, Timpl R (2001) Domain IVa of laminin alpha5 chain is cell-adhesive and binds beta1 and alphaVbeta3 integrins through Arg-Gly-Asp. FEBS Lett 509(2):181–185. https://doi.org/10.1016/s0014-5793(01)03167-2

Sasaki T, Göhring W, Mann K, Brakebusch C, Yamada Y, Fässler R, Timpl R (2001) Short arm region of laminin-5 gamma2 chain: structure, mechanism of processing and binding to heparin and proteins. J Mol Biol 314(4):751–763. https://doi.org/10.1006/jmbi.2001.5176

Sato H, Higashi S, Miyazaki K (2015) Amino-terminal fragments of laminin γ2 chain stimulate migration of metastatic breast cancer cells by interacting with CD44. Clin Exp Metastasis 32(5): 405–415. https://doi.org/10.1007/s10585-015-9705-6

Schneidman-Duhovny D, Inbar Y, Nussinov R, Wolfson HJ (2005) PatchDock and SymmDock: servers for rigid and symmetric docking. Nucleic Acids Res 33 (Web Server issue):W363–W367. https://doi.org/10.1093/nar/gki481

Screaton GR, Bell MV, Bell JI, Jackson DG (1993) The identification of a new alternative exon with highly restricted tissue expression in transcripts encoding the mouse Pgp-1 (CD44) homing receptor. Comparison of all 10 variable exons between mouse, human, and rat. J Biol Chem 268(17):12235–12238

Seiki M (2002) The cell surface: the stage for matrix metalloproteinase regulation of migration. Curr Opin Cell Biol 14(5):624–632. https://doi.org/10.1016/s0955-0674(02)00363-0

Senyürek I, Kempf WE, Klein G, Maurer A, Kalbacher H, Schäfer L, Wanke I, Christ C, Stevanovic S, Schaller M, Rousselle P, Garbe C, Biedermann T, Schittek B (2014) Processing of laminin α chains generates peptides involved in wound healing and host defense. J Innate Immun 6(4):467–484. https://doi.org/10.1159/000357032

Shay G, Lynch CC, Fingleton B (2015) Moving targets: emerging roles for MMPs in cancer progression and metastasis. Matrix Biol 44–46:200–206. https://doi.org/10.1016/j.matbio.2015.01.019

Siu MKY, Cheng CY (2004) Interactions of proteases, protease inhibitors, and the β1 integrin/laminin γ3 Protein complex in the regulation of ectoplasmic specialization dynamics in the rat testis1. Biol Reprod 70(4):945–964. https://doi.org/10.1095/biolreprod.103.023606

Smirnov SP, McDearmon EL, Li S, Ervasti JM, Tryggvason K, Yurchenco PD (2002) Contributions of the LG modules and furin processing to laminin-2 functions. J Biol Chem 277(21): 18928–18937. https://doi.org/10.1074/jbc.M201880200

Spessotto P, Yin Z, Magro G, Deutzmann R, Chiu A, Colombatti A, Perris R (2001) Laminin isoforms 8 and 10 are primary components of the subendothelial basement membrane promoting interaction with neoplastic lymphocytes. Cancer Res 61(1):339–347. https://cancerres.aacrjournals.org/content/61/1/339

Su L, Mruk DD, Lie PP, Silvestrini B, Cheng CY (2012) A peptide derived from laminin-γ3 reversibly impairs spermatogenesis in rats. Nat Commun 3:1185. https://doi.org/10.1038/ncomms2171

Suenaga N, Mori H, Itoh Y, Seiki M (2005) CD44 binding through the hemopexin-like domain is critical for its shedding by membrane-type 1 matrix metalloproteinase. Oncogene 24(5): 859–868. https://doi.org/10.1038/sj.onc.1208258

Suzuki N, Yokoyama F, Nomizu M (2005) Functional sites in the laminin alpha chains. Connect Tissue Res 46(3):142–152. https://doi.org/10.1080/03008200591008527

Sypecka J, Dragun-Szymczak P, Zalewska T, Domańska-Janik K (2009) Laminin promotes oligogliogenesis and increases MMPs activity in human neural stem cells of HUCB-NSC line. Acta Neurobiol Exp (Wars) 69(1):37–45

Takahashi K, Eto H, Tanabe KK (1999) Involvement of CD44 in matrix metalloproteinase-2 regulation in human melanoma cells. Int J Cancer 80(3):387–395. https://doi.org/10.1002/(sici)1097-0215(19990129)80:3<387::aid-ijc9>3.0.co;2-t

Takaishi S, Okumura T, Tu S, Wang SSW, Shibata W, Vigneshwaran R, Gordon SAK, Shimada Y, Wang TC (2009) Identification of gastric cancer stem cells using the cell surface marker CD44. Stem Cells 27(5):1006–1020. https://doi.org/10.1002/stem.30

Takizawa M, Arimori T, Taniguchi Y, Kitago Y, Yamashita E, Takagi J, Sekiguchi K (2017) Mechanistic basis for the recognition of laminin-511 by α6β1 integrin. Sci Adv 3(9):e1701497. https://doi.org/10.1126/sciadv.1701497

Talts JF, Andac Z, Göhring W, Brancaccio A, Timpl R (1999) Binding of the G domains of laminin alpha1 and alpha2 chains and perlecan to heparin, sulfatides, alpha-dystroglycan and several extracellular matrix proteins. EMBO J 18(4):863–870. https://doi.org/10.1093/emboj/18.4.863

Thapa R, Wilson GD (2016) The importance of CD44 as a stem cell biomarker and therapeutic target in cancer. Stem Cells Int 2016:2087204. https://doi.org/10.1155/2016/2087204

Timpl R, Tisi D, Talts JF, Andac Z, Sasaki T, Hohenester E (2000) Structure and function of laminin LG modules. Matrix Biol 19(4):309–317. https://doi.org/10.1016/S0945-053X(00)00072-X

Tran M, Rousselle P, Nokelainen P, Tallapragada S, Nguyen NT, Fincher EF, Marinkovich MP (2008) Targeting a tumor-specific laminin domain critical for human carcinogenesis. Cancer Res 68(8):2885–2894. https://doi.org/10.1158/0008-5472.Can-07-6160

Tsukita S, Oishi K, Sato N, Sagara J, Kawai A, Tsukita S (1994) ERM family members as molecular linkers between the cell surface glycoprotein CD44 and actin-based cytoskeletons. J Cell Biol 126(2):391–401. https://doi.org/10.1083/jcb.126.2.391

Tzu J, Marinkovich MP (2008) Bridging structure with function: structural, regulatory, and developmental role of laminins. Int J Biochem Cell Biol 40(2):199–214. https://doi.org/10.1016/j.biocel.2007.07.015

Udani M, Zen Q, Cottman M, Leonard N, Jefferson S, Daymont C, Truskey G, Telen MJ (1998) Basal cell adhesion molecule/lutheran protein. The receptor critical for sickle cell adhesion to laminin. J Clin Invest 101(11):2550–2558. https://doi.org/10.1172/JCI1204

Udayakumar TS, Chen ML, Bair EL, Von Bredow DC, Cress AE, Nagle RB, Bowden GT (2003) Membrane type-1-matrix metalloproteinase expressed by prostate carcinoma cells cleaves human laminin-5 beta3 chain and induces cell migration. Cancer Res 63(9):2292–2299

Utani A, Momota Y, Endo H, Kasuya Y, Beck K, Suzuki N, Nomizu M, Shinkai H (2003) Laminin alpha 3 LG4 module induces matrix metalloproteinase-1 through mitogen-activated protein kinase signaling. J Biol Chem 278(36):34483–34490. https://doi.org/10.1074/jbc.M304827200

Väänänen A, Srinivas R, Parikka M, Palosaari H, Bartlett JD, Iwata K, Grenman R, Stenman UH, Sorsa T, Salo T (2001) Expression and regulation of MMP-20 in human tongue carcinoma cells. J Dent Res 80(10):1884–1889. https://doi.org/10.1177/00220345010800100501

Vainionpää N, Kikkawa Y, Lounatmaa K, Miner JH, Rousselle P, Virtanen I (2006) Laminin-10 and Lutheran blood group glycoproteins in adhesion of human endothelial cells. Am J Physiol Cell Physiol 290(3):C764–C775. https://doi.org/10.1152/ajpcell.00285.2005

Van Goethem E, Guiet R, Balor S, Charrière GM, Poincloux R, Labrousse A, Maridonneau-Parini I, Le Cabec V (2011) Macrophage podosomes go 3D. Eur J Cell Biol 90(2-3):224–236. https://doi.org/10.1016/j.ejcb.2010.07.011

Vivinus-Nebot M, Rousselle P, Breittmayer JP, Cenciarini C, Berrih-Aknin S, Spong S, Nokelainen P, Cottrez F, Marinkovich MP, Bernard A (2004) Mature human thymocytes migrate on laminin-5 with activation of metalloproteinase-14 and cleavage of CD44. J Immunol 172(3):1397–1406. https://doi.org/10.4049/jimmunol.172.3.1397

Waterhouse A, Bertoni M, Bienert S, Studer G, Tauriello G, Gumienny R, Heer FT, de Beer TAP, Rempfer C, Bordoli L, Lepore R, Schwede T (2018) SWISS-MODEL: homology modelling of protein structures and complexes. Nucleic Acids Res 46(W1):W296–W303. https://doi.org/10.1093/nar/gky427

Wilson CL, Matrisian LM (1996) Matrilysin: an epithelial matrix metalloproteinase with potentially novel functions. Int J Biochem Cell Biol 28(2):123–136. https://doi.org/10.1016/1357-2725(95)00121-2

Winkler J, Abisoye-Ogunniyan A, Metcalf KJ, Werb Z (2020) Concepts of extracellular matrix remodelling in tumour progression and metastasis. Nat Commun 11(1):5120. https://doi.org/10.1038/s41467-020-18794-x

Wizemann H, Garbe JH, Friedrich MV, Timpl R, Sasaki T, Hohenester E (2003) Distinct requirements for heparin and alpha-dystroglycan binding revealed by structure-based mutagenesis of the laminin alpha2 LG4-LG5 domain pair. J Mol Biol 332(3):635–642. https://doi.org/10.1016/s0022-2836(03)00848-9

Xu H, Niu M, Yuan X, Wu K, Liu A (2020) CD44 as a tumor biomarker and therapeutic target. Exp Hematol Oncol 9(1):36. https://doi.org/10.1186/s40164-020-00192-0

Yaghobi Z, Movassaghpour A, Talebi M, Abdoli Shadbad M, Hajiasgharzadeh K, Pourvahdani S, Baradaran B (2021) The role of CD44 in cancer chemoresistance: a concise review. Eur J Pharmacol 903:174147. https://doi.org/10.1016/j.ejphar.2021.174147

Yamashita H, Beck K, Kitagawa Y (2004) Heparin binds to the laminin alpha4 chain LG4 domain at a site different from that found for other laminins. J Mol Biol 335(5):1145–1149. https://doi.org/10.1016/j.jmb.2003.11.047

Yan HHN, Cheng CY (2006) Laminin α 3 forms a complex with β3 and γ3 chains that serves as the ligand for α 6β1-integrin at the apical ectoplasmic specialization in adult rat testes. J Biol Chem 281(25):17286–17303. https://doi.org/10.1074/jbc.M513218200

Yan HHN, Mruk DD, Wong EWP, Lee WM, Cheng CY (2008) An autocrine axis in the testis that coordinates spermiation and blood–testis barrier restructuring during spermatogenesis. Proc Natl Acad Sci USA 105(26):8950. https://doi.org/10.1073/pnas.0711264105

Yao Y (2017) Laminin: loss-of-function studies. Cell Mol Life Sci 74(6):1095–1115. https://doi.org/10.1007/s00018-016-2381-0

Yap L, Tay HG, Nguyen MTX, Tjin MS, Tryggvason K (2019) Laminins in cellular differentiation. Trends Cell Biol 29(12):987–1000. https://doi.org/10.1016/j.tcb.2019.10.001

Yu Q, Stamenkovic I (1999) Localization of matrix metalloproteinase 9 to the cell surface provides a mechanism for CD44-mediated tumor invasion. Genes Dev 13(1):35–48. https://doi.org/10.1101/gad.13.1.35

Yu Q, Stamenkovic I (2000) Cell surface-localized matrix metalloproteinase-9 proteolytically activates TGF-beta and promotes tumor invasion and angiogenesis. Genes Dev 14(2):163–176. https://doi.org/10.1101/gad.14.2.163

Yu WH, Woessner JF Jr, McNeish JD, Stamenkovic I (2002) CD44 anchors the assembly of matrilysin/MMP-7 with heparin-binding epidermal growth factor precursor and ErbB4 and regulates female reproductive organ remodeling. Genes Dev 16(3):307–323. https://doi.org/10.1101/gad.925702

Yurchenco PD, Cheng YS, Colognato H (1992) Laminin forms an independent network in basement membranes. J Cell Biol 117(5):1119–1133. https://doi.org/10.1083/jcb.117.5.1119

Zöller M (2011) CD44: can a cancer-initiating cell profit from an abundantly expressed molecule? Nat Rev Cancer 11(4):254–267. https://doi.org/10.1038/nrc3023

Zöller M (2015) CD44, hyaluronan, the hematopoietic stem cell, and leukemia-initiating cells. Front Immunol 6:235. https://doi.org/10.3389/fimmu.2015.00235

Chapter 10
Basement Membrane, Collagen, and Fibronectin: Physical Interactions with Cancer Cells

Marco Franchi, Valentina Masola, Konstantinos-Athanasios Karamanos, Leonardo Franchi, Konstantina Kyriakopoulou, Maurizio Onisto, and Concettina Cappadone

Abstract Cancer cell invasion is regulated by extracellular matrix (ECM) chemical signaling and gene expression but it also consists in a mechanical process controlled by ECM's array. Scanning electron microscope analysis, invasion test, and real-time PCR demonstrated that Matrigel mimicking basement membrane (BM) doesn't promote epithelial–mesenchymal transition (EMT) in both low and very aggressive breast cancer cells (MCF-7 and MDA-MB-231). A loose network of type I collagen mimicking the sub-BM favors EMT in MCF-7 cells but physically limits their invasion ability vs. Matrigel, as collagen does not induce an increase of metalloproteases (MMPs) in cells following ameboid-invasion mode. Collagen doesn't change MDA-MB-231 phenotypes but further improves their invasion capability vs. Matrigel, by stimulating MMPs production. Concentrated type I collagen mimicking deeper ECM induces cells adhesion, further development of microvesicles, microvilli, long filopodia, and tunneling nanotubes (TNTs). Non-aligned fibronectin favors breast cancer cells adhesion, microvesicles, and TNTs development. Densely packed and parallel collagen fibers mimicking a collagen

M. Franchi (✉)
Department for Life Quality Studies, University of Bologna, Rimini, Italy
e-mail: marco.franchi3@unibo.it

V. Masola · M. Onisto
Department of Biomedical Sciences, University of Padova, Padova, Italy

K.-A. Karamanos
Department of Pharmacy and Industrial Pharmacy, University of Bologna, Bologna, Italy

L. Franchi
Department of Medicine, University of Bologna, Bologna, Italy

K. Kyriakopoulou
Biochemistry, Biochemical Analysis & Matrix Pathobiology Research Group, Laboratory of Biochemistry, Department of Chemistry, University of Patras, Patras, Greece

C. Cappadone
Dipartimento di Farmacia e Biotecnologie, Università of Bologna, Bologna, Italy

I. Kovalszky et al. (eds.), *The Extracellular Matrix and the Tumor Microenvironment*, Biology of Extracellular Matrix 11,
https://doi.org/10.1007/978-3-030-99708-3_10

array in mammary tumor progression oppose invasion. In colon cancer, LoVo-R cells, resistant to doxorubicin, concentrated collagen favors development of invaginating phenotypes with invadopodia. We suggest that collagen acts as a physical factor inducing EMT in breast cancer cells and drug resistance in LoVo-R cells.

10.1 The Biomechanical Role of ECM in Cancer Progression

In normal conditions, extracellular matrix (ECM) connects but also separates different tissues and provides the gaseous and metabolic support to most of them. Similarly, progression of most diseases depends on the physical status and chemical properties of ECM which interacts and responds to the biological changes of the pathological processes. This is what usually happens in wounds, foreign body granulomas, as well as in cancer (Schäfer and Werner 2008; Fouad and Aanei 2017; Foster et al. 2018; Paolillo and Schinelli 2019).

Malignant transformation is related to an uncontrolled cells growth, dissociation of cells with loss of intercellular adhesiveness and acquisition of migration capability. These changes allow cells to breach the basement membrane (BM), invade the peri-tumoral microenvironment, and then colonize distant anatomical sites by forming metastatic points. In addition, the peri-tumoral ECM undergoes significant structural alterations which can also induce the epithelial–mesenchymal transition (EMT) in cancer cells (Zolota et al. 2021; Karamanos et al. 2021). As a matter of fact, the uncontrolled tumor growth generates by itself compressive stresses within tumor which can stimulate cancer cells migration by developing of invading leader cells and improving cell–substrate adhesion for grouped cell invasion (Tse et al. 2012).

Cancer cell invasion is foremost a mechanical process, and many experimental studies and reviews concerning the interactions between cancer cells and surrounding tissues have mainly focused on gene regulation and signaling that lead to the development of the tumor. However, the knowledge on the role that physical properties of tumor microenvironment exert in affecting cancer cell behavior is relatively limited (Brabek et al. 2010; Baker et al. 2011; Balkwill et al. 2012; Kim et al. 2018; Wang et al. 2018; Malandrino et al. 2018; Scott et al. 2019). In particular, changes in structural and biomechanical properties of ECM which favor cancer cell adhesion ability and development of new cell phenotypes with an improved motility just exerted by generations of mechanical actions like contractile and tensional forces are undervalued (Friedl and Brocker 2000; Ridley et al. 2003; Zaman et al. 2006; Mierke et al. 2008; Zanotelli et al. 2021). When cancer cells migrate and travel through the ECM, they encounter a mechanical resistance which cells must overcome in order to reach blood and lymphatic vessels.

From a chemo-mechanical point of view, ECM consists in a meshwork of insoluble proteins like collagen, assembling in collagen fibrils and then in fibers, and in minor amount elastin which forms the elastic fibers; but ECM also contains a high molecular mass of anionic substances of predominant carbohydrate nature including proteoglycans (PGs) and glycosaminoglycans (GAGs), which are particularly able to sequester high amount of water. The insoluble collagen fibrils, mainly resisting to tensile stresses (Franchi et al. 2007, 2009, 2010; Pissarenko et al. 2019), are held together by GAGs and PGs which fill the interfiber spaces and acting like elastic strings in solution oppose compression. In ECM, these components interact with water and contribute to form a functional dynamic scaffold which can control the shape of tissues, organs, and whole body (Scott 1975, 2002, 2003; Yue 2014). ECM can bind cells of different tissues and offer anchorage sites for cell movement and migration, but also plays the role of a physical barrier separating different tissues or opposing to cell migration (Mould and Humphries 2004; Larsen et al. 2006; Lu et al. 2011; Yue 2014; Kular et al. 2014; Ringer et al. 2017).

ECM architecture has a significant effect on cell motility and the ability of cells to adhere and sense the mechano-chemical environment is particularly discussed. It is widely recognized that the chemical, mechanical, and topographical properties of the extracellular environment of peri-tumoral tissue can also play a possible primary role in affecting both tumor progression and cancer cells invasion (Lu et al. 2011; Baker et al. 2011). Collagen, the main component of ECM, is widely used as substrate in three-dimensional (3D) cultures of cancer cells. The chemical composition and physical characteristics of the ECM, such as geometry, macro-microporosity, collagen alignment, density, and stiffness, strongly contribute to cancer cell adhesion, migration, and invasion (Wolf and Friedl 2011). Moreover, both cancer and stromal cells can activate an enzymatic-dependent pericellular proteolysis which generates micro- and macrotracks delimited by relatively aligned collagen bundles which further favor cell adhesion and invasion (Gaggioli et al. 2007; Wolf et al. 2007; Provenzano et al. 2008).

Among other factors, also confinement and topology can affect modes of cell migration (Kurniawan et al. 2016; Pandya et al. 2017). Microporosity of ECM can depend on caliber, density, and array of collagen fibers, but also on the hydration level of tissue which is directly related to the content of GAGs like hyaluronic acid. Tissue mesh size, related to microporosity, represents a physical limit to cancer cell invasion and is related to deformation capability of the nucleus. Smaller than 7 μm^2 mesh sizes in ECM do not allow cancer cells to directly invade the peri-tumoral ECM (Wolf et al. 2013), whereas larger mesh sizes allow globular-shaped cells to migrate through the network by the so-called amoeboid movement. Even though collagen network plays a contrasting role to cancer cell invasion, some aggressive cancer cells can degrade the matrix architecture by releasing proteolytic enzymes named metalloproteases (MMPs) (Wolf et al. 2007; Friedl and Wolf 2010; Brabek et al. 2010; Scott et al. 2019).

Alterations of ECM mechanical properties can also affect cancer cell phenotypes as cells can sense the physical status and array of peri-tumoral ECM and respond by adapting their phenotypes to the new environment. For instance, an increase of ECM stiffness induces cancer cells to gain an elongated shape and re-enforce focal adhesions, cytoplasmic protrusions, and ability to exert high-traction forces (Peyton et al. 2008; Koch et al. 2012; Ringer et al. 2017). An increased collagen stiffness also affects cells in microenvironment surrounding tumor by modulating remodeling cytoskeleton of fibroblasts and cancer associated fibroblasts (CAFs) (Stylianou et al. 2019).

Particular attention should be given to the different invasion modes and cancer cells ability in switching them or "plasticity." The different invasion modes are related to different cell phenotypes and cell shape seems to predict ECM-driven 3D cell invasion in breast cancer (Baskaran et al. 2020). Both individual and collective invasions can occur: the first mode concerns cells loosing cell–cell contacts, whereas the second one involves cells which still maintain cell–cell junctions. Cells which invade in a collective mode can move as multicellular streams, budding, or larger clusters, whereas cells invading individually can develop elongated-mesenchymal phenotypes, contractility-driven rounded-amoeboid, and filopodial spike-mediated invading strategies. Cells adopting an elongated phenotype are characterized by actin-rich protrusions, focal adhesion formation, MMPs activity, and actomyosin contractility at the rear of the cells (Pandya et al. 2017). In general, these mesenchymal-like cells develop a highly motile phenotype and can easily cross basement membranes, interstitial spaces, and endothelial barriers in a single-cell migration. The strong cohesive intercellular forces favor the collective invasion and offer potential advantages for tumor survival such as covering different cell roles and improving resistance to immunologic attack (Christiansen and Rajasekaran 2006). Transition from collective to single-cell invasion or tumor plasticity may enhance a metastatic ability (Friedl et al. 2012). Anyway, all migration movement or invasion mode is related to intracellular actin polymerization with development of different cytoskeletons and new cell phenotypes.

To colonize distant organs, cancer cells must start a journey through different body tissues, at first by spreading from the body epithelial outer layer to the neighboring deeper tissues and then by invading blood and lymphatic vessels. In the first step of their journey, cancer cells have to cross ECM of connective tissues which contains different biological barriers opposing cell migration. ECM "biological barriers" can physically opposite but also favor migration/invasion of cancer cells: the first barrier comprises the *basement membrane* (BM) which serves as a scaffold for all epithelial cells; under the BM, the second one is represented by an *interstitial loose connective tissue* called papillary and subpapillary dermis in skin, lamina propria in hollow organs, or interstitial stroma in solid organs; a third biological barrier may be constituted by the *deeper dense connective tissue* which contains higher amount of collagen fibers of different size and array. As an example, in skin under the subpapillary dermis, we find the reticular dermis layer whose collagen accounts for about 70% of the weight of dried tissue. Both density and collagen array change from the papillary and subpapillary dermis, where collagen fibers composed of thin fibrils are sparse and show a low aggregation, to the deeper reticular dermis which contains thicker collagen fiber bundles with a horizontal laminar organization, densely distributed and including long and thick fibrils (Ribeiro et al. 2013; Breitkreutz et al. 2013; Limbert 2017; Ueda et al. 2019). As shown in Fig. 10.1a, b ECM just beneath the epithelium includes thin loose collagen

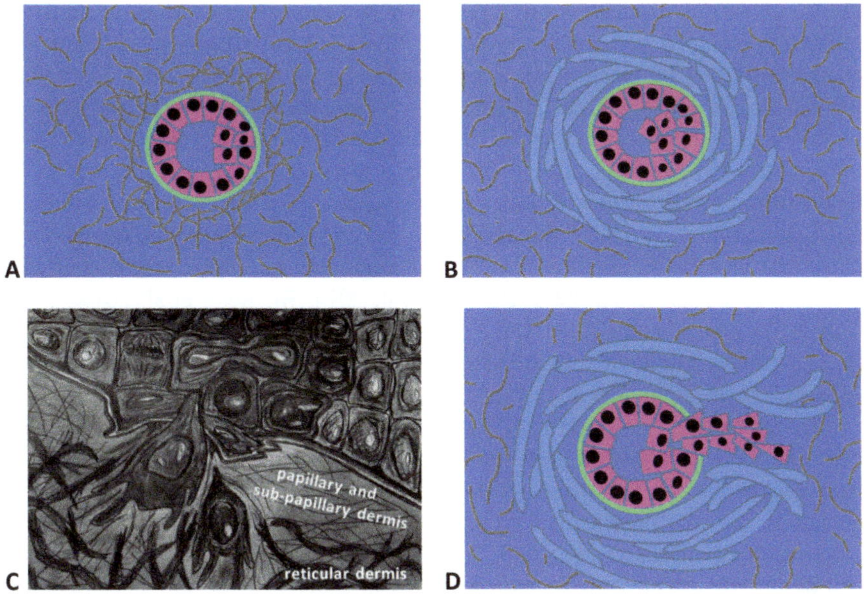

Fig. 10.1 Tumor-Associated Collagen Signatures (TACS). (**a**) In the first stage, a peri-tumoral deposition of collagen occurs (TACS-I). (**b**) When tumor grows, the collagen fibers around the pathological mass change their array and appear aligned and parallel to the tumor surface (TACS-II). (**c**) When cancer cells lose contact with each other and become able to penetrate the basement membrane, which is the first biological physical barrier opposing invasion, they can invade ECM of connective tissues. As an example, in skin cancer, cells pierce the basement membrane and then begin to migrate into the papillary and sub-papillary dermis containing sparse and thin collagen fibers. They further penetrate the depth of the tissue by invading the reticular dermis which contains thick collagen fibers. (**d**) With further development of the tumor mass, the collagen fibers develop a radial array that can favor cancer cell invasion (TACS-III)

fibrils, whereas the deep connective tissue layer comprises thicker collagen fibril bundles (Ottani et al. 1998).

The previously described biological barriers: (a) the BM, (b) the interstitial loose connective tissue/lamina propria/interstitial stroma, and (c) the deeper dense connective tissue, not only oppose cancer cells invasion but also offer anchorage and adhesion to MDA-MB-231 breast cancer cells which can better invade ECM by improving cell contraction and adapting their cytoskeleton (Byrne et al. 2021). In particular, type I and type III collagen assembled in fibrils and fibrillar fibronectin represent the main molecules composing physical scaffolds opposing and supporting the cells migratory movement. During tumor growth, changes of architecture of peri-tumoral collagen network are related to tumor progression and clinical prognosis (Egeblad et al. 2010; Conklin et al. 2011).

Markers of mammary carcinoma progression have been identified in a clinical-histological Tumor-Associated-Collagen-Signatures (TACS) classification. In a first step, only an increase in collagen deposition or desmoplasia around the tumor occurs

(TACS-I) and no metastasis are reported. When tumor mass grows in size, the randomly arranged peri-tumoral collagen fibers are substituted by straight and aligned collagen fibers showing a circumferential orientation (TACS-II), and at this histological stage, no clinical metastasis occurs. However, a further remodeling of the collagen fibers can occur, collagen fibers arrange running perpendicularly to the tumor boundary (TACS-III). Unfortunately, this radial collagen fiber array seems to be strongly related to a higher risk of cancer cell invasion into peri-tumoral microenvironment, because the stretched and radially arranged collagen fibers allow but also favor interfiber cancer cell migration and colonization in distant organs (Provenzano et al. 2008; Riching et al. 2014; Bredfeldt et al. 2014; Wang et al. 2018; Franchi et al. 2019). From a mechanical point of view, the remodeling of ECM promotes matrix alignment and compaction that favors higher tractions exerted by cancer cells (Anguiano et al. 2020) (Fig. 10.1a–c).

Besides collagen, fibronectin (FN) is one of the most abundant proteins mediating cell adhesion and migration in ECM. It can develop into fibrils that widely differ in their thickness, between 10 and 1000 nm (Kular et al. 2014). FN regulates collagen assembly, and the presence of collagen fibrils induces the formation of highly colocalized FN fibrils (Paten et al. 2019). In tumor stroma, both type I collagen and fibronectin are strongly increased and create a denser and mechanically stiffer tissue if compared to normal tissue (Tlsty and Coussens 2006; Rick et al. 2019). FN, arranged into a mesh of fibrils like collagen, but also connected to and within the (BM) of ECM, has been investigated as a possible ECM component which can favor cancer invasion and metastasis and is linked to cell surface receptors (integrins) (Kular et al. 2014). Tumor growth promotes compressive forces within tumor which are accomplished by a coordinated migration of invasive phenotypes in mammary carcinoma cells accompanied by fibronectin deposition and stronger cell–matrix adhesion (Tse et al. 2012).

Experimental mechanical stimulation applied to a collagen–fibronectin matrix highly enhanced invasion ability of HT1080 fibrosarcoma cancer cells only when FN was present. This suggests a primary potential role of this ECM component in governing the mechanical response of tumor cells and tumor progression (Menon and Beningo 2011). Even though its role in tumorigenesis and malignant progression has been highly controversial (Lin et al. 2019), it was reported that cancer-associated fibroblasts (CAFs), through increased contractility and traction forces, can assemble aligned FN matrix which seem to regulate directional migration of cancer cells (Erdogan et al. 2017). The stretching of fibronectin can activate a mechanical response pathway also in normal fibroblasts (Friedland et al. 2009), even though induced mechanical extension of single FN fibers seems to limit the spreading and migration of endothelial cells (Hubbard et al. 2015).

In this chapter, we present and critically discuss the changes of phenotypes and behavior of different breast cancer cell lines (MCF-7, MDA-MB-231) and colon cancer cells (LoVo-R) in 3D cultures mimicking the natural ECM barriers to cancer invasion. To mimic the natural BM, we prepared a layer of Matrigel at standard concentration (0.18 μg/mL) for PCR analysis (Albini et al. 1987), and at higher concentration (3.0 μg/mL) for scanning electron microscope (SEM) observations, as at lower concentration, we never observed a continuous layer of this coating at SEM (Franchi et al. 2019). To reproduce the loose collagen meshwork below the BM (subpapillary and papillary dermis in skin, lamina propria in hollow organs or interstitial stroma in solid organs), we prepared low concentrated type I collagen network (50 μg/mL), whereas to mimic the deeper dense connective tissue we prepared high concentrated (3000 μg/mL) similar collagen on Millipore filters. We also tested FN (130 μg/mL) which functioning as ECM "biological glue" is critical for the attachment and guides directional movement of breast cancer cells (Yue 2014; Oudin et al. 2016).

Finally, just to mimic ECM collagen array of peri-tumoral denser connective tissues as described in TACS-II (collagen fibers arrayed parallel to the tumor boundary, i.e., orthogonal to the cancer cell invasion), we cultivated breast cancer cells on commercial lyophilized and compressed collagen membranes of tightly packed type I collagen fibers (Franchi et al. 2019).

10.2 2D Polystyrene Flask Cultures Partially Transformed in 3D Cultures Induce EMT in MCF-7 Breast Cancer Cells

The role that the physical aspect of substrate surface plays in governing the cancer cell phenotypes was evident when we casually observed an occasional line corresponding to a physical defect of some polystyrene flasks surface. An irregular culture surface may represent by itself a 3D substrate for both normal and cancer cells which become able to move in all spatial directions. When we use flat and smooth polystyrene flasks to cultivate cells, we perform 2D cultures. Where the flask surface was regular and very smooth, the low aggressive MCF-7 breast cancer cells grew exhibiting an epithelial shape and characteristics: cobblestone or flattened polygonal cells showed cell–cell contacts by short filopodia and some of them looked melted together by intercellular tight junctions. Only few globular isolated cells rich in cytoplasmic microvesiscles were detectable (Fig. 10.2a–d). But when the flask surface where cells were seeded showed a line, about 10 μm deep, the 2D substrate transformed the 2D cultures to 3D cultures and the cells which were very next to the flask line completely changed their shape exhibiting a morphological aggressive phenotype: these cells assuming an elongated or fusiform shape developed a morphological epithelial–mesenchymal transition (EMT). Moreover, some of them partially lost intercellular adhesiveness and seemed also to migrate and even cross the line. We demonstrated that changes of the physical aspect of the flask surface, such as a line, seem to induce migration in the low aggressive MCF-7 cells (Fig. 10.2a–d). They seem to adapt to the new microenvironment: to cross the line the cells develop a long filopodia, which exhibiting lateral short filopodia runs from one edge of the line to the other one (Fig. 10.2b) or a lamellipodia (Fig. 10.2d).

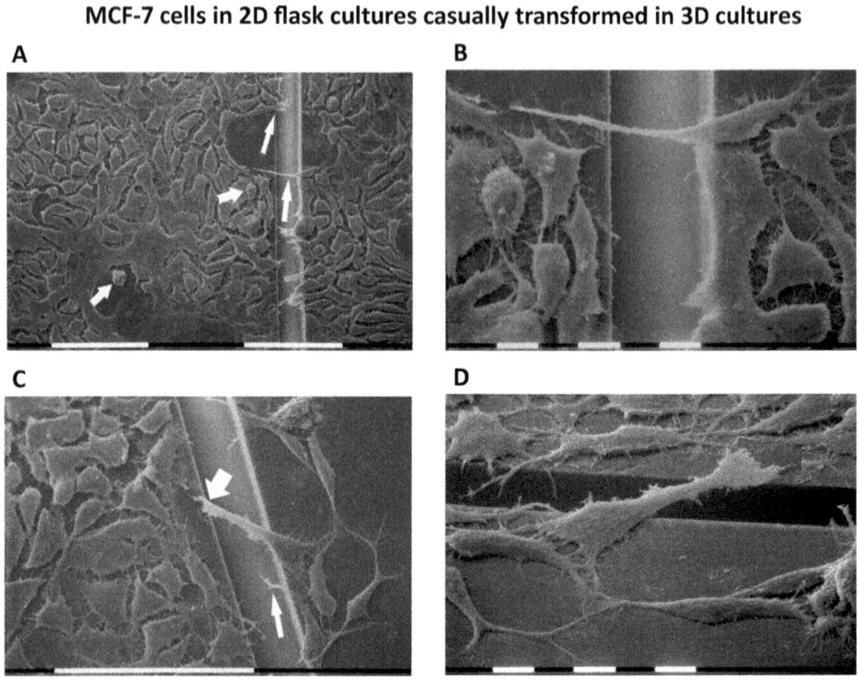

Fig. 10.2 MCF-7 cells on polystyrene flask 2D culture observed at SEM. (**a**) MCF-7 cells appear like flattened and smooth epithelial polygonal cells in contact with each other by short filopodia or in tight contact so to appear melted. Very few globular isolated cells with cytoplasmic microvesicles are detectable (large arrow). On the right of the picture, an occasional and regular line of the flask surface is present. MCF-7 cells very next to the line assume a mesenchymal elongate shape (thin arrows) to cross from one side to the other of the line edges. Note that cells crossing the line do not adhere to the bottom of the line (White bar = 100 μm). (**b**) A higher magnification of the previous picture shows a crossing cell developing a long filopodia. The surrounding cells appear like epithelial poligonal flattened cells (White bar = 10 μm). (**c**) The MCF-7 cells next to the flask line partially lose their cell–cell contact, exhibit an elongated and fusiform shape and develop lamellipodia (large arrow) or long filopodia (thin arrow) which demonstrate that they are moving. All the other cells show an epithelial phenotype and look in contact with each other (White bar = 100 μm). (**d**) A higher magnification of the previous picture shows elongated cells crossing the line and developing filopodia or lamellipodia (White bar = 10 μm)

Cancer invasion consists of a translocation of the cancer cell body and must be considered as a heterogeneous and adaptive process of cancer cells involving morphological changes and generation of cell polarity (Pandya et al. 2017). Physical aspect (line) of a culture substrate can affect phenotypes of low-aggressive breast cancer MCF-7 cells which develop more aggressive morphological shapes. In particular, we observed that these cancer cells cultivated on flasks in 2D cultures appeared isolated and developed EMT phenotypes when they casually grew on an irregular line which transformed the 2D culture in a 3D one (Balachander et al. 2015).

10.3 MCF-7 Breast Cancer Cells in 2D and 3D Cultures (Millipore Filter, Concentrated Matrigel-Covered Millipore, Low Concentrated Type I Collagen-Covered Millipore and Membrane of Densely Packed Collagen Fibers)

To investigate if ECM can affect phenotypes of a low aggressive breast cancer cell line (MCF-7), we reproduced in vitro the biological barriers which cancer cells must invade in order to colonize distant organs. The first biological barrier of outer ECM which cancer cells have to breach is the BM. To obtain continuous layer of culture substrate mimicking the BM and visible at SEM, we covered a Millipore filter with a thick layer of Matrigel. Then we also investigated how others deeper matrix biological barriers, which were mimicked with different concentrations and array of type I collagen fibrils or using a membrane of densely collagen fibers, could induce changes of MCF-7 cancer cell phenotypes.

The BM is a specialized form of ECM which underlies or surrounds most tissues, including epithelial, endothelial, muscle, and adipose tissues. BM is a sheet-like structure of more than 200 nm thickness composed by two independent polymeric networks, one of laminin and one of type IV collagen, which are probably linked by several additional ECM proteins, including nidogen and perlecan (Breitkreutz et al. 2013; Jayadev and Sherwood 2017; Scott et al. 2019). Glycoproteins and PGs, including fibulin, hemicentin, SPARC, agrin, and type XVIII collagen are also present in BMs, creating biochemically and biophysically distinct structures serving different functions (Jayadev and Sherwood 2017). BM was described at the transmission electron microscope (TEM) as composed of two layers: the lighter-stained layer closer to the epithelium and called lamina lucida, and the denser-stained layer one closer to the connective tissue and called lamina densa.

BM physically supports epithelia but also acts like a tissue scaffold separating but also connecting two different tissues: BM laminin interacts with both integrins on epithelial cell surfaces and type VII collagen of stromal anchoring fibrils which interact with types I and III collagen fibrils of ECM. At the same time, also BM provides essential functions of barrier as it separates epithelial cells from stromal connective tissues: usually cells cannot pass through the very thin pores (10–140 nm) of this barrier without degrading it. BM is an important regulator of cell behavior because it sends signals to epithelial cells about the external microenvironment. Moreover, BM is also a component of blood vessels and represents an extracellular microenvironment sensor for endothelial cells in intravasation events of cancer cells (Kalluri 2003a, b). To mimic the natural BM, we prepared a highly concentrated (3.0 μg/mL) layer of Matrigel because at standard concentration (0.18 μg/mL), we never observed at SEM the physical presence of this coating. Matrigel mimics the because it contains laminin, entactin/nidogen, type IV collagen, and heparan sulfate proteoglycans. Moreover, we prepared a layer of low concentrated type I collagen (50 μg/mL) as coating for a Millipore filter so to reproduce the collagen array in ECM of loose connective tissues (subpapillary and papillary dermis, lamina propria, interstitial stroma).

Both 2D and all 3D cultures were observed at the SEM to evaluate the changes of cancer cells phenotypes growing on ECM components. Always function is related to shape but at the same time structural morphology depends on function. The morphological investigations on the shape of cancer cells allow us to better understand their behavior during invasion of the surrounding tissues and the biological events which lead to metastasis.

MCF-7 cells cultivated on smooth polystyrene substrate displayed epithelial phenotypes: cobblestone flattened polygonal shape with cell–cell contacts through many thin and short filopodia. No microvesicles were detectable on their cytoplasmic surface (Fig. 10.3a).

When the same MCF-7 cells were cultivated in a 3D culture on a Millipore filter, they lost their flattened epithelial phenotype observed in 2D flask cultures. All cells still appeared gathered but exhibited a globular shape and developed a collective invasion ability through an ameboid movement when they crossed the Millipore holes (Fig. 10.3b).

This invasion mode was not directly observed when the same cells were seeded on a Millipore filter covered with concentrated Matrigel (3.0 μg/mL), which appeared like a relative thick and very smooth layer: MCF-7 cells still maintained strong cell–cell contacts but tried to invaginate the thick Matrigel layer and developed short invadopodia breaching the artificial basement membrane (Fig. 10.3c). It was reported that laminin receptors and collagen IV of BM can induce EMT thus favoring cancer progression (Scott et al. 2019).

When MCF-7 cells were cultivated on a Millipore filter covered by low concentrated (50 μg/mL) type I collagen, mimicking ECM of the loose interstitial connective tissues below the BM, the fibrils occluded all the Millipore filter pores thus partially limiting the invasion through an ameboid movement. MCF-7 cells completely changed their phenotypes as many of them looked like isolated cells exhibiting a mesenchymal elongated shape which suggested an EMT induced by collagen fibrils. Cells showed also large lamellipodia strongly adhering to collagen fibrils and binding to them, but other ones developed intercellular cytoplasmic connections morphologically corresponding to tunneling nanotubes (Fig. 10.3d).

If MCF-7 cells are cultivated on a membrane of densely packed fibers of type I collagen mimicking TACS-II, the stage of breast cancer with collagen fibers densely packed and parallel to the tumor boundary and no clinical metastasis, they show grouped globular and few polygonal-shaped cells with cell–cell contacts and short microvilli on their surface, but they seem unable to invade the collagen membrane (Fig. 10.3e, f) (Provenzano et al. 2008; Franchi et al. 2019).

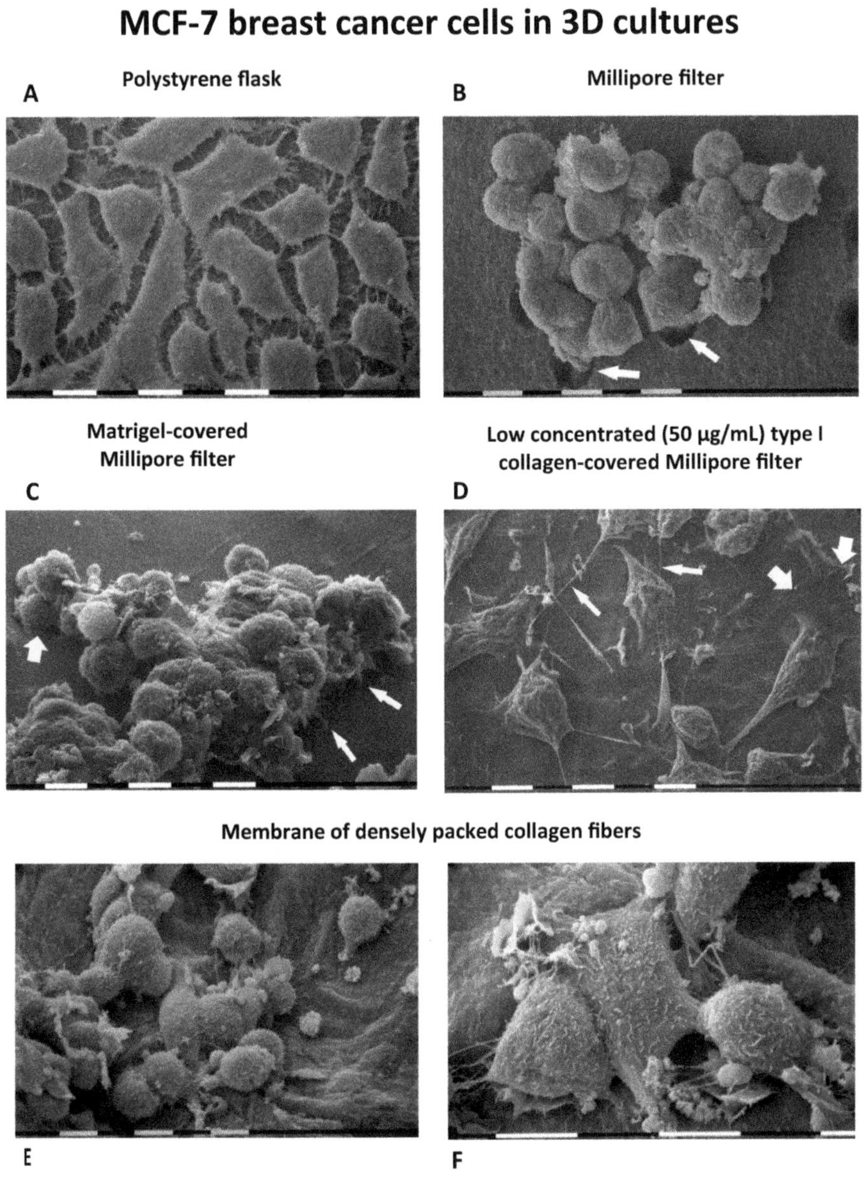

Fig. 10.3 SEM analysis of MCF-7 cells. (**a**) MCF-7 cells cultivated on polystyrene flasks in 2D cultures show epithelial phenotypes with cell–cell contact through many thin and short filopodia. (**b**) MCF-7 cells on the Millipore filter in 3D cultures show a globular shape and are grouped together. They can easily pass through the free Millipore pores (arrows) (White bar = 10 μm). (**c**) MCF-7 cells on Millipore filter covered by a concentrated Matrigel (3.0 μg/mL) in 3D cultures. Concentrated Matrigel appears like a relatively thick and very smooth layer on which very grouped globular MCF-7 cells array in tight contact with each other. They intimately adhere to Matrigel and it seems that they are invaginating the collagen fibrils (large arrow). They develop cytoplasmic protrusions which might morphologically correspond to invadopodia (thin arrows) (White bar = 10 μm). (**d**) The same MCF-7 cells cultivated on a Millipore filter covered by low

10.4 MDA-MB-231 Breast Cancer Cells in 2D and 3D Cultures (Millipore Filter, Concentrated Matrigel-Covered Millipore, FN, Low and High Concentrated Type I Collagen-Covered Millipore and Membrane of Densely Packed Collagen Fibers)

To explore whether ECM can affect phenotypes of a high aggressive breast cancer cell line (MDA-MB-231 cells), we reproduced in vitro the biological barriers which cancer cells have to break and invade in vivo to develop metastasis. We reproduced the first biological barrier of outer ECM, the BM, by covering a Millipore filter with a thick layer of Matrigel, but to mimic the other deeper matrix biological barriers, we covered a Millipore filter with different concentrations and array of type I collagen fibrils or we used a collagen membrane of densely packed collagen fibers.

Both 2D and 3D cultures were then observed at SEM to detect eventual changes of cancer cells phenotypes and cytoplasmic protrusions when cells grow on different ECM components. When MDA-MB-231 cells are cultivated on smooth polystyrene flasks in 2D cultures, they show different isolated phenotypes exhibiting cytoplasmic microvesicles: elongated fusiform-shaped cells, globular ones, and few flattened polygonal ones (Fig. 10.4a). When the same cells are cultivated on a Millipore filter in 3D cultures, they do not change their phenotypes but exhibit more microvesicles on their cytoplasmic surface. Single isolated cells can migrate through the Millipore holes developing filopodia or lamellipodia (Fig. 10.4b). It is interesting to observe that MDA-MB-231 cell phenotypes do not drastically change from 2D to 3D cultures. The thick layer of Matrigel covering the Millipore filter and mimicking the BM does not induce any phenotypes changes in MDA-MB-231 cells even though cells seem to better adhere to the substrate and some of them develop ventral cytoplasmic protrusions which are morphologically comparable to invadopodia (Fig. 10.4c). This data confirms the results of Anguiano et al. (2017, 2020) who demonstrated that at high concentrations, Matrigel slows down cancer cells migration, possibly due to excessive attachment, whereas at low concentration, it may facilitates migration. From a morphological point of view, at the standard concentration, Matrigel distribution does not covers all the Millipore pores so indirectly allows free cancer cell migration.

Fig. 10.3 (continued) concentrated (50 μg/mL) type I collagen fibrils occluding all the Millipore filter pores in a 3D culture. More flattened and isolated cells show a mesenchymal elongated shape with large lamellipodia binding collagen fibrils (large arrows), and intercellular cytoplasmic connections which morphologically correspond to tunneling nanotubes (thin arrows) (White bar = 10 μm). (**e**) MCF-7 cells cultured on a commercial membrane of densely packed collagen fibers appear grouped with cell–cell contacts and show a globular shape, but develop many microvilli on their surface (White bar = 10 μm). (**f**) At higher magnification, a polygonal and globular cell adhering to the membrane of densely collagen fibers are in tight contact and develop many cytoplasmic microvilli (White bar = 10 μm)

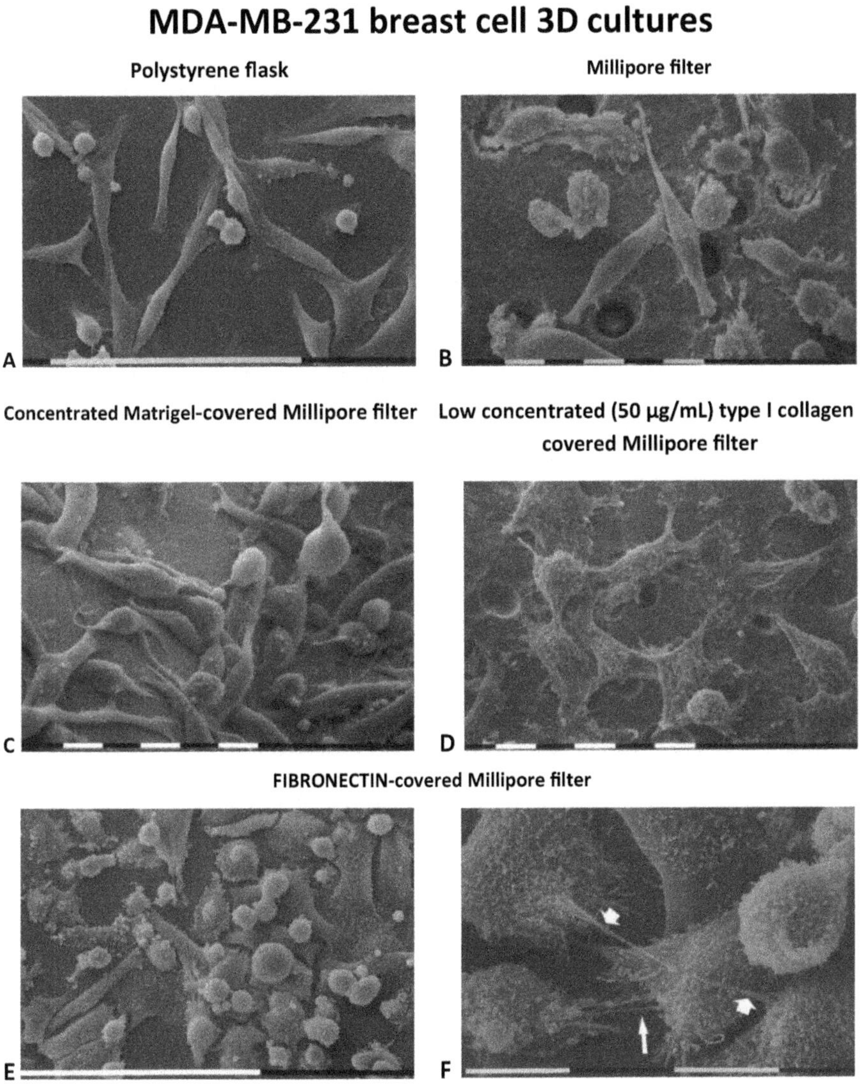

Fig. 10.4 SEM analysis of MDA-MB-231 cells. (**a**) MDA-MB-231 cells cultivated on polystyrene flasks in 2D cultures display different isolated phenotypes exhibiting cytoplasmic microvesicles: elongated-fusiform shaped cells, globular ones, and flattened polygonal ones (White bar = 100 µm). *From CELLS, 2020, 9, 2031,* Franchi M. et al. Extracellular Matrix-Mediated Breast Cancer Cells Morphological Alterations, Invasiveness, and Microvesicles/Exosomes Release. (**b**) MDA-MB-231 cells on Millipore filter in 3D cultures show the same phenotype described in flask cultures but microvesicles are more present. Individual cells are migrating through the free Millipore filter holes (White bar = 10 µm). (**c**) MDA-MB-231 cells in 3D cultures on concentrated Matrigel-covered Millipore. The three phenotypes observed in Millipore filter cultures are still visible: many isolated elongated-fusiform and globular cells and fewer flattened polygonal ones (White bar = 100 µm). (**d**) MDA-MB-231 cells in 3D cultures on low concentrated (50 µg/mL) type I collagen-covered Millipore. The three isolated previously described different phenotypes of MDA-MB-231 cells are still recognizable but show more microvesicles on their cytoplasmic surface. *From CELLS, 2020, 9, 2031,* Franchi M. et al. Extracellular Matrix-Mediated Breast Cancer Cells Morphological

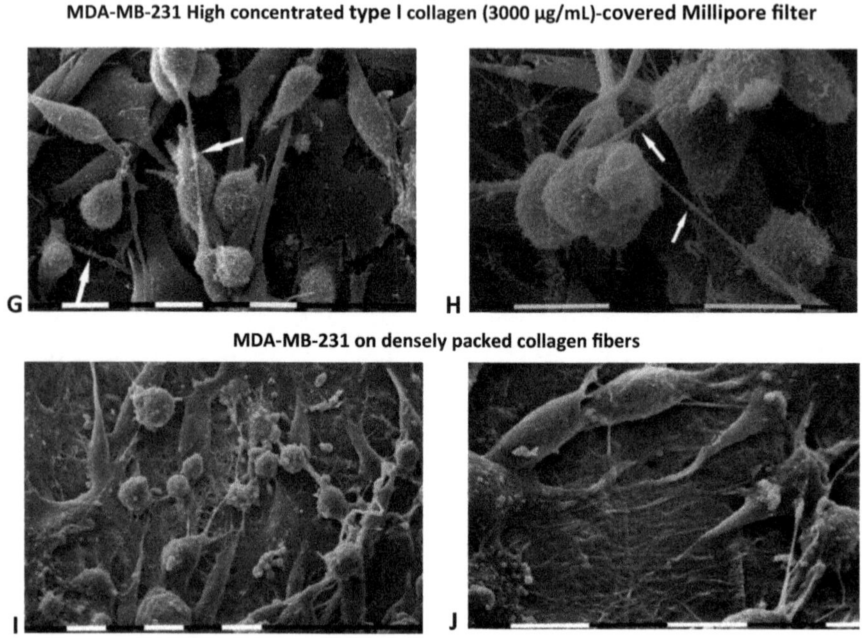

Fig. 10.4 (continued) Alterations, Invasiveness, and Microvesicles/Exosomes Release. (**e**) MDA-MB-231 cells in 3D cultures on Millipore filter covered by fibronectin (130 μg/mL) occluding all filter holes. MDA-MB-231 cells cultured on fibronectin seem to be more grouped and better adhere to the substrate. They show many "cobblestone" flattened polygonal cells and globular-shaped ones, but only a few elongated ones. All cells exhibit many short filopodia or microvilli and microvesicles (White bar = 100 μm). *From CELLS, 2020, 9, 2031,* Franchi M. et al. Extracellular Matrix-Mediated Breast Cancer Cells Morphological Alterations, Invasiveness, and Microvesicles/Exosomes Release. (**f**) MDA-MB-231 cells in 3D cultures on Millipore filter covered by fibronectin (130 μg/mL). Polygonal cells show microvilli and appear to be connected by thin single TNTs (thin arrow) and forming thicker ones composed of single thin TNTs tightly bundled together (large arrows). On the right, a globular cell exhibits many microvesicles on the cytoplasmic surface (White bar = 10 μm). *From CELLS, 2020, 9, 2031,* Franchi M. et al. Extracellular Matrix-Mediated Breast Cancer Cells Morphological Alterations, Invasiveness, and Microvesicles/Exosomes Release. (**g, h**) MDA-MB-231 cells in 3D cultures on Millipore filter covered by high concentrated type I collagen (3000 μg/mL). Breast cancer cells grow in more than one layer, but no tight cell–cell contact is visible. Only the flattened polygonal cells lay in direct contact with the collagen fibrils, whereas both the elongated or fusiform cells and globular ones grow over the flattened ones. All cells show both microvilli and microvesicles. TNTs were apparently distinguishable in all phenotypes and long and thin filopodia originating from elongated cells exhibit exosomes and microvesicles (arrows) on their surface (White bar = 10 μm). *From CELLS, 2020, 9, 2031,* Franchi M. et al. Extracellular Matrix-Mediated Breast Cancer Cells Morphological Alterations, Invasiveness, and Microvesicles/Exosomes Release. (**i**) The MDA-MB-231 cells attached to densely packed and parallel collagen fibers include a few isolated flattened polygonal cells, globular and elongated or also "squid"--shaped ones developing lamellipodia. All cells produce many microvesicles. Thin intercellular connections corresponding to TNTs are visible between adjacent cells. *From CELLS, 2020, 9, 2031,* Franchi M. et al. Extracellular Matrix-Mediated Breast Cancer Cells Morphological Alterations, Invasiveness, and Microvesicles/Exosomes Release. (**j**) In areas where fibrils of the collagen membrane are more exposed, elongated and fusiform shaped cells appear arrayed parallel to the fibrils axis (White bar = 10 μm). *From CELLS, 2020, 9, 2031,* Franchi M. et al. Extracellular Matrix-Mediated Breast Cancer Cells Morphological Alterations, Invasiveness, and Microvesicles/Exosomes Release

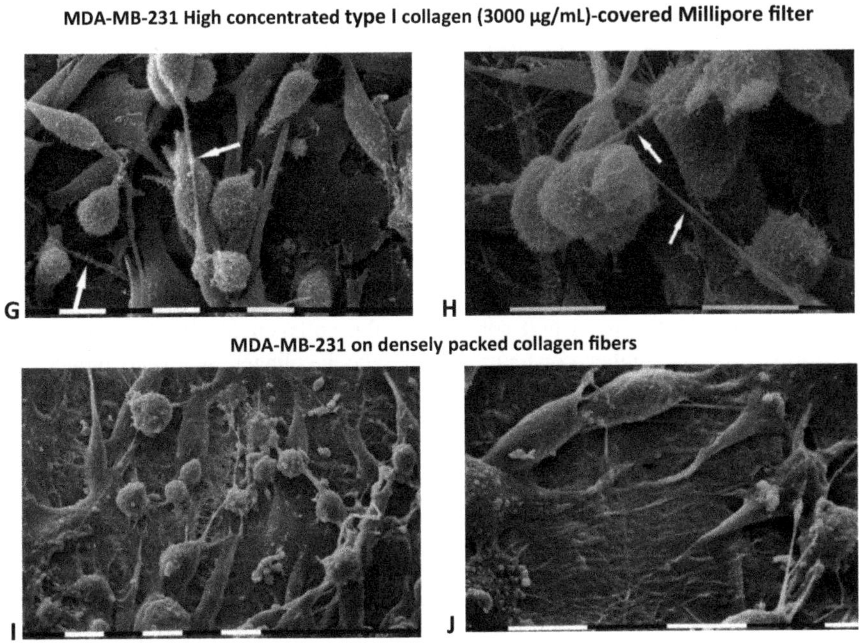

Fig. 10.4 (continued)

The MDA-MB-231 cells growing on low concentrated type I collagen (50 μg/mL) network still show the same phenotypes described in previous cultures, but cells display more evident and numerous cytoplasmic microvesicles (Fig. 10.4d). As both Matrigel and low concentrated collagen meshwork do not affect MDA-MB-231 phenotypes, we investigated in our experiments if FN, another ECM fibrillar component, could affect breast cancer phenotypes. Differently from previous samples, MDA-MB-231 cells cultured on fibronectin show many "cobblestone" flattened polygonal cells and globular shaped ones and only few elongated ones mainly growing on the previous ones. All cells appear more grouped with cell–cell contacts but exhibit many microvesicles and microvilli (Fig. 10.4e). Some of them are connected by thin single and thicker composed intercellular cytoplasmic protrusions which are morphologically comparable to tunneling nanotubes (TNTs) (Franchi et al. 2020a) (Fig. 10.4f). These data are explained considering that tested FN is arrayed in a network, so that the main effect of FN is just to favor cell adhesion. Only highly aligned FN is strongly related with EMT and high-speed migration (Wang et al. 2018). However, the presence of many microvesicles and the intercellular connections (TNTs) which allow a direct transfer of microvesicles and mitochondria between the connected cells suggest an increased metabolic activity in MDA-MB-231 cells. The observation of an increased number of polygonal cell phenotypes, which showing many microvesicles cannot be morphologically considered as low aggressive cells, suggest MDA-MB-231 cells develop a high adhesion to the FN substrate. Park and Helfman (2019) demonstrated that MDA-MB-231 cells develop a high FN expression in 3D suspension culture and an increase of FN facilitates cell-attachment via integrin β-5. FN matrix can affect both cancer cells phenotypes and guide migration via integrins which connecting FN fibrils to cell's cytoskeleton might act as "stretch sensors" representing a "mechanical continuity" between the inner and outer environment of the cells (Kular et al. 2014; Oudin et al. 2016).

To observe possible stronger phenotypic changes, we covered the Millipore filter with higher concentrated type I collagen (3000 μg/mL). In these cultures, the previously described phenotypes showing both microvilli and microvesicles are equally distributed and grow in more than one layer: both elongated cells showing a fusiform shape and globular ones lay on flattened and polygonal ones which are the only ones in direct contact and adhering to the collagen fibrils. In elongated or fusiform cells, the higher concentration of type I collagen seems to induce the development of long and thin filopodia with exosomes and microvesicles on their surface. All phenotypes show microvilli and microvesicles (Fig. 10.4g, h). The higher concentration of collagen fibrils seems to further favor the metabolic activity and adhesion of MDA-MB-231 cells.

To mimic the collagen fiber array described in TACS-II (collagen fibers run parallel to the tumor boundary, i.e., orthogonal to the direction of cancer cell invasion) and corresponding to patient with no metastasis, we cultivated breast cancer cells on a commercial lyophilized and compressed membrane of tightly packed type I collagen fibers. It was described that at this stage, the remodeled collagen arrangement does not allow breast cancer cell invasion both in vivo and in vitro (Provenzano et al. 2008; Franchi et al. 2019). Cells include very few isolated flattened polygonal cells adhering to collagen fibrils of the membrane, globular and many elongated-fusiform ones and "squid"-shaped ones showing lamellipodia. All cells show microvesicles and TNTs (Fig. 10.4i). Where collagen fibrils are more exposed the fusiform cells are arrayed parallel to the fibril axis (Fig. 10.4j) (Franchi et al. 2020a).

10.5 Migration Test and Real-Time PCR of MCF-7 and MDA-MB-231 Cells

The migration/invasion process of cancer cells is the expression of their invasiveness properties. MDA-MB-231 cells have a higher ability in migrating/invading vs. MCF-7 cells, thus confirming the worst clinical prognosis of some patients. BM (Matrigel at standard concentration: 0.18 μg/mL) seems not to strongly affect both MCF-7 and MDA-MB-231 breast cancer cells phenotypes because when they are cultivated on both Millipore filter and Matrigel-covered Millipore filter all cells show the same rate of migration. However, when cells grow on collagen fibrils network-covered Millipore filter, MDA-MB-231 cells show a higher invasion capacity, whereas MCF-7 cells even decrease their ability in penetrating the barrier of collagen fibrils (Fig. 10.5). This different behavior between the different cancer

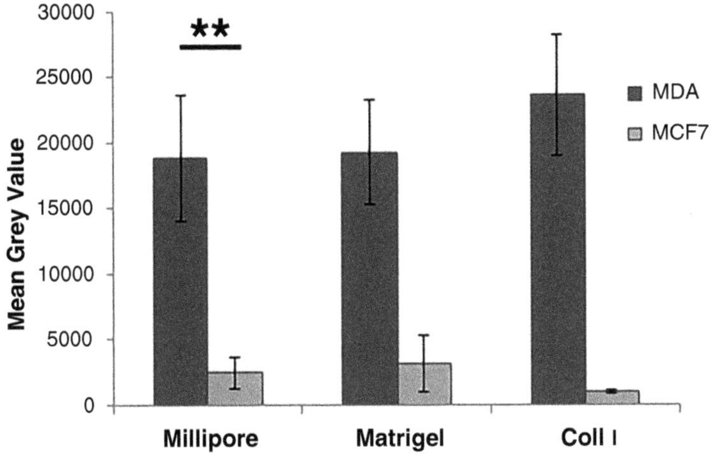

Fig. 10.5 Migration of breast cancer cells cultured on different substrates. MDA-MB-231 and MCF-7 cells were seeded on Millipore filters or Millipore filters covered with Matrigel (0.18 μg/mL) or collagen type I and incubated for 5 h. Double asterisks indicate statistically significant differences ($p \leq 0.01$). *From J. Clin. Med. 2019, 8, 213,* Franchi M. et al., *Collagen Fiber Array of Peritumoral Stroma Influences Epithelial-to-Mesenchymal Transition and Invasive Potential of Mammary Cancer Cells*

cell lines in contact with collagen might be explained by the strong and higher increase of MMP-2 and MMP-9 expression observed in MDA-MB-231 but not in MCF-7 (Fig. 10.6). It is well known that the release of MMPs from cancer cells cause an ECM digestion. This suggests an individual mesenchymal protease-dependent invasion mode for the MDA-MB-231 cultivated on a collagen network which instead represents a resistant biological barrier for the collective ameboid-dependent invasion of the MCF-7 cells (Bozzuto et al. 2015). In fact, if MCF-7 cells did not follow a protease-dependent invasion mode which creates wide spaces in ECM through MMPs of invadopodia, the very small pores size of the relative dense collagen network could not allow the crossing of cancer cell nuclei (Wolf et al. 2013). Anyway, collagen substrate seems to affect the MCF-7 cells phenotypes which show mesenchymal, elongated, or fusiform phenotypes (Fig. 10.3c): in fact, they display a VIM and FN-expression improvement and loss of cell–cell contacts as confirmed by a strong decrease of E-cadherin which was not observed in MDA-MB-231 cells collagen group (Khalil et al. 2017) (Fig. 10.6). It was reported that cadherin complexes act also as mechano-transducers that can sense changes in tension and trigger adaptive reinforcement of intercellular junctions (Leckband and de Rooij 2014).

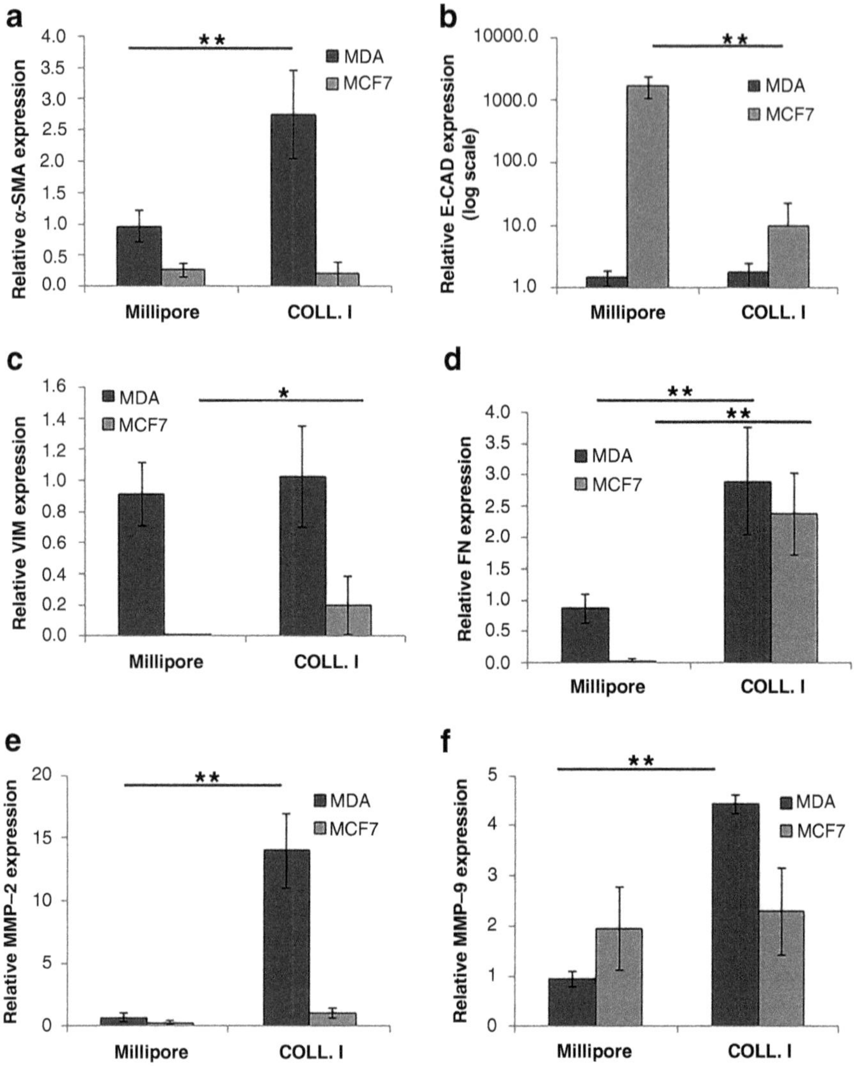

Fig. 10.6 Type I collagen induces striking changes in EMT markers and significant alterations in the expression of ECM components of breast cancer cells. Real-time PCR analysis of a-SMA, E-cadherin, vimentin, fibronectin, MMP-2 and MMP-9 on MDA-MB-231 and MCF-7 cells seeded on uncoated Millipore filter or filter covered with collagen type I and incubated for 24 h. Expression was normalized to GAPDH expression. Single and double asterisks indicate statistically significant differences ($p \leq 0.05$ and $p \leq 0.01$, respectively). *From J. Clin. Med. 2019, 8, 213,* Franchi M. et al., *Collagen Fiber Array of Peritumoral Stroma Influences Epithelial-to-Mesenchymal Transition and Invasive Potential of Mammary Cancer Cells*

10.6 Collagen as a Triggering Factor Inducing EMT in Breast Cancer Cells

The irregular surface of 3D cultures represents the first factor favoring the EMT of breast cancer cells, particularly evident in low aggressive (MCF-7) vs. higher aggressive (MDA-MB-231) cells, but also in 3D cultures, Matrigel (BM) does not affect so much the morphological phenotypes of all breast cancer cells. This data may be explained by the fact that even cancer cells like all normal epithelial cells lie, adhere, and grow in direct physical contact with a very flat and smooth scaffold, which epithelial cells sense and recognize as the BM (a smooth 2D surface). But when even some of the low aggressive MCF-7 cells lose their cell–cell contacts and breach, the BM they come in contact with a new unknown microenvironment represented by the ECM: very rough and porous collagen fibril loose meshwork, GAGs, PGs, glycoproteins, and water. They find a new and never sensed irregular and hostile physical microenvironment, so they adapt their phenotypes by developing cytoplasmic protrusions and phenotype changes which allow them to best invade ECM and reach vessels to provide nutrients. We can say that collagen itself induces morphological EMT in MCF-7 cells even though it could not enhance cell's ability to perform clinical metastasis, as it does not improve the production of MMP2 and MMP9 which allow cancer cells to penetrate lymphatic and blood vessels in metastatic process. This is confirmed by Invasion Test data which show that MCF-7 cells invasion decreases in collagen cultures, vs. Millipore and Matrigel substrates.

A molecular EMT in MCF-7 cultivated on type I collagen meshwork was, however, confirmed by a strong decrease of E-cadherin and an increase of both VIM and FN-expression. In conclusion, MCF-7, usually preferring a collective ameboid-dependent migration by squeezing through pliable barriers, could find a valid barrier in the irregular and resistant collagen network (Christiansen and Rajasekaran 2006; Bozzuto et al. 2015).

Deposition of collagen in human solid tumors is associated with higher incidence of metastasis (Ramaswamy et al. 2003). Even though collagen seems to not markedly promote morphological changes or EMT in MDA-MB-231 phenotypes, type I collagen fibrils even at low concentration are able to strongly improve MDA-MB-231 cells migration vs. the same cells cultivated in Millipore filter or in Matrigel-covered Millipore filter. Also, real-time PCR shows that type I collagen slightly promoted VIM-expression, but reveals a strong increase of α-SMA expression, FN-expression, MMP-2, and MMP-9 expression. The physical impact with the sub-BM collagen network could itself represent a "shock" for MDA-MB-231 cells which try to improve their invasiveness ability by increasing the production of MMPs. Moreover MDA-MB-231 cells showed at SEM a slight development of cytoplasmic microvesicles which contain also MMPs as reported in literature (Rilla et al. 2014).

Highly invasive MDA-MB-231 cells can induce metastasis in vivo so they are also capable of invading the deeper ECM including a higher concentration of collagen. In Millipore filter covered by high concentrated type I collagen

(3000 μg/mL) MDA-MB-231 flattened cells strongly adhere to the collagen network; on these cells, globular and elongated mesenchymal cells grow and develop microvilli, microvesicles, and TNTs which suggest an increased invasive potential (Franchi et al. 2020b). Also, when MDA-MB-231 cells were cultivated on FN-covered Millipore filter, they developed a higher number of flattened cells, expression of a good adhesion to FN, but showed a uniform and complete distribution of microvesicles on their cytoplasmic surface (Byrne et al. 2021). These different aspects in the same cells agree with the consideration that transition to an aggressive malignant phenotype is not an "all or nothing" event, so that we could not always observe morphological pure epithelial or pure mesenchymal phenotypes (Christiansen and Rajasekaran 2006).

Collagen and in particular the collagen fibrils/fibers array in tumor microenvironment plays a fundamental role in regulating tumor development and cancer cell invasion. Maller et al. (2013) found that collagen architecture in pregnancy can induce protection from breast cancer. They demonstrated that collagen organization, rather than density alone, is a fundamental contributor to induce an invasive phenotype. In fact, high density nonfibrillar type I collagen seems to suppress the tumor development by increasing junctional E-cadherin, upregulation of cell–cell junction genes, and downregulation of mesenchymal and metalloproteinase genes, whereas aligned fibrillar type I collagen favor cellular aggressive cancer cells phenotypes in mammary tumor cells and alters β1 integrin subcellular distribution.

Considering the particular peri-tumoral collagen array described in TACS-II and including the densely packed collagen fibers parallel to each other and to the tumor boundary, both MCF-7 and MDA-MB-231 cells seem to be unable to penetrate the collagen membrane used to mimic histological TACS-II stage. This ECM response is mediated by CAFs which seem to play an important role in changing the collagen array of tumor ECM: firstly opposing to cancer cell invasion (TACS-I and TACS-II) and then favoring the cancer cell invasion (TACS-III) (Provenzano et al. 2008; Egeblad et al. 2010; Conklin et al. 2011; Luhr et al. 2012; Bredfeldt et al. 2014; Costanza et al. 2017; Franchi et al. 2019, 2020a). Densely packed and parallel collagen fibers array observed in TACS-II are usually described in other anatomical situations and sites: the response of ECM in foreign body granuloma develops a tissue-encapsulation to limit diffusion of exogenous material by sheets of collagen fibers running parallel to the implant surface. Moreover, gingival ECM around titanium dental implants contains densely packed collagen fibers which following a circular array around the neck of the implants are presumably able to physically opposite virus and bacteria invasions from the oral plaque (Ruggeri et al. 1992, 1994). Which and how do the mechanical events which favor or also induce changes of collagen array in tumor ECM from TACS-II to TACS-III act? The densely packed, straight, and parallel collagen fibers array described in TACS-II are comparable to that of tendons. Tendon collagen fibers resist tension but when they are compressed, tenocytes change their phenotypes and ECM responds by developing a fibrocartilage, which protect both tenocytes and collagen fibers integrity. In particular, compressed tenocytes produce large amount of large PGs like aggrecan and versican which linking significant amount of water oppose compression (Docking et al. 2013). However, considering the peri-tumoral microenvironment in cancer, HA acts as a modulator of the tumor microenvironment through signal transduction mechanisms, but also regulating the hydration and osmotic balance in tumor microenvironment. We also know that a very hydrated peri-tumor ECM could create interfibers spaces where cancer cells enter to invade the surrounding deeper tissue and a HA synthesis is increased in different cancers, such as breast, prostate, lung, colon, and ovarian (Tavianatou et al. 2019, 2021). Moreover, HA can increase the integrin-mediated transduction thus favoring cancer cell adhesion and invasion (Chopra et al. 2014).

If epithelial cell dissociation or loss of cell–cell adhesion related to E-cadherin decrease and changes of ECM physical array are the main factors which favor cancer cell invasion, clinical data suggest that cancer progression and metastasis occur also without E-cadherin deregulation (Khalil et al. 2017). Thus, the collagen fiber array in peri-tumor microenvironment seems to be the first actor in affecting tumor development and cancer cell invasion. The epithelial E-cadherin-positive MCF-7 cells, lacking the expression of membrane type 1-matrix metalloproteinase 1 (MT1-MMP)/MMP14) and not possessing the ability to proteolytically remodel the ECM, are able to invade collectively the ECM when they find a low dense collagen network rich in relatively wide spaces. Interestingly, repressing E-cadherin MCF-7 cells loose cell–cell contacts but do not invade high-density collagen which they can cross only when the MT1-MMP/MMP14 is ectopically expressed (Ilina et al. 2020).

10.7 LoVo-R Colon Cancer Cells in 2D Cultures and 3D Cultures (Matrigel-Covered Millipore, High Concentrated Type I Collagen-Covered Millipore)

The tumor microenvironment and in particular collagen array play an "epigenetic" role in influencing primary tumor growth, migration, and metastatic potential, but also can favor the ability of cancer cells to resist chemotherapy (Provenzano et al. 2008; Conklin et al. 2011; Franchi et al. 2019; Montagner and Dupont 2020). ECM concentration is an important and crucial parameter to evaluate drug screening. An increase of collagen concentration in 3D microenvironment induces an increase of the intracellular stiffness and motor activity of cancer cells and seems to influence the intracellular fluctuations even in the presence of chemotherapeutic and anti-MMP drugs (Kim et al. 2018). To evaluate the possible role that collagen can play in influencing the drug resistance in cancer, we cultivated a colon cancer cell line which is resistant to doxorubicin (LoVo-R) in a Matrigel-covered Millipore filter (control) and in a concentrated type I collagen network (3000 μg/mL). LoVo-R cells cultivated on Matrigel mimicking the basement membrane appeared grouped with relative cell–cell contacts through intercellular thin and short cytoplasmic processes; they exhibited a smooth surface and a polygonal and flattened shape which resemble epithelial cells phenotype (Fig. 10.7a, b). The same cells cultivated on highly concentrated collagen showed a rounded shape and assembled in groups of few cells but some of them appeared isolated and showed a particular phenotype which resembles a "donut" shape (Fig. 10.7c, d). These cells displayed thin and short cytoplasmic protrusions which were in tight contact with the collagen fibrils or connected adjacent cells. We suggest that these cells were invading the collagen fibrils layer through a "funnel" invagination (Fig. 10.7e). From sectioned areas of the collagen-covered Millipore filter, we were able to detect thin and short cytoplasmic protrusions which developing from the ventral side of some cells penetrated the collagen fibrils network and morphologically corresponded to invadopodia (Fig. 10.7f). Moreover, a loss of collagen next to the cytoplasmic protrusions or invadodia suggests that a proteolitic activity next to the same protrusions occurred.

The ultrastructural data of LoVo-R cells cultivated in Matrigel or collagen substrates suggest that type I collagen induces EMT with drastic morphological changes of phenotypes observed in BM cultures: cells on collagen appeared more isolated and invading the collagen layer by developing invadopodia.

All these data suggest that collagen type I inducing an increase of cancer cells aggressiveness could also play a key role in favoring the development of a drug resistance. In fact, the previously described shock of MCF-7 and MDA-MB-231 cells when come in contact with the rough and irregular sub-BM microenvironment, could affect also LoVo-R cells, which developing a more aggressive phenotype could further increase their intrinsic resistance to therapeutic treatments. The development of drug resistance is a significant complication to successful of therapeutic procedures and seems partially favored by type I collagen composition of peri-tumoral ECM, which can significantly vary from patient to patient like clinical results suggest. The question is how can collagen improve cancer cells resistance to therapy? This could be partially explained because T-cell proliferation significantly reduces in a high-density matrix when compared to a low-density one (Kuczek et al. 2019). Other researchers suggest that tumor microenvironment could control the transport and thus efficacy of free doxorubicin in vitro (Brancato et al. 2018). Drug resistance is related to cancer cell lines but also tumor microenvironment: MDA-MB-231 cells displayed a stiffness-dependent response to the chemotherapeutic doxorubicin, whereas low aggressive MCF7 breast carcinoma cells cultured in the same conditions did not exhibit this stiffness-dependent resistance to the drug (Joyce et al. 2018). Anyway, the role of tumor ECM in modulating elasticity, adhesion, senescence, and EMT of MDA-MB-231 cells cancer cells seems associated with an improving of integrins upregulations (Poincloux et al. 2011; Mierke et al. 2011; Brandão-Costa et al. 2020; Byrne et al. 2021). Integrins are transmembrane proteins that physically connect cells to different proteins of ECM like fibronectin, vitronectin, collagens, and laminins but also to other cell surface receptors (Hynes 2002; Ringer et al. 2017). If the cadherin-based adherens junctions transduce the intercellular stresses (Leckband and de Rooij 2014), both chemical composition and all mechanical properties of the ECM are sensed by multimolecular complexes which are called focal adhesions (FAs) and include integrin receptors associated with the ECM (Geiger et al. 2009; Case and Waterman 2015). In

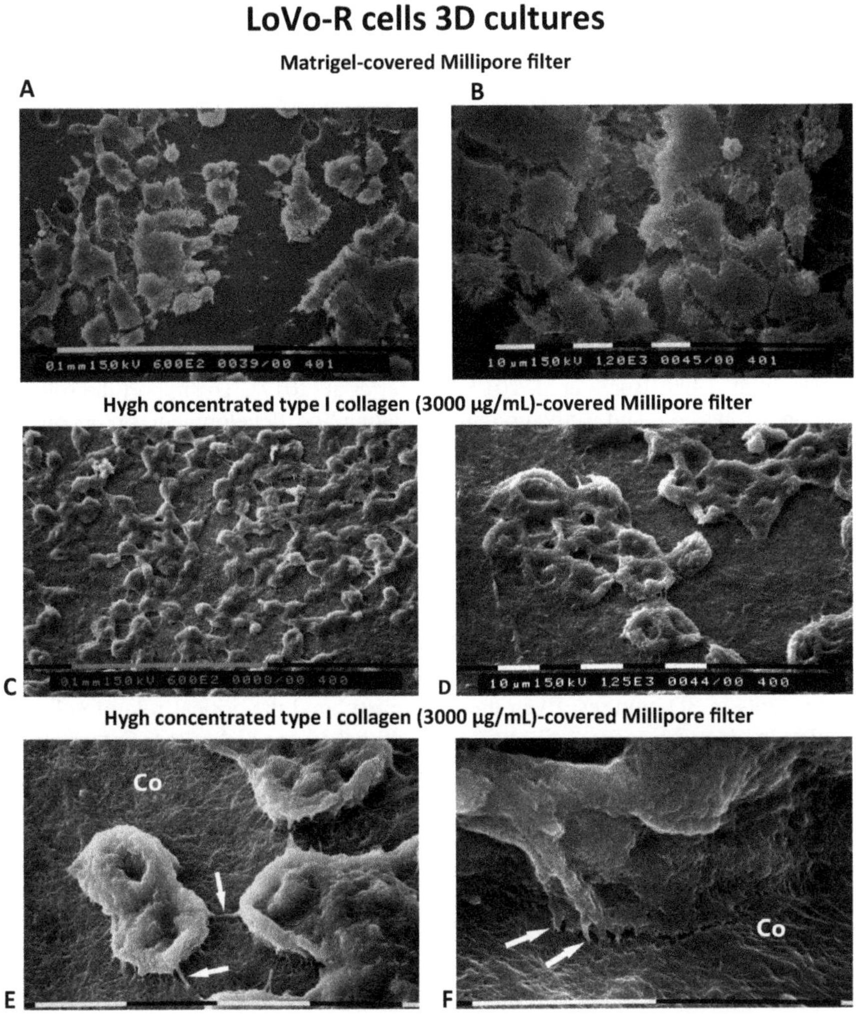

Fig. 10.7 SEM analysis of LoVo-R cells. (**a**) LoVo-R cells cultivated on a Matrigel-covered Millipore filter show a smooth surface and a polygonal-flattened shape like normal epithelial cells (White bar = 100 μm). (**b**) They appear distributed in a single cell layer and grouped with cell–cell contacts through intercellular thin and short cytoplasmic processes (White bar = 10 μm). (**c**) LoVo-R cells on high concentrated type I collagen (3000 μg/mL)-covered Millipore filter show rounded-shaped phenotypes assembled in small groups (White bar = 100 μm). (**d**) Some cells appear isolated and exhibit a particular phenotype that resembles a "donut" shape (White bar = 10 μm). (**e**) At a higher enlargement the "donut"-shaped cells show thin and short cytoplasmic protrusions in tight contact with the collagen fibrils or connecting adjacent cells (arrows). These cells are invading the collagen fibrils layer through a "funnel" invagination (Co = collagen) (White bar = 10 μm). (**f**) A lateral view of a sectioned collagen-covered Millipore filter shows that LoVo-R cells cultivated on high concentrated type I collagen (3000 μg/mL)-covered Millipore filter develop thin and short cytoplasmic protrusions from their ventral side which are penetrating the collagen fibrils network. These cytoplasmic protrusions correspond to invadopodia degrading adjacent collagen fibrils (arrows) (Co = collagen) (White bar = 10 μm)

particular, Ringer et al. (2017) described an outer FA layer including ECM-specific anchorage by integrin receptors, an intermediate layer where chemical and mechanical signals are processed, and an inner layer that is dominated by the actomyosin cytoskeleton of cells (Ringer et al. 2017). FAs complexes, linking integrins to the actin cytoskeleton either directly or indirectly by interacting with vinculin and α-actinin, act as mechanically elastic springs which enforce by talin unfolding when mechanical tension increases but remain small and short-lived when forces decrease in ECM. Therefore, they allow FAs to transmit the chemo-mechanical stimuli from ECM to the inner actomyosin cytoskeleton (Ringer et al. 2017). The main role of type I collagen in affecting the tumor progression is also supported by the fact that cells adhering to collagen type I by integrin α2β1 induce a very different signaling response compared to cells which bind collagen type IV with α1β1 integrin (Ivaska et al. 1999; Borza et al. 2012,; Ringer et al. 2017). To support the role for integrin expression and activation in regulating the development of oncogenic processes, it was reported that also hyaluronan, a soft polymeric glycosaminoglycan ECM component which is also considered a marker of breast cancer, can affect the cellular response by increasing the integrin mechano-transduction, thus suggesting the interaction of signaling between hyaluronan receptors and specific integrins. In particular, hyaluronan augments or overrides mechanical signaling by some classes of integrins to produce a spread morphology of cells otherwise observed only on very rigid substrates (Chopra et al. 2014). It is evident that in further experiments, it will be necessary to reproduce the in vivo conditions and include most of the ECM components in 3D cultures. Even though we explored the critical role of collagen and FN in affecting cancer cell phenotype and behavior vs. the BM, further investigations concerning the other molecules of ECM will help to better understand how tumor microenvironment could play a primary role in regulating cancer cell invasion. It is also plausible that only through evaluation and comparison of the interplay among all different ECM components, we will be able to suggest innovative therapeutic strategies to prevent or limit cancer invasion.

10.8 Conclusions: Take Home Message

1. Just a physically irregular surface of a culture substrate can transform 2D to 3D cultures simply because cells are free to move in all three spatial directions and may induce EMT in the low aggressive MCF-7 breast cancer cells.
2. In 3D cultures, a thick layer of Matrigel mimicking the BM does not promote changes in morphological phenotypes of both low aggressive MCF-7 and very aggressive MDA-MB-231 cells, probably because all epithelial cells recognize the smooth BM on which they usually grow.
3. In 3D cultures, a loose network of type I collagen fibrils mimicking ECM of sub-BM induces different effects on low aggressive MCF-7 cells vs. very aggressive MDA-MB-231 ones. Collagen favors the development of mesenchymal phenotypes morphology in MCF-7 cells but limits the invasion ability of the

same cells if compared to Matrigel cultures, as collagen induces a reduction in MMPs production in cells preferring an ameboid-invasion mode. On the contrary, type I collagen fibrils do not improve the EMT-related morphological changes in MDA-MB-231 cells which, however, in invasion test, show a strong increased ability to invade the collagen layer if compared to the same cells cultivated on Matrigel, as collagen seems to stimulate MMPs production.

4. In 3D cultures, nonaligned FN favors both MCF-7 and MDA-MB-231 cancer cells adhesion by developing more flattened and polygonal cells suggesting that nonaligned FN favors cancer cells adhesion. The increase of cytoplasmic microvesicles and the development of TNTs suggest an improve of cell–cell communications.
5. In 3D cultures, high concentrated type I collagen fibrils mimicking the deeper ECM induce breast cancer cells to grow in more than one layer, thus suggesting that concentrated collagen favors a further adhesion of cancer cells. The development of microvilli, microvesicles, long filopodia, and TNTs suggests that collagen may also stimulate communications among both adjacent and distant cells.
6. In 3D cultures, densely packed and parallel collagen fibers mimicking TACS-II stage in mammary tumor progression contrast both MCF-7 and MDA-MB-231 breast cancer cells invasion.
7. In 3D cultures of high concentrated type I collagen network, the LoVo-R colorectal cancer cells, resistant to doxorubicin, develop new morphological invaginating phenotypes and invadopodia. These morphological aspects suggest that the increased invasion ability of cancer cells could be related to the drug resistance.
8. Collagen, also in different arrays, could act as a physical factor inducing EMT in MCF-7 and MDA-MB-231 cells and drug resistance in LoVo-R cells.

References

Albini A, Iwamoto Y, Kleinman HK, Martin GR, Aaronson SA, Kozlowski JM, McEwan RN (1987) A rapid in vitro assay for quantitating the invasive potential of tumor cells. Cancer Res 47:3239–3245

Anguiano M, Castilla C, Maska M, Ederra C, Pelaez R, Morlaes X, Munoz-Arrieta G, Mujika M, Kozubek M, Munoz-Barrutia A, Rouzaut A, Arana S, Garcia-Aznar JM, Ortiz-De-Solorzano C (2017) Characterization of three-dimensional cancer cell migration in mixed collagen-Matrigel scaffolds using microfluidics and image analysis. PLoS One 12(2):e0171417. https://doi.org/10.1371/journal.pone.0171417

Anguiano M, Morales X, Castilla C, Pena AR, Ederra C, Martinez M, Ariz M, Esparza M, Amaveda H, Mora M, Movilla N, Garcia-Aznar JM, Cortes-Dominquez I, Ortiz-De-Solorzano C (2020) The use of mixed collagen-Matrigel matrices of increasing complexity recapitulates the biphasic role of cell adhesion in cancer cell migration: ECM sensing, remodeling and forces at the leading edge of cancer invasion. PLoS One 15(1):e0220019. https://doi.org/10.1371/journal.pone.0220019

Baker EL, Srivastava J, Yu D, Bonnecaze RT, Zaman MH (2011) Cancer cell migration: integrated roles of matrix mechanics and transforming potential. PLoS One 6(5):e20355. https://doi.org/10.1371/journal.pone.0020355. Epub 2011 May 27

Balachander GM, Balaji SA, Rangarajan A, Chatterjee K (2015) Enhanced metastatic potential in a 3D tissue scaffold toward a comprehensive in vitro model for breast cancer metastasis. ACS Appl Mater Interfaces 7(50):27810–27822. https://doi.org/10.1021/acsami.5b09064. Epub 2015 Dec 8

Balkwill FR, Capasso M, Hagemann T (2012) The tumor microenvironment at a glance. J Cell Sci 125:5591–5596

Baskaran JP, Weldy A, Guarin J, Munoz G, Shpilker PH, Kotlik M, Subbiah N, Wishart A, Peng Y, Miller MA, Cowen L, Oudin MJ (2020) Cell shape, and not 2D migration, predicts extracellular matrix-driven 3D cell invasion in breast cancer. APL Bioeng 4:026105. https://doi.org/10.1063/1.5143779

Borza CM, Su Y, Chen X, Yu L, Mont S, Chetyrkin S, Voziyan P, Hudson BG, Billings PC, Jo H, Bennett JS, Degrado WF, Eckes B, Zent R, Pozzi A (2012) Inhibition of integrin alpha2beta1 ameliorates glomerular injury. J Am Soc Nephrol 23(6):1027–1038

Bozzuto G, Condello M, Molinari A (2015) Migratory behaviour of tumour cells: a scanningelectron microscopy study. Ann Ist Super Sanita 51(2):139–147. https://doi.org/10.4415/ANN_15_02_12

Brabek J, Mierke CT, Rosel D, Vesely P, Fabry B (2010) The role of the tissue microenvironment in the regulation of cancer cell motility and invasion. Cell Commun Signal 8:22. http://www.biosignaling.com/content/8/1/22

Brandão-Costa RM, Helal-Neto E, Vieira AM, Barcellos-de-Souza P, Morgado-Diaz J, Barja-Fidalgo C (2020) Extracellular matrix derived from high metastatic human breast cancer triggers epithelial-mesenchymal transition in epithelial breast cancer cells through $\alpha v\beta 3$ integrin. Int J Mol Sci 21:2995. https://doi.org/10.3390/ijms21082995

Bredfeldt JS, Liu Y, Conklin MW, Keely PJ, Mackie TR, Eliceiri KW (2014) Automated quantification of aligned collagen for human breast carcinoma prognosis. J Pathol Inform 5:28. https://doi.org/10.4103/2153-3539.139707

Breitkreutz D, Koxolt I, Thiemann K, Nischt R (2013) Skin basement membrane: the foundation of epidermal integrity—BM functions and diverse roles of bridging molecules nidogen and Perlecan. BioMed Res Int 2013:179784. https://doi.org/10.1155/2013/179784

Byrne CE, Decombe JB, Bingham GC, Remont J, Miller LG, Khalif L, King T, Hamel K, Bunnell BA, Burow ME, MA E.C. (2021) Evaluation of extracellular matrix composition to improve breast cancer modeling. Tissue Eng Part A 27(7-8):500–511. https://doi.org/10.1089/ten.TEA.2020.0364

Case LB, Waterman CM (2015) Integration of actin dynamics and cell adhesion by a three-dimensional, mechanosensitive molecular clutch. Nat Cell Biol 17(8):955–963

Chopra A, Murray ME, Byfield FJ, Mendez MG, Halleluyan R, Restle DJ, Raz-Ben Aroush D, Galie PA, Pogoda K, Bucki R, Marcinkiewicz C, Prestwich GD, Zarembinski TI, Chen CS, Pure E, Kresh JY, Janmey PA (2014) Augmentation of integrin-mediated mechanotransduction by hyaluronic acid. Biomaterials 35(1):71–82

Christiansen JJ, Rajasekaran AK (2006) Reassessing epithelial to mesenchymal transition as a prerequisite for carcinoma invasion and metastasis. Cancer Res 66:8319–8326

Conklin MW, Eickhoff JC, Riching KM, Pehlke CA, Eliceiri KW, Provenzano PP, Friedl A, Keely PJ (2011) Aligned collagen is a prognostic signature for survival in human breast carcinoma. Am J Pathol 178:1221–1232. https://doi.org/10.1016/j.ajpath.2010.11.076

Costanza B, Umelo IA, Bellier J, Castronovo V, Turtoi A (2017) Stromal modulators of TGF-beta in cancer. J Clin Med 6:7. https://doi.org/10.3390/jcm6010007

Docking S, Samiric T, Scase E, Purdam C, Cook J (2013) Relationship between compressive loading and ECM changes in tendons. Muscles Ligaments Tendons J 3(1):7–11. https://doi.org/10.11138/mltj/2013.3.1.007. Published online 2013 May 21

Egeblad M, Rasch MG, Weaver VM (2010) Dynamic interplay between the collagen scaffold and tumor evolution. Curr Opin Cell Biol 22:697–706. https://doi.org/10.1016/j.ceb.2010.08.015

Erdogan B, Ao M, White LM, Means AL, Brewer BM, Yang L, Washington MK, Shi C, Franco OE, Weaver AM, Hayward SW, Li D, Webb J (2017) Cancer-associated fibroblasts promote directional cancer cell migration by aligning fibronectin. Cell Biol 216(11):3799–3816. https://doi.org/10.1083/jcb.201704053. Epub 2017 Oct 11

Foster DS, Jones RE, Ransom RC, Longaker MT, Norton JA (2018) The evolving relationship of wound healing and tumor stroma. JCI Insight 3:e99911

Fouad YA, Aanei C (2017) Revisiting the hallmarks of cancer. Am J Cancer Res 7:1016–1036

Franchi M, Fini M, Quaranta M, De Pasquale V, Raspanti M, Giavaresi G, Ottani V, Ruggeri A (2007) Crimp morphology in relaxed and stretched rat Achilles tendon. J Anat 210(1):1–7

Franchi M, Quaranta M, Macciocca M, De Pasquale V, Ottani V, Ruggeri A (2009) Structure relates to elastic recoil and functional role in quadriceps tendon and patellar ligament. Micron 40(3): 370–377

Franchi M, Ottani V, Stagni R, Ruggeri A (2010) Tendon and ligament fibrillar crimps give rise to left-handed helices of collagen fibrils in both planar and helical crimps. J Anat 216(3):301–309

Franchi M, Masola V, Bellin G, Onisto M, Karamanos KA, Piperigkou Z (2019) Collagen fiber array of peritumoral stroma influences epithelial-to-mesenchymal transition and invasive potential of mammary cancer cells. J Clin Med 8:213. https://doi.org/10.3390/jcm8020213

Franchi M, Piperigkou Z, Karamanos K-A, Franchi L, Masola V (2020a) Extracellular matrix-mediated breast cancer cells morphological alterations, invasiveness, and microvesicles/exosomes release. Cells 9(9):2031

Franchi M, Piperigkou Z, Riti E, Masola V, Onisto M, Karamanos NK (2020b) Long filopodia and tunneling nanotubes define new phenotypes of breast cancer cells in 3D cultures. Matrix Biol Plus 6-7:100026

Friedl P, Brocker EB (2000) The biology of cell locomotion within threedimensional extracellular matrix. Cell Mol Life Sci 57:41–64

Friedl P, Wolf K (2010) Plasticity of cell migration: a multiscale tuning model. J Cell Biol 188:11–19

Friedl P, Locker J, Sahai E, Segall JE (2012) Classifying collective cancer cell invasion. Nat Cell Biol 14:777–783

Friedland JC, Lee MH, Boettiger D (2009) Mechanically activated integrin switch controls alpha5beta1 function. Science 323:642–644

Geiger B, Spatz JP, Bershadsky AD (2009) Environmental sensing through focal adhesions. Nat Rev Mol Cell Biol 10(1):21–33

Hubbard B, Buczek-Thomas JA, Nugent MA, Michael L, Smith ML (2015) Fibronectin fiber extension decreases cell spreading and migration. J Cell Physiol 231(8):1728–1736. https://doi.org/10.1002/jcp.25271. Epub 2015 Dec 28

Hynes RO (2002) Integrins: bidirectional, allosteric signaling machines. Cell 110(6):673–687

Ilina O, Gritsenko PG, Syga S, Lippoldt J, La Porta CAM, Chepizhko O, Grosser S, Vullings M, Bakker G-J, Starruß J, Bult P, Zapperi S, Käs JA, Deutsch A, Friedl P (2020) Cell-cell adhesion and 3D matrix confinement determine jamming transitions in breast cancer invasion. Nat Cell Biol 22(9):1103–1115. https://doi.org/10.1038/s41556-020-0552-6. Epub 2020 Aug 24

Ivaska J, Reunanen H, Westermarck J, Koivisto L, Kahari VM, Heino J (1999) Integrin alpha2beta1 mediates isoformspecific activation of p38 and upregulation of collagen gene transcription by a mechanism involving the alpha2 cytoplasmic tail. J Cell Biol 147(2):401–416

Jayadev R, Sherwood DR (2017) Basement membranes. Curr Biol 27:R207–R211

Joyce MH, Lu C, James ER, Hegab R, Allen SC, Suggs LJ, Brock A (2018) Phenotypic basis for matrix stiffness-dependent chemoresistance of breast cancer cells to doxorubicin. Front Oncol 8: 337. https://doi.org/10.3389/fonc.2018.00337

Kalluri R (2003a) Basement membranes: structure, assembly and role in tumour angiogenesis. Nat Rev Cancer 3:422–433

Kalluri R (2003b) Basement membranes: structure, assembly and role in tumour angiogenesis. Nat Rev Cancer 3(6):422–433. https://doi.org/10.1038/nrc1094
Karamanos NK, Piperigkou Z, Passi A, Götte M, Rousselle P, Vlodavsky I (2021) Extracellular matrix-based cancer targeting. Trends Mol Med 27(10):1000–1013
Khalil AA, Ilina O, Gritsenko PG, Bult P, Span PN, Friedlc P (2017) Collective invasion in ductal and lobular breast cancer associates with distant metastasis. Clin Exp Metastasis 34:421–429. https://doi.org/10.1007/s10585-017-9858-6
Kim JE, Reynolds DS, Zaman MH, Mak M (2018) Characterization of the mechanical properties of cancer cells in 3D matrices in response to collagen concentration and cytoskeletal inhibitors. Integr Biol (Camb) 10(4):232–241. https://doi.org/10.1039/c8ib00044a
Koch TM, Munster S, Bonakdar N, Butler JP, Fabry B (2012) 3D traction forces in cancer cell invasion. PLoS One 7(3):e33476
Kuczek DE, Larsen AMH, Thorseth ML, Carretta M, Kalvisa A, Siersbæk MS, Simões AMC, Roslind A, Engelholm LH, Noessner E, Donia M, Svane IM, Straten PT, Grøntved L, Madsen DH (2019) Collagen density regulates the activity of tumor-infiltrating T cells. J Immunother Cancer 7(1):68. https://doi.org/10.1186/s40425-019-0556-6
Kular JK, Basu S, Sharma RI (2014) The extracellular matrix: structure, composition, age-related differences, tools for analysis and applications for tissue engineering. J Tissue Eng 5: 2041731414557112. https://doi.org/10.1177/2041731414557112
Kurniawan NA, Vos BE, Biebricher A, Wuite GJL, Peterman EJG, Koenderink GH (2016) Fibrin networks support recurring mechanical loads by adapting their structure across multiple scales. Biophys J 111(5):1026–1034. https://doi.org/10.1016/j.bpj.2016.06.034
Larsen M, Artym VV, Green JA et al (2006) The matrix reorganized: extracellular matrix remodeling and integrin signaling. Curr Opin Cell Biol 18:463–471
Leckband DE, de Rooij J (2014) Cadherin adhesion and mechanotransduction. Annu Rev Cell Dev Biol 30:291–315
Lin TC, Yang CH, Cheng LI, Chang WT, Lin YR, Cheng HC (2019) Fibronectin in cancer: friend or foe. Cell 9(1):27. https://doi.org/10.3390/cells9010027
Lu P, Takai K, Weaver VM, Werb Z (2011) Extracellular matrix degradation and remodeling in development and disease. Cold Spring Harb Perspect Biol 3:a005058
Luhr I, Friedl A, Overath T, Tholey A, Kunze T, Hilpert F, Sebens S, Arnold N, Rosel F, Oberg HH et al (2012) Mammary fibroblasts regulate morphogenesis of normal and tumorigenic breast epithelial cells by mechanical and paracrine signals. Cancer Lett 325:175–188. https://doi.org/10.1016/j.canlet.2012.06.014
Malandrino A, Kamm RD, Moeendarbary E (2018) In vitro modeling of mechanics in cancer metastasis. ACS Biomater Sci Eng 4:294–301. https://doi.org/10.1021/acsbiomaterials.7b00041
Maller O, Hansen KC, Lyons TR, Acerbi I, Weaver VM, Prekeris R, Tan AT, Schedin P (2013) Collagen architecture in pregnancy-induced protection from breast cancer. J Cell Sci 126(Pt 18): 4108–4110. https://doi.org/10.1242/jcs.121590. Epub 2013 Jul 10
Menon S, Beningo KA (2011) Cancer cell invasion is enhanced by applied mechanical stimulation. PLoS One 6(2):e17277. https://doi.org/10.1371/journal.pone.0017277
Mierke CT, Rosel D, Fabry B, Brabek J (2008) Contractile forces in tumor cell migration. Eur J Cell Biol 87:669–676
Mierke CT, Frey B, Fellner M, Herrmann M, Fabry B (2011) Integrin α5β1 facilitates cancer cell invasion through enhanced contractile forces. J Cell Sci 124:369–383
Montagner M, Dupont S (2020) Mechanical forces as determinants of disseminated metastatic cell fate. Rev Cells 9(1):250. https://doi.org/10.3390/cells9010250
Mould AP, Humphries MJ (2004) Regulation of integrin function through conformational complexity: not simply a kneejerk reaction? Curr Opin Cell Biol 16:544–551
Ottani V, Franchi M, De Pasquale V, Leonardi L, Morocutti M, Ruggeri A (1998) Collagen fibril arrangement and size distribution in monkey oral mucosa. J Anat 192(3):321–328

Oudin MJ, Jonas O, Kosciuk T, Broye LC, Guido BC, Wyckoff J, Riquelme D, Lamar JM, Asokan SB, Whittaker C, Ma D, Langer R, Cima MJ, Wisinski KB, Hynes RO, Lauffenburger DA, Keely PJ, Bear JE, Gertler FB (2016) Tumor cell-driven extracellular matrix remodeling drives haptotaxis during metastatic progression. Cancer Discov 6(5):516–531. https://doi.org/10.1158/2159-8290.CD-15-1183

Pandya P, Orgaz JL, Sanz-Moreno V (2017) Modes of invasion during tumour dissemination. Mol Oncol 11:5–27. https://doi.org/10.1002/1878-0261.12019

Paolillo M, Schinelli S (2019) Extracellular matrix alterations in metastatic processes. Int J Mol Sci 20:4947

Park HJ, Helfman DM (2019) Up-regulated fibronectin in 3D culture facilitates spreading of triple negative breast cancer cells on 2D through integrin β-5 and Src. Sci Rep 9(1):19950. https://doi.org/10.1038/s41598-019-56,276-3

Paten JA, Martin CA, Wanis JT, Siadat SM, Figueroa-Navedo AM, Ruberti JW, Deravi LF (2019) Molecular interactions between collagen and fibronectin: a reciprocal relationship that regulates de novo fibrillogenesis. Chem. 5(8):2126–2145. https://doi.org/10.1016/j.chempr.2019.05.011

Peyton SR, Kim PD, Ghajar CM, Seliktar D, Putnam AJ (2008) The effects of matrix stiffness and RhoA on the phenotypic plasticity of smooth muscle cells in a 3-D biosynthetic hydrogel system. Biomaterials 29:2597–2607

Pissarenko A, Yang W, Quan H, Brown KA, Williams A, Proud WG, Meyers MA (2019) Tensile behavior and structural characterization of pig dermis. Acta Biomaterialia 86:77–95. https://doi.org/10.1016/j.actbio.2019.01.023

Poincloux R, Collin O, Lizárraga F, Romao M, Debray M, Piel M et al (2011) Contractility of the cell rear drives invasion of breast tumor cells in 3D Matrigel. Proc Natl Acad Sci U S A 108: 1943–1948. https://doi.org/10.1073/pnas.1010396108. PMID: 21245302

Provenzano PP, Inman DR, Eliceiri KW, Knittel JG, Yan L, Rueden CT, White JG, Keely PJ (2008) Collagen density promotes mammary tumor initiation and progression. BMC Med 6:11. https://doi.org/10.1186/1741-7015-6-11

Ramaswamy S, Ross KN, Lander ES, Golub TR (2003) A molecular signature of metastasis in primary solid tumors. Nat Genet 33:49–64. https://doi.org/10.1038/ng1060. PMID: 12469122

Ribeiro JF, dos Anjos EHM, Mello MLS, de Campos Vidal B (2013) Skin collagen fiber molecular order: a pattern of distributional fiber orientation as assessed by optical anisotropy and image analysis. PLoS ONE 8:e54724. https://doi.org/10.1371/journal.pone.0054724

Riching KM, Cox BL, Salick MR, Pehlke C, Riching AS, Ponik SM, Bass BR, Crone WC, Jiang Y, Weaver AM et al (2014) 3D collagen alignment limits protrusions to enhance breast cancer cell persistence. Biophys J 107:2546–2558. https://doi.org/10.1016/j.bpj.2014.10.035

Rick JW, Chandra A, Ore CD, Nguyen AT, Yagnik G, Aghi MK (2019) Fibronectin in malignancy: cancer-specific alterations, protumoral effects, and therapeutic implications. Semin Oncol 46(3): 284–290. https://doi.org/10.1053/j.seminoncol.2019.08.002

Ridley AJ, Schwartz MA, Burridge K, Firtel RA, Ginsberg MH, Borisy G, Parsons JT, Horwitz AR (2003) Cell migration: integrating signals from front to back. Science 302:1704–1709

Rilla K, Siiskonen H, Tammi M, Tammi R (2014) Hyaluronan-coated extracellular vesicles-a novel link between hyaluronan and cancer. Adv Cancer Res 123:121–148. https://doi.org/10.1016/B978-0-12-800,092-2.00005-8

Ringer P, Colo G, Fässler R, Grashoff C (2017) Sensing the mechano-chemical properties of the extracellular matrix. Matrix Biol 64:6–16. https://doi.org/10.1016/j.matbio.2017.03.004

Ruggeri A, Franchi M, Marini N, Trisi P, Piattelli A (1992) Supracrestal circular collagen fiber network around osseointegrated nonsubmerged titanium implants. Clin Oral Implants Res 3: 169–175

Ruggeri A, Franchi M, Trisi P, Piattelli A (1994) Histologic and ultrastructural findings of gingival circular ligament surrounding osseointegrated nonsubmerged loaded titanium implants. Int J Oral Maxillofac Implants 9(6):1–15

Schäfer M, Werner S (2008) Cancer as an overhealing wound: an old hypothesis revisited. Nat Rev Mol Cell Biol 9:628–638

Scott JE (1975) Composition and structure of the pericellular environment. Physiological function and chemical composition of pericellular proteoglycan (an evolutionary view). Philos Trans R Soc Lond B Biol Sci 271(912):135–142. https://doi.org/10.1098/rstb.1975.0047

Scott JE (2002) The first and second 'laws' of chemical morphology, exemplified in mammalian extracellular matrices. Eur J Histochem 46(2):111–124. https://doi.org/10.4081/1660

Scott LE (2003) Elasticity in extracellular matrix 'shape modules' of tendon, cartilage, etc. A sliding proteoglycan-filament model. J Physiol 553(Pt 2):335–343. https://doi.org/10.1113/jphysiol.2003.050179. Published online 2003 Aug 15

Scott LE, Weinberg SH, Lemmon CA (2019) Mechanochemical signaling of the extracellular matrix in epithelial-mesenchymal transition. Front Cell Dev Biol 7:135. https://doi.org/10.3389/fcell.2019.00135

Stylianou A, Gkretsi V, Louca M, Zacharia LC, Stylianopoulos T (2019) Collagen content and extracellular matrix cause cytoskeletal remodelling in pancreatic fibroblasts. J R Soc Interface 16:20190226. https://doi.org/10.1098/rsif.2019.0226

Tavianatou AG, Piperigkou Z, Barbera C, Beninatto R, Masola V, Caon I, Onisto M, Franchi M, Galesso D, Karamanos NK (2019) Molecular Size-Dependent Specificity of Hyaluronan on Functional Properties, Morphology and Matrix Composition of Mammary Cancer Cells. Matrix Biol Plus 3:100008. https://doi.org/10.1016/j.mbplus.2019.100008

Tavianatou AG, Piperigkou Z, Koutsakis C, Barbera C, Beninatto R, Franchi M, Karamanos NK (2021) The action of hyaluronan in functional properties, morphology and expression of matrix effectors in mammary cancer cells depends on its molecular size. FEBS J 288(14):4291–4310. https://doi.org/10.1111/febs.15734

Tlsty TD, Coussens LM (2006) Tumor stroma and regulation of cancer development. Annu Rev Pathol 1:119–150

Tse JM, Cheng G, Tyrrell JA, Wilcox-Adelman SA, Boucher Y, Jain RK, Munn LL (2012) Mechanical compression drives cancer cells toward invasive phenotype. PNAS 109(3): 911–916. https://doi.org/10.1073/pnas.1118910109

Ueda M, Saito S, Murata T, Hirano T, Bise R, Kabashima K, Suzuki S (2019) Combined multiphoton imaging and biaxial tissue extension for quantitative analysis of geometric fiber organization in human reticular dermis. Sci Rep 9:10644. https://doi.org/10.1038/s41598-019-47,213-5

Wang WY, Pearson AT, Kutys ML, Choi CK, Wozniak MA, Baker BM, Chen CS (2018) Extracellular matrix alignment dictates the organization of focal adhesions and directs uniaxial cell migration. APL Bioeng 2:046107

Wolf K, Friedl P (2011) Extracellular matrix determinants of proteolytic and non-proteolytic cell migration. Trends Cell Biol 21(12):736–744. https://doi.org/10.1016/j.tcb.2011.09.006. Epub 2011 Oct 27

Wolf K, Wu YI, Liu Y, Geiger J, Tam E, Overall C, Stack MS, Friedl P (2007) Multistep pericellular proteolysis controls the transition from individual to collective cancer cell invasion. Nat Cell Biol 9:893–904

Wolf K, te Lindert M, Krause M, Alexander S, te Riet J, Willis AL, Hoffman RM, Figdor CG, Weiss SJ, Friedl P (2013) Physical limits of cell migration: control by ECM space and nuclear deformation and tuning by proteolysis and traction force. J Cell Biol 201(7):1069–1084. https://doi.org/10.1083/jcb.201210152

Yue B (2014) Biology of the extracellular matrix: an overview. J Glaucoma 23:S20–S23

Zaman MH, Trapani LM, Sieminski AL, Mackellar D, Gong H, Kamm RD, Wells A, Lauffenburger DA, Matsudaira P (2006) Migration of tumor cells in 3 D matrices is governed by matrix stiffness along with cell-matrix adhesion and proteolysis. Proc Natl Acad Sci USA 103:10889–10894

Zanotelli MR, Zhang J, Reinhart-King CA, Nancy E, Peter C (2021) Mechanoresponsive metabolism in cancer cell migration and metastasis. Cell Metab 33(7):1307–1321. https://doi.org/10.1016/j.cmet.2021.04.002

Zolota V, Tzelepi V, Piperigkou Z, Kourea H, Papakonstantinou E, Argentou MA, Karamanos NK (2021) Epigenetic alterations in triple-negative breast cancer-the critical role of extracellular matrix. Cancers (Basel) 13(4):713. https://doi.org/10.3390/cancers13040713

Chapter 11
Integrins in Cancer: Refocusing on the Tumor Microenvironment

Cédric Zeltz, Ning Lu, Ritva Heljasvaara, and Donald Gullberg

Abstract Despite having been characterized for several decades, the development of therapeutically useful compounds based on integrins has shown little success. In cancer biology, previous studies mainly focused on the roles of integrins in tumor cells and tumor vasculature, but currently, the role of integrins in cancer-associated fibroblasts (CAFs) in the tumor microenvironment (TME) has entered the limelight. The TME is a complex meshwork of extracellular matrix (ECM) macromolecules filled with a collection of cells, including CAFs, blood vessel-associated endothelial cells, smooth muscle cells, pericytes, mesenchymal stem cells, and a variety of immune cells. As a main producer of ECM and paracrine signals, the CAF is a central cell type during ECM remodeling, but CAFs are also receiving increased attention as key players of immunosuppression and chemoresistance. Whereas the paracrine signaling by CAFs has been extensively studied in the context of tumor stroma interactions, the nature of the numerous integrin-mediated cell-ECM interactions occurring in the TME has remained understudied. We will in this chapter discuss integrins with a focus on the TME and the interplay between cancer cells, immune cells, and CAFs in the context of three solid tumor types. As an example of the multifaceted mode of action of the stromal integrin α11β1, we will summarize our current understanding on the role of this CAF-expressed collagen-binding integrin in these three tumor types and discuss data suggesting that blocking

C. Zeltz · N. Lu · D. Gullberg (✉)
Department of Biomedicine and Centre for Cancer Biomarkers, University of Bergen, Bergen, Norway
e-mail: donald.gullberg@uib.no

R. Heljasvaara
Oulu Center for Cell-Matrix Research, Faculty of Biochemistry and Molecular Medicine, University of Oulu, Oulu, Finland

I. Kovalszky et al. (eds.), *The Extracellular Matrix and the Tumor Microenvironment*, Biology of Extracellular Matrix 11,
https://doi.org/10.1007/978-3-030-99708-3_11

desmoplasia might have opposite effects in different tumor types. The recent discoveries of CAF heterogeneity and tumor specific effects of desmoplasia suggest that much remains to be learned about tumor stroma interactions. We suggest that integrin-based targeting of cancer cells and angiogenesis should be shifted to targeting of tumor-supportive CAF subpopulations, which so far has been a relatively unexplored venue.

11.1 Introduction

In solid tumors, the activated fibroblasts, or cancer-associated fibroblasts (CAFs), constitute a major population of cells in the tumor microenvironment (TME). CAFs have been documented to affect tumor growth, formation of stem cell niches, immunosuppression, metastasis, and chemoresistance (Multhaupt et al. 2016; Su et al. 2018). We recently summarized the role of CAFs and integrins in three desmoplastic tumor types (Zeltz et al. 2020). In this chapter, we will update this information, but with more of a focus on CAF-expressed integrins and extend our discussion to the role of TME-mediated fibrosis in tumor progression and chemoresistance (Fig. 11.1). We will stick to the three desmoplastic tumor types we previously discussed (pancreatic cancer, breast cancer, and lung cancer) and retain a certain focus on integrin α11 (Zeltz et al. 2019). We regard this as a timely subject since it is likely that in the coming years, there will be an increased interest in cellular TME interactions.

The *fibroblast* is a cell type of primary significance for extracellular matrix (ECM) production and remodeling in interstitial tissues (Nagalingam et al. 2018). Activated fibroblasts, i.e., *myofibroblasts* or *CAFs,* are crucial in the processes of wound healing and tissue- and tumor fibrosis, respectively. In accordance with this, experimental data show that fibroblasts use analogous "toolkits" to reorganize the ECM in these different conditions (Rybinski et al. 2014; Gullberg et al. 2016; Kalluri 2016). The key characteristics of normal tissue fibroblasts, myofibroblasts, and CAFs are summarized in information Boxes 11.1 and 11.2.

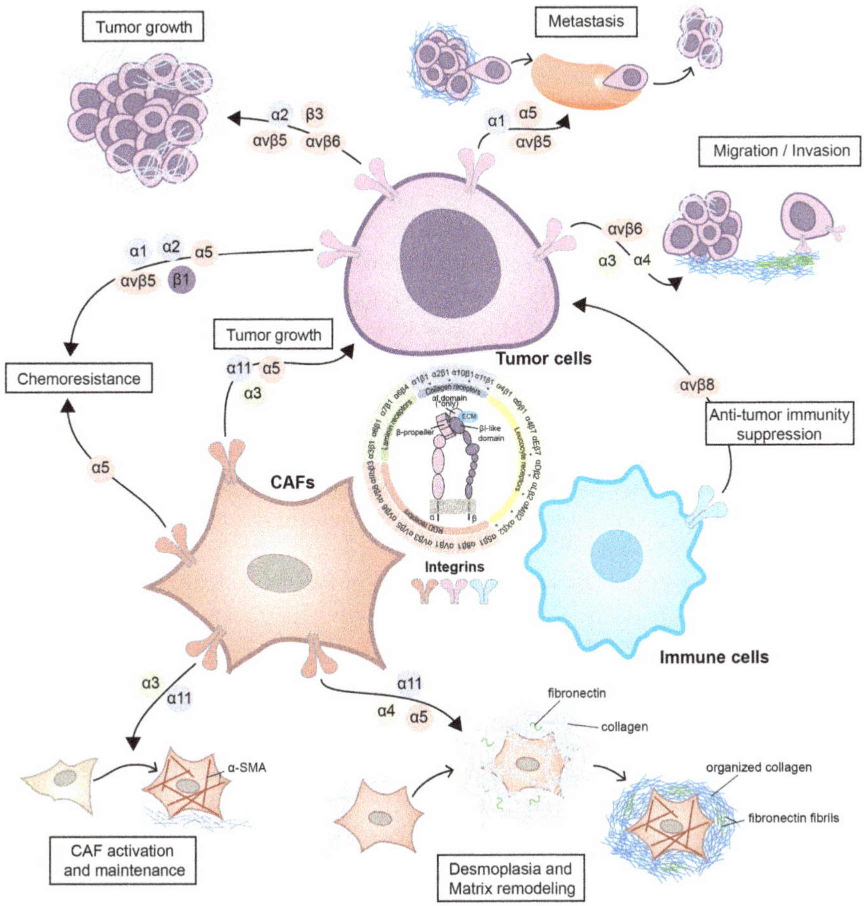

Fig. 11.1 Integrin functions in CAFs, tumor, and immune cells. The schematic summarizes integrin functions in pancreatic, lung, and breast cancer based on studies cited in Table 11.1. Integrins or integrin subunits involved in each process are indicated. Integrins are transmembrane receptors involved in extracellular matrix (ECM)–cell interactions. The integrin family is formed by 18 α and 8 β subunits that dimerize to form 24 integrins with distinct ligand specificities. Cancer-associated fibroblasts (CAFs) can differentiate into myofibroblastic CAFs (myCAFs), which show increased contractility due to alpha smooth muscle actin (α-SMA) expression. These cells, present in desmoplastic tumors, synthesize and remodel the ECM to increase stiffness, which in turn can affect tumor growth, invasion, and chemoresistance. A major CAF integrin involved in this process is α11β1. CAF integrins also act in a paracrine manner, via cytokine secretion, to promote tumorigenesis. Integrins expressed on tumor cells promote tumor growth, tumor invasion, and metastasis. Some tumor integrins confer chemoresistance and targeting αvβ5 integrin potentiates chemotherapy. Integrin αvβ8 on immune cells is responsible for antitumor immunity suppression

Box 11.1

The *fibroblast* is a cell type, of mesenchymal or neural-crest origin, that is nonepithelial, noninflammatory, and nonvascular. The fibroblast is a not only a key factor in processes involving synthesis of fibrillar collagens and other interstitial ECM components, but also an active cell type in integrin-mediated matrix remodeling as well as being a source of matrix metalloproteinases during tissue regeneration events (Multhaupt et al. 2016; Nagalingam et al. 2018). The transcriptional profile of fibroblasts varies with the anatomical location (Rinn et al. 2006). Cell lineage tracing has highlighted distinct origins of mouse fibroblasts in skin and cardiac tissue. Fibroblasts of mouse heart are derived from endocardium or epicardium (Moore-Morris et al. 2016). In mouse skin, a shared multipotent progenitor for papillary and reticular fibroblasts has been identified where subtypes of neonatal fibroblast are distinguished by a dynamic biomarker expression (Driskell et al. 2013; Guerrero-Juarez et al. 2019). Further heterogeneity of dermal fibroblasts arises from the complex cellular organization of hair follicles and due to the different embryonic origins of dermal fibroblasts (from dermomyotome in the posterior part of the body, from lateral plate mesoderm in the anterior part, and from the neural crest in the face). A dynamic expression of biomarkers also exists in human dermal fibroblasts with some notable differences compared to mouse skin (Philippeos et al. 2018). Recently, a comprehensive database was generated based on RNA-seq data from fibroblasts in mouse and human healthy and diseased tissues (fibroXplorer.com), and a summary of the data in this database was published (Buechler et al. 2021). The extensive analysis of the data generated a cross-tissue comparative atlas of fibroblasts gene expression and revealed a general organizing principle of the fibroblast lineage within and across organs. The study by Buechlet et al. describes two universal fibroblast lineages that express Dpt (dermatopontin) and Pi16 (peptidase inhibitor 16) or Dpt and Col15a1 (collagen XV) across tissues. The authors also demonstrate that these universal lineages give rise to eight types of tissue-specialized fibroblasts and in addition three types of activated fibroblast types during tissue injury.

Box 11.2

Myofibroblasts are activated fibroblasts considered to be contractile due to expression of the contractile isoform of actin, alpha smooth muscle actin (α-SMA) (Hinz et al. 2001; Tomasek et al. 2002). Myofibroblasts express αvβ1 integrin with a central role in transforming growth factor beta (TGF-β) activation in fibrotic conditions (Reed et al. 2015). Myofibroblasts were originally always thought to be depleted via apoptosis after completed wound healing (Tomasek et al. 2002; Schulz et al. 2018). However, mouse

(continued)

Box 11.2 (continued)
cardiac myofibroblasts have been observed to turn off α-SMA expression in the heart and form a cell type called matrifibrocyte with different properties than the undifferentiated premyofibroblasts (Fu et al. 2018). Emerging data thus suggest that myofibroblasts display more plasticity than previously thought. The finding that subsets of mouse skin myofibroblasts under certain conditions can differentiate into adipocytes further stresses the plasticity of myofibroblasts (Plikus et al. 2017).

Cancer-associated fibroblasts (*CAF*s) are fibroblast-like cells, of different origins, which are present in the TME. Sometimes used as abbreviation for carcinoma-associated fibroblasts, to specifically denote cells associated with epithelial-derived tumors. Demonstrated to be surprisingly heterogeneous. Several surprisingly heterogeneous CAF subtypes have been defined within tumor stroma. Pioneer work has defined three major types of fibroblasts in human and mouse pancreatic cancer ductal adenocarcinomas (PDACs): inflammatory CAFs (iCAFs), myofibroblastic CAFs (myCAFs) (Ohlund et al. 2017), and antigen presenting CAFs (apCAFs) (Elyada et al. 2019). Four major subclasses of CAFs in breast cancer (CAF-S1–S4), distinguished by different levels of α-SMA and fibroblasts activation protein (FAP) expression (Costa et al. 2018; Costa-Almeida et al. 2018) have also been described. Due to the plasticity and dynamic nature of the CAFs, it has been suggested that the CAF subtypes do not represent fixed cell types, but rather represent fibroblast "states" (Nurmik et al. 2019). Epigenetic changes do however result in more stable phenotypes (Marks et al. 2016; Eckert et al. 2019). Accumulating evidence indicates that some subpopulations of CAFs are tumor-supportive, whereas others are tumor-suppressive (Ireland and Mielgo 2018; Biffi and Tuveson 2018). In summary, CAFs have been demonstrated to act in a paracrine manner to affect different aspects of tumorigenesis, and via matrix synthesis and matrix remodeling to induce stiffness and hypoxia, which in turn also affect tumor growth. The comprehensive summary of fibroblast lineages described above also included data on transcriptional subtypes from human pancreatic ductal adenocarcinoma (PDAC) CAFs and nonsmall cell lung cancer (NSCLC) CAFs and demonstrated enrichment of an Lrrc15+ (leucine rich repeat containing 15) CAF subset (Buechler et al. 2021).

11.2 Tumor Fibrosis

Tumor desmoplasia is to a large part driven by paracrine and autocrine signaling of chemokines, cytokines, and growth factors. These factors regulate cell migration and cell proliferation, the secretion of ECM molecules, and the CAF-mediated crosslinking of fibrillar collagen matrices that often result in increased tissue stiffness and hypoxia (Kalluri 2016; Piersma et al. 2020; Zeltz et al. 2020). Additionally, ECM remodeling and reorganization are key factors that modify the properties of the

TME. The origin of CAFs varies with the tissue. Endogenous tissue fibroblasts, ADAM12$^+$ perivascular cells and pericytes, have all been described as sources of CAFs (Dulauroy et al. 2012; Kalluri 2016; Ohlund et al. 2014). Moreover, through cell lineage tracing methods, a platelet-derived growth factor receptor alpha negative (PDGFRα^-) and clusterin$^+$- breast cancer CAF subset have been demonstrated to emerge from mesenchymal, nonhematopoietic bone marrow cells in the polyoma middle T (PyMT) oncogene-driven mouse mammary carcinoma model (Raz et al. 2018). The epithelial-to-mesenchymal transition (EMT) as a source of CAF seems to be restricted but contributes to form an invasive mesenchymal tumor cell phenotype and to create a niche for the emergence of cancer stem cells (Nieto et al. 2016). Furthermore, EMT indirectly affects the stroma. For example, Pastushenko et al. have convincingly demonstrated that the stroma evolves (composition, immune cells) as tumor cells progressed toward EMT in a spontaneous EMT genetic model (*Kras* $^{LSL\text{-}12GD}$/*p53* $^{fl/fl}$/*Lgr5CreER*) of squamous cell carcinoma (SCC) (Pastushenko et al. 2018). Westcott and colleagues have characterized breast cancer cell invasion in vitro and identified a switch of tumor cell state into a mesenchymal invasive state that is described as not being a completed EMT. The invasive tumor cells, named trailblazer cells, which initiate the path of invasive migration, were identified by a mesenchymal seven gene signature including *ITGA11* and *PDGFRA* (Westcott et al. 2015). Thus, when discussing CAF heterogeneity, it is essential to distinguish CAFs from: (1) EMT-derived cells that express varying mesenchymal biomarkers, (2) resident mesenchymal stem cells, and (3) trailblazer cells, which display a mesenchymal signature. So far, we are lacking biomarkers that can distinctly distinguish these different cell types.

11.2.1 CAFs in Desmoplastic TME

In the tumor stroma, cellular interactions of CAFs with the ECM contribute to CAF activation, tumor cell growth, tumor cell invasion, and metastasis, and also affect immunosuppression, chemoresistance, and the formation of stem cell niches. Some of these aspects have been discussed in previous reviews [see (Östman and Augsten 2009; Kwa et al. 2019; Chen and Song 2019; Zeltz et al. 2020)]. In this chapter, we will update the role of CAF interactions in pancreatic, mammary, and lung fibrotic tumors.

Since the role of integrins in the TME is understudied, we foresee new molecular data in years to come, which will include pairing of paracrine mechanisms with active integrins in different interacting cell types, including CAFs, as well as a deeper understanding of integrin-dependent chemoresistance mechanisms in response to tyrosine kinase inhibitors (Cruz da Silva et al. 2019).

The availability of integrin-binding sites in tightly packed collagen fibrils in vivo has been suggested to limit the activity of collagen-binding integrins (Zeltz et al. 2014; Woltersdorf et al. 2017). An emerging picture suggests that integrin-binding sites are to some extent made available during remodeling of the collagen fibril surface thanks in part to proline-mediated flexibility (Chow et al. 2018; Zhu et al. 2018). However, this "microremodeling" is different from major matrix remodeling events including deposition of freshly synthesized collagens where the availability of integrin binding sites is not limited. In this context, CAFs in a growing tumor would be able to directly interact with the collagen matrix through collagen-binding integrins. In a fibronectin (FN)-rich matrix, a switch might occur to also include indirect linkages via α5β1 (Musiime et al. 2021).

Although several studies suggest that reorganization of collagen to a linearized and stiff matrix promotes tumor metastasis, the role of the ECM in tumorigenesis regarding its supportive or suppressive effects seems to vary with tumor type (Kai et al. 2019). A seminal study by Sahai and colleagues has shown that CAFs can pave the path for invading tumor cells, by piercing holes and reorganizing the matrix (Gaggioli et al. 2007). In the original study, α3β1 and α5β1 integrins were needed for vulval integrin-dependent CAF cell migration in an artificial-mixed collagen I/ laminin-111 matrix. A separate study of pancreatic ductal adenocarcinoma (PDAC) CAFs has demonstrated that α3β1 integrin in this experimental system could bind to laminin-332 to support CAF differentiation and maintenance and to promote PDAC cancer cell invasion (Cavaco et al. 2018).

We have below assembled and updated knowledge related to the role of CAFs in pancreatic, lung, and breast cancers. In this updated version, we have highlighted the importance of integrin-mediated cell–ECM interactions in the TME, including potential roles of these interactions in chemoresistance which as another aspect of TME biology where the role of integrins is severely underexplored.

11.3 Pancreatic Cancer

11.3.1 CAF Heterogeneity in Pancreatic Cancer

An increasing number of studies highlights the significance of the endogenous stroma as source of CAFs. The developmental origin of the tumor stroma varies depending on the tissue. In pancreatic cancer, resident fibroblasts and stellate cells are thought to be the two main stromal origins of CAFs. Although hepatic stellate cells (HSCs) have been found to be of mesothelial origin (Friedman 2000; Asahina et al. 2011), the origin of pancreatic stellate cells (PSCs) is still debated. PSCs have been suggested to have a neuroectodermal origin, but more recent data suggest that PSCs may arise from pancreatic mesenchyme (Harari et al. 2019; Wu et al. 2020). In the resting state, HSCs and PSCs can easily be recognized by storing lipid droplets. However, this does not remain an identifying criterion in PSC-derived CAFs though this is still an area of active research. A stromal cell population that is receiving attention in the pancreas are Gli1-positive cells. In a detailed study where Gli1 expression in pancreas was compared with a subpopulation of stromal cells expressing Hoxb6, only the Gli^{+} cells expanded under fibrotic conditions suggesting that one should be careful when analyzing stromal cell subpopulations, since dynamic changes during disease progression take place (Garcia et al. 2020). Another study suggests that only 10–15% of PDAC CAFs are derived from PSCs (Helms et al. 2021). In the literature, the nomenclature for the nonepithelial fibroblast-like cells in pancreas is confusing. It is unfortunate when the PSCs are called fibroblasts, whereas with the current nomenclature, it is unproblematic to call PSC-derived activated cells CAFs. When in doubt about the nature of nonepithelial cells in a tissue, the term "stromal cells" seems most appropriate.

Öhlund and colleagues (Ohlund et al. 2017) were among the first to characterize two major types of CAFs in the transgenic $\text{LSLS}^{\text{KrasG12D/+}}$; $\text{Trp53}^{\text{R172H/+}}$; $\text{Pdx}^{\text{cre/+}}$ (KPC) mouse model of PDAC and named these myofibroblastic CAFs (myCAFs) and inflammatory CAFs (iCAF). myCAFs are located peritumorally, express high levels of α-SMA and fibroblasts activation protein (FAP), and require cell–cell contact to be competent to differentiate into this subtype. In contrast, iCAFs are present at farther distance from tumor cells and express lower levels of FAP and α-SMA and secrete inflammatory cytokines such as IL-6. The authors furthermore convincingly show that CAFs could alternate from myCAFs to iCAFs and inversely, in a dynamic manner. Later studies identified a third subset of mouse CAFs characterized by expression of major histocompatibility complex (MHC) class II antigens suggested to be active in T-cell activation (Elyada et al. 2019). This third population, which is demonstrated to share transcriptional profile with mesothelial-like cells has been named antigen-presenting fibroblasts (apCAF).

In a separate study by Dominguez et al., careful comparisons of CAF subtypes in mouse PDAC tumor models and human PDAC tissues resulted in several noteworthy observations (Dominguez et al. 2020). In mouse PDAC stroma, five subclusters of stromal cells were identified, all characterized by podoplanin (PDPN) expression, with three of the subclusters corresponding to myCAFs, iCAFS, and apCAFs, respectively, and two subclusters with characteristics of normal fibroblasts. The apCAF subcluster was noted to have a mesothelial-like transcriptional profile. Comparison of the mouse and human pancreatic tumor tissues suggested the absence of specialized antigen-presenting apCAFs in human PDAC TME. Instead, the two major subsets of CAFs, iCAFs and myCAFs, expressed CD74 and HLA-DRA and were implicated in antigen presentation. Interestingly, only one integrin gene, *ITGA11*, was profiled to be highly expressed on the myCAFs. When comparing the data from PDAC TME tissues with other human tumor types, a "TGF-β CAF signature" consisting of an 11-gene-signature was identified. These data reveal a potential functional difference between human and mouse CAFs. From an integrin perspective, there are few integrins present in these gene lists, an observation made not only from the study by Dominguez et al. (Dominguez et al. 2020), but also from other studies. This is indeed intriguing since integrin are known to be central in matrix remodeling and integration of paracrine signaling. We can think of a few reasons to explain these observations:

1. In transcriptional analysis of a mixed cell population, potential changes in integrin regulation in an individual cell population may not be easily detected (up in some cell types, down in some other cell types, possibly neutralizing each

other). To analyze changes in integrins might require synchronized, more pure cell populations or using single cell profiling.
2. Integrins are known to be sensitive to growth factor levels, and TGF-β in general being a strong regulator of integrin levels. A failure to detect changes in integrin regulation in fairly homogenous sorted cell populations might thus reflect poor sensitivity of the assay for low abundant proteins (including membrane proteins such as integrins). Even though membrane protein levels are of low abundance in comparison to many intracellular and secreted proteins, they are often central in cell regulatory events. We predict that future analysis of molecular mechanism with more sensitive analyses will clarify the role of integrins in the TME. In the study of Dominguez et al., the presence ITGA11 on CAFs with a TGF-β regulated signature is interesting since the ITGA11 promoter contains a SMAD-binding element (Lu et al. 2006, 2010).

In the majority of models involving tumor-stroma interactions, detailed studies suggest that the stroma is tumor supportive (Alexander and Cukierman 2016; Han et al. 2015). In molecular terms, this corresponds to various types of cross-talk between tumor and stromal cells including both paracrine and integrin-mediated signaling. In pancreatic cancer, the stroma has been described to support tumor growth and metastasis as well as the development of chemoresistance (Pan et al. 2015). Other studies suggest that stroma might act as a restraining barrier preventing tumor expansion and tumor spread.

Several studies involving TME will have to be revisited in regard with the new knowledge about CAF heterogeneity.

New data generated in more targeted approaches to CAF subsets support data from broadly cited study from Ozdemir and colleagues, suggesting that conditional deletion of α-SMA-expressing fibroblasts in experimental PDAC mouse model aggravates tumor outcome (Ozdemir et al. 2014). In this study, the α-SMA-thymidine kinase mouse was crossed with two different models of PDAC, namely the KPC mouse and Ptf1a$^{cre/+}$; Kras$^{Gt2D/+}$;TGFbr2 $^{flox/flox}$ (PKT) mouse, allowing depletion of α-SMA-expressing cells. This approach led to decreased number of myofibroblasts and ensued more invasive, undifferentiated, and necrotic tumors. The cell population targeted in this approach was most likely the CAF subpopulation we now call myCAFs, but also α-SMA positive cells in the tumor vasculature are affected. In a more recent study, an advanced PDAC mouse model was used to analyze the effects of collagen I deletion in α-SMA-expressing cells and in the context of TME (Chen et al. 2021). The results indicate that reduced collagen I expression results in more proliferative tumor cells, supporting the concept that the collagen I-producing myCAFs are tumor re-restraining. But also in this model, there are caveats, since not only myCAF-sythesized collagen I was targeted, but also the CAF-independent collagen I that could also contribute to the observed phenotype. Interestingly, Chen et al. propose that myCAFs lacking collagen I synthesis stimulate the PDAC cells to secrete CXCL5 which in turn attracts neutrophils contributing to an immunosuppressive environment, which further explains the tumor-restraining phenotype. These results agree with recent data from a study showing a correlation between collagen content and disease severeness (Jiang et al. 2020). PDAC patients with more desmoplasia had a better prognosis, which led the authors to inhibit LOXL2 (lysyl oxidase-like 2) activity in a mouse tumor model. Their data thus also support the concept that a collagen matrix restrains tumor growth in the PDAC context.

Curiously, focal adhesion kinase (FAK) inhibition in tumor cells appears to reduce tumor desmoplasia, but without leading to increased tumor aggressiveness or tumor spread (Jiang et al. 2016). This study clearly demonstrates that the molecular mechanism whereby ECM amounts and properties are changed can have different effects on tumorigenesis (cf. collagen deletion in myCAFs above). Similar findings, where cell type specific features determine molecular mechanisms, have been observed for the role of αvβ6 in experimental PDAC models (Reader et al. 2019). Great care thus must be taken when analyzing different models where knowledge about, for example, receptor repertoires and mutations in signaling pathways must be considered.

11.3.2 Tumor Cell Integrins in Pancreatic Cancer

FAK plays an important role in integrin signaling. The importance of integrin signaling in PDAC is indirectly indicated in studies in which FAK inhibitors failed to inhibit tumor angiogenesis, apoptosis, or necrosis but decreased tumor size and the populations of CAFs and tumor-associated macrophages (TAMs) within tumors (Stokes et al. 2011). This work is further supported by studies in which blocking FAK in PDAC models where a FAK inhibitor reduced tumor cell proliferation, the secretion of pro-fibrogenic paracrine signals and fibrosis in the TME. An important finding in this study is that blocking FAK and reducing ECM of TME did not lead to increased PDAC tumor cell growth in these experimental systems, but instead had the opposite effect. This is thus in contrast to the recent study by Chen et al. (2021) in which reducing collagen I synthesis in myCAFs increased tumor invasiveness. Obviously, conditional deletion of FAK in different cell types would be required to understand the mechanism of how FAK works in the context of integrin adhesion and paracrine signaling.

The collagen-binding integrins α1β1 and α2β1 are reported to be expressed on PDAC cells. One study suggests that α1β1 can be used as a premalignant biomarker in PDAC, and it was also shown to promote metastatic potential of cancer cells and induce therapy resistance (Gharibi et al. 2017). α2β1 is expressed also in blood vessels and immune cells, and in epithelial and endothelial cells, it binds the basement membrane collagen IV (Barczyk et al. 2010). In this context, it is interesting to note that the secreted glycoprotein galectin-3, which is often found upregulated in PDAC, can, when secreted, form networks in the ECM which specifically influence α2β1 integrin to impede integrin clustering and avidity (Friedrichs et al. 2008). This could be one of many mechanisms by which TME composition affects integrin activity in different cell types expressing α2β1 in PDAC tumors (Table 11.1).

Table 11.1 Role of integrins in desmoplastic tumors

Tumor type	Cell type	Experimental model	Integrin/ integrin subunit	Role	Effects tumor microenvironment
Pancreatic ductal adenocarcinoma (PDAC)	Tumor cells	PDAC/CAF spheroids, orthotopic PDAC models in nude mice	αvβ5	Regulated by TGF-β in TME. Regulates tumor growth and metastasis, role in chemoresistance (Hurtado de Mendoza et al. 2021)	ND
	Tumor cells	PDAC cell lines in vitro, KDC mouse model in vivo	αvβ6	TGF-β-activating, pro-migratory via Eps8, acting as switch between migratory and TGF-β activating-role (Tod et al. 2017). Promotes tumor growth, and tumor desmoplasia in KDC mice (Reader et al. 2019)	Indirect via activation of TGF-β ++
	Tumor cells	Human tissues, PANC-1 cells in vitro	α1	Pre-malignant biomarker, promotes metastasis, chemoresistance (Gharibi et al. 2017)	ND
	Tumor cells	Human xenografts in mice, PDAC cells in vitro	α2	Promots growth and chemoresistance (Semba et al. 2004; Dangi-Garimella et al. 2011)	ND
	Tumor cells	In vitro studies	α5	RAB5A-mediated endocytosis of α5β1 to promote migration (He et al. 2019)	ND
	CAFs	In vitro and in vivo studies.	αvβ5	Regulates α5 integrin levels and activity by regulating endocytosis (Franco-Barraza et al. 2017)	++
	CAFs	In vitro studies	α3	Supports maintenance of CAFs and stimulates PDAC cell growth (Cavaco et al. 2018)	+
	CAFs	PDAC subcutaneous tumors in mouse models in vivo	α5	Stimulates tumor growth and tumor desmoplasia, affect chemotherapy efficacy (Kuninty et al. 2019)	+++

(continued)

Table 11.1 (continued)

Tumor type	Cell type	Experimental model	Integrin/ integrin subunit	Role	Effects tumor microenvironment
	CAFs	Cultured PSC. Primary and immortalized PDAC CAFs in vitro	α11	Present on PDAC CAFs in vitro and in vivo (Zeltz et al. 2019). Identified on human PDAC myCAFs by scRNA-seq as part of TGF-β CAF signature (Dominguez et al. 2020). Promotes PSC myofibroblast activation and cell migration in vitro (Schnittert et al. 2019). Functional in vivo data lacking	+++
Breast cancer	Tumor cells		αvβ6	Activate TGF-β, SOX4 activation and tumor cell resistance to cytotoxic T cells (Bagati et al. 2021). Expressed in subset myoepithelial cells (Allen et al. 2014)	+++, CAF activation and ECM production
	Tumor cells	$Itgb1^{-/-}$//$Itgb3^{-/-}$ mouse model	β1/β3	In ERbB2-dependent breast cancer integrins form a complex with insulin receptors (IRs) and activate Akt/mTORC1 signaling axis to stimulate growth and metastasis (Huck et al. 2010; Bui et al. 2019)	ND
	Tumor cells		β3	Takes part in extracellular vesicle uptake, shown to influence clonal expansion of tumor cells (Fuentes et al. 2020)	ND
	Tumor cells		α3	Represses Reelin expression to promote invasion (Ndoye et al. 2021)	ND
	Tumor cells		α4	Tumor cell invasion (Kwon et al. 2020).	ND
	Tumor cells		α5	FN fibrillogenesis in a stiff matrix in a Mena-dependent manner, facilitating α4β1-mediated tumor spread (Berger et al. 2020). Mediates metastasis to bone and osteolytic lesions (Pantano et al. 2021)	++++

	CAFs	Ddr2$^{-/-}$ mouse model	β1 (collagen-binding integrins)	Activated by DDR2 (Bayer et al. 2019), leads to remodeled collagen matrix	+++
	CAFs	Spin90$^{-/-}$ mouse model	α4	FN fibrillogenesis (Kwon et al. 2020) in SPIN90 knockout model, increased fibrillar matrix	+++
	CAFs	PyMT//Itga11$^{-/-}$ mouse model	α11	CAF activation, collagen remodeling, tumor growth, metastasis, expressed in subset of myoepithelial cells (Zeltz et al. 2020; Smeland et al. 2020; Primac et al. 2019)	+++
Non-small cell lung cancer	Tumor cells	NSCLC tumor cells, co-culture fibroblasts	αvβ6	Activates TGF-β in CAFs to stimulate tumorigenesis (Eberlein et al. 2015)	+++
	Tumor cells	PC9 tumor cells, PDX mouse tumor models	β1	Erlotinib resistance (Kanda et al. 2013), CAFs in inflammatory TME niche produce collagen, binding to β1 integrins on cancer cells, induce cell proliferation and tumor relapse (Cho et al. 2020)	CAFs are the cells that stimulate cancer cell proliferation
	Tumor cells	PC9 tumor cells	α2	Erlotinib resistance (Kanda et al. 2013)	ND
	Tumor cells	PC9 tumor cells	α5	Erlotinib resistance (Kanda et al. 2013)	ND
	CAFs	Mouse models in vivo, human NSCLC CAFs in vitro, gene signature	α11	Regulates stromal stiffness and metastasis in vivo, CAF migration in vitro (Navab et al. 2016; Iwai et al. 2021)	+++

A collection of selected references has been assembled to illustrate importance of integrins in tumor cells and cancer-associated fibroblasts (CAFs). We have also noted if integrins are implied in affecting the tumor microenvironment

FN fibronectin, ND not determined

In PDAC, αvβ3 integrin on tumor cells interacts with CAF-produced osteopontin to stimulate EMT and cancer stem cell-like properties by regulating the expression of FOXM1 at the tumor–stroma interface (Cao et al. 2018). The CAFs that secrete osteopontin probably correspond to the myCAFs. A study of human colon cancer has identified interactions between a specific subset of CAFs at tumor stroma with osteopontin, which in turn was suggested to influence cancer stem cells (Lenos et al. 2018). These studies that show CAF-mediated signaling activity at the stroma–tumor interface is clearly distinct from CAF interactions elsewhere in the tumor (Costa et al. 2018; Ohlund et al. 2017).

Two other αv-integrins that are relevant in the pancreatic PDAC cells are αvβ5 and αvβ6. High level of β5 in PDAC cancer correlates with reduced survival (Uhlen et al. 2017). In a careful study using KPC mice, β5 on PDAC cells was shown to promote tumor progression and metastasis (Hurtado de Mendoza et al. 2021). Interestingly, CAFs were shown to stimulate β5 levels on cancer cells via TGF-β signaling, again attesting to the importance of bidirectional paracrine signaling involving the TME.

Integrin αvβ6 has been found to be upregulated in a majority of PDAC cases (Brown and Marshall 2019). It has been reported to promote invasion and metastasis and is also the principal epithelial activator of TGF-β via binding to LAP (latency-associated peptide) (Munger et al. 1999). In this role as activator of TGF-β, αvβ6 indirectly influences the TME by providing TGF-β to CAFs and thus contributes to ECM synthesis and matrix remodeling (Reader et al. 2019). TGF-β can also act as tumor suppressor, and thus targeting TGF-β-activating integrins can be problematic, which highlights the need to dissect molecular mechanisms of action of αvβ6 and TGF-β in different tumors, tumor types, and tumor stages (Ahmed et al. 2017). Hence, for TGF-β-activating integrins, it is thus critical to choose the right animal or cell model and be aware if the function of TGF-β in the given model is growth-supportive or growth-suppressive. It is furthermore interesting to note that in addition to reduced tumor growth, the antibody blockade of αvβ6 led to reduced tumor desmoplasia (Reader et al. 2019), lending evidence to the indirect effect of epithelial αvβ6 on the CAF-produced ECM in the TME (Brown and Marshall 2019).

11.3.3 CAF Integrins in Pancreatic Cancer

We have generated a novel function-blocking antibody against integrin α11 that inhibits CAF interaction with collagen I, collagen reorganization, and CAF migration in a spheroid-based PDAC model (Zeltz et al. 2019) (Fig. 11.2). Schnittert and colleagues suggested that α11β1 on PDAC CAFs is indeed the main integrin that can drive PDAC cell invasion in an hetero-spheroid model (Schnittert et al. 2019). This study was based on use of a commercial polyclonal antibody to α11 with limited ability to distinguish specific reactivity on different stromal cells. Although in vitro data demonstrated a role of α11β1 in stellate cell myofibroblast differentiation, the lack of markers that can unequivocally separate PSC/pericytes from fibroblasts

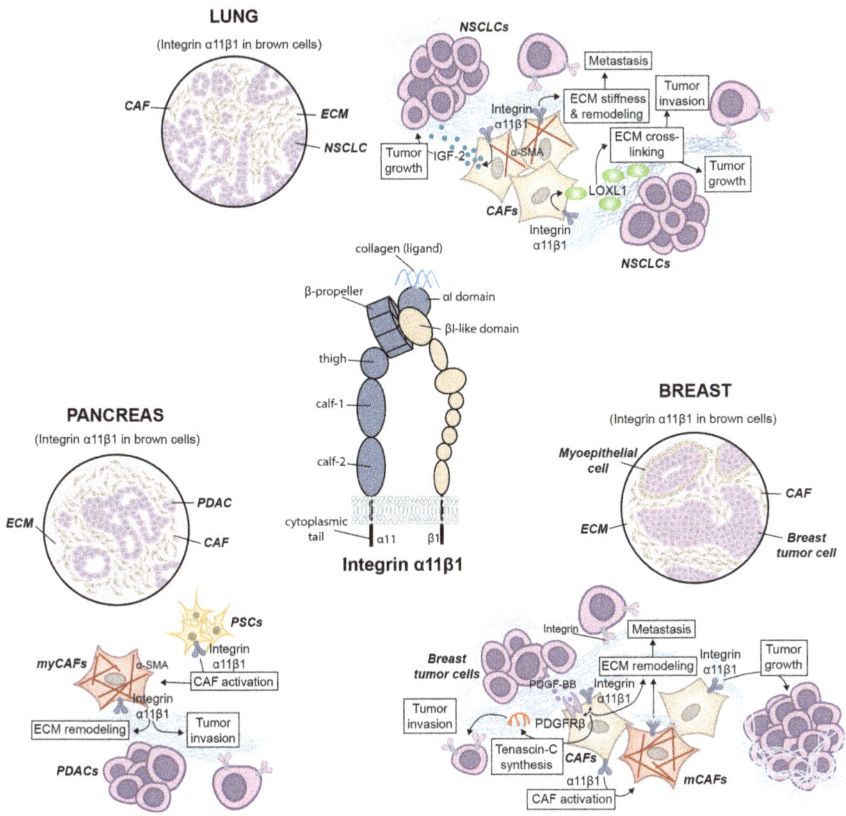

Fig. 11.2 Focus on integrin α11β1 in cancer. Integrin α11β1 is a collagen receptor expressed on a subpopulation of fibroblasts. The schematic summarizes its expression and functions in pancreatic, lung, and breast cancers. Integrin α11β1 is expressed on cancer-associated fibroblasts (CAFs) in the three cancer types and in myoepithelial cells surrounding duct in breast cancer. *Integrin α11β1 in lung tumor microenvironment (TME):* Integrin α11β1 expressed on CAFs controls ECM stiffness and remodeling and regulates the secretion of IGF-2, contributing to metastasis and tumor growth of nonsmall cell lung cancer (NSCLC), respectively. Moreover, integrin α11β1 modulates the expression of lysyl oxidase-like 1 (LOXL1), which is an ECM cross-linking enzyme implicated in tumor cell growth and invasion. *Integrin α11β1 in breast TME:* PDGF-BB is released by mammary tumor cells and binds to its receptor PDGFRβ on CAFs. PDGFRβ cross-talks with integrin α11β1 to support metastasis. Integrin α11β1 activate CAFs into myofibroblasts to remodel the extracellular matrix, which in turn affects tumor growth and metastasis. *Integrin α11β1 in pancreas TME:* TGF-β released in pancreatic ductal adenocarcinoma (PDAC) induces the formation of desmoplasia by CAFs. Integrin α11β1 in pancreatic stellate cells (PSCs) signals to differentiate into myofibroblastic CAFs (myCAFs). Integrin α11β1 controls ECM reorganization to mediate PDAC invasion

renders it difficult to judge on which cell type α11 is active in PDAC in an in vivo situation. Is it active on CAFs derived from PSC population or is it active on CAFs derived from a distinct fibroblast subpopulation? Whereas blockage of α11 on PSC in vitro inhibits myofibroblast differentiation, no blockage of α11 has so far been performed in a PDAC tumor model in vivo. We have confirmed that α11 is present on CAFs in PDAC tumor stroma in vivo using the novel monoclonal α11 antibodies we have generated, but not in NG2 (Nerve/Glial antigen 2)-positive CAFs (Zeltz et al. 2019). We have also failed to detect α11β1 in cells with pericyte characteristics (Zeltz et al. 2019). Thus, the determination of the source of α11-expressing CAFs using cell lineage tracing will be crucial to further study the role of α11β1 in PDAC, both to define the CAF subset with relation to other markers and to test effect of blocking or deleting α11 in an in vivo model. Careful analysis of the data by Dominguez et al. reveals that in PDAC, *ITGA11* is transcriptionally active in a subcluster of fibroblasts which correspond to a subset of myCAFs (Dominguez et al. 2020). In the pancreas, it is thus possible that α11-positive myCAFs are both of PSC and fibroblast origin. When analyzing integrins of different subsets of CAFs, it is important to remember that integrin upregulation in cell culture is a well-known fact. Integrin α11 is readily upregulated on cells in vitro that never have been found to express α11 in vivo. This might actually be the case for both mouse liver and pancreas where we have failed to detect α11 expression in either HSC or PSC in vivo, but α11 is clearly detected on both stellate cell subsets during in vitro culture (Martin et al. 2016; Schnittert et al. 2019).

Interestingly, PDAC cells can migrate on FN deposited on the CAF surface protrusions in a 3D collagen matrix using their α5β1 integrin (Miyazaki et al. 2019). It has also been demonstrated that α5β1 integrin endocytosis is regulated by αvβ5, indirectly affecting CAF activation (Franco-Barraza et al. 2017). In colon cancer, αvβ3 integrin expressed on CAF, together with integrin α5β1, contributes to FN fibrillogenesis and tumor cell invasion (Attieh et al. 2017). It will be interesting to establish whether αvβ3 has such a role also in PDAC CAFs.

Thus, present data imply that α5β1 and α11β1 integrins are important CAF receptors involved in matrix assembly and reorganization engaged in tumor cell growth and migration. Integrins αvβ3 and αvβ5 have both been found to assist or regulate integrin α5β1 activity while little is known about crosstalk of α11β1 with other integrins in pancreatic cancer. In tissue fibrosis, αvβ1 integrin has been demonstrated to orchestrate the activation of TGF-β on myofibroblasts and it will be thus interesting to have a comparable role on (a) particular PDAC CAF subtype (s). In addition, it is important to show which CAF subpopulations in PDAC are tumor-supportive and which are tumor-suppressive and integrate the relative contribution of specific CAF integrins in the paracrine signaling of iCAFs and myCAFs. In parallel, it will also be essential to understand whether different CAF subtypes hold prognostic value and have different functional properties (Table 11.1).

11.3.4 CAF/Integrin-Mediated Chemoresistance in Pancreatic Cancer

Integrin α1β1 is a receptor for the basement membrane collagen IV, and as already mentioned, one study suggests that it is associated with chemotherapy resistance in PDAC (Gharibi et al. 2017). The authors used mainly human PDAC cell lines to demonstrate that α1β1 is upregulated and coregulated with other TGF-β-responsive genes during EMT in PDAC cells. Interestingly, knocking down ITGA1 sensitized mesenchymal-like PDAC cells to the cytotoxic nucleoside analog gemcitabine.

αvβ5 is present on both PDAC cancer cells and stromal cells and is potentially a useful target to reduce desmoplasia. A recent study documented in vivo evidence for the role of αvβ5 in KPC mice and showed that iRGD (internalizing RGD peptides) targeting αvβ5 potentiated the effect of chemotherapy in this model (Hurtado de Mendoza et al. 2021). Although αvβ6 is present on PDAC cells, its central role in TGF-β activation indirectly affects CAF-mediated ECM synthesis and is therefore also an attractive target molecule that can be used to overcome CAF-mediated chemoresistance (Reader et al. 2019; Moore et al. 2020; Meecham and Marshall 2021). Tumor-associated macrophages and CAFs both contribute to TME-mediated chemoresistance in orthotopic genetic animal models like the KPC mouse and involvement of insulin-like growth factors (IGFs) in these processes has been suggested (Ireland et al. 2016; Ireland and Mielgo 2018). In this context, it would be interesting to determine the effect of IGFs on the integrin repertoire and on CAF-immune cell cross-talk.

11.4 Breast Cancer

11.4.1 CAF Heterogeneity in Breast Cancer

Breast is a complex organ, which undergoes hormonally regulated changes. In normal mouse mammary glands, the stroma largely determines the glandular epithelium development. Two subsets of fibroblasts have been identified in human mammary gland, lobular ($CD105^{high}/CD26^{low}$) and interlobular ($CD105^{low}/CD26^{high}$) fibroblasts (Morsing et al. 2016). CAF heterogeneity in breast cancer stroma has been observed both in human patient samples (Costa et al. 2018) and in mouse models (Bartoschek et al. 2018; Raz et al. 2018). In the human breast cancer, four different subsets of CAFs, CAFS1–CAFS4, have been identified, based on the expression profiles of six different markers (Costa et al. 2018). Notably, two of the subsets, CAF-S1 and CAF-S4 express high levels of α-SMA, but only CAF-S1 expresses fair amounts of FAP. CAF-S1 is found close to the tumor, attracts T-cells and contributes to immunosuppression. The immunosuppressive function of CAF-S1 partly depends on dipeptidylpeptidase 4 (DPP4, also known as CD26) and DPP4-mediated cleavage of CXCL10, leading to a reduction in T-cell recruitment to the tumor. The careful study by Costa et al. also indicates that within the four CAF subclasses there is probably even more heterogeneity. The CAF-S1 displays similar characteristics to those in myCAF population in pancreatic cancer (Costa et al. 2018).

Using single-cell RNA sequencing (scRNA-seq), the issue of CAF heterogeneity was independently addressed in the MMTV-PyMT (hereafter, PyMT) mouse model of mammary cancer at late stages of tumor progression (Bartoschek et al. 2018). This study identified four transcriptionally distinct subsets of CAFs with different functionalities and biophysical properties. They were termed as vCAFs (vascular CAFs), mCAFs (matrix CAFs), cCAFs (cell cycle CAFs), and dCAFs (developmental CAFs) and presented distinct spatial location within the tumor parenchyma. vCAFs were shown to originate from the perivascular compartment with cCAFs being a segment of proliferative vCAFs. Conversely, mCAFs mostly derived from resident fibroblasts, while dCAFs seemed to originate from the malignant epithelial compartment via EMT. Interestingly, PDGFRα was specifically expressed by mCAFs, whereas PDGFRβ was expressed by most CAF subsets, with exception of dCAFs. In contrast to the previously mentioned study (Costa et al. 2018), FAP and α-SMA markers were not specifically associated with a distinct subset of murine CAFs, but rather displayed a salt-and-pepper expression pattern in all four CAF subpopulations. These discrepancies are presumably related to differences between species, breast cancer subtypes and stages, and to detection methodologies.

As already mentioned, cell lineage tracing in PyMT tumors demonstrated the contribution of mesenchymal, nonhematopoietic bone marrow cells to a PDGFRα$^-$ and clusterin$^+$ CAF subpopulation (Raz et al. 2018). In vitro, bidirectional paracrine signaling between tumor cells and this CAF subpopulation had effects on both tumor cells and the CAFs. Clusterin, which has pleiotropic effects including stimulation of endothelial cell proliferation, was suggested to promote tumor growth mainly via enhancing angiogenesis. This further highlights the complexity of fibroblast heterogeneity in breast cancer and suggests that this challenging issue requires additional investigation with regards to biomarker expression, spatial localization, and functionality of CAFs in all the subtypes and stages of breast cancers.

11.4.2 Tumor Cell Integrins in Breast Cancer

The literature on integrins in breast cancer is quite comprehensive and until recently has focused on integrins on the tumor cells. We will summarize some of this data below before going to the relatively scarce literature on CAF integrins in the TME. Detailed studies reveal a redundancy of β1 and β3 integrins to promote ErbB2-driven breast cancer (Bui et al. 2019). When β1 integrin is inactivated in epithelial cells in an ErbB2-driven mouse mammary cancer model, the resulting phenotype is a modest delay in tumor induction, which has been suggested to be due to β3 integrin compensation (Huck et al. 2010). In a double *Itgb1/Itgb3* knockout model of ErbB2-dependent breast cancer, the mice demonstrate a more severe delay in tumor onset, thus supporting the previous suggestion that mild tumor phenotype in *Itgb1* null background was due to integrin β3 compensation. Mechanistically, it was shown that in wild-type tumor cells, integrins can form a complex with insulin receptors (IRs) and activate Akt/mTORC1 signaling axis to stimulate tumor growth and metastasis (Bui et al. 2019). IRs are degraded in soft matrices, but the integrin-mediated ECM stiffness seems to stabilize IR levels at the cell surface and increase cell signaling. Interestingly, β3 integrins on triple negative human MDA-MB-231 breast cancer cells have been demonstrated to be needed in a totally different context, namely for extracellular vesicle uptake allowing intercellular communication affecting metastasis and clonal growth capacity (Fuentes et al. 2020).

Regarding specific integrin α-chains and integrin heterodimers which have been implied on breast cancer cells, we have collected some recent data. In one example, α5 integrin chain is expressed in breast cancer cells and interacts with the intracellular an actin regulating protein called Mena. Increased stiffness and epidermal growth factor receptor 1 (EGFR) activation upregulate expression of Mena in breast cancer cells (Berger et al. 2020). Mena binds directly to α5 cytoplasmic tail to enhance α5-mediated outside-in integrin signaling and formation of FN-containing fibrillar adhesions (Gupton et al. 2012). The increased Mena levels stimulate α5β1 integrin-dependent EDA–FN organization to facilitate matrix metalloprotease (MMP)-dependent integrin α4β1- and α9β1-mediated breast cancer cell invasion. Independent studies suggest that α5β1 mediates metastasis to bone and osteolytic lesions (Pantano et al. 2021).

αvβ6 in breast cancer is involved in activating TGF-β with effects on both the tumor cells and CAFs in the TME such as increased collagen production and collagen remodeling by CAFs (Raab-Westphal et al. 2017; Brown and Marshall 2019). In one study, induction of αvβ6 in myoepithelial cells in preinvasive ductal carcinoma in situ was suggested as biomarker of disease progression (Allen et al. 2014). An elegant study furthermore demonstrated downstream effects of TGF-β activation in tumor cells leading to SOX4 activation and tumor cell resistance to cytotoxic T cells (Bagati et al. 2021). In this model antibody inhibition of αvβ6 induced T cell-mediated immunity in immunotherapy-resistant tumor models.

In addition to integrins that bind interstitial matrix components during invasion, integrins that adhere tumor cells to basement membranes have also been studied. These include laminin-binding integrin α3β1, which has been shown to repress secretion of Reelin, a secreted glycoprotein that inhibits invasion, with the end-effect of promoting invasion (Ndoye et al. 2021). Another study shows that ablation of α3β1 promotes tumor progression and invasion in Erb2-driven mouse mammary cancer model and in HER2-overexpressing human breast cancer cells, but not in triple-negative cancer cells (Ramovs et al. 2019). It should be also noted that α6β1 is a common marker of breast cancer stem cells (CSCs) and integrin-ECM interactions and FAK signaling are essential for maintenance of CSC pool and tumor progression (Guan 2010; Krebsbach and Villa-Diaz 2017).

11.4.3 CAF Integrins in Breast Cancer

One of the most notable features of tumor stroma interactions in breast cancer is the desmoplastic reaction. Extensive desmoplastic reaction in normal breast tissue in form of mammographic density is strongly correlated to an increased risk of breast cancer development and has been proposed as a diagnostic and prognostic marker. The ECM composition and architecture associated with this fibrotic reaction emerge from an intimate crosstalk between fibroblasts and epithelial cells in breast tissues. Desmoplasia has also been linked to increased activation of integrins in breast cancer. In a breast cancer model, tumor cell-secreted LOXL2 activates fibroblasts and promotes the expression of α-SMA in a FAK-dependent manner (Barker et al. 2013). A previous landmark paper has demonstrated that increased tumor stroma stiffness promotes tumor progression by β1 integrin signaling in a FAK and Rho-signaling dependent manner (Levental et al. 2009). Likewise, in a mouse model of breast cancer, FAK inhibition decreased tumor growth and reduced infiltration of leukocytes and macrophages (Walsh et al. 2010; Wendt and Schiemann 2009). Together, these studies support the notion that β1 integrin and FAK sustains the pro-tumor functions of CAFs. The cooperation between integrins and receptor tyrosine kinase (RTKs) in tumor and stromal cells regulates cell invasion during metastatic dissemination in breast cancer. In this context, we have recently shown that stromal integrin α11 displays a pro-tumorigenic and pro-metastatic activity in breast cancer and strongly associates with a PDGFRβ+ CAF subset (Primac et al. 2019) (Fig. 11.2). Integrin α11 expression is strongly upregulated in the stromal compartment during mammary tumor progression. Histological analyses revealed a significant association between integrin α11 and PDGFRβ, both in human breast cancer samples and in the preclinical transgenic PyMT model. Among several tested stromal markers (PDGFRα, PDGFRβ, α-SMA, FAP, FSP1, and NG2), this collagen-binding integrin was mostly associated with a PDGFRβ$^+$ CAF subpopulation at late stages of tumorigenesis. As both integrin α11 and PDGFRβ are well-known regulators of ECM, it is plausible that the identified CAF subset overlaps with myCAF subpopulation described in the aforementioned study (Bartoschek et al. 2018). Indeed, genetic ablation of integrin α11 in the PyMT model drastically reduced not only tumor growth and metastasis, but also the desmoplastic reaction in these tumors, further highlighting the contribution of this specific α11$^+$ CAF subset to tumor progression through ECM regulation. This is further supported by the fact that mCAFs are thought to derive from resident fibroblasts, as well as integrin α11/PDGFRβ$^+$ CAFs. Mechanistically, this study revealed that integrin α11/PDGFRβ crosstalk in CAFs endows mammary cancer cells with pro-invasive features through the deposition of tenascin-C. Tenascin-C was strongly expressed by the same subset of CAFs expressing integrin α11 and PDGFRβ in the late stage PyMT tumors, as well as in clinical samples of invasive breast cancer. Overall, this study discloses an example of a collaborative crosstalk between an integrin and a RTK in CAFs, which acts as a driver of tumor invasiveness in breast cancer. Similar molecular partnerships have been previously reported, although not on CAFs. Indeed, microenvironment-induced c-Met/β1 integrin complex formation was shown to sustain breast cancer metastasis via the promotion of c-Met phosphorylation, as well as an increase of integrin affinity for FN on the tumor cells (Jahangiri et al. 2017). Further examples of cooperation between integrins and growth factor receptors in the context of cancer are thoroughly discussed in previous reviews (Ivaska and Heino 2011; Schnittert et al. 2018). It is worth noting that α2 integrin subunit, which heterodimerizes with β1 subunit to form another fibrillar collagen-binding integrin, exerts opposite functions to α11β1 in a related mouse mammary cancer model (Table 11.1). In contrast to α11 integrin chain, integrin α2 chain is expressed not only by CAFs, but also by tumor cells and other stromal cells. Furthermore, unlike integrin α11, the α2 subunit is downregulated in human breast cancer and suppresses metastasis in vivo (Ramirez et al. 2011). Hence, α2-deficient MMTV-neu mice display increased metastasis, which is suggested to result from the increased capacity of tumor cells to intravasate into the bloodstream.

Discoidin domain receptor 2 (DDR2) is a cell surface tyrosine kinase activated by collagens (Multhaupt et al. 2016). The functions of DDR2 in both CAFs and mammary carcinoma cells have been studied by performing global and tumor cell specific deletion of Ddr2 in mice (Corsa et al. 2016). Global deletion of Ddr2 does not affect primary tumor growth but results in reduced metastasis. Closer examination reveals that *Ddr2*$^{-/-}$ stroma contains reduced amounts of fibrillar collagen, with reduced diameter and impaired organization, suggesting critical Ddr2 functions in CAFs. A more recent analysis of DDR2 in the breast cancer stroma suggests that the DDR2 effect is mediated via β1 integrins and the role of DDR2 is thus via inside out signaling to activate the integrins taking part in collagen remodeling (Bayer et al. 2019). Since the cross-talk between β1 integrins and DDR receptors supposedly is not restricted to the TME, it is possible that integrins, together with DDRs, are involved in tumor cell invasion and metastasis (Xu et al. 2012). In support of a role of DDR2 in metastasis, its small molecule inhibitor WRG-28 inhibits tumor cell–TME interactions and tumor cell invasion (Grither and Longmore 2018). It will be interesting to sort out a possible cooperation of DDR2 and integrins in tumor metastasis (cancer cell integrins) and collagen reorganization (CAF integrins) and if such a link exists, identify the specific integrin(s) involved.

In summary, an important cooperation between integrin and RTKs is detected in breast cancer CAFs, raising several interesting questions. Future studies will for example determine if one and the same integrin can cooperate with different RTKs in different tumor stroma contexts or if the cooperation is integrin-specific and limited to some kinases. This also extends to mechanism of TME-mediated chemoresistance where data are emerging on the role of integrins in mediating chemoresistance to drugs targeting RTKs (Cruz da Silva et al. 2019). These mechanisms have mainly been described in cancer cells, but similar mechanisms are likely to operate in CAFs. Finally, the role of αv integrins (TGF-β activating mechanisms), α5β1 integrin and FN matrix assembly, and relation to the collagen remodeling α11β1 integrin will be important to study more in detail in breast cancer, just as in other desmoplastic tumor types.

11.4.4 CAF-Mediated Chemoresistance in Breast Cancer

CSCs are largely responsible for the continuous tumor growth and development of drug resistance, and many different signaling pathways contribute to these processes. Some of these pathways involving CAFs are highlighted below. For example, Hedgehog (Hh) ligand activity is detected in one third of triple negative breast cancer (TNBC). In an animal model of TNBC with high Hh level, an upregulation of fibroblast growth factor 5 (FGF5) and an increased collagen remodeling activity was observed in CAFs (Cazet et al. 2018). The increased concentration of remodeled collagen at tumor–stroma interface correlated with increased pFAK levels as well as increased number of CSCs at the tumor–stroma interface. In this breast cancer model, treatment with an inhibitor of the Hh receptor Smoothened (SMOi) sensitized mice to chemotherapy. It will be interesting to determine what specific integrins are present and mediate the increased pFAK at the tumor–stroma interface.

Studies in breast cancer cell lines offer additional details as to how CSC formation via EMT may occur. Snail1, which is a central transcription factor in EMT, is an unstable protein that is ubiquinated. In experiments performed by Lambies et al., deubiquitination by a specific ubiquitinase (USP27X) contributes to Snail1 stability in turn contributing to increased EMT, increased numbers of CSCs and chemoresistance (Lambies et al. 2018). In addition, Snail1 stabilization in CAFs contributes to increased CAF activation. It will be interesting to see whether CAF Snail1 in this context integrates with integrin-dependent mechanoregulated signaling. Such mechanoregulation in myofibroblasts involving Snail1 has been shown to contribute to fibrosis and to depend on both YAP/TAZ and myocardin-related transcription factors (MRTFs) (Zhang et al. 2016).

In an impressive study of chemo-resistant patients of breast cancer, various CAF subsets were identified (Su et al. 2018). The $CD10^{+}GPR77^{+}$ CAF subset was found to sustain cancer stemness and to promote tumor resistance. Furthermore, targeting this subset of CAFs using antibodies to GPR77 restored chemosensitivity. The authors suggest that targeting the $CD10^{+}GPR77^{+}$ CAF subset could be an effective strategy against CSC-driven solid tumors. It will be interesting to relate this subset of CAFs with the different breast cancer CAF subsets identified by Costa et al., as well as with CAF subsets in other cancer forms (Costa et al. 2018). It will also be important to characterize the role of integrins in CAF-mediated chemoresistance in breast cancer. We believe this will be an important area of research in the coming years.

11.5 Lung Cancer

11.5.1 Lung Fibroblast Heterogeneity

The lung is also a complex organ where fibroblasts have several functions associated with normal lung function. Cell lineage tracing has been performed to identify and characterize the origin of fibroblast subsets (Li et al. 2018), but in mouse lung, single-cell transcriptional analysis has been even more instrumental and has resulted in the identification of five subsets of fibroblasts in healthy lung and six subsets in fibrotic lung, in addition to a mesothelial subtype (Xie et al. 2018). In normal lung, these were grouped as myofibroblasts (Acta2+), col3a1 matrix fibroblasts (Col3a1+; Itga8+), Col4a1 matrix fibroblasts (Col4a1+;dcn+), lipofibroblasts (Lp1+), and mesenchymal progenitors (CD52+) (Xie et al. 2018). In the fibrotic lung, a distinct fibroblast cell type with high PDGFRβ expression, distinct from pericytes, was identified (Xie et al. 2018). In a separate study, the focus was on identifying subsets of cells involved in collagen synthesis in normal and diseased lung. The scRNA-seq analysis identified several subclusters within alveolar, adventitial, and peribronchial fibroblasts (Tsukui et al. 2020). Some major findings in this careful study include the identification of collagen XIII as a marker for lipofibroblasts, collagen triple helix repeat containing 1 (CTHRC1) as a marker for a subset of fibrotic fibroblasts in fibrosis predicted to be derived from endogenous alveolar fibroblasts and showing poor correlation between α-SMA-positive myofibroblasts and high collagen I production. Whereas studies of pancreatic and breast cancer tumors have started to unravel different subsets of CAFs, less detailed information is published on CAF heterogeneity in non-small cell lung carcinoma (NSCLC). In one interesting recent study, the influence of vascular adventitial fibroblasts on A549 lung cancer cells in a xenograft model indicated a higher tumor-promoting activity of the adventitial fibroblasts compared to non-vessel-associated lung fibroblasts (Hoshino et al. 2011). Microarray analysis further demonstrated a high level of PDPN expression in these adventitial fibroblasts, which in combination with other studies suggest a role for PDPN in tumorigenesis and lymph node metastasis. The study by Hoshini et al. in addition to demonstrating CAF heterogeneity in the lung tumor stroma suggests that a perivascular environment in lung constitutes a specific niche for tumor progression in the lung. PDPN expression has recently been shown to regulate β1 integrin levels in keratinocytes (Shibuya et al. 2018), whether this activity also applies to NSCLC CAFs remains to be determined. Studies using antibodies to detect FAP as a biomarker have also provided information, which clearly indicate the existence of CAF subsets in the lung (Kilvaer et al. 2018). Immunostaining of human NSCLC tumor sections in studies by Kilvaer et al. typically showed α-SMA and FAP expression on different CAFs, suggesting that in the lung, FAP is not highly expressed on myofibroblastic CAFs. The study also points out a major weakness with FAP antibodies in the context of the TME namely, FAP antibodies also immunostain macrophages. In tumors from NSCLC patients with high levels CD3+ and CD8+ T cells, high FAP levels on CAFs were associated with better prognosis. The latter finding indicates that FAP-directed therapy as a general antistroma therapy needs to be performed with great caution, and as already mentioned might not be suitable as a general antistroma therapy, but rather be suitable for a subset of tumors. The general expression pattern of CAF markers needs great attention in therapy situations. This has also been illustrated by experiments with FAP-directed immunotherapy where a side-effect of the tumor-directed treatment was cachexia, due to expression of FAP in muscle (Roberts et al. 2013). Finally, as yet another example of the complex events that take place in the TME, a recent study of a cohort of NSCLC patients identified glutamin-fructose-6-phosphate transaminase 2 (GFPT2) in CAFs as being responsible for increased glucose uptake and metabolic reprogramming in the TME (Zhang et al. 2018b).

11.5.2 Tumor Cell Integrins in Lung Cancer

As in other solid tumors, integrin αvβ6 stands out again as an integrin on cancer cells, which when activated will have effects on the TME. Careful coculture studies with NSCLC cells and fibroblasts demonstrate activation of fibroblasts and stimulation of cancer cell growth by a mechanism dependent on αvβ6-mediated TGF-β activation (Eberlein et al. 2015). Please also see chemoresistance section below for more integrins implicated in tumor cells.

11.5.3 CAF Integrins; The Role of Integrin α11β1 in Lung Cancer

In 2002, the Tsao laboratory published a list of six novel candidate genes for lung adenocarcinoma (obtained by comparing pooled RNA from tumors with normal lung RNA), which included *Lc-19, HABP2, CRYM, CP, COL11A1, and ITGA11* (Wang et al. 2002). In 2011, Navab et al. published a molecular signature for the NSCLC stroma (Navab et al. 2011). Matched pairs of normal lung fibroblasts and lung CAFs were isolated from 15 patients and their transcription profiles established. This effort resulted in identification of 46 differentially expressed genes in CAFs that formed a prognostic gene expression signature. Interestingly, six of the identified genes could be induced by TGF-β in normal fibroblasts, including the collagen receptor α11β1, identified in the original gene set from 2002. Comparison of these CAF genes with tumor stroma genes indicated a shared upregulation of four genes; ITGA11, THB2, COL11a1, and CRTHRC1. In the same study, analyses of epigenetic changes identified limited methylation changes in tested genes. Following the identification of α11 as one of a limited set of genes being upregulated in the stroma of experimental NSCLC tumors, the role of α11 in lung cancer was explored further. In xenograft models, coimplantation of NSCLC tumor cells with mouse embryonic fibroblasts lacking α11 greatly reduced NSCLC tumor growth (Zhu et al. 2007), which in this xenograft model was correlated to α11-dependent secretion of IGF-2. To functionally further test the potential contribution of α11 to CAF function in NSCLC, the integrin $\alpha11^{-/-}$ mouse strain has been very helpful (Popova et al. 2007). Analysis of NSCLS tumor growth in the $\alpha11^{-/-}$ mice demonstrated that absence of α11 in the stroma impeded NSCLC tumor growth (Navab et al. 2016) and metastasis. Analysis of tumor stroma demonstrated reduced organization of the collagen stroma and reduced tumor stiffness. Analysis of signaling status demonstrated reduced FAK and ERK phosphorylation in the tumor stroma from $\alpha11^{-/-}$ mice and reduced expression of α-SMA in the tumor stroma. Interestingly, integrin α11 has recently been shown to regulate the lysyl oxidase-like 1 expression in lung CAF to mediate NSCLC cell invasion and tumor growth (Zeltz et al. 2019) (Fig. 11.2). Protein–protein interaction network analysis in addition identified several interactions affected in the $\alpha11^{-/-}$ stroma and previously identified in CAF differentiation and tumorigenesis, including latent transforming growth factor beta binding proteins 3 and 4 (LTBP3 and 4), WNT1-inducible signaling pathway protein 2 (WISP2), insulin-like growth factor binding protein 2 and 4 (IGFBP2 and 4), and syndecan-4. It will be interesting to determine whether, and how, these proteins take part in α11-mediated effects in CAFs and other types of stromal cells. That CAFs can regulate plasticity of lung cancer stemness via paracrine signaling was shown in experiments, which identified IGF-2 producing CAFs as inducers of Nanog expression in cancer cells and thus established that these lung CAFs constitute a supporting niche for cancer stemness (Chen et al. 2014). With regards to IGF-2, it is interesting to note that the integrin α11β-expressing fibroblasts in the xenograft model of lung cancer produce IGF-2 (Zhu et al. 2007), and part of the pro-tumorigenic action of α11β1 might thus be related to a stemness-stimulating activity.

11.5.4 Integrin-Mediated Chemoresistance in Lung Cancer

A study using PC9 cells demonstrated the involvement of β1 integrins and specifically α2β1 and α5β1 in Erlotinib chemoresistance by creating a Src bypass pathway to induce Akt signaling (Akt also activated by EGFR) (Kanda et al. 2013). A number of studies have addressed paracrine signaling at tumor–stroma interface. In the future, it will be important to correlate these with effects of dynamic changes in integrin repertoire in the TME cell types, including CAFs. Examples of such studies that would benefit from integrin analysis of TME-induced chemoresistance in the lung include a study where CAFs have been reported to produce IGF-2 as an inducer of the ABC transporter P-GP in A549 cells and to mediate drug resistance (Zhang et al. 2018a). In another study, lung adenocarcinoma CAFs treated with Cisplatin upregulated IL-11 and conferred chemoresistance to lung cancer cells by activating STAT3 antiapoptotic pathway (Tao et al. 2016). In agreement with these in vitro findings, patients with high levels of IL-11R display poor response to Cisplatin. Another careful study focused on the importance of activated CAFs in an inflammatory TME niche in relation to chemoresistance in dormant slow cycling cancer cells. In this study, combined in vitro experiments and in vivo mouse PDX (patient-derived xenograft) tumor model data suggest that CAFs produce collagen I that activate slow cycling cancer cells via β1 integrin and Src to become actively proliferating, leading to tumor relapse (Cho et al. 2020). Again, it would be interesting to identify the specific integrins involved in this mechanism.

11.6 Conclusions

With scRNA-seq-based transcriptional profiles of normal and fibrotic connective tissues accumulating, we will in the next few years see a more detailed stratification of fibroblast and CAF subtypes. The accumulated transcriptional profiles will have to be verified at protein level and these validated profiles will in next step inspire the development of RNA and antibody tools that can be used as "biomarker fingerprints" for the increasing number of fibroblast- and CAF-subsets. These subset-specific profile tools will replace the old concept of "one fibroblast biomarker" and will contribute to a better understanding of the functionally specialized fibroblasts and fibroblast-derived cells. The emerging data suggesting an increasing heterogeneity of CAFs will most likely also be reflected in tissue- and subtype-specific modes of integrin functions. In this context, it will be important to characterize cell- and tissue-specific integrin functions in various CAFs in order to understand how paracrine and direct integrin-mediated interactions cooperate to regulate TME functions. Finally, we foresee that in the coming years, function-blocking integrin antibodies, integrin-based small molecules, as well as integrin antibody-drug conjugates have the potential to increase the efficacy of antifibrosis therapy strategies and to lead to development of improved novel immunotherapy protocols in tumor fibrosis.

Acknowledgments This work was supported by Norwegian Centre of Excellence grant from Research Council of Norway (ID 223250), and MOTIF a Norway-North America grant from Norwegian Centre for International Cooperation in Education (SIU; PNA-2014/10057). Authors also thank Dr. Marion Kusche-Gullberg for stimulating discussions.

References

Ahmed S, Bradshaw AD, Gera S, Dewan MZ, Xu R (2017) The TGF-beta/Smad4 Signaling pathway in pancreatic carcinogenesis and its clinical significance. J Clin Med 6(1):5. https://doi.org/10.3390/jcm6010005

Alexander J, Cukierman E (2016) Stromal dynamic reciprocity in cancer: intricacies of fibroblastic-ECM interactions. Curr Opin Cell Biol 42:80–93. https://doi.org/10.1016/j.ceb.2016.05.002

Allen MD, Marshall JF, Jones JL (2014) alphavbeta6 expression in myoepithelial cells: a novel marker for predicting DCIS progression with therapeutic potential. Cancer Res 74(21): 5942–5947. https://doi.org/10.1158/0008-5472.CAN-14-1841

Asahina K, Zhou B, Pu WT, Tsukamoto H (2011) Septum transversum-derived mesothelium gives rise to hepatic stellate cells and perivascular mesenchymal cells in developing mouse liver. Hepatology 53(3):983–995. https://doi.org/10.1002/hep.24119

Attieh Y, Clark AG, Grass C, Richon S, Pocard M, Mariani P, Elkhatib N, Betz T, Gurchenkov B, Vignjevic DM (2017) Cancer-associated fibroblasts lead tumor invasion through integrin-beta3-dependent fibronectin assembly. J Cell Biol 216(11):3509–3520. https://doi.org/10.1083/jcb.201702033

Bagati A, Kumar S, Jiang P, Pyrdol J, Zou AE, Godicelj A, Mathewson ND, Cartwright ANR, Cejas P, Brown M, Giobbie-Hurder A, Dillon D, Agudo J, Mittendorf EA, Liu XS, Wucherpfennig KW (2021) Integrin alphavbeta6-TGFbeta-SOX4 pathway drives immune evasion in triple-negative breast cancer. Cancer Cell 39(1):54–67.e59. https://doi.org/10.1016/j.ccell.2020.12.001

Barczyk M, Carracedo S, Gullberg D (2010) Integrins. Cell Tissue Res 339(1):269–280. https://doi.org/10.1007/s00441-009-0834-6

Barker HE, Bird D, Lang G, Erler JT (2013) Tumor-secreted LOXL2 activates fibroblasts through FAK signaling. Mol Cancer Res 11(11):1425–1436. https://doi.org/10.1158/1541-7786.MCR-13-0033-T

Bartoschek M, Oskolkov N, Bocci M, Lovrot J, Larsson C, Sommarin M, Madsen CD, Lindgren D, Pekar G, Karlsson G, Ringner M, Bergh J, Bjorklund A, Pietras K (2018) Spatially and functionally distinct subclasses of breast cancer-associated fibroblasts revealed by single cell RNA sequencing. Nat Commun 9(1):5150. https://doi.org/10.1038/s41467-018-07582-3

Bayer SV, Grither WR, Brenot A, Hwang PY, Barcus CE, Ernst M, Pence P, Walter C, Pathak A, Longmore GD (2019) DDR2 controls breast tumor stiffness and metastasis by regulating integrin mediated mechanotransduction in CAFs. elife 8:e45508. https://doi.org/10.7554/eLife.45508

Berger AJ, Renner CM, Hale I, Yang X, Ponik SM, Weisman PS, Masters KS, Kreeger PK (2020) Scaffold stiffness influences breast cancer cell invasion via EGFR-linked Mena upregulation and matrix remodeling. Matrix Biol 85-86:80–93. https://doi.org/10.1016/j.matbio.2019.07.006

Biffi G, Tuveson DA (2018) Deciphering cancer fibroblasts. J Exp Med 215(12):2967–2968. https://doi.org/10.1084/jem.20182069

Brown NF, Marshall JF (2019) Integrin-mediated TGFbeta activation modulates the tumour microenvironment. Cancers (Basel) 11(9):1221. https://doi.org/10.3390/cancers11091221

Buechler MB, Pradhan RN, Krishnamurty AT, Cox C, Calviello AK, Wang AW, Yang YA, Tam L, Caothien R, Roose-Girma M, Modrusan Z, Arron JR, Bourgon R, Muller S, Turley SJ (2021) Cross-tissue organization of the fibroblast lineage. Nature 593(7860):575–579. https://doi.org/10.1038/s41586-021-03549-5

Bui T, Rennhack J, Mok S, Ling C, Perez M, Roccamo J, Andrechek ER, Moraes C, Muller WJ (2019) Functional redundancy between beta1 and beta3 integrin in activating the IR/Akt/mTORC1 signaling axis to promote ErbB2-driven breast cancer. Cell Rep 29(3):589–602.e586. https://doi.org/10.1016/j.celrep.2019.09.004

Cao J, Li J, Sun L, Qin T, Xiao Y, Chen K, Qian W, Duan W, Lei J, Ma J, Ma Q, Han L (2018) Hypoxia-driven paracrine osteopontin/integrin alphavbeta3 signaling promotes pancreatic cancer cell epithelial-mesenchymal transition and cancer stem cell-like properties by modulating FOXM1. Mol Oncol 13(2):228–245. https://doi.org/10.1002/1878-0261.12399

Cavaco ACM, Rezaei M, Caliandro MF, Lima AM, Stehling M, Dhayat SA, Haier J, Brakebusch C, Eble JA (2018) The interaction between laminin-332 and alpha3beta1 integrin determines differentiation and maintenance of CAFs, and supports invasion of pancreatic duct adenocarcinoma cells. Cancers (Basel) 11(1):14. https://doi.org/10.3390/cancers11010014

Cazet AS, Hui MN, Elsworth BL, Wu SZ, Roden D, Chan CL, Skhinas JN, Collot R, Yang J, Harvey K, Johan MZ, Cooper C, Nair R, Herrmann D, McFarland A, Deng N, Ruiz-Borrego M, Rojo F, Trigo JM, Bezares S, Caballero R, Lim E, Timpson P, O'Toole S, Watkins DN, Cox TR, Samuel MS, Martin M, Swarbrick A (2018) Targeting stromal remodeling and cancer stem cell plasticity overcomes chemoresistance in triple negative breast cancer. Nat Commun 9(1):2897. https://doi.org/10.1038/s41467-018-05220-6

Chen X, Song E (2019) Turning foes to friends: targeting cancer-associated fibroblasts. Nat Rev Drug Discov 18(2):99–115. https://doi.org/10.1038/s41573-018-0004-1

Chen WJ, Ho CC, Chang YL, Chen HY, Lin CA, Ling TY, Yu SL, Yuan SS, Chen YJ, Lin CY, Pan SH, Chou HY, Chen YJ, Chang GC, Chu WC, Lee YM, Lee JY, Lee PJ, Li KC, Chen HW, Yang PC (2014) Cancer-associated fibroblasts regulate the plasticity of lung cancer stemness via paracrine signalling. Nat Commun 5:3472. https://doi.org/10.1038/ncomms4472

Chen Y, Kim J, Yang S, Wang H, Wu CJ, Sugimoto H, LeBleu VS, Kalluri R (2021) Type I collagen deletion in alphaSMA(+) myofibroblasts augments immune suppression and accelerates progression of pancreatic cancer. Cancer Cell 39(4):548–565.e546. https://doi.org/10.1016/j.ccell.2021.02.007

Cho J, Lee HJ, Hwang SJ, Min HY, Kang HN, Park AY, Hyun SY, Sim JY, Lee HJ, Jang HJ, Suh YA, Hong S, Shin YK, Kim HR, Lee HY (2020) The interplay between slow-cycling,

chemoresistant cancer cells and fibroblasts creates a proinflammatory niche for tumor progression. Cancer Res 80(11):2257–2272. https://doi.org/10.1158/0008-5472.CAN-19-0631
Chow WY, Forman CJ, Bihan D, Puszkarska AM, Rajan R, Reid DG, Slatter DA, Colwell LJ, Wales DJ, Farndale RW, Duer MJ (2018) Proline provides site-specific flexibility for in vivo collagen. Sci Rep 8(1):13809. https://doi.org/10.1038/s41598-018-31937-x
Corsa CA, Brenot A, Grither WR, Van Hove S, Loza AJ, Zhang K, Ponik SM, Liu Y, DeNardo DG, Eliceiri KW, Keely PJ, Longmore GD (2016) The action of discoidin domain receptor 2 in basal tumor cells and stromal cancer-associated fibroblasts is critical for breast cancer metastasis. Cell Rep 15(11):2510–2523. https://doi.org/10.1016/j.celrep.2016.05.033
Costa A, Kieffer Y, Scholer-Dahirel A, Pelon F, Bourachot B, Cardon M, Sirven P, Magagna I, Fuhrmann L, Bernard C, Bonneau C, Kondratova M, Kuperstein I, Zinovyev A, Givel AM, Parrini MC, Soumelis V, Vincent-Salomon A, Mechta-Grigoriou F (2018) Fibroblast heterogeneity and immunosuppressive environment in human breast cancer. Cancer Cell 33(3):463–479.e10. https://doi.org/10.1016/j.ccell.2018.01.011
Costa-Almeida R, Soares R, Granja PL (2018) Fibroblasts as maestros orchestrating tissue regeneration. J Tissue Eng Regen Med 12(1):240–251. https://doi.org/10.1002/term.2405
Cruz da Silva E, Dontenwill M, Choulier L, Lehmann M (2019) Role of integrins in resistance to therapies targeting growth factor receptors in cancer. Cancers (Basel) 11(5):692. https://doi.org/10.3390/cancers11050692
Dangi-Garimella S, Krantz SB, Barron MR, Shields MA, Heiferman MJ, Grippo PJ, Bentrem DJ, Munshi HG (2011) Three-dimensional collagen I promotes gemcitabine resistance in pancreatic cancer through MT1-MMP-mediated expression of HMGA2. Cancer Res 71(3):1019–1028. https://doi.org/10.1158/0008-5472.CAN-10-1855
Dominguez CX, Muller S, Keerthivasan S, Koeppen H, Hung J, Gierke S, Breart B, Foreman O, Bainbridge TW, Castiglioni A, Senbabaoglu Y, Modrusan Z, Liang Y, Junttila MR, Klijn C, Bourgon R, Turley SJ (2020) Single-cell RNA sequencing reveals stromal evolution into LRRC15(+) myofibroblasts as a determinant of patient response to cancer immunotherapy. Cancer Discov 10(2):232–253. https://doi.org/10.1158/2159-8290.CD-19-0644
Driskell RR, Lichtenberger BM, Hoste E, Kretzschmar K, Simons BD, Charalambous M, Ferron SR, Herault Y, Pavlovic G, Ferguson-Smith AC, Watt FM (2013) Distinct fibroblast lineages determine dermal architecture in skin development and repair. Nature 504(7479):277–281. https://doi.org/10.1038/nature12783
Dulauroy S, Di Carlo SE, Langa F, Eberl G, Peduto L (2012) Lineage tracing and genetic ablation of ADAM12(+) perivascular cells identify a major source of profibrotic cells during acute tissue injury. Nat Med 18:1262–1270. https://doi.org/10.1038/nm.2848
Eberlein C, Rooney C, Ross SJ, Farren M, Weir HM, Barry ST (2015) E-Cadherin and EpCAM expression by NSCLC tumour cells associate with normal fibroblast activation through a pathway initiated by integrin alphavbeta6 and maintained through TGFbeta signalling. Oncogene 34(6):704–716. https://doi.org/10.1038/onc.2013.600
Eckert MA, Coscia F, Chryplewicz A, Chang JW, Hernandez KM, Pan S, Tienda SM, Nahotko DA, Li G, Blazenovic I, Lastra RR, Curtis M, Yamada SD, Perets R, McGregor SM, Andrade J, Fiehn O, Moellering RE, Mann M, Lengyel E (2019) Proteomics reveals NNMT as a master metabolic regulator of cancer-associated fibroblasts. Nature 569(7758):723–728. https://doi.org/10.1038/s41586-019-1173-8
Elyada E, Bolisetty M, Laise P, Flynn WF, Courtois ET, Burkhart RA, Teinor JA, Belleau P, Biffi G, Lucito MS, Sivajothi S, Armstrong TD, Engle DD, Yu KH, Hao Y, Wolfgang CL, Park Y, Preall J, Jaffee EM, Califano A, Robson P, Tuveson DA (2019) Cross-species single-cell analysis of pancreatic ductal adenocarcinoma reveals antigen-presenting cancer-associated fibroblasts. Cancer Discov 9(8):1102–1123. https://doi.org/10.1158/2159-8290.CD-19-0094
Franco-Barraza J, Francescone R, Luong T, Shah N, Madhani R, Cukierman G, Dulaimi E, Devarajan K, Egleston BL, Nicolas E, Katherine Alpaugh R, Malik R, Uzzo RG, Hoffman JP, Golemis EA, Cukierman E (2017) Matrix-regulated integrin alphavbeta5 maintains

alpha5beta1-dependent desmoplastic traits prognostic of neoplastic recurrence. elife 6:e20600. https://doi.org/10.7554/eLife.20600
Friedman SL (2000) Molecular regulation of hepatic fibrosis, an integrated cellular response to tissue injury. J Biol Chem 275(4):2247–2250. https://doi.org/10.1074/jbc.275.4.2247
Friedrichs J, Manninen A, Muller DJ, Helenius J (2008) Galectin-3 regulates integrin alpha2beta1-mediated adhesion to collagen-I and -IV. J Biol Chem 283(47):32264–32272. https://doi.org/10.1074/jbc.M803634200
Fu X, Khalil H, Kanisicak O, Boyer JG, Vagnozzi RJ, Maliken BD, Sargent MA, Prasad V, Valiente-Alandi I, Blaxall BC, Molkentin JD (2018) Specialized fibroblast differentiated states underlie scar formation in the infarcted mouse heart. J Clin Invest 128(5):2127–2143. https://doi.org/10.1172/JCI98215
Fuentes P, Sese M, Guijarro PJ, Emperador M, Sanchez-Redondo S, Peinado H, Hummer S, Ramon YCS (2020) ITGB3-mediated uptake of small extracellular vesicles facilitates intercellular communication in breast cancer cells. Nat Commun 11(1):4261. https://doi.org/10.1038/s41467-020-18081-9
Gaggioli C, Hooper S, Hidalgo-Carcedo C, Grosse R, Marshall JF, Harrington K, Sahai E (2007) Fibroblast-led collective invasion of carcinoma cells with differing roles for RhoGTPases in leading and following cells. Nat Cell Biol 9(12):1392–1400
Garcia PE, Adoumie M, Kim EC, Zhang Y, Scales MK, El-Tawil YS, Shaikh AZ, Wen HJ, Bednar F, Allen BL, Wellik DM, Crawford HC, Pasca di Magliano M (2020) Differential contribution of pancreatic fibroblast subsets to the pancreatic cancer stroma. Cell Mol Gastroenterol Hepatol 10(3):581–599. https://doi.org/10.1016/j.jcmgh.2020.05.004
Gharibi A, La Kim S, Molnar J, Brambilla D, Adamian Y, Hoover M, Hong J, Lin J, Wolfenden L, Kelber JA (2017) ITGA1 is a pre-malignant biomarker that promotes therapy resistance and metastatic potential in pancreatic cancer. Sci Rep 7(1):10060. https://doi.org/10.1038/s41598-017-09946-z
Grither WR, Longmore GD (2018) Inhibition of tumor-microenvironment interaction and tumor invasion by small-molecule allosteric inhibitor of DDR2 extracellular domain. Proc Natl Acad Sci U S A 115(33):E7786–E7794. https://doi.org/10.1073/pnas.1805020115
Guan JL (2010) Integrin signaling through FAK in the regulation of mammary stem cells and breast cancer. IUBMB Life 62(4):268–276. https://doi.org/10.1002/iub.303
Guerrero-Juarez CF, Dedhia PH, Jin S, Ruiz-Vega R, Ma D, Liu Y, Yamaga K, Shestova O, Gay DL, Yang Z, Kessenbrock K, Nie Q, Pear WS, Cotsarelis G, Plikus MV (2019) Single-cell analysis reveals fibroblast heterogeneity and myeloid-derived adipocyte progenitors in murine skin wounds. Nat Commun 10(1):650. https://doi.org/10.1038/s41467-018-08247-x
Gullberg D, Kletsas D, Pihlajaniemi T (2016) Editorial: wound healing and fibrosis-two sides of the same coin. Cell Tissue Res 365(3):449–451. https://doi.org/10.1007/s00441-016-2478-7
Gupton SL, Riquelme D, Hughes-Alford SK, Tadros J, Rudina SS, Hynes RO, Lauffenburger D, Gertler FB (2012) Mena binds alpha5 integrin directly and modulates alpha5beta1 function. J Cell Biol 198(4):657–676. https://doi.org/10.1083/jcb.201202079
Han Y, Zhang Y, Jia T, Sun Y (2015) Molecular mechanism underlying the tumor-promoting functions of carcinoma-associated fibroblasts. Tumour Biol 36(3):1385–1394. https://doi.org/10.1007/s13277-015-3230-8
Harari N, Sakhneny L, Khalifa-Malka L, Busch A, Hertel KJ, Hebrok M, Landsman L (2019) Pancreatic pericytes originate from the embryonic pancreatic mesenchyme. Dev Biol 449(1):14–20. https://doi.org/10.1016/j.ydbio.2019.01.020
He R, Wang M, Zhao C, Shen M, Yu Y, He L, Zhao Y, Chen H, Shi X, Zhou M, Pan S, Liu Y, Guo X, Li X, Qin R (2019) TFEB-driven autophagy potentiates TGF-beta induced migration in pancreatic cancer cells. J Exp Clin Cancer Res 38(1):340. https://doi.org/10.1186/s13046-019-1343-4
Helms EJ, Berry MW, Chaw RC, DuFort CC, Sun D, Onate MK, Oon C, Bhattacharyya S, Sanford-Crane H, Horton W, Finan JM, Sattler A, Makar R, Dawson DW, Xia Z, Hingorani SR, Sherman MH (2021) Mesenchymal lineage heterogeneity underlies non-redundant functions

of pancreatic cancer-associated fibroblasts. Cancer Discov 12(2):484–501. https://doi.org/10.1158/2159-8290.CD-21-0601
Hinz B, Celetta G, Tomasek JJ, Gabbiani G, Chaponnier C (2001) Alpha-smooth muscle actin expression upregulates fibroblast contractile activity. Mol Biol Cell 12(9):2730–2741
Hoshino A, Ishii G, Ito T, Aoyagi K, Ohtaki Y, Nagai K, Sasaki H, Ochiai A (2011) Podoplanin-positive fibroblasts enhance lung adenocarcinoma tumor formation: podoplanin in fibroblast functions for tumor progression. Cancer Res 71(14):4769–4779. https://doi.org/10.1158/0008-5472.CAN-10-3228
Huck L, Pontier SM, Zuo DM, Muller WJ (2010) beta1-integrin is dispensable for the induction of ErbB2 mammary tumors but plays a critical role in the metastatic phase of tumor progression. Proc Natl Acad Sci U S A 107(35):15559–15564. https://doi.org/10.1073/pnas.1003034107
Hurtado de Mendoza T, Mose ES, Botta GP, Braun GB, Kotamraju VR, French RP, Suzuki K, Miyamura N, Teesalu T, Ruoslahti E, Lowy AM, Sugahara KN (2021) Tumor-penetrating therapy for beta5 integrin-rich pancreas cancer. Nat Commun 12(1):1541. https://doi.org/10.1038/s41467-021-21858-1
Ireland LV, Mielgo A (2018) Macrophages and fibroblasts, key players in cancer chemoresistance. Front Cell Dev Biol 6:131. https://doi.org/10.3389/fcell.2018.00131
Ireland L, Santos A, Ahmed MS, Rainer C, Nielsen SR, Quaranta V, Weyer-Czernilofsky U, Engle DD, Perez-Mancera PA, Coupland SE, Taktak A, Bogenrieder T, Tuveson DA, Campbell F, Schmid MC, Mielgo A (2016) Chemoresistance in pancreatic cancer is driven by stroma-derived insulin-like growth factors. Cancer Res 76(23):6851–6863. https://doi.org/10.1158/0008-5472.CAN-16-1201
Ivaska J, Heino J (2011) Cooperation between integrins and growth factor receptors in signaling and endocytosis. Annu Rev Cell Dev Biol 27:291–320. https://doi.org/10.1146/annurev-cellbio-092910-154017
Iwai M, Tulafu M, Togo S, Kawaji H, Kadoya K, Namba Y, Jin J, Watanabe J, Okabe T, Hidayat M, Sumiyoshi I, Itoh M, Koyama Y, Ito Y, Orimo A, Takamochi K, Oh S, Suzuki K, Hayashizaki Y, Yoshida K, Takahashi K (2021) Cancer-associated fibroblast migration in non-small cell lung cancers is modulated by increased integrin alpha11 expression. Mol Oncol 15(5):1507–1527. https://doi.org/10.1002/1878-0261.12937
Jahangiri A, Nguyen A, Chandra A, Sidorov MK, Yagnik G, Rick J, Han SW, Chen W, Flanigan PM, Schneidman-Duhovny D, Mascharak S, De Lay M, Imber B, Park CC, Matsumoto K, Lu K, Bergers G, Sali A, Weiss WA, Aghi MK (2017) Cross-activating c-Met/beta1 integrin complex drives metastasis and invasive resistance in cancer. Proc Natl Acad Sci U S A 114(41): E8685–E8694. https://doi.org/10.1073/pnas.1701821114
Jiang H, Hegde S, Knolhoff BL, Zhu Y, Herndon JM, Meyer MA, Nywening TM, Hawkins WG, Shapiro IM, Weaver DT, Pachter JA, Wang-Gillam A, DeNardo DG (2016) Targeting focal adhesion kinase renders pancreatic cancers responsive to checkpoint immunotherapy. Nat Med 22(8):851–860. https://doi.org/10.1038/nm.4123
Jiang H, Torphy RJ, Steiger K, Hongo H, Ritchie AJ, Kriegsmann M, Horst D, Umetsu SE, Joseph NM, McGregor K, Pishvaian MJ, Blais EM, Lu B, Li M, Hollingsworth M, Stashko C, Volmar K, Yeh JJ, Weaver VM, Wang ZJ, Tempero MA, Weichert W, Collisson EA (2020) Pancreatic ductal adenocarcinoma progression is restrained by stromal matrix. J Clin Invest 130(9):4704–4709. https://doi.org/10.1172/JCI136760
Kai F, Drain AP, Weaver VM (2019) The extracellular matrix modulates the metastatic journey. Dev Cell 49(3):332–346. https://doi.org/10.1016/j.devcel.2019.03.026
Kalluri R (2016) The biology and function of fibroblasts in cancer. Nat Rev Cancer 16(9):582–598. https://doi.org/10.1038/nrc.2016.73
Kanda R, Kawahara A, Watari K, Murakami Y, Sonoda K, Maeda M, Fujita H, Kage M, Uramoto H, Costa C, Kuwano M, Ono M (2013) Erlotinib resistance in lung cancer cells mediated by integrin beta1/Src/Akt-driven bypass signaling. Cancer Res 73(20):6243–6253. https://doi.org/10.1158/0008-5472.CAN-12-4502

Kilvaer TK, Rakaee M, Hellevik T, Ostman A, Strell C, Bremnes RM, Busund LT, Donnem T, Martinez-Zubiaurre I (2018) Tissue analyses reveal a potential immune-adjuvant function of FAP-1 positive fibroblasts in non-small cell lung cancer. PLoS One 13(2):e0192157. https://doi.org/10.1371/journal.pone.0192157

Krebsbach PH, Villa-Diaz LG (2017) The role of integrin alpha6 (CD49f) in stem cells: more than a conserved biomarker. Stem Cells Dev 26(15):1090–1099. https://doi.org/10.1089/scd.2016.0319

Kuninty PR, Bansal R, De Geus SWL, Mardhian DF, Schnittert J, van Baarlen J, Storm G, Bijlsma MF, van Laarhoven HW, Metselaar JM, Kuppen PJK, Vahrmeijer AL, Ostman A, Sier CFM, Prakash J (2019) ITGA5 inhibition in pancreatic stellate cells attenuates desmoplasia and potentiates efficacy of chemotherapy in pancreatic cancer. Sci Adv 5(9):eaax2770. https://doi.org/10.1126/sciadv.aax2770

Kwa MQ, Herum KM, Brakebusch C (2019) Cancer-associated fibroblasts: how do they contribute to metastasis? Clin Exp Metastasis 36(2):71–86. https://doi.org/10.1007/s10585-019-09959-0

Kwon A, Chae IH, You E, Kim SH, Ahn SY, Lee OJ, Park ZY, Rhee S, Huh YH, Song WK (2020) Extra domain A-containing fibronectin expression in Spin90-deficient fibroblasts mediates cancer-stroma interaction and promotes breast cancer progression. J Cell Physiol 235(5):4494–4507. https://doi.org/10.1002/jcp.29326

Lambies G, Miceli M, Martinez-Guillamon C, Olivera-Salguero R, Pena R, Frias CP, Calderon I, Atanassov BS, Dent SYR, Arribas J, Garcia de Herreros A, Diaz VM (2018) TGFbeta-activated USP27X deubiquitinase regulates cell migration and chemoresistance via stabilization of Snail1. Cancer Res 79(1):33–46. https://doi.org/10.1158/0008-5472.CAN-18-0753

Lenos KJ, Miedema DM, Lodestijn SC, Nijman LE, van den Bosch T, Romero Ros X, Lourenco FC, Lecca MC, van der Heijden M, van Neerven SM, van Oort A, Leveille N, Adam RS, de Sousa EMF, Otten J, Veerman P, Hypolite G, Koens L, Lyons SK, Stassi G, Winton DJ, Medema JP, Morrissey E, Bijlsma MF, Vermeulen L (2018) Stem cell functionality is microenvironmentally defined during tumour expansion and therapy response in colon cancer. Nat Cell Biol 20(10):1193–1202. https://doi.org/10.1038/s41556-018-0179-z

Levental KR, Yu H, Kass L, Lakins JN, Egeblad M, Erler JT, Fong SF, Csiszar K, Giaccia A, Weninger W, Yamauchi M, Gasser DL, Weaver VM (2009) Matrix crosslinking forces tumor progression by enhancing integrin signaling. Cell 139(5):891–906. https://doi.org/10.1016/j.cell.2009.10.027

Li R, Bernau K, Sandbo N, Gu J, Preissl S, Sun X (2018) Pdgfra marks a cellular lineage with distinct contributions to myofibroblasts in lung maturation and injury response. elife 7:e36865. https://doi.org/10.7554/eLife.36865

Lu N, Heuchel R, Barczyk M, Zhang WM, Gullberg D (2006) Tandem Sp1/Sp3 sites together with an Ets-1 site cooperate to mediate alpha11 integrin chain expression in mesenchymal cells. Matrix Biol 25(2):118–129. https://doi.org/10.1016/j.matbio.2005.10.002

Lu N, Carracedo S, Ranta J, Heuchel R, Soininen R, Gullberg D (2010) The human alpha11 integrin promoter drives fibroblast-restricted expression in vivo and is regulated by TGF-beta1 in a Smad- and Sp1-dependent manner. Matrix Biol 29(3):166–176. https://doi.org/10.1016/j.matbio.2009.11.003

Marks DL, Olson RL, Fernandez-Zapico ME (2016) Epigenetic control of the tumor microenvironment. Epigenomics 8(12):1671–1687. https://doi.org/10.2217/epi-2016-0110

Martin K, Pritchett J, Llewellyn J, Mullan AF, Athwal VS, Dobie R, Harvey E, Zeef L, Farrow S, Streuli C, Henderson NC, Friedman SL, Hanley NA, Piper Hanley K (2016) PAK proteins and YAP-1 signalling downstream of integrin beta-1 in myofibroblasts promote liver fibrosis. Nat Commun 7:12502. https://doi.org/10.1038/ncomms12502

Meecham A, Marshall J (2021) Harnessing the power of foot-and-mouth-disease virus for targeting integrin alpha-v beta-6 for the therapy of cancer. Expert Opin Drug Discov. 16(7):737–744. https://doi.org/10.1080/17460441.2021.1878143

Miyazaki K, Oyanagi J, Hoshino D, Togo S, Kumagai H, Miyagi Y (2019) Cancer cell migration on elongate protrusions of fibroblasts in collagen matrix. Sci Rep 9(1):292. https://doi.org/10.1038/s41598-018-36646-z

Moore KM, Desai A, Delgado BL, Trabulo SMD, Reader C, Brown NF, Murray ER, Brentnall A, Howard P, Masterson L, Zammarchi F, Hartley JA, van Berkel PH, Marshall JF (2020) Integrin alphavbeta6-specific therapy for pancreatic cancer developed from foot-and-mouth-disease virus. Theranostics 10(7):2930–2942. https://doi.org/10.7150/thno.38702

Moore-Morris T, Cattaneo P, Puceat M, Evans SM (2016) Origins of cardiac fibroblasts. J Mol Cell Cardiol 91:1–5. https://doi.org/10.1016/j.yjmcc.2015.12.031

Morsing M, Klitgaard MC, Jafari A, Villadsen R, Kassem M, Petersen OW, Ronnov-Jessen L (2016) Evidence of two distinct functionally specialized fibroblast lineages in breast stroma. Breast Cancer Res 18(1):108. https://doi.org/10.1186/s13058-016-0769-2

Multhaupt HA, Leitinger B, Gullberg D, Couchman JR (2016) Extracellular matrix component signaling in cancer. Adv Drug Deliv Rev 97:28–40. https://doi.org/10.1016/j.addr.2015.10.013

Munger JS, Huang X, Kawakatsu H, Griffiths MJ, Dalton SL, Wu J, Pittet JF, Kaminski N, Garat C, Matthay MA, Rifkin DB, Sheppard D (1999) The integrin alpha v beta 6 binds and activates latent TGF beta 1: a mechanism for regulating pulmonary inflammation and fibrosis. Cell 96(3): 319–328

Musiime M, Chang J, Hansen U, Kadler KE, Zeltz C, Gullberg D (2021) Collagen assembly at the cell surface: dogmas revisited. Cells 10(3):662. https://doi.org/10.3390/cells10030662

Nagalingam RS, Al-Hattab DS, Czubryt MP (2018) What's in a name? On fibroblast phenotype and nomenclature. Can J Physiol Pharmacol 97(6):493–497. https://doi.org/10.1139/cjpp-2018-0555

Navab R, Strumpf D, Bandarchi B, Zhu CQ, Pintilie M, Ramnarine VR, Ibrahimov E, Radulovich N, Leung L, Barczyk M, Panchal D, To C, Yun JJ, Der S, Shepherd FA, Jurisica I, Tsao MS (2011) Prognostic gene-expression signature of carcinoma-associated fibroblasts in non-small cell lung cancer. Proc Natl Acad Sci U S A 108(17):7160–7165. https://doi.org/10.1073/pnas.1014506108

Navab R, Strumpf D, To C, Pasko E, Kim KS, Park CJ, Hai J, Liu J, Jonkman J, Barczyk M, Bandarchi B, Wang YH, Venkat K, Ibrahimov E, Pham NA, Ng C, Radulovich N, Zhu CQ, Pintilie M, Wang D, Lu A, Jurisica I, Walker GC, Gullberg D, Tsao MS (2016) Integrin alpha11beta1 regulates cancer stromal stiffness and promotes tumorigenicity and metastasis in non-small cell lung cancer. Oncogene 35(15):1899–1908. https://doi.org/10.1038/onc.2015.254

Ndoye A, Miskin RP, DiPersio CM (2021) Integrin alpha3beta1 represses reelin expression in breast cancer cells to promote invasion. Cancers (Basel) 13(2):344. https://doi.org/10.3390/cancers13020344

Nieto MA, Huang RY, Jackson RA, Thiery JP (2016) Emt: 2016. Cell 166(1):21–45. https://doi.org/10.1016/j.cell.2016.06.028

Nurmik M, Ullmann P, Rodriguez F, Haan S, Letellier E (2019) In search of definitions: cancer-associated fibroblasts and their markers. Int J Cancer 146(4):895–905. https://doi.org/10.1002/ijc.32193

Ohlund D, Elyada E, Tuveson D (2014) Fibroblast heterogeneity in the cancer wound. J Exp Med 211(8):1503–1523. https://doi.org/10.1084/jem.20140692

Ohlund D, Handly-Santana A, Biffi G, Elyada E, Almeida AS, Ponz-Sarvise M, Corbo V, Oni TE, Hearn SA, Lee EJ, Chio II, Hwang CI, Tiriac H, Baker LA, Engle DD, Feig C, Kultti A, Egeblad M, Fearon DT, Crawford JM, Clevers H, Park Y, Tuveson DA (2017) Distinct populations of inflammatory fibroblasts and myofibroblasts in pancreatic cancer. J Exp Med 214(3):579–596. https://doi.org/10.1084/jem.20162024

Östman A, Augsten M (2009) Cancer-associated fibroblasts and tumor growth--bystanders turning into key players. Curr Opin Genet Dev 19(1):67–73. https://doi.org/10.1016/j.gde.2009.01.003

Ozdemir BC, Pentcheva-Hoang T, Carstens JL, Zheng X, Wu CC, Simpson TR, Laklai H, Sugimoto H, Kahlert C, Novitskiy SV, De Jesus-Acosta A, Sharma P, Heidari P,

Mahmood U, Chin L, Moses HL, Weaver VM, Maitra A, Allison JP, LeBleu VS, Kalluri R (2014) Depletion of carcinoma-associated fibroblasts and fibrosis induces immunosuppression and accelerates pancreas cancer with reduced survival. Cancer Cell 25:719–734. https://doi.org/10.1016/j.ccr.2014.04.005

Pan B, Liao Q, Niu Z, Zhou L, Zhao Y (2015) Cancer-associated fibroblasts in pancreatic adenocarcinoma. Future Oncol 11(18):2603–2610. https://doi.org/10.2217/FON.15.176

Pantano F, Croset M, Driouch K, Bednarz-Knoll N, Iuliani M, Ribelli G, Bonnelye E, Wikman H, Geraci S, Bonin F, Simonetti S, Vincenzi B, Hong SS, Sousa S, Pantel K, Tonini G, Santini D, Clezardin P (2021) Integrin alpha5 in human breast cancer is a mediator of bone metastasis and a therapeutic target for the treatment of osteolytic lesions. Oncogene 40(7):1284–1299. https://doi.org/10.1038/s41388-020-01603-6

Pastushenko I, Brisebarre A, Sifrim A, Fioramonti M, Revenco T, Boumahdi S, Van Keymeulen A, Brown D, Moers V, Lemaire S, De Clercq S, Minguijon E, Balsat C, Sokolow Y, Dubois C, De Cock F, Scozzaro S, Sopena F, Lanas A, D'Haene N, Salmon I, Marine JC, Voet T, Sotiropoulou PA, Blanpain C (2018) Identification of the tumour transition states occurring during EMT. Nature 556(7702):463–468. https://doi.org/10.1038/s41586-018-0040-3

Philippeos C, Telerman SB, Oules B, Pisco AO, Shaw TJ, Elgueta R, Lombardi G, Driskell RR, Soldin M, Lynch MD, Watt FM (2018) Spatial and single-cell transcriptional profiling identifies functionally distinct human dermal fibroblast subpopulations. J Invest Dermatol 138(4):811–825. https://doi.org/10.1016/j.jid.2018.01.016

Piersma B, Hayward MK, Weaver VM (2020) Fibrosis and cancer: a strained relationship. Biochim Biophys Acta Rev Cancer 1873(2):188356. https://doi.org/10.1016/j.bbcan.2020.188356

Plikus MV, Guerrero-Juarez CF, Ito M, Li YR, Dedhia PH, Zheng Y, Shao M, Gay DL, Ramos R, Hsi TC, Oh JW, Wang X, Ramirez A, Konopelski SE, Elzein A, Wang A, Supapannachart RJ, Lee HL, Lim CH, Nace A, Guo A, Treffeisen E, Andl T, Ramirez RN, Murad R, Offermanns S, Metzger D, Chambon P, Widgerow AD, Tuan TL, Mortazavi A, Gupta RK, Hamilton BA, Millar SE, Seale P, Pear WS, Lazar MA, Cotsarelis G (2017) Regeneration of fat cells from myofibroblasts during wound healing. Science 355(6326):748–752. https://doi.org/10.1126/science.aai8792

Popova SN, Barczyk M, Tiger CF, Beertsen W, Zigrino P, Aszodi A, Miosge N, Forsberg E, Gullberg D (2007) Alpha11 beta1 integrin-dependent regulation of periodontal ligament function in the erupting mouse incisor. Mol Cell Biol 27(12):4306–4316. https://doi.org/10.1128/MCB.00041-07

Primac I, Maquoi E, Blacher S, Heljasvaara R, Van Deun J, Smeland HY, Canale A, Louis T, Stuhr L, Sounni NE, Cataldo D, Pihlajaniemi T, Pequeux C, De Wever O, Gullberg D, Noel A (2019) Stromal integrin alpha11 regulates PDGFR-beta signaling and promotes breast cancer progression. J Clin Invest 129:4609–4628. https://doi.org/10.1172/JCI125890

Raab-Westphal S, Marshall JF, Goodman SL (2017) Integrins as therapeutic targets: successes and cancers. Cancers (Basel) 9(9):110. https://doi.org/10.3390/cancers9090110

Ramirez NE, Zhang Z, Madamanchi A, Boyd KL, O'Rear LD, Nashabi A, Li Z, Dupont WD, Zijlstra A, Zutter MM (2011) The alpha(2)beta(1) integrin is a metastasis suppressor in mouse models and human cancer. J Clin Invest 121(1):226–237. https://doi.org/10.1172/jci42328

Ramovs V, Secades P, Song JY, Thijssen B, Kreft M, Sonnenberg A (2019) Absence of integrin alpha3beta1 promotes the progression of HER2-driven breast cancer in vivo. Breast Cancer Res 21(1):63. https://doi.org/10.1186/s13058-019-1146-8

Raz Y, Cohen N, Shani O, Bell RE, Novitskiy SV, Abramovitz L, Levy C, Milyavsky M, Leider-Trejo L, Moses HL, Grisaru D, Erez N (2018) Bone marrow-derived fibroblasts are a functionally distinct stromal cell population in breast cancer. J Exp Med 215(12):3075–3093. https://doi.org/10.1084/jem.20180818

Reader CS, Vallath S, Steele CW, Haider S, Brentnall A, Desai A, Moore KM, Jamieson NB, Chang D, Bailey P, Scarpa A, Lawlor R, Chelala C, Keyse SM, Biankin A, Morton JP, Evans TJ, Barry ST, Sansom OJ, Kocher HM, Marshall JF (2019) The integrin alphavbeta6 drives pancreatic cancer through diverse mechanisms and represents an effective target for therapy. J Pathol 249(3):332–342. https://doi.org/10.1002/path.5320

Reed NI, Jo H, Chen C, Tsujino K, Arnold TD, DeGrado WF, Sheppard D (2015) The alphavbeta1 integrin plays a critical in vivo role in tissue fibrosis. Sci Transl Med 7(288):288ra279. https://doi.org/10.1126/scitranslmed.aaa5094

Rinn JL, Bondre C, Gladstone HB, Brown PO, Chang HY (2006) Anatomic demarcation by positional variation in fibroblast gene expression programs. PLoS Genet 2(7):e119

Roberts EW, Deonarine A, Jones JO, Denton AE, Feig C, Lyons SK, Espeli M, Kraman M, McKenna B, Wells RJ, Zhao Q, Caballero OL, Larder R, Coll AP, O'Rahilly S, Brindle KM, Teichmann SA, Tuveson DA, Fearon DT (2013) Depletion of stromal cells expressing fibroblast activation protein-alpha from skeletal muscle and bone marrow results in cachexia and anemia. J Exp Med 210(6):1137–1151. https://doi.org/10.1084/jem.20122344

Rybinski B, Franco-Barraza J, Cukierman E (2014) The wound healing, chronic fibrosis, and cancer progression triad. Physiol Genomics 46(7):223–244. https://doi.org/10.1152/physiolgenomics.00158.2013

Schnittert J, Bansal R, Storm G, Prakash J (2018) Integrins in wound healing, fibrosis and tumor stroma: high potential targets for therapeutics and drug delivery. Adv Drug Deliv Rev 129:37–53. https://doi.org/10.1016/j.addr.2018.01.020

Schnittert J, Bansal R, Mardhian DF, van Baarlen J, Ostman A, Prakash J (2019) Integrin alpha11 in pancreatic stellate cells regulates tumor stroma interaction in pancreatic cancer. FASEB J 33(5):6609–6621. https://doi.org/10.1096/fj.201802336R

Schulz JN, Plomann M, Sengle G, Gullberg D, Krieg T, Eckes B (2018) New developments on skin fibrosis - essential signals emanating from the extracellular matrix for the control of myofibroblasts. Matrix Biol 68–69:522–532. https://doi.org/10.1016/j.matbio.2018.01.025

Semba T, Funahashi Y, Ono N, Yamamoto Y, Sugi NH, Asada M, Yoshimatsu K, Wakabayashi T (2004) An angiogenesis inhibitor E7820 shows broad-spectrum tumor growth inhibition in a xenograft model: possible value of integrin alpha2 on platelets as a biological marker. Clin Cancer Res 10(4):1430–1438. https://doi.org/10.1158/1078-0432.ccr-0109-03

Shibuya T, Honma M, Fujii M, Iinuma S, Ishida-Yamamoto A (2018) Podoplanin suppresses the cell adhesion of epidermal keratinocytes via functional regulation of beta1-integrin. Arch Dermatol Res 311(1):45–53. https://doi.org/10.1007/s00403-018-1878-9

Smeland HY, Askeland C, Wik E, Knutsvik G, Molven A, Edelmann RJ, Reed RK, Warren DJ, Gullberg D, Stuhr L, Akslen LA (2020) Integrin alpha11beta1 is expressed in breast cancer stroma and associates with aggressive tumor phenotypes. J Pathol Clin Res 6(1):69–82. https://doi.org/10.1002/cjp2.148

Stokes JB, Adair SJ, Slack-Davis JK, Walters DM, Tilghman RW, Hershey ED, Lowrey B, Thomas KS, Bouton AH, Hwang RF, Stelow EB, Parsons JT, Bauer TW (2011) Inhibition of focal adhesion kinase by PF-562,271 inhibits the growth and metastasis of pancreatic cancer concomitant with altering the tumor microenvironment. Mol Cancer Ther 10(11):2135–2145. https://doi.org/10.1158/1535-7163.MCT-11-0261

Su SC, Chen JN, Yao HR, Liu J, Yu SB, Lao LY, Wang MH, Luo ML, Xing Y, Chen F, Huang D, Zhao JH, Yang LB, Liao D, Su FX, Li MF, Liu Q, Song EW (2018) CD10(+) GPR77(+) cancer-associated fibroblasts promote cancer formation and chemoresistance by sustaining cancer stemness. Cell 172(4):841–856.e16. https://doi.org/10.1016/j.cell.2018.01.009

Tao L, Huang G, Wang R, Pan Y, He Z, Chu X, Song H, Chen L (2016) Cancer-associated fibroblasts treated with cisplatin facilitates chemoresistance of lung adenocarcinoma through IL-11/IL-11R/STAT3 signaling pathway. Sci Rep 6:38408. https://doi.org/10.1038/srep38408

Tod J, Hanley CJ, Morgan MR, Rucka M, Mellows T, Lopez MA, Kiely P, Moutasim KA, Frampton SJ, Sabnis D, Fine DR, Johnson C, Marshall JF, Scita G, Jenei V, Thomas GJ (2017) Pro-migratory and TGF-beta-activating functions of alphavbeta6 integrin in pancreatic cancer are differentially regulated via an Eps8-dependent GTPase switch. J Pathol 243(1):37–50. https://doi.org/10.1002/path.4923

Tomasek JJ, Gabbiani G, Hinz B, Chaponnier C, Brown RA (2002) Myofibroblasts and mechano-regulation of connective tissue remodelling. Nat Rev Mol Cell Biol 3(5):349–363

Tsukui T, Sun KH, Wetter JB, Wilson-Kanamori JR, Hazelwood LA, Henderson NC, Adams TS, Schupp JC, Poli SD, Rosas IO, Kaminski N, Matthay MA, Wolters PJ, Sheppard D (2020) Collagen-producing lung cell atlas identifies multiple subsets with distinct localization and relevance to fibrosis. Nat Commun 11(1):1920. https://doi.org/10.1038/s41467-020-15647-5

Uhlen M, Zhang C, Lee S, Sjostedt E, Fagerberg L, Bidkhori G, Benfeitas R, Arif M, Liu Z, Edfors F, Sanli K, von Feilitzen K, Oksvold P, Lundberg E, Hober S, Nilsson P, Mattsson J, Schwenk JM, Brunnstrom H, Glimelius B, Sjoblom T, Edqvist PH, Djureinovic D, Micke P, Lindskog C, Mardinoglu A, Ponten F (2017) A pathology atlas of the human cancer transcriptome. Science 357(6352):eaan2507. https://doi.org/10.1126/science.aan2507

Walsh C, Tanjoni I, Uryu S, Tomar A, Nam JO, Luo H, Phillips A, Patel N, Kwok C, McMahon G, Stupack DG, Schlaepfer DD (2010) Oral delivery of PND-1186 FAK inhibitor decreases tumor growth and spontaneous breast to lung metastasis in pre-clinical models. Cancer Biol Ther 9(10):778–790

Wang KK, Liu N, Radulovich N, Wigle DA, Johnston MR, Shepherd FA, Minden MD, Tsao MS (2002) Novel candidate tumor marker genes for lung adenocarcinoma. Oncogene 21(49): 7598–7604

Wendt MK, Schiemann WP (2009) Therapeutic targeting of the focal adhesion complex prevents oncogenic TGF-beta signaling and metastasis. Breast Cancer Res 11(5):R68. https://doi.org/10.1186/bcr2360

Westcott JM, Prechtl AM, Maine EA, Dang TT, Esparza MA, Sun H, Zhou Y, Xie Y, Pearson GW (2015) An epigenetically distinct breast cancer cell subpopulation promotes collective invasion. J Clin Invest 125(5):1927–1943. https://doi.org/10.1172/JCI77767

Woltersdorf C, Bonk M, Leitinger B, Huhtala M, Kapyla J, Heino J, Gil Girol C, Niland S, Eble JA, Bruckner P, Dreier R, Hansen U (2017) The binding capacity of alpha1beta1-, alpha2beta1- and alpha10beta1-integrins depends on non-collagenous surface macromolecules rather than the collagens in cartilage fibrils. Matrix Biol 63:91–105. https://doi.org/10.1016/j.matbio.2017.02.001

Wu Y, Zhang C, Jiang K, Werner J, Bazhin AV, D'Haese JG (2020) The role of stellate cells in pancreatic ductal adenocarcinoma: targeting perspectives. Front Oncol 10:621937. https://doi.org/10.3389/fonc.2020.621937

Xie T, Wang Y, Deng N, Huang G, Taghavifar F, Geng Y, Liu N, Kulur V, Yao C, Chen P, Liu Z, Stripp B, Tang J, Liang J, Noble PW, Jiang D (2018) Single-cell deconvolution of fibroblast heterogeneity in mouse pulmonary fibrosis. Cell Rep 22(13):3625–3640. https://doi.org/10.1016/j.celrep.2018.03.010

Xu H, Bihan D, Chang F, Huang PH, Farndale RW, Leitinger B (2012) Discoidin domain receptors promote alpha1beta1- and alpha2beta1-integrin mediated cell adhesion to collagen by enhancing integrin activation. PLoS One 7(12):e52209. https://doi.org/10.1371/journal.pone.0052209

Zeltz C, Orgel J, Gullberg D (2014) Molecular composition and function of integrin-based collagen glues-introducing COLINBRIs. Biochim Biophys Acta 1840(8):2533–2548. https://doi.org/10.1016/j.bbagen.2013.12.022

Zeltz C, Alam J, Liu H, Erusappan PM, Hoschuetzky H, Molven A, Parajuli H, Cukierman E, Costea DE, Lu N, Gullberg D (2019) alpha11beta1 integrin is induced in a subset of cancer-associated fibroblasts in desmoplastic tumor stroma and mediates in vitro cell migration. Cancers (Basel) 11(6):765. https://doi.org/10.3390/cancers11060765

Zeltz C, Primac I, Erusappan P, Alam J, Noel A, Gullberg D (2020) Cancer-associated fibroblasts in desmoplastic tumors: emerging role of integrins. Semin Cancer Biol 62:166–181. https://doi.org/10.1016/j.semcancer.2019.08.004

Zhang K, Grither WR, Van Hove S, Biswas H, Ponik SM, Eliceiri KW, Keely PJ, Longmore GD (2016) Mechanical signals regulate and activate SNAIL1 protein to control the fibrogenic response of cancer-associated fibroblasts. J Cell Sci 129(10):1989–2002. https://doi.org/10.1242/jcs.180539

Zhang Q, Yang J, Bai J, Ren J (2018a) Reverse of non-small cell lung cancer drug resistance induced by cancer-associated fibroblasts via a paracrine pathway. Cancer Sci 109(4):944–955. https://doi.org/10.1111/cas.13520

Zhang W, Bouchard G, Yu A, Shafiq M, Jamali M, Shrager JB, Ayers K, Bakr S, Gentles AJ, Diehn M, Quon A, West RB, Nair V, van de Rijn M, Napel S, Plevritis SK (2018b) GFPT2-expressing cancer-associated fibroblasts mediate metabolic reprogramming in human lung adenocarcinoma. Cancer Res 78(13):3445–3457. https://doi.org/10.1158/0008-5472.CAN-17-2928

Zhu CQ, Popova SN, Brown ER, Barsyte-Lovejoy D, Navab R, Shih W, Li M, Lu M, Jurisica I, Penn LZ, Gullberg D, Tsao MS (2007) Integrin alpha 11 regulates IGF2 expression in fibroblasts to enhance tumorigenicity of human non-small-cell lung cancer cells. Proc Natl Acad Sci U S A 104(28):11754–11759. https://doi.org/10.1073/pnas.0703040104

Zhu J, Hoop CL, Case DA, Baum J (2018) Cryptic binding sites become accessible through surface reconstruction of the type I collagen fibril. Sci Rep 8(1):16646. https://doi.org/10.1038/s41598-018-34616-z

Chapter 12
Adipose Compounds in Breast Tumor Extracellular Matrix

Flavia Piccioni and Paola De Luca

Abstract Adipocytes constitute the major cell population of adipose tissue (AT), which in turn is an important breast tissue component. Many epidemiological studies support that AT plays a significant role in breast cancer (BCa) development and progression. In fact, several conditions associated with an increase of AT, such as obesity or waist-to-hip ratio, impact on BCa risk, and outcomes. Besides, in the last years, it has been demonstrated that invasive BCa cells modify adjacent adipocytes, becoming cancer-associated adipocytes (CAAs). These cells differ phenotypic and functionally from normal adipocytes, and secrete factors such as adipokines, inflammatory molecules, and microRNAs (miRNAs) that promote a pro-tumoral microenvironment. Moreover, CAAs can regulate processes such as fat distribution, insulin secretion, energy consumption, and inflammatory reactions. Importantly, studies have revealed that a cyclic crosstalk between adipocytes and BCa cells is established, from which breast tumor feeds back in order to proliferate, invade, and metastasize. Additionally, the altered AT represents a barrier to resistance for BCa treatment. Thus, therapies targeting AT in the context of BCa are currently the subject of extensive study. In this chapter, we will focus on the signaling molecules and the feedback that is established between BCa cells and AT, which allow BCa progression.

Abbreviations

AKT	AKT serine/threonine kinase
AMPK	Protein kinase AMP-activated catalytic subunit alpha 1
AT	Adipose tissue
ATX	Autotaxin
BAT	Brown adipose tissue

F. Piccioni (✉) · P. De Luca
Laboratorio de Oncología Molecular y Nuevos Blancos Terapéuticos, Instituto de Biología y Medicina Experimental (IBYME), CONICET, Buenos Aires City, Argentina

I. Kovalszky et al. (eds.), *The Extracellular Matrix and the Tumor Microenvironment*, Biology of Extracellular Matrix 11,
https://doi.org/10.1007/978-3-030-99708-3_12

BCa	Breast cancer
BMI	Body mass index
C/EBPα	CCAAT/enhancer-binding protein α
CAAs	Cancer associated adipocytes
CAFs	Cancer associated fibroblast
CAP1	Adenylyl cyclase-associated protein 1
CCL2	Chemokine (C–C motif) ligand 2
CCL5	Chemokine (C–C motif) ligand 5
CCR2	C–C motif chemokine receptor 2
CCR4	C–C motif chemokine receptor 4
CCR5	C–C motif chemokine receptor 5
CSCs	Cancer stem cells
ECM	Extracellular matrix
EMT	Epithelial to mesenchymal transition
ER−	Estrogen receptor negative
ER	Estrogen receptor
ER+	Estrogen receptor positive
FABP4	Fatty-acid binding protein 4
FAs	Fatty acids
FFAs	Free fatty acids
FOXP4	Forkhead box P4
GLUT4	Glucose transporter type 4
HER2	Human epidermal growth factor receptor 2
HGF	Hepatocyte growth factor
IGF	Insulin growth factor
IGF-1	Insulin-like growth factor 1
IGF-1R	Insulin-like growth factor 1 receptor
IL-1	Interleukin-1
IL-1R	Interleukin-1 receptor
IL-1β	Interleukin-1 beta
IL-6	Interleukin-6
IL-8	Interleukin-8
JAK	Janus kinase
JAK2	Janus kinase 2
LDL	Low density lipoproteins
LPA	Lysophosphatidic acid
M1	Macrophage type 1
MCP-1	Monocyte chemoattractant protein-1
MCT1	MCTS1 re-initiation and release factor
MCT4	MCTS4 re-initiation and release factor
MeS	Metabolic syndrome
miRNAs	microRNAs
MMP11	Stromelysin-3
MMPs	Matrix metalloproteases

mTOR	Mechanistic target of rapamycin kinase
MVP	Major vault protein
NF-κB	Nuclear factor kappa-B
NGF	Nerve growth factor
NOTCH1	Notch receptor 1
ObR	Leptin receptor
p16INK4A	CDKN2A, cyclin dependent kinase inhibitor 2A
PAT	Pink adipose tissue
PDGF	Platelet-derived growth factor
PI3K	Phosphatidylinositol-4,5-bisphosphate 3-kinase
PPARγ	Peroxisome proliferator-activated receptor-γ
RANTES	Chemokine (C–C motif) ligand 5
S100B	S100 calcium binding protein B
SFRP1	Secreted Frizzled Related Protein 1
SPP1	Secreted phosphoprotein 1
STAT3	Signal transducer and activator of transcription 3
TNBC	Triple negative breast cancer
TNF-α	Tumor necrosis factor-alpha
UCP-1	Uncoupling protein 1
VEGF	Vascular endothelial growth factor
VEGFR	Vascular endothelial growth factor receptor
WAT	White adipose tissue
Wnt3a	Wnt family member 3A
Wnt5a	Wnt family member 5A

12.1 Introduction

According to GLOBOCAN 2018, BCa is the first cancer in incidence and mortality in women (Bray et al. 2018). Even when hereditary factors impact on BCa risk and progression, the nonhereditary are the main drivers of BCa disease (Czene et al. 2002; Willett 2002), some of which are also modifiable factors to be considered for cancer prevention and treatment. Many of these modifiable factors are related to AT content, for example, greater weight, weight gain during adulthood, or body fat distribution (Bray et al. 2018). Thus, obesity, body mass index (BMI), waist-to-hip ratio, central obesity, body fat content, and metabolic syndrome (MeS) influence BCa risk, development, and progression.

Extracellular matrix (ECM) composition has a key role in cancer development and progression and, conversely, it is well known that tumor cells modify ECM to enable their own survival, proliferation, and progression. In addition, ECM components determine the proper adipogenesis and AT integrity. Thus, by modifying ECM, tumor cells alter the surrounding adipocyte homeostasis and induce the transformation of normal adipocytes to CAAs, with low fat and mature adipocyte markers and overexpression of adipokines, inflammatory cytokines, and high energy metabolites, among other factors (Zhao et al. 2020). Considering AT represents around 56% of breast in nonlactating women (Zhao et al. 2020), this reciprocal interaction between cancer cells and adipocytes becomes more important and understanding the intricated molecular mechanism behind this phenomenon could determine new therapeutic approaches. Also, even in physiological conditions, the breast is a dynamic tissue that suffers several remodeling across women life and where the crosstalk between epithelial cells and adipocytes is fundamental in maintaining their proper development and modifications and conversely a disruption of this dialog could be cause of disease, including BCa (Zwick et al. 2018).

In this chapter, we will discuss the dynamic crosstalk between adipocytes and BCa cells with special emphasis in adipocyte secretions that modify ECM impacting on BCa survival, growth, progression, and resistance to therapy.

12.2 Breast Adipocytes

AT is composed by adipocytes (mature or preadipocytes), endothelial cells, fibroblast, pericytes, stem and progenitor cells, immune cells (macrophages, dendritic cells, mast cells, eosinophils, neutrophils, and lymphocytes), and ECM (Coelho et al. 2013; Miana and Prieto González 2018). According to its location, AT is subdivided in subcutaneous, visceral, intradermic, bone marrow-associated and intramuscular (Ibrahim 2010), been mammary AT an especial subtype of subcutaneous AT.

Adipocytes constitute the major population of AT (Coelho et al. 2013) and deserve special attention since they share a complex and dynamic crosstalk with epithelial cells in mammary gland both in physiological and pathological conditions.

Adipogenesis is a complex process with two master regulators, the nuclear receptor-peroxisome proliferator-activated receptor-γ (PPARγ) and the transcription coactivator CCAAT/enhancer-binding protein α (C/EBPα) (Rosen et al. 2000). Lipid accumulation, adipocyte fatty-acid binding protein 4 (FABP4), and insulin-responsive glucose transporter type 4 (GLUT4) expression determine early stage of differentiation, while adiponectin, leptin, adipose triglyceride lipase, lipoprotein lipase, and perilipin 1 are markers of mature adipocytes (Rybinska et al. 2020). Considering the presence of lipid droplets and mitochondrial content, adipocytes can be divided in white, brown, and beige or brown-like. White AT (WAT) is the most abundant in the body and, in particular, in mammary AT. It has a large lipid droplet and functions as an energy storage but also produce pro-inflammatory molecules and adipokines which gives it endocrine properties in response to inflammatory changes (Kothari et al. 2020). At molecular level, WAT is characterized by leptin and S100 calcium binding protein B (S100B) but not uncoupling protein 1 (UCP-1) expression. Brown AT (BAT) contains small lipid droplets and large spherical and packed iron-enriched mitochondria and contributes to the maintenance of body temperature through mitochondrial UCP-1 protein, which is its molecular marker. Finally, beige adipocytes retain both functions, energy storage and thermogenesis, and express

UCP-1, PPARγ, and leptin and has high mitochondrial content compared to WAT. No matter the subtype, AT is a complex and heterogeneous tissue that could present particular properties according to the specific location or to changes in the microenvironment. Thus, conversion of WAT to beige AT has been reported in response to cold and is called browning (Bartelt and Heeren 2014) and conversely, whitening is the conversion of BAT to beige and has been also observed (Kotzbeck et al. 2018). Also, WAT can transdifferentiate to BAT in response to cold (Himms-Hagen et al. 2000; Peres Valgas da Silva et al. 2019). In physiological conditions, mammary gland epithelium is a plastic tissue that undergoes various remodeling. After birth, mammary gland development and maturation continue during prepuberal to puberal life, and after menarque, respectively. Later in adulthood, remodeling could occur during pregnancy, lactation, and involution (postlactation and age-related). Remarkably, several studies showed that mammary AT interaction with epithelial cells is essential in all these processes (Zwick et al. 2018; Kothari et al. 2020). As well reviewed by Kothari et al., several studies demonstrated that dialogue between mammary adipocytes and epithelial cells is essential for the development of mammary gland (Howlett and Bissell 1993; Zangani et al. 1999). For example, mammary gland development comprises a prepuberal to puberal stage, where ductal branching and terminal end buds take place and, a maturation stage, where alveolar buds develop side branching that occurs upon menarque. It was reported that loss of mammary gland adipocytes causes a diminution of these processes (Landskroner-Eiger et al. 2010).

Pregnancy, lactation, and involution postlactation offer other examples of the dynamic interaction between adipose tissue and epithelial cells in mammary gland and were also well reviewed by Kothari et al. During pregnancy, alveolar development starts and WAT phenotype is modified. Morroni et al. demonstrated that WAT transdifferentiate to pink AT (PAT) with milk secretory capacity (Kothari et al. 2020; Morroni et al. 2004) in a process dependent on the transcription factor secreted phosphoprotein 1 (SPP1) (Prokesch et al. 2014). Also, it has been reported that WAT remains present during pregnancy but with a smaller size (Cinti 2018) but which is the predominant phenomenon is under discussion (Rybinska et al. 2020). Nonetheless, WAT tissue is required for normal alveolar formation and lactation, since it has been reported that deletion of WAT inhibits both processes (Vitali et al. 2012). It has been also demonstrated that BAT transdifferentiate to myoepithelial cells during pregnancy and lactation (L. Li et al. 2017). According to transdifferentiation model, during postlactation involution, PAT must transdifferentiate to WAT, which is dependent on PPARγ (Apostoli et al. 2014). Also, adipocytes with smaller size return to their normal shape by refilling their cytoplasm with lipids (Combs et al. 2004). On the other hand, Secreted Frizzled Related Protein 1 (SFRP1) secretion from adipocytes contributes to remodeling inducing apoptosis of epithelial cells from milk lobules (Clemenceau et al. 2020; Zheng et al. 2017).

Importantly, the aberrant expression of adipocyte markers associated to adipogenesis, breast development, pregnancy, lactation or involution of the mammary gland was also reported in BCa, indicating that AT altered function could be associated to tumor development. For example, it was reported that PPARγ and

C/EBPα are tumor suppressors for BCa development (Apostoli et al. 2014; Lourenço and Coffer 2017; Liu et al. 2019). Also, several transcription factors with functions in adipogenesis, either promoting or inhibiting it, play roles in tumor development and progression (Kothari et al. 2020). There was observed that high expression levels of BAT and beige AT markers in BCa xenografts favor tumor development (Singh et al. 2016).

12.3 Clinical Association Between BCa and Obesity

A clear demonstration that AT has a critical role in BCa development and progression comes from epidemiological studies regarding BCa risk and mortality in patients. Thus, several conditions associated with an increase in AT have an effect in BCa epidemiology (Burton and Foster 1985; Kolb and Zhang 2020; Dibaba et al. 2019) and deserve to be taken into account because they are also usually modifiable risk factors.

Association between overweight (BMI $= 25–30$ kg/m^2) and obesity (BMI ≥ 30 kg/m^2) and BCa was well studied (Burton and Foster 1985; Kolb and Zhang 2020; Cheraghi et al. 2012; Renehan et al. 2008). Thus, several studies demonstrated that BMI is a risk factor for BCa in postmenopausal women (Renehan et al. 2008; Lauby-Secretan et al. 2016). Importantly, a study with more than 87,000 women and another with about 250,000 postmenopausal women have shown that high body weight is associated with increased risk of estrogen receptor positive (ER +) BCa in postmenopausal women (Eliassen et al. 2006; McKenzie et al. 2015). In particular, body weight increasement after menopause raises BCa risk and this association is strong in obese women (Eliassen et al. 2006). When analyzed the role of BMI according cancer subtypes, increased BMI presented stronger association with hormone receptor positive BCa (Yang et al. 2011; Kolb and Zhang 2020; Vrieling et al. 2010).

Further studies should be carried out to determine BMI effect on premenopausal BCa risk. The fact that studies performed in both premenopausal and postmenopausal women show similar results compared to those carried out only in postmenopausal women suggest that overweight is an important factor for both populations (Renehan et al. 2008). However, different results about BMI effect were found in studies performed in premenopausal women. A meta-analysis study showed that BMI is associated with 20% lower risk for ER+ BCa with no association for ER- BCa in premenopausal women (Suzuki et al. 2009). On the other hand, a more recent meta-analysis shows that BMI increases the risk of triple negative BCa (TNBC) in premenopausal women (Pierobon and Frankenfeld 2013). In addition, several studies show that obesity is associated with a BCa worse outcome (Kolb and Zhang 2020). A meta-analyses study demonstrated that BMI increases both all-cause mortality and breast cancer-specific mortality (Chan et al. 2014). Also, obesity constitutes an independent prognostic factor for distant metastases and for death in BCa (Ewertz et al. 2011). A recent meta-analysis including 52,904 patients demonstrated that BMI increases the lymph node metastasis risk in BCa (Wang et al. 2020). The increase in body weight after diagnosis elevates overall mortality rate (Playdon et al. 2015). Finally, obesity is linked with an increased risk to develop secondary cancer (Druesne-Pecollo et al. 2012).

Considering visceral fat is associated with metabolic disease risk (Neeland et al. 2013) and woman with normal BMI could present visceral fat depots, recent clinical studies also evaluate waist-to-hip ratio. An interesting systematic review showed that while BMI has a neutral or an inverse effect in premenopausal women, central obesity increases their risk of BCa; and for postmenopausal women, obesity is a BCa risk factor but visceral obesity has no effect on BCa risk (Harvie et al. 2003). However, a recent meta-analyses review revealed that waist-to-hip-ratio increases BCa risk and this association is also significant when analyzed only postmenopausal women (Connolly et al. 2002). In concordance with these results, women with a normal BMI but relatively high levels of body fat measured by dual-energy x-ray absorptiometry present increased risk of invasive BCa (Iyengar et al. 2019). Also, elevated waist-to-hip ratio is associated with ER+ BCa mortality in post-menopausal women (Borugian et al. 2003). Finally, it was shown that weight gaining in adulthood elevates BCa risk in women with fattier breast suggesting that a minimum of breast fat is necessary for BCa development and the important role of mammary AT besides visceral fat in BCa (Soguel and Diorio 2016).

Interestingly, a study in 21,000 postmenopausal women investigating obesity in addition with MeS revealed that obesity is a risk factor for BCa but in combination with MeS the risk is even higher (Kabat et al. 2017). MeS is a group of pathophysiological disorders with visceral obesity as a component as comprise at least three of the following characteristics: abdominal obesity (circumference $\geq$ 35 inches in women) (Third Report of the National Cholesterol Education Program (NCEP) Expert Panel on Detection, Evaluation, and Treatment of High Blood Cholesterol in Adults (Adult Treatment Panel III) Final Report 2002), hyperglycemia ($\geq$110 mg/dL), blood triglycerides $\geq$150 mg/dL (in women), HDL-C cholesterol <50 mg/dL (in women), and hypertension ($\geq$130/85 mmHg) (Alberti et al. 2009). A systematic meta-analysis study demonstrated that MeS is a risk factor for BCa (Bhandari et al. 2014), being this association even greater in the subpopulation of postmenopausal women (Capasso et al. 2010). Also, evidences indicate that MeS is associated with a high grade of breast tumors (Reeves et al. 2007; Cleveland et al. 2007) and with a higher rate of recurrence and metastasis of BCa (Cao et al. 2015). Importantly, a recent study carried out with data from 94,555 women, demonstrated that MeS is significant associated with BCa mortality risk in postmenopausal women (Dibaba et al. 2019).

Altogether, epidemiological studies suggest that AT tissue; regardless being total, visceral or mammary AT; plays an important role in BCa development and progression.

12.4 CAAs and Crosstalk with BCa

Since adipocytes are the most abundant component of the breast tissue stroma, they make a relevant contribution to cancer progression. As described above, epidemiological evidence has demonstrated the strong relation between obesity and BCa. It is widely known the importance of cancer-adipocyte crosstalk in the modulation of tumor behavior. Consequently, it has been field of extensive study.

Importantly, a few years ago, a new concept has emerged since it was demonstrated that invasive cancer cells can modify adjacent adipocytes, currently called CAAs, which display different characteristics from normal adipocytes. These modified adipocytes, properly located at the invasive front of breast tumors, exhibit fibroblast-like phenotype, smaller size, dispersed small lipid droplets, overexpression of collagen VI, and low expression of adipokines, such as adiponectin, compared to normal adipocytes. Also, CAAs display a different secretome than mature adipocytes: they produce more chemokine (C–C motif) ligand 2 (CCL2) (Fletcher et al. 2017), chemokine (C–C motif) ligand 5 (CCL5) (Cha and Koo 2018), autotaxin (ATX), and pro-inflammatory cytokines such as interleukin-1 (IL-1), interleukin-1β (IL-1β), interleukin-6 (IL-6) (Bray et al. 2018), tumor necrosis factor-alpha (TNF-α), vascular endothelial growth factor (VEGF), leptin (Pallegar and Christian 2020), etc., all molecules that stimulate invasion and metastasis of BCa. Furthermore, CAAs present reduced expression of the adipocyte differentiation markers PPARγ and C/EBPα as well as their downstream targets FABP4, adiponectin, and hormone-sensitive lipase. The mechanism involved in dedifferentiation of mature adipocytes induced by tumor cells is still unknown, although TNF-α (Mauro et al. 2018), Wnt family member 3A (Wnt3a), Wnt family member 5A (Wnt5a) (Theriau et al. 2017), and stromelysin-3 (MMP11) (Ollberding et al. 2013), all molecules secreted by tumor cells have been related to this process. Interestingly, some authors have proposed that CAAs represent an intermediate form of cancer-associated fibroblast (CAFs) that are the main population of stromal cells in BCa (Bochet et al. 2013).

The interaction between BCa cells and CAAs fosters a pro-oncogenic tumor microenvironment, promoting proliferation, angiogenesis, invasion, and metastasis. Along diverse studies, it has been demonstrated that CAAs establish the communication with BCa cells by releasing soluble factors such as adipokines, impacting in tumor remodeling (Fig. 12.1).

12.4.1 Adipokines Produced by CAAs

As said before, crosstalk between BCa cells and adipocytes is established by soluble factors, and favors proliferation, survival, and metastasis of BCa. The whole set of factors produced by adipocytes, which include more than 600 metabolites,

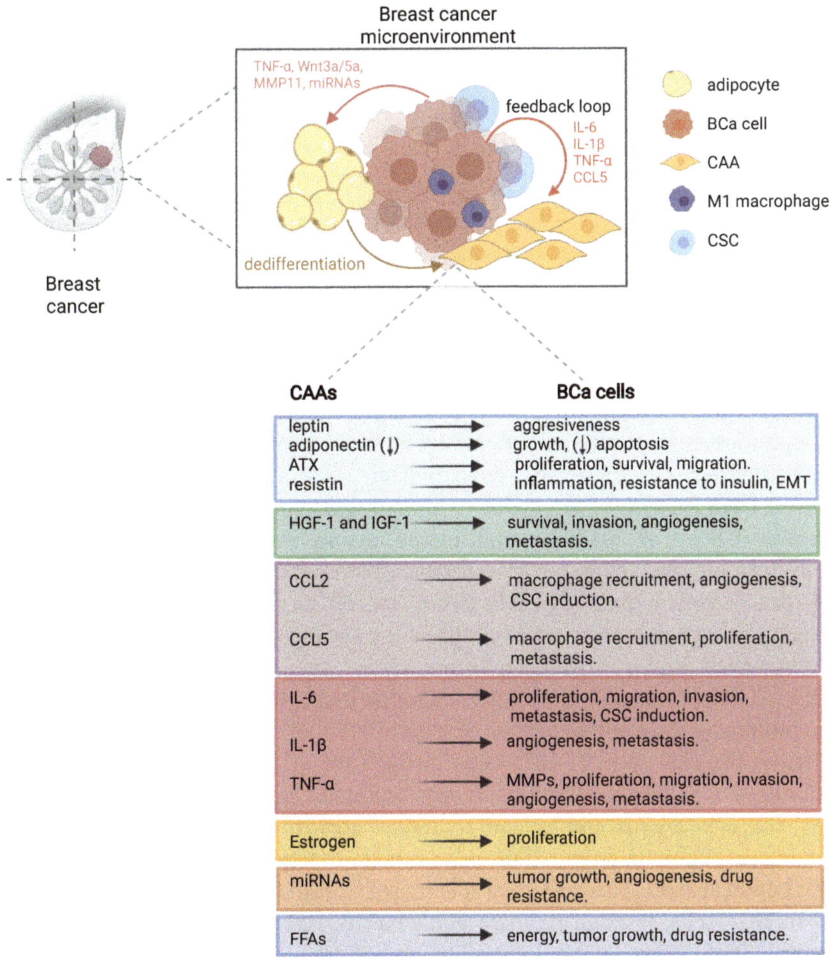

Fig. 12.1 Bidirectional adipocytes–BCa cells crosstalk is fundamental to support tumor growth and progression. BCa cells induce normal adipocytes to dedifferentiate to CAAs, in part through TNF-α, Wnt3a, Wnt5a, MMP11, and miRNAs. In addition, tumor cells induce a metabolic reprogramming of adipocytes in order to ensure high energy molecules for its own benefit. CAAs overexpress adipokines (except adiponectin), inflammatory cytokines, and high energy metabolites, among other factors. High levels of leptin secreted by CAAs promote BCa proliferation, EMT phenotype, progression, and resistance to targeted therapies. Diminished production of adiponectin by CAAs promotes BCa growth, invasion, and suppresses apoptosis. CAAs also secrete ATX (related with proliferation, survival, and migration) and resistin, which induces EMT, BCa cells stemness, and metastasis. Growth factors as IGF, VEGF, HGF, NGF, and PDGF promote survival, angiogenesis, invasion, and metastasis. In particular, HGF induces proliferation, metastasis, and angiogenesis, while IGF-1 induces BCa cells proliferation, EMT phenotype, and stemness. CAAs also produce a battery of inflammatory cytokines: CCL2, that induces BCa cells stemness and neovascularization; CCL5, that promotes invasion and metastasis; IL-6, that impacts on proliferation, EMT, migration, and invasion of BCa cells and is associated with resistance to therapy; IL-1β, associated with invasion, angiogenesis, and metastasis; TNF-α, that induces MMPs and dipeptidylpeptidases and increases BCa cells proliferation, chemoresistance, angiogenesis, and

hormones, and cytokines, are called adipokines. They regulate processes such as fat distribution, insulin secretion, energy consumption, and inflammatory reactions.

12.4.1.1 Leptin

Leptin is a hormone mainly synthesized and secreted by adipocytes. The binding to its receptor ObR induces proliferation, differentiation, inflammation, and nutrient absorption in breast cells (Sánchez-Jiménez et al. 2019). The main role of leptin is to regulate food intake and energy homeostasis leading to fat degradation in adipocytes (Kothari et al. 2020). Leptin circulates in plasma, and its levels increase according to total AT mass (Andò et al. 2014). Moreover, leptin expression has a role in several female cancers, including breast, cervical, endometrial, and ovarian. In particular, in BCa leptin levels in plasma of patients are higher compared to healthy volunteers and constitute a risk factor for BCa development (Niu et al. 2013). Also, leptin levels in plasma from BCa patients correlate with high-grade tumors, poor prognosis, and metastases (Choi et al. 2018). Several studies have shown that leptin promotes BCa growth. In a tumoral context, CAAs produce more leptin than mature adipocytes. Leptin can also be produced by CAFs (Choi et al. 2018). In vitro studies have shown that leptin promotes breast cancer proliferation through diverse pathways. Leptin activates janus kinase (JAK)/signal transducer and activator of transcription 3 (STAT3) and phosphatidylinositol 3-kinase pathway (PI3K)/AKT signaling pathways, promoting the proliferation of BCa cells (Choi et al. 2018). Moreover, leptin accelerates the cell cycle of BCa cells through the expression of cyclin D1 and cyclin-kinase 2 (Chen et al. 2006). It was reported that leptin, in the absence of ligand, can activate ER and human epidermal growth factor receptor 2 (HER2), which impacts in tumor progression and resistance to targeted therapies (Fiorio et al. 2008). Leptin also plays an important role in invasion and metastasis associated with BCa. In BCa cells, it could promote epithelial to mesenchymal transition (EMT) (Wei et al. 2016). Also, leptin produced by mature adipocytes in coculture with BCa cells together with IL-6 promotes local invasion and eventually metastasis. Adipocyte-derived leptin also drives self-renewal of BCa stem cells and chemoresistance in mouse breast tumors (Wang et al. 2018).

Additionally, leptin together with IL-1 favors angiogenesis by the induction of VEGF/VEGFR expression. It also upregulates IL-1-/IL-1R signaling, contributing to the angiogenic process (Zhou et al. 2011). On the other hand, leptin produced by

---

Fig. 12.1 (continued) metastasis. CAAs also exhibit augmented expression of aromatase, thus increasing estrogens production, which is highly relevant in BCa growth. Adipocytes suffer lipolysis driven by BCa cells. FFAs from adipocytes are transferred to BCa cells, where they can be used for newly synthesized membrane phospholipids from tumor cells. Also, miRNAs secreted by CAAs could be important in tumor growth, angiogenesis, and drug resistance Finally, BCa cells secrete inflammatory cytokines that induce CAAs and immune cells to secrete pro-inflammatory molecules in a cyclic feedback loop

adipocytes enables BCa progression by regulating immune cells as well. For example, IL-8 is produced by tumor-associated macrophages under leptin stimulus (Li et al. 2016). Further, leptin could weaken CD8+ T cell effector functions through activating STAT3- fatty acid β-oxidation and inhibiting glycolysis thus impairing breast tumor progression (Zhang et al. 2020).

12.4.1.2 Adiponectin

It is suggested that adiponectin works in the opposite way to leptin in the context of BCa progression, since it suppresses growth and invasion and induces apoptosis of BCa cells (Dieudonne et al. 2006; Körner et al. 2007) through protein kinase AMP-activated catalytic subunit alpha 1 (AMPK) activation (Dieudonne et al. 2006) and PI3K/AKT inhibition. It was reported that adiponectin secretion is reduced in CAAs, and is considered a protective factor against tumor progression (Wu et al. 2019). Several studies suggest that it is the leptin adiponectin ratio which determines the BCa risk and progression, and not leptin or adiponectin levels separately. Thus, high leptin to adiponectin ratio correlates with increased risk of postmenopausal BCa (Ollberding et al. 2013) and TNBC progression (Sultana et al. 2017).

12.4.1.3 Autotaxin and Resistin

ATX is a secreted glycoprotein produced by adipocytes but also platelets, endothelial cells, fibroblast, and cancer cells. Its function consists in the conversion of lysophosphatidylcholine to lipid signaling molecule lysophosphatidic acid mediating cells renewal, proliferation, survival, and migration (Lee et al. 2018). Besides, inflammation produced by tumor BCa cells triggers an increase of ATX expression in CAAs and fibroblasts, generating a feedback loop that favors tumor progression. ATX may be part of the connection between BCa and obesity since is augmented in subcutaneous AT and blood of obese patients (Ferry et al. 2003; Boucher et al. 2005).

Resistin is another adipokine observed in inflammatory zones. It can also be secreted by monocytes, macrophages, and bone marrow cells (McTernan et al. 2002; Patel et al. 2003). Among resistin functions, it is known that it increases inflammation, resistance to insulin and atherosclerosis as increases low-density lipoproteins (LDL) (Schwartz and Lazar 2011). It can also induce EMT and stemness in BCa, thus promoting the metastatic potential of BCa cells (Avtanski et al. 2019). As other adipokines circulating in blood, it can act systemically and locally. In fact, in circulation of obese patients is found in high levels (Makki et al. 2013). Interestingly, in BCa tissues, it can also be produced by CAAs, and its receptor adenylyl cyclase-associated protein 1 (CAP1) is expressed in primary human tumors. Moreover, its high expression in BCa tissue of patients was associated with aggressiveness and poor prognosis (Rosendahl et al. 2018).

12.4.1.4 Growth Factors

AT also secretes many growth factors that play known functions in several cancer, such as insulin growth factor (IGF), VEGF, hepatocyte growth factor (HGF), nerve growth factor (NGF), and platelet-derived growth factor (PDGF) (Peeraully et al. 2004; Pallua et al. 2009). In particular, they promote BCa survival, angiogenesis, invasion, and metastasis. These factors have also been studied in the majority of tumors, including the different stages of BCa progression. In addition, the increase of these factors in BCa milieu leads to resistance to therapies (Meyer et al. 2013; Voudouri et al. 2015; Linklater et al. 2016).

HGF and IGF-1

Both growth factors can be secreted by adipocytes or preadipocytes, also contributing with BCa progression.

HGF binds to its receptor, c-Met, located at tumor cells, and signals for proliferation, metastasis, and angiogenesis (Gallego et al. 2003). Interestingly, HGF was found to be higher both in adipocytes and serum from obese patients compared to lean individuals (Bell et al. 2006). Furthermore, c-Met expression was observed in the invasive front of BCa, which is located in direct contact with CAAs (Edakuni et al. 2001).

Insulin-like growth factor 1 (IGF-1), through binding to IGF-1R expressed in several subtypes of BCa (Law et al. 2008), activates PI3K/AKT and MAPK pathways, which impact in their proliferation. Importantly, insulin resistance, associated with obesity and other metabolic diseases, causes an increase in both insulin and IGF-1 levels, which constitute a risk factor for BCa (Esposito et al. 2013; Key et al. 2010). Moreover, IGF-1 was found to be twofold increased in CAAs from obese patients compared to lean individuals (D'Esposito et al. 2012; Nam et al. 1997). IGF-1R was described to have a role in cancer stem cells (CSCs) development and maintenance, EMT process and regulation of tumor microenvironment (Seccareccia and Brodt 2012). In addition, in hormone-dependent BCa, IGF-1R is coexpressed with ER-α, thus the cobinding of their ligands induce proliferation of tumor cells (Surmacz 2000).

12.4.1.5 Inflammatory Factors

It is well known that patients suffering from obesity and cancer display chronic inflammation and high levels of C-reactive protein (Dossus et al. 2014). Particularly in obesity, local inflammation is histologically recognized by the presence of "crown-like-structures," composed by dead/dying adipocytes surrounded by macrophages. These macrophages, which acquire an M1 pro-inflammatory phenotype (Engin 2017), are recruited in response to fat accumulation with the subsequent increase in chemokines CCL2 and CCL5 (Chen et al. 2005; Keophiphath et al. 2010). Free fatty acids (FFAs) and TNF-α signaling in macrophages and adipocytes, respectively, trigger the transcription of pro-inflammatory genes such as prostaglandin E2, TNF-α, IL-1, and IL-6, generating a feedback loop at the AT from breast and visceral fat of obese women (Rybinska et al. 2020).

CCL2 and CCL5 Chemokines

CCL2 is also known as monocyte chemoattractant protein-1 (MCP-1). It can be produced and secreted to extracellular environment by different cells composing the tumor microenvironment, such as tumor cells, fibroblasts, endothelial cells, and infiltrating monocytes (Melgarejo et al. 2009). CCL2 binds to the G-protein-coupled receptor C–C motif chemokine receptors 2 and 4 (CCR2 and CCR4), and is a chemoattractant for immune cells, especially monocytes, to the inflammatory site (Melgarejo et al. 2009). CCL2 also induces the activity of CSCs, by the induction of Notch receptor 1 (NOTCH1) expression (Tsuyada et al. 2012), and is significantly associated with neovascularization (Ueno and Zhang 2011). Noteworthy, it was suggested by Arendt et al. that crown-like structures, associated with malignant progression of BCa, are formed as consequence of macrophages recruitment by mammary epithelial cells-secreted CCL2 (Arendt et al. 2013).

CCL5 (previously called RANTES) can be produced by different types of cells in physiological processes. Its expression can also be found in BCa tissue at high levels, and can be produced by mesenchymal stem cells, as well (Khalid et al. 2015). CCL5, by binding to its receptor CCR5, induces invasion and metastasis of BCa (Velasco-Velázquez et al. 2012; Sax et al. 2016). This axis was found to be highly activated specifically in TNBC and HER2+ BCa. In fact, CCL5 expression was statistically associated with poor disease-free survival and overall cancer survival in patients with early HER2-positive BCa (Zazo et al. 2020), and the abundance of CCL5 in peritumoral AT of TNBC patients is also correlated with metastasis and poor overall survival (Song et al. 2018).

It is hypothesized that CCL5 is secreted by stromal cells into the tumor milieu and is recognized by CCR5, activating AKT/mTOR pathway, thus promoting tumor metastasis (Kim et al. 2018). On the other hand, CCL5 contributes to the inflammatory microenvironment since it could recruit macrophages and other immune cells (Keophiphath et al. 2010). Interestingly, a study showed that the invasiveness of MDA-MB-231 cells induced by human adipocytes was reduced by antagonizing CCL5 with specific peptides and antibodies (D'Esposito et al. 2016). Then, it can be suggested that antagonizing CCL2 and/or CCL5 in peritumoral adipocytes might be a potential target for BCa and its metastasis.

IL-6, IL-1(β) and TNF-α

IL-6 is a pleiotropic cytokine involved in various biological processes, such as immune regulation, hematopoiesis, and tumorigenesis. It can be secreted by adipocytes and acts locally and systemically. Locally, it suppresses lipogenesis and secretion of adiponectin (Bachelot et al. 2003). Under pathological conditions like obesity and cancer, the level of IL-6 secreted by adipocytes is significantly higher. IL-6 produced by adipocytes and BCa cells has been reported to impact on proliferation, migration, and invasion of BCa cells by modulating the expression of EMT-regulating genes (Gyamfi et al. 2019). Also, IL-6 increases aromatase expression by adipocytes and BCa cells, as well as estrogen synthesis, thus contributing with BCa progression (Purohit et al. 1995). It was demonstrated in vitro that upon the interaction of adipocytes and TNBC cells, the increment of IL-6 expression drove to higher migration and aggressiveness of these cancer cells (Nickel et al. 2018). Additionally, in clinical specimens from HER2-positive BCa, it was demonstrated that IL-6 induced the production and maintenance of breast CSCs through nuclear factor kappa-B (NF-κB) and STAT3 signaling pathways, then promoting BCa progression and resistance to therapy (Liu et al. 2018). Moreover, it is known that high levels of IL-6 in serum are correlated with poor prognosis of BCa (Rybinska et al. 2020).

IL-1β is involved in different steps of BCa progression. It is secreted into the extracellular microenvironment mainly by blood monocytes and tissue macrophages (Shen et al. 2017). It was observed that when adipocytes are cocultured with cancer cells, there is an increase in the expression of pro-inflammatory factors including IL-1β (Dirat et al. 2011). Importantly, this cytokine has been identified as a potential predictor of bone metastasis in BCa patients, since it increases the expression of osteoprotegerin, a promoter of invasion and metastasis of BCa. IL-1β can also promote tumor angiogenesis, by inducing the expression of VEGF and its receptors on endothelial cells (Tulotta and Ottewell 2018).

In addition, IL-1 has an important role since it is mainly secreted by AT in the tumor milieu. Moreover, IL-1 expression is increased in BCa, and is considered as poor prognosis, since this cytokine induces a secondary response by activating the secretion of other pro-inflammatory molecules and also by altering the expression of adhesion molecules strongly involved in metastasis. For these reasons, IL-1 is considered the most alarming factor in BCa evolution (Perrier et al. 2009).

TNF-α plays a physiological role in premenopausal non-obese women by regulating adipogenesis and lipid storage in adipocytes, thereby the total volume of AT (Cawthorn and Sethi 2008). TNF-α plays important functions in the tumor microenvironment, as well. It can be secreted by tumor and stromal cells, macrophages, or adipocytes. This cytokine increases tumorigenesis mainly by induction of matrix metalloproteases (MMPs) and dipeptidylpeptidases (Wolczyk et al. 2016). Noteworthy, in the serum of healthy women, TNF-α is generally not detected, while it is elevated in BCa patients (Alfano et al. 2017). TNF-α participates in BCa development, proliferation, chemoresistance, angiogenesis, and metastasis (Balkwill 2009).

In addition, it exerts a regulatory effect of estrogen synthesis (Liu et al. 2016) and enhances lipolysis in human adipocytes (Zhang et al. 2002).

12.4.1.6 Estrogens

Estrogens result from the aromatization of dihydrotestosterone that is catalyzed by aromatase. As well known, mammary epithelium is sensitive to estrogens and almost all BCa respond to this hormone either by canonical or noncanonical pathways acting in BCa cells or even by estrogens effect in the stromal cells (Péqueux et al. 2012). According to aromatase expression, the main sources of estrogens production are the ovaries in premenopausal women, but after menopause, AT becomes the principal site of its synthesis. Moreover, conditions of high AT, as obesity or high BMI, are linked to higher estrogen levels in circulation (Gérard and Brown 2018; Brown et al. 2017), hence they constitute one of the classical molecular explanations for increased risk of BCa in this subgroup of the population, and this issue has been deeply reviewed (Gérard and Brown 2018). Molecular mechanisms have been studied to explain the increased production of estrogens in AT of obese patients and have support on epidemiological studies (Gérard and Brown 2018). Thus, obesity induces overexpression of inflammatory factors such as TNF-α, IL-6 or CCL2, produced by adipocytes. Furthermore, other pro-inflammatory factors secreted by immune cells are also associated with increased expression of aromatase and, in consequence, of estrogens level (Purohit et al. 2002; Gérard and Brown 2018). Consistent with this, both aromatase expression and activity were found to be increased in AT of obese patients (Morris et al. 2011; Purohit et al. 2002; Lorincz and Sukumar 2006). However, in contrast, several studies have shown that circulating estrogens have a protective role against the development of breast cancer in obese women (Suba 2013; Key et al. 2003). Interestingly, it was demonstrated that in breast AT, the inflammatory factors associated with obesity induce a shift in the aromatase promoter PI.4 to the promoter PI.3/I, which is more sensitive to inflammatory factors, leading to an increase in estrogens level (To et al. 2015; Gérard and Brown 2018). Some authors suggested that estrogens produced locally in mammary AT could have more impact on BCa growth and progression associated with obesity, as estrogens levels are higher in the tumor microenvironment compared to circulation (van Landeghem et al. 1985; Rybinska et al. 2020).

12.4.1.7 microRNAs in AT and BCa Cells Crosstalk

MiRNAs are noncoding RNAs of 21–25 nucleotides considered global regulators of gene expression that play roles in multiple biological processes contributing to cellular homeostasis. Dysregulation of miRNA expression is associated with numerous pathologies, including cancer and metabolic disorders (Rottiers and Näär 2012; Takahashi et al. 2015). MiRNAs have functions in all steps of BCa development and progression (Kaboli et al. 2015). Also, it was demonstrated that they are master regulators of metabolism involved in processes such as cholesterol secretion, synthesis and degradation, fatty acid synthesis, lipoprotein degradation (Fernández-Hernando et al. 2013), maintenance of insulin, glucose and liver lipid homeostasis, and b-cell differentiation in pancreatic islets (Rottiers and Näär 2012). Several studies reported aberrant expression of circulating miRNAs in obesity and MeS (Deiuliis 2016; Karolina et al. 2012; Ramzan et al. 2020). These observations support the idea that miRNAs could also be part of AT and BCa crosstalk. In fact, several studies suggest that miRNAs could be part of the signal necessary for the transformation of normal adipocytes into CAAs and, conversely, miRNAs secreted by CAAs could be important in tumor growth, angiogenesis, and drug resistance (Bandini et al. 2019; Wu et al. 2019). A recent study showed that in vitro coculture of BCa cells with mature adipocytes induces proliferation, migration, and invasion and differential expression of 98 miRNAs in BCa cells where miR-3184-5p was the most increased while miR-181c-3p represent the most decreased miRNA whose direct targets were Forkhead box P4 (FOXP4) and PPARα, respectively (Rajarajan et al. 2019). Also, it was reported that miRNA-144, miRNA-126, and miRNA-155 secretion by BCa cells induce adipocytes to CAAs transformation (Sun et al. 2019). This study revealed that exosomal miR-144 induces beige/brown differentiation of adipocytes, and exosomal miRNA-126 is involved in metabolic reprogramming of adipocytes (Sun et al. 2019). Another study showed that exosomal miRNA-155 induces adipocyte lipolysis promoting aggressive phenotype of cancer cells (Wu et al. 2018). MiRNA-105 secreted by BCa cells induces metabolic reprogramming of CAAs and CAFs ensuring high energy conditions to support tumor cells growth (Yan et al. 2018). Interestingly, obesity downregulates p16INK4A protein in breast adipocytes increasing leptin secretion in a mechanism mediated by miR-141 and miR-146b-5p which is associated with EMT in breast ductal epithelial cells (Al-Khalaf et al. 2017). As showed above, several studies demonstrate a role of miRNAs secreted by tumor cells in adipocyte phenotype and transformation but further studies are needed to unveil the effect of miRNAs secreted by adipocytes in BCa growth and progression.

12.5 Metabolic Reprogramming

It is hypothesized that tumor cells, upon interaction with adipocytes, can induce a reprogramming of adipocytes metabolism in order to adapt it for tumor benefit. Since adipocytes are considered tremendous energy storage that provides high-energy metabolites (Reeves et al. 2007), their reprogramming could provide the tumors a high potential to proliferate. This reprogramming involves the metabolic regulation of almost all macronutrients, such as carbohydrates, lipids, and aminoacids (Choi et al. 2018).

It was observed in coculture assays with BCa cells and adipocytes that the latter suffer lipolysis driven by tumor cells. Besides, in this experiment, fatty acids (FAs) from adipocytes were transferred to BCa cells (Balaban et al. 2017) and this event favored BCa cells growth by increasing mitochondrial β-oxidation. This fact could explain the smaller size and less lipid droplets in CAAs (Dirat et al. 2011; Andarawewa et al. 2005). In turn, these transferred FAs can be used as substrates for newly synthesized membrane phospholipids from tumor cells (Menendez and Lupu 2007).

Importantly, FFAs secreted by CAAs can be used for the biosynthesis of lipid-signaling molecules that act as tumor promoters (Kotzbeck et al. 2018).

Interestingly, obese adipocytes provide more FAs to cancer cells than nonobese adipocytes (Balaban et al. 2017). Moreover, cell surface fatty acid translocase (CD36) is detected in most breast cancer tissues located adjacent to AT, and is recognized as a marker of cells initiating metastasis, even in other tumors (Pascual et al. 2017).

Adipocytes instructed by tumor cells rearrange their metabolism to glycolysis, thus releasing energy-rich metabolites such as lactate and pyruvate. Monocarboxylate transporters (MCT) also play an important role in the transport of these metabolites to BCa cells. In particular, MCT1 and MCT4 are overexpressed in BCa tissue (Baenke et al. 2015; Johnson et al. 2017). In vitro coculture experiments revealed that tumor cells induce MCT4 expression in adipocytes facilitating lactic acid efflux (Sun et al. 2019). A clinical study demonstrated that MCT1 and MCT4 expression in BCa tissue correlates with poor prognosis of patients. Moreover, MCT1 correlation with poor prognosis is stronger when is linked to MCT4 expression in adjacent AT. In addition, ER–BCa cells express high levels of MCT1 compared to ER+ tumors (Li et al. 2018).

Adipocytes that undergo glycolysis also produce ketone bodies, which serve as source for ATP production in invasive cancer cells. These ketone bodies induce mitochondrial biogenesis in BCa cells, increasing their ability to growth (Martinez-Outschoorn et al. 2012). Moreover, induction of ketone-specific gene signature was shown to be associated with worse outcomes in BCa patients (Argilés et al. 2014).

12.6 Therapeutic Approaches Focused on BCa Cells and AT-Crosstalk

There are a huge variety of approaches to treat the different subtypes of BCa. In this section, we will focus especially in those related to molecules differentially expressed in the AT under tight interaction with BCa tissue.

While inhibition of adipogenesis is considered as option, since a misbalance in AT volume fuels BCa progress, this strategy is not a first option since adipogenesis is a physiological process, and its inhibition could lead to adverse effects. Also, CAAs present a dedifferentiated phenotype, hence inhibition of adipogenesis could have controversial effects.

Adipokines and cytokines secreted by adipocytes are a more suggested target for therapy. Specifically, leptin depletion or treatment with leptin receptor antagonists is frequently employed in BCa treatment. Also, inhibition of ATX-LPA axis from AT is the focus of the latest strategies (Marshall et al. 2012).

Another option consists in reinforcing the effect of beneficial adipokines, as is the case of adiponectin. Peptide-based adiponectin receptor agonists, such as ADP-355, inhibit BCa growth in vitro and in xenografts (Otvos et al. 2011). In addition, (−) catechin flavonoid increases adiponectin expression (Si et al. 2007).

Blockade of inflammatory factors produced by CAAs is also a possible strategy. Some compounds are already employed in the clinical practice or are involved in clinical or preclinical studies, such as Tocilizumab (monoclonal antibody anti-IL-6R), Canakinumab (monoclonal antibody anti-IL-1β), Infliximab (monoclonal antibody anti- TNF-α), and AZD3695 (inhibitor of MCT1) (Polanski et al. 2014). In addition, it has been shown that inhibitors of CCR5 (receptor for the chemokine CCL5) Maraviroc and Vicriviroc can reduce infiltration and invasion of BCa cells (Levy 2009).

It is also possible targeting the metabolic interactions between BCa cells and adipocytes. An example of this case is Myricetin, which suppresses the accumulation of lipid droplets in the adipocytes (Chang et al. 2012). Suppression of FA oxidation constitutes a potential target, as FA is source of energy for BCa cells once transferred from adipocytes. Trimetazidine, which is an inhibitor of FA oxidation, induces apoptosis in cancer cells (Andela et al. 2005). Finally, another approach consists in inhibiting the transport of FA. BMS 309403, an inhibitor of FABP4, is a candidate assayed in preclinical studies (Nieman et al. 2011).

12.7 AT and Resistance in BCa Therapy

Secretome from adipocytes and cellular interactions not only can fuel BCa growth but also can promote resistance to multiple therapies for BCa, including chemotherapy, hormonal therapy, radiotherapy, and immunotherapy. Several studies have reported the negative influence of the adipose microenvironment both in resistance to chemo- and hormonal therapies, such as doxorubicin, tamoxifen, and fulvestrant, and a few examples will be shown in this section.

Major vault protein (MVP) is a transport-associated protein overexpressed in BCa cells as a result of interaction with adipocytes. In a coculture assay with adipocytes and BCa cells treated with doxorubicin, adipocytes induced an MVP-related multidrug-resistant phenotype, which could be relevant in obesity-related chemoresistance (Lehuédé et al. 2019). Another mechanism involving AT promoting resistance to therapy is the stimulation of CSCs, mediated through adipocyte-derived leptin (Delort et al. 2019). Besides, it was observed that leptin impaired the effect of hormonal tamoxifen therapy (Delort et al. 2019), and interestingly in overall and tamoxifen treated BCa patients, leptin expression correlated with poor prognosis (Chen et al. 2013).

These data reflect that adipokines could predict drug resistance to BCa therapy, evidencing the relevance of exploring more exhaustively the crosstalk between BCa cells and adipocytes.

12.8 Concluding Remarks

Mammary tissue constitutes a structure with high plasticity where the tight interaction between epithelial cells and AT is determinant in both physiological conditions and disease. Adipocytes comprise a key component of AT and strongly contribute to tumor development, growth and progression. Under unhealthy conditions such as obesity, adipocytes contribute to creation of a pro-inflammatory microenvironment that could impact on BCa development. Moreover, upon BCa development, these lipid-rich cells provide an important source of energy for the neighboring tumor cells, which establish a tight communication with adipocytes. This crosstalk promotes dedifferentiation of adipocytes into CAAs, which in turn produce several pro-tumoral stimuli, generating a cyclic feedback-loop.

The cooperation between adipocytes and BCa is emphasized in many epidemiologic studies, showing a clear connection between obesity and BCa outcome. For this reason, in the last years, AT became a challenging target for BCa therapies. Moreover, AT is also involved in resistance to radio-, chemo-, and immunotherapy. Thus, a deep comprehension about the crosstalk between BCa and adipocytes is under the focus of several studies. This knowledge would provide a very valuable tool for developing novel therapies in order to tackle BCa.

Acknowledgments and Funding This research was supported by the Argentinean Agency of Science and Technology (ANPCyT PICT 2018-2625 and PICT 2018-2264. We thank Fundación Williams (Argentina) and Fundación Rene Barón (Argentina) for their support.

Conflicts of Interest The authors declare no conflicts of interest.

References

Alberti KGMM, Eckel RH, Grundy SM, Zimmet PZ, Cleeman JI, Donato KA, Fruchart JC, Philip W, James T, Loria CM, Smith SC (2009) Harmonizing the metabolic syndrome: a joint interim statement of the International Diabetes Federation Task Force on Epidemiology and Prevention; National Heart, Lung, and Blood Institute; American Heart Association; World Heart Federation; International Atherosclerosis Society; And International Association for the Study of Obesity. Circulation 120. https://doi.org/10.1161/CIRCULATIONAHA.109.192644

Alfano CM, Peng J, Andridge RR et al (2017) Inflammatory cytokines and comorbidity development in breast cancer survivors versus noncancer controls: evidence for accelerated aging? J Clin Oncol 35. https://doi.org/10.1200/JCO.2016.67.1883

Al-Khalaf HH, Amir M, Al-Mohanna F, Tulbah A, Al-Sayed A, Aboussekhra A (2017) Obesity and P16 INK4A downregulation activate breast adipocytes and promote their protumorigenicity. Mol Cell Biol 37. https://doi.org/10.1128/mcb.00101-17

Andarawewa KL, Motrescu ER, Chenard MP, Gansmuller A, Stoll I, Tomasetto C, Rio MC (2005) Stromelysin-3 is a potent negative regulator of adipogenesis participating to cancer cell-adipocyte interaction/crosstalk at the tumor invasive front. Cancer Res 65. https://doi.org/10.1158/0008-5472.CAN-05-1231

Andela VB, Altuwaijri S, Wood J, Rosier RN (2005) Inhibition of β-oxidative respiration is a therapeutic window associated with the cancer chemo-preventive activity of PPARγ agonists. FEBS Lett 579. https://doi.org/10.1016/j.febslet.2005.01.082

Andò S, Barone I, Giordano C et al (2014) The multifaceted mechanism of leptin signaling within tumor microenvironment in driving breast cancer growth and progression. Front Oncol 4. https://doi.org/10.3389/fonc.2014.00340

Apostoli AJ, Skelhorne-Gross GEA, Rubino RE, Peterson NT, di Lena MA, Schneider MM, Sengupta SK, Nicol CJB (2014) Loss of PPARγ expression in mammary secretory epithelial cells creates a pro-breast tumorigenic environment. Int J Cancer 134. https://doi.org/10.1002/ijc.28432

Arendt LM, McCready J, Keller PJ et al (2013) Obesity promotes breast cancer by CCL2-mediated macrophage recruitment and angiogenesis. Cancer Res 73. https://doi.org/10.1158/0008-5472.CAN-13-0926

Argilés JM, Busquets S, Stemmler B, López-Soriano FJ (2014) Cancer cachexia: understanding the molecular basis. Nat Rev Cancer 14. https://doi.org/10.1038/nrc3829

Avtanski D, Garcia A, Caraballo B et al (2019) Resistin induces breast cancer cells epithelial to mesenchymal transition (EMT) and stemness through both adenylyl cyclase-associated protein 1 (CAP1)-dependent and CAP1-independent mechanisms. Cytokine 120. https://doi.org/10.1016/j.cyto.2019.04.016

Bachelot T, Ray-Coquard I, Menetrier-Caux C et al (2003) Prognostic value of serum levels of interleukin 6 and of serum and plasma levels of vascular endothelial growth factor in hormone-refractory metastatic breast cancer patients. Br J Cancer 88. https://doi.org/10.1038/sj.bjc.6600956

Baenke F, Dubuis S, Brault C, Weigelt B, Dankworth B, Griffiths B, Jiang M et al (2015) Functional screening identifies MCT4 as a key regulator of breast cancer cell metabolism and survival. J Pathol 237. https://doi.org/10.1002/path.4562

Balaban S, Shearer RF, Lee LS et al (2017) Adipocyte lipolysis links obesity to breast cancer growth: adipocyte-derived fatty acids drive breast cancer cell proliferation and migration. Cancer Metab 5. https://doi.org/10.1186/s40170-016-0163-7

Balkwill F (2009) Tumour necrosis factor and cancer. Nat Rev Cancer 9. https://doi.org/10.1038/nrc2628

Bandini E, Rossi T, Gallerani G, Fabbri F (2019) Adipocytes and micrornas crosstalk: a key tile in the mosaic of breast cancer microenvironment. Cancers 11. https://doi.org/10.3390/cancers11101451

Bartelt A, Heeren J (2014) Adipose tissue browning and metabolic health. Nat Rev Endocrinol 10. https://doi.org/10.1038/nrendo.2013.204

Bell LN, Ward JL, Degawa-Yamauchi M, Bovenkerk JE, Jones RM, Cacucci BM, Gupta CE et al (2006) Adipose tissue production of hepatocyte growth factor contributes to elevated serum HGF in obesity. Am J Physiol Endocrinol Metab 291. https://doi.org/10.1152/ajpendo.00174.2006

Bhandari R, Kelley GA, Hartley TA, Rockett IRH (2014) Metabolic syndrome is associated with increased breast cancer risk: a systematic review with meta-analysis. Int J Breast Cancer 2014. https://doi.org/10.1155/2014/189384

Bochet L, Lehuédé C, Dauvillier S, Wang YY, Dirat B, Laurent V, Dray C et al (2013) Adipocyte-derived fibroblasts promote tumor progression and contribute to the desmoplastic reaction in breast cancer. Cancer Res 73. https://doi.org/10.1158/0008-5472.CAN-13-0530

Borugian MJ, Sheps SB, Kim-Sing C, Olivotto IA, van Patten C, Dunn BP, Coldman AJ, Potter JD, Gallagher RP, Gregory Hislop T (2003) Waist-to-hip ratio and breast cancer mortality. Am J Epidemiol 158. https://doi.org/10.1093/aje/kwg236

Boucher J, Quilliot D, Pradères JP, Simon MF, Grès S, Guigné C, Prévot D et al (2005) Potential involvement of adipocyte insulin resistance in obesity-associated up-regulation of adipocyte lysophospholipase D/autotaxin expression. Diabetologia 48. https://doi.org/10.1007/s00125-004-1660-8

Bray F, Ferlay J, Soerjomataram I et al (2018) Global cancer statistics 2018: GLOBOCAN estimates of incidence and mortality worldwide for 36 cancers in 185 countries. CA Cancer J Clin 68. https://doi.org/10.3322/caac.21492

Brown KA, Iyengar NM, Zhou XK, Gucalp A, Subbaramaiah K, Wang H, Giri DD et al (2017) Menopause is a determinant of breast aromatase expression and its associations with BMI, inflammation, and systemic markers. J Clin Endocrinol Metabol 102. https://doi.org/10.1210/jc.2016-3606

Burton BT, Foster WR (1985) Health implications of obesity: an NIH consensus development conference. J Am Diet Assoc:85

Cao L, Yao G, Hu X, Chen L, Ye C (2015) Advances in studies on metabolic syndrome and breast cancer. Zhonghua Wai Ke Za Zhi [Chin J Surg] 53

Capasso I, Esposito E, Pentimalli F, Crispo A, Montella M, Grimaldi M, de Marco M et al (2010) Metabolic syndrome affects breast-cancer risk in postmenopausal women: national cancer institute of Naples experience. Cancer Biol Ther 10. https://doi.org/10.4161/cbt.10.12.13473

Cawthorn WP, Sethi JK (2008) TNF-alpha and adipocyte biology. FEBS Lett 582. https://doi.org/10.1016/j.febslet.2007.11.051

Cha YJ, Koo JS (2018) Adipokines as therapeutic targets in breast cancer treatment. Expert Opin Ther Targets 22. https://doi.org/10.1080/14728222.2018.1538356

Chan DSM, Vieira AR, Aune D, Bandera EV, Greenwood DC, McTiernan A, Navarro Rosenblatt D, Thune I, Vieira R, Norat T (2014) Body mass index and survival in women with breast cancer—systematic literature review and meta-analysis of 82 follow-up studies. Ann Oncol 25. https://doi.org/10.1093/annonc/mdu042

Chang CJ, Tzeng TF, Liou SS, Chang YS, Min Liu I (2012) Myricetin increases hepatic peroxisome proliferator-activated receptor protein expression and decreases plasma lipids and adiposity in rats. Evid Based Complement Alternat Med 2012. https://doi.org/10.1155/2012/787152

Chen A, Mumick S, Zhang C et al (2005) Diet induction of monocyte chemoattractant protein-1 and its impact on obesity. Obes Res 13. https://doi.org/10.1038/oby.2005.159

Chen C, Chang YC, Liu CL et al (2006) Leptin-induced growth of human ZR-75-1 breast cancer cells is associated with up-regulation of cyclin D1 and c-Myc and down-regulation of tumor suppressor p53 and p21WAF1/CIP1. Breast Cancer Res Treat 98. https://doi.org/10.1007/s10549-005-9139-y

Chen X, Zha X, Chen W, Zhu T, Qiu J, Røe OD, Li J, Wang Z, Yin Y (2013) Leptin attenuates the anti-estrogen effect of tamoxifen in breast cancer. Biomed Pharmacother 67. https://doi.org/10.1016/j.biopha.2012.10.001

Cheraghi Z, Poorolajal J, Hashem T, Esmailnasab N, Irani AD (2012) Effect of body mass index on breast cancer during premenopausal and postmenopausal periods: a meta-analysis. PLoS One 7. https://doi.org/10.1371/journal.pone.0051446

Choi J, Cha YJ, Koo JS (2018) Adipocyte biology in breast cancer: from silent bystander to active facilitator. Prog Lipid Res 69. https://doi.org/10.1016/j.plipres.2017.11.002

Cinti S (2018) Pink adipocytes. Trends Endocrinol Metab 29. https://doi.org/10.1016/j.tem.2018.05.007

Clemenceau A, Diorio C, Durocher F (2020) Role of secreted frizzled-related protein 1 in early mammary gland tumorigenesis and its regulation in breast microenvironment. Cells 9. https://doi.org/10.3390/cells9010208

Cleveland RJ, Eng SM, Abrahamson PE, Britton JA, Teitelbaum SL, Neugut AI, Gammon MD (2007) Weight gain prior to diagnosis and survival from breast cancer. Cancer Epidemiol Biomarkers Prev 16. https://doi.org/10.1158/1055-9965.EPI-06-0889

Coelho M, Oliveira T, Fernandes R (2013) Biochemistry of adipose tissue: an endocrine organ. Arch Med Sci 9. https://doi.org/10.5114/aoms.2013.33181

Combs TP, Pajvani UB, Berg AH, Lin Y, Jelicks LA, Laplante M, Nawrocki AR et al (2004) A transgenic mouse with a deletion in the collagenous domain of adiponectin displays elevated circulating adiponectin and improved insulin sensitivity. Endocrinology 145. https://doi.org/10.1210/en.2003-1068

Connolly BS, Barnett C, Vogt KN, Li T, Stone J, Boyd NF (2002) A meta-analysis of published literature on waist-to-hip ratio and risk of breast cancer. Nutr Cancer 44. https://doi.org/10.1207/S15327914NC4402_02

Czene K, Lichtenstein P, Hemminki K (2002) Environmental and heritable causes of cancer among 9.6 million individuals in the Swedish family-cancer database. Int J Cancer 99. https://doi.org/10.1002/ijc.10332

D'Esposito V, Passaretti F, Hammarstedt A, Liguoro D, Terracciano D, Molea G, Canta L et al (2012) Adipocyte-released insulin-like growth factor-1 is regulated by glucose and fatty acids and controls breast cancer cell growth in vitro. Diabetologia 55. https://doi.org/10.1007/s00125-012-2629-7

D'Esposito VD, Liguoro D, Ambrosio MR et al (2016) Adipose microenvironment promotes triple negative breast cancer cell invasiveness and dissemination by producing CCL5. Oncotarget 7. https://doi.org/10.18632/oncotarget.8336

Deiuliis JA (2016) MicroRNAs as regulators of metabolic disease: pathophysiologic significance and emerging role as biomarkers and therapeutics. Int J Obes (Lond) 40. https://doi.org/10.1038/ijo.2015.170

Delort L, Bougaret L, Cholet J, Vermerie M, Billard H, Decombat C, Bourgne C, Berger M, Dumontet C, Caldefie-Chezet F (2019) Hormonal therapy resistance and breast cancer: involvement of adipocytes and leptin. Nutrients 11. https://doi.org/10.3390/nu11122839

Dibaba DT, Ogunsina K, Braithwaite D, Akinyemiju T (2019) Metabolic syndrome and risk of breast cancer mortality by menopause, obesity, and subtype. Breast Cancer Res Treat 174. https://doi.org/10.1007/s10549-018-5056-8

Dieudonne MN, Bussiere M, Dos Santos E et al (2006) Adiponectin mediates antiproliferative and apoptotic responses in human MCF7 breast cancer cells. Biochem Biophys Res Commun 345. https://doi.org/10.1016/j.bbrc.2006.04.076

Dirat B, Bochet L, Dabek M et al (2011) Cancer-associated adipocytes exhibit an activated phenotype and contribute to breast cancer invasion. Cancer Res 71. https://doi.org/10.1158/0008-5472.CAN-10-3323

Dossus L, Jimenez-Corona A, Romieu I et al (2014) C-reactive protein and postmenopausal breast cancer risk: results from the E3N cohort study. Cancer Causes Control 25. https://doi.org/10.1007/s10552-014-0355-9

Druesne-Pecollo N, Touvier M, Barrandon E, Chan DSM, Norat T, Zelek L, Hercberg S, Latino-Martel P (2012) Excess body weight and second primary cancer risk after breast cancer: a systematic review and meta-analysis of prospective studies. Breast Cancer Res Treat 135. https://doi.org/10.1007/s10549-012-2187-1

Edakuni G, Sasatomi E, Satoh T, Tokunaga O, Miyazaki K (2001) Expression of the hepatocyte growth factor/c-Met pathway is increased at the cancer front in breast carcinoma. Pathol Int 51. https://doi.org/10.1046/j.1440-1827.2001.01182.x

Eliassen AH, Colditz GA, Rosner B, Willett WC, Hankinson SE (2006) Adult weight change and risk of postmenopausal breast cancer. JAMA 296. https://doi.org/10.1001/jama.296.2.193

Engin AB (2017) Adipocyte-macrophage cross-talk in obesity. Adv Exp Med Biol 960. https://doi.org/10.1007/978-3-319-48382-5_14

Esposito K, Chiodini P, Capuano A, Bellastella G, Maiorino MI, Rafaniello C, Giugliano D (2013) Metabolic syndrome and postmenopausal breast cancer: systematic review and meta-analysis. Menopause 20. https://doi.org/10.1097/GME.0b013e31828ce95d

Ewertz M, Jensen MB, Gunnarsdóttir KÁ, Højris I, Jakobsen EH, Nielsen D, Stenbygaard LE, Tange UB, Cold S (2011) Effect of obesity on prognosis after early-stage breast cancer. J Clin Oncol 29. https://doi.org/10.1200/JCO.2010.29.7614

Fernández-Hernando C, Ramírez CM, Goedeke L, Suárez Y (2013) MicroRNAs in metabolic disease. Arterioscler Thromb Vasc Biol 33. https://doi.org/10.1161/ATVBAHA.112.300144

Ferry G, Tellier E, Try A, Grés S, Naime I, Simon MF, Rodriguez M et al (2003) Autotaxin is released from adipocytes, catalyzes lysophosphatidic acid synthesis, and activates preadipocyte proliferation. up-regulated expression with adipocyte differentiation and obesity. J Biol Chem 278. https://doi.org/10.1074/jbc.M301158200

Fiorio E, Mercanti A, Terrasi M et al (2008) Leptin/HER2 crosstalk in breast cancer: in vitro study and preliminary in vivo analysis. BMC Cancer 8. https://doi.org/10.1186/1471-2407-8-305

Fletcher SJ, Sacca PA, Pistone-Creydt M et al (2017) Human breast adipose tissue: characterization of factors that change during tumor progression in human breast cancer. J Exp Clin Cancer Res 36. https://doi.org/10.1186/s13046-017-0494-4

Gallego MI, Bierie B, Hennighausen L (2003) Targeted expression of HGF/SF in mouse mammary epithelium leads to metastatic adenosquamous carcinomas through the activation of multiple signal transduction pathways. Oncogene 22. https://doi.org/10.1038/sj.onc.1207063

Gérard C, Brown KA (2018) Obesity and breast cancer – role of estrogens and the molecular underpinnings of aromatase regulation in breast adipose tissue. Mol Cell Endocrinol 466. https://doi.org/10.1016/j.mce.2017.09.014

Gyamfi J, Lee YH, Min BS, Choi J (2019) Niclosamide reverses adipocyte induced epithelial-mesenchymal transition in breast cancer cells via suppression of the interleukin-6/STAT3 signalling axis. Sci Rep 9. https://doi.org/10.1038/s41598-019-47707-2

Harvie M, Hooper L, Howell AH (2003) Central obesity and breast cancer risk: a systematic review. Obes Res 4. https://doi.org/10.1046/j.1467-789x.2003.00108.x

Himms-Hagen J, Melnyk A, Zingaretti MC, Ceresi E, Barbatelli G, Cinti S (2000) Multilocular fat cells in WAT of CL-316243-treated rats derive directly from white adipocytes. Am J Physiol Cell Physiol 279. https://doi.org/10.1152/ajpcell.2000.279.3.c670

Howlett AR, Bissell MJ (1993) The influence of tissue microenvironment (stroma and extracellular matrix) on the development and function of mammary epithelium. Epithelial Cell Biol 2

Ibrahim MM (2010) Subcutaneous and visceral adipose tissue: structural and functional differences. Obes Rev 11. https://doi.org/10.1111/j.1467-789X.2009.00623.x

Iyengar NM, Arthur R, Manson JE, Chlebowski RT, Kroenke CH, Peterson L, Cheng TYD et al (2019) Association of body fat and risk of breast cancer in postmenopausal women with normal body mass index: a secondary analysis of a randomized clinical trial and observational study. JAMA Oncol 5. https://doi.org/10.1001/jamaoncol.2018.5327

Johnson JM, Cotzia P, Fratamico R, Mikkilineni L, Chen J, Colombo D, Mollaee M et al (2017) MCT1 in invasive ductal carcinoma: monocarboxylate metabolism and aggressive breast cancer. Front Cell Dev Biol 5. https://doi.org/10.3389/fcell.2017.00027

Kabat GC, Kim MY, Lee JS, Ho GY, Going SB, Beebe-Dimmer J, Manson JAE, Chlebowski RT, Rohan TE (2017) Metabolic obesity phenotypes and risk of breast cancer in postmenopausal women. Cancer Epidemiol Biomarkers Prev 26. https://doi.org/10.1158/1055-9965.EPI-17-0495

Kaboli PJ, Rahmat A, Ismail P, Ling KH (2015) MicroRNA-based therapy and breast cancer: a comprehensive review of novel therapeutic strategies from diagnosis to treatment. Pharmacol Res 97. https://doi.org/10.1016/j.phrs.2015.04.015

Karolina DS, Tavintharan S, Armugam A, Sepramaniam S, Pek SLT, Wong MTK, Lim SC, Sum CF, Jeyaseelan K (2012) Circulating MiRNA profiles in patients with metabolic syndrome. J Clin Endocrinol Metabol 97. https://doi.org/10.1210/jc.2012-1996

Keophiphath M, Rouault C, Divoux A et al (2010) CCL5 promotes macrophage recruitment and survival in human adipose tissue. Arterioscler Thromb Vasc Biol 30. https://doi.org/10.1161/ATVBAHA.109.197442

Key TJ, Appleby PN, Reeves GK, Roddam A, Dorgan JF, Longcope C, Stanczyk FZ et al (2003) Body mass index, serum sex hormones, and breast cancer risk in postmenopausal women. J Natl Cancer Inst 95. https://doi.org/10.1093/jnci/djg022

Key TJ, Appleby PN, Reeves GK, Roddam AW, Helzlsouer KJ, Alberg AJ, Rollison DE et al (2010) Insulin-like growth factor 1 (IGF1), IGF binding protein 3 (IGFBP3), and breast cancer risk: pooled individual data analysis of 17 prospective studies. Lancet Oncol 11. https://doi.org/10.1016/S1470-2045(10)70095-4

Khalid A, Wolfram J, Ferrari I et al (2015) Recent advances in discovering the role of CCL5 in metastatic breast cancer. Mini Rev Med Chem 15. https://doi.org/10.2174/138955751513150923094709

Kim EJ, Kim YK, Kim S et al (2018) Adipochemokines induced by ultraviolet irradiation contribute to impaired fat metabolism in subcutaneous fat cells. Br J Dermatol 178. https://doi.org/10.1111/bjd.15907

Kolb R, Zhang W (2020) Obesity and breast cancer: a case of inflamed adipose tissue. Cancers 12. https://doi.org/10.3390/cancers12061686

Körner A, Pazaitou-Panayiotou K, Kelesidis T et al (2007) Total and high-molecular-weight adiponectin in breast cancer: in vitro and in vivo studies. J Clin Endocrinol Metab 92. https://doi.org/10.1210/jc.2006-1858

Kothari C, Diorio C, Durocher F (2020) The importance of breast adipose tissue in breast cancer. Int J Mol Sci 21. https://doi.org/10.3390/ijms21165760

Kotzbeck P, Giordano A, Mondini E, Murano I, Severi I, Venema W, Cecchini MP et al (2018) Brown adipose tissue whitening leads to brown adipocyte death and adipose tissue inflammation. J Lipid Res 59. https://doi.org/10.1194/jlr.M079665

Landskroner-Eiger S, Park J, Israel D, Pollard JW, Scherer PE (2010) Morphogenesis of the developing mammary gland: stage-dependent impact of adipocytes. Dev Biol 344. https://doi.org/10.1016/j.ydbio.2010.06.019

Lauby-Secretan B, Scoccianti C, Loomis D, Grosse Y, Bianchini F, Straif K (2016) Body fatness and cancer — viewpoint of the IARC Working Group. N Engl J Med 375. https://doi.org/10.1056/nejmsr1606602

Law JH, Habibi G, Hu K et al (2008) Phosphorylated insulin-like growth factor-I/insulin receptor is present in all breast cancer subtypes and is related to poor survival. Cancer Res 68. https://doi.org/10.1158/0008-5472.CAN-08-2755

Lee D, Suh DS, Lee SC et al (2018) Role of autotaxin in cancer stem cells. Cancer Metastasis Rev 37. https://doi.org/10.1007/s10555-018-9745-x

Lehuédé C, Li X, Dauvillier S, Vaysse C, Franchet C, Clement E, Esteve D et al (2019) Adipocytes promote breast cancer resistance to chemotherapy, a process amplified by obesity: role of the major vault protein (MVP). Breast Cancer Res 21. https://doi.org/10.1186/s13058-018-1088-6

Levy JA (2009) HIV pathogenesis: 25 years of progress and persistent challenges. AIDS 23. https://doi.org/10.1097/QAD.0b013e3283217f9f

Li K, Wei L, Huang Y et al (2016) Leptin promotes breast cancer cell migration and invasion via IL-18 expression and secretion. Int J Oncol 48. https://doi.org/10.3892/ijo.2016.3483

Li L, Li B, Li M, Niu C, Wang G, Li T, Król E, Jin W, Speakman JR (2017) Brown adipocytes can display a mammary basal myoepithelial cell phenotype in vivo. Mol Metab 6. https://doi.org/10.1016/j.molmet.2017.07.015

Li Z, Wu Q, Sun S, Wu J, Li J, Zhang Y, Wang C, Yuan J, Sun S (2018) Monocarboxylate transporters in breast cancer and adipose tissue are novel biomarkers and potential therapeutic targets. Biochem Biophys Res Commun 501. https://doi.org/10.1016/j.bbrc.2018.05.091

Linklater ES, Tovar EA, Essenburg CJ, Turner L, Madaj Z, Winn ME, Melnik MK et al (2016) Targeting MET and EGFR crosstalk signaling in triple-negative breast cancers. Oncotarget 7. https://doi.org/10.18632/oncotarget.12065

Liu D, Wang X, Chen Z (2016) Tumor necrosis factor-α, a regulator and therapeutic agent on breast cancer. Curr Pharm Biotechnol 17. https://doi.org/10.2174/1389201017666160301102713

Liu S, Lee JS, Jie C et al (2018) HER2 overexpression triggers an IL1a proinflammatory circuit to drive tumorigenesis and promote chemotherapy resistance. Cancer Res 78. https://doi.org/10.1158/0008-5472.CAN-17-2761

Liu LM, Sun WZ, Fan XZ, Ya Li X, Cheng MB, Zhang Y (2019) Methylation of C/EBPa by PRMT1 inhibits its tumor-suppressive function in breast cancer. Cancer Res 79. https://doi.org/10.1158/0008-5472.CAN-18-3211

Lorincz AM, Sukumar S (2006) Molecular links between obesity and breast cancer. Endocr Relat Cancer 13. https://doi.org/10.1677/erc.1.00729

Lourenço AR, Coffer PJ (2017) A tumor suppressor role for C/EBPα in solid tumors: more than fat and blood. Oncogene 36. https://doi.org/10.1038/onc.2017.151

Makki K, Froguel P, Wolowczuk I (2013) Adipose tissue in obesity-related inflammation and insulin resistance: cells, cytokines, and chemokines. ISRN Inflamm 2013. https://doi.org/10.1155/2013/139239

Marshall JCA, Collins JW, Nakayama J, Horak CE, Liewehr DJ, Steinberg SM, Albaugh M et al (2012) Effect of inhibition of the lysophosphatidic acid receptor 1 on metastasis and metastatic dormancy in breast cancer. J Natl Cancer Inst 104. https://doi.org/10.1093/jnci/djs319

Martinez-Outschoorn UE, Lin Z, Whitaker-Menezes D, Howell A, Lisanti MP, Sotgia F (2012) Ketone bodies and two-compartment tumor metabolism: stromal ketone production fuels mitochondrial biogenesis in epithelial cancer cells. Cell Cycle 11. https://doi.org/10.4161/cc.22136

Mauro L, Naimo GD, Gelsomino L et al (2018) Uncoupling effects of estrogen receptor a on LKB1/AMPK interaction upon adiponectin exposure in breast cancer. FASEB J 32. https://doi.org/10.1096/fj.201701315R

McKenzie F, Ferrari P, Freisling H, Chajès V, Rinaldi S, de Batlle J, Dahm CC et al (2015) Healthy lifestyle and risk of breast cancer among postmenopausal women in the European prospective investigation into cancer and nutrition cohort study. Int J Cancer 136. https://doi.org/10.1002/ijc.29315

McTernan CL, McTernan PG, Harte AL, Levick PL, Barnett AH, Kumar S (2002) Resistin, central obesity, and type 2 diabetes. Lancet 359. https://doi.org/10.1016/S0140-6736(02)07281-1

Melgarejo E, Medina MÁ, Sánchez-Jiménez F, Urdiales JL (2009) Monocyte chemoattractant protein-1: a key mediator in inflammatory processes. Int J Biochem Cell Biol 41. https://doi.org/10.1016/j.biocel.2008.07.018

Menendez JA, Lupu R (2007) Fatty acid synthase and the lipogenic phenotype in cancer pathogenesis. Nat Rev Cancer 7. https://doi.org/10.1038/nrc2222

Meyer AS, Miller MA, Gertler FB, Lauffenburger DA (2013) The receptor AXL diversifies EGFR signaling and limits the response to EGFR-targeted inhibitors in triple-negative breast cancer cells. Sci Signal 6. https://doi.org/10.1126/scisignal.2004155

Miana VV, Prieto González EA (2018) Adipose tissue stem cells in regenerative medicine. Ecancermedicalscience 12. https://doi.org/10.3332/ecancer.2018.822

Morris PG, Hudis CA, Giri D, Morrow M, Falcone DJ, Zhou XK, Baoheng D et al (2011) Inflammation and increased aromatase expression occur in the breast tissue of obese women with breast cancer. Cancer Prev Res 4. https://doi.org/10.1158/1940-6207.CAPR-11-0110

Morroni M, Giordano A, Zingaretti MC, Boiani R, De Matteis R, Kahn BB, Nisoli E et al (2004) Reversible transdifferentiation of secretory epithelial cells into adipocytes in the mammary gland. Proc Natl Acad Sci U S A 101. https://doi.org/10.1073/pnas.0407647101

Nam SY, Lee EJ, Kim KR, Cha BS, Song YD, Lim SK, Lee HC, Huh KB (1997) Effect of obesity on total and free insulin-like growth factor (IGF)-1, and their relationship to IGF-binding protein (BP)-1, IGFBP-2, IGFBP-3, insulin, and growth hormone. Int J Obes (Lond) 21. https://doi.org/10.1038/sj.ijo.0800412

Neeland IJ, Ayers CR, Rohatgi AK, Turer AT, Berry JD, Das SR, Vega GL et al (2013) Associations of visceral and abdominal subcutaneous adipose tissue with markers of cardiac and metabolic risk in obese adults. Obesity 21. https://doi.org/10.1002/oby.20135

Nickel A, Blücher C, Al KO et al (2018) Adipocytes induce distinct gene expression profiles in mammary tumor cells and enhance inflammatory signaling in invasive breast cancer cells. Sci Rep 8. https://doi.org/10.1038/s41598-018-27210-w

Nieman KM, Kenny HA, Penicka CV et al (2011) Adipocytes promote ovarian cancer metastasis and provide energy for rapid tumor growth. Nat Med 17. https://doi.org/10.1038/nm.2492

Niu J, Jiang L, Guo W, Shao L, Liu Y, Wang L (2013) The association between leptin level and breast cancer: a meta-analysis. PLoS One 8. https://doi.org/10.1371/journal.pone.0067349

Ollberding NJ, Kim Y, Shvetsov YB et al (2013) Prediagnostic leptin, adiponectin, C-reactive protein, and the risk of postmenopausal breast cancer. Cancer Prev Res 6. https://doi.org/10.1158/1940-6207.CAPR-12-0374

Otvos L, Haspinger E, la Russa F, Maspero F, Graziano P, Kovalszky I, Lovas S et al (2011) Design and development of a peptide-based adiponectin receptor agonist for cancer treatment. BMC Biotechnol 11. https://doi.org/10.1186/1472-6750-11-90

Pallegar NK, Christian SL (2020) Adipocytes in the tumour microenvironment. Adv Exp Med Biol 1234. https://doi.org/10.1007/978-3-030-37184-5_1

Pallua N, Pulsfort AK, Suschek C, Wolter TP (2009) Content of the growth factors BFGF, IGF-1, VEGF, and PDGF-BB in freshly harvested lipoaspirate after centrifugation and incubation. Plast Reconstr Surg 123. https://doi.org/10.1097/PRS.0b013e318199ef31

Pascual G, Avgustinova A, Mejetta S et al (2017) Targeting metastasis-initiating cells through the fatty acid receptor CD36. Nature 541. https://doi.org/10.1038/nature20791

Patel L, Buckels AC, Kinghorn IJ, Murdock PR, Holbrook JD, Plumpton C, Macphee CH, Smith SA (2003) Resistin is expressed in human macrophages and directly regulated by PPARγ activators. Biochem Biophys Res Commun 300. https://doi.org/10.1016/S0006-291X(02)02841-3

Peeraully MR, Jenkins JR, Trayhurn P (2004) NGF gene expression and secretion in white adipose tissue: regulation in 3T3-L1 adipocytes by hormones and inflammatory cytokines. Am J Physiol Endocrinol Metab 287. https://doi.org/10.1152/ajpendo.00076.2004

Péqueux C, Raymond-Letron I, Blacher S, Boudou F, Adlanmerini M, Fouque MJ, Rochaix P et al (2012) Stromal estrogen receptor-α promotes tumor growth by normalizing an increased angiogenesis. Cancer Res 72. https://doi.org/10.1158/0008-5472.CAN-11-3768

Peres Valgas da Silva C, Hernández-Saavedra D, White JD, Stanford KI (2019) Cold and exercise: therapeutic tools to activate brown adipose tissue and combat obesity. Biology 8. https://doi.org/10.3390/biology8010009

Perrier S, Caldefie-Chézet F, Vasson MP (2009) IL-1 family in breast cancer: potential interplay with leptin and other adipocytokines. FEBS Lett 583. https://doi.org/10.1016/j.febslet.2008.12.030

Pierobon M, Frankenfeld CL (2013) Obesity as a risk factor for triple-negative breast cancers: a systematic review and meta-analysis. Breast Cancer Res Treat 137. https://doi.org/10.1007/s10549-012-2339-3

Playdon MC, Bracken MB, Sanft TB, Ligibel JA, Harrigan M, Irwin ML (2015) Weight gain after breast cancer diagnosis and all-cause mortality: systematic review and meta-analysis. J Natl Cancer Inst 107. https://doi.org/10.1093/jnci/djv275

Polanski R, Hodgkinson CL, Fusi A et al (2014) Activity of the monocarboxylate transporter 1 inhibitor azd3965 in small cell lung cancer. Clin Cancer Res 20. https://doi.org/10.1158/1078-0432.CCR-13-2270

Prokesch A, Smorlesi A, Perugini J, Manieri M, Ciarmela P, Mondini E, Trajanoski Z et al (2014) Molecular aspects of adipoepithelial transdifferentiation in mouse mammary gland. Stem Cells 32. https://doi.org/10.1002/stem.1756

Purohit A, Ghilchik MW, Duncan L et al (1995) Aromatase activity and interleukin-6 production by normal and malignant breast tissues. J Clin Endocrinol Metab 80. https://doi.org/10.1210/jcem.80.10.7559896

Purohit A, Newman SP, Reed MJ (2002) The role of cytokines in regulating estrogen synthesis: implications for the etiology of breast cancer. Breast Cancer Res 4. https://doi.org/10.1186/bcr425

Rajarajan D, Selvarajan S, Charan MR, Raja SK, Mahapatra, and Ravi Kasiappan. (2019) Genome-wide analysis reveals MiR-3184-5p and MiR-181c-3p as a critical regulator for adipocytes-associated breast cancer. J Cell Physiol 234. https://doi.org/10.1002/jcp.28428

Ramzan F, D'Souza RF, Durainayagam BR, Milan AM, Markworth JF, Miranda-Soberanis V, Sequeira IR et al (2020) Circulatory MiRNA biomarkers of metabolic syndrome. Acta Diabetol 57. https://doi.org/10.1007/s00592-019-01406-6

Reeves GK, Pirie K, Beral V, Green J, Spencer E, Bull D (2007) Cancer incidence and mortality in relation to body mass index in the million women study: cohort study. Br Med J 335. https://doi.org/10.1136/bmj.39367.495995.AE

Renehan AG, Tyson M, Egger M, Heller RF, Zwahlen M (2008) Body-mass index and incidence of cancer: a systematic review and meta-analysis of prospective observational studies. Lancet 371. https://doi.org/10.1016/S0140-6736(08)60269-X

Rosen ED, Walkey CJ, Puigserver P, Spiegelman BM (2000) Transcriptional regulation of adipogenesis. Genes Dev 14. https://doi.org/10.1101/gad.14.11.1293

Rosendahl AH, Bergqvist M, Lettiero B, Kimbung S, Borgquist S (2018) Adipocytes and obesity-related conditions jointly promote breast cancer cell growth and motility: associations with CAP1 for prognosis. Front Endocrinol 9. https://doi.org/10.3389/fendo.2018.00689

Rottiers V, Näär AM (2012) MicroRNAs in metabolism and metabolic disorders. Nat Rev Mol Cell Biol 13. https://doi.org/10.1038/nrm3313

Rybinska I, Agresti R, Trapani A et al (2020) Adipocytes in breast cancer, the thick and the thin. Cells 9. https://doi.org/10.3390/cells9030560

Sánchez-Jiménez F, Pérez-Pérez A, de la Cruz-Merino L, Sánchez-Margalet V (2019) Obesity and breast cancer: role of leptin. Front Oncol 9. https://doi.org/10.3389/fonc.2019.00596

Sax MJ, Gasch C, Athota VR et al (2016) Cancer cell CCL5 mediates bone marrow independent angiogenesis in breast cancer. Oncotarget 7. https://doi.org/10.18632/oncotarget.13387

Schwartz DR, Lazar MA (2011) Human resistin: found in translation from mouse to man. Trends Endocrinol Metab 22. https://doi.org/10.1016/j.tem.2011.03.005

Seccareccia E, Brodt P (2012) The role of the insulin-like growth factor-I receptor in malignancy: an update. Growth Horm IGF Res 22. https://doi.org/10.1016/j.ghir.2012.09.003

Shen Q, Cohen B, Zheng W et al (2017) Notch shapes the innate immunophenotype in breast cancer. Cancer Discov 7. https://doi.org/10.1158/2159-8290.CD-17-0037

Si YC, Pil JP, Hyun JS, Kim YK, Dong WS, Eui SS, Hyoung HL, Byeong GL, Baik JH, Lee TR (2007) (-)-Catechin suppresses expression of Kruppel-like factor 7 and increases expression and secretion of adiponectin protein in 3T3-L1 cells. Am J Physiol Endocrinol Metab 292. https://doi.org/10.1152/ajpendo.00436.2006

Singh R, Parveen M, Basgen JM, Fazel S, Meshesha MF, Thames EC, Moore B et al (2016) Increased expression of beige/brown adipose markers from host and breast cancer cells influence xenograft formation in mice. Mol Cancer Res 14. https://doi.org/10.1158/1541-7786.MCR-15-0151

Soguel L, Diorio C (2016) Anthropometric factors, adult weight gain, and mammographic features. Cancer Causes Control 27. https://doi.org/10.1007/s10552-015-0706-1

Song X, Zhou X, Qin Y et al (2018) Emodin inhibits epithelial-mesenchymal transition and metastasis of triple negative breast cancer via antagonism of CC-chemokine ligand 5 secreted from adipocytes. Int J Mol Med 42. https://doi.org/10.3892/ijmm.2018.3638

Suba Z (2013) Circulatory estrogen level protects against breast cancer in obese women. Recent Pat Anticancer Drug Discov 8. https://doi.org/10.2174/1574892811308020004

Sultana R, Kataki AC, Borthakur BB, Basumatary TK, Bose S (2017) Imbalance in leptin-adiponectin levels and leptin receptor expression as chief contributors to triple negative breast cancer progression in Northeast India. Gene 621. https://doi.org/10.1016/j.gene.2017.04.021

Sun S, Wu Q, Li J et al (2019) Exosomes from the tumour-adipocyte interplay stimulate beige/brown differentiation and reprogram metabolism in stromal adipocytes to promote tumour progression. J Exp Clin Cancer Res 38. https://doi.org/10.1186/s13046-019-1210-3

Surmacz E (2000) Function of the IGF-I receptor in breast cancer. J Mammary Gland Biol Neoplasia 5. https://doi.org/10.1023/a:1009523501499

Suzuki R, Orsini N, Saji S, Key TJ, Wolk A (2009) Body weight and incidence of breast cancer defined by estrogen and progesterone receptor status-a meta-analysis. Int J Cancer 124. https://doi.org/10.1002/ijc.23943

Takahashi RU, Miyazaki H, Ochiya T (2015) The roles of microRNAs in breast cancer. Cancers 7. https://doi.org/10.3390/cancers7020598

Theriau CF, Sauvé OS, Beaudoin MS et al (2017) Proliferative endocrine effects of adipose tissue from obese animals on MCF7 cells are ameliorated by resveratrol supplementation. PLoS One 12. https://doi.org/10.1371/journal.pone.0183897

Third Report of the National Cholesterol Education Program (NCEP) Expert Panel on Detection, Evaluation, and Treatment of High Blood Cholesterol in Adults (Adult Treatment Panel III) Final Report (2002) Circulation:106. https://doi.org/10.1161/circ.106.25.3143

To SQ, Knower KC, Cheung V, Simpson ER, Clyne CD (2015) Transcriptional control of local estrogen formation by aromatase in the breast. J Steroid Biochem Mol Biol 145. https://doi.org/10.1016/j.jsbmb.2014.05.004

Tsuyada A, Chow A, Wu J et al (2012) CCL2 mediates cross-talk between cancer cells and stromal fibroblasts that regulates breast cancer stem cells. Cancer Res 72. https://doi.org/10.1158/0008-5472.CAN-11-3567

Tulotta C, Ottewell P (2018) The role of IL-1B in breast cancer bone metastasis. Endocr Relat Cancer 25. https://doi.org/10.1530/ERC-17-0309

Ueno NT, Zhang D (2011) Targeting EGFR in triple negative breast cancer. J Cancer 2. https://doi.org/10.7150/jca.2.324

van Landeghem AA, Poortman J, Nabuurs M, Thijssen JHH (1985) Endogenous concentration and subcellular distribution of estrogens in normal and malignant human breast tissue. Cancer Res:45

Velasco-Velázquez M, Jiao X, De La Fuente M et al (2012) CCR5 antagonist blocks metastasis of basal breast cancer cells. Cancer Res 72. https://doi.org/10.1158/0008-5472.CAN-11-3917

Vitali A, Murano I, Zingaretti MC, Frontini A, Ricquier D, Cinti S (2012) The adipose organ of obesity-prone C57BL/6J mice is composed of mixed white and brown adipocytes. J Lipid Res 53. https://doi.org/10.1194/jlr.M018846

Voudouri K, Berdiaki A, Tzardi M, Tzanakakis GN, Nikitovic D (2015) Insulin-like growth factor and epidermal growth factor signaling in breast cancer cell growth: focus on endocrine resistant disease. Anal Cell Pathol 2015. https://doi.org/10.1155/2015/975495

Vrieling A, Buck K, Kaaks R, Chang-Claude J (2010) Adult weight gain in relation to breast cancer risk by estrogen and progesterone receptor status: a meta-analysis. Breast Cancer Res Treat 123. https://doi.org/10.1007/s10549-010-1116-4

Wang T, Fahrmann JF, Lee H et al (2018) JAK/STAT3-regulated fatty acid β-oxidation is critical for breast cancer stem cell self-renewal and chemoresistance. Cell Metab 27. https://doi.org/10.1016/j.cmet.2017.11.001

Wang J, Cai Y, Fangfang Y, Ping Z, Liu L (2020) Body mass index increases the lymph node metastasis risk of breast cancer: a dose-response meta-analysis with 52904 subjects from 20 cohort studies. BMC Cancer 20. https://doi.org/10.1186/s12885-020-07064-0

Wei L, Li K, Pang X et al (2016) Leptin promotes epithelial-mesenchymal transition of breast cancer via the upregulation of pyruvate kinase M2. J Exp Clin Cancer Res 35. https://doi.org/10.1186/s13046-016-0446-4

Willett WC (2002) Balancing life-style and genomics research for disease prevention. Science 296. https://doi.org/10.1126/science.1071055

Wolczyk D, Zaremba-Czogalla M, Hryniewicz-Jankowska A et al (2016) TNF-α promotes breast cancer cell migration and enhances the concentration of membrane-associated proteases in lipid rafts. Cell Oncol 39. https://doi.org/10.1007/s13402-016-0280-x

Wu Q, Sun S, Li Z, Yang Q, Li B, Zhu S, Wang L, Wu J, Yuan J, Yang C, Li J, Sun S (2018) Tumour-originated exosomal miR-155 triggers cancer-associated cachexia to promote tumour progression. Mol Cancer 17:155

Wu Q, Li B, Li Z et al (2019) Cancer-associated adipocytes: key players in breast cancer progression. J Hematol Oncol 12. https://doi.org/10.1186/s13045-019-0778-6

Yan W, Wu X, Zhou W, Fong MY, Cao M, Liu J, Liu X et al (2018) Cancer-cell-secreted exosomal MiR-105 promotes tumour growth through the MYC-dependent metabolic reprogramming of stromal cells. Nat Cell Biol 20. https://doi.org/10.1038/s41556-018-0083-6

Yang XR, Chang-Claude J, Goode EL, Couch FJ, Nevanlinna H, Milne RL, Gaudet M et al (2011) Associations of breast cancer risk factors with tumor subtypes: a pooled analysis from the breast cancer association consortium studies. J Natl Cancer Inst 103. https://doi.org/10.1093/jnci/djq526

Zangani D, Darcy KM, Shoemaker S, Ip MM (1999) Adipocyte-epithelial interactions regulate the in vitro development of normal mammary epithelial cells. Exp Cell Res 247. https://doi.org/10.1006/excr.1998.4373

Zazo S, González-Alonso P, Martín-Aparicio E et al (2020) Autocrine CCL5 effect mediates trastuzumab resistance by ERK pathway activation in HER2-positive breast cancer. Mol Cancer Ther 19. https://doi.org/10.1158/1535-7163.MCT-19-1172

Zhang HH, Halbleib M, Ahmad F et al (2002) Tumor necrosis factor-α stimulates lipolysis in differentiated human adipocytes through activation of extracellular signal-related kinase and elevation of intracellular cAMP. Diabetes 51. https://doi.org/10.2337/diabetes.51.10.2929

Zhang C, Yue C, Herrmann A et al (2020) STAT3 activation-induced fatty acid oxidation in CD8+ T effector cells is critical for obesity-promoted breast tumor growth. Cell Metab 31. https://doi.org/10.1016/j.cmet.2019.10.013

Zhao C, Wu M, Zeng N, Xiong M, Hu W, Lv W, Yi Y, Zhang Q, Yiping W (2020) Cancer-associated adipocytes: emerging supporters in breast cancer. J Exp Clin Cancer Res 39. https://doi.org/10.1186/s13046-020-01666-z

Zheng X, Ning C, Dong Y, Zhao P, Li J, Fan Z, Li J, Yu Y, Mrode R, Liu JF (2017) Quantitative proteome analysis of bovine mammary gland reveals protein dynamic changes involved in peak and late lactation stages. Biochem Biophys Res Commun 494. https://doi.org/10.1016/j.bbrc.2017.10.038

Zhou W, Guo S, Gonzalez-Perez RR (2011) Leptin pro-angiogenic signature in breast cancer is linked to IL-1 signalling. Br J Cancer 104. https://doi.org/10.1038/sj.bjc.6606013

Zwick RK, Guerrero-Juarez CF, Horsley V, Plikus MV (2018) Anatomical, physiological, and functional diversity of adipose tissue. Cell Metab 27. https://doi.org/10.1016/j.cmet.2017.12.002

Chapter 13
Extracellular Matrix as a Metabolic Niche in Cancer

Anna Sebestyén, Titanilla Dankó, Dániel Sztankovics, Dorottya Moldvai, Ildikó Krencz, Regina Raffay, and Gábor Petővári

Abstract Heterogeneity in tumor mass with an altered tumor microenvironment is a hallmark of cancer metabolism, which provides a metabolic niche for cells. Tumor cells with their higher metabolic plasticity have competitive advantage in tumor tissue evolution. The high glucose consumption, in parallel with impaired vascularization related to tumor growth, usually results in lactate production, extracellular glutamine depletion, and increasing acidification. These alterations completely rewire the activity of non-tumor cells (cancer-associated fibroblasts, inflammatory cells, adipocytes, endothelial cells, etc.). Additionally, amino acids, lipids, or other structural matrix elements and their degradation products—as nutrients—have critical importance in building the biomass for cancer growth. Changes in the extracellular matrix influence and regulate metabolite composition of this niche; moreover, these modifications can induce metabolic alterations in tumor cells (e.g., Warburg effect, OXPHOS phenotype, and autophagy), alter metabolic crosstalk between tumor and non-tumor cells, generate metabolic heterogeneity in tumor tissue, and finally affect the homeostasis of the organ-body, as well. This metabolic symbiosis can initiate and/or promote the aggressiveness and progression of cancer.

13.1 Metabolic Rewiring and Altered Cellular Bioenergetics

Growing cancer has many well-known hallmarks—including metabolic rewiring and altered cellular bioenergetics which are highly influenced by the special environment, nutrients, and growth factors—and requires anabolic properties to facilitate cell growth. During the adaptation to the permanently changing conditions in tumor tissues, cells need to rearrange the activity of metabolic pathways (catabolic/anabolic

---

A. Sebestyén (✉) · T. Dankó · D. Sztankovics · D. Moldvai · I. Krencz · R. Raffay · G. Petővári
1st Department of Pathology and Experimental Cancer Research – Tumor Metabolism Research Group, Semmelweis University, Budapest, Hungary
e-mail: sebestyen.anna@med.semmelweis-univ.hu

I. Kovalszky et al. (eds.), *The Extracellular Matrix and the Tumor Microenvironment*, Biology of Extracellular Matrix 11,
https://doi.org/10.1007/978-3-030-99708-3_13

processes) to achieve nutrient supplements, bioenergetic forces regarding tumor cell growth and survival. This feature—metabolic rewiring—has an important function in tumor development and progression. Stromal components can support or inhibit tumor cell proliferation, survival, and adaptation (Hanahan and Weinberg 2011). Accordingly, the complex stromal microenvironment with its matrix elements and the non-tumorous cells deserves special interest (Schwörer et al. 2019). The altered metabolic by-products of tumor cells influence the functions and even the extracellular matrix (ECM) composition of the surrounding cells, and also respond to the tumor cell-derived paracrine factors. Oncogenic alterations in both tumorous and parental cells disrupt normal organ and tissue ECM structure (Egeblad et al. 2010; Li et al. 2018; Elia and Haigis 2021). These alterations are able to reprogram non-malignant cells, and influence their produced ECM proteins and other metabolic by-products (as potential nutrients). These drive the supporting functions of ECM to force tumor cell adaptation and cancer progression.

Proliferating cancer cells reorganize the tissue architecture and the adjacent stroma; furthermore, tumor growth becomes dependent on vascularization and stromal supports. Several studies suggest that tumor/stroma ratio, and the structure of the stroma (e.g., compact, desmoplastic stroma) may correlate with the prognosis of cancers (e.g., colon and breast cancers) subtype dependently (Kramer et al. 2019; van Pelt et al. 2018). However, to use tumor/stroma ratio as a reproducible prognostic factor needs robust clinical validation in future prospective studies. The special roles of certain cell types—e.g., cancer-associated fibroblasts (CAFs) and immune regulatory cells—were described in the progression and formation of several tumors in the last three decades (Hanahan and Coussens 2012; Casazza et al. 2014; Louault et al. 2020; Karta et al. 2021). Alterations in ECM—including matrix proteins, metabolites, and acidification—are intensively studied in parallel with the non-tumorous stromal cells (mesenchymal, immune, and endothelial cells) in the microenvironment (Fig. 13.1). It was proposed that *heterogeneity of metabolic pathway alterations in tumor mass with altered tumor microenvironment is a hallmark in cancer* metabolism, which provides a metabolic niche in cancer (Elia and Haigis 2021; Pavlova and Thompson 2016). It was highlighted that ECM nutrients produced by several different cells provide building blocks for tumor cell growth. In addition, the accumulation of metabolic "waste"/by-products has several regulatory functions that influence the behavior of surrounding cells and the remodeling in ECM (Baltazar et al. 2020). The secreted metabolic products (as lactate or different amino acids) by certain cells can be utilized in a competition among various cell types (metabolic symbiosis) where tumor cells can orchestrate the consumption and gain several advantages. Metabolites have both self-mediated paracrine effects in the microenvironment and/or induced actions via signaling mechanisms in stromal cells or the related matrix elements. These effects could have special roles both in the matrix and in the cellular elements of tumor tissues. In this chapter, we summarize the roles of ECM in metabolic symbiosis and review how these affect tumor microenvironment and growth in a complex manner (Yoshida 2021).

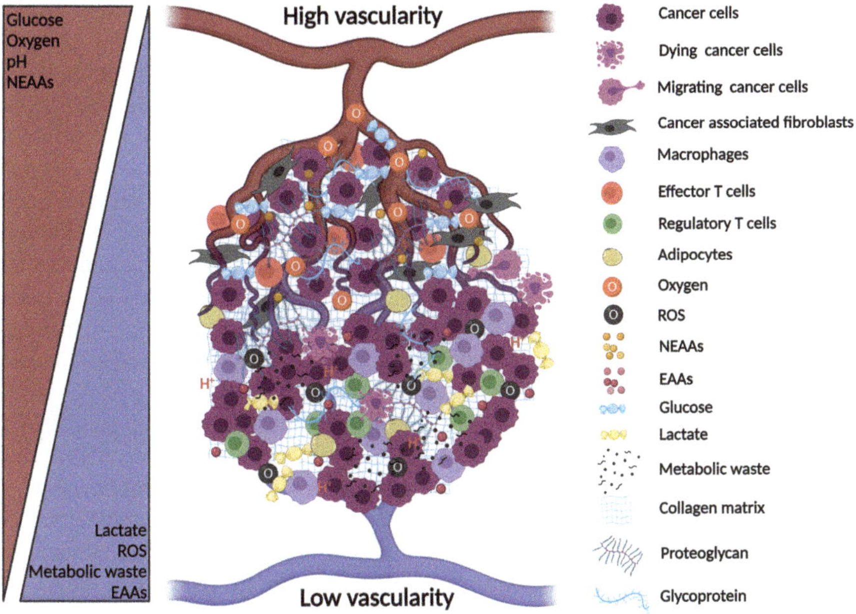

Fig. 13.1 Altering metabolic compartments including metabolic microenvironment in tumor tissue. Several elements of the microenvironment and their altered metabolic activity contribute to tissue and metabolic complexity. Interactions/communications and the symbiosis of different cell populations include the altered concentration of macromolecules and metabolites. Altering nutrients, amino acids (essential and non-essential amino acids—EAAs/NEAAs), glucose, lactate, waste, and the developing pH gradients, oxygen concentration differences, reactive oxygen pressures (ROS—reactive oxygen species), and ECM element deposition rewire the tissue microenvironment and increase cellular and metabolic heterogeneity of the tumor tissue in correlation with vascularization. Gradients, altering conditions were labeled in blue/red triangles on the left side of the figure. Cellular elements (proliferating, surviving, and dying cancer cells, effector and regulator T cells, macrophages, adipocytes, and fibroblast—cancer-associated fibroblasts) and extracellular proteins, proteoglycans, and metabolites were also indicated

13.2 Extracellular Matrix Elements Compensate Starving Conditions During Tumor Cell Growth and Survival in Tissue Microenvironment

Tumor growth and survival need both bioenergetic supply and building blocks for new macromolecule biosynthesis at starving conditions. These processes can be maintained by the uptake of ECM-derived metabolites. Glucose and amino acid consumption are the main sources of carbon required for proliferating cell mass in tumor tissues. *Glucose consumption* is high in correlation with the glycolytic activity of cancer cells; however, there are other metabolic pathways that require glucose (including pentose phosphate pathway—generating pentose phosphates for nucleotide synthesis and NADPH, hexosamine pathway—requiring for protein glycosylation, gluconeogenesis—generating glycogen in glucose storage, amino acid synthesis, and one-carbon metabolism—producing NADPH, purines, glutathione, methylation, etc.). Compared with normal cells, the glucose influx and glycolysis are accelerated in tumor cells due to the overexpression of transporters and glycolytic enzymes. It is also well known, a general statement that "anaerobe glycolysis is enhanced and OXPHOS capacity is reduced in many cancer cells". The well-known Warburg effect, the lactate producing anaerobe glycolysis, is the more suitable and fast way to gain energy for cancer growth, but it yields less ATP/glucose. Therefore, growing new cells need other metabolic intermediates for maintaining macromolecule biosynthesis, as well. These demands can be supplied by truncated TCA cycle and other TCA fulfilling processes. These mechanisms highly reduce the concentrations of glucose and some additional metabolites (including glutamine or other amino acids, lipids) in the ECM.

Glutamine utilization and TCA cycle fulfilling mechanisms, the anaplerosis, support tumor growth/survival in hypoxic-pseudohypoxic conditions (Wise and Thompson 2010). Additionally, glutamine uptake facilitates the neutralization of reactive oxygen species (ROS) and nucleotide biosynthesis. As glutathione plays a role in the exchange of non-essential amino acids (NEAAs) such as glutamine, cysteine, and serine, glutamine and glutamate levels can be reduced in the ECM. Consequently, the ammonia level is elevated due to amino acid consumption intra- and extracellularly. Moreover, the inappropriate vascularization, the constant delay in nutrient and oxygen supply—further decreases the level of building block metabolites in the ECM (Lane et al. 2020; Tímár et al. 2021). Acetate, as acetyl-CoA source, could also be used to support lipid/fatty acid synthesis by ACSS2 (acetyl coenzyme synthetase) overexpression or Ras-transformed cells scavenge unsaturated fatty acids from phospholipids in several cancers (Lovinfosse et al. 2016; Huang et al. 2018; Currie et al. 2013). These alterations also support the growth and survival of tumor cells; however, the non-malignant cells have no such adaptation possibility or plasticity in this nutrient- and O_2-depleted microenvironment. Thus, tumor cells with their higher metabolic plasticity have competitive advantages in selection and tissue evolution. The most important metabolic pathways and some key elements in nutrient utilization regarding cellular metabolic adaptation were summarized in Fig. 13.2.

Additionally, growing tumors invade the microenvironment, cause injury, and inflammation as permanent stress, which damage cellular and matrix components. These result in degradation-derived supplements, where ECM elements and matrix proteins provide nutrient sources for tumor cells in case of poor vascularization (Johnson et al. 2016). The matrix could compensate starving conditions at protein, amino acid, and lipid metabolism levels (Commisso et al. 2013; Olivares et al. 2017; Guillaume et al. 2019). The affected and actively participating stroma of the developing tumor has many features similar to wound healing which influence tumor cell growth and survival (Dvorak 1986, 2015; Singer and Clark 1999; Balkwill and Mantovani 2001). Cytokine and other factors producing tumor cells alter the cellular and metabolite compositions. These changes are also accompanied by the accumulation and/or decrease of different metabolites and alter the non-tumorous cell

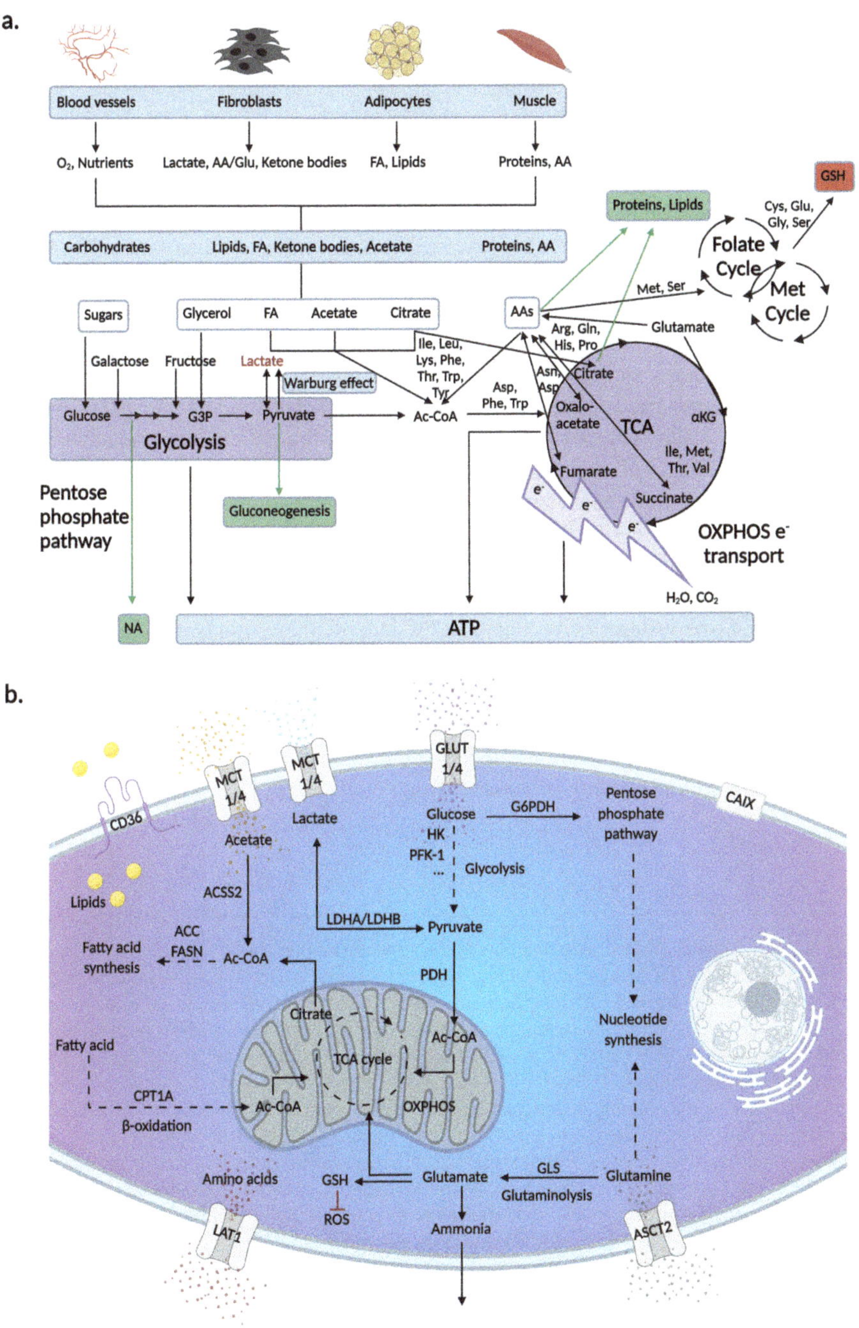

Fig. 13.2 Metabolic pathways in nutrients and building block cycles. (**a**) Cancer cells can undergo metabolic changes to adapt to altering microenvironments. Metabolic plasticity helps to use available metabolites which come from different sources (vessels, fibroblasts, adipocytes, muscle, or other cellular elements). Metabolic compartments O_2, nutrients, lactate, ketone bodies, lipids, proteins, and amino acids (AA) allow supply flexibility by using multiple metabolites for growing and surviving tumor cell requirements. Carbohydrates (e.g., sugars), lipids, fatty acids (FA), or

composition in the microenvironment. Tumor cells facilitate further reorganization of the surrounding cellular milieu, including immune, mesenchymal cells, fibroblasts, and other cells with non-tumor origin (e.g., adipocytes). Stromal cells participate in the coordination of tissue regeneration; therefore, these produce several regulatory factors (cytokines, growth factors) to adjust tissue remodeling (Hynes and Naba 2012). Monocytes release many small molecules, as danger-associated-molecular patterns, which are usually paracrine metabolites, nucleotides, ATP, pro- and/or anti-inflammatory cytokines. The increased amount of immunosuppressive regulatory T (T_{reg}) cells, M2 macrophages, and myofibroblasts contributes to the production of tissue regenerating cytokines (TGFβ; PDGF) and chemotactic factors initiating ECM remodeling. The produced cytokines and other factors promote the trans-differentiation of myofibroblasts to *cancer-associated fibroblasts (CAFs)*. CAFs are characteristic for many solid cancers, and support the proliferation, adaptation, and survival mechanisms as well as facilitate the metastatic capacity of tumor cells. Cancer cell metabolism itself has a direct effect on the establishment of tumor-promoting microenvironment. In addition with tumor growth-related high nutrient consumption, the nutrient pool of tumor tissues could be completely exhausted, as a consequence of metabolic waste release into the microenvironment (Guillaume et al. 2019).

In 1986, *Dvorak* described cancer as a "wound that does not heal". This statement was confirmed by several observations since many characteristics of growing tumors are similar to healing wounds: such as *high rate of glucose consumption, lactate production, and extracellular glutamine depletion* after trauma and hypoxia (Dvorak 1986, 2015). Based on these, the metabolic alterations occurring in tumor tissues resemble the wound healing processes (Byun et al. 2018; Dvorak 1986).

Fig. 13.2 (continued) other metabolites can fuel the metabolic pathways (glycolysis, pentose phosphate pathway, TCA cycle, and ATP generation). Moreover, several nodes in these pathways (their metabolites) present different building blocks for the newly produced cancer cells to synthetize their macromolecule requirements (green arrows, nucleic acid—NA, protein, and lipid synthetic processes). This metabolic plasticity and flexibility contribute to the production of certain important building blocks (green background) and energy for growing tumor cells (gray-blue background) and maintaining redox homeostasis (red background, glutathione—GSH) help to survive starved, injured situations, e.g., in dormant state (glycerol 3 phosphate—G3P; αKG—alpha-ketoglutarate; Ac-CoA—acetyl-coenzyme A; amino acids were indicated using their common three-letter abbreviations). (**b**) Schematic figure of the above summarized main pathways in the cell which help to orientate the associations of several metabolic pathways during metabolic rewiring. The figure does not contain all metabolic enzymes and processes but we tried to present the ones which were referred in the main text (ACSS2—acetyl coenzyme synthetase, ACC—acetyl coenzyme A carboxylase, Ac-CoA—acetyl coenzyme A, LDHA/B—lactate dehydrogenase A/B, HK—hexokinase, PFK-1—phospho-fructokinase-1, G6PDH—glucose-6-phosphate dehydrogenase, PDH—pyruvate dehydrogenase, OXPHOS—oxidative phosphorylation, CPT1A—carnitine palmitoyltransferase 1A, GSH—glutathione, GLS—glutaminase, CAIX—carbonic anhydrase IX, GLUT1/4—glucose transporter 1/4, MCT1/4—monocarboxylate transporter 1/4, LAT1—L-type amino acid transporters, ASCT2—alanine, serine, cysteine transporter 2)

13.3 The Effects of Lowering Glucose Concentration and High Lactate Level (Tissue Acidification)

Tumor cell proliferation is supported by several oncogenic mutations and hyperactivated growth signals. These induce the expression of certain genes to serve bioenergetic demands. Increase in the transcription of myc-regulated glucose transporters and glycolytic enzymes speeds up glucose uptake and catabolism. Since glucose concentration has several regulatory functions in ECM homeostasis, the reduced glucose level—as an important metabolic checkpoint—exerts numerous effects in the tumorous ECM. The high glucose consumption, in parallel with impaired vascularization, results in about *90% decrease of glucose concentration* in interstitial tissue fluids during intensive tumor growth (Gullino et al. 1964). Simultaneously, the immune response activating cells, inflammatory T cells, macrophages, and fibroblasts are mainly glycolytic and require glucose for their ATP consumption. High ^{18}F-flourodeoxiglucose (FDG) uptake of the tumor and other non-tumor cells (as CAFs or activated immune cells) in the cancer microenvironment is a well-known problem of routine analyses in PET-CT (e.g., the FDG positivity of immune cells). This phenomenon is associated with both the Warburg phenotype of rewired non-tumor cells and the metabolic symbiosis (Sugita et al. 2021).

The increased aerobe and/or anaerobe (pseudohypoxia) glycolysis (Warburg effect) were described in tumor cells by *Otto Warburg* about 100 years ago (Warburg et al. 1927). The elevated intracellular level of lactate initiates the lactate/H^+ transport accomplished by the monocarboxylate transporters (MCTs) which increase *tissue acidosis*. The consequences of acidic microenvironment and its tumor-promoting effects were clarified in the last decade (Payen et al. 2020). O_2 consumption and the resulted CO_2 release in tumor mass also contribute to the decrease of the pH induced by the HCO_3^- transports of carbonic anhydrases (CAs). CAIX, together with other bicarbonate transporters, promote the extracellular H^+ accumulation and acidification (Benej et al. 2020; Gillies 2021; Lee and Griffiths 2020). CAIX inhibitors, which are already studied in clinical trials, may help to inhibit the metabolic adaptation in the future (McDonald et al. 2020; Jamali et al. 2015; Aldera and Govender 2021; Lau et al. 2017).

13.3.1 Immunosuppressive Effects of Increased Lactate Level in the Extracellular Matrix

Effector T cells have high glucose demand in immune defense, and the lowering glucose impairs many immune effector functions in the microenvironment such as IFNγ production and T helper 1 differentiation. In addition, the depleted glucose shifts T_{reg} differentiation and function. T_{reg} cells are less sensitive to lowered glucose concentration than other effector T cells. In T_{reg} cells, the lowered glycose concentration and as a consequence the decreasing ATP level elevate AMPK activity and reduce glucose oxidation. In parallel, however, lipid/fatty acid oxidation can be increased to compensate glucose depletion in T_{reg} cells. Based on these, the available glucose concentration could be an important metabolic checkpoint to regulate anti-tumor immune response of T cells in tumor mass (Siska et al. 2020; Angelin et al. 2017). It was described that PD-1/PD-L1 interactions and their signs in disrupting CD28-mediated co-stimulation inhibit glucose transport and consumption in activated T cells. Regarding this, it was detected that PDL1 immune checkpoint therapy upregulates GLUT1 expression of effector T-cell population to uptake glucose more effectively in competition with tumor cells in their microenvironment. The regulatory role and the potential competition for available glucose in the microenvironment were confirmed by other findings since adaptive T-cell therapy is less effective in highly glycolytic tumors (Chang et al. 2015). There are many publications about glucose demand and Warburg phenotype of *inflammatory M1 macrophages*, as well. All these underline that enhancing glycolytic flux promotes M1 macrophage differentiation and low glucose concentration reverses this and promotes the polarization of anti-inflammatory M2 macrophage in the microenvironment (Bader et al. 2020). Tumor-promoting capabilities are characteristic for M2 macrophages involving immunosuppressive and angiogenic effects, as well as stromal remodeling.

It was summarized that increasing lactate level inhibits the functions of NK and NKT cells during cancer progression. It was documented that lactate induces the apoptosis of NK cells (Harmon et al. 2019) and the silenced lactate dehydrogenase (LDH) caused the better cytolytic activity of NK cells in a pancreatic cancer xenograft model (Husain et al. 2013). In correlation with these, a recently published work described that lactate could inhibit mTOR signaling, block IFNγ, and IL-4 in NKT cells (Xie et al. 2016).

Increasing lactate can influence cytokine production of immune regulatory cells, which can prevent the final differentiation of dendritic cells (remaining tolerogenic) leading to an increase in immunosuppressive IL-10 production (Nasi et al. 2013). In addition, lactate promotes the overexpression of IL-23 in different types of tumor cells (e.g., melanoma, colon, breast, and gastric cancers) (Langowski et al. 2006), which enhances the expression of IL-17, matrix metalloprotease 9 (MMP-9), induces angiogenesis, and reduces the number of cytotoxic T cells, as immunosuppressive effects (Shime et al. 2008).

The above-described examples are evidences which indicate that lactate helps to avoid the recognition of tumor cells and promotes tumor progression through its diverse immunosuppressive effects in tissue microenvironment.

13.3.2 The Effects of Lactate in Metabolic Symbiosis (Including Several Functions of Fibroblasts and Endothelial Cells)

Tumorigenic alterations lead to PI3K/Akt/mTOR pathway hyperactivation in tumor cells and increase glucose uptake as metabolic driving forces. The lowered oxygen level, the hypoxic conditions, or the pseudohypoxic regulatory changes induce the Warburg effect and the accumulation of lactate in the microenvironment (de la Cruz-López et al. 2019; Koppenol et al. 2011). However, in oxygenated environment both tumor and surrounding other cells (mainly fibroblasts) can uptake and catabolize lactate and other metabolic intermediates (e.g., carbohydrates, ribose, certain lipids, fatty acids, acetate, and amino acids) which ensure the requirements of their growth and survival. These situations can contribute to varied metabolite concentrations in the extracellular environment and alter metabolic milieu (Thompson and Bielska 2019). Furthermore, tumor cells have special adaptation processes and advantages in nutrient- and O_2-depleted ECM supported by the rearrangement of catabolic and anabolic processes, and force CAFs to initiate nutrient production and release for themselves (as nutrient supplies for cancer progression). Local metabolic pathway redistributions of nutrient utilization and building block cycles as well as cellular *metabolic symbiosis* serve tumor cell adaptation.

Decreasing glucose concentration resulting from tumoral consumption lowers intracellular ATP level in starving stromal fibroblasts. This activates AMPK, consequently inactivates mTOR kinase, and switches on autophagy for bioenergetic survival mechanism in fibroblasts. These processes and the autophagy-dependent mechanisms guide amino acids into the ECM to replenish the consumed nutrients and selectively promote tumor cell growth (Sousa et al. 2016). Further effects of low glucose in the growing tumors create a *circulus vitiosus* in tissue vascularization. Low glucose level decreases phospho-fructokinase-1 (PFK-1), and consequently the proliferation and migration of sprouting *endothelial cells*. This impaired vascularization initiates the decrease of nutrient and glucose supply in certain parts of the tumor mass (Cantelmo et al. 2016). All these effects of lowered glucose level in the ECM promote the transformation of tumor-associated stromal myofibroblasts to CAFs, and alter the distribution of immune cells participating in the matrix remodeling of growing tumor mass. The above-described situation with intensive aerobe glycolysis and hypovascularization causes *lactate accumulation* in the ECM (García-Cañaveras et al. 2019). Additionally, the high intratumoral and vascular concentrations of lactate increase the risk of metastasis and death.

Tissue/tumor heterogeneity and the adaptation to oxygen gradient (related to different vascularization) lead to rational energy consumption which is in correlation with lactate transport. This promotes the so-called *"waste" (e.g., lactate) use in oxygen-rich microenvironment* near the vessels. For directing this shuttle, MCTs help the lactate flux. MCT1 is involved in lactate influx and efflux in all cells. In addition, MCT4 expression is characteristic for tumor cells in hypo-oxygenated microenvironment, where effective lactate efflux is necessary (in highly glycolytic cells). Recent studies emphasize the importance of *both glucose and lactate*—as metabolic substrates. These two metabolites can be nutrients for cells depending on their microenvironment and oxygen supply (Payen et al. 2020).

This implies that *lactate is not a real waste* in the ECM of tumor mass. Lactate can be taken up and oxidized in cells with mainly oxidative phosphorylation (OXPHOS) metabolism or in chemotherapy-resistant cells (Taddei et al. 2020). For example, in spheroid-forming breast cancer cells (mammospheres), lactate oxidation can promote long-term survival and late tumor initiation (Lamb et al. 2015). Several new results described that lactate among other alternative nutrient substrates can be oxidized in quiescent “stem”-like cancer cells with OXPHOS phenotype (Farnie et al. 2015; Ózsvári et al. 2020; Intlekofer and Finley 2019). It was also suggested that the therapy-resistant cells are also able to use lactate in the course of tumor evolution (Ippolito et al. 2016). Regarding these, the inhibition of lactate uptake and/or oxidation can disturb the survival and adaptation of cancer cells in disease progression, especially in surviving “dormant” cells (Akkoc et al. 2021).

Lactate released from hypo-oxygenated cells can be taken up either by benign or malignant highly oxygenated cells in the so-called “two- or three-compartment models” in different cancers (breast, pancreas, lung cancer, etc.). This *metabolic symbiosis* can promote or initiate the aggressiveness of cancer and additionally, the altered metabolite levels could have several indirect tumor growth-promoting effects both on malignant and on non-malignant cells. Based on these, *lactate is an important oncometabolite* in this symbiosis and its consequence—tissue acidosis—has further extracellular and intracellular signaling effects. These effects influence the secretion of growth factors both by the tumor cells and by other cellular elements of the tumor stroma, including *fibroblasts, endothelial cells, and adipocytes*. CAFs start to secrete HGF, ILs, other cytokines, and growth factors (TGFβ, IFNs, FGFs, VEGF, etc.) and in addition, release regulatory proteins, metabolites, and/or epigenetic regulators (e.g., regulatory miRs) including exosomal transport (Gorchs and Kaipe 2021). These microenvironmental rewirings promote the motility, migratory, and invasive alterations of tumor cells.

Intracellularly lactate can function as a direct inducer of hypoxia response in hypoxia-related protein stabilization. Lactate can bind and stabilize oxygen level-regulated NDGR3 protein (this prevents its PHD2/VHL-dependent degradation) (Lee et al. 2015). Therefore, NDRG3 protein is able to bind c-Raf, activate Raf-Erk pathway, and induce NF-kB, which promotes, for example, VEGF and bFGF production in cancer cells and growth factor receptor expression in the surrounding endothelial cells tumor type dependently (Sonveaux et al. 2012).

13.3.3 Direct Receptor-Ligand Signaling Effects of Lactate on Tumor Cells

It was described that several G-protein-coupled receptors can be activated by metabolites responding to secreted products of metabolism (fatty acids, mono- and disaccharides, amino acids, or metabolic intermediates ketone bodies, lactate, succinate, etc.). In the extracellular niche, the direct receptor-ligand effect works through lactate-activated G-protein-coupled receptor GPR81, also known as hydroxycarboxylic acid receptor 1 (HCAR1) (Parks et al. 2020). This receptor can be found in the plasma membrane and other intracellular membrane organelles; its activation lowers cAMP level and inhibits PKA signaling mechanisms. The receptor function could be regulated by direct lactate receptor interaction or conformation modification in acidic microenvironment. Recent publications reported the elevated expression of GPR81 in many cancers (e.g., breast, cervical, and liver) and cancer cell lines (e.g., colon, breast, lung, and pancreatic cancer cell lines). These alterations showed association with the aggressiveness, therapy resistance, and survival of cancer cells in model systems.

Furthermore, proton-sensitive lactate sensors (GPR4/65/68/132) were also described which are influenced by acidic tissue microenvironment (low pH) in macrophages (Parks et al. 2020). The H^+-sensing mechanisms can activate intracellular Ca^{2+}, cAMP, ROS, and these alterations lead to CREB-mediated alterations in different cells through cellular signaling network including MAPK pathways and protein kinase activations. In addition, these further increase matrix MMP expressions and epithelial-mesenchymal transitions (Parks and Pouysségur 2017) influencing tissue remodeling and cancer progression.

13.3.4 Lactate Induces Tissue Remodeling

High lactate concentration and its consequence—acidosis—in the microenvironment (the acidic niche) reorganize tissue elements which stimulate cancer cell overgrowth and invasion into normal tissues (Gillies and Gatenby 2015). Acidic microenvironment is linked to proteolysis of ECM elements, as a result of the synthesis and the activation of MMPs. These proteins can derive from tumor and stromal cells; moreover, macrophages and fibroblasts redistribute their lysosomes to the cell periphery in acidic microenvironment. Besides, cathepsinB is secreted in that way directly into the microenvironment. *Kobayashi* suggested first that this membrane cathepsinB-enhanced proteolytic network could have a role in metastatic matrix remodeling (Kobayashi et al. 1993; Sameni et al. 1995; Glunde et al. 2003).

Hypoxia itself and its consequences—e.g., the lactate production and the acidic microenvironment—could activate certain factors as latent TGFβ and increase the expression and/or activity of other lysosomal degradation enzymes. These alter the architecture of ECM, degrade collagens (e.g., collagen IV), increase enolase-1, and plasminogen activation (Taddei et al. 2013). Finally, the unbalanced ECM degradation, the remodeling of basement membrane, and stiffness contribute to the cancer cell invasion, as well. These alterations could help matrix remodeling which assist progression, alter the polarity and adhesion capacity of the tumor cells, and additionally, propagate invasion and metastasis. For example, collagen deposits are elevating in the early progression of breast cancer which increase tissue stiffness with their cellular consequences (Najafi et al. 2019). These result in the expression changes of lysyl oxidase and hydroxylases catalyzing intramolecular crosslinks between collagens and elastins (Qi and Xu 2018). Both enzymes are overexpressed and could hydroxylate collagens in response to hypoxic conditions. The elasticity and strength of the matrix, the increase/modification of crosslinking the matrix elements were described in many cancers, these usually negatively influenced the progression (e.g., breast, head and neck, pancreatic, bronchial cancers glioblastomas, and melanomas) (Deville and Cordes 2019). Additionally, non-enzymatic crosslinking by glycanation/transglutamination and fibronectin-mediated collagen reorganization could also occur in the later stages of tumor progression.

Based on the previously summarized alterations in the exchange of glucose and lactate of the tumor microenvironment, the increasing amount of lactate in the ECM contributes to the progression and therapy resistance of different cancers (Fig. 13.3). These in situ alterations favor migration and metastasis with the help of altered matrix-specific intracellular molecule distribution. Therefore, quantification of lactate and pH alterations has an increased interest in cancer research (Lau and Heiden 2020).

13.4 Other Altering Non-cellular Elements as Metabolic Factors in the Extracellular Matrix

During nutrient, glucose, and/or amino acid deprivations, the uptake and the catabolic degradation of ECM elements can also help to maintain bioenergetic homeostasis (Muranen et al. 2017; Olivares et al. 2017). In this situation, non-tumorous cells increase MAPK phosphorylation, decrease mTOR activity, and induce autophagy to cover the energy demand for survival. In this situation, CAFs could be forced by tumor cells to produce metabolites as nutrients through autophagy activation. Based on several studies, both nutrient deprivation and oxidative stress activate autophagy and stabilize HIF1α in CAFs and stromal fibroblast in cancers (Linares et al. 2017; Martinez-Outschoorn et al. 2010; Valencia et al. 2014; Ono et al. 2009).

Amino acids, as nutrients, and even their uptake from the extracellular environment have critical importance in cell growth for building the biomass of proliferating tumor cells (Hosios et al. 2016). These processes deplete the amount of NEAAs (Eagle 1955) in the ECM. As it was previously mentioned, glutamine is highly consumed by tumor cells promoting intensive cell growth in several ways, e.g., it can fulfill TCA cycle intermediates (anaplerosis). Additionally, glutamine—similarly to serine—is important for nucleotide biosynthesis. Glutamine, cysteine, and glycine have a role in maintaining redox homeostasis through glutathione exchange and/or precursors for glutathione production (Valencia et al. 2014). Tumorous and normal cells are in competition for amino acids in nutrient-depleted conditions (fast lowering amino acid level). At starving situation, the oncogene-driven increased expression of transporters and certain other mechanisms, as macropinocytosis of ECM proteins, can give some advantages for fast proliferating tumor mass (Su et al. 2021). Nutrients are used in a well-organized and optimized symbiosis in a heterocellular microenvironment. In a cooperative metabolic way, all cells consume, survive, and additionally, tumor cells mainly proliferate. In metabolic stress forced by the environment, CAFs can utilize glutamate and lactate; furthermore, they provide cysteine

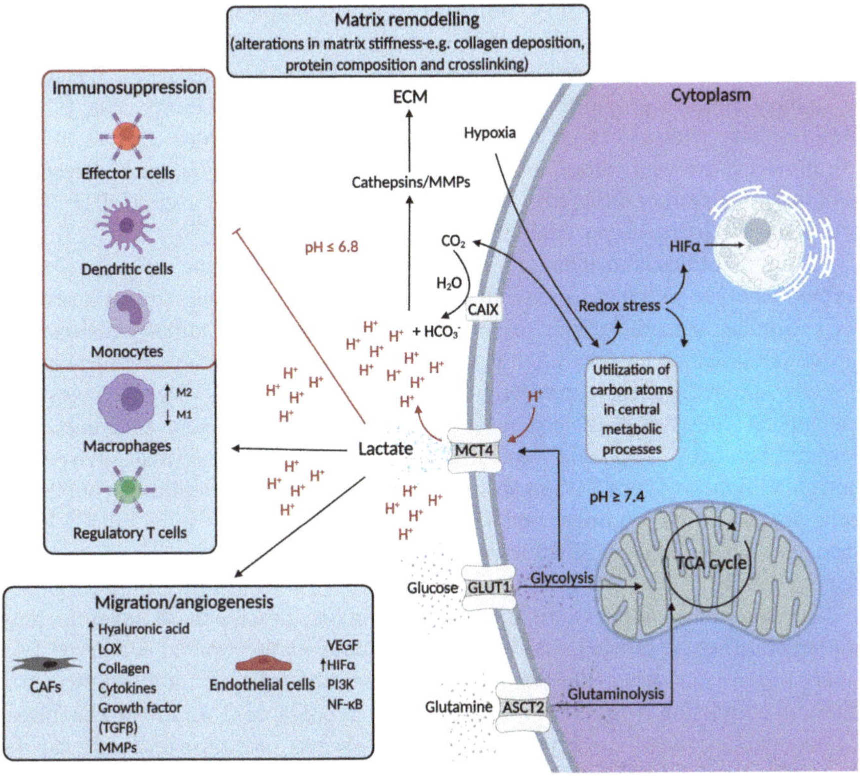

Fig. 13.3 Altering lactate concentration and pH, the role of acidic microenvironment in tumor mass. Both tumor and stromal cells can produce lactate related to their oxygen supply and/or tumorigenic alterations (e.g., pseudohypoxia). The accumulation of lactate and its consequence, the acidic pH, are associated with increased aggressiveness, which inhibit anti-tumor immune response and induce resistance to therapy, invasion, and metastasis. Lactate can be used as an alternative nutrient supply for well-oxygenated tumor cells; negatively regulates innate and adaptive tumor-infiltrating immune cells (e.g., impairs monocyte differentiation, decreases antigen presentation; inhibits immune effector cells, natural killer cells—NK and cytotoxic T cells; promotes immunosuppressive cells—M2 macrophage differentiation and myeloid-derived suppressor cells, and regulatory T cells), these contribute to immune escape mechanisms. The expression profile of metabolite transporters (CAIX—carbonic anhydrase IX, GLUT1—glucose transporter 1, MCT4—monocarboxylate transporter 4, ASCT2—alanine, serine, cysteine transporter 2) can also change at the surface of tumor cells to facilitate uptake and release of several nutrients from the ECM to serve and give selective growing source for the tumor cell proliferation. The tumor cells and the recruited cancer-associated fibroblasts (CAFs) start to produce cytokines, growth factors, and metabolites for the growing demand of tumor cells. Additionally, the acidic microenvironment activates several matrix-embedded enzymes (e.g., matrix metalloproteases—MMPs, cathepsins) for matrix degradation and remodeling, liberating tumor cells from tissues and promoting migration and metastasis. Based on these changes, the cancer-generated lactic acidic microenvironment through reprogrammed metabolism represents a critical immunosuppressive and tumor-promoting niche. Lactate is rather a metabolite than a "waste" product in these situations

and glutamine for cancer cells. Moreover, this adaptation mechanism—especially cysteine and glutamine uptake and their use in cancer cells—can be inhibited by effector T cells (Yang et al. 2016; Sousa et al. 2016). In other cases, the forced autophagy in CAFs provides alanine and several dipeptides for tumor cells, as well. Besides, fatty acids and adipocytes can provide glutamine-arginine conversion from citrulline for growing tumor cells. These feeding effects may support cancer cell proliferation in tumor mass, e.g., in correlation with obesity (Incio et al. 2016; Tajan and Vousden 2016; Meyer et al. 2016).

It was described that many amino acids (glutamine, arginine, cysteine, serine, tryptophane) are required for effective T-cell response in the microenvironment, as well. Similar to acidification, the high glutamine uptake of tumor cells and the lowering glutamine concentration in the ECM suppress the glutamine utilization of T cells and reduce their proliferation (Nakaya et al. 2014). Macrophages and endothelial cells are also very sensitive to depleted amino acid situations. The lowered amount of available cysteine and serine, which are essential for NADPH and glutathione, induces ROS to accumulate at a toxic level. In addition, endothelial cells depend on NEAA-supported TCA cycle anaplerosis, and some other VEGF-mediated angiogenic proliferations which could also be affected in nutrient-deprived ECM.

The three-dimensional matrix of extracellular molecules, the ECM, gives not only the structure but the biochemical, bioenergetic source for the growing tumor mass. In tumor progression and evolution, the components and the formation of this matrix alter in correlation with the consequence of hypoxia and acidosis. The major components—*collagens, fibronectin, laminin, elastin, and proteoglycans* can be degraded by low pH-activated MMPs, and other proteases. Moreover, this remodeling and the acidic microenvironment activate endoglycosidase to digest heparan sulfate glycosaminoglycans (HS-GAG) (Hammond et al. 2014). In addition, other released molecules such as lipoprotein- and exosome-delivered substances can influence the functions of proteoglycans (PGs) (Fuster and Esko 2005). The acidosis alters ligand-receptor interactions among cancer cells and ECM through the induced proteolytic activity, heparinase, tumor-promoting ligand, growth factor, cytokine, and chemokine, as well as lipoprotein productions and exosome secretions. Furthermore, the negatively charged GAGs attract protons in the ECM locally, and at the membrane surfaces (EV or cell membranes). The metabolite release-influenced pH alterations can elaborate matrix and several proteoglycan-bounded cytokines to circulation to help the re-localization of these factors to distant sites and form pre-metastatic niche for migrated, metastatic cells. Tumorigenic and hypoxic alterations (as glycolytic drivers) can be influenced by hyaluronan-mediated degradation (hyaluronidase) in correlation with metabolic rewiring of cancer cells. Additionally, hyaluronidase promotes glycolysis by receptor tyrosine kinase-mediated way which is required for the elevation of cell migration in tumorigenesis (e.g., in breast cancer models) (Sullivan et al. 2018).

Proteoglycans (PGs) are proteins which are conjugated with glycosaminoglycan polysaccharides. Localized either at intra- or extracellular site, or as co-receptors on the cell surface, PGs avidly bind basic proteins and other positively charged molecules as a consequence of their highly anionic character. The structures and the diverse functions of these molecules influence both tumor development and progression (Fuster and Esko 2005; Iozzo and Schaefer 2015). In addition, the negative charge and presence of PGs in the matrix can influence proton (H^+) distribution and alter local pH (Maroudas et al. 1988). The GAG chains and the related charge-mediated bindings have further dynamic alterations in the matrix composition and tissue homeostasis. Cytokines and growth factors can bind to these molecules and stored or released depending on the actual pH. Moreover, several activated proteolytic enzymes (metalloproteases, heparanase) can liberate these factors and alter the activity of growth factor/cytokine-receptor signaling network both in malignant and in non-malignant cells. It is also suggested that cell membrane-bounded HS proteoglycans can interact with lipoproteins, low-density lipoproteins (LDLs) and efficiently help in lipoprotein internalizations as receptors, and even as co-receptors in classical LDL and/or lipoprotein receptor-mediated pathways (Menard et al. 2018; Christianson and Belting 2014). In acidic microenvironment, the alterations in these functions have to be investigated further.

In addition, lipid droplet (LD) accumulation is general in the acidic ECM of chronic inflammation (which usually also occurs in tumor growth) (Shyu Jr. et al. 2018). This condition significantly facilitates the PG-mediated lipid uptake (Oörni and Kovanen 2006). It was reported that increased lipoprotein uptake assisted by surface heparan sulfate proteoglycans (HSPGs), and the followed cellular storage of LD are in correlation with increased survival and metastatic capacity of hypoxic glioblastomas and osteosarcoma cells (Menard et al. 2016; Cortini et al. 2021). Regarding cellular metabolism, new functions were recently described related to some PGs in the ECM. New structural functions have been described in correlation with biglycan and decorin. Biglycan can be synthesized de novo and released from the ECM during several stress responses including proteolytic matrix degradation. Upon this, it can act as Toll-like receptor ligand (as a potent danger signal) and induces inflammatory reactions by inflammatory cytokine secretions (IL1β, TNFα, and chemokines) (Schaefer et al. 2017). Additionally, biglycans can stabilize and bind HIF2α to influence erythropoiesis and stimulate NOX enzymes, ROS, and VEGF which are contributing to inflammation-induced genetic alterations and genomic instability. Decorin also affects the cellular milieu influencing secretion and the release of several angiogenic and growth factors. Moreover, the newly developed function of this PG is to influence metabolic activity of certain cells. Decorin, as an agonist of VEGFR2, facilitates Peg3-dependent accumulation of autophagic vacuoles in endothelial cells, which means that this ECM-localized PG can regulate cellular autophagy (Buraschi et al. 2019). It was also published that decorin accumulation is developing during nutrient deprivation in parallel with autophagy and mitophagy induction. This regulatory function of an ECM PG was underlined in in vivo KO experiments, where the autophagy flux and the maturation of autophagosomes were impaired in Dcn−/− mice (Gubbiotti et al. 2015).

Based on these, decorin as an ECM-localized PG can regulate cellular autophagy in metabolic processes (Buraschi et al. 2019) (Gubbiotti et al. 2015).

It is also known that *exogenous lipids, fatty acids, lipid vesicles, and endogenous lipids* produced by cells using fatty acid synthase activity are stored in LDs (Corbet et al. 2016). LD accumulation was shown in a number of different cancers (Bensaad et al. 2014; Cotte et al. 2018), and it was also suggested that this phenotype may promote chemoresistance, as well. In case cholesteryl ester accumulation is targeted, the migratory capacity of several highly metastatic tumor cells (e.g., pancreas cancer or glioma) can be reduced (Li et al. 2016; Bemlih et al. 2010). In triple-negative breast cancers, however, the LD accumulation showed no correlation with metastatic potential; the accumulated LDs could have a prognostic role in metastatic properties tumor type dependently (Wright et al. 2017). Hypoxia and the related acidosis activate LIPIN-dependent mechanisms, with SREBP-mediated lipogenic programs, and induce lipoprotein internalizations enhancing Erk signaling (Chen et al. 2018). It was also proposed that fatty acid oxidation and synthesis are mainly balanced in healthy tissues, whereas cells decrease acetyl CoA carboxylase (which prevents lipid production from fatty acid oxidation) during adaptation in acidic microenvironment. Usually, glutamine utilization and reductive glutamine metabolism could balance lipid metabolism with de novo lipogenesis (Metallo et al. 2011). These processes are very important and promoted by active mitochondrial functions and TCA cycle. The exogenous lipids were also considered to influence metastasis based on some new publications; CD36 and/or FABP4 are hypoxia-induced by the extracellular lipid receptors at the cell surface. Using these mechanisms, cholesterol and fatty acid uptake can participate in progression and initiate tumor spreading (Ladanyi et al. 2018; Gharpure et al. 2018). As it was previously mentioned, obesity influences both normal and tumorous ECM and their homeostasis, as well. In correlation with these effects, the prevalence and the risk of several cancers can be increased. Obesity-related endocrine, inflammatory, and metabolic factors assist in tumorigenesis, which can alter nutrient-dependent metabolic regulation and cellular signaling network such as IGF and PI3K/Akt/mTOR axis activity, respectively. Obesity-associated cytokines (e.g., IL-6 and TNFα) and adipokines (e.g., adiponectin and leptin) increase estrogen synthesis (e.g., in breast cancers) and promote tumor growth (Rajesh and Sarkar 2021). These factors, their release, and local accumulation in the tumor matrix are very important metabolic players of the microenvironment.

Inorganic polyphosphates (polyP, ATP/ADP)—linear orthophosphate residues with high energy bonds of phosphoanhydrides—are able to form and function to store biochemically useful energy both intra- and extracellularly. Phosphotransfer for energy-consuming also needs to be considered in ECM; alkaline phosphatases and adenylate kinases function both in cells and in ECM and are also crucial in ATP/ADP generation (Müller et al. 2019). ATP has a role in providing energy and molecule transfer channeling maintained by ATP-pumps, contraction of muscle, and building new cellular compartments which require this energy source in many cells. Nutrients are metabolized to liberate energy, which can be stored in phosphoanhydride bonds of ATP. The intracellular ATP concentration (~100 mM) is usually higher than in the ECM (10 nM), and its level varies in the range of 20–100 nM in human blood. Measuring tissue ATP concentration is very difficult.

The ECM does not contain mitochondria, thus its energy level can be fueled without extracellular mitochondria. ATP cannot diffuse through the plasma membrane, but ATP-permeable and ATP-export channels can transport it to the ECM. ATP and polyP are associated with binding proteins in the extracellular space, but no specific ATP/ADP carrier has been identified so far. The roles of extracellular ATP and polyPs were highlighted by the following facts: (a) ATP can be a signaling molecule which is linked to autocrine signaling loop through receptors, such as purinergic receptors (Di Virgilio and Adinolfi 2017); (b) ATP, as a nucleotide, feeds metabolic energy-requiring mechanism in the ECM (e.g., secreted tyrosine kinase in platelet degranulation, phosphorylation by ECM kinases, packaging collagen maturation—peptidyl-trans-isomerases: cyclophylins, FK506-binding proteins, and parvulins)—(Bordoli et al. 2014; Fanghänel and Fischer 2004); (c) ATP can help heat shock protein clustering (e.g., HSP70 is supposed to bind ATP in extracellular space) (Trcka et al. 2019); (d) sol-gel transition theory suggests that this conversion requires free energy (e.g., claudins in tight junctions—(Zhao et al. 2018)); (e) endothelial cell-released ATP can function as a chemoattractant for migrating ring forming cells (Müller et al. 2018a, b); (f) ECM-stored polyPs can be the source for ATP production both intra- and extracellularly. Additionally, there are more evidences that microenvironmental polyPs affect fibroblasts influencing FGF-induced adhesion (Segawa et al. 2011; Shiba et al. 2003). PolyP synthesis is linked to mitochondria that act as a reservoir for metabolic energy, and its degradation is related to alkaline phosphatase (ALP) which can act as polyP transferase. ALPs are membrane-bounded glycoproteins. ALP and adenylate kinase (ADK) inhibitors can block cell migratory behavior, especially in fibroblasts and endothelial cells in different in vitro experiments. Moreover, polyP addition increases and speeds up the migration of endothelial cells demonstrated in scratch tests (Müller et al. 2018a). In these assays, the increased extracellular ATP level is highly modified the migratory behavior of the studied cells. ALP and ADK can phospho-transfer intra- and extracellularly; additionally, polyPs can be transported by exosomes, as well. ECM influences the stability and the degradation of polyPs to charged fibrous proteins (collagen, proteoglycan, and GAGs electrostatic interactions) and their transport mechanisms, but the role of ECM as an energy reservoir has not been described well yet (Müller et al. 2019).

In some studies, organelle transfer can also be detected in cancer cells. In parallel, mitochondrial DNA and mitochondria transfer were confirmed from stromal cells to mitochondrion-deficient cancer cells; these help OXPHOS-dependent survival and adaptations in cancers (Spees et al. 2006; Tan et al. 2015). Based on these transfers, cancer metabolism can be completely rewired depending on the necessities of tumor cell growth.

The increased release of exosomes was also documented in different cancer cells (King et al. 2012; Svensson et al. 2011). It was also suggested that *exosomes* can be enriched in PG-bonded lipoproteins, and in addition, their ligands (e.g., apolipoproteins) could influence cellular communications and *vesicle cargo* (Boussadia et al. 2018; Shao et al. 2018; Parolini et al. 2009). The metabolite composition of exosomes could also modulate cancer cell functions. Hypoxic adipocyte-derived exosomes and other hypoxic vesicles showed increased accumulations of triglycerides in the tumor microenvironment. These showed higher level of lipogenic enzymes which can be transferred to the neighboring or other distant cells (Sano et al. 2014). It was described that acidic exosome fractions can positively influence the invasive potential of melanoma cells (Peppicelli et al. 2014; Kucharzewska et al. 2013). Additionally, hypoxic vesicles can transfer and mimic hypoxic response in exosome-treated non-hypoxic tumor cells influencing other cellular compartments (Zhao et al. 2016). Based on these, proteins, nucleic acids, and metabolites transported by exosomes through the ECM have to be considered in tumor metabolism and development, as well.

13.5 Complex Metabolic Regulation in the Extracellular Matrix

The continuous changes in ECM influence and regulate the molecule composition of the metabolic niche at different levels: (a) metabolic alterations in tumor cells; (b) metabolic crosstalk between tumor and non-tumor cells (completion and collaboration); (c) metabolic heterogeneity in tumor tissue (e.g., anatomical location of the tumor, tissue-specific metabolites, local hypoxia, and metabolite gradients); (d) homeostasis of the organ, the whole body, and the personal metabolic state (e.g., obesity, diet, systematic glucose-insulin homeostasis, diverse metabolic niches in the body, and gut microbiome-released metabolites) (Elia and Haigis 2021).

The heterocellular collaborations, fibrotic elements, ECM proteins, metabolites, and their metabolic consequences in tissue microenvironment promote tumor growth. Accordingly, tumor cells force many metabolic pathways among normal metabolic processes along with their demand. This summary underlines the consequences of metabolic adaptations, alterations in the constitution of ECM elements. These alterations provide site for metabolic symbiosis, where many players and molecules communicate, compete, cross-feed each other in a way guided by tumor cells to promote tumor cell survival, growth, and if it is necessary to overgrowth non-tumor cell populations, migrate, and metastasize (Figs. 13.1 and 13.2).

These changes together with the exhausting of reserved nutrients and energy forces lead to metabolic collapse in the whole organ and body (additionally, could result in cachexia in the patients). In this multi-stepped evolution, the tumor cells have to proliferate in an organ-specific manner in a special microenvironment. As a consequence, cells undergo adaptation mechanisms and further alterations (e.g., epithelial-mesenchymal transition and degradation of adhesion molecules) to gain migratory and metastatic properties. Metastatic cells have to survive in the blood circulation, under oxidative stress situations, and nutrient depletion. Finally, these cells need to adapt to the new place or reprogram the new microenvironment for maintaining their own growth and survival demand. In these complex processes, especially in tissue environment, cells use matrix elements as frameworks, solid base, nutrient source, growth cytokine depo, and intercellular communication niche. In this symbiosis, the ECM elements have both passive and active roles including autophagy and exocytosis-mediated feeding; small metabolite exchanges between hypoxic and normoxic cells; LDs and lipoprotein uptake and release of growth factor and cytokine production. In this tumor orchestrated milieu, tumor cells have several advantages (Boedtkjer and Pedersen 2020). Targeting metabolic adaptation mechanisms, metabolic communications and this metabolic symbiosis are in the focus of many recent studies to increase metabolic stress in cancer cells and induce metabolic catastrophe in tumor mass (Anderson et al. 2021; Petővári et al. 2020; Holloway and Marignani 2021). Targeting metabolic vulnerabilities of cancer cells reduce metabolite-derived immunosuppression by enhancing both metabolic fitness and anti-tumor immune functions (Bonglack et al. 2021; Shriwas et al. 2021; Georganaki et al. 2018). The problem is that tissue composition and the ECM are depending on cellular origin, where the local microenvironment influences the metabolic adaptation mechanisms and rewires cancer cell bioenergetics. Moreover, several metabolites and metabolic effects could have certain genetic and epigenetic effects which can modify tumor plasticity. To target metabolic communications in tissue microenvironment, we need to consider many different factors including tumor type, location, stage, and potential metabolic compensatory mechanisms represented in tumor heterogeneity.

However, several recently used drugs with various primary targets affect the ECM as a by-product of the drug actions and metabolites. Until recently, ECM has been less considered in drug discovery. Moreover, the ECM, its composition and elements could also influence drug transport, drug accumulations/concentrations, drug delivery to tumor cells, and finally, the therapeutic effects. Certain recent therapeutics specifically modify the composition of ECM elements: immunomodulatory drugs, chemotherapeutics—e.g., methotrexate and nucleotide synthesis targeting drugs; statins, which alter collagen and proteoglycan expression; non-steroidal anti-inflammatory drugs and mTOR/calcineurin inhibitors modify fibronectin and chondroitin sulfate PGs; antimicrobial antibiotics reduce MMPs production; TK inhibitors reduce the expression of TGFβ-enhanced matrix elements (Järveläinen et al. 2009). Drugs and their metabolites can bind to ECM proteins themselves, and it is also well known that the structure of ECM can often be disrupted as a consequence of therapeutic treatments. It is an alternative strategy to influence the delivery and/or deposition of anti-tumor drugs and their metabolites in the tumoral extracellular matrix. Furthermore, the by-stander effects of these drugs have to be considered to eliminate both tumorous and non-tumorous cells. Adipose compartments, lipids can solve and continuously release lipophyl drugs (e.g., antipsychotic drugs—fluphenazine, olanzapine). Moreover, lipid compartments and their ratio can influence the delivery/blood-brain barrier crossing (e.g., Temozolomide) and the effect of such treatments, as well. There are some drugs that direct target cellular and structural (e.g., Natalizumab, Vedolizumab) elements of the ECM in certain immunological diseases (Schön 2008; Scribano 2018). In conclusion, pharmacological targeting of ECM is a promising way, but more studies are required to evaluate its potential translation to the clinical trials, especially in cancer.

13.6 Concluding Remarks

Cancer metabolism studies have made great progress in describing alterations of cancer and other cells as well as the importance of metabolic niche in tissue microenvironment. Understanding the effects of local metabolite availability and metabolic dependencies in tumor and stromal cells could provide new options in individual therapy using known drugs with metabolic targets. Considering the complexity of tissue heterogeneity and its players including ECM elements, the experimental cancer models and their limits have to be revised in future studies (Lau and Heiden 2020). Hopefully, we will characterize better the metabolic communications, metabolic phenotype of cancer cells, and their important molecular mechanisms in the ECM. In addition, we could target these tumor-promoting alterations more precisely and turn this knowledge to find more benefits for the development of recent personalized therapies.

References

Akkoc Y, Peker N, Akcay A, Gozuacik D (2021) Autophagy and cancer dormancy. Front Oncol 11: 627023. https://doi.org/10.3389/fonc.2021.627023

Aldera AP, Govender D (2021) Carbonic anhydrase IX: a regulator of pH and participant in carcinogenesis. J Clin Pathol. https://doi.org/10.1136/jclinpath-2020-207073

Anderson NM, Qin X, Finan JM, Lam A, Athoe J, Missiaen R, Skuli N, Kennedy A, Saini AS, Tao T, Zhu S, Nissim I, Look AT, Qing G, Simon MC, Feng H (2021) Metabolic enzyme DLST promotes tumor aggression and reveals a vulnerability to OXPHOS inhibition in high-risk neuroblastoma. Cancer Res. https://doi.org/10.1158/0008-5472.can-20-2153

Angelin A, Gil-de-Gómez L, Dahiya S, Jiao J, Guo L, Levine MH, Wang Z, Quinn WJ 3rd, Kopinski PK, Wang L, Akimova T, Liu Y, Bhatti TR, Han R, Laskin BL, Baur JA, Blair IA, Wallace DC, Hancock WW, Beier UH (2017) Foxp3 reprograms T cell metabolism to function in low-glucose, high-lactate environments. Cell Metab 25(6):1282–1293.e1287. https://doi.org/10.1016/j.cmet.2016.12.018

Bader JE, Voss K, Rathmell JC (2020) Targeting metabolism to improve the tumor microenvironment for cancer immunotherapy. Mol Cell 78(6):1019–1033. https://doi.org/10.1016/j.molcel.2020.05.034

Balkwill F, Mantovani A (2001) Inflammation and cancer: back to Virchow? Lancet 357(9255): 539–545. https://doi.org/10.1016/s0140-6736(00)04046-0

Baltazar F, Afonso J, Costa M, Granja S (2020) Lactate beyond a waste metabolite: metabolic affairs and signaling in malignancy. Front Oncol 10:231. https://doi.org/10.3389/fonc.2020.00231

Bemlih S, Poirier MD, El Andaloussi A (2010) Acyl-coenzyme A: cholesterol acyltransferase inhibitor Avasimibe affect survival and proliferation of glioma tumor cell lines. Cancer Biol Ther 9(12):1025–1032. https://doi.org/10.4161/cbt.9.12.11875

Benej M, Svastova E, Banova R, Kopacek J, Gibadulinova A, Kery M, Arena S, Scaloni A, Vitale M, Zambrano N, Papandreou I, Denko NC, Pastorekova S (2020) CA IX stabilizes intracellular pH to maintain metabolic reprogramming and proliferation in hypoxia. Front Oncol 10:1462. https://doi.org/10.3389/fonc.2020.01462

Bensaad K, Favaro E, Lewis CA, Peck B, Lord S, Collins JM, Pinnick KE, Wigfield S, Buffa FM, Li JL, Zhang Q, Wakelam MJO, Karpe F, Schulze A, Harris AL (2014) Fatty acid uptake and

lipid storage induced by HIF-1α contribute to cell growth and survival after hypoxia-reoxygenation. Cell Rep 9(1):349–365. https://doi.org/10.1016/j.celrep.2014.08.056
Boedtkjer E, Pedersen SF (2020) The acidic tumor microenvironment as a driver of cancer. Annu Rev Physiol 82:103–126. https://doi.org/10.1146/annurev-physiol-021119-034627
Bonglack EN, Messinger JE, Cable JM, Ch'ng J, Parnell KM, Reinoso-Vizcaíno NM, Barry AP, Russell VS, Dave SS, Christofk HR, Luftig MA (2021) Monocarboxylate transporter antagonism reveals metabolic vulnerabilities of viral-driven lymphomas. Proc Natl Acad Sci U S A 118(25). https://doi.org/10.1073/pnas.2022495118
Bordoli MR, Yum J, Breitkopf SB, Thon JN, Italiano JE Jr, Xiao J, Worby C, Wong SK, Lin G, Edenius M, Keller TL, Asara JM, Dixon JE, Yeo CY, Whitman M (2014) A secreted tyrosine kinase acts in the extracellular environment. Cell 158(5):1033–1044. https://doi.org/10.1016/j.cell.2014.06.048
Boussadia Z, Lamberti J, Mattei F, Pizzi E, Puglisi R, Zanetti C, Pasquini L, Fratini F, Fantozzi L, Felicetti F, Fecchi K, Raggi C, Sanchez M, D'Atri S, Carè A, Sargiacomo M, Parolini I (2018) Acidic microenvironment plays a key role in human melanoma progression through a sustained exosome mediated transfer of clinically relevant metastatic molecules. J Exp Clin Cancer Res 37(1):245. https://doi.org/10.1186/s13046-018-0915-z
Buraschi S, Neill T, Iozzo RV (2019) Decorin is a devouring proteoglycan: remodeling of intracellular catabolism via autophagy and mitophagy. Matrix Biol 75-76:260–270. https://doi.org/10.1016/j.matbio.2017.10.005
Byun JS, Park S, Caban A, Jones A, Gardner K (2018) Linking race, cancer outcomes, and tissue repair. Am J Pathol 188(2):317–328. https://doi.org/10.1016/j.ajpath.2017.10.009
Cantelmo AR, Conradi LC, Brajic A, Goveia J, Kalucka J, Pircher A, Chaturvedi P, Hol J, Thienpont B, Teuwen LA, Schoors S, Boeckx B, Vriens J, Kuchnio A, Veys K, Cruys B, Finotto L, Treps L, Stav-Noraas TE, Bifari F, Stapor P, Decimo I, Kampen K, De Bock K, Haraldsen G, Schoonjans L, Rabelink T, Eelen G, Ghesquière B, Rehman J, Lambrechts D, Malik AB, Dewerchin M, Carmeliet P (2016) Inhibition of the glycolytic activator PFKFB3 in endothelium induces tumor vessel normalization, impairs metastasis, and improves chemotherapy. Cancer Cell 30(6):968–985. https://doi.org/10.1016/j.ccell.2016.10.006
Casazza A, Di Conza G, Wenes M, Finisguerra V, Deschoemaeker S, Mazzone M (2014) Tumor stroma: a complexity dictated by the hypoxic tumor microenvironment. Oncogene 33(14): 1743–1754. https://doi.org/10.1038/onc.2013.121
Chang CH, Qiu J, O'Sullivan D, Buck MD, Noguchi T, Curtis JD, Chen Q, Gindin M, Gubin MM, van der Windt GJ, Tonc E, Schreiber RD, Pearce EJ, Pearce EL (2015) Metabolic competition in the tumor microenvironment is a driver of cancer progression. Cell 162(6):1229–1241. https://doi.org/10.1016/j.cell.2015.08.016
Chen M, Zhang J, Sampieri K, Clohessy JG, Mendez L, Gonzalez-Billalabeitia E, Liu XS, Lee YR, Fung J, Katon JM, Menon AV, Webster KA, Ng C, Palumbieri MD, Diolombi MS, Breitkopf SB, Teruya-Feldstein J, Signoretti S, Bronson RT, Asara JM, Castillo-Martin M, Cordon-Cardo C, Pandolfi PP (2018) An aberrant SREBP-dependent lipogenic program promotes metastatic prostate cancer. Nat Genet 50(2):206–218. https://doi.org/10.1038/s41588-017-0027-2
Christianson HC, Belting M (2014) Heparan sulfate proteoglycan as a cell-surface endocytosis receptor. Matrix Biol 35:51–55. https://doi.org/10.1016/j.matbio.2013.10.004
Commisso C, Davidson SM, Soydaner-Azeloglu RG, Parker SJ, Kamphorst JJ, Hackett S, Grabocka E, Nofal M, Drebin JA, Thompson CB, Rabinowitz JD, Metallo CM, Vander Heiden MG, Bar-Sagi D (2013) Macropinocytosis of protein is an amino acid supply route in Ras-transformed cells. Nature 497(7451):633–637. https://doi.org/10.1038/nature12138
Corbet C, Pinto A, Martherus R, Santiago de Jesus JP, Polet F, Feron O (2016) Acidosis drives the reprogramming of fatty acid metabolism in cancer cells through changes in mitochondrial and histone acetylation. Cell Metab 24(2):311–323. https://doi.org/10.1016/j.cmet.2016.07.003
Cortini M, Armirotti A, Columbaro M, Longo DL, Di Pompo G, Cannas E, Maresca A, Errani C, Longhi A, Righi A, Carelli V, Baldini N, Avnet S (2021) Exploring metabolic adaptations to the

acidic microenvironment of osteosarcoma cells unveils sphingosine 1-phosphate as a valuable therapeutic target. Cancers (Basel) 13(2). https://doi.org/10.3390/cancers13020311

Cotte AK, Aires V, Fredon M, Limagne E, Derangère V, Thibaudin M, Humblin E, Scagliarini A, de Barros JP, Hillon P, Ghiringhelli F, Delmas D (2018) Lysophosphatidylcholine acyltransferase 2-mediated lipid droplet production supports colorectal cancer chemoresistance. Nat Commun 9(1):322. https://doi.org/10.1038/s41467-017-02732-5

Currie E, Schulze A, Zechner R, Walther TC, Farese RV Jr (2013) Cellular fatty acid metabolism and cancer. Cell Metab 18(2):153–161. https://doi.org/10.1016/j.cmet.2013.05.017

de la Cruz-López KG, Castro-Muñoz LJ, Reyes-Hernández DO, García-Carrancá A, Manzo-Merino J (2019) Lactate in the regulation of tumor microenvironment and therapeutic approaches. Front Oncol 9:1143. https://doi.org/10.3389/fonc.2019.01143

Deville SS, Cordes N (2019) The extracellular, cellular, and nuclear stiffness, a trinity in the cancer resistome—a review. Front Oncol 9(1376). https://doi.org/10.3389/fonc.2019.01376

Di Virgilio F, Adinolfi E (2017) Extracellular purines, purinergic receptors and tumor growth. Oncogene 36(3):293–303. https://doi.org/10.1038/onc.2016.206

Dvorak HF (1986) Tumors: wounds that do not heal. Similarities between tumor stroma generation and wound healing. N Engl J Med 315(26):1650–1659. https://doi.org/10.1056/nejm198612253152606

Dvorak HF (2015) Tumors: wounds that do not heal-redux. Cancer Immunol Res 3(1):1–11. https://doi.org/10.1158/2326-6066.Cir-14-0209

Eagle H (1955) Nutrition needs of mammalian cells in tissue culture. Science 122(3168):501–514. https://doi.org/10.1126/science.122.3168.501

Egeblad M, Nakasone ES, Werb Z (2010) Tumors as organs: complex tissues that interface with the entire organism. Dev Cell 18(6):884–901. https://doi.org/10.1016/j.devcel.2010.05.012

Elia I, Haigis MC (2021) Metabolites and the tumour microenvironment: from cellular mechanisms to systemic metabolism. Nat Metab 3(1):21–32. https://doi.org/10.1038/s42255-020-00317-z

Fanghänel J, Fischer G (2004) Insights into the catalytic mechanism of peptidyl prolyl cis/trans isomerases. Front Biosci 9:3453–3478. https://doi.org/10.2741/1494

Farnie G, Sotgia F, Lisanti MP (2015) High mitochondrial mass identifies a sub-population of stem-like cancer cells that are chemo-resistant. Oncotarget 6(31):30472–30486. https://doi.org/10.18632/oncotarget.5401

Fuster MM, Esko JD (2005) The sweet and sour of cancer: glycans as novel therapeutic targets. Nat Rev Cancer 5(7):526–542. https://doi.org/10.1038/nrc1649

García-Cañaveras JC, Chen L, Rabinowitz JD (2019) The tumor metabolic microenvironment: lessons from lactate. Cancer Res 79(13):3155–3162. https://doi.org/10.1158/0008-5472.Can-18-3726

Georganaki M, van Hooren L, Dimberg A (2018) Vascular targeting to increase the efficiency of immune checkpoint blockade in cancer. Front Immunol 9:3081. https://doi.org/10.3389/fimmu.2018.03081

Gharpure KM, Pradeep S, Sans M, Rupaimoole R, Ivan C, Wu SY, Bayraktar E, Nagaraja AS, Mangala LS, Zhang X, Haemmerle M, Hu W, Rodriguez-Aguayo C, McGuire M, Mak CSL, Chen X, Tran MA, Villar-Prados A, Pena GA, Kondetimmanahalli R, Nini R, Koppula P, Ram P, Liu J, Lopez-Berestein G, Baggerly K, Eberlin LS, Sood AK (2018) FABP4 as a key determinant of metastatic potential of ovarian cancer. Nat Commun 9(1):2923. https://doi.org/10.1038/s41467-018-04987-y

Gillies RJ (2021) Cancer heterogeneity and metastasis: life at the edge. Clin Exp Metastasis. https://doi.org/10.1007/s10585-021-10101-2

Gillies RJ, Gatenby RA (2015) Metabolism and its sequelae in cancer evolution and therapy. Cancer J 21(2):88–96. https://doi.org/10.1097/ppo.0000000000000102

Glunde K, Guggino SE, Solaiyappan M, Pathak AP, Ichikawa Y, Bhujwalla ZM (2003) Extracellular acidification alters lysosomal trafficking in human breast cancer cells. Neoplasia 5(6): 533–545. https://doi.org/10.1016/s1476-5586(03)80037-4

Gorchs L, Kaipe H (2021) Interactions between cancer-associated fibroblasts and T cells in the pancreatic tumor microenvironment and the role of chemokines. Cancers (Basel) 13(12). https://doi.org/10.3390/cancers13122995

Gubbiotti MA, Neill T, Frey H, Schaefer L, Iozzo RV (2015) Decorin is an autophagy-inducible proteoglycan and is required for proper in vivo autophagy. Matrix Biol 48:14–25. https://doi.org/10.1016/j.matbio.2015.09.001

Guillaume JD, Celano SL, Martin KR, MacKeigan JP (2019) Determining the impact of metabolic nutrients on autophagy. Methods Mol Biol 1862:151–162. https://doi.org/10.1007/978-1-4939-8769-6_11

Gullino PM, Clark SH, Grantham FH (1964) The interstitial fluid of solid tumors. Cancer Res 24: 780–794

Hammond E, Khurana A, Shridhar V, Dredge K (2014) The role of heparanase and sulfatases in the modification of heparan Sulfate proteoglycans within the tumor microenvironment and opportunities for novel cancer therapeutics. Front Oncol 4:195. https://doi.org/10.3389/fonc.2014.00195

Hanahan D, Coussens LM (2012) Accessories to the crime: functions of cells recruited to the tumor microenvironment. Cancer Cell 21(3):309–322. https://doi.org/10.1016/j.ccr.2012.02.022

Hanahan D, Weinberg RA (2011) Hallmarks of cancer: the next generation. Cell 144(5):646–674. https://doi.org/10.1016/j.cell.2011.02.013

Harmon C, Robinson MW, Hand F, Almuaili D, Mentor K, Houlihan DD, Hoti E, Lynch L, Geoghegan J, O'Farrelly C (2019) Lactate-mediated acidification of tumor microenvironment induces apoptosis of liver-resident NK cells in colorectal liver metastasis. Cancer Immunol Res 7(2):335–346. https://doi.org/10.1158/2326-6066.Cir-18-0481

Holloway RW, Marignani PA (2021) Targeting mTOR and glycolysis in HER2-positive breast cancer. Cancers (Basel) 13(12):2922

Hosios AM, Hecht VC, Danai LV, Johnson MO, Rathmell JC, Steinhauser ML, Manalis SR, Vander Heiden MG (2016) Amino acids rather than glucose account for the majority of cell mass in proliferating mammalian cells. Dev Cell 36(5):540–549. https://doi.org/10.1016/j.devcel.2016.02.012

Huang Z, Zhang M, Plec AA, Estill SJ, Cai L, Repa JJ, McKnight SL, Tu BP (2018) ACSS2 promotes systemic fat storage and utilization through selective regulation of genes involved in lipid metabolism. Proc Natl Acad Sci U S A 115(40):E9499–e9506. https://doi.org/10.1073/pnas.1806635115

Husain Z, Huang Y, Seth P, Sukhatme VP (2013) Tumor-derived lactate modifies antitumor immune response: effect on myeloid-derived suppressor cells and NK cells. J Immunol 191(3):1486–1495. https://doi.org/10.4049/jimmunol.1202702

Hynes RO, Naba A (2012) Overview of the matrisome--an inventory of extracellular matrix constituents and functions. Cold Spring Harb Perspect Biol 4(1):a004903. https://doi.org/10.1101/cshperspect.a004903

Incio J, Liu H, Suboj P, Chin SM, Chen IX, Pinter M, Ng MR, Nia HT, Grahovac J, Kao S, Babykutty S, Huang Y, Jung K, Rahbari NN, Han X, Chauhan VP, Martin JD, Kahn J, Huang P, Desphande V, Michaelson J, Michelakos TP, Ferrone CR, Soares R, Boucher Y, Fukumura D, Jain RK (2016) Obesity-induced inflammation and desmoplasia promote pancreatic cancer progression and resistance to chemotherapy. Cancer Discov 6(8):852–869. https://doi.org/10.1158/2159-8290.Cd-15-1177

Intlekofer AM, Finley LWS (2019) Metabolic signatures of cancer cells and stem cells. Nat Metab 1(2):177–188. https://doi.org/10.1038/s42255-019-0032-0

Iozzo RV, Schaefer L (2015) Proteoglycan form and function: a comprehensive nomenclature of proteoglycans. Matrix Biol 42:11–55. https://doi.org/10.1016/j.matbio.2015.02.003

Ippolito L, Marini A, Cavallini L, Morandi A, Pietrovito L, Pintus G, Giannoni E, Schrader T, Puhr M, Chiarugi P, Taddei ML (2016) Metabolic shift toward oxidative phosphorylation in docetaxel resistant prostate cancer cells. Oncotarget 7(38):61890–61904. https://doi.org/10.18632/oncotarget.11301

Jamali S, Klier M, Ames S, Barros LF, McKenna R, Deitmer JW, Becker HM (2015) Hypoxia-induced carbonic anhydrase IX facilitates lactate flux in human breast cancer cells by non-catalytic function. Sci Rep 5:13605. https://doi.org/10.1038/srep13605

Järveläinen H, Sainio A, Koulu M, Wight TN, Penttinen R (2009) Extracellular matrix molecules: potential targets in pharmacotherapy. Pharmacol Rev 61(2):198–223. https://doi.org/10.1124/pr.109.001289

Johnson MO, Siska PJ, Contreras DC, Rathmell JC (2016) Nutrients and the microenvironment to feed a T cell army. Semin Immunol 28(5):505–513. https://doi.org/10.1016/j.smim.2016.09.003

Karta J, Bossicard Y, Kotzamanis K, Dolznig H, Letellier E (2021) Mapping the metabolic networks of tumor cells and cancer-associated fibroblasts. Cell 10(2). https://doi.org/10.3390/cells10020304

King HW, Michael MZ, Gleadle JM (2012) Hypoxic enhancement of exosome release by breast cancer cells. BMC Cancer 12:421. https://doi.org/10.1186/1471-2407-12-421

Kobayashi H, Moniwa N, Sugimura M, Shinohara H, Ohi H, Terao T (1993) Effects of membrane-associated cathepsin B on the activation of receptor-bound prourokinase and subsequent invasion of reconstituted basement membranes. Biochim Biophys Acta 1178(1):55–62. https://doi.org/10.1016/0167-4889(93)90109-3

Koppenol WH, Bounds PL, Dang CV (2011) Otto Warburg's contributions to current concepts of cancer metabolism. Nat Rev Cancer 11(5):325–337. https://doi.org/10.1038/nrc3038

Kramer CJH, Vangangelt KMH, van Pelt GW, Dekker TJA, Tollenaar R, Mesker WE (2019) The prognostic value of tumour-stroma ratio in primary breast cancer with special attention to triple-negative tumours: a review. Breast Cancer Res Treat 173(1):55–64. https://doi.org/10.1007/s10549-018-4987-4

Kucharzewska P, Christianson HC, Welch JE, Svensson KJ, Fredlund E, Ringnér M, Mörgelin M, Bourseau-Guilmain E, Bengzon J, Belting M (2013) Exosomes reflect the hypoxic status of glioma cells and mediate hypoxia-dependent activation of vascular cells during tumor development. Proc Natl Acad Sci U S A 110(18):7312–7317. https://doi.org/10.1073/pnas.1220998110

Ladanyi A, Mukherjee A, Kenny HA, Johnson A, Mitra AK, Sundaresan S, Nieman KM, Pascual G, Benitah SA, Montag A, Yamada SD, Abumrad NA, Lengyel E (2018) Adipocyte-induced CD36 expression drives ovarian cancer progression and metastasis. Oncogene 37(17): 2285–2301. https://doi.org/10.1038/s41388-017-0093-z

Lamb R, Harrison H, Smith DL, Townsend PA, Jackson T, Ozsvari B, Martinez-Outschoorn UE, Pestell RG, Howell A, Lisanti MP, Sotgia F (2015) Targeting tumor-initiating cells: eliminating anabolic cancer stem cells with inhibitors of protein synthesis or by mimicking caloric restriction. Oncotarget 6(7):4585–4601. https://doi.org/10.18632/oncotarget.3278

Lane AN, Higashi RM, Fan TW (2020) Metabolic reprogramming in tumors: contributions of the tumor microenvironment. Genes Dis 7(2):185–198. https://doi.org/10.1016/j.gendis.2019.10.007

Langowski JL, Zhang X, Wu L, Mattson JD, Chen T, Smith K, Basham B, McClanahan T, Kastelein RA, Oft M (2006) IL-23 promotes tumour incidence and growth. Nature 442(7101):461–465. https://doi.org/10.1038/nature04808

Lau AN, Heiden MGV (2020) Metabolism in the tumor microenvironment. Annu Rev Cancer Biol 4(1):17–40. https://doi.org/10.1146/annurev-cancerbio-030419-033333

Lau J, Lin KS, Bénard F (2017) Past, present, and future: development of theranostic agents targeting carbonic anhydrase IX. Theranostics 7(17):4322–4339. https://doi.org/10.7150/thno.21848

Lee SH, Griffiths JR (2020) How and why are cancers acidic? Carbonic anhydrase IX and the homeostatic control of tumour extracellular pH. Cancers (Basel) 12(6). https://doi.org/10.3390/cancers12061616

Lee DC, Sohn HA, Park ZY, Oh S, Kang YK, Lee KM, Kang M, Jang YJ, Yang SJ, Hong YK, Noh H, Kim JA, Kim DJ, Bae KH, Kim DM, Chung SJ, Yoo HS, Yu DY, Park KC, Yeom YI (2015) A lactate-induced response to hypoxia. Cell 161(3):595–609. https://doi.org/10.1016/j.cell.2015.03.011

Li J, Gu D, Lee SS, Song B, Bandyopadhyay S, Chen S, Konieczny SF, Ratliff TL, Liu X, Xie J, Cheng JX (2016) Abrogating cholesterol esterification suppresses growth and metastasis of pancreatic cancer. Oncogene 35(50):6378–6388. https://doi.org/10.1038/onc.2016.168

Li W, Ng JM, Wong CC, Ng EKW, Yu J (2018) Molecular alterations of cancer cell and tumour microenvironment in metastatic gastric cancer. Oncogene 37(36):4903–4920. https://doi.org/10.1038/s41388-018-0341-x

Linares JF, Cordes T, Duran A, Reina-Campos M, Valencia T, Ahn CS, Castilla EA, Moscat J, Metallo CM, Diaz-Meco MT (2017) ATF4-induced metabolic reprograming is a synthetic vulnerability of the p62-deficient tumor stroma. Cell Metab 26(6):817–829.e816. https://doi.org/10.1016/j.cmet.2017.09.001

Louault K, Li RR, DeClerck YA (2020) Cancer-associated fibroblasts: understanding their heterogeneity. Cancers (Basel) 12(11). https://doi.org/10.3390/cancers12113108

Lovinfosse P, Koopmansch B, Lambert F, Jodogne S, Kustermans G, Hatt M, Visvikis D, Seidel L, Polus M, Albert A, Delvenne P, Hustinx R (2016) (18)F-FDG PET/CT imaging in rectal cancer: relationship with the RAS mutational status. Br J Radiol 89(1063):20160212. https://doi.org/10.1259/bjr.20160212

Maroudas A, Weinberg PD, Parker KH, Winlove CP (1988) The distributions and diffusivities of small ions in chondroitin sulphate, hyaluronate and some proteoglycan solutions. Biophys Chem 32(2–3):257–270. https://doi.org/10.1016/0301-4622(88)87012-1

Martinez-Outschoorn UE, Trimmer C, Lin Z, Whitaker-Menezes D, Chiavarina B, Zhou J, Wang C, Pavlides S, Martinez-Cantarin MP, Capozza F, Witkiewicz AK, Flomenberg N, Howell A, Pestell RG, Caro J, Lisanti MP, Sotgia F (2010) Autophagy in cancer associated fibroblasts promotes tumor cell survival: role of hypoxia, HIF1 induction and NFκB activation in the tumor stromal microenvironment. Cell Cycle 9(17):3515–3533. https://doi.org/10.4161/cc.9.17.12928

McDonald PC, Chia S, Bedard PL, Chu Q, Lyle M, Tang L, Singh M, Zhang Z, Supuran CT, Renouf DJ, Dedhar S (2020) A phase 1 study of SLC-0111, a novel inhibitor of carbonic anhydrase IX, in patients with advanced solid tumors. Am J Clin Oncol 43(7):484–490. https://doi.org/10.1097/coc.0000000000000691

Menard JA, Christianson HC, Kucharzewska P, Bourseau-Guilmain E, Svensson KJ, Lindqvist E, Indira Chandran V, Kjellén L, Welinder C, Bengzon J, Johansson MC, Belting M (2016) Metastasis stimulation by hypoxia and acidosis-induced extracellular lipid uptake is mediated by proteoglycan-dependent endocytosis. Cancer Res 76(16):4828–4840. https://doi.org/10.1158/0008-5472.Can-15-2831

Menard JA, Cerezo-Magaña M, Belting M (2018) Functional role of extracellular vesicles and lipoproteins in the tumour microenvironment. Philos Trans R Soc Lond Ser B Biol Sci 373(1737). https://doi.org/10.1098/rstb.2016.0480

Metallo CM, Gameiro PA, Bell EL, Mattaini KR, Yang J, Hiller K, Jewell CM, Johnson ZR, Irvine DJ, Guarente L, Kelleher JK, Vander Heiden MG, Iliopoulos O, Stephanopoulos G (2011) Reductive glutamine metabolism by IDH1 mediates lipogenesis under hypoxia. Nature 481(7381):380–384. https://doi.org/10.1038/nature10602

Meyer KA, Neeley CK, Baker NA, Washabaugh AR, Flesher CG, Nelson BS, Frankel TL, Lumeng CN, Lyssiotis CA, Wynn ML, Rhim AD, O'Rourke RW (2016) Adipocytes promote pancreatic cancer cell proliferation via glutamine transfer. Biochem Biophys Rep 7:144–149. https://doi.org/10.1016/j.bbrep.2016.06.004

Müller WEG, Ackermann M, Tolba E, Neufurth M, Ivetac I, Kokkinopoulou M, Schröder HC, Wang X (2018a) Role of ATP during the initiation of microvascularization: acceleration of an autocrine sensing mechanism facilitating chemotaxis by inorganic polyphosphate. Biochem J 475(20):3255–3273. https://doi.org/10.1042/bcj20180535

Müller WEG, Ackermann M, Wang S, Neufurth M, Muñoz-Espí R, Feng Q, Schröder HC, Wang X (2018b) Inorganic polyphosphate induces accelerated tube formation of HUVEC endothelial cells. Cell Mol Life Sci 75(1):21–32. https://doi.org/10.1007/s00018-017-2601-2

Müller WEG, Schröder HC, Wang X (2019) Inorganic polyphosphates as storage for and generator of metabolic energy in the extracellular matrix. Chem Rev 119(24):12337–12374. https://doi.org/10.1021/acs.chemrev.9b00460

Muranen T, Iwanicki MP, Curry NL, Hwang J, DuBois CD, Coloff JL, Hitchcock DS, Clish CB, Brugge JS, Kalaany NY (2017) Starved epithelial cells uptake extracellular matrix for survival. Nat Commun 8:13989. https://doi.org/10.1038/ncomms13989

Najafi M, Farhood B, Mortezaee K (2019) Extracellular matrix (ECM) stiffness and degradation as cancer drivers. J Cell Biochem 120(3):2782–2790. https://doi.org/10.1002/jcb.27681

Nakaya M, Xiao Y, Zhou X, Chang JH, Chang M, Cheng X, Blonska M, Lin X, Sun SC (2014) Inflammatory T cell responses rely on amino acid transporter ASCT2 facilitation of glutamine uptake and mTORC1 kinase activation. Immunity 40(5):692–705. https://doi.org/10.1016/j.immuni.2014.04.007

Nasi A, Fekete T, Krishnamurthy A, Snowden S, Rajnavölgyi E, Catrina AI, Wheelock CE, Vivar N, Rethi B (2013) Dendritic cell reprogramming by endogenously produced lactic acid. J Immunol 191(6):3090–3099. https://doi.org/10.4049/jimmunol.1300772

Olivares O, Mayers JR, Gouirand V, Torrence ME, Gicquel T, Borge L, Lac S, Roques J, Lavaut MN, Berthezène P, Rubis M, Secq V, Garcia S, Moutardier V, Lombardo D, Iovanna JL, Tomasini R, Guillaumond F, Vander Heiden MG, Vasseur S (2017) Collagen-derived proline promotes pancreatic ductal adenocarcinoma cell survival under nutrient limited conditions. Nat Commun 8:16031. https://doi.org/10.1038/ncomms16031

Ono A, Utsugi M, Masubuchi K, Ishizuka T, Kawata T, Shimizu Y, Hisada T, Hamuro J, Mori M, Dobashi K (2009) Glutathione redox regulates TGF-beta-induced fibrogenic effects through Smad3 activation. FEBS Lett 583(2):357–362. https://doi.org/10.1016/j.febslet.2008.12.021

Oörni K, Kovanen PT (2006) Enhanced extracellular lipid accumulation in acidic environments. Curr Opin Lipidol 17(5):534–540. https://doi.org/10.1097/01.mol.0000245259.63505.c2

Ózsvári B, Sotgia F, Lisanti MP (2020) First-in-class candidate therapeutics that target mitochondria and effectively prevent cancer cell metastasis: mitoriboscins and TPP compounds. Aging (Albany NY) 12(11):10162–10179. https://doi.org/10.18632/aging.103336

Parks SK, Pouysségur J (2017) Targeting pH regulating proteins for cancer therapy-progress and limitations. Semin Cancer Biol 43:66–73. https://doi.org/10.1016/j.semcancer.2017.01.007

Parks SK, Mueller-Klieser W, Pouysségur J (2020) Lactate and acidity in the cancer microenvironment. Annu Rev Cancer Biol 4(1):141–158. https://doi.org/10.1146/annurev-cancerbio-030419-033556

Parolini I, Federici C, Raggi C, Lugini L, Palleschi S, De Milito A, Coscia C, Iessi E, Logozzi M, Molinari A, Colone M, Tatti M, Sargiacomo M, Fais S (2009) Microenvironmental pH is a key factor for exosome traffic in tumor cells. J Biol Chem 284(49):34211–34222. https://doi.org/10.1074/jbc.M109.041152

Pavlova NN, Thompson CB (2016) The emerging hallmarks of cancer metabolism. Cell Metab 23(1):27–47. https://doi.org/10.1016/j.cmet.2015.12.006

Payen VL, Mina E, Van Hée VF, Porporato PE, Sonveaux P (2020) Monocarboxylate transporters in cancer. Mol Metab 33:48–66. https://doi.org/10.1016/j.molmet.2019.07.006

Peppicelli S, Bianchini F, Torre E, Calorini L (2014) Contribution of acidic melanoma cells undergoing epithelial-to-mesenchymal transition to aggressiveness of non-acidic melanoma cells. Clin Exp Metastasis 31(4):423–433. https://doi.org/10.1007/s10585-014-9637-6

Petővári G, Dankó T, Krencz I, Hujber Z, Rajnai H, Vetlényi E, Raffay R, Pápay J, Jeney A, Sebestyén A (2020) Inhibition of metabolic shift can decrease therapy resistance in human high-grade glioma cells. Pathol Oncol Res 26(1):23–33. https://doi.org/10.1007/s12253-019-00677-2

Qi Y, Xu R (2018) Roles of PLODs in collagen synthesis and cancer progression. Front Cell Dev Biol 6:66. https://doi.org/10.3389/fcell.2018.00066

Rajesh Y, Sarkar D (2021) Association of adipose tissue and adipokines with development of obesity-induced liver cancer. Int J Mol Sci 22(4). https://doi.org/10.3390/ijms22042163

Sameni M, Elliott E, Ziegler G, Fortgens PH, Dennison C, Sloane BF (1995) Cathepsin B and D are localized at the surface of human breast cancer cells. Pathol Oncol Res 1(1):43–53. https://doi.org/10.1007/bf02893583
Sano S, Izumi Y, Yamaguchi T, Yamazaki T, Tanaka M, Shiota M, Osada-Oka M, Nakamura Y, Wei M, Wanibuchi H, Iwao H, Yoshiyama M (2014) Lipid synthesis is promoted by hypoxic adipocyte-derived exosomes in 3T3-L1 cells. Biochem Biophys Res Commun 445(2):327–333. https://doi.org/10.1016/j.bbrc.2014.01.183
Schaefer L, Tredup C, Gubbiotti MA, Iozzo RV (2017) Proteoglycan neofunctions: regulation of inflammation and autophagy in cancer biology. FEBS J 284(1):10–26. https://doi.org/10.1111/febs.13963
Schön MP (2008) Treatment of psoriasis: a journey from empiricism to evidence. Clin Dermatol 26(5):417–418. https://doi.org/10.1016/j.clindermatol.2008.07.002
Schwörer S, Vardhana SA, Thompson CB (2019) Cancer metabolism drives a stromal regenerative response. Cell Metab 29(3):576–591. https://doi.org/10.1016/j.cmet.2019.01.015
Scribano ML (2018) Vedolizumab for inflammatory bowel disease: from randomized controlled trials to real-life evidence. World J Gastroenterol 24(23):2457–2467. https://doi.org/10.3748/wjg.v24.i23.2457
Segawa S, Fujiya M, Konishi H, Ueno N, Kobayashi N, Shigyo T, Kohgo Y (2011) Probiotic-derived polyphosphate enhances the epithelial barrier function and maintains intestinal homeostasis through integrin-p38 MAPK pathway. PLoS One 6(8):e23278. https://doi.org/10.1371/journal.pone.0023278
Shao M, Yang Q, Zhu W, Jin H, Wang J, Song J, Kong Y, Lv X (2018) LncHOXA10 drives liver TICs self-renewal and tumorigenesis via HOXA10 transcription activation. Mol Cancer 17(1): 173. https://doi.org/10.1186/s12943-018-0921-y
Shiba T, Nishimura D, Kawazoe Y, Onodera Y, Tsutsumi K, Nakamura R, Ohshiro M (2003) Modulation of mitogenic activity of fibroblast growth factors by inorganic polyphosphate. J Biol Chem 278(29):26788–26792. https://doi.org/10.1074/jbc.M303468200
Shime H, Yabu M, Akazawa T, Kodama K, Matsumoto M, Seya T, Inoue N (2008) Tumor-secreted lactic acid promotes IL-23/IL-17 proinflammatory pathway. J Immunol 180(11):7175–7183. https://doi.org/10.4049/jimmunol.180.11.7175
Shriwas P, Roberts D, Li Y, Wang L, Qian Y, Bergmeier S, Hines J, Adhicary S, Nielsen C, Chen X (2021) A small-molecule pan-class I glucose transporter inhibitor reduces cancer cell proliferation in vitro and tumor growth in vivo by targeting glucose-based metabolism. Cancer Metab 9(1):14. https://doi.org/10.1186/s40170-021-00248-7
Shyu P Jr, Wong XFA, Crasta K, Thibault G (2018) Dropping in on lipid droplets: insights into cellular stress and cancer. Biosci Rep 38(5). https://doi.org/10.1042/bsr20180764
Singer AJ, Clark RA (1999) Cutaneous wound healing. N Engl J Med 341(10):738–746. https://doi.org/10.1056/nejm199909023411006
Siska PJ, Singer K, Evert K, Renner K, Kreutz M (2020) The immunological Warburg effect: can a metabolic-tumor-stroma score (MeTS) guide cancer immunotherapy? Immunol Rev 295(1): 187–202. https://doi.org/10.1111/imr.12846
Sonveaux P, Copetti T, De Saedeleer CJ, Végran F, Verrax J, Kennedy KM, Moon EJ, Dhup S, Danhier P, Frérart F, Gallez B, Ribeiro A, Michiels C, Dewhirst MW, Feron O (2012) Targeting the lactate transporter MCT1 in endothelial cells inhibits lactate-induced HIF-1 activation and tumor angiogenesis. PLoS One 7(3):e33418. https://doi.org/10.1371/journal.pone.0033418
Sousa CM, Biancur DE, Wang X, Halbrook CJ, Sherman MH, Zhang L, Kremer D, Hwang RF, Witkiewicz AK, Ying H, Asara JM, Evans RM, Cantley LC, Lyssiotis CA, Kimmelman AC (2016) Pancreatic stellate cells support tumour metabolism through autophagic alanine secretion. Nature 536(7617):479–483. https://doi.org/10.1038/nature19084
Spees JL, Olson SD, Whitney MJ, Prockop DJ (2006) Mitochondrial transfer between cells can rescue aerobic respiration. Proc Natl Acad Sci U S A 103(5):1283–1288. https://doi.org/10.1073/pnas.0510511103

Su H, Yang F, Sun B, Karin M (2021) Macropinocytosis: the big drinker behind cancer cell self-consumption. Autophagy 17(5):1290–1291. https://doi.org/10.1080/15548627.2021.1919969

Sugita S, Yamato M, Hatabu T, Kataoka Y (2021) Involvement of cancer-derived EMT cells in the accumulation of 18F-fluorodeoxyglucose in the hypoxic cancer microenvironment. Sci Rep 11(1):9668. https://doi.org/10.1038/s41598-021-88414-1

Sullivan WJ, Mullen PJ, Schmid EW, Flores A, Momcilovic M, Sharpley MS, Jelinek D, Whiteley AE, Maxwell MB, Wilde BR, Banerjee U, Coller HA, Shackelford DB, Braas D, Ayer DE, de Aguiar Vallim TQ, Lowry WE, Christofk HR (2018) Extracellular matrix remodeling regulates glucose metabolism through TXNIP destabilization. Cell 175(1):117–132.e121. https://doi.org/10.1016/j.cell.2018.08.017

Svensson KJ, Kucharzewska P, Christianson HC, Sköld S, Löfstedt T, Johansson MC, Mörgelin M, Bengzon J, Ruf W, Belting M (2011) Hypoxia triggers a proangiogenic pathway involving cancer cell microvesicles and PAR-2-mediated heparin-binding EGF signaling in endothelial cells. Proc Natl Acad Sci U S A 108(32):13147–13152. https://doi.org/10.1073/pnas.1104261108

Taddei ML, Giannoni E, Comito G, Chiarugi P (2013) Microenvironment and tumor cell plasticity: an easy way out. Cancer Lett 341(1):80–96. https://doi.org/10.1016/j.canlet.2013.01.042

Taddei ML, Pietrovito L, Leo A, Chiarugi P (2020) Lactate in sarcoma microenvironment: much more than just a waste product. Cell 9(2). https://doi.org/10.3390/cells9020510

Tajan M, Vousden KH (2016) The quid pro quo of the tumor/stromal interaction. Cell Metab 24(5):645–646. https://doi.org/10.1016/j.cmet.2016.10.017

Tan AS, Baty JW, Dong LF, Bezawork-Geleta A, Endaya B, Goodwin J, Bajzikova M, Kovarova J, Peterka M, Yan B, Pesdar EA, Sobol M, Filimonenko A, Stuart S, Vondrusova M, Kluckova K, Sachaphibulkij K, Rohlena J, Hozak P, Truksa J, Eccles D, Haupt LM, Griffiths LR, Neuzil J, Berridge MV (2015) Mitochondrial genome acquisition restores respiratory function and tumorigenic potential of cancer cells without mitochondrial DNA. Cell Metab 21(1):81–94. https://doi.org/10.1016/j.cmet.2014.12.003

Thompson CB, Bielska AA (2019) Growth factors stimulate anabolic metabolism by directing nutrient uptake. J Biol Chem 294(47):17883–17888. https://doi.org/10.1074/jbc.AW119.008146

Tímár J, Sebestyén A, Kopper L, Dankó T (2021) Hypoxia signaling in cancer: from basics to clinical practice. Pathol Oncol Res 27(100). https://doi.org/10.3389/pore.2021.1609802

Trcka F, Durech M, Vankova P, Chmelik J, Martinkova V, Hausner J, Kadek A, Marcoux J, Klumpler T, Vojtesek B, Muller P, Man P (2019) Human stress-inducible Hsp70 has a high propensity to form ATP-dependent antiparallel dimers that are differentially regulated by cochaperone binding. Mol Cell Proteomics 18(2):320–337. https://doi.org/10.1074/mcp.RA118.001044

Valencia T, Kim JY, Abu-Baker S, Moscat-Pardos J, Ahn CS, Reina-Campos M, Duran A, Castilla EA, Metallo CM, Diaz-Meco MT, Moscat J (2014) Metabolic reprogramming of stromal fibroblasts through p62-mTORC1 signaling promotes inflammation and tumorigenesis. Cancer Cell 26(1):121–135. https://doi.org/10.1016/j.ccr.2014.05.004

van Pelt GW, Sandberg TP, Morreau H, Gelderblom H, van Krieken J, Tollenaar R, Mesker WE (2018) The tumour-stroma ratio in colon cancer: the biological role and its prognostic impact. Histopathology 73(2):197–206. https://doi.org/10.1111/his.13489

Warburg O, Wind F, Negelein E (1927) The metabolism of tumors in the body. J Gen Physiol 8(6):519–530. https://doi.org/10.1085/jgp.8.6.519

Wise DR, Thompson CB (2010) Glutamine addiction: a new therapeutic target in cancer. Trends Biochem Sci 35(8):427–433. https://doi.org/10.1016/j.tibs.2010.05.003

Wright HJ, Hou J, Xu B, Cortez M, Potma EO, Tromberg BJ, Razorenova OV (2017) CDCP1 drives triple-negative breast cancer metastasis through reduction of lipid-droplet abundance and stimulation of fatty acid oxidation. Proc Natl Acad Sci U S A 114(32):E6556–e6565. https://doi.org/10.1073/pnas.1703791114

Xie D, Zhu S, Bai L (2016) Lactic acid in tumor microenvironments causes dysfunction of NKT cells by interfering with mTOR signaling. Sci China Life Sci 59(12):1290–1296

Yang L, Achreja A, Yeung TL, Mangala LS, Jiang D, Han C, Baddour J, Marini JC, Ni J, Nakahara R, Wahlig S, Chiba L, Kim SH, Morse J, Pradeep S, Nagaraja AS, Haemmerle M, Kyunghee N, Derichsweiler M, Plackemeier T, Mercado-Uribe I, Lopez-Berestein G, Moss T, Ram PT, Liu J, Lu X, Mok SC, Sood AK, Nagrath D (2016) Targeting stromal glutamine synthetase in tumors disrupts tumor microenvironment-regulated cancer cell growth. Cell Metab 24(5):685–700. https://doi.org/10.1016/j.cmet.2016.10.011

Yoshida GJ (2021) The harmonious interplay of amino acid and monocarboxylate transporters induces the robustness of cancer cells. Meta 11(1). https://doi.org/10.3390/metabo11010027

Zhao H, Yang L, Baddour J, Achreja A, Bernard V, Moss T, Marini JC, Tudawe T, Seviour EG, San Lucas FA, Alvarez H, Gupta S, Maiti SN, Cooper L, Peehl D, Ram PT, Maitra A, Nagrath D (2016) Tumor microenvironment derived exosomes pleiotropically modulate cancer cell metabolism. elife 5:e10250. https://doi.org/10.7554/eLife.10250

Zhao J, Krystofiak ES, Ballesteros A, Cui R, Van Itallie CM, Anderson JM, Fenollar-Ferrer C, Kachar B (2018) Multiple claudin-claudin cis interfaces are required for tight junction strand formation and inherent flexibility. Commun Biol 1:50. https://doi.org/10.1038/s42003-018-0051-5

Chapter 14
The Role of Inflammatory Cells in Tumor Angiogenesis

Roberto Tamma, Tiziana Annese, and Domenico Ribatti

Abstract Tumor growth depends on angiogenesis. The complex tissue environment surrounding tumor cells, which is composed of a variety of resident and infiltrating host cells, secreted factors and extracellular matrix proteins, influences tumor angiogenesis and progression. Moreover, the tumor microenvironment contributes to determining therapeutic responses and resistance to therapy. The ability to block tumor resistance is related to the understanding of the cellular and molecular pathways activated in the tumor microenvironment. Novel emerging targeted therapeutic strategies are based on the combination of different antitumor approaches with the aim of resolving refractory tumors and improving cancer treatment efficiency.

14.1 Tumor Angiogenesis

Healthy and pathologic tissue homeostasis requires an adequate supply of oxygen and nutrients that is connected to efficient development of the vascular system. Additionally, tumor cells to survive and proliferate need oxygen and nutrients and consequently the closeness to blood vessels. Angiogenesis is the physiological process through which new blood vessels form from pre-existing vessels (Carmeliet and Jain 2011). Generally, tumor development is an angiogenesis-dependent process, and the angiogenetic process depends on the temporal coordination of factors and related pathways needed for the establishment of stable channels to provide a supply to tumor cells (Weis and Cheresh 2011). It has been well established that during cancer progression, the interactions between tumor cells and inflammatory cells are closely associated with each other and with angiogenesis (Wang et al. 2019a).

R. Tamma · T. Annese · D. Ribatti (✉)
Department of Basic Medical Sciences, Neurosciences and Sensory Organs, University of Bari Medical School, Bari, Italy
e-mail: roberto.tamma@uniba.it; tiziana.annese@uniba.it; domenico.ribatti@uniba.it

I. Kovalszky et al. (eds.), *The Extracellular Matrix and the Tumor Microenvironment*, Biology of Extracellular Matrix 11,
https://doi.org/10.1007/978-3-030-99708-3_14

The growth of solid tumor mass, its progression and the metastatic process, how it is widely described, are strongly influenced by angiogenesis (Folkman 1971). In 1966, Warren and collaborators implanted melanoma nodules in experimental animals and observed a rapid vessel sprout toward the mass, the formation of new capillaries, their penetration into the tumor, and the establishment of blood flow. This phenomenon was more evident during tumor growth than in inflammation processes (Warren and Shubik 1966). Research conducted by Folkman showed that without appropriate vascularization and therefore oxygen and nutrient supply, a tumor can grow limitedly to a size of a few millimeters and a cell content of approximately a few thousand cells (Folkman 1971; Nishida et al. 2006). Under these conditions, tumors induce a process recognized as an angiogenic switch in which tumor cells acquire angiogenic properties, leading to the transition from a quiescent to active endothelium and consequently the vascularization of the growing cell mass (Baeriswyl and Christofori 2009; Ribatti et al. 2007). In tumor murine models, this switch coincides with malignant transition of the growing mass and is needed for malignant tumor progression (Lin et al. 2006; Folkman et al. 1989). It became evident that some soluble factors released by the tumor induced the activation of angiogenesis. Folkman hypothesized that until the appropriate blood flow is created, the tumor mass stops its growth and enters a dormant state (Folkman et al. 1971). On this basis, in the last 50 years, research on mechanisms related to tumor angiogenesis has intensified to discover molecules usable as new targets in anticancer therapy. Tumor angiogenesis is a multiphasic process initiated directly by the tumor when it reaches a size that makes it hypoxic, which further leads to cancer development.

14.2 Tumor Microenvironment

It is well known that tumor cells develop in a complex tissue environment, the so-called tumor microenvironment (TME), which includes cancer cells, stromal cells, blood vessels, nerve fibers, extracellular matrix, and acellular components. The TME is involved in tumor initiation as well as during tumor progression and metastasis; furthermore, it also has important effects on therapeutic efficacy (Tamma et al. 2019a). It is believed that although cancer initiation is due to the acquisition of oncogenic mutations in cells, its progression depends on the surrounding cells that are recruited and subsequently release many cytokines and chemokines (Tysnes and Bjerkvig 2007). In 1863, Rudolf Virchow postulated the crosstalk between inflammation and cancer (Virchow 1989), and 20 years after Stephen Paget illustrated the "seed and soil" theory assuming that the choice of the target organ depends on the interactions between metastatic tumor cells (the "seed") and their organ microenvironment (the "soil") (Paget 1989). One hundred years later, Hanahan and Weinberg expanded from six to ten hallmarks of cancer and recognized the important role of the TME in cancer development (Hanahan and Weinberg 2011). The main cytokines and chemokines secreted by cells of the TME are involved in the regulation of angiogenesis, including proangiogenic factors, such as the vascular endothelial growth factor (VEGF) family, fibroblast growth factors (FGFs), platelet-derived growth factor (PDGF), angiopoietins (Ang), and hypoxia-inducible factor (HIF), and angiostatic factors, such as angiostatin, endostatin, platelet factor 4 (PF4), and thrombospondin-1 (TSP1) (Ucuzian et al. 2010).

14.3 Pro-Angiogenic Factors

VEGF The human VEGF family includes VEGF-A, VEGF-B, VEGF-C, VEGF-D, and placental growth factor (PlGF) originating from different genes (Melincovici et al. 2018). The VEGF family exerts its function by binding three transmembrane tyrosine kinase receptors (RTKs), VEGFR-1 (FLT1), VEGFR-2 (KDR, FLK1), and VEGFR-3 (FLT4). VEGFR-1 is expressed in monocytes, macrophages, hematopoietic stem cells, vascular smooth cells, and leukemic cells. VEGFR-2 is expressed in vascular endothelial cells, endothelial progenitor cells, and megakaryocytes, whereas VEGFR-3 is expressed in lymphatic endothelial cells. VEGFs can also interact with other proteins, integrins, cadherins, heparan sulfate proteoglycans, and with the coreceptors neuropilin-1 and -2 (NRP-1 and NRP-2), which enhance VEGFR-1 and VEGFR-2 action (Stuttfeld and Ballmer-Hofer 2009). VEGF-A is the main component of the VEGF family and is produced by endothelial and vascular smooth muscle cells, activated platelets, fibroblasts, lymphocytes, macrophages, and tumor cells. It is considered a crucial angiogenic stimulator involved in numerous pleiotropic effects, including the proliferation and inhibition of apoptosis of vascular endothelial cells (Ferrara and Davis-Smyth 1997; Gerber et al. 1998), permeability, chemotaxis and activation of monocytes and hematopoietic stem cells, and exerts neurotrophic and neuroprotective action (Storkebaum and Carmeliet 2004). Through alternative splicing, the VEGF-A transcript produces several isoforms with proangiogenic or antiangiogenic activities, including VEGF-A_{121}, VEGF-A_{145}, VEGF-A_{165}, VEGF-A_{189}, and VEGF-A_{206} (Yang et al. 2018a; Logue et al. 2016; Dehghani et al. 2018). $VEGFA_{165}$ is the most important both quantitatively and qualitatively. VEGF-B is involved in pulmonary angiogenesis after chronic hypoxia and has been found in cardiac and skeletal muscle. VEGF-C and VEGF-D are important lymphangiogenesis regulators (Rauniyar et al. 2018; Stacker and Achen 2018). PlGF, discovered in the human placenta, is highly expressed in trophoblast cells (Hang et al. 2013) and has also been found in the thyroid, lungs, heart, and skeletal muscle (Maglione et al. 1991). It includes four different subtypes that bind VEGFR-1, and the PlGF isoform also binds NRP-1 and NRP-2. PlGF regulates the growth, migration, and survival of endothelial cells directly through VEGFR-1 or indirectly through VEGFR-2/VEGF-A-mediated activation or formation of a PlGF/VEGF-A heterodimer (Autiero et al. 2003).

Fibroblast Growth Factors (FGFs) The human FGF family includes 22 members involved in the regulation of endothelial cell differentiation, proliferation, migration, survival, and vessel maturation (Yun et al. 2010). FGF-1 and FGF-2, the first known as acid FGF and the latter as basic FGF, mostly mediate the angiogenic response (Motomura et al. 2008). FGF receptors (FGFRs) belong to the RTK superfamily. Upon activation, they undergo dimerization and internalization and initiate large-scale tyrosine phosphorylation responses and signaling cascades activating the Ras/MAP-kinase pathway (Mathew et al. 2016).

Platelet-Derived Growth Factor (PDGF) The PDGF family comprises four PDGF homodimers, namely, PDGF-AA, PDGF-BB, PDGF-CC, and PDGF-DD, and one heterodimer, PDGF-AB (Fredriksson et al. 2004). PDGF was originally isolated from platelets, but it has been expressed by numerous other cell types, including epithelial and endothelial cells. PDGF receptors (PDGFRs) belong to the family of RTKs and include PDGFRα and PDGFRβ, which are encoded by two different genes (Gao et al. 2018). These receptors are expressed by fibroblasts, pericytes, vascular smooth muscle cells, monocytes, macrophages, lymphocytes, and mast cells and stimulate their proliferation and motility. PDGFs participate in vascular development by acting on the proliferation and survival of vascular mural cells (Olson and Soriano 2011).

Angiopoietins (Angs) The Ang protein family includes four members: Ang-1, Ang-2, Ang-3, and Ang-4 (Lee et al. 2004); the first two are the major members involved in vasculogenesis and vascular repair (Akwii et al. 2019). Angs bind to two receptors belonging to the family of RTKs named Tie1 and Tie2. Tie2 is expressed by endothelial and myeloid cells (Patan 1998). Tie1 is an orphan poorly characterized receptor that seems to be involved in the modulation of Ang/Tie-2 through the formation of heterodimers with Tie-2 (Eklund et al. 2017). Ang-1 is expressed by both mural cells and other nonvascular stromal and tumor cells. It is involved in the regulation of vessel stabilization during embryonic development, vessel remodeling, and maintenance of the normal vasculature (Brindle et al. 2006). Ang-2 is produced by the VEGF-stimulated endothelium, hypoxia, and shear stress, promoting blood vessel wall destabilization through competitive inhibition of Tie-2 and integrin activation. Furthermore, Ang-2 stimulates pericyte detachment, permeability, vascular regression, and lymphangiogenesis (Akwii et al. 2019).

Hypoxia-Inducible Factors (HIFs) HIFs are DNA-binding transcription factors that associate with specific nuclear cofactors under hypoxia (Palazon et al. 2014). They are heterodimers that include both the constitutively expressed HIF-1β subunit and oxygen-regulated HIF-1α or HIF-2α subunit (Hu et al. 2003). In humans, HIF-1α is ubiquitously expressed, while HIF-2α, although it is expressed mainly in the endothelium, in hypoxic conditions, is also expressed in the kidney, pancreas, brain, liver, intestine, and myocardium. When cells are in a hypoxic environment, the hydroxylation process is inhibited, and HIF-α escapes proteasomal degradation, dimerizes with HIF-1β, and associates with transcriptional coactivators (Berra et al. 2001). The latter recognizes hypoxia-responsive genes, resulting in physiological adaptation to hypoxia. Other stimuli, such as nitric oxide and reactive oxygen species (ROS), can also activate HIFs (Wellman et al. 2004).

Many human cancers are characterized by increased levels of HIF, and its expression correlates with mortality (Zhong et al. 1999; Talks et al. 2000). Hypoxic conditions contribute to increased HIF activity, which translates into the regulation of genes involved in angiogenesis, cell survival, metabolism, invasion, and metastasis. In solid tumors, the rapid proliferation of cancer cells limits oxygen diffusion within the tumor, decreasing its concentrations under physiological conditions. This leads to increased expression and activity of HIF, contributing to tumor angiogenesis (Huang et al. 2017; Shi and Fang 2004).

14.4 Angiogenic Inhibitors

Angiostatin Angiostatin is a 38 kDa internal fragment of plasminogen (Cao and Xue 2004; Ji et al. 1998). Angiostatin inhibits endothelial cell proliferation, migration, and tube formation (Pozzi et al. 2000) and induces apoptosis of endothelial cells (Ramirez-Moreno et al. 2020). Moreover, angiostatin inhibits the signaling induced by FGF-2 and VEGF in human microvascular endothelial cells (Redlitz et al. 1999) and inhibits primary tumor growth as well as angiogenesis-dependent growth of metastases (Dell'Eva et al. 2002).

Endostatin Endostatin is an angiostatic 20 kDa internal type XVIII collagen fragment released by proteolytic activity (Wenzel et al. 2006). The hinge region of endostatin contains several proteolytic cleavage sites where matrix metalloproteinases (MMPs), cathepsins, and elastases induce its release and consequently the interaction with cell membrane receptors, including $\alpha5\beta1$, $\alpha v\beta3$, and $\alpha v\beta5$ integrin receptors, on endothelial cells (Zatterstrom et al. 2000). Endostatin inhibits the mitogen-activated protein kinase pathway in endothelial cells, leading to the inhibition of angiogenesis (Wickstrom et al. 2005). Endostatin affects VEGF to VEGFR-2 binding and tyrosine phosphorylation (Jia et al. 2004) and inhibits the activities of matrix metalloproteinases-2, -9, and -13 (MMP-2, MMP-9, and MMP-13) (Kim et al. 2000).

Platelet Factor 4 (PF4) PF4 is the most abundant chemokine member of the C-X-C family found in platelets and megakaryocytes. It exhibits antiangiogenic effects both in vivo and in vitro and directly interacts with VEGF-A_{165} (Hang et al. 2013; Maurer et al. 2006).

Thrombospondin-1 (TSP-1) TSP-1 belongs to a family of extracellular matrix (ECM) glycoproteins. TSP-1, initially discovered in platelet granules, is also produced by endothelial cells, monocytes/macrophages, and smooth muscle cells. TSP-1 interacts with numerous ECM proteins, modulates extracellular protease levels, and activates transforming growth factor beta (TGF-β) (Lawler 2002). TSP-1 inhibits angiogenesis by inhibiting the growth, sprouting, and motility of endothelial cells. High concentrations of TSP-1 have the opposite effect, promoting angiogenesis (Lawler and Lawler 2012).

14.5 TME Infiltrating Cells

Macrophages Tumor-associated macrophages (TAMs) are one of the major tumor-infiltrating innate immune cells and play an important role in the TME because they are involved in promoting tumor growth, invasion, metastasis, and therapeutic resistance (Chanmee et al. 2014). TAMs are described in two different polarization states: M1 CD68-positive and M2, CD-163 and CD-206-positive (Medbury et al. 2013). It is generally believed that M1 macrophages are involved in proinflammatory processes by migrating to inflamed tissues and targeting pathogens directly or activating cells of the adaptive immune system. It has been demonstrated that the M1 subpopulation has antitumor function because of its ability to kill tumor cells and recruit cytotoxic T lymphocytes to activate adaptive immune responses (Chanmee et al. 2014). The M2 subpopulation, on the other hand, has the functions of debris removal, angiogenesis stimulation, and tissue reconstruction and promotes tumorigenesis. They induce immune tolerance and attract T regulatory cells and Th2 T cells. It is believed that M2 TAMs have protumor activity because they stimulate angiogenesis and tumor growth (Jayasingam et al. 2019). Usually, TAM recruitment is correlated with the induction of angiogenic switching and is associated with a poor prognosis in most cancer types. Many cytokines and chemokines are secreted by vascular and perivascular cells, stromal cells, and cancer cells that recruit TAMs in the TME and include C-C motif ligand 2 (CCL2), CCL5, CCL7, Ang-2, colony-stimulating factor-1 (CSF1), VEGF, interleukin-33 (IL-33), semaphorin 3D (Sema 3D), endothelial monocyte-activating polypeptide-II (EMAP-II), endothelin (ET)-1 and 2, stromal cell-derived factor 1α (SDF1α/CXCL12), eotaxin, and oncostatin (Wang et al. 2019a).

TAMs can transdifferentiate into vessel-like structures by vasculogenic mimicry. In gliomas, the areas where vascular mimicry is found are characterized by high TAM infiltration and correlated with M2 density (Rong et al. 2016). The angiogenic factors secreted by TAMs include EGF-A, TGF-β, FGF-2, CCL18, Sema4D, adrenomedullin (ADM), and PlGF (Riabov et al. 2014). TAMs express the MCT1-lactate transporter. Furthermore, TAMs express VEGF-A when exposed to hypoxia or in the presence of lactate produced by tumor cells following aerobic or anaerobic glycolysis (Zhang et al. 2020). This effect is mediated by HIF1α, and lactate seems to lead to M2-like polarization of TAMs (Colegio et al. 2014). TAMs have been found to localize frequently in avascular and hypoxic areas of invasive carcinoma of the breast, where the expression of VEGF-A is upregulated (Lewis and Pollard 2006). Fra-1 and the IL-6/JAK/Stat3 signaling pathway in TAMs are involved in the secretion of proangiogenic factors (Choi et al. 2018). TAMs produce CCL18, which stimulates angiogenesis in synergy with VEGF-A (Lin et al. 2015). ADM is a potent vasodilator belonging to the calcitonin superfamily whose secretion by macrophages is upregulated by inflammatory factors and hypoxia. In melanoma, TAM-derived ADM induces angiogenesis in a paracrine manner via the endothelial nitric oxide synthase (eNOS) signaling pathway (Chen et al. 2011). MMP-9, which is highly expressed by M2 macrophages, triggers the angiogenic switch during carcinogenesis by the release of VEGF-A from the ECM in colorectal cancer (Deryugina and Quigley 2015; Yahaya et al. 2019).

Mast Cells (MCs) MCs are involved in a large spectrum of biological processes, ranging from inflammation and immune modulation to angiogenesis, tissue repair, remodeling, and cancer (Welker et al. 2000). MC precursors complete their differentiation and maturation in target tissues under the control of local growth factors, including IL-9, IL-10, IL-3, IL-4, IL-33, CXCL12, nerve growth factor (NGF), and TGF-β (Hu et al. 2007). MCs are traditionally classified based on the production of tryptase and chymase, and resident MCs of various organs are characterized by the expression and release of peculiar factors related to their tissue-specific functions (Krystel-Whittemore et al. 2015). MCs can be recruited to the tumor microenvironment by tumor cell-released chemoattractants, including stem cell factor (SCF) or CCL-15 (Yu et al. 2018). In the TME, MCs release proangiogenic factors such as FGF2, VEGFA, tumor necrosis factor alpha (TNFα), and CXCL8 (Norrby 2002). Furthermore, they produce MMPs and chymase, and tryptase activates pro-MMPs (Kanbe et al. 1999; Johnson et al. 1998). The localization of MCs in the TME is determined by interactions of CCR2, CXCR2, and CXCR3 with their respective ligands CCL2, CXCL1, and CXCL10. In this way, MCs facilitate tumor angiogenesis and promote tumor invasiveness (Ramirez-Moreno et al. 2020; Komi and Redegeld 2020). On the other hand, numerous cytokines released by MCs contribute to inflammation, inhibiting tumor cell growth and inducing tumor cell apoptosis (Ribatti and Crivellato 2012). MC tryptase activates the Ang-1 pathway and induces endothelial cell proliferation in pancreatic cancer (Guo et al. 2016). MC inactivation delayed the angiogenic switch and malignant progression in early preneoplastic lesion experimental squamous epithelial, intestinal, and pancreatic islet cancer models (Maciel et al. 2015).

Neutrophils Neutrophils release large amounts of soluble factors, including cytokines and chemokines, through which they recruit and activate other immune cells (Malech et al. 2014). Moreover, they are involved in chronic inflammation regulation and in various steps of tumor progression and angiogenesis, exerting both pro-(tumor-associated neutrophil, TAN-N2) and antitumor (TAN-N1) roles. Normal density neutrophils (NDNs) have been associated with cytotoxic antitumor activities, while immature low-density neutrophils (LDNs) exert immunosuppressive protumor activities (Cerecedo et al. 2021). The TME is infiltrated with $CD66b^+$ neutrophils, and their number is correlated with poor clinical outcome (Carus et al. 2013). TGFβ reduces endothelial adhesiveness of neutrophils and neutrophil transmigration through the endothelium as well as the number of antitumor neutrophils in the TME (Granot 2019). In a Nod Scid mouse model of human prostate cancer, TANs are the major source of MMP-9 (Li et al. 2020a). In gliomas, the high TME infiltration of neutrophils was correlated with the tumor grade as well as resistance to anti-VEGF therapy (Liang et al. 2014). Neutrophils produce low amounts of tissue inhibitors of metalloproteinases-1 (TIMP-1), thus enhancing the angiogenic effect of MMP-9 (Wang et al. 2019b). In a RIP-Tag murine model, granulocyte-CSF (G-CSF) stimulates neutrophils to release the proangiogenic molecule Bv8, which is critical for VEGF-independent tumor angiogenesis (Bjornmalm et al. 2017). Resistance to anti-VEGF therapy in tumors has been correlated with the infiltration of neutrophils and associated with Bv8 neutrophil expression (Shojaei et al. 2008). On the other hand, neutrophils are also involved in the inhibition of angiogenesis through the release of antiangiogenic factors, such as affecting neutrophil migration toward CXCL1 and CXCL8 (Jeronimo et al. 2017).

Lymphocytes The role of lymphocytes in tumor progression and angiogenesis remains to be further explored, and conflicting data about their function in the TME are emerging (Paijens et al. 2021). B cells are often present in the TME, and it is believed that they may contribute to tumor angiogenesis via STAT3 activation. STAT3 activation in cancer promotes tumor cell survival and proliferation, and a positive correlation has been established between its expression and VEGF release (Yang et al. 2013). It is thought that although only a subset of B cells infiltrating the tumor express STAT3, this might be enough to potentiate and maintain persistent STAT3 activation. Transplantation of STAT3-expressing B cells in tumor mouse models increased tumor growth and angiogenesis through the production of VEGF (Wang et al. 2019c). Another way by which B cells contribute to tumor angiogenesis is the antibody-mediated activation of Fcγ receptors on TAMs. This mechanism induces the secretion of IL-1, leading to the recruitment of myofibroblasts and promotion of tumor angiogenesis (Voronov et al. 2014). Tumor-infiltrating T cells play an important role in the antitumor response by the production of many cytokines, such as TNF-α, interferon gamma (IFN-γ), IL-2, IL-17, IL-22, and IL-36. TAMs inhibit CD8$^+$ T-cell infiltration and antitumor function (de Ruiter et al. 2017; Lan et al. 2021). Regulatory T (Treg) cells are immunosuppressive cells that affect the specialization and function of antigen-presenting cells (APCs), decrease their interactions with T cells, and subsequently inhibit effector T-cell function (Maimela et al. 2019). In addition, Tregs suppress natural killer (NK) cell activities (Li et al. 2020b). Cytotoxic T cells in the TME release IL-2, IL-12, and IFN-γ, improving the cytotoxic functions of CD8$^+$ T cells through the production of TNF-related apoptosis-inducing ligands (TRAILs), ROS, and perforin (Grossman et al. 2004). Tumor cells express coinhibitory receptors such as programmed death ligand-1 (PD-L1) and CD80 that interact with the inhibitory molecules programmed death-1 (PD-1) and cytotoxic T lymphocyte antigen-4 (CTLA-4) expressed by CD8$^+$ T cells. These interactions may inhibit CD8$^+$ T-cell activation and function (Cai et al. 2019). CD4$^+$ and CD8$^+$ T cells produce FGF-2 and heparin-binding epidermal-like growth factor (HB-EGF), which are both proangiogenic factors (Blotnick et al. 1994). On the other hand, T cells are also involved in the antiangiogenic response through TNFα, TGFβ, and IFNs. IFNs induce the expression of CXCL-9, CXCL-10, and CXCL-11 with angiostatic activities that can directly bind CXCR3 on endothelial cells (Blotnick et al. 1994; Beatty and Paterson 2001). NK cells are able to control tumor growth through their cytotoxic activity (Wu and Lanier 2003). Intratumor NK cells display phenotypic and/or functional alterations compared with peripheral NK cells depending on the influence of local factors and/or the interaction with other cell types of the TME (Larsen et al. 2014). The presence of TGF-β inhibits CD16, perforins, granzymes, and IFN-γ secretion, reverting NK cells to a proangiogenic phenotype characterized by the secretion of VEGF. Furthermore, the interaction between the immunoregulatory class I MHC molecule HLA-G and the KIR2DL4, ILT-4, and ILT-2 inhibitory NK cell receptors induces NK cells to acquire proangiogenic activities. Prostaglandin E2 (PGE2) is believed to contribute to the NK cell angiogenic switch (Bassani et al. 2019). Tumor-infiltrating NK cells express high levels of CD56, but low levels or none of CD16 produce several factors, such as VEGF, angiogenin, Ang-1, PIGF, CXCL8, and MMPs, which stimulate endothelial cell growth and angiogenesis (Bruno et al. 2018).

Cancer-Associated Fibroblasts (CAFs) CAFs are able to interact with tumor cells and form a myofibroblastic microenvironment that supports tumor progression and angiogenesis via secretion of various growth factors, cytokines, chemokines, and the degradation of ECM (Liu et al. 2019). A significant percentage of CAFs may share endothelial markers such as PECAM/CD31, and this allows us to suppose that they originate from an endothelial subpopulation through endothelial-to-mesenchymal transition (Potenta et al. 2008). Regarding their influence on angiogenesis, several studies have shown that their secretome is rich in several cytokines with proangiogenic effects, including VEGF, CXCL-8, and FGFs (Linares et al. 2020). Furthermore, CAF release is able to form capillary-like structures through vasculogenic mimicry by TGF-β and SDF-1 paracrine action (Yang et al. 2016a). Moreover, SDF-1 recruits endothelial precursor cells (EPCs), which may transdifferentiate into endothelial cells and stimulate the formation of novel vasculature at the tumor-host cell interface (Orimo et al. 2005). CAFs express podoplanin, which promotes angiogenesis in breast cancer via upregulation of VEGF-C rather than VEGF-A or VEGF-D (Kubouchi et al. 2018). The galectin family of glycan-binding proteins displays important functions in cancer development and progression. In gastric cancer, CAF expression of Galectin-1 is upregulated, leading to enhanced VEGF expression. Under hypoxic conditions, G-protein-coupled estrogen receptor (GPER) downregulation in CAFs reduces VEGF expression (Ham et al. 2019). In human pancreatic adenocarcinoma, VEGF expression by CAFs may be regulated by fibroblast activation protein α (FAP α), which is involved in affecting the balance of pro- and anti-angiogenic mediators (Higashino et al. 2019).

14.6 TME Inflammatory Cells and Angiogenesis. Our Experience in the Study of Human Lymphomas

We studied the inflammatory cell infiltrate and its role in tumor angiogenesis in diffuse large B-cell lymphoma (DLBCL) by comparing activated B-cell-like (ABC) patients to germinal center B-cell-like (GCB) patients. We demonstrated that increased ABC expression of STAT3 was correlated with poor prognosis in DLBCL and was associated with higher M2 TAM (Fig. 14.1a, b) and $CD8^+$ (Fig. 14.1c, d) cell infiltration into the TME, which, in turn, induced a strong

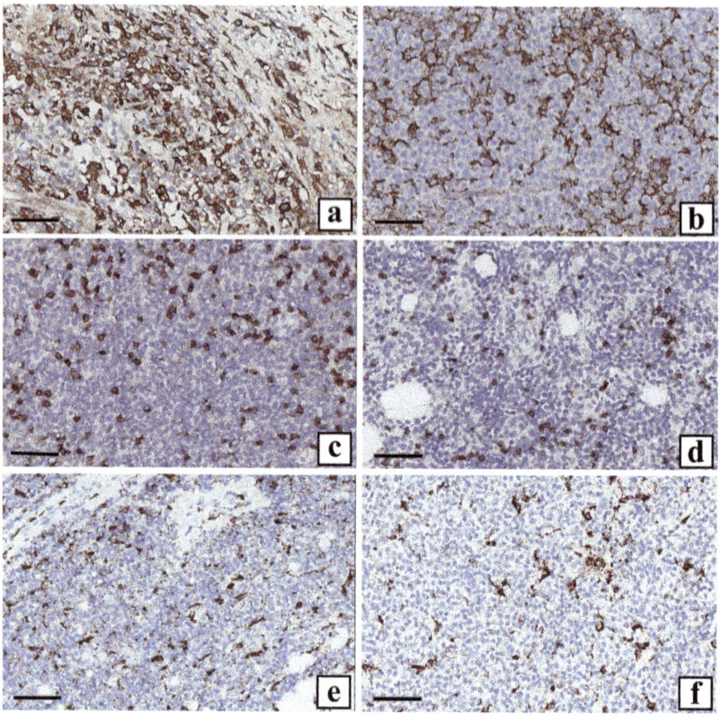

Fig. 14.1 Immunohistochemical staining of CD163^{+} macrophages in ABC (**a**) and GCB (**b**) DLBCL samples; CD8^{+} T cells in ABC (**c**) and GCB (**d**) DLBCL samples; CD68^{+} macrophages in MALT lymphoma (**e**) and control (**f**) samples. Scale bar 60 mm

angiogenic response in the ABC group (Tamma et al. 2020). Moreover, tumor vessels appeared lined by endothelial cells expressing both FVIII and STAT3 (Tamma et al. 2019b). Regarding the morphological distribution of the different TME cells in DLBCL, we established that cell patterns generated by CD4^{+}, CD8^{+}, CD68^{+}, CD163^{+}, and tryptase^{+} mast cell profiles have a higher uniformity index in the ABC, indicating a tendency of the cells to assume a more uniform distribution in the tissues in this more aggressive DLBCL subtype (Guidolin et al. 2021). Recently, Laddaga and coworkers suggested that the number of tumor infiltrating lymphocytes in the DLBCL TME is connected to a pre-existing antitumor immune response and then to an improved therapy response (Laddaga et al. 2021).

In a further study, we demonstrated that mucosa-associated lymphoid tissue (MALT)-type lymphoma and the tumor inflammatory TME included a high number of CD3^{+}, CD4^{+} and CD8^{+} lymphocytes, CD68^{+} (Fig. 14.1e, f), CD163^{+} macrophages, and tryptase^{+} mast cells. Interestingly, CD8^{+} cell content positively correlated with both CD34^{+} vessels, remarking on the important role of these cells in tumor angiogenesis and with CD163^{+} TAMs. Moreover, tryptase^{+} mast cells correlated with CD4^{+} lymphocytes (Tamma et al. 2021).

14.7 Targeting Angiogenesis and Inflammatory Cells in TME

Chemotherapy associated with surgery and/or radiotherapy is the principal cancer therapy worldwide (Bjornmalm et al. 2017). The TME has been gradually recognized as a crucial contributor to cancer progression and drug resistance (Heinrich et al. 2012), so the study of the components of the TME was deepened to identify new therapeutic targets.

Targeting Angiogenesis Bevacizumab was the first anti-VEGF antibody Food and Drug Administration (FDA) approved and actually used in different cancers, including metastatic colorectal cancer, lung cancer, kidney cancer, glioblastoma metastasis, and HER2-negative breast cancer, with response rates and durations highly variable (Jang et al. 2017). The addition of bevacizumab to chemotherapy has shown improvements in progression-free and overall survival with respect to chemotherapy alone (Jang et al. 2017; Yang et al. 2017). Another strategy consists of the inhibition of VEGF binding to its receptors by soluble decoy receptors (Holash et al. 2002). Aflibercept is a recombinant fusion protein containing portions of human VEGFR-1 and VEGFR-2 extracellular domains fused to the Fc portion of human immunoglobulin G1 able to bind with high-affinity VEGF and PlGF, inhibiting the activation of cognate VEGFRs (Holash et al. 2002). Experimental data about the use of aflibercept in cancer xenograft models demonstrated greater antitumor activity than bevacizumab (Chiron et al. 2014). Ramucirumab is a monoclonal anti-VEGFR-2 antibody used as monotherapy or in combination with paclitaxel for the treatment of advanced gastric or gastroesophageal junction adenocarcinoma, metastatic non-small cell lung cancer (NSCLC), and colorectal cancer (Singh and Parmar 2015; Aprile et al. 2014). Tyrosine kinase inhibitors (TKIs) are used for the inhibition of VEGFRs, PDGF-A and PDGF-BRs, and c-Kit (Hamberg et al. 2010; Wang et al. 2016). Among TKIs, pazopanib is commonly used for the treatment of advanced renal cell carcinoma and soft tissue sarcoma (Hamberg et al. 2010; Nakano et al. 2019) and sunitinib is used in metastatic renal cell carcinoma (Roma-Rodrigues et al. 2019). Sunitinib has more benefits than sorafenib as a first-line therapy, although sunitinib has higher toxicity than sorafenib (Deng et al. 2019). M-TOR inhibitors decrease endothelial cell proliferation through the mTOR/AP-1/VEGF pathway, among which everolimus (Wang et al. 2016). Patients treated with antiangiogenic agents have a reduced response to therapies for the acquisition of drug resistance. Two mechanisms of this resistance are the activation of alternative signaling pathways and the upregulation of alternative angiogenic factors and cytokines. Deepening these pathways would allow us to elaborate new treatments and the development of combination regimens with more durable clinical benefits (Philips and Atkins 2014). Anti-VEGF treatment in pancreatic cancer induces increased expression of FGF-1 and -2 and Ang-1 (Zhuang et al. 2010). In patients affected by colorectal cancer treated with bevacizumab, high levels of Ang-2 were detectable (Goede et al. 2010). In glioblastoma multiforme, anti-VEGFR therapy leads to increased levels of FGF-2 and SDF-1. Similar results have been found in lung cancer models resistant to angiogenesis inhibitors in which epidermal growth factor receptors (EGFRs) and FGFRs are overexpressed (Cascone et al. 2011). In colorectal and renal cancer patients treated with TKIs, increased levels of PIGF and VEGF were detectable (Motzer and Bukowski 2006). Vanucizumab, a bispecific anti-Ang-2/anti-VEGF-A antibody, revealed an acceptable safety profile and promising antitumor activity (Hidalgo et al. 2018). FGFR inhibitors restore the sensitivity to bevacizumab in tumor mouse models (Gyanchandani et al. 2013), but further research failed to determine the relevance of this association (Norden et al. 2015; Semrad et al. 2017). The VEGFR, FGFR, and PDGFR multiple receptor TKI lenvatinib showed promising effects in several tumors and should be considered for counteracting resistance to antiangiogenic agents (Suyama and Iwase 2018).

Anti-angiogenic therapies induce the production of cytokines, such as SDF1, IL-8, and G-CSF, involved in the recruitment of bone marrow-derived cells (BMDCs), which contributes negatively to the anti-angiogenic effect (Montemagno and Pages 2020). An increase in $CD11b^+$ $Gr1^+$ myeloid-derived suppressor cells (MDSCs) has been observed in tumors not sensitive to anti-VEGF-A treatment (Shojaei et al. 2007). Th-17 cells induce the expression of G-CSF by CAFs and consequently the recruitment of MDSCs (Shojaei et al. 2009). Hypoxia has been related to sunitinib resistance in glioblastoma and breast and metastatic renal cell carcinoma as a consequence of the increased recruitment of MDSCs to the tumor niche (Piao et al. 2012).

Vessel co-option is believed to be correlated with refractoriness to anti-VEGF drug treatment of colorectal cancer liver metastases (Frentzas et al. 2016) and has been observed following anti-VEGFR-2 inhibition in cerebral melanoma metastases (Frentzas et al. 2016). Moreover, vessel co-option has been evidenced in human breast cancer liver metastases, NSCLC, and lung metastases (Kuczynski et al. 2016). The blockade of both VEGF-A and ARP2/3, VEGFA and c-MET or VEGF-A and ZEB2 suppresses vessel co-option and tumor invasion (Sennino et al. 2012; Depner et al. 2016). Vasculogenic mimicry is deeply associated with poor patient survival (Sun et al. 2004). In ovarian cancer models, bevacizumab may induce the progression of metastatic disease, which would correlate with a hypoxic response and vasculogenic mimicry (Xu et al. 2012). Studies on the TME in everolimus-resistant renal carcinoma demonstrated that the antiangiogenic drug stimulates vasculogenic mimicry by differentiating tumor cells into endothelial-like cells (Serova et al. 2016). Moreover, everolimus induces triple-negative breast cancer invasion via vasculogenic mimicry; thus, its evaluation could be helpful in predicting the efficacy of antiangiogenic therapy in these patients (Sun et al. 2017).

Targeting TAMs Targeting TAM-recruiting mediators, which include chemokines, complement components, CSF-1, and VEGF, is being studied (Liu et al. 2020). It has been reported that the inhibition of CSF1R in glioblastoma and cervical and breast cancer murine models induces a dramatic reduction in tumor volume and survival of mice (Pyonteck et al. 2013). This inhibition seems to reprogram TAMs by GM-CSF to induce their repolarization to an antitumoral state (Quail and Joyce 2013; DeNardo et al. 2011). The monoclonal antibody RG7155 in human patients led to a remarkable reduction in CSF-1R$^+$ CD163$^+$ macrophages in diffuse-type giant cell tumor patients (Ackermann et al. 2013). TAM reduction improves antiangiogenic treatments. Treatment with vascular-disrupting agents such as combretastatin-A4-phosphate has been reported to markedly increase its efficacy when TIE2$^+$ TAM recruitment is blocked (Welford et al. 2011). The reduction in TAMs augmented the effects of sorafenib (Zhang et al. 2010). In addition, TAMs improved the antiangiogenic and antitumor effects of VEGF/VEGFR2 antibodies in subcutaneous tumor models (Priceman et al. 2010). TAMs limit the cytotoxic activity of CD8+ cytotoxic T cells during tumor progression, mainly in the M2 polarization state. Inhibiting TAM recruitment or blocking TAM polarization to the M2 phenotype may enhance T-cell-mediated antitumor responses and improve the efficacy of immunotherapies (Coussens et al. 2013). Moreover, some immunotherapies may also depend on the reprogramming of TAMs toward an M1 phenotype. One method used to reprogram TAMs is histidine-rich glycoprotein (HRG) treatment, which induces macrophage downregulation of PlGF and stimulates the normalization of blood vessels and the efficiency of chemotherapy in mouse tumor models (Rolny et al. 2011). Other strategies include the suppression of nuclear factor-kB signaling (Hagemann et al. 2008) or exposure to anti-IL-10R antibodies combined with the TLR9 ligand CpG (Guiducci et al. 2005).

Targeting TANs Inhibition of the protumor functions of TANs (Hsu et al. 2020) may be combined with conventional or new anticancer therapies to improve the antitumor effects (Khan et al. 2020). CXCR2 inhibitors are also used in combination with other therapies in clinical evaluation in patients with different tumors (Li et al. 2019; Timaxian et al. 2021; Groth et al. 2021; Cabrero-de Las Heras and Martinez-Balibrea 2018). The neutrophil-derived enzyme elastase promotes tumor growth and invasiveness. The elastase inhibitor ONO-5046 reduced tumor growth in NSCLC (Houghton et al. 2010). Another approach has been to reprogram neutrophil function in the TME through the inhibition of TGFβ (Qin et al. 2020). The inhibition of angiotensin-converting enzyme and the angiotensin II type 1 receptor nicotinamide phosphoribosyltransferase (NAMPT) or CXCR4 is another approach to reprogram neutrophils to an antitumor state (Shrestha et al. 2016; Yang et al. 2018b).

Targeting CAFs The protein FAP is considered a candidate for targeting CAFs because it is expressed in tumors but not in healthy tissues and is considered a predictor of poor survival (Liao et al. 2013). Nevertheless, both sibrotuzumab, an antibody targeting FAP, and inhibitors of FAP activity induced lower survival rates (Liu et al. 2019; Yang et al. 2016b). An IL-2 variant targeting FAP, RO6874281, is under investigation (Joshi 2020; Koustoulidou et al. 2021).

14.8 Concluding Remarks

Cytokines and chemokines secreted by cells of the TME are involved in the regulation of tumor angiogenesis based on the balance of pro- and antiangiogenic factors. Deepening the mechanisms underlying the crosstalk between the TME and tumor cells has allowed the discovery of numerous molecular-targeted drugs that control diverse elements of the TME. Different approaches varying from traditional and emerging inhibitors of angiogenic cytokines and their receptors to the modulation of TME cell activities and novel immune checkpoint inhibitors proved to be promising in tumor progression. Despite the promising results of these new therapeutic approaches, their efficacy is often limited by evasion, and resistance mechanisms have emerged. Overcoming resistance to antitumor therapies is a great challenge but might lead to the improvement of the clinical outcome of patients and, for this reason, currently constitutes a major focus of research.

Acknowledgments This work was supported by Associazione "Il Sorriso di Antonio," Corato, Italy.

References

Ackermann M, Tsuda A, Secomb TW, Mentzer SJ, Konerding MA (2013) Intussusceptive remodeling of vascular branch angles in chemically-induced murine colitis. Microvasc Res 87:75–82

Akwii RG, Sajib MS, Zahra FT, Mikelis CM (2019) Role of angiopoietin-2 in vascular physiology and pathophysiology. Cell 8(5):471

Aprile G, Rijavec E, Fontanella C, Rihawi K, Grossi F (2014) Ramucirumab: preclinical research and clinical development. Onco Targets Ther 7:1997–2006

Autiero M, Luttun A, Tjwa M, Carmeliet P (2003) Placental growth factor and its receptor, vascular endothelial growth factor receptor-1: novel targets for stimulation of ischemic tissue revascularization and inhibition of angiogenic and inflammatory disorders. J Thromb Haemost 1(7): 1356–1370

Baeriswyl V, Christofori G (2009) The angiogenic switch in carcinogenesis. Semin Cancer Biol 19(5):329–337

Bassani B, Baci D, Gallazzi M, Poggi A, Bruno A, Mortara L (2019) Natural killer cells as key players of tumor progression and angiogenesis: old and novel tools to divert their pro-tumor activities into potent anti-tumor effects. Cancers (Basel) 11(4):461

Beatty G, Paterson Y (2001) IFN-gamma-dependent inhibition of tumor angiogenesis by tumor-infiltrating CD4+ T cells requires tumor responsiveness to IFN-gamma. J Immunol 166(4): 2276–2282

Berra E, Roux D, Richard DE, Pouyssegur J (2001) Hypoxia-inducible factor-1 alpha (HIF-1 alpha) escapes O(2)-driven proteasomal degradation irrespective of its subcellular localization: nucleus or cytoplasm. EMBO Rep 2(7):615–620

Bjornmalm M, Thurecht KJ, Michael M, Scott AM, Caruso F (2017) Bridging bio-Nano science and cancer nanomedicine. ACS Nano 11(10):9594–9613

Blotnick S, Peoples GE, Freeman MR, Eberlein TJ, Klagsbrun M (1994) T lymphocytes synthesize and export heparin-binding epidermal growth factor-like growth factor and basic fibroblast

growth factor, mitogens for vascular cells and fibroblasts: differential production and release by CD4+ and CD8+ T cells. Proc Natl Acad Sci U S A 91(8):2890–2894
Brindle NP, Saharinen P, Alitalo K (2006) Signaling and functions of angiopoietin-1 in vascular protection. Circ Res 98(8):1014–1023
Bruno A, Bassani B, D'Urso DG, Pitaku I, Cassinotti E, Pelosi G et al (2018) Angiogenin and the MMP9-TIMP2 axis are up-regulated in proangiogenic, decidual NK-like cells from patients with colorectal cancer. FASEB J 32(10):5365–5377
Cabrero-de Las Heras S, Martinez-Balibrea E (2018) CXC family of chemokines as prognostic or predictive biomarkers and possible drug targets in colorectal cancer. World J Gastroenterol 24(42):4738–4749
Cai J, Wang D, Zhang G, Guo X (2019) The role of PD-1/PD-L1 axis in Treg development and function: implications for cancer immunotherapy. Onco Targets Ther 12:8437–8445
Cao Y, Xue L (2004) Angiostatin. Semin Thromb Hemost 30(1):83–93
Carmeliet P, Jain RK (2011) Molecular mechanisms and clinical applications of angiogenesis. Nature 473(7347):298–307
Carus A, Ladekarl M, Hager H, Nedergaard BS, Donskov F (2013) Tumour-associated CD66b+ neutrophil count is an independent prognostic factor for recurrence in localised cervical cancer. Br J Cancer 108(10):2116–2122
Cascone T, Herynk MH, Xu L, Du Z, Kadara H, Nilsson MB et al (2011) Upregulated stromal EGFR and vascular remodeling in mouse xenograft models of angiogenesis inhibitor-resistant human lung adenocarcinoma. J Clin Invest 121(4):1313–1328
Cerecedo D, Martinez-Vieyra I, Lopez-Villegas EO, Hernandez-Cruz A, Loza-Huerta ADC (2021) Heterogeneity of neutrophils in arterial hypertension. Exp Cell Res 402(2):112577
Chanmee T, Ontong P, Konno K, Itano N (2014) Tumor-associated macrophages as major players in the tumor microenvironment. Cancers (Basel) 6(3):1670–1690
Chen P, Huang Y, Bong R, Ding Y, Song N, Wang X et al (2011) Tumor-associated macrophages promote angiogenesis and melanoma growth via adrenomedullin in a paracrine and autocrine manner. Clin Cancer Res 17(23):7230–7239
Chiron M, Bagley RG, Pollard J, Mankoo PK, Henry C, Vincent L et al (2014) Differential antitumor activity of aflibercept and bevacizumab in patient-derived xenograft models of colorectal cancer. Mol Cancer Ther 13(6):1636–1644
Choi J, Gyamfi J, Jang H, Koo JS (2018) The role of tumor-associated macrophage in breast cancer biology. Histol Histopathol 33(2):133–145
Colegio OR, Chu NQ, Szabo AL, Chu T, Rhebergen AM, Jairam V et al (2014) Functional polarization of tumour-associated macrophages by tumour-derived lactic acid. Nature 513(7519):559–563
Coussens LM, Zitvogel L, Palucka AK (2013) Neutralizing tumor-promoting chronic inflammation: a magic bullet? Science 339(6117):286–291
de Ruiter EJ, Ooft ML, Devriese LA, Willems SM (2017) The prognostic role of tumor infiltrating T-lymphocytes in squamous cell carcinoma of the head and neck: a systematic review and meta-analysis. Onco Targets Ther 6(11):e1356148
Dehghani S, Nosrati R, Yousefi M, Nezami A, Soltani F, Taghdisi SM et al (2018) Aptamer-based biosensors and nanosensors for the detection of vascular endothelial growth factor (VEGF): a review. Biosens Bioelectron 110:23–37
Dell'Eva R, Pfeffer U, Indraccolo S, Albini A, Noonan D (2002) Inhibition of tumor angiogenesis by angiostatin: from recombinant protein to gene therapy. Endothelium 9(1):3–10
DeNardo DG, Brennan DJ, Rexhepaj E, Ruffell B, Shiao SL, Madden SF et al (2011) Leukocyte complexity predicts breast cancer survival and functionally regulates response to chemotherapy. Cancer Discov 1(1):54–67
Deng H, Liu W, He T, Hong Z, Yi F, Wei Y et al (2019) Comparative efficacy, safety, and costs of sorafenib vs. sunitinib as first-line therapy for metastatic renal cell carcinoma: a systematic review and meta-analysis. Front Oncol. 9:479

Depner C, Zum Buttel H, Bogurcu N, Cuesta AM, Aburto MR, Seidel S et al (2016) EphrinB2 repression through ZEB2 mediates tumour invasion and anti-angiogenic resistance. Nat Commun 7:12329

Deryugina EI, Quigley JP (2015) Tumor angiogenesis: MMP-mediated induction of intravasation- and metastasis-sustaining neovasculature. Matrix Biol 44-46:94–112

Eklund L, Kangas J, Saharinen P (2017) Angiopoietin-Tie signalling in the cardiovascular and lymphatic systems. Clin Sci (Lond) 131(1):87–103

Ferrara N, Davis-Smyth T (1997) The biology of vascular endothelial growth factor. Endocr Rev 18(1):4–25

Folkman J (1971) Tumor angiogenesis: therapeutic implications. N Engl J Med 285(21):1182–1186

Folkman J, Merler E, Abernathy C, Williams G (1971) Isolation of a tumor factor responsible for angiogenesis. J Exp Med 133(2):275–288

Folkman J, Watson K, Ingber D, Hanahan D (1989) Induction of angiogenesis during the transition from hyperplasia to neoplasia. Nature 339(6219):58–61

Fredriksson L, Li H, Eriksson U (2004) The PDGF family: four gene products form five dimeric isoforms. Cytokine Growth Factor Rev 15(4):197–204

Frentzas S, Simoneau E, Bridgeman VL, Vermeulen PB, Foo S, Kostaras E et al (2016) Vessel co-option mediates resistance to anti-angiogenic therapy in liver metastases. Nat Med 22(11): 1294–1302

Gao Z, Daquinag AC, Su F, Snyder B, Kolonin MG (2018) PDGFRalpha/PDGFRbeta signaling balance modulates progenitor cell differentiation into white and beige adipocytes. Development 145(1):dev155861

Gerber HP, McMurtrey A, Kowalski J, Yan M, Keyt BA, Dixit V et al (1998) Vascular endothelial growth factor regulates endothelial cell survival through the phosphatidylinositol 3′-kinase/Akt signal transduction pathway. Requirement for Flk-1/KDR activation. J Biol Chem 273(46): 30336–30343

Goede V, Coutelle O, Neuneier J, Reinacher-Schick A, Schnell R, Koslowsky TC et al (2010) Identification of serum angiopoietin-2 as a biomarker for clinical outcome of colorectal cancer patients treated with bevacizumab-containing therapy. Br J Cancer 103(9):1407–1414

Granot Z (2019) Neutrophils as a therapeutic target in cancer. Front Immunol 10:1710

Grossman WJ, Verbsky JW, Barchet W, Colonna M, Atkinson JP, Ley TJ (2004) Human T regulatory cells can use the perforin pathway to cause autologous target cell death. Immunity 21(4):589–601

Groth C, Arpinati L, Shaul ME, Winkler N, Diester K, Gengenbacher N et al (2021) Blocking migration of polymorphonuclear myeloid-derived suppressor cells inhibits mouse melanoma progression. Cancers (Basel) 13(4):726

Guidolin D, Tamma R, Annese T, Tortorella C, Ingravallo G, Gaudio F et al (2021) Different spatial distribution of inflammatory cells in the tumor microenvironment of ABC and GBC subgroups of diffuse large B cell lymphoma. Clin Exp Med 21(4):573–578

Guiducci C, Vicari AP, Sangaletti S, Trinchieri G, Colombo MP (2005) Redirecting in vivo elicited tumor infiltrating macrophages and dendritic cells towards tumor rejection. Cancer Res 65(8): 3437–3446

Guo X, Zhai L, Xue R, Shi J, Zeng Q, Gao C (2016) Mast cell Tryptase contributes to pancreatic cancer growth through promoting angiogenesis via activation of angiopoietin-1. Int J Mol Sci 17(6):834

Gyanchandani R, Ortega Alves MV, Myers JN, Kim S (2013) A proangiogenic signature is revealed in FGF-mediated bevacizumab-resistant head and neck squamous cell carcinoma. Mol Cancer Res 11(12):1585–1596

Hagemann T, Lawrence T, McNeish I, Charles KA, Kulbe H, Thompson RG et al (2008) "Re-educating" tumor-associated macrophages by targeting NF-kappaB. J Exp Med 205(6): 1261–1268

Ham IH, Lee D, Hur H (2019) Role of cancer-associated fibroblast in gastric cancer progression and resistance to treatments. J Oncol 2019:6270784

Hamberg P, Verweij J, Sleijfer S (2010) (Pre-)clinical pharmacology and activity of pazopanib, a novel multikinase angiogenesis inhibitor. Oncologist 15(6):539–547

Hanahan D, Weinberg RA (2011) Hallmarks of cancer: the next generation. Cell 144(5):646–674

Hang TC, Tedford NC, Reddy RJ, Rimchala T, Wells A, White FM et al (2013) Vascular endothelial growth factor (VEGF) and platelet (PF-4) factor 4 inputs modulate human microvascular endothelial signaling in a three-dimensional matrix migration context. Mol Cell Proteomics 12(12):3704–3718

Heinrich EL, Walser TC, Krysan K, Liclican EL, Grant JL, Rodriguez NL et al (2012) The inflammatory tumor microenvironment, epithelial mesenchymal transition and lung carcinogenesis. Cancer Microenviron 5(1):5–18

Hidalgo M, Martinez-Garcia M, Le Tourneau C, Massard C, Garralda E, Boni V et al (2018) First-in-human phase I study of single-agent vanucizumab, a first-in-class bispecific anti-angiopoietin-2/anti-VEGF-A antibody, in adult patients with advanced solid tumors. Clin Cancer Res 24(7):1536–1545

Higashino N, Koma YI, Hosono M, Takase N, Okamoto M, Kodaira H et al (2019) Fibroblast activation protein-positive fibroblasts promote tumor progression through secretion of CCL2 and interleukin-6 in esophageal squamous cell carcinoma. Lab Investig 99(6):777–792

Holash J, Davis S, Papadopoulos N, Croll SD, Ho L, Russell M et al (2002) VEGF-Trap: a VEGF blocker with potent antitumor effects. Proc Natl Acad Sci U S A 99(17):11393–11398

Houghton AM, Rzymkiewicz DM, Ji H, Gregory AD, Egea EE, Metz HE et al (2010) Neutrophil elastase-mediated degradation of IRS-1 accelerates lung tumor growth. Nat Med 16(2):219–223

Hsu BE, Shen Y, Siegel PM (2020) Neutrophils: orchestrators of the malignant phenotype. Front Immunol 11:1778

Hu CJ, Wang LY, Chodosh LA, Keith B, Simon MC (2003) Differential roles of hypoxia-inducible factor 1alpha (HIF-1alpha) and HIF-2alpha in hypoxic gene regulation. Mol Cell Biol 23(24): 9361–9374

Hu ZQ, Zhao WH, Shimamura T (2007) Regulation of mast cell development by inflammatory factors. Curr Med Chem 14(28):3044–3050

Huang Y, Lin D, Taniguchi CM (2017) Hypoxia inducible factor (HIF) in the tumor microenvironment: friend or foe? Sci China Life Sci 60(10):1114–1124

Jang HJ, Kim BJ, Kim JH, Kim HS (2017) The addition of bevacizumab in the first-line treatment for metastatic colorectal cancer: an updated meta-analysis of randomized trials. Oncotarget 8(42):73009–73016

Jayasingam SD, Citartan M, Thang TH, Mat Zin AA, Ang KC, Ch'ng ES (2019) Evaluating the polarization of tumor-associated macrophages into M1 and M2 phenotypes in human cancer tissue: technicalities and challenges in routine clinical practice. Front Oncol 9:1512

Jeronimo A, Rodrigues G, Vilas-Boas F, Martins GG, Bagulho A, Real C (2017) Hydrogen peroxide regulates angiogenesis-related factors in tumor cells. Biochem Cell Biol 95(6): 679–685

Ji WR, Castellino FJ, Chang Y, Deford ME, Gray H, Villarreal X et al (1998) Characterization of Kringle domains of angiostatin as antagonists of endothelial cell migration, an important process in angiogenesis. FASEB J 12(15):1731–1738

Jia YH, Dong XS, Wang XS (2004) Effects of endostatin on expression of vascular endothelial growth factor and its receptors and neovascularization in colonic carcinoma implanted in nude mice. World J Gastroenterol 10(22):3361–3364

Johnson JL, Jackson CL, Angelini GD, George SJ (1998) Activation of matrix-degrading metalloproteinases by mast cell proteases in atherosclerotic plaques. Arterioscler Thromb Vasc Biol 18(11):1707–1715

Joshi S (2020) Targeting the tumor microenvironment in neuroblastoma: recent advances and future directions. Cancers (Basel) 12(8):2057

Kanbe N, Tanaka A, Kanbe M, Itakura A, Kurosawa M, Matsuda H (1999) Human mast cells produce matrix metalloproteinase 9. Eur J Immunol 29(8):2645–2649

Khan S, Mittal S, McGee K, Alfaro-Munoz KD, Majd N, Balasubramaniyan V et al (2020) Role of neutrophils and myeloid-derived suppressor cells in glioma progression and treatment resistance. Int J Mol Sci 21(6):1954

Kim YM, Jang JW, Lee OH, Yeon J, Choi EY, Kim KW et al (2000) Endostatin inhibits endothelial and tumor cellular invasion by blocking the activation and catalytic activity of matrix metalloproteinase. Cancer Res 60(19):5410–5413

Komi DEA, Redegeld FA (2020) Role of mast cells in shaping the tumor microenvironment. Clin Rev Allergy Immunol 58(3):313–325

Koustoulidou S, Hoorens MWH, Dalm SU, Mahajan S, Debets R, Seimbille Y et al (2021) Cancer-associated fibroblasts as players in cancer development and progression and their role in targeted radionuclide imaging and therapy. Cancers (Basel) 13(5):1100

Krystel-Whittemore M, Dileepan KN, Wood JG (2015) Mast cell: a multi-functional master cell. Front Immunol 6:620

Kubouchi Y, Yurugi Y, Wakahara M, Sakabe T, Haruki T, Nosaka K et al (2018) Podoplanin expression in cancer-associated fibroblasts predicts unfavourable prognosis in patients with pathological stage IA lung adenocarcinoma. Histopathology 72(3):490–499

Kuczynski EA, Yin M, Bar-Zion A, Lee CR, Butz H, Man S et al (2016) Co-option of liver vessels and not sprouting angiogenesis drives acquired sorafenib resistance in hepatocellular carcinoma. J Natl Cancer Inst 108(8):djw030

Laddaga FE, Ingravallo G, Mestice A, Tamma R, Perrone T, Maiorano E et al (2021) Correlation between circulating blood and microenvironment T lymphocytes in diffuse large B-cell lymphomas. J Clin Pathol. https://doi.org/10.1136/jclinpath-2020-207048

Lan HR, Du WL, Liu Y, Mao CS, Jin KT, Yang X (2021) Role of immune regulatory cells in breast cancer: foe or friend? Int Immunopharmacol 96:107627

Larsen SK, Gao Y, Basse PH (2014) NK cells in the tumor microenvironment. Crit Rev Oncog 19(1–2):91–105

Lawler J (2002) Thrombospondin-1 as an endogenous inhibitor of angiogenesis and tumor growth. J Cell Mol Med 6(1):1–12

Lawler PR, Lawler J (2012) Molecular basis for the regulation of angiogenesis by thrombospondin-1 and -2. Cold Spring Harb Perspect Med 2(5):a006627

Lee HJ, Cho CH, Hwang SJ, Choi HH, Kim KT, Ahn SY et al (2004) Biological characterization of angiopoietin-3 and angiopoietin-4. FASEB J 18(11):1200–1208

Lewis CE, Pollard JW (2006) Distinct role of macrophages in different tumor microenvironments. Cancer Res 66(2):605–612

Li Y, He Y, Butler W, Xu L, Chang Y, Lei K et al (2019) Targeting cellular heterogeneity with CXCR2 blockade for the treatment of therapy-resistant prostate cancer. Sci Transl Med 11(521): eaax0428

Li P, Lu M, Shi J, Hua L, Gong Z, Li Q et al (2020a) Dual roles of neutrophils in metastatic colonization are governed by the host NK cell status. Nat Commun 11(1):4387

Li C, Jiang P, Wei S, Xu X, Wang J (2020b) Regulatory T cells in tumor microenvironment: new mechanisms, potential therapeutic strategies and future prospects. Mol Cancer 19(1):116

Liang J, Piao Y, Holmes L, Fuller GN, Henry V, Tiao N et al (2014) Neutrophils promote the malignant glioma phenotype through S100A4. Clin Cancer Res 20(1):187–198

Liao Y, Ni Y, He R, Liu W, Du J (2013) Clinical implications of fibroblast activation protein-alpha in non-small cell lung cancer after curative resection: a new predictor for prognosis. J Cancer Res Clin Oncol 139(9):1523–1528

Lin EY, Li JF, Gnatovskiy L, Deng Y, Zhu L, Grzesik DA et al (2006) Macrophages regulate the angiogenic switch in a mouse model of breast cancer. Cancer Res 66(23):11238–11246

Lin L, Chen YS, Yao YD, Chen JQ, Chen JN, Huang SY et al (2015) CCL18 from tumor-associated macrophages promotes angiogenesis in breast cancer. Oncotarget 6(33):34758–34773

Linares J, Marin-Jimenez JA, Badia-Ramentol J, Calon A (2020) Determinants and functions of CAFs secretome during cancer progression and therapy. Front Cell Dev Biol 8:621070

Liu T, Zhou L, Li D, Andl T, Zhang Y (2019) Cancer-associated fibroblasts build and secure the tumor microenvironment. Front Cell Dev Biol 7:60

Liu H, Yang Z, Lu W, Chen Z, Chen L, Han S et al (2020) Chemokines and chemokine receptors: a new strategy for breast cancer therapy. Cancer Med 9(11):3786–3799

Logue OC, McGowan JW, George EM, Bidwell GL 3rd. (2016) Therapeutic angiogenesis by vascular endothelial growth factor supplementation for treatment of renal disease. Curr Opin Nephrol Hypertens 25(5):404–409

Maciel TT, Moura IC, Hermine O (2015) The role of mast cells in cancers. F1000Prime Rep 7:09

Maglione D, Guerriero V, Viglietto G, Delli-Bovi P, Persico MG (1991) Isolation of a human placenta cDNA coding for a protein related to the vascular permeability factor. Proc Natl Acad Sci U S A 88(20):9267–9271

Maimela NR, Liu S, Zhang Y (2019) Fates of CD8+ T cells in tumor microenvironment. Comput Struct Biotechnol J 17:1–13

Malech HL, Deleo FR, Quinn MT (2014) The role of neutrophils in the immune system: an overview. Methods Mol Biol 1124:3–10

Mathew G, Hannan A, Hertzler-Schaefer K, Wang F, Feng GS, Zhong J et al (2016) Targeting of Ras-mediated FGF signaling suppresses Pten-deficient skin tumor. Proc Natl Acad Sci U S A 113(46):13156–13161

Maurer AM, Zhou B, Han ZC (2006) Roles of platelet factor 4 in hematopoiesis and angiogenesis. Growth Factors 24(4):242–252

Medbury HJ, James V, Ngo J, Hitos K, Wang Y, Harris DC et al (2013) Differing association of macrophage subsets with atherosclerotic plaque stability. Int Angiol 32(1):74–84

Melincovici CS, Bosca AB, Susman S, Marginean M, Mihu C, Istrate M et al (2018) Vascular endothelial growth factor (VEGF) - key factor in normal and pathological angiogenesis. Romanian J Morphol Embryol 59(2):455–467

Montemagno C, Pages G (2020) Resistance to anti-angiogenic therapies: a mechanism depending on the time of exposure to the drugs. Front Cell Dev Biol 8:584

Motomura K, Hagiwara A, Komi-Kuramochi A, Hanyu Y, Honda E, Suzuki M et al (2008) An FGF1:FGF2 chimeric growth factor exhibits universal FGF receptor specificity, enhanced stability and augmented activity useful for epithelial proliferation and radioprotection. Biochim Biophys Acta 1780(12):1432–1440

Motzer RJ, Bukowski RM (2006) Targeted therapy for metastatic renal cell carcinoma. J Clin Oncol 24(35):5601–5608

Nakano K, Funauchi Y, Hayakawa K, Tanizawa T, Ae K, Matsumoto S et al (2019) Relative dose intensity of induction-phase pazopanib treatment of soft tissue sarcoma: its relationship with prognoses of pazopanib responders. J Clin Med 8(1):60

Nishida N, Yano H, Nishida T, Kamura T, Kojiro M (2006) Angiogenesis in cancer. Vasc Health Risk Manag 2(3):213–219

Norden AD, Schiff D, Ahluwalia MS, Lesser GJ, Nayak L, Lee EQ et al (2015) Phase II trial of triple tyrosine kinase receptor inhibitor nintedanib in recurrent high-grade gliomas. J Neuro-Oncol 121(2):297–302

Norrby K (2002) Mast cells and angiogenesis. APMIS 110(5):355–371

Olson LE, Soriano P (2011) PDGFRbeta signaling regulates mural cell plasticity and inhibits fat development. Dev Cell 20(6):815–826

Orimo A, Gupta PB, Sgroi DC, Arenzana-Seisdedos F, Delaunay T, Naeem R et al (2005) Stromal fibroblasts present in invasive human breast carcinomas promote tumor growth and angiogenesis through elevated SDF-1/CXCL12 secretion. Cell 121(3):335–348

Paget S (1989) The distribution of secondary growths in cancer of the breast. 1889. Cancer Metastasis Rev 8(2):98–101

Paijens ST, Vledder A, de Bruyn M, Nijman HW (2021) Tumor-infiltrating lymphocytes in the immunotherapy era. Cell Mol Immunol 18(4):842–859

Palazon A, Goldrath AW, Nizet V, Johnson RS (2014) HIF transcription factors, inflammation, and immunity. Immunity 41(4):518–528

Patan S (1998) TIE1 and TIE2 receptor tyrosine kinases inversely regulate embryonic angiogenesis by the mechanism of intussusceptive microvascular growth. Microvasc Res 56(1):1–21

Philips GK, Atkins MB (2014) New agents and new targets for renal cell carcinoma. Am Soc Clin Oncol Educ Book:e222–e227

Piao Y, Liang J, Holmes L, Zurita AJ, Henry V, Heymach JV et al (2012) Glioblastoma resistance to anti-VEGF therapy is associated with myeloid cell infiltration, stem cell accumulation, and a mesenchymal phenotype. Neuro-Oncology 14(11):1379–1392

Potenta S, Zeisberg E, Kalluri R (2008) The role of endothelial-to-mesenchymal transition in cancer progression. Br J Cancer 99(9):1375–1379

Pozzi A, Moberg PE, Miles LA, Wagner S, Soloway P, Gardner HA (2000) Elevated matrix metalloprotease and angiostatin levels in integrin alpha 1 knockout mice cause reduced tumor vascularization. Proc Natl Acad Sci U S A 97(5):2202–2207

Priceman SJ, Sung JL, Shaposhnik Z, Burton JB, Torres-Collado AX, Moughon DL et al (2010) Targeting distinct tumor-infiltrating myeloid cells by inhibiting CSF-1 receptor: combating tumor evasion of antiangiogenic therapy. Blood 115(7):1461–1471

Pyonteck SM, Akkari L, Schuhmacher AJ, Bowman RL, Sevenich L, Quail DF et al (2013) CSF-1R inhibition alters macrophage polarization and blocks glioma progression. Nat Med 19(10): 1264–1272

Qin F, Liu X, Chen J, Huang S, Wei W, Zou Y et al (2020) Anti-TGF-beta attenuates tumor growth via polarization of tumor associated neutrophils towards an anti-tumor phenotype in colorectal cancer. J Cancer 11(9):2580–2592

Quail DF, Joyce JA (2013) Microenvironmental regulation of tumor progression and metastasis. Nat Med 19(11):1423–1437

Ramirez-Moreno IG, Ibarra-Sanchez A, Castillo-Arellano JI, Blank U, Gonzalez-Espinosa C (2020) Mast cells localize in hypoxic zones of tumors and secrete CCL-2 under hypoxia through activation of L-type calcium channels. J Immunol 204(4):1056–1068

Rauniyar K, Jha SK, Jeltsch M (2018) Biology of vascular endothelial growth factor C in the morphogenesis of lymphatic vessels. Front Bioeng Biotechnol 6:7

Redlitz A, Daum G, Sage EH (1999) Angiostatin diminishes activation of the mitogen-activated protein kinases ERK-1 and ERK-2 in human dermal microvascular endothelial cells. J Vasc Res 36(1):28–34

Riabov V, Gudima A, Wang N, Mickley A, Orekhov A, Kzhyshkowska J (2014) Role of tumor associated macrophages in tumor angiogenesis and lymphangiogenesis. Front Physiol 5:75

Ribatti D, Crivellato E (2012) Mast cells, angiogenesis, and tumour growth. Biochim Biophys Acta 1822(1):2–8

Ribatti D, Nico B, Crivellato E, Roccaro AM, Vacca A (2007) The history of the angiogenic switch concept. Leukemia 21(1):44–52

Rolny C, Mazzone M, Tugues S, Laoui D, Johansson I, Coulon C et al (2011) HRG inhibits tumor growth and metastasis by inducing macrophage polarization and vessel normalization through downregulation of PlGF. Cancer Cell 19(1):31–44

Roma-Rodrigues C, Mendes R, Baptista PV, Fernandes AR (2019) Targeting tumor microenvironment for cancer therapy. Int J Mol Sci 20(4):840

Rong X, Huang B, Qiu S, Li X, He L, Peng Y (2016) Tumor-associated macrophages induce vasculogenic mimicry of glioblastoma multiforme through cyclooxygenase-2 activation. Oncotarget 7(51):83976–83986

Semrad TJ, Kim EJ, Tanaka MS, Sands J, Roberts C, Burich RA et al (2017) Phase II study of Dovitinib in patients progressing on anti-vascular endothelial growth factor therapy. Cancer Treat Res Commun 10:21–26

Sennino B, Ishiguro-Oonuma T, Wei Y, Naylor RM, Williamson CW, Bhagwandin V et al (2012) Suppression of tumor invasion and metastasis by concurrent inhibition of c-Met and VEGF signaling in pancreatic neuroendocrine tumors. Cancer Discov 2(3):270–287

Serova M, Tijeras-Raballand A, Dos Santos C, Martinet M, Neuzillet C, Lopez A et al (2016) Everolimus affects vasculogenic mimicry in renal carcinoma resistant to sunitinib. Oncotarget 7(25):38467–38486
Shi YH, Fang WG (2004) Hypoxia-inducible factor-1 in tumour angiogenesis. World J Gastroenterol 10(8):1082–1087
Shojaei F, Wu X, Malik AK, Zhong C, Baldwin ME, Schanz S et al (2007) Tumor refractoriness to anti-VEGF treatment is mediated by CD11b+Gr1+ myeloid cells. Nat Biotechnol 25(8): 911–920
Shojaei F, Singh M, Thompson JD, Ferrara N (2008) Role of Bv8 in neutrophil-dependent angiogenesis in a transgenic model of cancer progression. Proc Natl Acad Sci U S A 105(7): 2640–2645
Shojaei F, Wu X, Qu X, Kowanetz M, Yu L, Tan M et al (2009) G-CSF-initiated myeloid cell mobilization and angiogenesis mediate tumor refractoriness to anti-VEGF therapy in mouse models. Proc Natl Acad Sci U S A 106(16):6742–6747
Shrestha S, Noh JM, Kim SY, Ham HY, Kim YJ, Yun YJ et al (2016) Angiotensin converting enzyme inhibitors and angiotensin II receptor antagonist attenuate tumor growth via polarization of neutrophils toward an antitumor phenotype. Onco Targets Ther 5(1):e1067744
Singh AD, Parmar S (2015) Ramucirumab (Cyramza): a breakthrough treatment for gastric cancer. P T 40(7):430–468
Stacker SA, Achen MG (2018) Emerging roles for VEGF-D in human disease. Biomol Ther 8(1):1
Storkebaum E, Carmeliet P (2004) VEGF: a critical player in neurodegeneration. J Clin Invest 113(1):14–18
Stuttfeld E, Ballmer-Hofer K (2009) Structure and function of VEGF receptors. IUBMB Life 61(9): 915–922
Sun B, Zhang S, Zhao X, Zhang W, Hao X (2004) Vasculogenic mimicry is associated with poor survival in patients with mesothelial sarcomas and alveolar rhabdomyosarcomas. Int J Oncol 25(6):1609–1614
Sun H, Zhang D, Yao Z, Lin X, Liu J, Gu Q et al (2017) Anti-angiogenic treatment promotes triple-negative breast cancer invasion via vasculogenic mimicry. Cancer Biol Ther 18(4):205–213
Suyama K, Iwase H (2018) Lenvatinib: a promising molecular targeted agent for multiple cancers. Cancer Control 25(1):1073274818789361
Talks KL, Turley H, Gatter KC, Maxwell PH, Pugh CW, Ratcliffe PJ et al (2000) The expression and distribution of the hypoxia-inducible factors HIF-1alpha and HIF-2alpha in normal human tissues, cancers, and tumor-associated macrophages. Am J Pathol 157(2):411–421
Tamma R, Annese T, Ruggieri S, Brunetti O, Longo V, Cascardi E et al (2019a) Inflammatory cells infiltrate and angiogenesis in locally advanced and metastatic cholangiocarcinoma. Eur J Clin Investig 49(5):e13087
Tamma R, Ingravallo G, Albano F, Gaudio F, Annese T, Ruggieri S et al (2019b) STAT-3 RNAscope determination in human diffuse large B-cell lymphoma. Transl Oncol 12(3): 545–549
Tamma R, Ingravallo G, Gaudio F, Annese T, Albano F, Ruggieri S et al (2020) STAT3, tumor microenvironment, and microvessel density in diffuse large B cell lymphomas. Leuk Lymphoma 61(3):567–574
Tamma R, Ingravallo G, Annese T, Giorgis MDE, DI Giovanni F, Gaudio F et al (2021) Tumor cell microenvironment and microvessel density analysis in MALT type lymphoma. Anticancer Res 41(3):1291–1297
Timaxian C, Vogel CFA, Orcel C, Vetter D, Durochat C, Chinal C et al (2021) Pivotal role for Cxcr2 in regulating tumor-associated neutrophil in breast cancer. Cancers (Basel) 13(11):2584
Tysnes BB, Bjerkvig R (2007) Cancer initiation and progression: involvement of stem cells and the microenvironment. Biochim Biophys Acta 1775(2):283–297
Ucuzian AA, Gassman AA, East AT, Greisler HP (2010) Molecular mediators of angiogenesis. J Burn Care Res 31(1):158–175

Virchow R (1989) Cellular pathology. As based upon physiological and pathological histology. Lecture XVI--Atheromatous affection of arteries. 1858. Nutr Rev 47(1):23–25

Voronov E, Carmi Y, Apte RN (2014) The role IL-1 in tumor-mediated angiogenesis. Front Physiol 5:114

Wang S, Lu J, You Q, Huang H, Chen Y, Liu K (2016) The mTOR/AP-1/VEGF signaling pathway regulates vascular endothelial cell growth. Oncotarget 7(33):53269–53276

Wang J, Li D, Cang H, Guo B (2019a) Crosstalk between cancer and immune cells: role of tumor-associated macrophages in the tumor microenvironment. Cancer Med 8(10):4709–4721

Wang X, Rojas-Quintero J, Wilder J, Tesfaigzi Y, Zhang D, Owen CA (2019b) Tissue inhibitor of metalloproteinase-1 promotes polymorphonuclear neutrophil (PMN) pericellular proteolysis by anchoring matrix metalloproteinase-8 and -9 to PMN surfaces. J Immunol 202(11):3267–3281

Wang SS, Liu W, Ly D, Xu H, Qu L, Zhang L (2019c) Tumor-infiltrating B cells: their role and application in anti-tumor immunity in lung cancer. Cell Mol Immunol 16(1):6–18

Warren BA, Shubik P (1966) The growth of the blood supply to melanoma transplants in the hamster cheek pouch. Lab Investig 15(2):464–478

Weis SM, Cheresh DA (2011) Tumor angiogenesis: molecular pathways and therapeutic targets. Nat Med 17(11):1359–1370

Welford AF, Biziato D, Coffelt SB, Nucera S, Fisher M, Pucci F et al (2011) TIE2-expressing macrophages limit the therapeutic efficacy of the vascular-disrupting agent combretastatin A4 phosphate in mice. J Clin Invest 121(5):1969–1973

Welker P, Grabbe J, Zuberbier T, Guhl S, Henz BM (2000) Mast cell and myeloid marker expression during early in vitro mast cell differentiation from human peripheral blood mononuclear cells. J Invest Dermatol 114(1):44–50

Wellman TL, Jenkins J, Penar PL, Tranmer B, Zahr R, Lounsbury KM (2004) Nitric oxide and reactive oxygen species exert opposing effects on the stability of hypoxia-inducible factor-1alpha (HIF-1alpha) in explants of human pial arteries. FASEB J 18(2):379–381

Wenzel D, Schmidt A, Reimann K, Hescheler J, Pfitzer G, Bloch W et al (2006) Endostatin, the proteolytic fragment of collagen XVIII, induces vasorelaxation. Circ Res 98(9):1203–1211

Wickstrom SA, Alitalo K, Keski-Oja J (2005) Endostatin signaling and regulation of endothelial cell-matrix interactions. Adv Cancer Res 94:197–229

Wu J, Lanier LL (2003) Natural killer cells and cancer. Adv Cancer Res 90:127–156

Xu Y, Li Q, Li XY, Yang QY, Xu WW, Liu GL (2012) Short-term anti-vascular endothelial growth factor treatment elicits vasculogenic mimicry formation of tumors to accelerate metastasis. J Exp Clin Cancer Res 31:16

Yahaya MAF, Lila MAM, Ismail S, Zainol M, Afizan N (2019) Tumour-associated macrophages (TAMs) in colon cancer and how to reeducate them. J Immunol Res 2019:2368249

Yang C, Lee H, Pal S, Jove V, Deng J, Zhang W et al (2013) B cells promote tumor progression via STAT3 regulated-angiogenesis. PLoS One 8(5):e64159

Yang J, Lu Y, Lin YY, Zheng ZY, Fang JH, He S et al (2016a) Vascular mimicry formation is promoted by paracrine TGF-beta and SDF1 of cancer-associated fibroblasts and inhibited by miR-101 in hepatocellular carcinoma. Cancer Lett 383(1):18–27

Yang X, Lin Y, Shi Y, Li B, Liu W, Yin W et al (2016b) FAP promotes immunosuppression by cancer-associated fibroblasts in the tumor microenvironment via STAT3-CCL2 signaling. Cancer Res 76(14):4124–4135

Yang SB, Gao KD, Jiang T, Cheng SJ, Li WB (2017) Bevacizumab combined with chemotherapy for glioblastoma: a meta-analysis of randomized controlled trials. Oncotarget 8(34): 57337–57344

Yang JG, Wang LL, Ma DC (2018a) Effects of vascular endothelial growth factors and their receptors on megakaryocytes and platelets and related diseases. Br J Haematol 180(3):321–334

Yang J, Kumar A, Vilgelm AE, Chen SC, Ayers GD, Novitskiy SV et al (2018b) Loss of CXCR4 in myeloid cells enhances antitumor immunity and reduces melanoma growth through NK cell and FASL mechanisms. Cancer Immunol Res 6(10):1186–1198

Yu Y, Blokhuis B, Derks Y, Kumari S, Garssen J, Redegeld F (2018) Human mast cells promote colon cancer growth via bidirectional crosstalk: studies in 2D and 3D coculture models. Onco Targets Ther 7(11):e1504729

Yun YR, Won JE, Jeon E, Lee S, Kang W, Jo H et al (2010) Fibroblast growth factors: biology, function, and application for tissue regeneration. J Tissue Eng 2010:218142

Zatterstrom UK, Felbor U, Fukai N, Olsen BR (2000) Collagen XVIII/endostatin structure and functional role in angiogenesis. Cell Struct Funct 25(2):97–101

Zhang W, Zhu XD, Sun HC, Xiong YQ, Zhuang PY, Xu HX et al (2010) Depletion of tumor-associated macrophages enhances the effect of sorafenib in metastatic liver cancer models by antimetastatic and antiangiogenic effects. Clin Cancer Res 16(13):3420–3430

Zhang J, Muri J, Fitzgerald G, Gorski T, Gianni-Barrera R, Masschelein E et al (2020) Endothelial lactate controls muscle regeneration from ischemia by inducing M2-like macrophage polarization. Cell Metab 31(6):1136–53 e7

Zhong H, De Marzo AM, Laughner E, Lim M, Hilton DA, Zagzag D et al (1999) Overexpression of hypoxia-inducible factor 1alpha in common human cancers and their metastases. Cancer Res 59(22):5830–5835

Zhuang G, Brantley-Sieders DM, Vaught D, Yu J, Xie L, Wells S et al (2010) Elevation of receptor tyrosine kinase EphA2 mediates resistance to trastuzumab therapy. Cancer Res 70(1):299–308

Chapter 15
Cancer Angiogenesis and Its Master Regulator Perlecan

Ilona Kovalszky, Loránd Váncza, Andrea Reszegi, Péter Tátrai, and Kornélia Baghy

Abstract Angiogenesis and vasculogenesis are not only indispensable for normal tissue growth and regeneration but also particularly vital for tumors. As without sufficient blood supply that provides nutrients, tumor cells cannot proliferate. Tumors may employ several strategies to receive sufficient blood supply: they can promote the development of new blood vessels, engulf already existing arteries, or induce the splitting or glomeruli-like proliferation of pre-existing blood vessels. The outcome depends on the equilibrium of pro- and antiangiogenic factors that regulate these events. The most potent stimulatory factor is VEGF, but several other growth factors, cytokines, chemokines, and miRNAs have a stimulatory potential, whereas specific proteolytic breakdown products and other miRNAs can inhibit angiogenesis. Like tumor cells, newly generated tumor vasculature also displays deregulated behavior: blood vessels created upon the angiogenic switch differ from their normal counterparts in both structure and function. *Perlecan*, a large extracellular matrix heparan sulfate/chondroitin sulfate (HS/CS) proteoglycan, is a ubiquitous component of basement membranes, including vascular basement membranes of both normal and tumorous blood vessels. Besides its structural role, perlecan regulates angiogenesis by interacting with a large number of partner molecules, mainly with growth factors and cytokines bound to its HS chains. Whereas intact perlecan promotes angiogenesis, a tryptic fragment of its domain V called endorepellin exerts anti-angiogenic action and triggers tumor autophagy, whose potentials may be exploited in the future management of cancer.

I. Kovalszky (✉) · L. Váncza · A. Reszegi · K. Baghy
Department of Pathology & Experimental Cancer Research, Semmelweis University, Budapest, Hungary
e-mail: kovalszky.ilona@med.semmelweis-univ.hu

P. Tátrai
Solvo Biotechnology, Budapest, Hungary

I. Kovalszky et al. (eds.), *The Extracellular Matrix and the Tumor Microenvironment*, Biology of Extracellular Matrix 11,
https://doi.org/10.1007/978-3-030-99708-3_15

15.1 Cancer Angiogenesis

Since the first publications on cancer angiogenesis appeared more than 20 years ago, anti-angiogenic therapies, with the aim to deteriorate tumor blood supply, have emerged as leading approaches toward conquering cancer. To obtain sufficient nutrition and oxygen, rapidly proliferating cancer cells require an increased blood supply (Folkman 1971). However, no trivial physiological mechanism is available to fulfill this demand. Therefore, tumor cells need to deploy different strategies depending on their type and localization and utilize multiple factors to trick stromal components into forming new blood vessels. The last decades have seen intensive efforts to clarify the critical steps of tumor angiogenesis, a key prerequisite to cancer cell proliferation and invasion (Pluda 1997). Research has elucidated a number of typical pathways that characterize tumor angiogenesis (Döme et al. 2007). (*1*) Tumors can facilitate angiogenesis by promoting the development of new capillary buds from already available blood vessels. This process, also referred to as sprouting, is supported by a series of stimulatory factors like vascular endothelial growth factor (VEGF), platelet-derived growth factor (PDGF), fibroblast growth factor (FGF), transforming growth factor (TGF) beta, matrix metalloproteases (MMPs), and integrins (Bergers and Benjamin 2003; Mehrzadi et al. 2021). (*2*) Tumor tissue can surround and engulf preexisting blood vessels, a process typically observed in liver metastases (Latacz et al. 2020). (*3*) A further approach to increase the number of tumor-supplying blood vessels is the so-called intussusception, whereby preexisting blood vessels split up via ingrowth of a transmural pillar (De Spiegelaere et al. 2012). (*4*) A special form of angiogenesis characteristic of glioblastoma multiform is glomeruloid angiogenesis, whereby preexisting blood vessels (Brat and Van Meir 2001) form convoluted loops in response to the pulling force from proliferating tumor cells (Lugano et al. 2020).

15.2 Critical Participants of Tumor Angiogenesis

The development of blood vessels requires a set of structural and regulatory factors. This is also true for the different forms of tumor angiogenesis discussed above (Fig. 15.1).

15.3 Structural Compartments of Tumor Blood Vessels

(a) *Endothelial cells*. Contrary to the historical view of tumors, abnormal phenotype is not restricted to cancer cells but also characterizes stromal cells including vascular endothelial cells (Hida et al. 2018, 2010, 2016). Upon modulation by the tumor microenvironment, these cells change their phenotype, express new

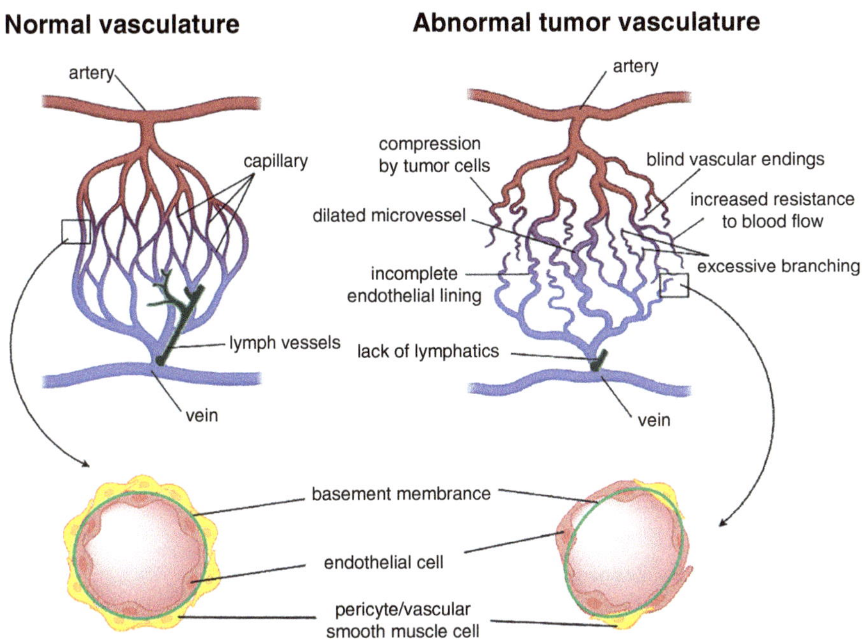

Fig. 15.1 Tumor vasculature is characterized by a disorganized arrangement which involves loss of the well-structured fashion of gradually decreasing diameter of the arteries ending in the branching capillary system followed by the collection of deoxygenized blood in the venous side, always detectable in normal situation. Instead, leaky blood vessels with uneven diameters, insufficient to withstand the vascular pressure, small arteries with dead ends, incomplete endothelial linings, uneven arrangement of pericytes, and incomplete basement membranes can be observed

cell surface markers (St Croix et al. 2000; Høye et al. 2018), and display altered response to regulatory factors (Amin et al. 2006; Matsuda et al. 2010). Changes may encompass increased angiogenic capacity and drug resistance (Xiong et al. 2009).

(b) *Pericytes* in tumor microvessels are arranged abnormally loosely on the abluminal side of endothelial cells, display irregular shape, and possess aberrant cytoplasmic processes. Therefore, they fail to provide adequate support to the tumor neovasculature (Raza et al. 2010; Armulik et al. 2005).

(c) *Basement membrane* is also irregular, with varying thickness and loose association to the endothelial cells. These abnormalities result in leaky blood vessels surrounded by small pools of extravasated blood (Baluk et al. 2003; Iozzo et al. 2009; Zuazo-Gaztelu and Casanovas 2018).

15.4 Regulatory Factors of Tumor Angiogenesis

15.4.1 Tumor-Associated Stromal Cells

Bone marrow-derived stromal cells are implicated in the production of regulatory factors that promote or inhibit blood vessel generation. Neutrophils and eosinophil granulocytes, monocytes, and tumor-associated macrophages (TAMs) migrate to the tumor and secrete angiogenesis modulators like VEGFA, FGF2, and MMPs. TAMs also produce chemokines with pro- or antiangiogenic potential and interleukins with proinflammatory effect (Cassetta and Pollard 2020; Chen et al. 2019; Stockmann et al. 2014; Lin et al. 2006; Owen and Mohamadzadeh 2013). Lymphocytes actively participate in local immune reactions and may inhibit angiogenesis and tumor proliferation by interfering with the effect of PD-L1 (Asadzadeh et al. 2017; Tian et al. 2017; Thienpont and Lambrechts 2017).

15.4.2 Tumor-Associated Fibroblasts (TAFs)

Whereas normal fibrocytes do not synthesize angiogenic factors, TAFs transformed from normal fibrocytes as well as from bone marrow stem cells engage in the production of VEGF upon stimulation from cancer cells (Wang et al. 2019); VEGF, in turn, may promote either vasculogenesis or angiogenesis. This interaction is likely to be bidirectional, as the conditioned medium of TAFs facilitates VEGF production of ovarian cancer cells in culture (Xu et al. 2013), and IL-6, the major stimulator of VEGF secretion, is produced by both the tumor and TAF cells. Fibrocytes can also gain TAF phenotype as an effect of IL-6. In addition to VEGF, TAFs can synthesize FGF2, PDGF, and CXCL12, all of which promote tumor angiogenesis (Wang et al. 1999; Cao et al. 2008) (Table 15.1).

Table 15.1 Representative regulatory factors of angiogenesis

Pro-angiogenic factors	Anti-angiogenic factors
VEGF	Angiostatin
FGF	Endostatin
PDGF	Endorepellin
EGF	Canstatin
HGF	Tumstatin
Angiopoietin	Thrombospondin
TGF-beta	
CXCL12	

15.4.3 Proangiogenic Factors

Under physiological conditions, a delicate balance of pro- and antiangiogenic factors results in quiescence of blood vessels. This equilibrium breaks down in tumors as rapidly proliferating cancer cells require more blood and O_2 supply, thus together with stromal cells start to secrete angiogenesis-stimulating factors. While VEGF is the best known among them, there are many more mediators that trigger the proliferation of tumor blood vessels.

The VEGF family contains four members: A, B, C, and D; among these, VEGFA is the most important. VEGFA binds to its cell surface tyrosine kinase (TK) receptors VEGFR1 and VEGFR2 to stimulate the proliferation of endothelial cells (Holmes and Zachary 2005; Leone et al. 2019). Among the members of the FGF family, basic FGF (bFGF, FGF2) is the most potent stimulator of angiogenesis both in normal and tumor-induced blood vessels. bFGF is secreted by both tumor and stromal cells. In cooperation with VEGF, they collectively promote the secretion of proteases that facilitate the degradation of ECM and support the invasion of cancer cells (Turner and Grose 2010; Lugano et al. 2020). A plethora of data suggests that in addition to VEGF and FGF, PDGF is also implicated in both normal and tumor-associated angiogenesis. In tumors, PDGF signals through the PDGF receptor. Endothelial cells of tumor blood vessels secrete PDGF, while pericytes express PDGF receptor, thus promoting their own recruitment to the tumor blood vessels. PDGF stimulates both angiogenesis and cancer cell proliferation, and decreases intratumoral pressure (Raica and Cimpean 2010).

By signaling through the TK receptor Tie2, angiopoietin 2 also acts as a potent driver of tumor angiogenesis and its increased expression generally indicates poor tumor prognosis. Angiopoietin 2 mobilizes myeloid cells which, once transformed into stromal cells, promote tumor progression and downregulate tumor immunity (Yu and Ye 2020). Furthermore, by signaling via integrinβ1, angiopoietin 2 induces the expression of MMPs, which can destabilize the vasculature (Etoh et al. 2001). MMP2, MMP9, and MMP14 represent all active contributors to the angiogenic switch as matrix degradation increases the bioavailability of angiogenic factors (Bergers et al. 2000). At the same time, the degradation products of matrix components may exert anti-angiogenic effects, like in the case of endostatin and angiostatin (O'Reilly et al. 1999).

Hypoxia induced by insufficient blood supply inhibits the proteasomal degradation of hypoxia-inducible factor1α (HIF1α). HIF1α enters the nucleus and by associating with HIF1β it forms a heterodimeric transcriptional complex that binds to hypoxia-responsive elements. This initiates the synthesis of a plethora of proteins involved in angiogenesis including VEGF, nitric oxide synthase (NOS), leptin, erythropoietin (EPO), and many others (Masoud and Li 2015). Chemokines and their receptors have also been shown to play an active role in angiogenesis and also several members of the CXC and CXCL families possess an angiogenesis-stimulating potential. CXCL2, 3, 5, 6, 7, and 8 stimulate angiogenesis by binding to the endothelial cell surface receptor CXCR2. CXCL8 is one of the most potent chemokines that facilitates angiogenesis in ovarian, lung, and colon carcinoma, as well as in melanoma (Singh et al. 2007). Also other chemokines can exert an angiostatic effect (Sozzani et al. 2015; Belperio et al. 2000; Strieter et al. 2006).

Tumor cells may shed small cytoplasmic vesicles called exosomes into the bloodstream. Among other cargoes, tumor-derived exosomes may carry miRNAs and long noncoding RNAs. In the past 10 years, it has been increasingly recognized that these miRNAs may affect the action of tumor angiogenesis regulators, e.g., pro-angiogenic potential has been reported for miR155 (VHL), miR566 (VHL), miR210 (EFNA3), miR21(PTEN), miR182, miR296, miR17-92, miR378, miR221/222, and let-7b/-7f (Arcucci et al. 2021).

15.5 Natural Inhibitors of Angiogenesis

15.5.1 Proteolytic Breakdown Products

Endostatin, angiostatin, and endorepellin exemplify a peculiar antiangiogenic mechanism. All three molecules are degradation products of proteins that in their native form fulfill a completely different function. Angiostatin is derived from plasminogen, endostatin is the tryptic fragment of collagen XVIII, while endorepellin corresponds to the domain V of perlecan. Interestingly, angiostatin on its own has only a modest effect on endothelial cells; instead, it targets the innate immune system and utilizes the cytokine IL-12 to mediate its inhibitory effect (Albini et al. 2009). The beneficial effect of IL-12 was confirmed in a model of pathologic angiogenesis as well (Zhou et al. 2016). In contrast with angiostatin, endostatin directly targets endothelial cells by binding to Frizzled and VEGF receptors, as well as to glypicans (Poluzzi et al. 2016). The binding of endostatin triggers diverse downstream events mediated by intracellular pathways involved in angiogenesis such as MAPK, Hif1α, NFκB, STAT, etc. (Abdollahi et al. 2004); consequently, endostatin inhibits angiogenesis and induces autophagy. The activity of endorepellin overlaps with that of endostatin; it also utilizes VEGF receptors, whereas it interacts with different integrins (Kapoor et al. 2020).

Five members of the *thrombospondin* (TSP) family are all large multidomain proteins of the ECM. Their major representative is TSP1 with several functions including anti-angiogenesis. It inhibits the proliferation, migration and survival of endothelial cells, induces their apoptosis and antagonizes VEGF (Lawler and Lawler 2012). This effect is based on the interaction of its so-called TSR (thrombospondin type I repeat) domain with the cell surface proteins CD36, CD47(IAP), and integrins on endothelial cells. The TSR domain also sequesters active MMP2 and MMP9, thereby hindering the release of VEGF from the ECM (Bein and Simons 2000; Rodríguez-Manzaneque et al. 2001). Whereas low levels of TSP were originally thought to be associated with poor tumor survival, and TSP overexpression was linked with delayed angiogenesis, recent publications have reported contradictory data (Huang et al. 2017). Thus, it seems that the effect of TSP1 also depends on other microenvironmental factors. Besides TSP1, the potential of TSP2 to inhibit tumor angiogenesis has been also reported, which effect could be reversed by miR-221-3p (Wu et al. 2019).

Tumstatin is a 28-kDa cleavage product of type IV collagen, a major constituent of basement membranes. Via signaling through αvβ3-integrin, tumstatin induces apoptosis of vascular endothelial cells, inhibits the synthesis of mTOR pathway proteins, and prevents the dissociation of eIF4E from the 4E binding protein (Maeshima et al. 2002). Similar to tumstatin, *arresten* and *canstatin* are also derived from type IV collagen (Mundel and Kalluri 2007). They also suppress tube formation, proliferation, and invasion of endothelial cells, and promote their apoptosis. Both signal via integrin receptors (Aikio et al. 2012; Panka and Mier 2003) and canstatin induce FAS-dependent apoptosis of endothelial cells.

15.5.2 Noncoding RNAs

As mentioned above, certain noncoding RNAs may promote or inhibit angiogenesis. Some may even behave in a context-dependent fashion, e.g., miR221/222 can act in either direction depending on the tissue (Khella et al. 2015; Nicoli et al. 2012). These miRNAs most commonly target VEGF (Kong et al. 2016; Xue et al. 2016) (mir121, 122, 126, 29b, 206, 140-5p, 497, 377, 134), but effects have been reported on several other molecules implicated in angiogenesis such as VEGFR1,2, HIF1α, MMP2, ETS1, STAT3, STAT5A, CD44, and the Akt-mTOR pathway (Fang et al. 2011; Jeyapalan et al. 2011; Chen et al. 2016).

15.5.3 Heparin and Heparan Sulfate in the Regulation of Angiogenesis

Glycosaminoglycans (GAGs) play pivotal roles in the regulation of angiogenesis. GAG sugar chains are constructed of repeating glucuronic acid/*N*-acetyl glucosamine disaccharide units. This basic structure is modified by deacetylation and sulfation through the action of *N*-deacetylase/*N*-sulfotransferase and subsequently by epimerase that converts glucuronic acid to iduronic acid as well as three different *O*-sulfotransferases (Annaval et al. 2020). The ratio of *N*-sulfation to *N*-acetylation greatly determines the actual behavior of GAG chains, lending either stimulatory or inhibitory potential to the molecule. GAGs may exist as free sugar chains in the ECM, or may be attached to a core protein like syndecan and glypican on the cell surface or perlecan in the basement membrane (BM). Crucially to our topic, almost all factors involved in angiogenesis—close to 60 molecules we know of—can establish interactions with heparin or heparan sulfate (HS). Accordingly, GAGs may exert a wide variety of effects, the most prominent of which are the following:

(1) Heparin and HS may sequester angiogenic factors and thereby inhibit their binding to endothelial cells. (2) Proangiogenic growth factors may bind to the HS chains of perlecan and thus accumulate in the vicinity of endothelial cells. (3) HSPGs may present angiogenic signals to their respective cell surface receptors. (4) These signals, in turn, may facilitate the expression of cell surface HSPG or (5) Promote the synthesis of proteases. (6) Proteases may cleave HSPGs from the cell surface. (7) Free HS chains can bind to antiangiogenic factors and promote their effect (Chiodelli et al. 2015).

15.5.4 Discovery and Structure of Perlecan

Perlecan is a large proteoglycan that typically resides in BMs. Its protein core, sized over 470 kDa, is built up of five domains that show conspicuous homologies with other ECM proteins (Fig. 15.2) (Murdoch et al. 1992). The complete sequence of mouse perlecan was published in 1991 (Noonan et al. 1991), while the human sequence including the regulatory region was completed in 1993 (Cohen et al.

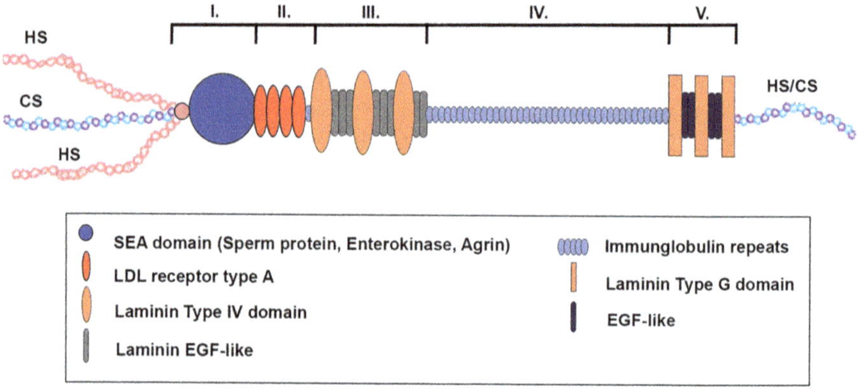

Interactive proteins with roles in matrix organisation and stabilisation.

Domain I.	Domain II.	Domain III.	Domain IV.	Domain V.
Laminin-1	Fibrillin-1	WARP	Nidogen-1	Nidogen-1
Collagen IV, V,	VLDL	Collagen VI	Nidogen-2	Fibulin-2
VI, VII, XI,	LDL	Tropoelastin	Collagen IV, VI,	ECM-1
XII, XIV.	Wnt		Fibulin-2	Collagen VI
Col99, ColQ			Tropoelastin	Fibulin-2
Fibronectin				Tropoelastin
PRELP, WARP				α-dystoglycan
Fibrillin-1, TSP-1				matriglycan
Histone H1				

Fig. 15.2 Domain structure of perlecan showing its most important interacting partners. Glycosaminoglycans binding critical regulators of angiogenesis like VEGF, FGF, and PDGF are attached to domain I of the molecule. In contrast to the proangiogenic action of perlecan, domain V, if liberated by protease activity renamed endorepellin, gains the potential to inhibit angiogenesis. *Int. J. Mol. Sci. 2021*, *22*(5), 2716

1993). Generally, perlecan carries HS GAG chains attached to the first domain, but its glycanation by chondroitin sulfate has also been reported (Lord et al. 2014). Domain I is also designated as SEA for its homology with sperm protein, enterokinase, and agrin. The other four domains (II-V) are built up from several modules and display homologies with LDL receptors, laminin, and the immunoglobulin superfamily (Farach-Carson et al. 2014; Melrose et al. 2006). This multidomain structure enables perlecan to establish interactions with a multitude of ECM molecules including BMs components like laminins and type IV collagen, interstitial matrix components like fibronectin, thrombospondin, and fibrillin, as well as growth factors, integrins, and others. (Bix and Iozzo 2008; McCarthy 2015; Mongiat et al. 2003a). This implies that besides its structural role perlecan possesses regulatory functions as well (Melrose 2020).

15.5.5 Perlecan Is an Indispensable Component of Basement Membranes

Type IV collagen and laminins are the principal building blocks of BMs that form a mat-like support for epithelial cells (Hohenester and Yurchenco 2013) or, as atypical BM, surround individual cells of mesenchymal origin (Pastor-Pareja 2020). Perlecan and nidogen are interspersed irregularly in this structure and connect collagen and laminin layers. Interactions of perlecan with other components of the BM both via its protein core and its HS chains adds stability to the structure. Besides acting as a molecular sieve, perlecan stores active molecules such as growth factors and cytokines, creates connections between growth factors and their cognate cell surface receptors, and promotes interactions between extracellular matrix molecules and their integrin receptors on the cell surface (Bix and Iozzo 2008; Gubbiotti et al. 2017). The importance of perlecan was clearly supported by the fact that its complete lack results in embryonic lethality in both mice and humans owing to inadequate vasculogenesis (Arikawa-Hirasawa et al. 1999; Iozzo et al. 2009).

15.5.6 Physiological Functions of Perlecan

Its large size and structural diversity endow perlecan with a plethora of physiological functions. As a component of the cartilaginous pericellular matrix it supports the interaction between chondrocytes (Wijeratne et al. 2016) and stimulates their proliferation in the growth plate via binding FGF 18; also, its domain I facilitates chondrogenic differentiation and supports the bone-forming activity of BMPII (Yang et al. 2006; Decarlo et al. 2012). Hence, perlecan is indispensable for proper skeletal development (Smith et al. 2007) and is a key component of all weight-bearing connective tissues including the articular *cartilage*, intervertebral disk, meniscus, ligaments, and tendons (Guilak et al. 2021). Since perlecan is a prominent player of vasculogenesis and angiogenesis both in physiology and pathology (Bix and Iozzo 2008), its lack impairs cardiovascular development (Sasse et al. 2008) and hinders wound healing (Lord et al. 2017). Growth factors and cytokines attached to the HS chains of domain I can modulate cellular functions such as proliferation, migration, and angiogenesis (Lord et al. 2014). Perlecan regulates embryonic development by interacting with sonic hedgehog action (Palma et al. 2011). Via its domain V perlecan supports the function of pericytes in the maintenance and repair of the blood-brain barrier (Roberts et al. 2012; Nakamura et al. 2019). Conversely, the same perlecan domain V when disengaged by protease cleavage is called endorepellin and displays antiangiogenic properties (Mongiat et al. 2003b).

15.5.7 Perlecan in Pathology

Its modular structure and the great number of its interaction partners explain why perlecan also participates in so many pathological processes. The low-density lipoprotein (LDL) receptor-like domain II binds LDL and very low-density lipoprotein (VLDL), thereby interfering with lipid metabolism. This may impair muscle tissue through lipid deposition and facilitate the development of fatty liver, arteriosclerosis, and metabolic syndrome. Experiments utilizing perlecan-deficient mice rescued from perinatal lethality revealed that these animals have less and smaller white fat adipocytes. This cell type secretes adiponectin, a cytokine that prevents fat deposition into other organs (Yamashita et al. 2018). These results drew attention to a potentially unfavorable impact of perlecan on lipid metabolism (Vikramadithyan et al. 2004). Perlecan has also been shown to promote arteriosclerosis and osteophyte production in knee osteoarthritis (Kaneko et al. 2013). As a reservoir of cytokines and growth factors perlecan participates in wound healing by attracting fibroblasts and supporting the proliferation of capillaries. However, the same mechanisms can result in pathological fibrosis in chronic injuries or inflammation (Zhou et al. 2004).

In spite of these negative effects, perlecan is indispensable for embryonic development and tissue organization (Hassell et al. 2002). Various mutations of perlecan including missense, deletion, and splicing mutations may result in Silverman-Handmaker-type dyssegmental dysplasia (DDSH) (Hassell et al. 2002), skeletal dysplasias, and a combination of chondrodysplasia and myotonia characterized by disrupted neuromuscular junction called Schwartz-Jampel syndrome (Nicole et al. 2000; Arikawa-Hirasawa et al. 2002).

15.5.8 Perlecan and Cancer

Perlecan is thought to promote cancer progression by attracting ECM proteins that support tumor growth and via its HS chains that bind growth factors and facilitate

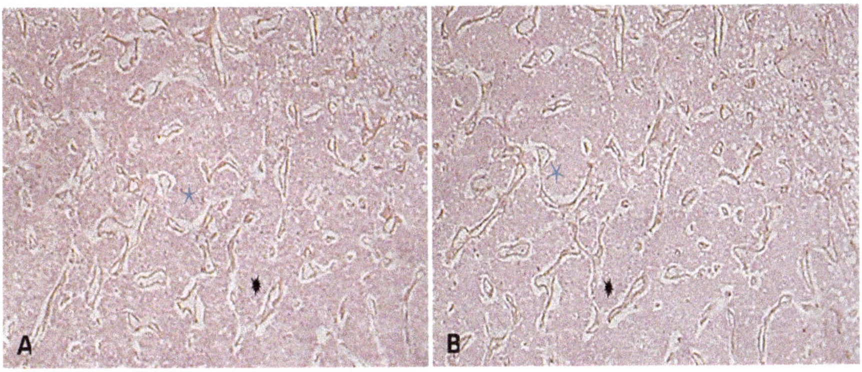

Fig. 15.3 Colocalization of perlecan and bFGF in the tumorous blood vessels of hepatocellular carcinoma. (**a**) perlecan, (**b**) bFGF. Original magnification x10. Asterisks show the regions positive for both perlecan and bFGF

their interaction with their cognate receptors on the surface of tumor cells. In experiments, p53-mutant pancreatic cancer cells could instruct their TAFs to upregulate the expression of perlecan and thus enhance the chemotherapy resistance and invasiveness of the tumor (Vennin et al. 2019; Ritchie et al. 2020) (Fig. 15.3).

In prostate cancer patients, increased serum levels of MMP7 and a tryptic fragment of perlecan domain IV are markers of invasiveness (Grindel et al. 2016); conversely, downregulation of perlecan reversed the invasive phenotype of prostate cancer (Savorè et al. 2005). Perlecan maintains SHH signaling in prostate cancer (Datta et al. 2006). Knockdown of agrin and perlecan inhibits the migration and adhesion of oral cancer cells and sensitizes them toward cisplatin (Correction: Agrin and perlecan mediate tumorigenic processes in oral squamous cell carcinoma 2015; Kawahara et al. 2014).

Besides supporting local spread, perlecan also promotes invasion and distant metastasis in multiple tumor types. The metastatic cascade is a multistep process that begins with the degradation of the basement membrane in the originating tissue and continues with migration through the extracellular matrix, intravasation into blood or lymphatic vessels and extravasation at a distant site where the tumor cells colonize. To successfully proceed through all these steps, tumor cells must permanently destroy epithelial and vascular BMs where they establish contacts with perlecan which in turn promotes colonization in the host tissue via its multivalent interactions (Elgundi et al. 2020). Upregulation of perlecan has been described in melanoma, lung, breast, liver, cholangiocellular, prostate, ovarian, pancreatic and oral carcinomas, as well as in glioblastoma (Cohen et al. 1994; Ilhan-Mutlu et al. 2016; Nackaerts et al. 1997; Nerlich et al. 1998; Hagedorn et al. 2001; Roskams et al. 1998; Sabit et al. 2001; Ida-Yonemochi et al. 2002; Kazanskaya et al. 2018; Warren et al. 2014; Sharma et al. 1998).

15.5.9 Perlecan and Tumor Angiogenesis

As perlecan is an indispensable resident of both epithelial and vascular BMs, its role in tumor angiogenesis comes as no surprise. Mechanistically, perlecan is thought to facilitate angiogenesis via its HS chains attached to the first, N-terminal domain. These negatively charged GAG chains bind several factors with angiogenic potential and present them to their receptors on the surface of endothelial cells. As discussed before, the best known and most potent of these factors is VEGFA that binds avidly to VEGFR 1 and 2 (Jiang and Couchman 2003); however, a host of other factors are induced by the hypoxic internal environment within fast-growing tumors. Basic FGF, PDGF, HGF, cytokines, and microRNAs—essentially the same factors detected in physiological angiogenesis—bind to perlecan, mostly to HS chains or occasionally to domain III, and stimulate endothelial cells to produce new blood vessels. What distinguishes tumor angiogenesis is dysregulation. Experimental evidence indicates that tumor-associated perlecan is not restricted into the BMs, but can also be detected in the tumor stroma, where it is synthesized both by stromal fibroblasts and tumor cells (Sabit et al. 2001). At the same time, deposition of perlecan diminishes in the fragmented basement membranes of the tumor tissue (Nackaerts et al. 1997). Tumor blood vessels fail to differentiate into arterial and venous arms, and their structure is abnormal with a decreased number of pericytes (Fig. 15.1) (Cao 2009).

15.5.10 Endorepellin, the Angiostatic Fragment of Perlecan

Similar to other basement membrane proteins like collagen type IV and XVIII, a fragment of perlecan can be shed via protease cleavage. Perlecan domain V, once released, is called endorepellin and it is known to inhibit angiogenesis and vasculogenesis by binding to VEGFR2 in regions different from the VEGFA binding sites. This interaction initiates signaling through VEGFR2 and α2β1-integrin that antagonizes the effect of angiogenic factors. The fact that endorepellin needs both VEGFR2 and α2β1-integrin receptors to exert its angiostatic effects confers specificity to its action, as only endothelial cells express both receptors on their surface (Goyal et al. 2011). Allosteric inhibition of VEGFR2 triggers internalization of the receptor, as well as downregulation of PI3K- and PLCγ-related signaling required for the transcription of VEGFA (Goyal et al. 2012). Although both angiostatin and endorepellin bind to VEGFR2, they utilize different integrins and different pathways: endostatin competes with VEGF for VEGFR2 and engages glypican/Wnt signaling and α5β1-integrin, whereas endorepellin binds to its own binding site on VEGFR2 and engages α2β1-integrin (Poluzzi et al. 2016).

15.5.11 Autophagy and Tumor Angiogenesis

Autophagy is a physiological process aimed to clear damaged, degraded, or hazardous molecules from the cytoplasm by delivering them to the lysosomes via autophagosomes. Autophagy helps to maintain intracellular homeostasis and assists adaptation to restricted nutrient supply (Schaaf et al. 2019). Thus, autophagy supports normal cells under starvation, prevents the accumulation of defective proteins, and participates in the rejuvenation of organelles (Oliva Trejo et al. 2020). Fast-proliferating malignant tumors with poor blood supply are even more dependent on autophagy for survival, as self-degradation provides tumor cells with nutrients under hypoxic conditions (Douglass et al. 2015; White 2012). On the other hand, when endorepellin binds to VEGFR2 on endothelial cells of tumor blood vessels it promotes not only autophagy but also cell death, which further potentiates its angiostatic effect (Poluzzi et al. 2014). Again, perlecan, and endorepellin have opposing roles as intact perlecan was shown to inhibit autophagy in muscles and thus maintain homeostasis (Ning et al. 2015; Gubbiotti et al. 2017). Of note, although perlecan and endorepellin have been studied most thoroughly in the context of tumor angiogenesis and autophagy, other ECM proteoglycans like decorin, biglycan, endostatin, and lumican have recently been found to display such activities (Chen and Iozzo 2020).

References

Abdollahi A, Hahnfeldt P, Maercker C, Gröne H-J, Debus J, Ansorge W, Folkman J, Hlatky L, Huber PE (2004) Endostatin's antiangiogenic signaling network. Mol Cell 13(5):649–663. https://doi.org/10.1016/S1097-2765(04)00102-9

Aikio M, Alahuhta I, Nurmenniemi S, Suojanen J, Palovuori R, Teppo S, Sorsa T, López-Otín C, Pihlajaniemi T, Salo T, Heljasvaara R, Nyberg P (2012) Arresten, a collagen-derived angiogenesis inhibitor, suppresses invasion of squamous cell carcinoma. PLoS One 7(12):e51044. https://doi.org/10.1371/journal.pone.0051044

Albini A, Brigati C, Ventura A, Lorusso G, Pinter M, Morini M, Mancino A, Sica A, Noonan DM (2009) Angiostatin anti-angiogenesis requires IL-12: the innate immune system as a key target. J Transl Med 7(1):5. https://doi.org/10.1186/1479-5876-7-5

Amin DN, Hida K, Bielenberg DR, Klagsbrun M (2006) Tumor endothelial cells express epidermal growth factor receptor (EGFR) but not ErbB3 and are responsive to EGF and to EGFR kinase inhibitors. Cancer Res 66(4):2173–2180. https://doi.org/10.1158/0008-5472.Can-05-3387

Annaval T, Wild R, Crétinon Y, Sadir R, Vivès RR, Lortat-Jacob H (2020) Heparan sulfate proteoglycans biosynthesis and post synthesis mechanisms combine few enzymes and few Core proteins to generate extensive structural and functional diversity. Molecules 25(18): 4215. https://doi.org/10.3390/molecules25184215

Arcucci V, Stacker SA, Achen MG (2021) Control of gene expression by exosome-derived non-coding RNAs in cancer angiogenesis and lymphangiogenesis. Biomol Ther 11(2):249. https://doi.org/10.3390/biom11020249

Arikawa-Hirasawa E, Watanabe H, Takami H, Hassell JR, Yamada Y (1999) Perlecan is essential for cartilage and cephalic development. Nat Genet 23(3):354–358. https://doi.org/10.1038/15537

Arikawa-Hirasawa E, Le AH, Nishino I, Nonaka I, Ho NC, Francomano CA, Govindraj P, Hassell JR, Devaney JM, Spranger J, Stevenson RE, Iannaccone S, Dalakas MC, Yamada Y (2002) Structural and functional mutations of the perlecan gene cause Schwartz-Jampel syndrome, with myotonic myopathy and chondrodysplasia. Am J Hum Genet 70(5):1368–1375. https://doi.org/10.1086/340390

Armulik A, Abramsson A, Betsholtz C (2005) Endothelial/pericyte interactions. Circ Res 97(6): 512–523. https://doi.org/10.1161/01.RES.0000182903.16652.d7

Asadzadeh Z, Mohammadi H, Safarzadeh E, Hemmatzadeh M, Mahdian-Shakib A, Jadidi-Niaragh-F, Azizi G, Baradaran B (2017) The paradox of Th17 cell functions in tumor immunity. Cell Immunol 322:15–25. https://doi.org/10.1016/j.cellimm.2017.10.015

Baluk P, Morikawa S, Haskell A, Mancuso M, McDonald DM (2003) Abnormalities of basement membrane on blood vessels and endothelial sprouts in tumors. Am J Pathol 163(5):1801–1815. https://doi.org/10.1016/s0002-9440(10)63540-7

Bein K, Simons M (2000) Thrombospondin type 1 repeats interact with matrix metalloproteinase. 2: Regulation of metalloproteinase activity. J Biol Chem 275(41):32167–32173. https://doi.org/10.1074/jbc.M003834200

Belperio JA, Keane MP, Arenberg DA, Addison CL, Ehlert JE, Burdick MD, Strieter RM (2000) CXC chemokines in angiogenesis. J Leukoc Biol 68(1):1–8

Bergers G, Benjamin LE (2003) Tumorigenesis and the angiogenic switch. Nat Rev Cancer 3(6): 401–410. https://doi.org/10.1038/nrc1093

Bergers G, Brekken R, McMahon G, Vu TH, Itoh T, Tamaki K, Tanzawa K, Thorpe P, Itohara S, Werb Z, Hanahan D (2000) Matrix metalloproteinase-9 triggers the angiogenic switch during carcinogenesis. Nat Cell Biol 2(10):737–744. https://doi.org/10.1038/35036374

Bix G, Iozzo RV (2008) Novel interactions of perlecan: unraveling perlecan's role in angiogenesis. Microsc Res Tech 71(5):339–348. https://doi.org/10.1002/jemt.20562

Brat DJ, Van Meir EG (2001) Glomeruloid microvascular proliferation orchestrated by VPF/VEGF: a new world of angiogenesis research. Am J Pathol 158(3):789–796. https://doi.org/10.1016/s0002-9440(10)64025-4

Cao Y (2009) Tumor angiogenesis and molecular targets for therapy. Front Biosci (Landmark Ed) 14:3962–3973. https://doi.org/10.2741/3504

Cao Y, Cao R, Hedlund E-M (2008) R regulation of tumor angiogenesis and metastasis by FGF and PDGF signaling pathways. J Mol Med 86(7):785–789. https://doi.org/10.1007/s00109-008-0337-z

Cassetta L, Pollard JW (2020) Tumor-associated macrophages. Curr Biol 30(6):R246–r248. https://doi.org/10.1016/j.cub.2020.01.031

Chen CG, Iozzo RV (2020) Angiostatic cues from the matrix: endothelial cell autophagy meets hyaluronan biology. J Biol Chem 295(49):16797–16812. https://doi.org/10.1074/jbc.REV120.014391

Chen QY, Jiao DM, Wu YQ, Chen J, Wang J, Tang XL, Mou H, Hu HZ, Song J, Yan J, Wu LJ, Chen J, Wang Z (2016) MiR-206 inhibits HGF-induced epithelial-mesenchymal transition and angiogenesis in non-small cell lung cancer via c-Met/PI3k/Akt/mTOR pathway. Oncotarget 7(14):18247–18261. https://doi.org/10.18632/oncotarget.7570

Chen Y, Song Y, Du W, Gong L, Chang H, Zou Z (2019) Tumor-associated macrophages: an accomplice in solid tumor progression. J Biomed Sci 26(1):78. https://doi.org/10.1186/s12929-019-0568-z

Chiodelli P, Bugatti A, Urbinati C, Rusnati M (2015) Heparin/Heparan sulfate proteoglycans glycomic interactome in angiogenesis: biological implications and therapeutical use. Molecules 20(4):6342–6388. https://doi.org/10.3390/molecules20046342

Cohen IR, Grässel S, Murdoch AD, Iozzo RV (1993) Structural characterization of the complete human perlecan gene and its promoter. Proc Natl Acad Sci U S A 90(21):10404–10408. https://doi.org/10.1073/pnas.90.21.10404

Cohen IR, Murdoch AD, Naso MF, Marchetti D, Berd D, Iozzo RV (1994) Abnormal expression of perlecan proteoglycan in metastatic melanomas. Cancer Res 54(22):5771–5774

Correction: Agrin and perlecan mediate tumorigenic processes in oral squamous cell carcinoma (2015). PLoS One 10(3):e0119247. https://doi.org/10.1371/journal.pone.0119247

Datta MW, Hernandez AM, Schlicht MJ, Kahler AJ, DeGueme AM, Dhir R, Shah RB, Farach-Carson C, Barrett A, Datta S (2006) Perlecan, a candidate gene for the CAPB locus, regulates prostate cancer cell growth via the Sonic Hedgehog pathway. Mol Cancer 5(1):9. https://doi.org/10.1186/1476-4598-5-9

De Spiegelaere W, Casteleyn C, Van den Broeck W, Plendl J, Bahramsoltani M, Simoens P, Djonov V, Cornillie P (2012) Intussusceptive angiogenesis: a biologically relevant form of angiogenesis. J Vasc Res 49(5):390–404. https://doi.org/10.1159/000338278

Decarlo AA, Belousova M, Ellis AL, Petersen D, Grenett H, Hardigan P, O'Grady R, Lord M, Whitelock JM (2012) Perlecan domain 1 recombinant proteoglycan augments BMP-2 activity and osteogenesis. BMC Biotechnol 12:60. https://doi.org/10.1186/1472-6750-12-60

Döme B, Hendrix MJC, Paku S, Tóvári J, Tímár J (2007) Alternative vascularization mechanisms in cancer: pathology and therapeutic implications. Am J Pathol 170(1):1–15. https://doi.org/10.2353/ajpath.2007.060302

Douglass S, Goyal A, Iozzo RV (2015) The role of perlecan and endorepellin in the control of tumor angiogenesis and endothelial cell autophagy. Connect Tissue Res 56(5):381–391. https://doi.org/10.3109/03008207.2015.1045297

Elgundi Z, Papanicolaou M, Major G, Cox TR, Melrose J, Whitelock JM, Farrugia BL (2020) Cancer metastasis: the role of the extracellular matrix and the heparan sulfate proteoglycan perlecan. Front Oncol 9:1482. https://doi.org/10.3389/fonc.2019.01482

Etoh T, Inoue H, Tanaka S, Barnard GF, Kitano S, Mori M (2001) Angiopoietin-2 is related to tumor angiogenesis in gastric carcinoma: possible in vivo regulation via induction of proteases. Cancer Res 61(5):2145–2153

Fang JH, Zhou HC, Zeng C, Yang J, Liu Y, Huang X, Zhang JP, Guan XY, Zhuang SM (2011) MicroRNA-29b suppresses tumor angiogenesis, invasion, and metastasis by regulating matrix metalloproteinase 2 expression. Hepatology 54(5):1729–1740. https://doi.org/10.1002/hep.24577

Farach-Carson MC, Warren CR, Harrington DA, Carson DD (2014) Border patrol: insights into the unique role of perlecan/heparan sulfate proteoglycan 2 at cell and tissue borders. Matrix Biol 34:64–79. https://doi.org/10.1016/j.matbio.2013.08.004

Folkman J (1971) Tumor angiogenesis: therapeutic implications. N Engl J Med 285(21):1182–1186. https://doi.org/10.1056/nejm197111182852108

Goyal A, Pal N, Concannon M, Paul M, Doran M, Poluzzi C, Sekiguchi K, Whitelock JM, Neill T, Iozzo RV (2011) Endorepellin, the angiostatic module of perlecan, interacts with both the α2β1 integrin and vascular endothelial growth factor receptor 2 (VEGFR2): a dual receptor antagonism. J Biol Chem 286(29):25947–25962. https://doi.org/10.1074/jbc.M111.243626

Goyal A, Poluzzi C, Willis CD, Smythies J, Shellard A, Neill T, Iozzo RV (2012) Endorepellin affects angiogenesis by antagonizing diverse vascular endothelial growth factor receptor 2 (VEGFR2)-evoked signaling pathways: transcriptional repression of hypoxia-inducible factor 1α and VEGFA and concurrent inhibition of nuclear factor of activated T cell 1 (NFAT1) activation. J Biol Chem 287(52):43543–43556. https://doi.org/10.1074/jbc.M112.401786

Grindel B, Li Q, Arnold R, Petros J, Zayzafoon M, Muldoon M, Stave J, Chung LW, Farach-Carson MC (2016) Perlecan/HSPG2 and matrilysin/MMP-7 as indices of tissue invasion: tissue localization and circulating perlecan fragments in a cohort of 288 radical prostatectomy patients. Oncotarget 7(9):10433–10447. https://doi.org/10.18632/oncotarget.7197

Gubbiotti MA, Neill T, Iozzo RV (2017) A current view of perlecan in physiology and pathology: a mosaic of functions. Matrix Biol 57-58:285–298. https://doi.org/10.1016/j.matbio.2016.09.003

Guilak F, Hayes AJ, Melrose J (2021) Perlecan in pericellular mechanosensory cell-matrix communication, extracellular matrix stabilisation and mechanoregulation of load-bearing connective tissues. Int J Mol Sci 22(5):2716. https://doi.org/10.3390/ijms22052716

Hagedorn HG, Sauer U, Schleicher ED, Nerlich AG (2001) Divergence in distribution and prognostic significance of major basement components in laryngeal carcinomas. Int J Oncol 18(5):1045–1051. https://doi.org/10.3892/ijo.18.5.1045
Hassell J, Yamada Y, Arikawa-Hirasawa E (2002) Role of perlecan in skeletal development and diseases. Glycoconj J 19(4):263–267. https://doi.org/10.1023/A:1025340215261
Hida K, Ohga N, Hida Y, Shindoh M (2010) Significance of anti-angiogenic therapy in head and neck cancer—heterogeneity of tumor endothelium. Jpn Dent Sci Rev 46(1):26–32. https://doi.org/10.1016/j.jdsr.2009.10.001
Hida K, Maishi N, Torii C, Hida Y (2016) Tumor angiogenesis--characteristics of tumor endothelial cells. Int J Clin Oncol 21(2):206–212. https://doi.org/10.1007/s10147-016-0957-1
Hida K, Maishi N, Annan DA, Hida Y (2018) Contribution of tumor endothelial cells in cancer progression. Int J Mol Sci 19(5):1272. https://doi.org/10.3390/ijms19051272
Hohenester E, Yurchenco PD (2013) Laminins in basement membrane assembly. Cell Adhes Migr 7(1):56–63. https://doi.org/10.4161/cam.21831
Holmes DI, Zachary I (2005) The vascular endothelial growth factor (VEGF) family: angiogenic factors in health and disease. Genome Biol 6(2):209. https://doi.org/10.1186/gb-2005-6-2-209
Høye AM, Tolstrup SD, Horton ER, Nicolau M, Frost H, Woo JH, Mauldin JP, Frankel AE, Cox TR, Erler JT (2018) Tumor endothelial marker 8 promotes cancer progression and metastasis. Oncotarget 9(53):30173–30188. https://doi.org/10.18632/oncotarget.25734
Huang T, Sun L, Yuan X, Qiu H (2017) Thrombospondin-1 is a multifaceted player in tumor progression. Oncotarget 8(48):84546–84558. https://doi.org/10.18632/oncotarget.19165
Ida-Yonemochi H, Ikarashi T, Nagata M, Hoshina H, Takagi R, Saku T (2002) The basement membrane-type heparan sulfate proteoglycan (perlecan) in ameloblastomas: its intercellular localization in stellate reticulum-like foci and biosynthesis by tumor cells in culture. Virchows Arch 441(2):165–173. https://doi.org/10.1007/s00428-001-0556-y
Ilhan-Mutlu A, Siehs C, Berghoff AS, Ricken G, Widhalm G, Wagner L, Preusser M (2016) Expression profiling of angiogenesis-related genes in brain metastases of lung cancer and melanoma. Tumour Biol 37(1):1173–1182. https://doi.org/10.1007/s13277-015-3790-7
Iozzo RV, Zoeller JJ, Nyström A (2009) Basement membrane proteoglycans: modulators par excellence of cancer growth and angiogenesis. Mol Cells 27(5):503–513. https://doi.org/10.1007/s10059-009-0069-0
Jeyapalan Z, Deng Z, Shatseva T, Fang L, He C, Yang BB (2011) Expression of CD44 3-′-untranslated region regulates endogenous microRNA functions in tumorigenesis and angiogenesis. Nucleic Acids Res 39(8):3026–3041. https://doi.org/10.1093/nar/gkq1003
Jiang X, Couchman JR (2003) Perlecan and tumor angiogenesis. J Histochem Cytochem 51(11): 1393–1410. https://doi.org/10.1177/002215540305101101
Kaneko H, Ishijima M, Futami I, Tomikawa-Ichikawa N, Kosaki K, Sadatsuki R, Yamada Y, Kurosawa H, Kaneko K, Arikawa-Hirasawa E (2013) Synovial perlecan is required for osteophyte formation in knee osteoarthritis. Matrix Biol 32(3-4):178–187. https://doi.org/10.1016/j.matbio.2013.01.004
Kapoor A, Chen CG, Iozzo RV (2020) Endorepellin evokes an angiostatic stress signaling cascade in endothelial cells. J Biol Chem 295(19):6344–6356. https://doi.org/10.1074/jbc.RA120.012525
Kawahara R, Granato DC, Carnielli CM, Cervigne NK, Oliveria CE, Rivera C et al (2014) Agrin and Perlecan mediate tumorigenic processes in oral squamous cell carcinoma. PLoS One 9(12): e115004. https://doi.org/10.1371/journal.pone.0115004. PMID: 255069
Kazanskaya GM, Tsidulko AY, Volkov AM, Kiselev RS, Suhovskih AV, Kobozev VV, Gaytan AS, Aidagulova SV, Krivoshapkin AL, Grigorieva EV (2018) Heparan sulfate accumulation and perlecan/HSPG2 up-regulation in tumour tissue predict low relapse-free survival for patients with glioblastoma. Histochem Cell Biol 149(3):235–244. https://doi.org/10.1007/s00418-018-1631-7
Khella HWZ, Butz H, Ding Q, Rotondo F, Evans KR, Kupchak P, Dharsee M, Latif A, Pasic MD, Lianidou E, Bjarnason GA, Yousef GM (2015) miR-221/222 are involved in response to

sunitinib treatment in metastatic renal cell carcinoma. Mol Ther 23(11):1748–1758. https://doi.org/10.1038/mt.2015.129

Kong R, Ma Y, Feng J, Li S, Zhang W, Jiang J, Zhang J, Qiao Z, Yang X, Zhou B (2016) The crucial role of miR-126 on suppressing progression of esophageal cancer by targeting VEGF-A. Cell Mol Biol Lett 21:3. https://doi.org/10.1186/s11658-016-0004-2

Latacz E, Caspani E, Barnhill R, Lugassy C, Verhoef C, Grünhagen D, Van Laere S, Fernández Moro C, Gerling M, Dirix M, Dirix LY, Vermeulen PB (2020) Pathological features of vessel co-option versus sprouting angiogenesis. Angiogenesis 23(1):43–54. https://doi.org/10.1007/s10456-019-09690-0

Lawler PR, Lawler J (2012) Molecular basis for the regulation of angiogenesis by thrombospondin-1 and -2. Cold Spring Harb Perspect Med 2(5):a006627. https://doi.org/10.1101/cshperspect.a006627

Leone P, Buonavoglia A, Fasano R, Solimando AG, De Re V, Cicco S, Vacca A, Racanelli V (2019) Insights into the regulation of tumor angiogenesis by micro-RNAs. J Clin Med 8(12):2030. https://doi.org/10.3390/jcm8122030

Lin EY, Li JF, Gnatovskiy L, Deng Y, Zhu L, Grzesik DA, Qian H, Xue XN, Pollard JW (2006) Macrophages regulate the angiogenic switch in a mouse model of breast cancer. Cancer Res 66(23):11238–11246. https://doi.org/10.1158/0008-5472.Can-06-1278

Lord MS, Chuang CY, Melrose J, Davies MJ, Iozzo RV, Whitelock JM (2014) The role of vascular-derived perlecan in modulating cell adhesion, proliferation and growth factor signaling. Matrix Biol 35:112–122. https://doi.org/10.1016/j.matbio.2014.01.016

Lord MS, Ellis AL, Farrugia BL, Whitelock JM, Grenett H, Li C, O'Grady RL, DeCarlo AA (2017) Perlecan and vascular endothelial growth factor-encoding DNA-loaded chitosan scaffolds promote angiogenesis and wound healing. J Control Release 250:48–61. https://doi.org/10.1016/j.jconrel.2017.02.009

Lugano R, Ramachandran M, Dimberg A (2020) Tumor angiogenesis: causes, consequences, challenges and opportunities. Cell Mol Life Sci 77(9):1745–1770. https://doi.org/10.1007/s00018-019-03351-7

Maeshima Y, Sudhakar A, Lively JC, Ueki K, Kharbanda S, Kahn CR, Sonenberg N, Hynes RO, Kalluri R (2002) Tumstatin, an endothelial cell-specific inhibitor of protein synthesis. Science 295(5552):140–143. https://doi.org/10.1126/science.1065298

Masoud GN, Li W (2015) HIF-1α pathway: role, regulation and intervention for cancer therapy. Acta Pharm Sin B 5(5):378–389. https://doi.org/10.1016/j.apsb.2015.05.007

Matsuda K, Ohga N, Hida Y, Muraki C, Tsuchiya K, Kurosu T, Akino T, Shih SC, Totsuka Y, Klagsbrun M, Shindoh M, Hida K (2010) Isolated tumor endothelial cells maintain specific character during long-term culture. Biochem Biophys Res Commun 394(4):947–954. https://doi.org/10.1016/j.bbrc.2010.03.089

McCarthy KJ (2015) The basement membrane proteoglycans perlecan and agrin: something old, something new. Curr Top Membr 76:255–303. https://doi.org/10.1016/bs.ctm.2015.09.001

Mehrzadi S, Pourhanifeh MH, Mirzaei A, Moradian F, Hosseinzadeh A (2021) An updated review of mechanistic potentials of melatonin against cancer: pivotal roles in angiogenesis, apoptosis, autophagy, endoplasmic reticulum stress and oxidative stress. Cancer Cell Int 21(1):188. https://doi.org/10.1186/s12935-021-01892-1

Melrose J (2020) Perlecan, a modular instructive proteoglycan with diverse functional properties. Int J Biochem Cell Biol 128:105849. https://doi.org/10.1016/j.biocel.2020.105849

Melrose J, Roughley P, Knox S, Smith S, Lord M, Whitelock J (2006) The structure, location, and function of perlecan, a prominent pericellular proteoglycan of fetal, postnatal, and mature hyaline cartilages. J Biol Chem 281(48):36905–36914. https://doi.org/10.1074/jbc.M608462200

Mongiat M, Fu J, Oldershaw R, Greenhalgh R, Gown AM, Iozzo RV (2003a) Perlecan protein core interacts with extracellular matrix protein 1 (ECM1), a glycoprotein involved in bone formation and angiogenesis. J Biol Chem 278(19):17491–17499. https://doi.org/10.1074/jbc.M210529200

Mongiat M, Sweeney SM, San Antonio JD, Fu J, Iozzo RV (2003b) Endorepellin, a novel inhibitor of angiogenesis derived from the C terminus of perlecan. J Biol Chem 278(6):4238–4249. https://doi.org/10.1074/jbc.M210445200

Mundel TM, Kalluri R (2007) Type IV collagen-derived angiogenesis inhibitors. Microvasc Res 74(2-3):85–89. https://doi.org/10.1016/j.mvr.2007.05.005

Murdoch AD, Dodge GR, Cohen I, Tuan RS, Iozzo RV (1992) Primary structure of the human heparan sulfate proteoglycan from basement membrane (HSPG2/perlecan). A chimeric molecule with multiple domains homologous to the low density lipoprotein receptor, laminin, neural cell adhesion molecules, and epidermal growth factor. J Biol Chem 267(12):8544–8557

Nackaerts K, Verbeken E, Deneffe G, Vanderschueren B, Demedts M, David G (1997) Heparan sulfate proteoglycan expression in human lung-cancer cells. Int J Cancer 74(3):335–345. https://doi.org/10.1002/(sici)1097-0215(19970620)74:3<335::aid-ijc18>3.0.co;2-a

Nakamura K, Ikeuchi T, Nara K, Rhodes CS, Zhang P, Chiba Y, Kazuno S, Miura Y, Ago T, Arikawa-Hirasawa E, Mukouyama Y-S, Yamada Y (2019) Perlecan regulates pericyte dynamics in the maintenance and repair of the blood-brain barrier. J Cell Biol 218(10):3506–3525. https://doi.org/10.1083/jcb.201807178

Nerlich AG, Lebeau A, Hagedorn HG, Sauer U, Schleicher ED (1998) Morphological aspects of altered basement membrane metabolism in invasive carcinomas of the breast and the larynx. Anticancer Res 18(5a):3515–3520

Nicole S, Davoine CS, Topaloglu H, Cattolico L, Barral D, Beighton P, Hamida CB, Hammouda H, Cruaud C, White PS, Samson D, Urtizberea JA, Lehmann-Horn F, Weissenbach J, Hentati F, Fontaine B (2000) Perlecan, the major proteoglycan of basement membranes, is altered in patients with Schwartz-Jampel syndrome (chondrodystrophic myotonia). Nat Genet 26(4): 480–483. https://doi.org/10.1038/82638

Nicoli S, Knyphausen CP, Zhu LJ, Lakshmanan A, Lawson ND (2012) miR-221 is required for endothelial tip cell behaviors during vascular development. Dev Cell 22(2):418–429. https://doi.org/10.1016/j.devcel.2012.01.008

Ning L, Xu Z, Furuya N, Nonaka R, Yamada Y, Arikawa-Hirasawa E (2015) Perlecan inhibits autophagy to maintain muscle homeostasis in mouse soleus muscle. Matrix Biol 48:26–35. https://doi.org/10.1016/j.matbio.2015.08.002

Noonan DM, Fulle A, Valente P, Cai S, Horigan E, Sasaki M, Yamada Y, Hassell JR (1991) The complete sequence of perlecan, a basement membrane heparan sulfate proteoglycan, reveals extensive similarity with laminin A chain, low density lipoprotein-receptor, and the neural cell adhesion molecule. J Biol Chem 266(34):22939–22947

Oliva Trejo JA, Tanida I, Suzuki C, Kakuta S, Tada N, Uchiyama Y (2020) Characterization of starvation-induced autophagy in cerebellar Purkinje cells of pHluorin-mKate2-human LC3B transgenic mice. Sci Rep 10(1):9643. https://doi.org/10.1038/s41598-020-66370-6

O'Reilly MS, Wiederschain D, Stetler-Stevenson WG, Folkman J, Moses MA (1999) Regulation of angiostatin production by matrix metalloproteinase-2 in a model of concomitant resistance. J Biol Chem 274(41):29568–29571. https://doi.org/10.1074/jbc.274.41.29568

Owen JL, Mohamadzadeh M (2013) Macrophages and chemokines as mediators of angiogenesis. Front Physiol 4:159. https://doi.org/10.3389/fphys.2013.00159

Palma V, Carrasco H, Reinchisi G, Olivares G, Faunes F, Larraín J (2011) SHh activity and localization is regulated by perlecan. Biol Res 44:63–67

Panka DJ, Mier JW (2003) Canstatin inhibits Akt activation and induces Fas-dependent apoptosis in endothelial cells. J Biol Chem 278(39):37632–37636. https://doi.org/10.1074/jbc.M307339200

Pastor-Pareja JC (2020) Atypical basement membranes and basement membrane diversity - what is normal anyway? J Cell Sci 133(8):jcs241794. https://doi.org/10.1242/jcs.241794

Pluda JM (1997) Tumor-associated angiogenesis: mechanisms, clinical implications, and therapeutic strategies. Semin Oncol 24(2):203–218

Poluzzi C, Casulli J, Goyal A, Mercer TJ, Neill T, Iozzo RV (2014) Endorepellin evokes autophagy in endothelial cells. J Biol Chem 289(23):16114–16128. https://doi.org/10.1074/jbc.M114.556530

Poluzzi C, Iozzo RV, Schaefer L (2016) Endostatin and endorepellin: a common route of action for similar angiostatic cancer avengers. Adv Drug Deliv Rev 97:156–173. https://doi.org/10.1016/j.addr.2015.10.012

Raica M, Cimpean AM (2010) Platelet-derived growth factor (PDGF)/PDGF receptors (PDGFR) axis as target for antitumor and antiangiogenic therapy. Pharmaceuticals (Basel) 3(3):572–599. https://doi.org/10.3390/ph3030572

Raza A, Franklin MJ, Dudek AZ (2010) Pericytes and vessel maturation during tumor angiogenesis and metastasis. Am J Hematol 85(8):593–598. https://doi.org/10.1002/ajh.21745

Ritchie S, Pereira BA, Vennin C, Timpson P (2020) Targeting genetically-tuned CAFs in pancreatic cancer via perlecan manipulation. Expert Opin Ther Targets 24(3):171–174. https://doi.org/10.1080/14728222.2020.1727887

Roberts J, Kahle M, Bix G (2012) Perlecan and the blood-brain barrier: beneficial proteolysis? Front Pharmacol 3:155. https://doi.org/10.3389/fphar.2012.00155

Rodríguez-Manzaneque JC, Lane TF, Ortega MA, Hynes RO, Lawler J, Iruela-Arispe ML (2001) Thrombospondin-1 suppresses spontaneous tumor growth and inhibits activation of matrix metalloproteinase-9 and mobilization of vascular endothelial growth factor. Proc Natl Acad Sci 98(22):12485. https://doi.org/10.1073/pnas.171460498

Roskams T, De Vos R, David G, Van Damme B, Desmet V (1998) Heparan sulphate proteoglycan expression in human primary liver tumours. J Pathol 185(3):290–297. https://doi.org/10.1002/(sici)1096-9896(199807)185:3<290::Aid-path91>3.0.Co;2-i

Sabit H, Tsuneyama K, Shimonishi T, Harada K, Cheng J, Ida H, Saku T, Saito K, Nakanuma Y (2001) Enhanced expression of basement-membrane-type heparan sulfate proteoglycan in tumor fibro-myxoid stroma of intrahepatic cholangiocarcinoma. Pathol Int 51(4):248–256. https://doi.org/10.1046/j.1440-1827.2001.01201.x

Sasse P, Malan D, Fleischmann M, Roell W, Gustafsson E, Bostani T, Fan Y, Kolbe T, Breitbach M, Addicks K, Welz A, Brem G, Hescheler J, Aszodi A, Costell M, Bloch W, Fleischmann BK (2008) Perlecan is critical for heart stability. Cardiovasc Res 80(3):435–444. https://doi.org/10.1093/cvr/cvn225

Savorè C, Zhang C, Muir C, Liu R, Wyrwa J, Shu J, Zhau HE, Chung LW, Carson DD, Farach-Carson MC (2005) Perlecan knockdown in metastatic prostate cancer cells reduces heparin-binding growth factor responses in vitro and tumor growth in vivo. Clin Exp Metastasis 22(5):377–390. https://doi.org/10.1007/s10585-005-2339-3

Schaaf MB, Houbaert D, Meçe O, Agostinis P (2019) Autophagy in endothelial cells and tumor angiogenesis. Cell Death Differ 26(4):665–679. https://doi.org/10.1038/s41418-019-0287-8

Sharma B, Handler M, Eichstetter I, Whitelock JM, Nugent MA, Iozzo RV (1998) Antisense targeting of perlecan blocks tumor growth and angiogenesis in vivo. J Clin Invest 102(8):1599–1608. https://doi.org/10.1172/JCI3793

Singh S, Sadanandam A, Singh RK (2007) Chemokines in tumor angiogenesis and metastasis. Cancer Metastasis Rev 26(3-4):453–467. https://doi.org/10.1007/s10555-007-9068-9

Smith SM, West LA, Hassell JR (2007) The core protein of growth plate perlecan binds FGF-18 and alters its mitogenic effect on chondrocytes. Arch Biochem Biophys 468(2):244–251. https://doi.org/10.1016/j.abb.2007.10.006

Sozzani S, Del Prete A, Bonecchi R, Locati M (2015) Chemokines as effector and target molecules in vascular biology. Cardiovasc Res 107(3):364–372. https://doi.org/10.1093/cvr/cvv150

St Croix B, Rago C, Velculescu V, Traverso G, Romans KE, Montgomery E, Lal A, Riggins GJ, Lengauer C, Vogelstein B, Kinzler KW (2000) Genes expressed in human tumor endothelium. Science 289(5482):1197–1202. https://doi.org/10.1126/science.289.5482.1197

Stockmann C, Schadendorf D, Klose R, Helfrich I (2014) The impact of the immune system on tumor: angiogenesis and vascular remodeling. Front Oncol 4:69. https://doi.org/10.3389/fonc.2014.00069

Strieter RM, Burdick MD, Mestas J, Gomperts B, Keane MP, Belperio JA (2006) Cancer CXC chemokine networks and tumour angiogenesis. Eur J Cancer 42(6):768–778. https://doi.org/10.1016/j.ejca.2006.01.006

Thienpont B, Lambrechts D (2017) It's T time for normal blood vessels. Dev Cell 41(2):125–126. https://doi.org/10.1016/j.devcel.2017.04.005

Tian L, Goldstein A, Wang H, Ching Lo H, Sun Kim I, Welte T, Sheng K, Dobrolecki LE, Zhang X, Putluri N, Phung TL, Mani SA, Stossi F, Sreekumar A, Mancini MA, Decker WK, Zong C, Lewis MT, Zhang XH (2017) Mutual regulation of tumour vessel normalization and immunostimulatory reprogramming. Nature 544(7649):250–254. https://doi.org/10.1038/nature21724

Turner N, Grose R (2010) Fibroblast growth factor signalling: from development to cancer. Nat Rev Cancer 10(2):116–129. https://doi.org/10.1038/nrc2780

Vennin C, Mélénec P, Rouet R, Nobis M, Cazet AS, Murphy KJ, Herrmann D, Reed DA, Lucas MC, Warren SC, Elgundi Z, Pinese M, Kalna G, Roden D, Samuel M, Zaratzian A, Grey ST, Da Silva A, Leung W, Mathivanan S, Wang Y, Braithwaite AW, Christ D, Benda A, Parkin A, Phillips PA, Whitelock JM, Gill AJ, Sansom OJ, Croucher DR, Parker BL, Pajic M, Morton JP, Cox TR, Timpson P (2019) CAF hierarchy driven by pancreatic cancer cell p53-status creates a pro-metastatic and chemoresistant environment via perlecan. Nat Commun 10(1):3637. https://doi.org/10.1038/s41467-019-10968-6

Vikramadithyan RK, Kako Y, Chen G, Hu Y, Arikawa-Hirasawa E, Yamada Y, Goldberg IJ (2004) Atherosclerosis in perlecan heterozygous mice. J Lipid Res 45(10):1806–1812. https://doi.org/10.1194/jlr.M400019-JLR200

Wang D, Huang H-JS, Kazlauskas A, Cavenee WK (1999) Induction of vascular endothelial growth factor expression in endothelial cells by platelet-derived growth factor through the activation of phosphatidylinositol 3-kinase. Cancer Res 59(7):1464–1472

Wang FT, Sun W, Zhang JT, Fan YZ (2019) Cancer-associated fibroblast regulation of tumor neo-angiogenesis as a therapeutic target in cancer. Oncol Lett 17(3):3055–3065. https://doi.org/10.3892/ol.2019.9973

Warren CR, Grindel BJ, Francis L, Carson DD, Farach-Carson MC (2014) Transcriptional activation by NFκB increases perlecan/HSPG2 expression in the desmoplastic prostate tumor microenvironment. J Cell Biochem 115(7):1322–1333. https://doi.org/10.1002/jcb.24788

White E (2012) Deconvoluting the context-dependent role for autophagy in cancer. Nat Rev Cancer 12(6):401–410. https://doi.org/10.1038/nrc3262

Wijeratne SS, Martinez JR, Grindel BJ, Frey EW, Li J, Wang L, Farach-Carson MC, Kiang CH (2016) Single molecule force measurements of perlecan/HSPG2: a key component of the osteocyte pericellular matrix. Matrix Biol 50:27–38. https://doi.org/10.1016/j.matbio.2015.11.001

Wu XG, Zhou CF, Zhang YM, Yan RM, Wei WF, Chen XJ, Yi HY, Liang LJ, Fan LS, Liang L, Wu S, Wang W (2019) Cancer-derived exosomal miR-221-3p promotes angiogenesis by targeting THBS2 in cervical squamous cell carcinoma. Angiogenesis 22(3):397–410. https://doi.org/10.1007/s10456-019-09665-1

Xiong YQ, Sun HC, Zhang W, Zhu XD, Zhuang PY, Zhang JB, Wang L, Wu WZ, Qin LX, Tang ZY (2009) Human hepatocellular carcinoma tumor-derived endothelial cells manifest increased angiogenesis capability and drug resistance compared with normal endothelial cells. Clin Cancer Res 15(15):4838–4846. https://doi.org/10.1158/1078-0432.Ccr-08-2780

Xu LN, Xu BN, Cai J, Yang JB, Lin N (2013) Tumor-associated fibroblast-conditioned medium promotes tumor cell proliferation and angiogenesis. Genet Mol Res 12(4):5863–5871. https://doi.org/10.4238/2013.November.22.14

Xue D, Yang Y, Liu Y, Wang P, Dai Y, Liu Q, Chen L, Shen J, Ju H, Li Y, Tan Z (2016) MicroRNA-206 attenuates the growth and angiogenesis in non-small cell lung cancer cells by blocking the 14-3-3ζ/STAT3/HIF-1α/VEGF signaling. Oncotarget 7(48):79805–79813. https://doi.org/10.18632/oncotarget.12972

Yamashita Y, Nakada S, Yoshihara T, Nara T, Furuya N, Miida T, Hattori N, Arikawa-Hirasawa E (2018) Perlecan, a heparan sulfate proteoglycan, regulates systemic metabolism with dynamic changes in adipose tissue and skeletal muscle. Sci Rep 8(1):7766. https://doi.org/10.1038/s41598-018-25635-x

Yang W, Gomes RR, Brown AJ, Burdett AR, Alicknavitch M, Farach-Carson MC, Carson DD (2006) Chondrogenic differentiation on perlecan domain I, collagen II, and bone morphogenetic protein-2-based matrices. Tissue Eng 12(7):2009–2024. https://doi.org/10.1089/ten.2006.12.2009

Yu X, Ye F (2020) Role of angiopoietins in development of cancer and neoplasia associated with viral infection. Cell 9(2):457. https://doi.org/10.3390/cells9020457

Zhou Z, Wang J, Cao R, Morita H, Soininen R, Chan KM, Liu B, Cao Y, Tryggvason K (2004) Impaired angiogenesis, delayed wound healing and retarded tumor growth in perlecan heparan sulfate-deficient mice. Cancer Res 64(14):4699–4702. https://doi.org/10.1158/0008-5472.Can-04-0810

Zhou Y, Yoshida S, Kubo Y, Kobayashi Y, Nakama T, Yamaguchi M, Ishikawa K, Nakao S, Ikeda Y, Ishibashi T, Sonoda K-H (2016) Interleukin-12 inhibits pathological neovascularization in mouse model of oxygen-induced retinopathy. Sci Rep 6(1):28140. https://doi.org/10.1038/srep28140

Zuazo-Gaztelu I, Casanovas O (2018) Unraveling the role of angiogenesis in cancer ecosystems. Front Oncol 8:248. https://doi.org/10.3389/fonc.2018.00248

Chapter 16
The microRNA-Extracellular Matrix Interplay in Breast Cancer

Zoi Piperigkou, Dimitra Manou, Dimitra Bainantzou, Vasiliki Zolota, Efthymia Papakonstantinou, Achilleas D. Theocharis, and Nikos K. Karamanos

Abstract The subclonal evolution of breast cancer is closely related to epigenetic regulation. Secreted microRNAs (miRNAs) spotted within the complex extracellular matrix (ECM) network are responsible for post-transcriptional and functional alterations to matrix constituents affecting vital cell processes for the initiation of metastasis, such as cell proliferation, migration, and invasion. The focus of this chapter is to highlight how the two-way relationship between miRNAs and ECM affects breast cancer pathogenesis and progression. Future investigation on the epigenetic regulation of matrix biomolecules focusing on miRNAs will improve current approaches to target tumor microenvironment and may expand our perspective in the mechanistic aspects of this pathology, contributing to a more effective breast cancer patient management.

Z. Piperigkou (✉) · N. K. Karamanos (✉)
Biochemistry, Biochemical Analysis & Matrix Pathobiology Research Group, Laboratory of Biochemistry, Department of Chemistry, University of Patras, Patras, Greece

Foundation for Research and Technology-Hellas (FORTH)/Institute of Chemical Engineering Sciences (ICE-HT), Patras, Greece
e-mail: zoipip@upatras.gr; n.k.karamanos@upatras.gr

D. Manou · D. Bainantzou · A. D. Theocharis
Biochemistry, Biochemical Analysis & Matrix Pathobiology Research Group, Laboratory of Biochemistry, Department of Chemistry, University of Patras, Patras, Greece

V. Zolota
Department of Pathology, School of Medicine, University of Patras, Patras, Greece

E. Papakonstantinou
Department of Gynecology and Obstetrics, University Hospital of Patras, Patras, Greece

I. Kovalszky et al. (eds.), *The Extracellular Matrix and the Tumor Microenvironment*, Biology of Extracellular Matrix 11,
https://doi.org/10.1007/978-3-030-99708-3_16

16.1 Introduction

MicroRNAs (miRNAs) hold an important role as post-transcriptional regulatory non-coding RNAs that target 3′ untranslated region (UTR) of one or multiple mRNAs (De Rie et al. 2017). Primary miRNAs are processed to precursor miRNAs (pre-miRNAs) by a specific microprocessor complex. Pre-miRNAs are then cleaved by the RNase III Dicer to generate mature miRNAs that direct the RNA-induced silencing complex (RISC) to mRNAs with complementary sequence (target-mRNAs) (Chendrimada et al. 2005).

MiRNAs are considered as powerful breast cancer biomarkers as they may govern several signaling cascades in (patho)physiological conditions including breast cancer, since more than 60% of protein-coding mRNAs may be targets of miRNAs as indicated by bioinformatics predictions (Cui et al. 2019). It is plausible to suggest that differential miRNA expression in a specific tissue or cell line affects the biogenesis and turnover of extracellular matrix (ECM), which constitutes a dynamic, multifunctional 3D network of macromolecules (Karamanos et al. 2021, 2018). Its constituents including proteoglycans (PGs), glycosaminoglycans (GAGs), collagen and elastin fibrils, glycoproteins, glycosidases and proteases, interconnect to form a regulatory niche that affect vital cancer cell functions, such as epithelial-to-mesenchymal transition (EMT), cell migration and invasion (Manou et al. 2020; Piperigkou et al. 2021). The miRNA-evoked control of the ECM has emerged as a novel mechanism of mediating matrix-dependent cancer cell processes, including cell proliferation, migration, adhesion, differentiation, apoptosis, and stem cell properties (Table 16.1) (Zolota et al. 2021; Piperigkou et al. 2018; Rutnam et al. 2013).

Opposing miRNA functions have been identified in breast cancer, as specific miRNAs may serve either as tumor suppressors or as oncogenes, depending on target-mRNA functions (Svoronos et al. 2016). MiR-21 and let-7 are among the most well-known breast cancer oncogenes and their plasma levels are crucial novel biomarkers for breast cancer monitoring (Khalighfard et al. 2018). Moreover, miR-155 is an oncogenic miRNA which has been mainly explored in leukemia (Ferrajoli et al. 2013); it also acts as an oncogene in breast cancer by targeting the suppressor of cytokine signaling 1 gene and promotes cell proliferation (Zhang et al. 2019). In breast carcinoma cells, miR-10b can directly suppress the translation (Ma et al. 2007). Moreover, the oncogenic axis of miR-10b involves the direct targeting of syndecan-1 to induce breast cancer cell migration and invasion (Piperigkou et al. 2016; Ibrahim et al. 2012). On the other hand, important tumor suppressors in breast cancer include miR-145(Wang et al. 2016), miR-205 (Chao et al. 2014), miR-335 (Heyn et al. 2011) and miR-372 (Zhao et al. 2017).

Targeting ECM and its cellular receptors through miRNAs is considered a novel mechanism to control matrix-dependent cellular processes that guide breast cancer progression. It is well established that miRNAs modulate the expression of matrix components, thus mediating cancer cell properties (Piperigkou and Karamanos 2019); however, there is a lack of evidence regarding the mechanistic aspects of

Table 16.1 MicroRNA-mediated regulation of extracellular matrix molecules in breast cancer

miRNA	ECM-related target	Cell line	Theranostic value	Reference
Breast cancer				
miR-10b	SDC-1	MDA-MB-231, MCF-7	– MiR-10b has a reported clinical correlation in preinvasive breast cancer – Emerging biomarker for BC diagnosis (Bertoli et al. 2015)	Ibrahim et al. (2012)
miR-140-3p	SDC-4	MDA-MB-231, SKBR3, MCF-7	Inverse relation between the impact of miR-140-3p and syndecan-4 on patient survival	Onyeisi et al. (2021)
miR-142-3p	ITGAV	MDA-MB-231, MDA-MB-468, MCF-7	Dysregulation of miR-142-3p in clinical samples of breast cancer and breast cancer cell lines compared to normal breast tissue	Schwickert et al. (2015)
miR-200b	LOX, SDC-2	MDA-MB-231, MCF-7	– Estrogen receptors mediate miR-200b expression in mammary cancer cells – Lower miR-200b levels have been correlated with poor disease-free survival (Piperigkou et al. 2020)	Sun et al. (2014)
miR-206	VEGF	MDA-MB-231, MCF-7	TNBC tissues express prominently lower levels of miR-206 compared to non-TNBC tissue samples and normal breast tissues	Liang et al. (2016)
ERα-positive				
miR-181a-5p	MMP14	MCF-7	miR-181a-5p is a critical regulator for MMP-14 expression and can affect MMP-14-mediated cancer cell migration, invasion, and angiogenesis	Li et al. (2015)
miR-221/222	ITGB4, ADAM17	MCF-7	– Integrin β4 expression was associated with poor primary luminal Lum-IC differentiation (G3 tumors) and with low miR-221/222 expression – An inverse correlation of miR-221/222 with Ki67 is also observed	Dentelli et al. (2014)
miR-373	CD44	MCF-7	Metastasis-promoting micro-RNA	Negrini and Calin (2008)
miR-520c	CD44	MCF-7	Metastasis-promoting micro-RNA	(Negrini and Calin (2008)

(continued)

Table 16.1 (continued)

miRNA	ECM-related target	Cell line	Theranostic value	Reference
ERα-negative				
miR-124	ITGB1	MDA-MB-231	MiR-124 is related to blockage of metastasis in ER- cell lines compared to ER+ cell line	Lv et al. (2011)
miR-1258	HPSE	SUM-149	MiR-1258 levels inversely correlate with metastatic ability of human breast cell lines and patient tissues	Zhang et al. (2011)
miR-143	CD44	SKBR3	MiR-143 functions as a tumor suppressor in breast cancer and that low miR-143 expression in breast cancer tissues may be an unfavorable prognostic factor	Yang et al. (2016)
miR-145	MMP11	HCC1937, MDA-MB-231	MiR-145 expression was downregulated in a graded manner in TNBC, with higher grades showing lowest expression	Tang et al. (2016)
miR-181a miR-193a/b	uPA	MDA-MB-231, MDA-MB-436	MiR-193a/b and miR-181a inhibit in vitro invasion of breast cancer cells mainly by downregulating uPA expression	Noh et al. (2011)
miR-193b	uPA	MDA-MB-231	MiR-193b is closely associated with clinical metastasis	Li et al. (2009)
miR-205	LamC1	MDA-MB-231, BT-549, SUM-149	MiR-205 is a novel transcriptional target of p53, and it exerts a role as oncosuppressor in triple-negative breast cancer, and in particular in a model representative of the most undifferentiated and mesenchymal subgroup	Piovan et al. (2012)
miR-205	VEGF-A	MDA-MB-231	MiR-205 is specifically downregulated in breast cancer	Wu et al. (2009)
miR-205	ITGA5	SUM-159	MiR-205 expression level is drastically lower in basal mesenchymal-like TNBC cells than other subtype breast cancer cells	Xiao et al. (2018)
miR-21	TIMP-3	MDA-MB-231	– Correlation of miR-21 with lymph node metastasis – Promising of therapy with	Song et al. (2010)

(continued)

Table 16.1 (continued)

miRNA	ECM-related target	Cell line	Theranostic value	Reference
			miRNA-based or non-miRNA treatments (Bertoli et al. 2015)	
miR-29b	ITGA6 (CD49f), LOX, VEGF-A	MDA-MB-231	MiR-29 expression correlates with more favorable outcomes, more differentiated phenotypes in normal and cancer cells and reduced metastatic potential	Chou et al. (2013)
miR-301	Col2A1	MDA-MB-231	Patients with higher miR-301 expression level have a worse DFS survival compared to those with lower miR-301 expression, while the relapses associated with miR-301 expression were nodal or distant metastases, not local recurrences	Shi et al. (2011)
miR-301a-3p from MDA-MB-231	TIMP-2 of astrocytes when they internalize EVs from BCC	MDA-MB-231	Higher levels of miR-301a-3p were significantly associated with reduced survival	Morad et al. (2020)
miR-30a-5p	ITGB3	MDA-MB-468, MDA-MB-231	– MiR-30a-5p expression strongly correlated with histological grade and survival status – High β3 integrin expression was markedly associated with reduced overall survival in TNBC patient subgroup	Li et al. (2016)
miR-31	ITGA2, ITGA5, ITGAV, ITGB3	MDA-MB-231	MiR-31 regulates cancer invasion-metastasis cascade	Augoff et al. (2011)
miR-335	TNC	MDA-MB-231	Expression of miR-126 and miR-335 is lost in the majority of primary breast tumors from patients who relapse, and the loss of expression of either microRNA is associated with poor distal metastasis-free survival TNC is involved in CSC immune evasion pathways	Tavazoie et al. (2008) and Graham et al. (2017)
miR-33a	ADAM9	MDA-MB-231	– MiR-33a expression is reduced in breast cancer tissues compared with matched normal tissues. – A correlation is observed between lower miR-33a	Zhang et al. (2015)

(continued)

Table 16.1 (continued)

miRNA	ECM-related target	Cell line	Theranostic value	Reference
			expression and increased lymph node metastasis	
miR-373	ITGA2	MDA-MB-468, MDA-MB-231	MiR-373^{high}/ITGA2^{low} may be a prognosis biomarker for breast cancer patients	Ding et al. (2015)
miR-4443	TIMP2	MDA-MB-231	MiR-4443 probably added to the chemoresistant capability of breast cancer cells by targeting TIMP2	Chen et al. (2016)
miR-506	CD151	MDA-MB-231	MiR-506 played a role as a master suppressor of EMT in breast cancer	Arora et al. (2013)
miR-520a-3p	CD44	Hs578T, BT20	MiR-520a-3p may play a suppressing role in breast cancer	Li et al. (2017)
miR-539	LamA4	BT-549	LAMA4 plays an important role in tumor progression and may be an important target in the treatment of TNBC	Yang et al. (2018)
miR-645	uPA	MDA-MB-231	uPA downregulation by miR-645 results in decrease of invasion in vitro and in vivo	Meng et al. (2018)
miR-9 from MDA-MB-468	ECM glycoprotein fibulin-3 EFEMP1 of normal fibroblasts	MDA-MB-468	– MiR-9 expression in TNBC correlates with different CAF subsets and resistance to cisplatin – EFEMP1 downmodulation is linked to the acquisition of a malignant phenotype in tumor-associated fibroblasts, which seems to be particularly relevant in TNBC subtype. – Emerging biomarker for BC diagnosis (Bertoli et al. 2015)	Cosentino et al. (2020)

Abbreviations: ADMA a disintegrin and metalloproteinase, EFEMP epidermal growth factor-containing fibulin-like extracellular matrix protein, HPSE heparanase, ITG integrin, Lam laminin, LOX lysyl oxidase, MMP matrix metalloproteinase, SDC syndecan, TIMP tissue inhibitor for matrix metalloproteinases, TNC tenascin C, uPA urokinase-type plasminogen activator, VEGF vascular endothelial growth factor

matrix-mediated miRNAs in breast cancer progression. Recent reports indicate that estrogen receptors (ERs) are responsible for the epigenetic targeting of miRNAs with established effects on breast cancer progression, such as miR-10b, miR-21, miR-145, and miR-200b in hormone- and non-hormone-dependent breast cancer cells (Piperigkou et al. 2017, 2020). In the aspect of matrix receptors, integrin β4 suppresses miR-92ab, miR-99ab, and miR-100, than in turn stimulate integrin β4-mediated cell motility in breast cancer (Gerson et al. 2012a). Moreover, circulating miRNAs loaded in tumor-derived exosomes or traveling within Argonaute (AGO) protein-positive ribonucleoprotein multivesicular bodies, functionally affect recipient cells, thus serving as critical biomarkers to control vital steps during breast cancer progression (Sun et al. 2018; Turchinovich et al. 2016).

This chapter focuses on the functional relationship among miRNAs and ECM biomolecules in the projection of breast cancer pathogenesis and development. Specifically, major examples of the well thought out effects of miRNAs on ECM components will be discussed. More focus is given to the functional relationship of matrix-mediated miRNAs in breast cancer, which is less elucidated.

16.2 ECM Regulates Breast Cancer Cell Progression Through miRNA Modulation

Altered miRNA expression in human cancers indicates that miRNAs contribute to tumor development (Chou and Werb 2012; Melo and Kalluri 2013). miRNAs are known regulators of matrix expression and through this pathway play an important role in cancer cell invasion and metastasis (Table 16.1). Recent reports indicate that ECM biomolecules have the ability to modify the expression and function of miRNAs (Fig. 16.1) (Bourguignon 2019). In the next sections, we review published data concerning the effect of ECM components on miRNA modulation in breast cancer cells. Investigating and understanding the mechanisms of interaction of the extracellular substance with miRNAs will allow the emergence of therapeutic targets related to these biological processes.

Most studies have focused on the effect of major ECM macromolecules, such as integrins, CD44, and hyaluronan (HA) on miRNA expression. An association between β4 expression and poor prognosis in "basal-like" breast cancer (Lu et al. 2008) has been proposed and the role of this integrin in breast cancer progression has been highlighted. First, Gerson et al. concluded that β4 integrin status affects the expression of specific families of miRNAs (Gerson et al. 2012a). In particular, through the performance of a novel miRNA microarray analysis (qNPA) they identified two miRNA families, miR-25/32/92abc/363/363-3p/367 and miR-99ab/100, that are downregulated by β4 integrin expression. Analysis of Affymetrix GeneChip data identified 54 common targets of miR-92ab and miR-99ab/100 within the subset of β4-regulated mRNAs, revealing several genes known to be key components of β4- regulated signaling cascades and effectors of cell motility. A novel function for the β4 integrin has been also reported in downregulating miR-29a expression which targets secreted protein acidic and rich in cysteine (SPARC), a glycoprotein that plays an important role in ECM remodeling and cell invasion. Thus, miR-29a repression by β4 integrin facilitated tumor invasion (Gerson et al. 2012b). The above results show that the integrin not only activates multiple

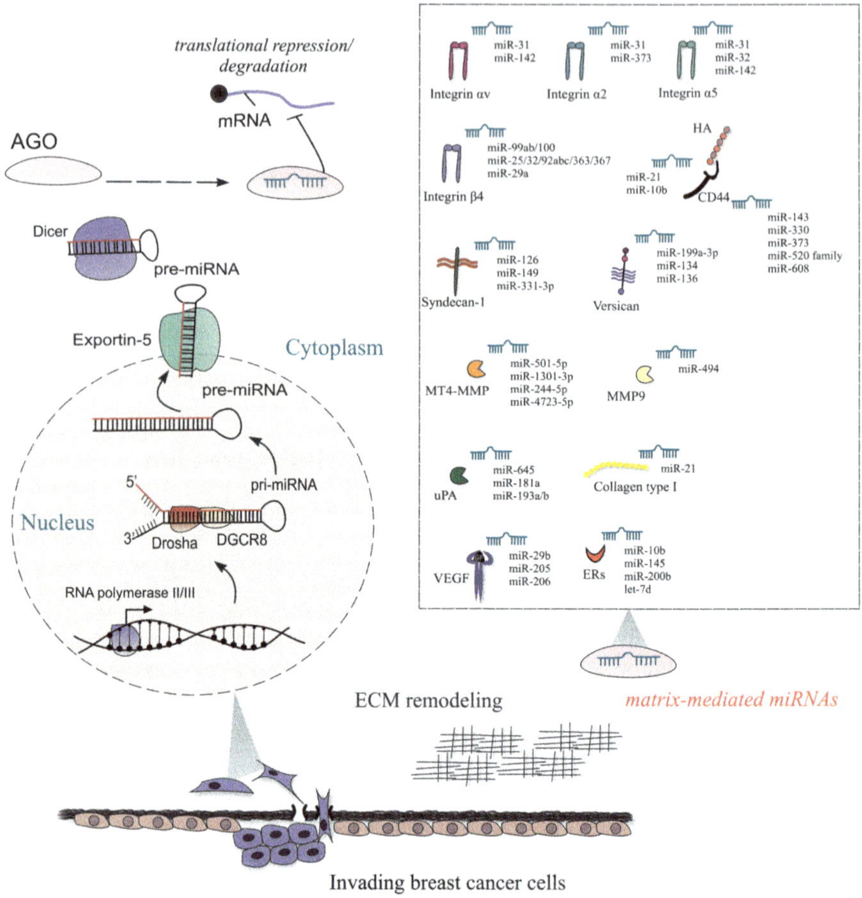

Fig. 16.1 MiRNA biogenesis and matrix regulation in breast cancer. The canonical miRNA biogenesis cascade produces mature miRNA sequences loaded to Argonaute (AGO) proteins to target the 3′ untranslated regions (UTR) of one or multiple mRNAs. Depending on target-mRNA functions, miRNAs act either as oncogenes or as tumor suppressors. During breast cancer progression, cancer cells-ECM affects the expression of certain miRNAs that in turn are involved in several steps of metastasis. Matrix-mediated miRNAs mediate cell-cell and cell-matrix interactions, thus regulating breast cancer cell behavior. Major ECM constituents, including integrins, estrogen receptors (ERs), syndecans, hyaluronan (HA)-CD44 interactions, matrix metalloproteinases (MMPs), urokinase-type plasminogen activator (uPA), growth factor receptors (GFRs), and collagen type I, regulate the expression of several miRNAs that are involved in breast cancer progression

downstream signaling pathways through phosphorylation, but it also has the ability to regulate specific miRNAs which can influence β4-mediated migration and invasion.

HA and its binding receptor, CD44, have been studied in cancer and have been associated with tumor survival, proliferation, and migration (Tavianatou et al. 2021). It has been reported that HA-CD44 interaction exerts its carcinogenic effect through miRNA signaling. Bourguignon et al. focused their research on detecting the mechanisms by which hyaluronic acid affects the expression miRNA in cancer and proposing novel drug targets for sensitizing tumor cell death. HA binding to CD44 promotes Nanog phosphorylation, a known stem cell marker, in the breast cancer cell line MCF-7. Phosphorylated Nanog through association with RNase III Drosha and RNA helicase p68 induces miR-21 production and program cell death 4 (PDCD4) reduction (Bourguignon et al. 2009). The final result of these events is inhibition of apoptosis, chemotherapy resistance, and blocking HA-CD44-mediated tumor cell behavior.

Moreover, it has been reported that HA binding to CD44 promotes c-Src kinase activation, which, in turn, increases Twist phosphorylation, leading to the nuclear translocation of Twist and transcriptional activation in MDA-MB-231 breast cancer cells (Bourguignon et al. 2010). Twist, which is one of c-Src substrates, is generally considered as a putative oncogene for its role in promoting a variety of tumor-related properties (e.g., EMT, invasion, and drug resistance). During HA/CD44 signaling, Twist induces miR-10b gene expression/production in tumor cells through interacting with miR-10 promoter (with E-box domain). This process results in the reduction of a tumor suppressor protein (HOXD10), RhoA/RhoC upregulation, Rho-kinase (ROK) activation, and breast tumor cell invasion. Src inhibitor, PP2, or Twist siRNA when applied to cancer cells significantly suspended miR-10 expression and downstream RhoGTPase (RhoC)-ROK effector functions (Bourguignon et al. 2010).

Therefore, it is suggested that the HA-CD44-mediated signaling events offer the potential for new therapeutic approaches that will target CD44 using anti-CD44 specific antibodies, inhibitors, anti-sense molecules, or HA-based nanoparticles containing therapeutic drugs.

With respect to the interaction with miRNA of additional single ECM macromolecules, recent data suggest that the metalloproteinase MT4-MMP (or MMP17) contributes in breast cancer growth and metastasis through regulation of miRNAs, including miR-501-5p, miR-1301-3p, miR-224-5p, and miR-4723-5p (Cervantes-Garduno et al. 2018). The targets of the most frequently altered (upregulated or downregulated) miRNAs affected several pathways that cooperate in hallmarks of cancer, including EMT and cell growth, and regulate the expression of transforming growth factor beta (TGF-β), mitogen activation protein kinase (MAPK), ErbB, and Wnt signaling, as well as several cell adhesion molecules.

ECM as a whole possesses biomechanical properties that affect cellular function (Manou et al. 2020). Composition and rigidity of ECM influences epithelial cell behavior and has recently been associated with miRNA regulation (Manou et al. 2020). Deposition of collagen type I promotes loss of epithelial polarity and tumor progression. Recently, studies using 3D cultures using Matrigel, which recapitulate the basement membrane matrix, explored a link between microRNAs and collagen type I in MCF-7 breast cancer cells (Li et al. 2011). Treatment of MCF-7 cells with collagen type I disrupted acini and upregulated the expression of miR-21, a well-documented oncogenic miRNA, via a post-transcriptional mechanism. The same group in two subsequent papers investigated the expression and function of miRNAs in Matrigel 3D culture using tumor cells (Li et al. 2012; Nguyen et al. 2012). Comparing the expression of miRNAs in rBM 3D and 2D cultures of MCF-7 and MDA-MB231 cells, a profound difference has been revealed in miRNA profiles between the two cultures within each cell type, which also correlated with distinct mass morphogenesis. Collagen type I in rBM 3D culture substantially altered the miRNA signature of mass morphogenesis of MCF-7 cells. Overexpression of miR-200 family members, which are potent suppressors of EMT, in MCF-7 cells, correlated with strong cell-cell adhesion. In contrast, the silenced expression of miR-200 family members in MDA-MB-231 cells correlated with the stellate, mesenchymal, and highly invasive morphology (Piperigkou et al. 2017, 2020). Recently, the first miRNA signatures of morphogenesis of human breast cancer cells have been provided in rBM 3D culture and warrant further utilization of rBM 3-D culture in the investigation of miRNAs in breast cancer (Nguyen et al. 2012).

Using a similar approach identified significant differential expression of miRNAs between cancer cells cultured in Matrigel and those in 2D cultures (Price et al. 2012).

A common Matrigel-induced miRNA signature comprised of upregulated miR-1290 and miR-210 and downregulated miR-29b and miR-32 was identified using RT-qPCR across five epithelial cancer cell lines (SW480, SW620, HT-29, A549, and MDA-MB-231). Integrin α5 was identified as a novel putative target of miR-32 that may facilitate cancer cell interactions with the ECM. The authors proposed Matrigel culture as a valuable approach to the in vitro study of miRNAs.

All these results highlight that dysregulated ECM deposition quality may affect miRNA expression and 3D culture is a principal investigational approach in this field. It has been shown that the gene expression signature of 3D culture of breast cancer cells has great prognostic value for breast cancer patients (Martin et al. 2008). However, the molecular mechanism by which 3D culture affects miRNA expression has not yet been unravelled. Matrigel has a heterogeneous composition, and it should be important to establish whether miRNA expression signatures in 3D vs. 2D cultures are affected exclusively by the stiffness of Matrigel or also by its components.

All previous studies partially explain how ECM components interact with miRNA and promote tumor progression (Fig. 16.1). However, further research is necessary to highlight the mechanisms by which other ECM molecules regulate miRNA biogenesis.

It is a common knowledge that the 3′ UTR contains both binding sites for regulatory proteins as well as miRNAs. It has been proposed that the 3′ UTR of ECM molecule mRNAs, containing miRNA-binding sites, can regulate miRNA levels and function (Lee et al. 2010, 2011; Jeyapalan et al. 2011). Data suggest that 3′ UTRs of mRNAs act as 'endogenous decoys' for miRNAs and subsequently inhibit miRNA functions, by blocking target mRNA suppression (Salmena et al. 2011). This function has been extended to versican, CD44, and nephronectin mRNAs. Lee et al. showed that expression of versican 3′ UTR induced cell, tissue, and organ adhesion through competing miR-199a-3p levels and function (Lee et al. 2010). They demonstrated that the overexpression of versican 3′ UTR interacted with endogenous miR-199a-3p, deliberating versican and fibronectin mRNA for translation. Overexpression of versican 3′ UTR also reduced breast cancer cell growth by decreasing miR-199a-3p, miR-134, and miR-136 levels and therefore restoring phosphatase and tensin homolog (PTEN) function (Lee et al. 2010).

Furthermore, research demonstrated that the 3′ UTR of CD44 interacts with miR-216a, miR-330, and miR-608 and inhibits tumor growth in a human breast cancer cell line, partly through modulation of angiogenesis (Jeyapalan et al. 2011). Moreover, it has been shown that the overexpression of the CD44 3′ UTR results in enhanced cell motility, invasion, and cell adhesion in human breast carcinoma cell line MDA-MB-231 cells, as well as metastasis in vivo. Using computational analysis, it has been indicated that miRNAs that interact with CD44 3′ UTR also have binding sites in other matrix-encoding mRNA 3′ UTRs, including collagen type I $\alpha 1$ repressed by miR-328 and fibronectin type 1 repressed by miR-512-3p, miR-491, and miR-671 (Rutnam and Yang 2012). Expression of CD44, collagen type I $\alpha 1$, and fibronectin type 1 was synergistically upregulated in vitro and in vivo upon transfection of the CD44 3′ UTR (Rutnam and Yang 2012).

Finally, it was shown that overexpression of the 3′ UTR of nephronectin through binding to miR-378 stimulates cell differentiation in the osteoblast progenitor cells MC3T3-E1. 3′ UTR-transfected cells had enhanced levels of β-catenin and GSK3 which contributed to cell differentiation, through minimizing EGFR and ERK phosphorylation (Lee et al. 2011; Kahai et al. 2010).

Evidence from recent research data suggests that a reciprocal and harmonious collaboration is taking place between miRNAs and the ECM that facilitates the balance of the cell with its environment.

As more and more mechanisms and actions come to light concerning ECM and miRNA interplay, a great challenge is to detect which constituents are most important for breast cancer cell survival and metastasis among them. A better understanding of the role and function of ECM-dependent regulation of specific miRNAs may lead to new treatment design with targeted therapies.

16.3 Conclusions and Perspectives

The miRNA biogenesis machinery is dysregulated in human cancer through complex mechanisms, such as abnormal transcriptional control of miRNA genes and defective epigenetic changes. Depending on the biological function of target mRNAs, miRNAs serve either as a novel class of oncogenes or tumor suppressors; thus, they directly orchestrate multifaceted signaling factors that in turn regulate breast cancer phenotype and progression. Focusing on the endogenous processing machinery by targeting the ECM-evoked miRNAs in breast cancer may help to better control cancer cell behavior that benefits patient management.

Acknowledgements ZP, ADT, and NKK acknowledge support by the Action for the Strategic Development on the Research and Technological Sector (MIS5045810 and MIS5033644), funded by the Operational Programme “Competitiveness, Entrepreneurship and Innovation” (NSRF

2014-2020) and co-financed by Greece and the European Union (European Regional Development Fund).

References

Arora H, Qureshi R, Park WY (2013) miR-506 regulates epithelial mesenchymal transition in breast cancer cell lines. PLoS One 8:e64273

Augoff K, Das M, Bialkowska K, Mccue B, Plow EF, Sossey-Alaoui K (2011) miR-31 is a broad regulator of β1-integrin expression and function in cancer cells. Mol Cancer Res 9:1500–1508

Bertoli G, Cava C, Castiglioni I (2015) MicroRNAs: new biomarkers for diagnosis, prognosis, therapy prediction and therapeutic tools for breast cancer. Theranostics 5:1122–1143

Bourguignon LYW (2019) Matrix hyaluronan-CD44 interaction activates MicroRNA and LncRNA signaling associated with chemoresistance, invasion, and tumor progression. Front Oncol 9:492

Bourguignon LY, Spevak CC, Wong G, Xia W, Gilad E (2009) Hyaluronan-CD44 interaction with protein kinase C(epsilon) promotes oncogenic signaling by the stem cell marker Nanog and the production of microRNA-21, leading to down-regulation of the tumor suppressor protein PDCD4, anti-apoptosis, and chemotherapy resistance in breast tumor cells. J Biol Chem 284: 26533–26546

Bourguignon LY, Wong G, Earle C, Krueger K, Spevak CC (2010) Hyaluronan-CD44 interaction promotes c-Src-mediated twist signaling, microRNA-10b expression, and RhoA/RhoC up-regulation, leading to Rho-kinase-associated cytoskeleton activation and breast tumor cell invasion. J Biol Chem 285:36721–36735

Cervantes-Garduno A, Zampedri C, Espinosa M, Maldonado V, Melendez-Zajgla J, Ceballos-Cancino G (2018) MT4-MMP modulates the expression of miRNAs in breast cancer cells. Arch Med Res 49:471–478

Chao CH, Chang CC, Wu MJ, Ko HW, Wang D, Hung MC, Yang JY, Chang CJ (2014) MicroRNA-205 signaling regulates mammary stem cell fate and tumorigenesis. J Clin Invest 124:3093–3106

Chen X, Zhong S-L, Lu P, Wang D-D, Zhou S-Y, Yang S-J, Shen H-Y, Zhang L, Zhang X-H, Zhao J-H, Tang J-H (2016) miR-4443 participates in the malignancy of breast cancer. PLoS One 11: e0160780

Chendrimada TP, Gregory RI, Kumaraswamy E, Norman J, Cooch N, Nishikura K, Shiekhattar R (2005) TRBP recruits the Dicer complex to Ago2 for microRNA processing and gene silencing. Nature 436:740–744

Chou J, Werb Z (2012) MicroRNAs play a big role in regulating ovarian cancer-associated fibroblasts and the tumor microenvironment. Cancer Discov 2:1078–1080

Chou J, Lin JH, Brenot A, Kim J-W, Provot S, Werb Z (2013) GATA3 suppresses metastasis and modulates the tumour microenvironment by regulating microRNA-29b expression. Nat Cell Biol 15:201–213

Cosentino G, Romero-Cordoba S, Plantamura I, Cataldo A (2020) miR-9-mediated inhibition of EFEMP1 contributes to the acquisition of pro-tumoral properties in normal fibroblasts. Cells 9: 2143

Cui M, Wang H, Yao X, Zhang D, Xie Y, Cui R, Zhang X (2019) Circulating microRNAs in cancer: potential and challenge. Front Genet 10:626

De Rie D, Abugessaisa I, Alam T, Arner E, Arner P, Ashoor H, Astrom G, Babina M, Bertin N, Burroughs AM, Carlisle AJ, Daub CO, Detmar M, Deviatiiarov R, Fort A, Gebhard C, Goldowitz D, Guhl S, Ha TJ, Harshbarger J, Hasegawa A, Hashimoto K, Herlyn M, Heutink P, Hitchens KJ, Hon CC, Huang E, Ishizu Y, Kai C, Kasukawa T, Klinken P, Lassmann T, Lecellier CH, Lee W, Lizio M, Makeev V, Mathelier A, Medvedeva YA, Mejhert N, Mungall CJ, Noma S, Ohshima M, Okada-Hatakeyama M, Persson H, Rizzu P,

Roudnicky F, Saetrom P, Sato H, Severin J, Shin JW, Swoboda RK, Tarui H, Toyoda H, Vitting-Seerup K, Winteringham L, Yamaguchi Y, Yasuzawa K, Yoneda M, Yumoto N, Zabierowski S, Zhang PG, Wells CA, Summers KM, Kawaji H, Sandelin A, Rehli M, Consortium F, Hayashizaki Y, Carninci P, Forrest ARR, De Hoon MJL (2017) An integrated expression atlas of miRNAs and their promoters in human and mouse. Nat Biotechnol 35:872–878

Dentelli P, Traversa M, Rosso A, Togliatto G, Olgasi C, Marchiò C, Provero P, Lembo A, Bon G, Annaratone L, Sapino A, Falcioni R, Brizzi MF (2014) miR-221/222 control luminal breast cancer tumor progression by regulating different targets. Cell Cycle 13:1811–1826

Ding W, Fan XL, Xu X, Huang JZ, Xu SH, Geng Q, Li R, Chen D, Yan GR (2015) Epigenetic silencing of ITGA2 by MiR-373 promotes cell migration in breast cancer. PLoS One 10: e0135128

Ferrajoli A, Shanafelt TD, Ivan C, Shimizu M, Rabe KG, Nouraee N, Ikuo M, Ghosh AK, Lerner S, Rassenti LZ, Xiao L, Hu J, Reuben JM, Calin S, You MJ, Manning JT, Wierda WG, Estrov Z, O'brien S, Kipps TJ, Keating MJ, Kay NE, Calin GA (2013) Prognostic value of miR-155 in individuals with monoclonal B-cell lymphocytosis and patients with B chronic lymphocytic leukemia. Blood 122:1891–1899

Gerson KD, Maddula VS, Seligmann BE, Shearstone JR, Khan A, Mercurio AM (2012a) Effects of beta4 integrin expression on microRNA patterns in breast cancer. Biol Open 1:658–666

Gerson KD, Shearstone JR, Maddula V, Seligmann BE, Mercurio AM (2012b) Integrin beta4 regulates SPARC protein to promote invasion. J Biol Chem 287:9835–9844

Graham ÉA, Mallet JF, Jambi M, Nishioka H, Homma K, Matar C (2017) MicroRNA signature in the chemoprevention of functionally-enriched stem and progenitor pools (FESPP) by active hexose correlated compound (AHCC). Cancer Biol Ther 18:765–774

Heyn H, Engelmann M, Schreek S, Ahrens P, Lehmann U, Kreipe H, Schlegelberger B, Beger C (2011) MicroRNA miR-335 is crucial for the BRCA1 regulatory cascade in breast cancer development. Int J Cancer 129:2797–2806

Ibrahim SA, Yip GW, Stock C, Pan JW, Neubauer C, Poeter M, Pupjalis D, Koo CY, Kelsch R, Schule R, Rescher U, Kiesel L, Gotte M (2012) Targeting of syndecan-1 by microRNA miR-10b promotes breast cancer cell motility and invasiveness via a Rho-GTPase- and E-cadherin-dependent mechanism. Int J Cancer 131:E884–E896

Jeyapalan Z, Deng Z, Shatseva T, Fang L, He C, Yang BB (2011) Expression of CD44 3′-untranslated region regulates endogenous microRNA functions in tumorigenesis and angiogenesis. Nucleic Acids Res 39:3026–3041

Kahai S, Lee SC, Seth A, Yang BB (2010) Nephronectin promotes osteoblast differentiation via the epidermal growth factor-like repeats. FEBS Lett 584:233–238

Karamanos NK, Piperigkou Z, Theocharis AD, Watanabe H, Franchi M, Baud S, Brezillon S, Gotte M, Passi A, Vigetti D, Ricard-Blum S, Sanderson RD, Neill T, Iozzo RV (2018) Proteoglycan chemical diversity drives multifunctional cell regulation and therapeutics. Chem Rev 118:9152–9232

Karamanos NK, Theocharis AD, Piperigkou Z, Manou D, Passi A, Skandalis SS, Vynios DH, Orian-Rousseau V, Ricard-Blum S, Schmelzer CEH, Duca L, Durbee M, Afratis NA, Troeberg L, Franchi M, Masola V, Onisto M (2021) A guide to the composition and functions of the extracellular matrix. FEBS J 288:6850–6912

Khalighfard S, Alizadeh AM, Irani S, Omranipour R (2018) Plasma miR-21, miR-155, miR-10b, and let-7a as the potential biomarkers for the monitoring of breast cancer patients. Sci Rep 8: 17981

Lee DY, Jeyapalan Z, Fang L, Yang J, Zhang Y, Yee AY, Li M, Du WW, Shatseva T, Yang BB (2010) Expression of versican 3′-untranslated region modulates endogenous microRNA functions. PLoS One 5:e13599

Lee SC, Fang L, Wang CH, Kahai S, Deng Z, Yang BB (2011) A non-coding transcript of nephronectin promotes osteoblast differentiation by modulating microRNA functions. FEBS Lett 585:2610–2616

Li XF, Yan PJ, Shao ZM (2009) Downregulation of miR-193b contributes to enhance urokinase-type plasminogen activator (uPA) expression and tumor progression and invasion in human breast cancer. Oncogene 28:3937–3948

Li C, Nguyen HT, Zhuang Y, Lin Y, Flemington EK, Guo W, Guenther J, Burow ME, Morris GF, Sullivan D, Shan B (2011) Post-transcriptional up-regulation of miR-21 by type I collagen. Mol Carcinog 50:563–570

Li C, Nguyen HT, Zhuang Y, Lin Z, Flemington EK, Zhuo Y, Kantrow SP, Morris GF, Sullivan DE, Shan B (2012) Comparative profiling of miRNA expression of lung adenocarcinoma cells in two-dimensional and three-dimensional cultures. Gene 511:143–150

Li Y, Kuscu C, Banach A, Zhang Q, Pulkoski-Gross A, Kim D, Liu J, Roth E, Li E, Shroyer KR, Denoya PI, Zhu X, Chen L, Cao J (2015) miR-181a-5p inhibits cancer cell migration and angiogenesis via downregulation of matrix metalloproteinase-14. Cancer Res 75:2674–2685

Li W, Liu C, Zhao C, Zhai L, Lv S (2016) Downregulation of β3 integrin by miR-30a-5p modulates cell adhesion and invasion by interrupting Erk/Ets-1 network in triple-negative breast cancer. Int J Oncol 48:1155–1164

Li J, Wei J, Mei Z, Yin Y, Li Y, Lu M, Jin S (2017) Suppressing role of miR-520a-3p in breast cancer through CCND1 and CD44. Am J Transl Res 9:146–154

Liang Z, Bian X, Shim H (2016) Downregulation of microRNA-206 promotes invasion and angiogenesis of triple negative breast cancer. Biochem Biophys Res Commun 477:461–466

Lu S, Simin K, Khan A, Mercurio AM (2008) Analysis of integrin beta4 expression in human breast cancer: association with basal-like tumors and prognostic significance. Clin Cancer Res 14: 1050–1058

Lv X-B, Jiao Y, Qing Y, Hu H, Cui X, Lin T, Song E, Yu F (2011) miR-124 suppresses multiple steps of breast cancer metastasis by targeting a cohort of pro-metastatic genes in vitro. Chin J Cancer 30:821–830

Ma L, Teruya-Feldstein J, Weinberg RA (2007) Tumour invasion and metastasis initiated by microRNA-10b in breast cancer. Nature 449:682–688

Manou D, Karamanos NK, Theocharis AD (2020) Tumorigenic functions of serglycin: regulatory roles in epithelial to mesenchymal transition and oncogenic signaling. Semin Cancer Biol 62: 108–115

Martin KJ, Patrick DR, Bissell MJ, Fournier MV (2008) Prognostic breast cancer signature identified from 3D culture model accurately predicts clinical outcome across independent datasets. PLoS One 3:e2994

Melo SA, Kalluri R (2013) miR-29b moulds the tumour microenvironment to repress metastasis. Nat Cell Biol 15:139–140

Meng D, Lei M, Han Y, Zhao D, Zhang X, Yang Y, Liu R (2018) MicroRNA-645 targets urokinase plasminogen activator and decreases the invasive growth of MDA-MB-231 triple-negative breast cancer cells. Onco Targets Ther 11:7733–7743

Morad G, Daisy CC, Otu HH, Libermann TA, Dillon ST, Moses MA (2020) Cdc42-dependent transfer of mir301 from breast cancer-derived extracellular vesicles regulates the matrix modulating ability of astrocytes at the blood-brain barrier. Int J Mol Sci 21:3851

Negrini M, Calin GA (2008) Breast cancer metastasis: a microRNA story. Breast Cancer Res 10: 303

Nguyen HT, Li C, Lin Z, Zhuang Y, Flemington EK, Burow ME, Lin YI, Shan B (2012) The microRNA expression associated with morphogenesis of breast cancer cells in three-dimensional organotypic culture. Oncol Rep 28:117–126

Noh H, Hong S, Dong Z, Pan ZK, Jing Q, Huang S (2011) Impaired microRNA processing facilitates breast cancer cell invasion by upregulating urokinase-type plasminogen activator expression. Genes Cancer 2:140–150

Onyeisi JOS, Greve B, Espinoza-Sánchez NA, Kiesel L, Lopes CC, Götte M (2021) microRNA-140-3p modulates invasiveness, motility, and extracellular matrix adhesion of breast cancer cells by targeting syndecan-4. J Cell Biochem 122:1491–1505

Piovan C, Palmieri D, Di Leva G, Braccioli L, Casalini P, Nuovo G, Tortoreto M, Sasso M, Plantamura I, Triulzi T, Taccioli C, Tagliabue E, Iorio MV, Croce CM (2012) Oncosuppressive role of p53-induced miR-205 in triple negative breast cancer. Mol Oncol 6:458–472

Piperigkou Z, Karamanos NK (2019) Dynamic interplay between miRNAs and the extracellular matrix influences the tumor microenvironment. Trends Biochem Sci 44:1076–1088

Piperigkou Z, Mohr B, Karamanos N, Gotte M (2016) Shed proteoglycans in tumor stroma. Cell Tissue Res 365:643–655

Piperigkou Z, Franchi M, Gotte M, Karamanos NK (2017) Estrogen receptor beta as epigenetic mediator of miR-10b and miR-145 in mammary cancer. Matrix Biol 64:94–111

Piperigkou Z, Gotte M, Theocharis AD, Karamanos NK (2018) Insights into the key roles of epigenetics in matrix macromolecules-associated wound healing. Adv Drug Deliv Rev 129:16–36

Piperigkou Z, Franchi M, Riethmüller C, Götte M, Karamanos NK (2020) miR-200b restrains EMT and aggressiveness and regulates matrix composition depending on ER status and signaling in mammary cancer. Matrix Biol Plus 6-7:100024

Piperigkou Z, Kyriakopoulou K, Koutsakis C, Mastronikolis S, Karamanos NK (2021) Key matrix remodeling enzymes: functions and targeting in cancer. Cancers (Basel) 13:1441

Price KJ, Tsykin A, Giles KM, Sladic RT, Epis MR, Ganss R, Goodall GJ, Leedman PJ (2012) Matrigel basement membrane matrix influences expression of microRNAs in cancer cell lines. Biochem Biophys Res Commun 427:343–348

Rutnam ZJ, Yang BB (2012) The non-coding 3′ UTR of CD44 induces metastasis by regulating extracellular matrix functions. J Cell Sci 125:2075–2085

Rutnam ZJ, Wight TN, Yang BB (2013) miRNAs regulate expression and function of extracellular matrix molecules. Matrix Biol 32:74–85

Salmena L, Poliseno L, Tay Y, Kats L, Pandolfi PP (2011) A ceRNA hypothesis: the Rosetta Stone of a hidden RNA language? Cell 146:353–358

Schwickert A, Weghake E, Brüggemann K, Engbers A, Brinkmann BF, Kemper B, Seggewiß J, Stock C, Ebnet K, Kiesel L, Riethmüller C, Götte M (2015) microRNA miR-142-3p inhibits breast cancer cell invasiveness by synchronous targeting of WASL, integrin alpha V, and additional cytoskeletal elements. PLoS One 10:e0143993

Shi W, Gerster K, Alajez NM, Tsang J, Waldron L, Pintilie M, Hui AB, Sykes J, P'ng C, Miller N, Mccready D, Fyles A, Liu F-F (2011) MicroRNA-301 mediates proliferation and invasion in human breast cancer. Cancer Res 71:2926–2937

Song B, Wang C, Liu J, Wang X, Lv L, Wei L, Xie L, Zheng Y, Song X (2010) MicroRNA-21 regulates breast cancer invasion partly by targeting tissue inhibitor of metalloproteinase 3 expression. J Exp Clin Cancer Res 29:29

Sun M, Gomes S, Chen P, Frankenberger CA, Sankarasharma D, Chung CH, Chada KK, Rosner MR (2014) RKIP and HMGA2 regulate breast tumor survival and metastasis through lysyl oxidase and syndecan-2. Oncogene 33:3528–3537

Sun Z, Shi K, Yang S, Liu J, Zhou Q, Wang G, Song J, Li Z, Zhang Z, Yuan W (2018) Effect of exosomal miRNA on cancer biology and clinical applications. Mol Cancer 17:147

Svoronos AA, Engelman DM, Slack FJ (2016) OncomiR or tumor suppressor? The duplicity of microRNAs in cancer. Cancer Res 76:3666–3670

Tang L, Wei D, Yan F (2016) MicroRNA-145 functions as a tumor suppressor by targeting matrix metalloproteinase 11 and Rab GTPase family 27a in triple-negative breast cancer. Cancer Gene Ther 23:258–265

Tavazoie SF, Alarcón C, Oskarsson T, Padua D, Wang Q, Bos PD, Gerald WL, Massagué J (2008) Endogenous human microRNAs that suppress breast cancer metastasis. Nature 451:147–152

Tavianatou AG, Piperigkou Z, Koutsakis C, Barbera C, Beninatto R, Franchi M, Karamanos NK (2021) The action of hyaluronan in functional properties, morphology and expression of matrix effectors in mammary cancer cells depends on its molecular size. FEBS J 288:4291–4310

Turchinovich A, Tonevitsky AG, Burwinkel B (2016) Extracellular miRNA: a collision of two paradigms. Trends Biochem Sci 41:883–892

Wang H, Hang C, Ou XL, Nie JS, Ding YT, Xue SG, Gao H, Zhu JX (2016) MiR-145 functions as a tumor suppressor via regulating angiopoietin-2 in pancreatic cancer cells. Cancer Cell Int 16:65

Wu H, Zhu S, Mo Y-Y (2009) Suppression of cell growth and invasion by miR-205 in breast cancer. Cell Res 19:439–448

Xiao Y, Li Y, Tao H, Humphries B, Li A, Jiang Y, Yang C, Luo R, Wang Z (2018) Integrin α5 down-regulation by miR-205 suppresses triple negative breast cancer stemness and metastasis by inhibiting the Src/Vav2/Rac1 pathway. Cancer Lett 433:199–209

Yang Z, Chen D, Nie J, Zhou S, Wang J, Tang Q, Yang X (2016) MicroRNA-143 targets CD44 to inhibit breast cancer progression and stem cell-like properties. Mol Med Rep 13:5193–5199

Yang Z-X, Zhang B, Wei J, Jiang G-Q, Wu Y-L, Leng B-J, Xing C-G (2018) MiR-539 inhibits proliferation and migration of triple-negative breast cancer cells by down-regulating LAMA4 expression. Cancer Cell Int 18:16–16

Zhang L, Sullivan PS, Goodman JC, Gunaratne PH, Marchetti D (2011) MicroRNA-1258 suppresses breast cancer brain metastasis by targeting heparanase. Cancer Res 71:645–654

Zhang C, Zhang Y, Ding W, Lin Y, Huang Z, Luo Q (2015) MiR-33a suppresses breast cancer cell proliferation and metastasis by targeting ADAM9 and ROS1. Protein Cell 6:881–889

Zhang L, Chen T, Yan L, Xu H, Wang Y, Li Y, Wang H, Chen S, Wang W, Chen C, Yang Q (2019) MiR-155-3p acts as a tumor suppressor and reverses paclitaxel resistance via negative regulation of MYD88 in human breast cancer. Gene 700:85–95

Zhao YX, Liu HC, Ying WY, Wang CY, Yu YJ, Sun WJ, Liu JF (2017) microRNA372 inhibits proliferation and induces apoptosis in human breast cancer cells by directly targeting E2F1. Mol Med Rep 16:8069–8075

Zolota V, Tzelepi V, Piperigkou Z, Kourea H, Papakonstantinou E, Argentou M, I. & Karamanos, N. K. (2021) Epigenetic alterations in triple-negative breast cancer-the critical role of extracellular matrix. Cancers (Basel) 13:713

Chapter 17
The Impact of the Extracellular Matrix on Immunotherapy Success

Manglio M. Rizzo, Mariel A. Fusco, and Mariana Malvicini

Abstract Biological therapies to promote and enhance the immune response against cancer have gained much relevance in the past recent decade. Since the first approval of immune checkpoints inhibitor immunotherapies by the American Food and Drugs Administration (FDA), the outcomes obtained in patients with a variety of solid tumors have stimulated a revolution regarding this type of target therapy. Therefore, therapeutic options and developments aimed at enhancing the immune response in patients are currently increasing. The evasion mechanisms mediated both by tumors and by the tumor microenvironment (TME), however, remain important barriers to acquiring a satisfactory antitumor response. The dynamic extracellular matrix (ECM) as a part of the TME plays a key role in this scenario. In this chapter, we addressed how the interplay between cancer cells and the TME components, including soluble factors, endothelial cells, tumor-associated macrophages (TAMs), T cells, cancer-associated fibroblasts, and a perturbed ECM, could contribute to cancer progression. Importantly, how the TME remodeling, especially ECM changes, are implicated in the success of cancer immunotherapy.

17.1 Introduction

Conventional strategies for advanced cancer patients with unresectable tumor include chemotherapy and radiation. Developing novel strategies such as antiangiogenic or immunotherapy, on the other hand, consists of targeted therapies. Even with the wide range of therapeutic options, the improvement of clinical response and/or survival of patients remains limited. Currently, there is a vast experience regarding the application of immunotherapy in tumors, and it is widely known that immune cells are involved in the control of cancer. During the last two

M. M. Rizzo · M. A. Fusco · M. Malvicini (✉)
Cancer Immunobiology Laboratory, Instituto de Investigaciones en Medicina Traslacional (IIMT, Universidad Austral-CONICET), Pilar, Buenos Aires, Argentina
e-mail: mmalvicini@austral.edu.ar

I. Kovalszky et al. (eds.), *The Extracellular Matrix and the Tumor Microenvironment*, Biology of Extracellular Matrix 11,
https://doi.org/10.1007/978-3-030-99708-3_17

decades, a growing area of research focused on the novel strategies, acting in many phases of the immune response regulating the innate and the adaptive immunity. In addition, many studies indicate that conventional chemotherapy and/or radiotherapy could act in synergy to generate immunity against many tumors. However, many immune evasion mechanisms and events, such as cancer cells' uncontrolled growth, resistance to apoptosis, or metabolic shift toward anaerobic glycolysis, trigger an adjustment of the extracellular matrix (ECM), modifying the response from stromal and immune cells and impacting on the efficacy of immunotherapies. Then, ECM composition and dysregulation become key drivers for both cancer progression and response to antitumor therapies. In this chapter, we will discuss how the ECM remodeling influences the immune response against cancer and how it can be harnessed to achieve the benefit of immunotherapy for eradicating tumors.

17.2 Extracellular Matrix Remodeling and Immune Tumor Microenvironment

17.2.1 ECM Components

The ECM is a dynamic network of macromolecules such as fibrillary collagen, elastin, glycosaminoglycans (GAGs), proteoglycans (PGs), and enzymes with essential support and biochemical function for cells and tissues. These extracellular macromolecules, produced mostly by fibroblast, are in constant and highly controlled remodeling. Abnormal changes or dysregulation in the conformation of ECM components could alter chemical and physical ECM properties and promote tumor progression (Bonnans et al. 2014).

1. *Fibroblasts and Cancer-associated Fibroblasts (CAFs):* Fibroblasts synthesize collagen and other fibrillar and non-fibrillar factors maintaining the ECM homeostasis. They produce the network to support tissue structures and integrity. Fibroblasts can acquire an immune regulatory phenotype and in cancer could play an important role as predominant cell type from the tumor stroma (Pankova et al. 2016). This subpopulation widely identified as cancer-associated fibroblasts (CAFs) are present in many types of solid tumors. CAFs are recruited from resident fibroblasts, bone marrow-derived progenitor cells, or by cancer cells (McDonald et al. 2015). They are activated in the TME, by factors such as the transforming growth factor-beta (TGF-β), secrete by tumor to become myofibroblasts (Calon et al. 2014). In addition to cancer cells, inflammatory and immune cells also can produce mediators to activate fibroblasts. CAF interactions within tumors are complex, and it is depending on specific tumor types and the cells of their TME but generally, CAFs and TME cells interrelate contributing to tumor growth (Takahashi et al. 2017). For instance CAFs isolated from breast cancer triple-negative tumors expressed the monocyte attractant chemokine CXCL16, stimulated by myeloid suppressor cells to, in turns, recruits

more myeloid cells and more fibroblasts (Allaoui et al. 2016). Myeloid-derived suppressor cells (MDSCs), as we will describe later, are an immunosuppressive population that impairs T-cell response. MDSCs and CAFs accumulation then promotes stroma activation, thickening of ECM, immune evasion, and cell metastasis, and is associated with poor prognosis in this aggressive phenotype of breast tumor (Takai et al. 2016). The accumulation of CAFs was also associated with the numbers of CD163-positive tumor-associated macrophages (TAMs) in oral squamous cell carcinoma (OSCC). For instance CAFs release IL6, CXCL8, and TGF-β determining M2 TAM polarization, influencing the TME to an immunosuppressive milieu and inducing protumoral phenotype of TAMs in OSCC (Takahashi et al. 2017).

2. *Collagens:* Type I fibrillar collagen as well as other types of collagens are secreted by fibroblasts to form macromolecules by cross-linking, conforming the most percentage of biomolecules of the ECM (Netti et al. 2000). The cross-linked collagen provides ECM flexibility but also complexity. On the one hand, type I collagen acts as a regulator of matrix stiffening, but also is implicated in growth factors and cytokines bioavailability across the ECM, impairing tumor immunity (Netti et al. 2000). It has been reported that aged fibroblast-derived collagen promoted biophysical changes in the ECM, inhibiting the motility of T cells and promoting the accumulation of myeloid cells, which results in an overall change in the immune profile through the ECM (Kaur et al. 2019). These events could have implications, for example, in the efficacy of current therapies for melanoma. In human breast cancer, collagen deposition and stiffness of ECM correlate with tumor invasion. In addition, it was observed that the number of infiltrating TGF-β-producing TAMs was highest in more aggressive tumor subtypes (Maller et al. 2021). Also, TGF-β secreted by TAMs activates collagen crosslinking, which is implicated, as we mentioned, in elasticity and complexity of the ECM and finally, in immune cells trafficking.
3. *Glycosaminoglycans:* the most abundant glycosaminoglycan (GAG) in the ECM is hyaluronan (HA) (Chanmee et al. 2016). HA is a high-molecular-weight GAG, synthesized by cell membrane hyaluronan synthases 1, 2, and 3 (HAS1–3), by alternative addition of glucuronic acid and N-acetylglucosamine. HA is involved in the elasticity and viscosity of the ECM because acts buffering ion exchanges, water, and also maintaining the osmotic balance. Then, HA also plays a key role in cell mobility, invasion, proliferation, and inflammation (Chanmee et al. 2016; Nikitovic et al. 2015).

   Heparan-, chondroitin-, and keratin-sulfate are other GAGs with viscoelastic effects on ECM. The addition of a sulfated site in this group allows GAGs to bind to proteins by covalent bonds. For example, chondroitin sulfate can bind to matrix proteins, soluble growth factors, chemokines, and cytokines interacting with water and then modulating ECM viscoelasticity (Berdiaki et al. 2021).
4. *Proteoglycans:* Proteoglycans (PGs) include a core protein with one or more covalently attached GAG side chains (Tzanakakis et al. 2019). PGs are important regulators in our body. Depending on their types they localize on the cell surface, in the ECM, and some of them is detected intracellular. Owing to their GAG

chains they are implicated in several biological and pathological events, including inflammation and cancer. Biglycan is the residence of the ECM, with diverse functions. Among those it is highly implicated in the development of inflammatory cancer milieu (Tzanakakis et al. 2019; Yamauchi et al. 2018). Secreted by the macrophages, it can stimulate their TRL2 and 4 receptors ending up by increased cytokine (TNFalfa, CCL2, CCL5, IL1beta, IFNgamma, CXCL13, etc.) synthesis. Its action concluding in the recruitment of neutrophils, macrophages, B and T lymphocytes promoting anti-apoptosis and tumor growth (Tzanakakais et al. 2019). Versican, another type of ECM PG, binds molecules such as GAGs (particularly HA) (Wight et al. 2020), the inflammation-associated protein tumor necrosis factor-stimulated gene-6 (TSG-6), and is also involved in leukocytes infiltration and cancer progression. In the TME, Versican interacts with cells both via HA or via receptors such as toll-like receptors (TLRs) or CD44 present on immune and non-immune cells. Versican is also involved in the availability of cytokines such as tumor necrosis factor alpha (TNFα), IL-6, and other soluble factors influencing on the phenotype of the immune cells in the TME (Wight et al. 2020). In this sense, it has been reported that Versican produced by CAFs lead to a pro-inflammatory milieu, while Versican expressed by MDSCs can lead to an anti-inflammatory and immunosuppressive TME (Hope et al. 2016).

5. *Matrikines and MMPs:* Matrikines are fragments produced from ECM macromolecules, i.e., collagen, elastin, and Versican partial degradation (Wells et al. 2015). Matrikines are also able to regulate many cellular processes such as proliferation and migration and have been described as regulators of physiological or pathological processes such as tissue repair, inflammation, and tumor invasion (Wells et al. 2015). Recently, it has been reported that Versikine—a fragment derived from Versican proteolysis—promotes CD8+ T-cell infiltration and generation of CD103 + CD11c^{hi}MHCII^{hi} conventional dendritic cells (DCs), a population of antigen-presenting cells implicated in T-cell activation in colorectal cancer (CRC) (Hope et al. 2017). On the other hand, matrix metalloproteinases (MMPs) are a family of 28 zinc-dependent endopeptidases that principally contribute to fibrillar and non-fibrillar collagens degradation. Remodeling of ECM by MMPs supports tumor dissemination and invasion (Wells et al. 2015). In this sense, it has been demonstrated that adhesion molecules CD44 and integrin alphavbeta3 enable that activated MMPs binds on the cell surface of melanoma cells promoting ECM degradation, tumor growth, and cell invasion (Yu and Stamenkovic 1999).

17.2.2 Immunomodulatory Roles of ECM: Matrix Components as Promoters of Immune Response

Despite ECM usually being considered as a passive architectural structure, the ECM is now being accepted as a dynamic red, source of damage-associated molecular patterns (DAMPs) with the ability to activate the immune response. Those DAMPs could be, for example, matrikines mentioned above (Shay et al. 2015). In homeostasis, the composition of the ECM is continuously adjusted by a sequence of synthesis and degradation events. Instead, during cancer progression, this dynamic balance could be disrupted and lead to the accumulation of ECM components and chronic inflammation. ECM deposition and transformation during tumorigenesis can result in the release or formation of ECM-derived DAMPs which can stimulate local inflammatory response and immune cell recruitment. ECM-derived DAMPs are capable of modulating immunity both directly via interaction with immune receptors such as CD44 and TLRs or indirectly through the release of cytokines and chemokines. Cells from innate and adaptive immune systems attracted by these soluble factors can be found within the ECM and the same immune cell type may delay or stimulate tumor progression. For example, the bioactive fragment Versikine, generated from Versican proteolysis by a-disintegrin-and-metalloproteinase-with-thrombospondin-motifs (ADAMTS) protease, promotes immunogenicity and antagonizes DCs dysfunction promoted by whole Versican. In addition, Versikine promotes T-cell infiltration through regulation of a particular DC subset that is critical for CD8+ T-cell trafficking and antitumor immunity (Hope et al. 2017). On the other hand, matrikine Val-Gly-Val-Ala-Pro-Gly (VGVAPG), an elastin peptide, specially attracts monocytes to the TME while impairing neutrophil reactivity (Senior et al. 1984). Other elastin fragments are also released to the ECM during tumor progression. For instance, elastin-derived nonapeptide AG-9 stimulates OSCC cells' invasive properties by increasing membrane type 1-matrix metalloproteinase (MT1-MMP, also known as MMP-14 expression and MMP-2 secretion (Bretaudeau et al. 2020). Overexpression of MMP-14 in the TME could become critical since it has been reported that MMP-14 modulates inflammatory signaling networks for controlling macrophage-mediated immune response. Additionally, upregulated MMP14 levels correlated with a lack of CD8+ cytotoxic T cells in CRC (Claesson-Welsh 2020). In this manner, ECM can directly contribute to recruit specific immune cells but also to the lack of infiltrating antitumor immune populations in the TME.

17.3 Immune Response in Tumors

17.3.1 The Cancer-Immunity Cycle

Immune -mediate response to cancer is a cyclic process that can be self-propagating, primary to an increase of immune-stimulatory factors such as genetic and cellular alterations presents in cancer cells (Chen and Mellman 2013). These characteristics that define tumors provide the immune system with the resources to generate T-cell responses to recognize and eradicate malignant cells. This cycle can be distributed into many steps, from the release of tumor cell antigens to the cancer cells' elimination. First, neoantigens generated by tumorigenesis are released and taken by antigen-presenting DCs. In this step, many signals are crucial to achieve a suitable T-cell-mediated immunity. These signals might include cytokines secretion and

DAMPs released by dying tumor cells. Then, DCs present the captured antigens on class I major histocompatibility complex (MHCI) and class II MHC (MHCII) to the priming and activation of effector T cells. Next, CD4+ and CD8+ T cells leave secondary lymphoid organs adjacent to the tumor and return to tumor tissue to support a response against the cancer-specific antigens. The landscape of the immune response is determined at this step, with an important balance between CD4+ and CD8+ T effector cells versus principally CD4+ but also CD8+ T regulatory cells. Elimination of the cancer cell at the tumor site releases additional antigens, increasing the scale and complexity of the response in successive reiterations of the cycle. In addition, this sequence is characterized by inhibitory factors that produce regulatory mechanisms, which can restrict immunity. The presence of an immunosuppressive milieu may explain the partial efficacy observed in immune-based cancer therapies and why these therapies may be more active in combination with agents that goal other steps of the cycle (Chen and Mellman 2013).

17.3.2 Immune-Cell Trafficking in the Stroma

1. *Tumor-infiltrating regulatory T cells:* naturally occurring regulatory T cells (Tregs) are originated in the thymus as a subtype of $CD4^+$ T cells that constitutively express the CD25 molecule and the transcription factor forkhead box P3 (Foxp3) which is essential to suppress $CD4^+$ and $CD8^+$ T cells effector responses in vivo (Sakaguchi 2004). A sub-population has been also identified as *induced* Tregs, characterized by their ability to inhibit the effector T-cell response by the secretion of soluble factors. Tregs inhibit tumor-infiltrating lymphocytes (TILs) through immunosuppressive cytokines such as interleukin 10 (IL-10) and tumor growth factor beta (TGFβ) and by the surface expression of cytotoxic T lymphocyte antigen 4 (CTLA-4). In addition, Tregs block anti-tumor immunity impairing NK cells cytokine production, inducing tolerant DCs, and increasing the activity of 2,3 indoleamine oxygenase (IDO) which is responsible for tryptophan degradation resulting in $CD4^+$ and $CD8^+$ T cells apoptosis (Thompson and Powrie 2004)

   It has been reported that patients with lung, skin, breast, ovarian, and pancreatic cancers showed an increase in the number of $CD4^+$ $CD25^+$ $Foxp3^+$ cells. Currently, monoclonal antibodies (mAbs) directed against CTLA-4 are used in patients. However, systemic depletion of Tregs by checkpoint inhibition may induce autoimmune responses (Tanaka and Sakaguchi 2017). With the aim to induce antitumor immunity without the induction of autoimmunity is possible to target other cell type that enables immune evasion in the TME such as CAFs. For instance, recently it has been observed that CD70-positive CAFs significantly increased the frequency of Tregs in CRC. Then, consideration of CD70-targeting antibodies to revert the lack of anti-tumor immune response promoted by Tregs could be an interesting approach for CRC (Jacobs et al. 2018).

2. *Tumor-infiltrating Regulatory Myeloid Cells:* Myeloid-derived suppressor cells (MDSCs) are immature cells with immunosuppressive capability (Gabrilovich and Nagaraj 2009). In humans, MDSCs are characterized by $CD11b^{+}$ $CD14^{-}$ HLA-$DR^{-/low}$ $CD33^{+}$ $CD15^{+}$ markers, and their presence in cancer patients has been associated with poor outcomes. MDSCs can promote T cells anergy. In addition, MDSCs express the inhibitory ligand, PD-L1, causing an exhausted phenotype of effector T cells (Noman et al. 2014). Additionally, MDSCs can release nitric oxide (NO) and peroxynitrite inhibiting T-cell activation, induce the increase of Tregs, and differentiate macrophages to a M2 phenotype. Modulation of the number of MDSCs and or inhibition of MDSCs activity are currently areas of great interest (Deng et al. 2017).
3. *Tumor-infiltrating Cytotoxic T Lymphocytes (TILs):* The success of an adequate antitumor immune response depends on preceding recruitment of tumor-infiltrating CD8+ cytotoxic T lymphocytes (TILs) in the TME (Chen and Mellman 2013). Chemokines of CCL and CXCL family have been related to TILs enrollment in many solid tumors (Harlin et al. 2009). The frequency of TILs is a favorable prognostic in many cancers such as melanoma, esophageal, breast, ovarian, head and neck, and non-small-cell lung cancer (NSCLC). Diverse mechanisms, as direct tumor cell killing, have been described for the antitumor activity exerted by TILs (Boon et al. 1994).

   ECM dynamic changes and remodeling are critical for immune cell trafficking, migration, spatial distribution, and activation. DCs and T cells are able to migrate along type 1 collagen self-regulating adhesion molecules whereas cancer cells use MMPs and integrin-dependent migration to penetrate collagen fibrils. It has been observed that in lung cancer, T-cell infiltration arises in low-density fibronectin and collagen regions while it is decreased in dense matrix fibers.

17.3.3 The Role of the Matrisome in Tumor Inflammation and Cancer

The set of proteins with different functions from cellular adhesion and motility to cell signaling that conformed the ECM has been identified as the *matrisome* (Socovich and Naba 2019; Yuzhalin et al. 2018). Modifications in the matrisome because of tumor progression could cause reciprocal alterations in the cancer cells and activate pathways responsible for many hallmarks of cancer such as inhibition of apoptosis and uncontrolled proliferation. Recently, Yuzhalin et al. have identified a nine-gene matrisome signature common to a range of solid cancers, which predicts poor prognosis in several tumor types. Matrisome gene signatures from CRC, gastric, lung, and ovarian cancers were significantly enriched in hypoxia, neovascularization, epithelial-mesenchymal transition, and inflammation (Yuzhalin et al. 2018).

17.4 Cancer Immunotherapy

17.4.1 Improving Immune-Mediated Antitumor Response

Immunotherapy purposes control the growth and spreading of malignant cells by the activation of a specific immune response (Rosenberg et al. 2004). In this sense, a vast number of strategies to achieve an effective immune response against tumors have been tested such as cytokine or immune checkpoint blockade mAbs administration, adoptive T-cell therapy, dendritic cells-based vaccines, etc. Some of them are under evaluation and actually practice in the clinic, particularly the use of immune checkpoint inhibitors (Perez-Gracia et al. 2014).

These strategies have been demonstrated to be potent in animal models, but it was not until a few years ago with the use of a DCs-based vaccine in hormone-refractory prostate cancer (*sipuleucel*) or immunostimulatory mAbs (*ipilimumab, nivolumab, pembrolizumab, tremelimumab*) that clinical results were more satisfactory (Hodi et al. 2010). An elucidation for the unsatisfying clinical results is based on immunosuppressive mechanisms used by tumors cells to escape from the host immune response. This has led to the design of new approaches to inhibit immunosuppressive factors derived from the TME, in part responsible for the lack of activation of an efficient antitumor response.

In the design of a therapeutic strategy, the need to implement multiple approaches to block immune suppressor mechanisms must be taken into account. In this context, protocols of combined therapy with focusing on modulate matrix and TME components might act in synergy.

17.4.2 Immune Checkpoint Inhibitors

The immune response is regulated by a highly complex equilibrium of signals transmitted by stimulatory and inhibitory receptors (Tanaka and Sakaguchi 2017). Immunostimulatory monoclonal antibodies (mAbs) have the ability to enhance ongoing immune responses. A number of resistant metastatic cancers (e.g., melanoma, renal cell carcinoma, or lung cancer) now have the possibility of being successfully controlled using mAbs with long-term clinical responses (Tanaka and Sakaguchi 2017). The main concept involves the block of negative regulatory cell surface molecules, which inhibit T-cell activation. Immune checkpoint inhibitors (ICI) antibody therapy has demonstrated that the CTLA-4 inhibitor *(ipilimumab, tremelimumab*) has favorable antitumor activity in patients with advanced solid tumors (Hodi et al. 2010). Lately, fully human (IgG4) monoclonal anti-PD-1 (*nivolumab*, *pembrolizumab*) and programmed death receptor ligand-1 (PD-L1, *atezolizumab,* durvalumab) antibodies have demonstrated clinical benefit in melanoma, refractory non-small cell lung cancer, advanced renal cell carcinoma, Hodgkin lymphoma, head and neck squamous cell carcinoma, urothelial carcinoma and hepatocellular carcinoma (Alsaab et al. 2017). Remarkably, anti-PD-1 therapy demonstrated longstanding survival, robust responses, and safety profiles in these tumor types. Moreover, for many types of solid tumors, clinical responses occurred regardless of PD-L1 expression on tumor cells. Re-establishing anti-tumor immunity seems to be feasible and the results with ICI are favorable; however, the efficacy of strategies based on immunostimulatory mAbs is vulnerable to the hostile TME that suppresses the effector response (Alsaab et al. 2017). In this sense, the role of ECM components might reveal potential future synergism for cancer treatment.

17.4.3 ECM Components as Emerging Players in Immunotherapy

As we mentioned, a successfully immune response rests on a tight balance between stimulatory factors and inhibitory factors. Regardless of immunotherapy with ICIs block checkpoints to induce anti-tumor response, stromal cells, including CAFs, ECM macromolecules (collagens, GAGs, and PGs), matrikines and cytokines are involved in the delivery and polarization of immune cells and regulates generation and proliferation of the sustained host immune response (Fig. 17.1a). Particularly, CAFs that secrete matrix-associated proteins, cytokines, and pro-inflammatory chemokines orchestrate the access of leukocytes in the TME (Freeman and Mielgo 2020). It has been reported that CAFs correlate with an immune-suppressive phenotype, with decreased IFN-gamma and granzyme B expression, supporting resistance to ICI therapy in experimental CRC models. In addition, CAFs derived from melanoma biopsies showed an upregulation of both programmed death ligands 1 and 2 (PD-L1 and PD-L2), abolishing CD8+ T-cell function (Li et al. 2019). PD-L1 and PD-L2 have also been upregulated in CAFs from pancreatic cancer patients. As well, CAFs promote the expression of other co-inhibitory immune checkpoint receptors such as LAG-3 and T-cell immunoglobulin and mucin-domain containing-3 (TIM-3) impairing effector T-cell function (Goehrig et al. 2019). It has been proposed that the upregulation of inhibitory checkpoints is mediated by secretion of soluble factors such as CXCL5, which enhanced the expression of PD-L1 in melanoma and CRC cell lines by activating PI3K-AKT pathway (Li et al. 2019). HCC-associated fibroblasts also were capable to upregulate the expression of PD-L1 on the surface of neutrophils, by IL-6 secretion. Then, neutrophils were then able to suppress T-cell activation in this HCC model (Cheng et al. 2016).

On the other hand, as we describe, matrikines are formed from collagen, laminin, elastin, Versican, and HA by MMPs and a family of matrix enzymes that include a-disintegrin-and-metalloproteinases (ADAM) and ADAMTS (Wight et al. 2020). Matrikines can directly contribute to recruit specific immune cells. To potentiate the efficacy of ICI, matrikines could be proposed to regulate T-cell infiltration and the availability of interleukins, promoting TME inflammation and contributing to

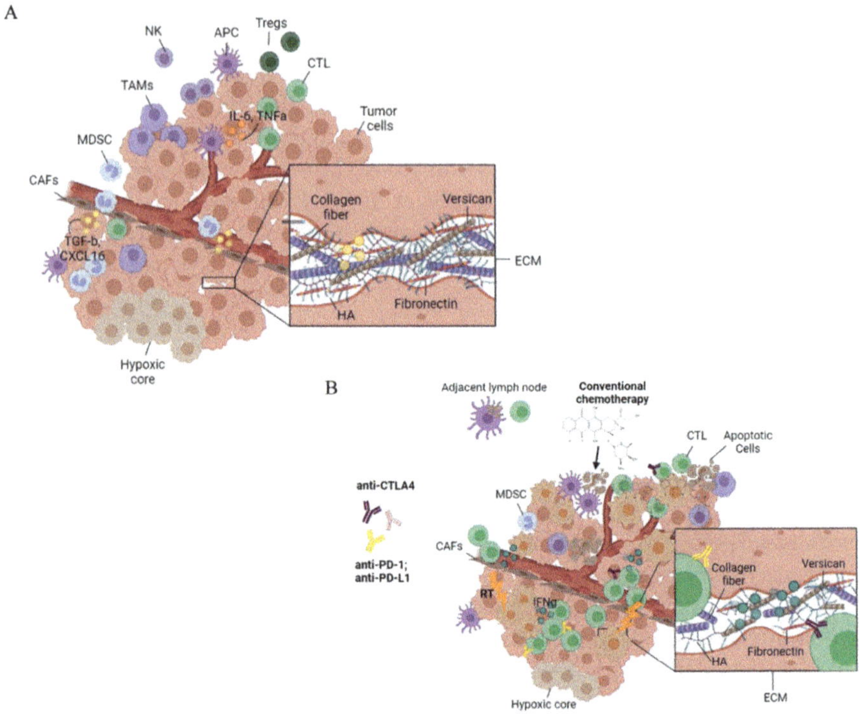

Fig. 17.1 The extracellular matrix plays a key role in the activation of TME immune response. (**a**) The dynamic network of stromal cells and macromolecules in constant remodeling supports the infiltration and polarization of immune cells. Fibroblasts and cancer-associated fibroblasts (CAFs) release soluble factors and matrix proteases degrade ECM proteins to produce matrikines (versikine, elastin-derived peptides) regulating the host immune response. Antigen-presenting cells (APC) such as dendritic cells (DCs) and T cells (cytotoxic T lymphocytes; CTL) migrate along collagen and fibronectin fibrils. ECM also is involved in the efficacy of immunotherapy: whereas matrikines can regulate T-cell infiltration and the availability of interleukins contributing to overcome resistance, hypoxia in the TME can promote immune evasion by upregulation of immunosuppressive factors (transforming growth factor-beta; TGF-b, CXCL16) and the recruiting of myeloid-derived suppressor cells and regulatory T cells (MDSCs; Tregs). (**b**) In spite of the demonstrated efficacy of immune checkpoint inhibitors (ICI) in advanced cancers, local radiotherapy (RT) and/or conventional chemotherapy can transform tumors and overcome some of the evasion mechanism. RT can induce an inflammatory response in the TME mediated by IL-1 and IL-6, the release of neoantigens, DCs presentation and T CD8+ cell activation with Interferon gamma (INF-g) secretion. In addition, chemotherapy could lead to tumor cell death by apoptotic and/or non-apoptotic mechanisms impacting on the cross-presentation of tumor-derived antigen mediated by DCs. Then, conventional chemotherapy and RT could promote immunity against many tumors and improve the efficacy of immunotherapy with immune checkpoint inhibitors such as anti-CTLA4, anti-PD-1, or anti-PD-L1

overcome resistance that suppresses the effector immune response pursued with immunotherapy.

17.5 Therapeutic Combinations to Enhance Immunotherapy Achievement

17.5.1 Influence of Radiotherapy on ECM and the Stroma

In spite of the demonstrated efficacy of ICI in advanced cancers, many patients do not experience therapeutic benefit, maybe by a lack of specific recognition, or to the presence of immunosuppressive mechanisms in the TME (Pitt et al. 2016). Recent advances have discovered that local radiotherapy (RT) can transform tumors and overcome some of the evasion mechanisms reported (Menon et al. 2019).

RT is a key tool to treat patients with cancer. Its main way of action is associated with its capacity to induce double strain DNA damage (DSD) triggering cell apoptosis. RT technics has been changing in parallel with technologies development. Hyper-fractionated RT has been classically used but, nowadays, there are a lot of different technics which administrate large doses of radiotherapy in few fractions.

These kinds of treatment could impact ECM affecting their cellular compartments like CAFs, MDSCs, and TAMs, but also induces changes in ECM density, composition, and stiffness likely modifying malignant cell invasion, survival, and proliferation (Menon et al. 2019).

In this sense, it has been shown that RT induces changes in CAFs modulating their genomic expression and regulating their pro-tumorigenic capability. They induce radioresistance by secreting CXCL1 and it was associated with worse prognostic in patients receiving RT (Zhang et al. 2017); in addition, it has been observed that STAT3 phosphorylation induced by RT promotes MDSCs infiltration and proliferation (Grinde et al. 2017; Allaoui et al. 2016; Oweida et al. 2021). However, the inhibition of this path in association with RT improves tumor response in pancreatic cancer (Oweida et al. 2021).

Thence, to maximize the therapeutic benefit, it is important to define the doses and also combinations to highlighting the pathways to promote anti-tumor immunity and effector T cell function while limiting pathways that mediate an immunosuppressive TME.

It has been reported that RT has the ability to generate immunogenic signals and release DAMPs (Bernier et al. 2004; Wu et al. 2017). In this context, RT is able to induce an inflammatory response in the TME mediated by cytokines such as IL-1 and IL-6. In addition, the release of DAMPs and tumor neoantigens induced by RT allows DCs presentation and T CD8+ cell activation (Garnett et al. 2004; Burnette et al. 2011).

RT also has been shown to promote the stimulator of interferon genes (STING) pathway activation and PD-L1 upregulation. When RT induces DNA damage and tumor cell apoptosis, DNA from dying tumor cells is uptake by DCs. This event induces the sensing of DNA by cyclic-GMP-AMP with activation of STING pathway that finally induces expression of Type I IFNs. Additionally, in response to RT-induced tissue damage, CAFs can secrete chemoattractant molecules such as CXCL9 and CXCL10, recruiting CD8+ T cells in the TME (Chen et al. 2016) (Fig. 17.1b).

Although it has been postulated that RT increases the number and activity of CAFs and promotes the accumulation of ECM macromolecules that physically block immune cells trafficking causing an immunosuppressive TME, it was recently demonstrated the benefit of RT strategy in the PACIFIC trial (NCT02125461) (Antonia et al. 2017). In this studio, as we will comment later, RT and immunotherapy are driven synergistically. The lack of recurrences and the improved ability of immune cells to infiltrate and eradicate metastases resulted in extended progression-free survival in patients. Then, RT can modify ECM toward a sketched, reduced tumor stroma and sensitize cancer cells to immune therapies (Menon et al. 2019).

17.5.2 Chemotherapy Induces ECM Remodeling and Promotes the Immune Response

Chemotherapy still represents the core of cancer treatment nowadays. It has been described that this strategy leads to tumor cell death by apoptotic and/or non-apoptotic mechanisms such as autophagy or necrosis (Kerr et al. 1972). For drugs such as cyclophosphamide, gemcitabine, doxorubicin, and cisplatin, the principal mechanism of action described is DNA damage followed by apoptosis (Binotto et al. 2003; Casares et al. 2005). Although apoptosis has been considered as a non-immunogenic cell death, it is now more clear that immunity can be triggered by apoptosis (Nowak et al. 2003). Doxorubicin induces immunogenic apoptosis mediated by the release of the histone HMGB1, which, in turn, activates TLR-4. Also, doxorubicin induces apoptosis by up-regulation of FAS-L in cancer cells (Casares et al. 2005). One possible consequence of chemotherapy on the promotion of immune response has been attributed to alkylating agents. In fact, cyclophosphamide induces the expansion of CD8+ DC, the main subset involved in the cross-presentation of cell-derived antigens (Nowak et al. 2003).

There is experimental evidence that reducing the dose of conventional chemotherapy could act in synergy to generate immunity against many tumors. In this sense, it has been demonstrated that low-dose cyclophosphamide leads to a significant reduction in Tregs population (Malvicini et al. 2009) and paclitaxel reduces the number of MDSCs in a murine model of melanoma (Sevko et al. 2013). Gemcitabine and 5-FU can also selectively deplete MDSCs. It has been described that 5-FU induces MDSC depletion and promotes IFN-γ production by tumor-specific CD8+ T cells in a thymoma model (Vincent et al. 2010). Blidner et al. characterized the effect of indomethacin (IND) on MDSCs and showed that IND inhibited the suppressive activity exerted by MDSCs on CD8 (+) T cells in lung adenocarcinoma-bearing mice (Blidner et al. 2015). In order to achieve tumor elimination, the immune response should include the activation of cytotoxic T cells that lyse tumor cells. To this end, chemotherapy could work in synergy with ICI therapy.

Focusing on ECM and other factors of the TME, it has been reported that HA production appears to be enhanced in different tumors including ovarian, CRC, breast cancer, and HCC (Freeman and Mielgo 2020). In addition, the presence of high HA levels in TME correlates with poor clinical outcomes. The natural compound 4-methylumbelliferone (4Mu) has been reported to inhibit HA synthesis with antitumor effects in vitro and in vivo (Piccioni et al. 2015). Also, 4Mu therapy reduces HA accumulation and increases the homing of specific antitumor T lymphocytes in CRC-bearing mice (Malvicini et al. 2015).

In addition to HA and GAGs, fibrillar and non-fibrillar ECM components may impact chemotherapy response. The accumulation of ECM proteins is associated with tumor progression and the progressive stiffening of ECM, modifying the sensitivity of cancer cells to chemotherapy. It has been observed that MDA-MB-231 triple-negative breast carcinoma cells exhibited stiffness-dependent resistance response to doxorubicin when were cultured in a 3D alginate-based hydrogel system, and the stiffness-dependent response was lost when cells were removed from 3D to monolayer cultures. These results indicate that ECM conditions are able to define the response to chemotherapy (Joyce et al. 2018). The effects of ECM stiffness on paclitaxel and gemcitabine response were also reported in pancreatic cancer (Amrutkar et al. 2019). The ECM rigidity induces chemoresistance suggesting that TME stiffness would be modulated to overcome resistance and acquire chemotherapy success.

17.5.3 Combined Strategies: Where Are We and Where Are We Going

Chemotherapy in combination with immunotherapy has shown clinical benefit in several tumor types and it is used in clinical practice (Lake and Robinson 2005). Patients with NSCLC were the first ones who benefit from this kind of treatment. The association of pembrolizumab with chemotherapy is used in squamous and non-squamous NSCLC based in the KeyNote 189 and the KeyNote 407 studies, respectively (https://clinicaltrials.gov/). Another interesting combination used in the clinic in this group of patients is chemo plus an anti-PDL1, *atezolizumab*, and an anti-VEGF (*bevacizumab*). All these combinations with platinum-based chemotherapy have shown significant overall survival benefits. The anti-PDL1 has reached its place in the group of patients with metastatic SCLC where *atezolizumab or durvalumab* are being used in combination with platinum plus etoposide. This fact is relevant because SCLC is such a refractory disease.

Another scenario where the chemo-immune combination is used is in esophageal or gastric cancer. The Keynote-590 (https://clinicaltrials.gov/) study showed the combination of cisplatin and 5-fluorouracil plus an anti-PD1 (*pembrolizumab*) improved overall survival over chemotherapy alone. Another anti-PD1 which showed a clinical benefit in combination with chemotherapy in CPS greater than five gastric cancers was *nivolumab*. The OS was 11 vs. 14 months favoring the combination arm. Certainly, there are other studies without so clear results in favor of the combination and these differences could be explained at least in part by the populate selection.

One step forward was done in inoperable stage III NSCLC. The triple combination of chemo-radiotherapy followed by immunotherapy has been approved in this group of patients based on the PACIFIC trial (Antonia et al. 2017). This trial was the first one in demonstrating a significant overall survival and progression-free survival benefit with the addition of immunotherapy after chemo-radiotherapy versus chemo-radiotherapy alone. The overall survival was significantly improved (42.9% vs 33.4%; HR 0.72) as well as progression-free survival (33.1 vs 19%; 0.55) after 5 years of follow-up (America Society of Clinical Oncology meeting, 2021).

There are registered more than 20 phase III clinical trials which combine chemo-, radio-, and immunotherapy. Interestingly, they are proving this triple combination in different ways, like sequential or concurrent; in different pathologies like NSCLC, nasopharyngeal carcinoma, pancreatic cancer, etc.; and in different cancer stages, metastatic or localized ones (https://clinicaltrials.gov/).

17.6 Concluding Comments

A number of immunotherapy strategies have been opened for clinical applications. Nevertheless, the success of these diverse approaches in experimental models is different to the limited outcomes achieved in advanced cancer patients. Immunotherapy could fail or have a slight effect on patient survival for many reasons. The shared scenario for solid tumors often involves patients with advanced diseases that impedes, or at least decreases, the possibility of immunotherapy success. ECM dynamic changes and remodeling are critical for immune cell trafficking, migration, spatial distribution and activation. In the design of an immunotherapeutic approach, the necessity to implement multiple treatments to inhibit TME suppressor mechanisms has to be taken into account. In this context, protocols of combined therapy with focusing in modulate ECM and TME components might act in synergy. Accordingly, conventional chemotherapy and radiotherapy could promote immunity against many tumors. Currently, different forms of immunotherapy including the use of immune checkpoint inhibitors are evaluated in combination along with conventional treatment regimens for overall clinical benefits evidencing improvements in patient survival.

References

Allaoui R, Bergenfelz C, Mohlin S, Hagerling C, Salari K, Werb Z, Anderson RL, Ethier SP, Jirstrom K, Pahlman S, Bexell D, Tahin B, Johansson ME, Larsson C, Leandersson K (2016) Cancer-associated fibroblast-secreted CXCL16 attracts monocytes to promote stroma activation in triple-negative breast cancers. Nat Commun 7:13050. https://doi.org/10.1038/ncomms13050

Alsaab H, Sau S, Alzhrani R, Tatiparti K, Bhise K, Kashaw S, Iyer A (2017) PD-1 and PD-1L checkpoint signaling inhibition for cancer immunotherapy: mechanisms, combinations and clinical outcome. Front Pharmacol 8:561. https://doi.org/10.3389/fphar.2017.00561

Amrutkar M, Aasrum M, Verbeke CS, Gladhaug IP (2019) Secretion of fibronectin by human pancreatic stellate cells promotes chemoresistance to gemcitabine in pancreatic cancer cells. BMC Cancer 19(1):596. https://doi.org/10.1186/s12885-019-5803-1

Antonia SJ, Villegas A, Daniel D, Vicente D, Murakami S, Hui R, Yokoi T, Chiappori A, Lee KH, de Wit M, Cho BC, Bourhaba M, Quantin X, Tokito T, Mekhail T, Planchard D, Kim YC, Karapetis CS, Hiret S, Ostoros G, Kubota K, Gray JE, Paz-Ares L, de Castro CJ, Wadsworth C, Melillo G, Jiang H, Huang Y, Dennis PA, Ozguroglu M, Investigators P (2017) Durvalumab after chemoradiotherapy in stage III non-small-cell lung cancer. N Engl J Med 377(20): 1919–1929. https://doi.org/10.1056/NEJMoa1709937

Berdiaki A, Neagu M, Giatagana EM, Kuskov A, Tsatsakis AM, Tzanakakis GN, Nikitovic D (2021) Glycosaminoglycans: carriers and targets for tailored anti-cancer therapy. Biomol Ther 11(3):395. https://doi.org/10.3390/biom11030395

Bernier J, Hall EJ, Giaccia A (2004) Radiation oncology: a century of achievements. Nat Rev Cancer 4(9):737–747. https://doi.org/10.1038/nrc1451

Binotto G, Trentin L, Semenzato G (2003) Ifosfamide and cyclophosphamide: effects on immunosurveillance. Oncology 65(Suppl 2):17–20. https://doi.org/10.1159/000073353

Blidner AG, Salatino M, Mascanfroni ID, Diament MJ, de Kier B, Joffe E, Jasnis MA, Klein SM, Rabinovich GA (2015) Differential response of myeloid-derived suppressor cells to the non-steroidal anti-inflammatory agent indomethacin in tumor-associated and tumor-free microenvironments. J Immunol 194(7):3452–3462. https://doi.org/10.4049/jimmunol.1401144

Bonnans C, Chou J, Werb Z (2014) Remodelling the extracellular matrix in development and disease. Nat Rev Mol Cell Biol 15(12):786–801. https://doi.org/10.1038/nrm3904

Boon T, Cerottini JC, Van den Eynde B, van der Bruggen P, Van Pel A (1994) Tumor antigens recognized by T lymphocytes. Annu Rev Immunol 12:337–365. https://doi.org/10.1146/annurev.iy.12.040194.002005

Bretaudeau C, Baud S, Dupont-Deshorgue A, Cousin R, Brassart B, Brassart-Pasco S (2020) AG-9, an elastin-derived peptide, increases in vitro oral tongue carcinoma cell invasion, through an increase in MMP-2 secretion and MT1-MMP expression, in a RPSA-dependent manner. Biomol Ther 11(1):39. https://doi.org/10.3390/biom11010039

Burnette BC, Liang H, Lee Y, Chlewicki L, Khodarev NN, Weichselbaum RR, Fu YX, Auh SL (2011) The efficacy of radiotherapy relies upon induction of type i interferon-dependent innate and adaptive immunity. Cancer Res 71(7):2488–2496. https://doi.org/10.1158/0008-5472.CAN-10-2820

Calon A, Tauriello DV, Batlle E (2014) TGF-beta in CAF-mediated tumor growth and metastasis. Semin Cancer Biol 25:15–22. https://doi.org/10.1016/j.semcancer.2013.12.008

Casares N, Pequignot MO, Tesniere A, Ghiringhelli F, Roux S, Chaput N, Schmitt E, Hamai A, Hervas-Stubbs S, Obeid M, Coutant F, Metivier D, Pichard E, Aucouturier P, Pierron G, Garrido C, Zitvogel L, Kroemer G (2005) Caspase-dependent immunogenicity of doxorubicin-induced tumor cell death. J Exp Med 202(12):1691–1701. https://doi.org/10.1084/jem.20050915

Chanmee T, Ontong P, Itano N (2016) Hyaluronan: a modulator of the tumor microenvironment. Cancer Lett 375(1):20–30. https://doi.org/10.1016/j.canlet.2016.02.031

Chen DS, Mellman I (2013) Oncology meets immunology: the cancer-immunity cycle. Immunity 39(1):1–10. https://doi.org/10.1016/j.immuni.2013.07.012

Chen Q, Sun L, Chen ZJ (2016) Regulation and function of the cGAS-STING pathway of cytosolic DNA sensing. Nat Immunol 17(10):1142–1149. https://doi.org/10.1038/ni.3558

Cheng JT, Deng YN, Yi HM, Wang GY, Fu BS, Chen WJ, Liu W, Tai Y, Peng YW, Zhang Q (2016) Hepatic carcinoma-associated fibroblasts induce IDO-producing regulatory dendritic cells through IL-6-mediated STAT3 activation. Oncogenesis 5:e198. https://doi.org/10.1038/oncsis.2016.7

Claesson-Welsh L (2020) How the matrix metalloproteinase MMP14 contributes to the progression of colorectal cancer. J Clin Invest 130(3):1093–1095. https://doi.org/10.1172/JCI135239

Deng Z, Rong Y, Teng Y, Zhuang X, Samykutty A, Mu J, Zhang L, Cao P, Yan J, Miller D, Zhang HG (2017) Exosomes miR-126a released from MDSC induced by DOX treatment promotes lung metastasis. Oncogene 36(5):639–651. https://doi.org/10.1038/onc.2016.229

Freeman P, Mielgo A (2020) Cancer associated fibroblast mediated inhibition of CD8+ cytotoxic T cell accumulation in tumours: mechanisms and therapeutic opportunities. Cancers (Basel) 12(9): 2687. https://doi.org/10.3390/cancers12092687

Gabrilovich DI, Nagaraj S (2009) Myeloid-derived suppressor cells as regulators of the immune system. Nat Rev Immunol 9(3):162–174. https://doi.org/10.1038/nri2506

Garnett CT, Palena C, Chakraborty M, Tsang KY, Schlom J, Hodge JW (2004) Sublethal irradiation of human tumor cells modulates phenotype resulting in enhanced killing by cytotoxic T lymphocytes. Cancer Res 64(21):7985–7994. https://doi.org/10.1158/0008-5472.CAN-04-1525

Goehrig D, Nigri J, Samain R, Wu Z, Cappello P, Gabiane G, Zhang X, Zhao Y, Kim IS, Chanal M, Curto R, Hervieu V, de La Fouchardiere C, Novelli F, Milani P, Tomasini R, Bousquet C, Bertolino P, Hennino A (2019) Stromal protein betaig-h3 reprogrammes tumour microenvironment in pancreatic cancer. Gut 68(4):693–707. https://doi.org/10.1136/gutjnl-2018-317570

Grinde MT, Vik J, Camilio KA, Martinez-Zubiaurre I, Hellevik T (2017) Ionizing radiation abrogates the pro-tumorigenic capacity of cancer-associated fibroblasts co-implanted in xenografts. Sci Rep 7:46714. https://doi.org/10.1038/srep46714

Harlin H, Meng Y, Peterson AC, Zha Y, Tretiakova M, Slingluff C, McKee M, Gajewski TF (2009) Chemokine expression in melanoma metastases associated with CD8+ T-cell recruitment. Cancer Res 69(7):3077–3085. https://doi.org/10.1158/0008-5472.CAN-08-2281

Hodi F, O'Day S, Mc Dermott D, Weber R, Sosman J, Haanen J, Gonzalez R, Robert C, Urba W (2010) Improved survival with ipilimumab in patients with metastatic melanoma. N Engl J Med 8:711–723. https://doi.org/10.1056/NEJMoa1033466

Hope C, Foulcer S, Jagodinsky J, Chen SX, Jensen JL, Patel S, Leith C, Maroulakou I, Callander N, Miyamoto S, Hematti P, Apte SS, Asimakopoulos F (2016) Immunoregulatory roles of versican proteolysis in the myeloma microenvironment. Blood 128(5):680–685. https://doi.org/10.1182/blood-2016-03-705780

Hope C, Emmerich PB, Papadas A, Pagenkopf A, Matkowskyj KA, Van De Hey DR, Payne SN, Clipson L, Callander NS, Hematti P, Miyamoto S, Johnson MG, Deming DA, Asimakopoulos F (2017) Versican-derived matrikines regulate Batf3-dendritic cell differentiation and promote T cell infiltration in colorectal cancer. J Immunol 199(5):1933–1941. https://doi.org/10.4049/jimmunol.1700529

Jacobs J, Deschoolmeester V, Zwaenepoel K, Flieswasser T, Deben C, Van den Bossche J, Hermans C, Rolfo C, Peeters M, De Wever O, Lardon F, Siozopoulou V, Smits E, Pauwels P (2018) Unveiling a CD70-positive subset of cancer-associated fibroblasts marked by pro-migratory activity and thriving regulatory T cell accumulation. Onco Targets Ther 7(7): e1440167. https://doi.org/10.1080/2162402X.2018.1440167

Joyce MH, Lu C, James ER, Hegab R, Allen SC, Suggs LJ, Brock A (2018) Phenotypic basis for matrix stiffness-dependent chemoresistance of breast cancer cells to doxorubicin. Front Oncol 8: 337. https://doi.org/10.3389/fonc.2018.00337

Kaur A, Ecker BL, Douglass SM, Kugel CH 3rd, Webster MR, Almeida FV, Somasundaram R, Hayden J, Ban E, Ahmadzadeh H, Franco-Barraza J, Shah N, Mellis IA, Keeney F, Kossenkov A, Tang HY, Yin X, Liu Q, Xu X, Fane M, Brafford P, Herlyn M, Speicher DW,

Wargo JA, Tetzlaff MT, Haydu LE, Raj A, Shenoy V, Cukierman E, Weeraratna AT (2019) Remodeling of the collagen matrix in aging skin promotes melanoma metastasis and affects immune cell motility. Cancer Discov 9(1):64–81. https://doi.org/10.1158/2159-8290.CD-18-0193

Kerr JF, Wyllie AH, Currie AR (1972) Apoptosis: a basic biological phenomenon with wide-ranging implications in tissue kinetics. Br J Cancer 26(4):239–257. https://doi.org/10.1038/bjc.1972.33

Lake RA, Robinson BW (2005) Immunotherapy and chemotherapy--a practical partnership. Nat Rev Cancer 5(5):397–405. https://doi.org/10.1038/nrc1613

Li Z, Zhou J, Zhang J, Li S, Wang H, Du J (2019) Cancer-associated fibroblasts promote PD-L1 expression in mice cancer cells via secreting CXCL5. Int J Cancer 145(7):1946–1957. https://doi.org/10.1002/ijc.32278

Maller O, Drain AP, Barrett AS, Borgquist S, Ruffell B, Zakharevich I, Pham TT, Gruosso T, Kuasne H, Lakins JN, Acerbi I, Barnes JM, Nemkov T, Chauhan A, Gruenberg J, Nasir A, Bjarnadottir O, Werb Z, Kabos P, Chen YY, Hwang ES, Park M, Coussens LM, Nelson AC, Hansen KC, Weaver VM (2021) Tumour-associated macrophages drive stromal cell-dependent collagen crosslinking and stiffening to promote breast cancer aggression. Nat Mater 20(4): 548–559. https://doi.org/10.1038/s41563-020-00849-5

Malvicini M, Rizzo M, Alaniz L, Pinero F, Garcia M, Atorrasagasti C, Aquino JB, Rozados V, Scharovsky OG, Matar P, Mazzolini G (2009) A novel synergistic combination of cyclophosphamide and gene transfer of interleukin-12 eradicates colorectal carcinoma in mice. Clin Cancer Res 15(23):7256–7265. https://doi.org/10.1158/1078-0432.CCR-09-1861

Malvicini M, Fiore E, Ghiaccio V, Piccioni F, Rizzo M, Olmedo Bonadeo L, Garcia M, Rodriguez M, Bayo J, Peixoto E, Atorrasagasti C, Alaniz L, Aquino J, Matar P, Mazzolini G (2015) Tumor microenvironment remodeling by 4-methylumbelliferone boosts the antitumor effect of combined immunotherapy in murine colorectal carcinoma. Mol Ther 23(9): 1444–1455. https://doi.org/10.1038/mt.2015.112

McDonald LT, Russell DL, Kelly RR, Xiong Y, Motamarry A, Patel RK, Jones JA, Watson PM, Turner DP, Watson DK, Soloff AC, Findlay VJ, LaRue AC (2015) Hematopoietic stem cell-derived cancer-associated fibroblasts are novel contributors to the pro-tumorigenic microenvironment. Neoplasia 17(5):434–448. https://doi.org/10.1016/j.neo.2015.04.004

Menon H, Ramapriyan R, Cushman TR, Verma V, Kim HH, Schoenhals JE, Atalar C, Selek U, Chun SG, Chang JY, Barsoumian HB, Nguyen QN, Altan M, Cortez MA, Hahn SM, Welsh JW (2019) Role of radiation therapy in modulation of the tumor stroma and microenvironment. Front Immunol 10:193. https://doi.org/10.3389/fimmu.2019.00193

Netti PA, Berk DA, Swartz MA, Grodzinsky AJ, Jain RK (2000) Role of extracellular matrix assembly in interstitial transport in solid tumors. Cancer Res 60(9):2497–2503

Nikitovic D, Tzardi M, Berdiaki A, Tsatsakis A, Tzanakakis GN (2015) Cancer microenvironment and inflammation: role of hyaluronan. Front Immunol 6:169. https://doi.org/10.3389/fimmu.2015.00169

Noman MZ, Desantis G, Janji B, Hasmim M, Karray S, Dessen P, Bronte V, Chouaib S (2014) PD-L1 is a novel direct target of HIF-1alpha, and its blockade under hypoxia enhanced MDSC-mediated T cell activation. J Exp Med 211(5):781–790. https://doi.org/10.1084/jem.20131916

Nowak AK, Lake RA, Marzo AL, Scott B, Heath WR, Collins EJ, Frelinger JA, Robinson BW (2003) Induction of tumor cell apoptosis in vivo increases tumor antigen cross-presentation, cross-priming rather than cross-tolerizing host tumor-specific CD8 T cells. J Immunol 170(10): 4905–4913. https://doi.org/10.4049/jimmunol.170.10.4905

Oweida AJ, Mueller AC, Piper M, Milner D, Van Court B, Bhatia S, Phan A, Bickett T, Jordan K, Proia T, Schulick R, Messersmith WA, Del Chiaro M, Clambey E, Gough MJ, Williams J, Hansen K, Goodman K, Karam SD (2021) Response to radiotherapy in pancreatic ductal adenocarcinoma is enhanced by inhibition of myeloid-derived suppressor cells using STAT3 anti-sense oligonucleotide. Cancer Immunol Immunother 70(4):989–1000. https://doi.org/10.1007/s00262-020-02701-w

Pankova D, Chen Y, Terajima M, Schliekelman MJ, Baird BN, Fahrenholtz M, Sun L, Gill BJ, Vadakkan TJ, Kim MP, Ahn YH, Roybal JD, Liu X, Parra Cuentas ER, Rodriguez J, Wistuba II, Creighton CJ, Gibbons DL, Hicks JM, Dickinson ME, West JL, Grande-Allen KJ, Hanash SM, Yamauchi M, Kurie JM (2016) Cancer-associated fibroblasts induce a collagen cross-link switch in tumor stroma. Mol Cancer Res 14(3):287–295. https://doi.org/10.1158/1541-7786.MCR-15-0307

Perez-Gracia J, Labiano S, Rodriguez-Ruiz M, Sanmamed M, Melero I (2014) Orchestanting immune check point blockade for cancer immunotherapy in combinations. Curr Opin Immunol 27:89–97. https://doi.org/10.1016/j.coi.2014.01.002

Piccioni F, Fiore E, Bayo J, Atorrasagasti C, Peixoto E, Rizzo M, Malvicini M, Tirado-Gonzalez I, Garcia MG, Alaniz L, Mazzolini G (2015) 4-methylumbelliferone inhibits hepatocellular carcinoma growth by decreasing IL-6 production and angiogenesis. Glycobiology 25(8): 825–835. https://doi.org/10.1093/glycob/cwv023

Pitt JM, Vetizou M, Daillere R, Roberti MP, Yamazaki T, Routy B, Lepage P, Boneca IG, Chamaillard M, Kroemer G, Zitvogel L (2016) Resistance mechanisms to immune-checkpoint blockade in cancer: tumor-intrinsic and -extrinsic factors. Immunity 44(6):1255–1269. https://doi.org/10.1016/j.immuni.2016.06.001

Rosenberg S, Yang J, Restifo N (2004) Cancer Immunotherapy: moving beyond current vaccines. Nat Med 10(9):909–915. https://doi.org/10.1038/nm1100

Sakaguchi S (2004) Naturally arising CD4+ regulatory t cells for immunologic self-tolerance and negative control of immune responses. Annu Rev Immunol 22:531–562. https://doi.org/10.1146/annurev.immunol.21.120601.141122

Senior RM, Griffin GL, Mecham RP, Wrenn DS, Prasad KU, Urry DW (1984) Val-Gly-Val-Ala-Pro-Gly, a repeating peptide in elastin, is chemotactic for fibroblasts and monocytes. J Cell Biol 99(3):870–874. https://doi.org/10.1083/jcb.99.3.870

Sevko A, Michels T, Vrohlings M, Umansky L, Beckhove P, Kato M, Shurin GV, Shurin MR, Umansky V (2013) Antitumor effect of paclitaxel is mediated by inhibition of myeloid-derived suppressor cells and chronic inflammation in the spontaneous melanoma model. J Immunol 190(5):2464–2471. https://doi.org/10.4049/jimmunol.1202781

Shay G, Lynch CC, Fingleton B (2015) Moving targets: emerging roles for MMPs in cancer progression and metastasis. Matrix Biol 44-46:200–206. https://doi.org/10.1016/j.matbio.2015.01.019

Socovich AM, Naba A (2019) The cancer matrisome: from comprehensive characterization to biomarker discovery. Semin Cell Dev Biol 89:157–166. https://doi.org/10.1016/j.semcdb.2018.06.005

Takahashi H, Sakakura K, Kudo T, Toyoda M, Kaira K, Oyama T, Chikamatsu K (2017) Cancer-associated fibroblasts promote an immunosuppressive microenvironment through the induction and accumulation of protumoral macrophages. Oncotarget 8(5):8633–8647. https://doi.org/10.18632/oncotarget.14374

Takai K, Le A, Weaver VM, Werb Z (2016) Targeting the cancer-associated fibroblasts as a treatment in triple-negative breast cancer. Oncotarget 7(50):82889–82901. https://doi.org/10.18632/oncotarget.12658

Tanaka A, Sakaguchi S (2017) Regulatory T cells in cancer immunotherapy. Cell Res 27(1): 109–118. https://doi.org/10.1038/cr.2016.151

Thompson C, Powrie F (2004) Regulatory T cells. Curr Opin Pharmacol 4(4):408–414. https://doi.org/10.1016/j.coph.2004.05.001

Tzanakakis G, Neagu M, Tsatsakis A, Nikitovic D (2019) Proteoglycans and immunobiology of cancer-therapeutic implications. Front Immunol 10:875. https://doi.org/10.3389/fimmu.2019.00875

Vincent J, Mignot G, Chalmin F, Ladoire S, Bruchard M, Chevriaux A, Martin F, Apetoh L, Rebe C, Ghiringhelli F (2010) 5-Fluorouracil selectively kills tumor-associated myeloid-derived suppressor cells resulting in enhanced T cell-dependent antitumor immunity. Cancer Res 70(8): 3052–3061. https://doi.org/10.1158/0008-5472.CAN-09-3690

Wells JM, Gaggar A, Blalock JE (2015) MMP generated matrikines. Matrix Biol 44-46:122–129. https://doi.org/10.1016/j.matbio.2015.01.016

Wight TN, Kang I, Evanko SP, Harten IA, Chang MY, Pearce OMT, Allen CE, Frevert CW (2020) Versican-A critical extracellular matrix regulator of immunity and inflammation. Front Immunol 11:512. https://doi.org/10.3389/fimmu.2020.00512

Wu Q, Allouch A, Martins I, Brenner C, Modjtahedi N, Deutsch E, Perfettini JL (2017) Modulating both tumor cell death and innate immunity is essential for improving radiation therapy effectiveness. Front Immunol 8:613. https://doi.org/10.3389/fimmu.2017.00613

Yamauchi M, Barker TH, Gibbons DL, Kurie JM (2018) The fibrotic tumor stroma. J Clin Invest 128(1):16–25. https://doi.org/10.1172/JCI93554

Yu Q, Stamenkovic I (1999) Localization of matrix metalloproteinase 9 to the cell surface provides a mechanism for CD44-mediated tumor invasion. Genes Dev 13(1):35–48. https://doi.org/10.1101/gad.13.1.35

Yuzhalin AE, Urbonas T, Silva MA, Muschel RJ, Gordon-Weeks AN (2018) A core matrisome gene signature predicts cancer outcome. Br J Cancer 118(3):435–440. https://doi.org/10.1038/bjc.2017.458

Zhang H, Yue J, Jiang Z, Zhou R, Xie R, Xu Y, Wu S (2017) CAF-secreted CXCL1 conferred radioresistance by regulating DNA damage response in a ROS-dependent manner in esophageal squamous cell carcinoma. Cell Death Dis 8(5):e2790. https://doi.org/10.1038/cddis.2017.180

Chapter 18
Exploiting Hyaluronan-CD44 Network in Tumor Therapy

Theodoros T. Karalis and Spyros S. Skandalis

Abstract Cancer is one of the leading causes of death worldwide. During tumor development and progression, extracellular matrix is being intensively re-organized. One of the most abundant molecules in the extracellular matrix is the polysaccharide hyaluronan. Hyaluronan accumulation and high CD44 expression—the major hyaluronan cellular receptor—correlate with higher malignant states of cancer cells, increased incidence of metastases and poor prognosis of the patients in a wide array of tumor types. Thus, hyaluronan interaction with CD44 emerges as an important target for cancer treatment. In this chapter, recent efforts to exploit hyaluronan/CD44 network for tumor therapy are being discussed. Overall, there is a wide variety of tools available to target this system like anti-CD44 antibodies and peptides, gene therapies against CD44, nanotechnology and modified-hyaluronan. Initial in vivo evidence shows promising results, nominating hyaluronan/CD44 network targeting as a potent candidate to be introduced into clinical settings for tumor therapy.

18.1 Introduction

Cancer is one of the most common-occurring diseases and one of the leading causes of death worldwide. During cancer development, progression, and metastasis, the extracellular matrix is extensively re-organized (Theocharis et al. 2016; Larsen et al. 2006). One of the most abundant molecules in the extracellular matrix is the polysaccharide hyaluronan, which is differentially regulated during tumorigenesis. Physiologically, hyaluronan is found in many tissues such as the skin and the cartilage, where owing to its viscoelastic properties regulates several physiological characteristics of the tissues as well as proper cell function. In cancer, hyaluronan

T. T. Karalis · S. S. Skandalis (✉)
Biochemistry, Biochemical Analysis & Matrix Pathobiology Res. Group, Laboratory of Biochemistry, Department of Chemistry, University of Patras, Patras, Greece
e-mail: skandalis@upatras.gr

I. Kovalszky et al. (eds.), *The Extracellular Matrix and the Tumor Microenvironment*, Biology of Extracellular Matrix 11,
https://doi.org/10.1007/978-3-030-99708-3_18

metabolism is altered compared to normal tissues and contributes to acquirement of tumor cell malignant properties (Skandalis et al. 2020; Karousou et al. 2017; Heldin et al. 2019). One of the main hyaluronan receptors, CD44, is considered a tumor promoter for most malignancies and plays significant roles in the regulation of cancer cell survival, proliferation, migration, invasion, metastasis, differentiation and drug resistance (Skandalis et al. 2019; Misra et al. 2011; Morath et al. 2016). Given the important roles of hyaluronan and its receptor CD44 in cancer, targeting this system opens a new avenue for the treatment of malignant diseases. Thereby, here we summarize the latest developments in the field of hyaluronan/CD44 network targeting for tumor therapy.

18.1.1 Hyaluronan Synthesis and Catabolism

Hyaluronan is a glycosaminoglycan consisting of repeating disaccharide units of *N*-acetyl-glucosamine (GlcNAc) and D-glucuronic acid (GlcUA), bound through alternate β1–3 and β1–4 linkages (Weigel 2015). Hyaluronan is synthesized in the plasma membrane by specific enzymes termed hyaluronan synthases (HASes). Three hyaluronan synthases have been described in the human genome, *HAS1–3*, which are encoded by distinct genes. *HAS1* is found in 19q13.41 (www.genecards.org, GCID: GC19M054887), *HAS2* in 8q24.13 (www.genecards.org, GCID: GC08M121594) and *HAS3* in 16q22.1 (www.genecards.org, GCID: GC16P069105). Currently there is no information regarding their protein structure, but it is known that they have six transmembrane domains as well as UDP-sugar binding regions (Weigel 2015). Although the three hyaluronan synthases are structurally related they display different spatio-temporal expression and activities. For example, higher expression of HAS1 can be found in adipose tissue and the ovary, HAS2 is highly expressed in the adipose tissue, while HAS3 is expressed in urinary bladder, esophagus, and the lung (proteinatlas.org). Hyaluronan synthases synthesize hyaluronan of different molecular weights. Specifically, HAS2 synthesizes hyaluronan with molecular weight higher than 2×10^6 Da, in contrast to HAS1 and HAS3 which produce hyaluronan ranging from 2×10^5 to 2×10^6 Da (Itano et al. 1999). Furthermore, HAS1 displays the lowest catalytic activity compared to the other HASes, while HAS2 is less active than HAS3. Moreover, HAS1 exhibits higher Km values for both UDP-GlcUA and UDP-GlcNAc compared to HAS2 and HAS3 (Itano and Kimata 2002).

These membrane-embedded glycosyl-transferases utilize as substrates UDP-GlcUA and UDP-GlcNAc which are derived from various metabolic pathways inside the cells (Vigetti et al. 2012; Flores-Diaz et al. 1997). Accordingly, the availability of UDP-sugars is a major factor that regulates hyaluronan biosynthesis (Hascall et al. 2014; Rilla et al. 2013). This in turn suggests that specific microenvironment conditions regulate hyaluronan production, since it has been shown that high glucose upregulates hyaluronan synthesis (Wang et al. 2014). Moreover, the presence of Mg^{2+} is crucial for hyaluronan production (Weigel 2015). The enzymatic activity of HASes is also regulated by their sub-cellular localization and trafficking from and toward the plasma membrane, where hyaluronan is normally synthesized. HAS1 is mainly localized in Golgi, HAS3 in Golgi and membrane protrusions, while HAS2 is found in the endoplasmic reticulum. The fact that HASes mainly reside inside the cells suggests that there is a reservoir of enzymes ready to translocate in plasma membrane and produce hyaluronan upon stimulation (Torronen et al. 2014). Indicative of this notion is the finding that HAS3 synthesizes a pericellular hyaluronan stroma after its translocation to the plasma membrane (Deen et al. 2014). Furthermore, hyaluronan synthase activity is also regulated by post-translational modifications such as *O*-GlcNAcylation (Vigetti et al. 2012), poly- and mono-ubiquitinylation (Karousou et al. 2010; Mehic et al. 2017) and phosphorylation (Vigetti et al. 2011), while homo- and hetero-dimerization of HAS2 with any of the other synthases leads to concomitant increase in hyaluronan production (Karousou et al. 2010; Bart et al. 2015).

HAS gene regulation is another crucial factor that controls hyaluronan production by the cells in different tissues. Different transcription factors bind to the promoter of each *HAS* gene and control their expression. Specifically, *HAS1* gene contains binding elements for SP1/3, SMAD, and E2F-myc. *HAS2* is regulated by the binding of CREB, NF-κB, RAR, STAT3, YY1, ZEB1, E2F-myc, and SP1, while HAS2-AS1 encoding—a long non-coding RNA that also regulates hyaluronan production by HAS2—is controlled by NF-κB, SP1/3, SMAD and HIF1α. Finally, *HAS3* expression is controlled by binding of ΔN-p63, NF-κB, C/EBP and SP1 in its promoter regions (Heldin et al. 2019).

Furthermore, hyaluronan synthesis is tightly regulated by several growth factors that by inducing intracellular signaling pathways control the expression of several HAS isoforms. The effect of each growth factor on hyaluronan synthesis is cell- and tissue-type-specific. Such growth factors include PDGF-BB, TGF-β, TNFα and IL-1, among others (Heldin et al. 2019).

Hyaluronan amount in the tissues is also modulated by catabolism from specific hyaluronan-degrading enzymes termed hyaluronidases (HYALs). In human, hyaluronan is recycled with high rates. Almost one-third of the total hyaluronan amount can be found in the skin and its half-life ranges from one to one and a half day (Pandey et al. 2008). Hyaluronan in the tissues has initial size about 1000–10,000 kDa and is degraded in the extracellular space in smaller fragments (10–100 kDa) (Fraser et al. 1997). Next, most of these generated fragments are drained through the lymphatic system and degraded in the lymph nodes. The remaining fragments enter the bloodstream and are finally removed by the liver, kidneys, and spleen (Pandey et al. 2008).

In the human genome, several hyaluronidase genes have been found and are encoded by different genes. HYAL-1 and HYAL-2 are widely expressed in several tissues.

HYAL-2 bears a glycosyl-phosphatidyl-inositol (GPI) tail which anchors the protein to the outside of plasma membrane mainly in lipid rafts, together with hyaluronan receptor CD44. HYAL-3 function despite its wide expression has not been fully deciphered (Shuttleworth et al. 2002; Flannery et al. 1998), HYAL-4 degrades chondroitin sulfate (CS) chains, while PHYAL-1 is a pseudogene and is not expressed in human. PH-20/SPAM1 is a hyaluronidase that is expressed in the testis and displays significant hyaluronan degrading activity (Cherr et al. 2001; Baba et al. 2002). Until recently, the widely recognized model for hyaluronan degradation suggested that hyaluronan residing in the extracellular space is degraded initially by HYAL-2 in 20 kDa fragments which then enter the cell by endocytosis through caveolae pathways in endosomes and transported to lysosomes for further degradation to disaccharides by HYAL-1 and exoglycosidases (Montanari et al. 2018). Recently, two new hyaluronidases that need to fit into the scheme of hyaluronan catabolism were discovered, HYBID/CEMIP/KIAA1199 and TMEM2. The transcription and translation of *CEMIP/KIAA1199* gene produces a 153 kDa protein that contains a 30 amino acid N-terminal domain that is required for hyaluronan degradation (Yoshida et al. 2013b). Moreover, it contains seven *N*-glycosylation sites, one G8 domain, two GG, and four PbH1 regions. The GG regions seem to be implicated in the process of hyaluronan degradation, G8 domain in interaction with other proteins and PbH1 in poly-saccharide hydrolysis (Guo et al. 2006; He et al. 2006; Birkenkamp-Demtroder et al. 2011; Yoshida et al. 2013a). Hyaluronan degradation by HYBID is performed by endocytosis in clathrin-coated vesicles with acidic pH, while the resulting hyaluronan fragments are released in the extracellular space (Yoshida and Okada 2019). The *TMEM2* gene product is a transmembrane protein of 154 kDa. The hyaluronidase TMEM2 contains one G8, one GG, and three PbH1 domains in the extracellular region, a transmembrane region, and a cytoplasmic tail (Yamaguchi et al. 2019). In contrast to HYAL-1/-2 which degrade hyaluronan in acidic pH, TMEM2 has optimal pH of enzymatic activity at 6–7. TMEM2 is located in plasma membrane and degrades extracellular hyaluronan in fragments of intermediate size, which then are endocytosed and degraded further in the lysosomes. The enzymatic activity of TMEM2 requires Ca^{2+} as a co-factor (Yamaguchi et al. 2019; Yamamoto et al. 2017). Finally, hyaluronan can be degraded by non-enzymatic ways by the action of reactive oxygen species (ROS) produced by diverse cellular metabolic pathways (Soltes et al. 2006; Agren et al. 1997).

18.1.2 CD44

Hyaluronan synthesized in the plasma membranes is subsequently extruded to the extracellular space where it can interact with several extracellular proteins, such as the proteoglycan family hyalectans, or plasma membrane receptors, and thus regulate several cell functional properties. The hyaluronan receptors discovered so far include CD44 (Cluster of Differentiation 44), RHAMM (Receptor for Hyaluronan-Mediated Motility), HARE/STAB2 (Hyaluronic Acid Receptor for Endocytosis/Stabilin-2), Laylin, Stabilin-1 and LYVE-1 (Lymphatic Vessel Endothelial Hyaluronic Acid Receptor 1).

The major and best characterized receptor for hyaluronan is CD44, which is expressed in various cell types and tissues. *CD44* gene is located in 11p13 (genecards.org) and its transcript mRNA is subjected to alternative splicing which after translation leads to encoding of several different isoforms of the receptor. Human *CD44* gene contains 19 exons, 10 standard (S1–10) and 9 variants (V2–10). The S1–10 exons are retained in all CD44 isoforms. CD44s (CD44 standard) isoform does not contain any of the V2–10 exons, while alternative splicing of V2–10 gives rise to CD44v (CD44 variants). From the mRNA translation the final product is a transmembrane protein with an extracellular region containing a LINK domain—which is responsible for hyaluronan/CD44 interactions—and a stalk-like domain—in which the exons V2–V10 are introduced—a transmembrane region, and a small cytoplasmic tail. The extracellular domain is modified by *O*- and *N*-glycosylations, while specific CD44 isoforms, such as CD44v3, bear covalently bound glycosaminoglycan chains (chondroitin sulfate or heparan sulfate chains), contributing further diversity to the resulting proteins. The cytoplasmic tail of CD44 despite not containing intrinsic kinase activity regulates several signaling pathways through interactions with cytoplasmic proteins such as Src, ERM (Ezrin, Radixin, Moesin) and IQGAP1 (Zoller 2011; Skandalis et al. 2010). Moreover, CD44 intracellular domain can be cleaved by γ-secretase and translocated to the nucleus where it regulates the expression of several genes (e.g., MMP9) (Miletti-Gonzalez et al. 2012). Apart from cellular signaling, CD44 can also participate in the endocytosis of hyaluronan, leading to its degradation (Thankamony and Knudson 2006; Skandalis et al. 2020).

18.2 Roles of Hyaluronan-CD44 Network in Tumors

Apart from the diverse roles of hyaluronan and its receptor CD44 in physiological processes like embryogenesis and cartilage function they play significant roles in tumor development and progression. In tumors, hyaluronan creates a highly hydrated extracellular matrix with specific physicochemical properties which allows cancer cells to proliferate and migrate. Moreover, hyaluronan synthesized by tumor stromal cells or cancer cells themselves, engages CD44 on the surface of cancer cells to regulate biological processes like growth/survival, epithelial-to-mesenchymal transition (EMT), differentiation, invasion, metastasis, drug resistance and cancer stem cell properties.

18.2.1 Growth/Survival

For tumors to successfully form, cancer cells need to deal with several stressful events, such as anchorage-independent growth, hypoxia and limited nutrient availability. Therefore, it is critical for cancer cells to take advantage of physiological molecular mechanisms and pathways allowing them to cope with such stressful events. One of the major cellular receptors correlated with survival and anti-apoptotic signaling in cancer cells is the hyaluronan receptor CD44. CD44 regulates the expression and activation of proteins involved in resistance to apoptosis and cell growth like Fas, caspase 3/9, Bcl-xl/Bak, Akt, pRb and Bcl-2 (Lakshman et al. 2004; Yasuda et al. 2001; Park et al. 2012b). For example, overexpression of CD44s, CD44v3–10, and CD44v8–10 in human colon cancer cells successfully attenuated etoposide-induced cell death (Lakshman et al. 2004). On the other hand, inhibition of CD44 expression in colon carcinoma cells reduced the expression of Bcl-2, Bcl-xL, while simultaneously increased the expression of apoptosis proteins Bax and caspase-3/8/9 (Park et al. 2012b). Furthermore, CD44 controls downstream activation of Akt, a major survival pathway, and cell cycle-regulating proteins p21 and pRb in many types of cancer cells, like breast, colon and lung cancer cells (Lakshman et al. 2004). In lung cancer cells, interaction of hyaluronan with CD44 reduced Fas expression and subsequent Fas-mediated apoptosis (Yasuda et al. 2001). In chronic lymphocytic leukemia (CLL) patients CD44 was found to promote cancer cell survival. The crucial importance of CD44 in CLL was further certified by the fact that CD44 knock-down reduced survival even in Akt-overexpressing cells. In that model, CD44 regulated the expression of MCL1 anti-apoptotic protein through Akt and Erk pathways (Fedorchenko et al. 2013). Hyaluronan engaged to CD44 also induced phosphorylation of FAK, which associates with PI3K to protect against apoptosis (Fujita et al. 2002). Furthermore, hyaluronan-CD44 interaction activated ErbB2 signaling through Hsp90, cdc37, p110 and p85 proteins (Chanmee et al. 2015).

18.2.2 Epithelial-to-Mesenchymal Transition (EMT) and Differentiation

Epithelial-to-mesenchymal transition is a dynamic process that cancer cells utilize in order to metastasize. During this process, hyaluronan synthase expression is induced and hyaluronan is synthesized to large amounts. Moreover, CD44 has been found to be overexpressed in mesenchymal cancer cells and its high expression correlates with a more undifferentiated phenotype (Misra et al. 2011; Heldin et al. 2014). In breast cancer, EMT correlates with poor prognosis and intriguingly, breast cancer cells with mesenchymal and more malignant phenotype express higher amounts of CD44, which also displays high hyaluronan-binding capacity (Bernert et al. 2011; Heldin et al. 1996). In breast epithelial cultures, TGF-β-induced EMT depends on the expression of HAS2. Specifically, TGF-β induces Smad and p38 MAPK pathways to upregulate HAS2 expression (Porsch et al. 2013). Moreover, HAS2 overexpression has been shown to promote the malignant phenotype via suppression of E-cadherin and translocation of β-catenin to the nucleus, signaling events that take place during EMT (Zoltan-Jones et al. 2003; Koyama et al. 2007). Interestingly, switching between CD44v isoforms to CD44s through alternative splicing promoted EMT by suppression of E-cadherin through PI3K/Akt pathways. Inhibition of splicing activity that produced CD44v occurred through downregulation of epithelial splicing regulatory protein 1 and 2 (ESRP1 and 2) by transcription factors Snail1, Zeb1, and Zeb2 (Reinke et al. 2012). Expression of CD44 and its interaction with hyaluronan also regulate differentiation of aggressive cancer cells like acute myeloid leukemia cells, thus downregulating their aggressive properties (Solis et al. 2012).

18.2.3 Invasion/Metastasis

Accumulation of hyaluronan as well as high CD44 expression in the cancerous tissues has been widely correlated with advanced incidence of invasion and metastasis of various types of cancer cells. CD44 interacts with several growth factor receptors like ErbB2 and PDGFR and its interaction with hyaluronan regulates their signaling activity (Bourguignon et al. 1997; Ghatak et al. 2005; Li et al. 2006). Binding of hyaluronan induces CD44 clustering, which activates downstream signaling pathways in a cell- and tissue-dependent manner. This clustering was critical for MMP9 activation and subsequent activation of TGF-β, which in turn induced cancer cell invasion and metastasis (Yu and Stamenkovic 1999). Moreover, CD44 interacted with MT1-MMP—a major metalloproteinase responsible for extracellular matrix degradation during invasion and metastasis—to enhance its activity. In turn, MT1-MMP enhanced the shedding of CD44 variants and promoted cancer cell invasiveness (Kajita et al. 2001; Stamenkovic and Yu 2009; Mori et al. 2002). Hyaluronan fragments also have a functional role during invasion and metastasis. Specifically, hyaluronan dodecasaccharides by engaging CD44 induced secretion of CXCL1, to enhance endothelial cell sprouting, which could be critical for tumor angiogenesis, a key process in metastasis (Takahashi et al. 2005). Hyaluronan is also synthesized by stromal cells. This stromal cell-derived hyaluronan engaged CD44v6 in the surface of colon tumor cells to sustain PI3K signaling in a positive CD44v6/PI3K loop and promote invasion (Misra et al. 2011).

18.2.4 Drug Resistance

One of the main properties that cancer cells acquire during tumor development is resistance to several drugs. Drug resistance in the tumors is thought to be acquired through different mechanisms. First, a subpopulation of cells with pre-existing potential for drug resistance could be present inside a tumor. Second, drugs utilized for cancer therapy can induce resistance by cancer cells. In both cases, cancer cells upregulate the expression of proteins, like multi-drug resistance proteins (MDRs) that allow them to evade drug-induced apoptosis. Intriguingly, there is extensive evidence correlating CD44 expression with resistance to radiotherapy or chemotherapy of various types of cancer cells (Yaghobi et al. 2021). CD44 physically interacts with P-glycoprotein to enhance drug resistance in cancer cells (Miletti-Gonzalez et al. 2005). In malignant cells, decrease in CD44 expression ameliorated drug resistance (Xu et al. 2015). Specifically, in hepatocellular carcinoma, inhibition of CD44 sensitized cancer cells to sorafenib (Fernando et al. 2015). Paclitaxel-resistant ovarian cells also showed increased expression of CD44 (Gao et al. 2015). It is important to note that interaction of high molecular weight hyaluronan with CD44 increased MDR expression and subsequently led to drug resistance, while low molecular weight hyaluronan-CD44 interaction led to MDR internalization, signifying that not only the presence of hyaluronan but also its size is critical for this process (Zoller 2011). Moreover, hyaluronan through CD44 binding activated PI3K signaling, which in turn induced MDR protein expression (Misra et al. 2005). In the context of CD44v, CD44v3 targeting in head and neck squamous cell carcinoma cells reduced resistance to cisplatin (Wang et al. 2007). Hyaluronan-CD44v3 interaction upregulates miRNA-302, enhanced Oct4-Sox2-NANOG signaling and increased expression of MDR1, leading to development of chemo-resistance (Bourguignon et al. 2012). CD44v3-hyaluronan interaction also regulated expression and activity of P300 which can in turn acetylate β-catenin and NF-κB-p65, resulting in upregulation of MDR1 expression (Zoller 2015). Interestingly, in prostate cancer, knocking down the expression of CD44v6 enhanced chemotherapy sensitivity (Ni et al. 2014).

18.2.5 Tumor Stem Cell Properties

In tumors, sub-populations of cancer cells, termed cancer stem cells (CSCs), have been shown to be responsible for tumor recurrence after chemotherapy or radiotherapy. Hyaluronan, like in normal stem cell niches, creates an ideal micro-environment for CSCs survival, self-renewal, and maintenance. Importantly, hyaluronan receptor CD44 is widely recognized as a stem cell marker (Skandalis et al. 2019). $CD44^{High}CD24^{Low}$ populations isolated from tumors of the breast showed stem cell properties, like self-renewal and tumor-initiating capacity (Shao et al. 2016; Li et al. 2017b; Wei et al. 2012). Mechanistically, ΔNp63 induced the expression of HAS3, HYAL-1, and CD44 to create a hyaluronan-rich environment that favored stemness of breast cancer cells (Gatti et al. 2018). The importance of hyaluronan and CD44 was further solidified by experiments in hyaluronan-based multilayer nanofilms, where pancreatic cells grown in such conditions upregulated CD44v6 and form colonies (Lee et al. 2018). In pancreatic tumor cells, the tumor suppressor KFL4 bound to CD44 promoter to block its expression and ameliorated cancer stem cell properties and metastasis (Yan et al. 2016). On the other hand, targeting hyaluronan with 4-methyl-umbelliferone—a widely utilized hyaluronan synthesis inhibitor—promoted phagocytosis of hepatocellular CSCs (Rodriguez et al. 2018). CD44 also activated Wnt/β-catenin pathway to upregulate FoxM1 and Twist, leading to enhanced stemness of lung adenocarcinoma cells (Su et al. 2016). Circulating oral squamous carcinoma cells that display CD44 expression were able to form spheres and displayed chemoresistance and self-renewal (Patel et al. 2016). For a more comprehensive analysis of hyaluronan/CD44 roles in cancer stem cell properties, refer to our previous review (Skandalis et al. 2019).

18.3 Hyaluronan/CD44 Network Targeting

Given the important roles of hyaluronan/CD44 network in tumor development, progression, and metastasis, this system emerges as a potent pharmacological target. Therefore, several efforts have been concentrated toward this purpose, such as incorporation of hyaluronan on nanoparticles for specific delivery of drugs to tumor cells, antibodies or peptides blocking CD44 actions and interactions, chemically modified hyaluronan, and utilizing gene therapies (CRISPR/Cas9 or sh/siRNAs) against CD44 (Fig. 18.1). Below, we describe recent advances in pharmacological targeting of this system for cancer treatment (summarized in Table 18.1).

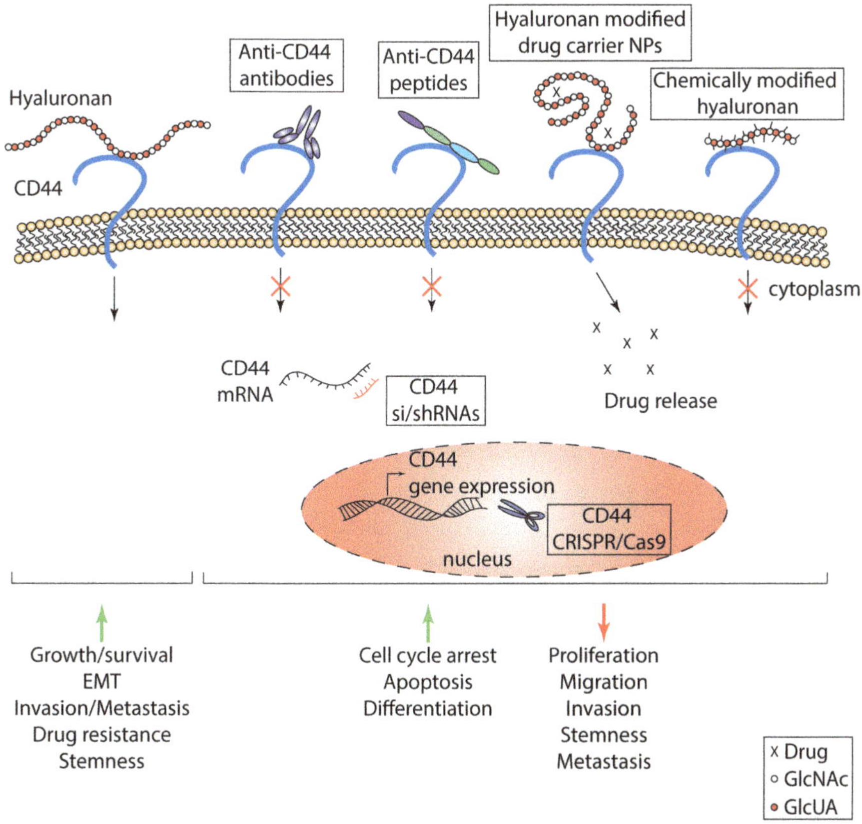

Fig. 18.1 Hyaluronan/CD44 network targeting approaches. Incorporation of hyaluronan on nanoparticles for specific delivery of drugs to tumor cells, antibodies or peptides blocking CD44 actions and interactions, chemically modified hyaluronan and gene therapies targeting CD44 mRNA (CRISPR/Cas9 or sh/siRNAs against CD44)

Table 18.1 Overview of hyaluronan/CD44 targeting tools (for further details, see text)

Approach	Tool	Tumor	Function	References
Nanomedicine	HA-PEI-PLGA-docetaxel NPs	Lung	↓ Proliferation	Maiolino et al. (2015)
	HA-PLGA-docetaxel NPs	Lung	↓ Tumor growth	Wu et al. (2017)
	HA-Dextran-modified quercetin-Cu(II) NPs with PARP inhibitor	Breast	↑ Animal survival	Cheng et al. (2021)
	HA-Resverastrol/chalcone NPs	Breast	↑ Drug cytoplasmic release	Shah et al. (2017)
	HA-Doxorubicin-super paramagnetic iron oxide NPs	Breast	↑ Drug uptake and release, apoptosis	Vyas et al. (2015)
	HA-Berberine derivative/doxorubicin NPs	Breast	↑ Mitochondrial apoptotic pathways ↓ Proliferation, migration, lung metastasis	Lin et al. (2021)
	HA-Chitosan-docetaxel NPs	Breast	↓ Cell viability	Shabani Ravari et al. (2016)
	HA-Curcumin/celecoxib NPs	Breast	↑ Apoptosis, animal survival ↓ Tumor growth, cell viability, lung metastasis	Liu et al. (2020)
	HA-Paclitaxel/curcumin PAAVB NPs	Breast	↑ Adaptive anti-tumor immunogenicity ↓ Immunosuppression, lung metastasis	Wang et al. (2021)
	HA-(PANi)-imiquimod NPs	Breast	↑ Immune responses against tumor	Yasothamani et al. (2021)
	HA-Salinomycin/curcumin NPs	Breast (CSCs)	↑ G1 cell cycle arrest ↓ EMT	Zhao et al. (2020)
	HA-NPs with saporin	Breast	↑ Apoptosis	Ding et al. (2018)
	HA-Hyaluronidase/Ce6 NPs	Breast	↑ Uptake, efficacy	Gong et al. (2016)
	HA-Gold-nanoclustered verteporfin NPs	Breast	↑ Blood stream stability, drug delivery ↓ Tumor growth	Han et al. (2016)
	Multifunctionalized iron oxide magnetic nanoparticles (MNPs) with anti-CD44 antibody and gemcitabine derivatives	Breast and pancreatic	↑ Cancer cell killing	Aires et al. (2016)
	HA-Cisplatin green tea catechin micelles	Ovarian	↑ Drug efficacy ↓ Off-target effects	Bae et al. (2017)

	Chitosan polylactic acid- coated with IM7 anti-CD44 antibody	Ovarian	↓ Toxicity	Yang et al. (2017)
	HA-Serum albumin- conjugated NPs	Ovarian	↑ Drug uptake, cytotoxicity against cancer cells	Edelman et al. (2017)
	HA-Paclitaxel NPs	Colon	↑ Animal survival	Zhu et al. (2018)
	HA-PTC209 NPs	Colon (CSCs)	↓ Tumor growth	Xu et al. (2019)
	HA-Ce6 phosphosensitizer NPs	Colon	↓ Tumor growth	Gao et al. (2017)
	HA-poly(hexamethylene biguanide) and chitosan NPS with KRAS si/shRNA	Colorectal	↑ Uptake and KRAS silencing	Tirella et al. (2019)
	HA-Doxorubicin lipoic-acid encapsulated NPs	Myeloma and acute myeloid leukemia	↓ Tumor growth	Zhong et al. (2017)
	A6 anti-CD44 peptide polymersomal epirubicin	Multiple myeloma	↑ Uptake, anti-cancer efficacy	Gu et al. (2019)
	HA-Cabazitaxel/Silibinin liposomes	Prostate (CSCs)	↑ Apoptosis ↓ Migration	Mahira et al. (2019)
	CDDP NPs	Prostate (CSCs)	↓ Stem cell properties	Jafari Malek et al. (2014)
	HA-Diethyldithiocarbamate- copper NPs	Pancreatic (CSCs)	↑ ROS generation ↓ Sphere formation	Marengo et al. (2019)
	HA-chitosan NPs with Bcl- 2 siRNA	Bladder	↓ Bcl-2 expression	Liang et al. (2021)
	HA-HPLR- with EGFR targeted peptides	Liver	↓ Tumor growth	Liang et al. (2019)
	HA NPs with Gli1 siRNA	Gastric (CSCs)	↓ Migration, invasion, tumor spheroid, and colony formation	Yao et al. (2020)
	HA-docetaxel NPs with AS1411 aptamer	Glioblastoma	↑ Blood-brain barrier penetration, drug uptake ↓ Spheroid formation, tumor growth	Wang et al. (2019)
	HA-Doxorubicin liposome NPs	Glioblastoma	↑ Anti-neoplastic function, doxorubicin efficacy	Hayward et al. (2016)
	HA-Zinc protoporphyrin nanoprobes	Sarcoma	↑ Anti-cancer effect ↓ Side effects	Gao et al. (2021)

(continued)

Table 18.1 (continued)

Approach	Tool	Tumor	Function	References
Antibodies	RG7356	Chronic lymphocytic leukemia	↑ Apoptosis	D'arena et al. (2014) and Zhang et al. (2013)
		Breast	↑ Secretion of immune cell chemo-attractants, antibody-dependent cellular phagocytosis ↓ Tumor growth	Weigand et al. (2012) and Maisel et al. (2016)
	A3D8	Acute myeloid leukemia	↑ G0/G1 cell cycle arrest	Zada et al. (2003), Gadhoum et al. (2004a, b), and Qian et al. (2012)
		Chronic lymphocytic leukemia	↑ Caspase activation ↓ Cell viability	Fedorchenko et al. (2013)
		Erythroleukem ia	↑ Caspase-independent apoptosis ↓ Cell growth	Artus et al. (2006)
		Ovarian (CSCs)	↑ S phase cell cycle arrest, apoptosis	Du et al. (2013)
	F77	Prostate	↑ Apoptosis	Chen et al. (2018b)
	Hermes-1	Colorectal carcinoma	↓ Binding of hyaluronan, laminin and collagen	Ishii et al. (1993)
	IM7	Colorectal carcinoma	↓ Liver metastasis	Ogoshi et al. (1998)
		Glioblastoma	↑ Apoptosis ↓ Hyaluronan biosynthesis	Wiranowska et al. (2010)
		Ovarian	↓ Proliferation, tumor development, and progression	Yang et al. (2017)
		Bladder	↓ NF-κB activation	Fitzgerald et al. (2000)
		Melanoma	↓ Adhesion of melanoma cells on endo-thelial cells	Ota et al. (1995)
	IM7 and S5	Natural killer cells	↑ Natural killer cell activity	Tan et al. (1993)
	IM7 conjugated to saporin	Prostate	↑ Apoptosis	Bostad et al. (2014)
	J173 and F10442	Natural killer cells	↑ CD16-mediated lysis, intracellular Ca^{2+}	Galandrini et al. (1994)

	J173	Burkitt's lymphoma and chronic myelogenous leukemia	↑ Killing activity of peripheral mononuclear cells against cancer cells	Ishizuka et al. (2008)
	HI44α	Acute myeloid leukemia	↑ Differentiation, apoptosis	Song et al. (2004)
	KMP1	Bladder	↓ Cell proliferation, migration and adhesion, tumor growth	Chen et al. (2018c)
	aCD44O^{2}LNCs with paclitaxel	Pancreatic (CSCs)	↑ Anti-tumor efficacy	Navarro-Marchal et al. (2021)
	5-mG2a-f	Oral squamous cell carcinoma	↓ Tumor development	Takei et al. (2020)
	P3D2	Breast	↓ Tumorigenesis	Lusche et al. (2021)
	Anti-CD44v9 mAb 44-IV	Colon	↓ Cancer cell capillary adhesion, liver metastasis	Seki et al. (1997)
Peptides	RP-1	Gastric	Detection of CD44$^+$ cancer cells, prognosis prediction	Zhang et al. (2015) and Li et al. (2017a)
	CV-1	Gastric	Detection of CD44v3–10$^+$ cancer cells	Zhang et al. (2016)
	PFT	Prostate	Detection of CD44v6$^+$ cancer cells	Peng et al. (2017)
	PDPP	Breast	Detection of CD44$^+$ CSCs	Cho et al. (2015)
	A6	Chronic lymphocytic leukemia	↑ Toxicity	Finlayson (2015)
		Ovarian and breast	↓ Migration	Piotrowicz et al. (2011)
		Melanoma	↓ Lung metastasis	Piotrowicz et al. (2011)
	A6 with reduction-sensitive polymersomal vincristine sulfate	Acute myeloid leukemia	↑ Animal survival ↓ Leukemia burden in the circulation, bone marrow, liver, and spleen	Gu et al. (2021)
	v6	Pancreatic	↑ Animal survival ↓ Tumor growth	Matzke-Ogi et al. (2016)
	NLN and NEW	Breast and pancreatic	↑ CD44v6 internalization ↓ c-Met/Erk activation, migration, invasion, metastasis	Khan et al. (2021)

(continued)

Table 18.1 (continued)

Approach	Tool	Tumor	Function	References
	NLN and NEW with KLA pro-apoptotic peptides	Breast and pancreatic	↑ Tumor cell killing ↓ Tumor growth, metastasis, side effects	Khan et al. (2021)
	FKBPL and AD-01	Breast and prostate	↑ Anti-angiogenic effects ↓ Tumor growth	Valentine et al. (2011)
		Breast	↓ Migration	Yakkundi et al. (2013)
	AD-01	Breast (CSCs)	↑ Differentiation ↓ Mammosphere formation, tumor initiation, chemotherapy, and radiotherapy resistance when combined with DAPT	McClements et al. (2013)
	Peptides derived from collagen type IV with liposomes and doxorubicin	Melanoma	↓ Tumor size	Ndinguri et al. (2012)
	P3 and P9	Chronic lymphocytic leukemia	↓ Adhesion to pro-MMP9, chemotaxis, transendothelial migration	Ugarte-Berzal et al. (2014)
	PCK3145	Fibrosarcoma	↓ Adhesion to hyaluronan	Annabi et al. (2005)
	A5G27	Melanoma	↓ Metastasis	Hibino et al. (2005)
	A5G27 with siRNA particles	Lung and ovarian	↓ Tumor growth	Golan et al. (2016)
	CD44BP	Breast	↓ Tumor sphere formation	Yang et al. (2013)
	CD44 cytoplasmic tail peptides bearing phosphor- $_{Ser}$325	Melanoma	↓ Migration	Peck and Isacke (1998)
	CD44-derived peptides as immunogens	Prostate (CSCs)	↑ Sensitize dendritic cells, anti-tumor activities	Wang et al. (2020)
Chemically modified hyaluronan	Modification with Au-Ag alloy in NPs	Breast	↑ Sensitivity to radiotherapy	Chong et al. (2020)
	Modification with a hydrazide group and bisphosphonate (BP)	Colon	↑ Targeting	Varghese et al. (2009)
	Modification with sulfate groups	Pancreatic	↑ Targeting of CD44 and P-selectin	Bhattacharya et al. (2020)

	Modification with poly(lactic-*co*-glycolic acid) and loaded to transferrin- targeted nanoformulated AUY922 NPs	Brain	↑ Uptake, apoptosis, side effects ↓ Tumor growth	Debele et al. (2021)
	Modification with β-cyclodextrin and drug conjugates	Lung and prostate	↑ Toxicity against cancer cells ↓ Toxicity against normal cells	Bai et al. (2020)
Gene therapies	CD44 siRNA	Lung	↓ Stem cell gene expression	Nurwidya et al. (2017)
	CD44 siRNA in combination with doxorubicin	Breast	↑ Apoptosis ↓ Tumor metastasis, proliferation, invasion, migration	Vahidian et al. (2020)
	CD44 siRNA in liposomes with Apt1 CD44-targeted aptamer	Breast	↓ CD44 expression	Alshaer et al. (2018)
	CD44 shRNA	Breast	↓ Proliferation, colony formation, invasion	Zhou et al. (2018)
	PLGANPs with FAK and CD44 siRNAs	Ovarian	↑ Apoptosis ↓ Tumor size, angiogenesis, proliferation	Zou et al. (2013)
	CD44 shRNA adenovirus	Colon	↑ Apoptosis ↓ Proliferation, migration, invasion	Lee et al. (2017)
	CD44v6 shRNA	Colon	↓ Adenoma growth	Misra et al. (2009)
	CD44 CRISPR/Cas9	Osteosarcoma	↓ Proliferation, spheroid formation, migration, invasion	Liu et al. (2018)
		Osteosarcoma	↑ Drug sensitivity of drug- resistant cells	Xiao et al. (2018)
		Liver (CSCs)	↑ Tumor differentiation ↓ Tumor malignancy	Han et al. (2015)
	CD44v6 CRISPR/Cas9	Gastric	↑ Cisplatin sensitization ↓ Self-renewal	Lobo et al. (2020)
	Cosmc CRISPR/Cas9	Breast	↓ CD44 O-glycosylation and expression, proliferation	Du et al. (2020a)

HA hyaluronan, NPs nanoparticles, CSCs cancer stem cells

18.3.1 Nanomedicine

Nanomedicine therapies utilize targeted drug delivery approaches to specifically target tumor cells and avoid adverse effects in normal tissues. Accordingly, there are two types of targeted drug delivery. First, passive targeting, which takes advantage of enhanced permeability and retention effect (EPR) in tumors. The EPR effect allows for accumulation of drugs and nanoparticles in tumors that have distinct architectural features from normal tissues, such as hyper-vasculogenesis, impaired lymphatic drainage, and different abnormal interstitial pressures (Fang et al. 2011). Second, active targeting, follows passive targeting to selectively deliver specific drugs to cancer cells overexpressing the desired receptor, taking advantage of receptor-substrate interactions (Danhier et al. 2010). Due to the important roles of hyaluronan/CD44 interactions and the correlation between CD44 expression and higher states of malignancy, hyaluronan is widely utilized in nanomedicine to specifically target CD44-expressing cancer cells and cancer stem cells. In blood circulation drug bioavailability and activity is altered by the body's defense, through chemical modification or binding to serum factors. In the last years, a "3S" transition concept has emerged, which incorporates stability transition, surface transition and size transition to overcome barriers in the delivery process. Hyaluronan therefore is an excellent candidate to be incorporated to nanoparticles and drug carriers, since it fulfills all three criteria due to its biocompatibility, biodegradability, and specific targeting of CD44-expressing cancer cells (Zhong et al. 2020).

One of the main purposes of hyaluronan-coated nanoparticles is to specifically deliver encapsulated drugs in cancer cells to enhance anti-tumor efficacy and avoid adverse effects. Hyaluronan can decorate polyetheleneimine (PEI) biodegradable nanoparticles of poly(lactic-*co*-glycolic) acid (PLGA) to deliver docetaxel specifically to lung cancer cells and inhibit their proliferation (Maiolino et al. 2015). Tumor growth inhibition of lung cancer cells in vivo was also observed by using PLGA hyaluronan-coated docetaxel nanoparticles without PEI, verifying the anti-cancer efficacy of this strategy (Wu et al. 2017). In another study, docetaxel loaded in chitosan-coated hyaluronan nanoparticles was more effective than free docetaxel against $CD44^+$ breast cancer cells (Shabani Ravari et al. 2016). Hyaluronan nanoparticles composed of branched cell-penetrating peptide B-mR9 could also successfully deliver methotrexate to $CD44^+$ cells and exerted antitumor activity (Yoo et al. 2020). Curcumin and celecoxib loaded in hyaluronan-coated nanoparticles displayed in vitro toxicity and inhibited tumor growth, enhanced survival of mice, induced apoptosis, and abrogated the formation of lung metastasis of breast cancer cells (Liu et al. 2020). Tumor cells display higher levels of glutathione. Taking advantage of this, hyaluronan-coated redox-sensitive micelles were developed to aim CD44-expressing tumor cells and release the drugs inside the target cells (Du et al. 2020b). Furthermore, dextran-modified quercetin-Cu(II)/hyaluronan nanoparticles with a natural PARP inhibitor could induce synthetic lethality in triple-negative breast cancer cells and extended animal survival, without displaying any adverse effects on normal organs (Cheng et al. 2021). Similarly, in triple-negative breast cancer cells conjugation of resveratrol and a chalcone to hyaluronan enhanced their uptake (Shah et al. 2017). Doxorubicin was loaded in hyaluronan-super paramagnetic iron oxide nanoparticles to achieve better uptake and cytoplasmic release of the drug and subsequently enhanced apoptosis in triple-negative breast cancer cells (Vyas et al. 2015). In addition, hyaluronan could facilitate lipoplexes delivery in breast cancer cells that express CD44 (Surace et al. 2009). Recently, targeting mitochondrial pathways in cancer cells has emerged as a promising strategy for tumor therapy. Indeed, hyaluronan nanomedicine utilizing a berberine derivative and doxorubicin was able to target mitochondria and subsequently inhibited proliferation, migration, and enhanced apoptosis of triple-negative breast cancer cells. Importantly, lung metastasis in vivo was abrogated after this treatment. These effects were mediated by suppression of MMP-2/-9 activities and induction of mitochondrial apoptotic pathways (Lin et al. 2021). Cisplatin can be also loaded to hyaluronan-green tea catechin micellar nanocomplexes, to enhance its efficacy against ovarian cancer cells, without causing side effects in vivo (Bae et al. 2017). In glioblastoma cells, hyaluronan-conjugated liposome nanoparticles effectively delivered doxorubicin inside the cells and enhanced its anti-neoplastic function (Hayward et al. 2016). Paclitaxel is another drug that can be also delivered with hyaluronan-nanoparticle complexes to treat CD44-positive colon cancer cells in orthotopic mouse models (Zhu et al. 2018). Paclitaxel and curcumin encapsulated in poly (acrylamide-*co*-acrylonitrile-*co*-vinylimidazole-*co*-bis(2-methacryloyl) oxyethyl disulfide) (PAAVB) polymer-based intelligent platform coated with hyaluronan, were able to promote adaptive anti-tumor immunogenicity and inhibit immunosuppression of CD44-overexpressing breast tumor cells while simultaneously abrogating lung metastasis (Wang et al. 2021). Hyaluronan can also decorate serum albumin conjugate-based nanoparticles to enhance drug delivery and their cytotoxic effects on cancer cells expressing CD44 (Edelman et al. 2017). In addition, promising results have been obtained by using hyaluronan-modified nanoparticles for treatment of hematological tumors, where CD44 plays significant roles. Specifically, doxorubicin encapsulated in lipoic acid-crosslinked hyaluronic acid nanoparticles was used to inhibit tumor growth of multiple myeloma and acute myeloid leukemic cells in vivo (Zhong et al. 2017).

The size of hyaluronan that may be used to coat nanoparticles is apparently important. Specifically, hyaluronan nanoparticles coated with low molecular weight hyaluronan displayed low CD44-binding activity, while high binding affinity was displayed by high molecular weight hyaluronan-coated nanoparticles, suggesting that high molecular weight hyaluronan is more suitable as a coating substance (Mizrahy et al. 2011). Furthermore, high molecular weight hyaluronan-coated nanoparticles displayed enhanced circulation time and better tumor targeting specificity than low molecular weight hyaluronan-coated nanoparticles (Mizrahy et al. 2014).

As already mentioned, CD44 serves as a major cancer stem cell marker. Therefore, hyaluronan can be used to guide nanoparticle-drug conjugates to cancer stem cells. To target prostate cancer stem cells, hyaluronan was used to decorate cabazitaxel and silibinin co-encapsulated cationic liposomes to induce apoptosis and decrease cell migration (Mahira et al. 2019). Cis-dichlorodiamminoplatium (II) (CDDP) glyconanoparticles were also coated with hyaluronan to deliver the drug in vivo and suppress stem cell properties of prostate cancer cells (Jafari Malek et al. 2014). For breast cancer treatment, co-delivery of salinomycin and curcumin in hyaluronan-coated nanoparticles was achieved in CD44-expressing breast cancer stem cells, inducing cell cycle arrest in G1 and inhibiting EMT (Zhao et al. 2020). To target pancreatic cancer stem cells, hyaluronan was utilized to coat diethyldithiocarbamate-copper complex nanoparticles. These nanoparticles were able to impair sphere formation of pancreatic cancer stem cells, probably by inducing the generation of ROS (Marengo et al. 2019). Nanoparticles can be delivered also orally. Specifically, hyaluronan-decorated nanoparticles containing PTC209, a BMI-1 inhibitor, were able to target colon cancer stem cells and abrogate tumor growth in vivo (Xu et al. 2019).

Hyaluronan-coated nanoparticles can encapsulate phosphosensitizers, like Ce6, to deliver them in human colon cancer cells in vivo. After photodynamic therapy, tumor growth was attenuated, while no side effects were observed (Gao et al. 2017). Furthermore, co-administration of hyaluronidase with such nanoparticles induced their uptake from tumor cells and enhanced their anti-cancer efficacy, probably by increasing the enhanced permeability and retention (EPR) effect inside the tumor (Gong et al. 2016). For triple-negative breast cancer treatment, photothermally targeted hyaluronan-polyaniline (PANi)-imiquimod (R837, a TLR7 agonist) nanoparticles were used to induce immune responses against the tumor (Yasothamani et al. 2021). Interestingly, hyaluronan-decorated nanoparticles could be effectively used to encapsulate and deliver hyaluronidase enzymes specifically to the tumors together with doxorubicin to enhance its efficacy (Chen et al. 2018a). Gold-nanoclustered hyaluronan nano-assemblies were also used as a platform for photodynamic or photothermal cancer therapies, by delivering the photodynamic therapy agent verteporfin in cancer cells. These nanoparticles displayed excellent stability in the blood and were able to completely inhibit tumorigenesis while displaying 100% survival rate (Han et al. 2016). Hyaluronan-conjugated zinc protoporphyrin nanoprobes were also used for photodynamic therapy against colon cancer and fibrosarcoma cells (Gao et al. 2021).

Among others, hyaluronan-coated nanoparticles can carry siRNAs or shRNAs against specific proteins expressed in cancer cells. KRAS is a known oncogene with no available small molecule inhibitors. To target KRAS hyaluronan-decorated nanocarriers [poly(hexamethylene biguanide) and chitosan] bearing KRAS-si/shRNA have been designed to specifically silence KRAS expression in CD44-expressing cancer cells (Tirella et al. 2019). Moreover, chitosan nanoparticles were conjugated with hyaluronan to target bladder cancer cells that expressed high levels of CD44, to deliver siRNA against the Bcl-2 oncogene. These particles were able to target bladder cancer cells in vivo and successfully interfere with the Bcl-2 expression (Liang et al. 2021). A new anti-cancer siRNA delivery system named by the authors HPLR was manufactured with a siRNA-peptide core surrounded by lipid bilayer, thin hyaluronan coating, and EGFR-targeted peptides to target EGFR and CD44 overexpressing cells. This system was able to successfully deliver the siRNAs in subcutaneous liver tumors and inhibit their growth without displaying any significant toxicity (Liang et al. 2019). Additionally, hyaluronan-conjugated nanoparticles were utilized to deliver Gli1 siRNA in gastric cancer stem cells and subsequently reduced migration, invasion as well as tumor spheroid and colony formation (Yao et al. 2020).

Notably, nanoparticles can also be decorated with CD44-targeting peptides. Indeed, A6 anti-CD44 peptide conjugated to polymersomal epirubicin enhanced uptake and anti-cancer efficacy of epirubicin against multiple myeloma cells in vivo (Gu et al. 2019). Furthermore, nanoparticles can be decorated with anti-CD44 antibodies. Multifunctionalized iron oxide magnetic nanoparticles with CD44 antibodies were able to selectively target CD44-positive cancer cells (Aires et al. 2016). In breast cancer cells, conjugation of saporin, a ribosome-inactivating protein, with hyaluronan modified nanoparticles achieved intracellular release of saporin leading to enhanced apoptosis (Ding et al. 2018). Hyaluronan-coated nanoparticles have been also conjugated with AS1411 aptamer, to penetrate blood-brain barrier and deliver docetaxel to CD44- overexpressing glioma cells. Delivery of docetaxel through this pathway significantly attenuated the formation of spheroids and tumor growth in vivo (Wang et al. 2019).

18.3.2 Antibodies

Different monoclonal antibodies have been raised against CD44 standard and variant isoforms to be utilized for cancer treatment. One of the best-studied anti-CD44 monoclonal antibodies is RG7356 which binds to the constant region of CD44, abrogating hyaluronan binding. In chronic lymphocytic leukemia (CLL), interruption of hyaluronan/CD44 interaction induced caspase-dependent apoptosis, with the strongest effects being observed in ZAP-70$^+$ CLL cells both in vitro and in vivo (D'arena et al. 2014). Mechanistically, RG7356 engages CD44 to induce its internalization in CLL cells, and subsequently decrease ZAP-70, which was found to be complexed with CD44 (Zhang et al. 2013). RG7356 is also effective against triple-negative breast cancer cells in mice xenografts by modifying the MAPK pathway (Weigand et al. 2012). Response to treatment with RG7356 in xenograft models and colorectal cancer patients depended on the presence of CD44s isoform, suggesting that only treatment in patients with CD44s expression could be effective (Birzele et al. 2015). RG7356 mechanism of function also involves activation of the immune system. Specifically, RG7356 treatment induced secretion of chemo-attractants responsible for recruitment of immune cells, such as macrophages, to the tumor, through activation of MAPK. Moreover, RG7356 activated antibody-dependent cellular phagocytosis (ADCP) of triple-negative breast cancer cells by macrophages (Maisel et al. 2016). In clinical trials, ^{89}Zirconium-labeled RG7356 was taken up by several tissues such as the spleen, liver, bone marrow, lung, and kidney in a dose-dependent manner in patients (Jauw et al. 2018), while selectively targeting CD44$^+$ breast and pancreatic cancer cells in monkeys (Vugts et al. 2014). Moreover, phase

1 clinical trials with RG7356 in patients with advanced, CD44-expressing solid tumors showed that it was well-tolerated, with most side effects being mild such as fever, headache, and fatigue. Its clinical efficacy was modest with 21% of patients experiencing disease stabilization (Menke-Van Der Houven Van Oordt et al. 2016). Furthermore, phase 1 studies in acute myeloid leukemia patients verified the mild adverse effects, although only 2 out of 44 patients showed response in the treatment suggesting that this antibody cannot be utilized as monotherapy (Vey et al. 2016).

Another popular anti-CD44 antibody is A3D8. On human acute myeloid leukemia (AML) cells, A3D8 treatment caused G0/G1 cell cycle arrest through induction of p21, p27 and reduction of pRb and CdK2/4 activities. Furthermore, JNK protein expression was reduced leading to reduction in c-Jun phosphorylation after A3D8 treatment (Zada et al. 2003; Gadhoum et al. 2004a, b; Li et al. 2016). A3D8, apart from inducing apoptosis and inhibiting cell proliferation, induced differentiation of the AML cells (Gadhoum et al. 2004a). This apoptotic effect was caused by A3D8-induced CD44s lipid raft clustering leading to Fas aggregation and subsequent caspase-8 activation (Qian et al. 2012). It is important to note that bone marrow stromal cells could protect acute myeloid leukemia cells from A3D8-induced apoptosis through activation of PI3K/Akt signaling to down-regulate p27 (Chen et al. 2015). In these studies, the effects of A3D8 were verified by another anti-CD44 antibody, H90. In chronic lymphocytic leukemia, A3D8 abrogated CLL cell viability and in vivo caused reduction of MCL1 protein and activation of caspases (Fedorchenko et al. 2013). In human erythroleukemia cells, A3D8 also evoked cell growth inhibition and caspase-independent apoptosis-like cell death, through disruption of mitochondrial membrane potential and release of AIF but not cytochrome c, which is typical for caspase-dependent apoptosis. This type of cell death involves activation of PARP and calpain, since their inhibition ameliorated A3D8-induced cell death (Artus et al. 2006). CD44 ligation by A3D8 was also able to arrest ovarian cancer stem cells in S phase and induced apoptosis (Du et al. 2013).

The monoclonal antibody F77 was developed to recognize glycolipids and O-glycosylation on prostate cancer cell proteins. F77 antigen was finally identified to be glycosylated CD44v10 isoform. On functional level, F77 induced apoptosis in prostate cancer cells. Moreover, this antibody can be utilized in ELISA assays to identify this glycosylated CD44v10 isoform in prostate cancer cells and serum of patients with prostate cancer (Chen et al. 2018b).

Hermes-1 is an anti-CD44-specific monoclonal antibody that specifically interrupts hyaluronan/CD44 interactions. When compared with other anti-CD44 antibodies, like Hermes-3, J173, and 50B4, Hermes-1 was the only one capable to completely inhibit binding of hyaluronan by colorectal carcinoma cells. It is important to note that Hermes-1, Hermes-3, and J173 inhibited the adhesion of the cells on laminin and collagen, while 50B4 did not have any effect (Ishii et al. 1993). Hermes-1 can be also utilized to investigate differences between metastatic cells and cells from the original tumor. Unexpectedly, utilizing Hermes-1, it was found that primary human colon carcinoma cells expressed higher amounts of CD44 on their cell surface, than the cells in lymph node metastases (Kubens and Zanker 1998).

Inhibition of hyaluronan/CD44 interactions has been also achieved with IM7 monoclonal antibody. Despite not displaying any effect on colon carcinoma cell growth in vitro, IM7 was able to inhibit liver metastasis in vivo, although tumor cell colonies were detected in all of the livers even in mice free of nodules, suggesting that IM7 probably delayed formation of metastases, rather than completely inhibiting them (Ogoshi et al. 1998). In glioma cells, which synthesize high levels of hyaluronan, IM7 decreased hyaluronan biosynthesis, and induced apoptosis (Wiranowska et al. 2010). IM7 can be also delivered with chitosan polylactic acid-coated nanoparticles to reduce its toxicity. This modified IM7 antibody was able to reduce proliferation of ovarian cancer cells and control the development and progression of ovarian cancer in vivo (Yang et al. 2017). It is known that hyaluronan fragments that engage CD44 can activate downstream NF-κB signaling in various tumor types, which in turn promotes inflammation and tumorigenesis. IM7 by inhibiting these interactions successfully reduced NF-κB activation in bladder cancer cells (Fitzgerald et al. 2000). IM7 was also able to block adhesion of melanoma cells on endothelial cells, a critical step during metastasis formation (Ota et al. 1995). IM7 can be also conjugated to ribosome inactivating protein saporin, to specifically target CD44-expressing cancer cells. These modified antibodies can release saporin in the target cell cytoplasm, which can then display its cytotoxic effects only on prostate cancer cells with high CD44 expression (Bostad et al. 2014). IM7 also enhanced natural killer cell activity (Tan et al. 1993), suggesting that CD44 targeting with antibodies can enhance activation of immune system against cancer cells. In the same study, different anti-CD44 antibodies were also studied, Hermes-1, S3 and S5, but the same effect was only observed with S5 antibody. The effect of CD44 ligation was also investigated with J173 and F10442 antibodies, where CD44 cross-linking upregulated CD16-mediated lysis. Moreover, CD44 blocking with these antibodies led to rapid increase of intracellular Ca^{2+} (Galandrini et al. 1994). Accordingly, J173 activated MAPK signaling to enhance the killing activity of peripheral mononuclear cells against cancer cells from Burkitt's lymphoma and chronic myelogenous leukemia (Ishizuka et al. 2008).

Several other less studied anti-CD44 monoclonal antibodies have been developed. The HI44α antibody effects were investigated in acute myeloid leukemia cells derived from patients. Ligation of CD44 by HI44α was able to induce differentiation and apoptosis of AML cells, probably through inhibition of c-Myc expression (Song et al. 2004). In colon cancer cells, engagement of CD44 by a specific antibody reversed the resistance to anti-integrin antibody, altered cell morphology, and enhanced apoptosis (Bates et al. 1998). U36 is another antibody that has been raised to recognize CD44v6- expressing squamous cell-carcinomas (Van Hal et al. 1996), but whether it can be used for treatment of these tumors remains to be investigated. KMP1 is a CD44-specific antibody that shows significant anti-tumor effects against bladder cancer cells. Specifically, KMP1 inhibited the bladder cancer cell proliferation, migration, and adhesion in vivo, while suppressed tumor growth in xenograft models. Importantly, expression of the KMP1 epitope correlated with clinical severity and prognosis of bladder cancer (Chen et al. 2018c). Anti-CD44 antibody conjugated to oil liquid nanocapsules, aCD44O^{2}LNCs, to target pancreatic cancer stem cells, has also been developed. This technology displayed high uptake of these nanoparticles by pancreatic cancer stem cells in vivo. Moreover, these nanoparticles when coupled with paclitaxel were able to enhance their anti-tumor efficacy (Navarro-Marchal et al. 2021). A defucosylated anti-CD44 antibody, 5-mG2a-f, also significantly reduced tumor development in oral squamous cell carcinoma xenograft models (Takei et al. 2020). In another study, the authors developed four anti-CD44 monoclonal antibodies, namely, P4G9, P3D2, P3A7 and P3G4 that recognized unglycosylated and conserved regions of CD44 ectodomain. P3D2 was able to inhibit breast cancer tumorigenesis in animal models (Lusche et al. 2021). Since CD44v7 is expressed in many human cancers, an antibody raised against CD44v7 is to be utilized as a diagnostic and therapeutic tool in clinical settings (Borgya et al. 1995). Whether this antibody is useful remains to be investigated. An antibody raised against CD44v9, mAb 44-IV was able to inhibit liver metastasis of human colon cancer cells by blocking their adhesion to the capillaries, a critical step in the metastatic cascade (Seki et al. 1997).

18.3.3 Peptides

Apart from antibodies, peptides have been also used to target CD44 and regulate hyaluronan binding capacity. Peptides can be used for detection of CD44-expressing tumor cells inside the tissues. Several peptides against CD44 have been described utilizing phage display libraries (Park et al. 2012a). For example, RP-1, a 12-mer peptide isolated from a phage peptide library, binds to CD44^{+} gastric cancer cells and allows their detection inside the tissues (Zhang et al. 2015). Importantly, RP1 peptide binding can predict prognosis of gastric cancer patients (Li et al. 2017a). Another 7-mer peptide also isolated from phage display library, termed CV-1, was able to detect CD44v3-v10 protein expression in gastric cancer cells and tissues (Zhang et al. 2016). Similarly, a phage displays 15 amino acid peptide PFT marked CD44v6-expressing prostate cancer stem cells (Peng et al. 2017). Polyvalent-directed peptide polymer (PDPP) specifically traced CD44-expressing breast cancer stem cells (Cho et al. 2015).

A widely studied peptide against CD44 is A6. A6 peptide is derived from uPA, but it does not bind to uPA receptor (uPAR) nor interferes with uPA/uPAR interactions. Importantly A6 did not cause any significant toxicity in animals. Moreover, A6 showed efficacy and exceptionally good safety profile in Phase 1a, 1b and 2 clinical trials. In chronic lymphocytic leukemia, A6 displayed significant toxicity against B-lymphocytes expressing ZAP-70 (Finlayson 2015). A6 peptides loaded with reduction-sensitive polymersomal vincristine sulfate targeted CD44-expressing acute myeloid leukemia cells and reduced the leukemia burden in the circulation, bone marrow, liver, and spleen, while extending survival of mice (Gu et al. 2021). A6 also inhibited migration of ovarian and breast cancer cells. Mechanistically, A6 regulated CD44-mediated adhesion to hyaluronan and the activation of downstream

FAK and MAP/ERK signaling pathways. In vivo treating mice with A6 inhibited the metastasis of melanoma cells to the lung (Piotrowicz et al. 2011).

Peptides that bind CD44v have been also developed. A peptide mimicking a specific extracellular motif of CD44v6 reduced CD44v6-mediated activation of c-Met and VEGFR-2. Treatment with this peptide in vivo reduced angiogenesis in tumors, suggesting that targeting the co-receptor functions of CD44 is another viable possibility (Tremmel et al. 2009). Following the same notion, v6 peptide that interfered with CD44v6 co-receptor functions with these receptors, blocked tumor growth and metastasis in pancreatic tumors in mice, while extended their survival time. Of note, the v6 peptide achieved higher inhibition than c-Met or VEGFR-2 inhibitors (Matzke-Ogi et al. 2016). NLN and NEW are peptides that bind CD44v6, inducing its internalization to inhibit c-Met/Erk pathways, leading to reduction of cell migration, invasion, and metastasis. In vivo these peptides after conjugation with KLA pro-apoptotic peptides successfully killed tumor cells and impaired tumor growth and metastasis without displaying systemic side effects (Khan et al. 2021).

The FK506-binding protein-like (FKBPL) and a 24-amino acid-derived peptide AD-01 exerted anti-angiogenic effects and reduced tumor growth in vivo, effects that are dependent on the presence of CD44 (Valentine et al. 2011). Moreover, FKBPL and AD-01 bound CD44 and reduced breast cancer cell migration, through inhibition of Rac-1 activity, upregulation of RhoA and the actin-interacting proteins profilin and vinculin (Yakkundi et al. 2013). Moreover, AD-01 treatment inhibited mammosphere-forming capacity of breast cancer stem cells in vitro and reduced tumor initiation in vivo. Specifically, AD-01 induced differentiation of breast cancer stem cells, while reducing the expression of stem cell markers Nanog, Oct4, and Sox2. Importantly when combined with DAPT—a Notch inhibitor—it caused significant reduction of chemotherapy and radiotherapy resistance in breast cancer stem cells (McClements et al. 2013).

CD44, apart from binding hyaluronan, can also interact with other proteins like collagen. Consequently, there are also peptides taking advantage of collagen/CD44 interactions. Peptides derived from collagen type IV incorporated to liposomes bearing doxorubicin targeted $CD44^+$ melanoma cells and reduced tumor size (Ndinguri et al. 2012). Moreover, taking advantage of pro-MMP9 and CD44 interactions, P3 and P6 peptide were prepared from PEX9 domain of pro-MMP9, and were shown to reduce chronic lymphocytic leukemia cell adhesion to pro-MMP9 and thus abrogated chemotaxis and transendothelial migration (Ugarte-Berzal et al. 2014). The peptide PCK3145 induced CD44 shedding by increasing MT1-MMP while decreasing MMP9 secretion. PCK3145 also abrogated adhesion of fibrosarcoma cells on hyaluronan, thus antagonizing tumor metastatic processes (Annabi et al. 2005). A5G27 peptides derived from laminin could also bind CD44v3 and CD44v6, and could be conjugated to particles bearing specific siRNAs. Introduction of such particles in vivo inhibited tumor growth by lung adenocarcinoma or ovarian carcinoma cells (Golan et al. 2016). A5G27 peptide blocked melanoma metastasis by inhibiting the binding of FGF2 in the heparan sulfate chains of CD44v3, and thus reducing bioactivity of FGF2 (Hibino et al. 2005). C21, a C-terminal peptide of thrombospondin-4 competed for osteopontin and hyaluronan binding to CD44, but its effect on tumor cells remains to be investigated (Sadvakassova et al. 2009). Another CD44-binding peptide (CD44BP) added to an engineered matrix showed significant anti-tumorigenic effect, as evidenced by reductions in tumor sphere formation in vitro (Yang et al. 2013).

Another interesting approach was synthesizing CD44 cytoplasmic tail peptides bearing phosphor-Ser^{325} which were conjugated to penetration sequences in order to enhance plasma membrane translocation. Such peptides blocked CD44-mediated cell migration without affecting hyaluronan binding or CD44 expression (Peck and Isacke 1998). CD44 could be used also as an antigen to be recognized by immune cells for cancer treatment. Specifically, CD44-derived peptides could function as immunogens to sensitize dendritic cells and enhance their anti-tumor activities against prostate cancer stem cells that expressed high CD44 levels in vitro and in vivo (Wang et al. 2020).

18.3.4 Chemically Modified Hyaluronan

Hyaluronan can be modified in several different chemical groups to inhibit binding to CD44, affect tissue architecture and to be utilized in hydrogels and as a scaffold for drug carriers. Sulfhydryl (-SH) modified hyaluronan is utilized to form hydrogels to be used as drug-loaded implant for chemotherapeutics, photosensitizer and photothermal reagent in chemotherapy, photodynamic and photothermal therapy against tumors (Xu et al. 2021). Hyaluronan modification with Au-Ag alloy can be used in nanoparticles that allow for the sensitization of breast cancer cells in radiotherapy, since ionizing radiation releases toxic Ag^+ and enhances production of OH^- in tumor sites (Chong et al. 2020). High molecular weight hyaluronan can be also modified with a hydrazide group and bisphosphonate (BP) for selective targeting of CD44 expressing cancer cells (Varghese et al. 2009). Notably, hyaluronan—a non-sulfated glycosaminoglycan—has been modified with sulfate groups which allowed simultaneous targeting of CD44 and P-selectin, to effectively target $CD44^+$P-$selectin^+$ cancer cells (Bhattacharya et al. 2020). Hyaluronan can be also modified with poly(lactic-*co*-glycolic acid) to produce highly stable and GSH sensitive micelles. Loading of these complexes with transferrin-targeted nanoformulated AUY922 eased their uptake by brain tumor cells and induced caspase-dependent cleavage of the apoptosis marker PARP followed by upregulation of p53. Moreover, this complex was drastic against tumor growth in vivo, without displaying any toxicity to other major organs (Debele et al. 2021). β-Cyclodextrin-modified hyaluronan with drug conjugates was able to exert significant toxicity against lung and prostate cancer cells with high expression of CD44, while on the other hand showed no toxicity against normal cells with low CD44 expression (Bai et al. 2020).

18.3.5 Gene Therapies

A promising strategy to target hyaluronan/CD44 network in tumors is to silence the expression of CD44 gene, with specific siRNA and/or shRNA or the most recently developed CRISPR/Cas9 system.

Introduction of a CD44-specific siRNA in lung cancer cells reduced the expression of the stem cell-related genes CXCR4 and POU5F genes after TGF-β1/TNFα treatments but failed to reverse EMT gene signature (Nurwidya et al. 2017). A recombinant adenovirus bearing CD44 shRNA was also able to inhibit cell proliferation, migration and invasion, while induced apoptosis of colon cancer cells. These adenoviral particles inhibited Akt and GSK-3β signaling pathways. Furthermore, the expression levels of anti-apoptotic proteins Bcl-2 and Bcl-xL were reduced, while the apoptotic proteins Bax, cleaved caspase-3/−9 and PARP were increased in colon cancer cells treated with the adenovirus (Lee et al. 2017). Targeted deletion of CD44v6 with specific shRNA in colon cancer cells also reduced the adenoma growth in vivo through interruption of hyaluronan/CD44v6/pErbB2/Cox-2 pathway (Misra et al. 2009). In triple-negative breast cancer cells, knock-down of CD44 with shRNA vectors significantly suppressed proliferation, colony formation, and invasion (Zhou et al. 2018). Doxorubicin resistance is a major caveat in the treatment of breast cancer. Targeting of breast cancer cells with CD44 siRNA in combination with doxorubicin treatment effectively limited tumor metastasis, proliferation, invasion, migration, and induced apoptosis in triple-negative breast cancer cells, signifying the importance of targeting CD44 to overcome drug resistance (Vahidian et al. 2020). Targeting of CD44-expressing breast cancer cells can be also performed with CD44-targeted aptamer Apt1. This aptamer conjugated to liposomes with encapsulated CD44 siRNA was able to successfully target CD44-expressing breast cancer cells to silence CD44 in vitro and in vivo (Alshaer et al. 2018). In another study, the authors utilized biodegradable poly D,L-lacticied-*co*-glycolide acid nanoparticles (PLGANPs) to deliver FAK and CD44 shRNAs in ovarian cancer cells in vivo, since both FAK and CD44 have active roles in tumor angiogenesis and cancer metastatic processes. Double knock-down of both FAK and CD44 reduced tumor size, inhibited angiogenesis, reduced proliferation and induced apoptosis (Zou et al. 2013).

CRISP/Cas9 technology in contrast to siRNA or shRNA systems offers silencing capacity of a specific gene by completely removing the target gene from the host genome. In osteosarcomas, high expression of CD44 predicts poor survival and higher incidence of metastases in patients. Targeting CD44 with CRISPR/Cas9 system in metastatic osteosarcoma cells abrogated the proliferation and spheroid formation in 3D cultures as well as migration and invasion (Liu et al. 2018). Furthermore, in multi-drug-resistant osteosarcoma cells, silencing of CD44 with CRISPR/Cas9 system could also enhance drug sensitivity (Xiao et al. 2018). Although CD44 seems to play an important role in tumor growth, CD44 silencing in vivo with CRISP/Cas9 in HAS3-overexpressing stromal fibroblasts or CD44 knock-down with shRNA in breast cancer cells was not able to inhibit tumor growth, in contrast with the reductions that were observed when hyaluronan production or accumulation was inhibited (Zhao et al. 2019). This rather contradicting study points out the context-dependent effects of hyaluronan and CD44 in several tumors. Given the fact that in most breast tumors HAS2 has been most extensively studied, while the role of HAS3 remains obscure, further studies are needed to clarify this discrepancy. CD44 can be also indirectly targeted, as evidenced by CRISPR/Cas9 silencing of Cosmc, an endoplasmic reticulum-localized chaperone that regulates protein O-glycosylation. Cosmc knock-out inhibited protein expression of CD44, confirming the notion that O-glycosylation is important for proper CD44 expression. Of note, reconstitution of CD44 reversed the effects of Cosmc disruption on MAPK signaling and breast cancer cell proliferation, verifying the important role of CD44 for breast tumor growth (Du et al. 2020a). In liver cancer stem cells, knock-out of CD44 utilizing CRISPR/Cas9 system resulted in less malignant and more differentiated tumors. Unexpectedly, CD44 silencing increased the expression of stem cell markers Oct4, Sox2, and Nanog. This contradictory finding warrants further investigation as in these cells CD44 was predominantly nuclear and bound in promoter regions of c-Myc and Sox2 (Han et al. 2015). The CRISPR/Cas9 system can be also utilized to target specific variants of CD44. In gastric cells, CD44 exon v6 was deleted, allowing for the rest of CD44 gene to be expressed. Removal of v6 exon from CD44 sensitized gastric cancer cells to cisplatin and abrogated their self-renewal (Lobo et al. 2020). In another study, utilizing CRISPR/Cas9 and overexpression approaches, chimeric antigen receptors (CARs) that specifically target CD44v6 from head and neck squamous cell carcinoma were expressed on T cells. Targeting CD44v6 demonstrated a direct correlation of CD44v6 expression and cytotoxic effects mediated from CAR T cells (Haist et al. 2021).

18.4 Conclusions

Conclusively, hyaluronan/CD44 network appears to offer an important target for translation into the clinic to treat tumors. However, given the omnipresent localization of hyaluronan in the body and ubiquitous expression of CD44 as well as their importance in physiological processes, targeting this system needs to be approached with caution. On the bright side, many of the hyaluronan effects depend on the expression of CD44v isoforms that are expressed specifically by tumor cells. Moreover, while many of the processes controlled by hyaluronan/CD44 network overlap with features of immune and inflammatory pathways, there are clear differences between tumor cells and physiologic events that can be exploited, like the size and interactions of hyaluronan found in cancerous tissues. Certainly, further research is needed to clarify the exact roles of hyaluronan and CD44 in tumorigenesis, but the evidence presented so far suggests that targeting of this system is a promising avenue for the development of safe and effective cancer treatment regiments in the future.

References

Agren UM, Tammi RH, Tammi MI (1997) Reactive oxygen species contribute to epidermal hyaluronan catabolism in human skin organ culture. Free Radic Biol Med 23:996–1001

Aires A, Ocampo SM, Simoes BM, Josefa Rodriguez M, Cadenas JF, Couleaud P, Spence K, Latorre A, Miranda R, Somoza A, Clarke RB, Carrascosa JL, Cortajarena AL (2016) Multifunctionalized iron oxide nanoparticles for selective drug delivery to CD44- positive cancer cells. Nanotechnology 27:065103

Alshaer W, Hillaireau H, Vergnaud J, Mura S, Delomenie C, Sauvage F, Ismail S, Fattal E (2018) Aptamer-guided siRNA-loaded nanomedicines for systemic gene silencing in CD-44 expressing murine triple-negative breast cancer model. J Control Release 271:98–106

Annabi B, Bouzeghrane M, Currie JC, Hawkins R, Dulude H, Daigneault L, Ruiz M, Wisniewski J, Garde S, Rabbani SA, Panchal C, Wu JJ, Beliveau R (2005) A PSP94-derived peptide PCK3145 inhibits MMP-9 secretion and triggers CD44 cell surface shedding: implication in tumor metastasis. Clin Exp Metastasis 22:429–439

Artus C, Maquarre E, Moubarak RS, Delettre C, Jasmin C, Susin SA, Robert-Lezenes J (2006) CD44 ligation induces caspase-independent cell death via a novel calpain/AIF pathway in human erythroleukemia cells. Oncogene 25:5741–5751

Baba D, Kashiwabara S, Honda A, Yamagata K, Wu Q, Ikawa M, Okabe M, Baba T (2002) Mouse sperm lacking cell surface hyaluronidase PH-20 can pass through the layer of cumulus cells and fertilize the egg. J Biol Chem 277:30310–30314

Bae KH, Tan S, Yamashita A, Ang WX, Gao SJ, Wang S, Chung JE, Kurisawa M (2017) Hyaluronic acid-green tea catechin micellar nanocomplexes: fail-safe cisplatin nanomedicine for the treatment of ovarian cancer without off-target toxicity. Biomaterials 148:41–53

Bai Y, Liu CP, Chen D, Liu CF, Zhuo LH, Li H, Wang C, Bu HT, Tian W (2020) beta-Cyclodextrin-modified hyaluronic acid-based supramolecular self-assemblies for pH- and esterase- dual-responsive drug delivery. Carbohydr Polym 246:116654

Bart G, Vico NO, Hassinen A, Pujol FM, Deen AJ, Ruusala A, Tammi RH, Squire A, Heldin P, Kellokumpu S, Tammi MI (2015) Fluorescence resonance energy transfer (FRET) and proximity ligation assays reveal functionally relevant homo- and heteromeric complexes among hyaluronan synthases HAS1, HAS2, and HAS3. J Biol Chem 290:11479–11490

Bates RC, Elith CA, Thorne RF, Burns GF (1998) Engagement of variant CD44 confers resistance to anti-integrin antibody-mediated apoptosis in a colon carcinoma cell line. Cell Adhes Commun 6:21–38

Bernert B, Porsch H, Heldin P (2011) Hyaluronan synthase 2 (HAS2) promotes breast cancer cell invasion by suppression of tissue metalloproteinase inhibitor 1 (TIMP-1). J Biol Chem 286: 42349–42359

Bhattacharya DS, Svechkarev D, Bapat A, Patil P, Hollingsworth MA, Mohs AM (2020) Sulfation modulates the targeting properties of hyaluronic acid to P-selectin and CD44. ACS Biomater Sci Eng 6:3585–3598

Birkenkamp-Demtroder K, Maghnouj A, Mansilla F, Thorsen K, Andersen CL, Oster B, Hahn S, Orntoft TF (2011) Repression of KIAA1199 attenuates Wnt-signalling and decreases the proliferation of colon cancer cells. Br J Cancer 105:552–561

Birzele F, Voss E, Nopora A, Honold K, Heil F, Lohmann S, Verheul H, Le Tourneau C, Delord JP, Van Herpen C, Mahalingam D, Coveler AL, Meresse V, Weigand S, Runza V, Cannarile M (2015) CD44 isoform status predicts response to treatment with anti-CD44 antibody in cancer patients. Clin Cancer Res 21:2753–2762

Borgya A, Woodman A, Sugiyama M, Donie F, Kopetzki E, Matsumura Y, Tarin D (1995) Isolation and characterisation of antibodies which specifically recognise the peptide encoded by exon 7 (v2) of the human CD44 gene. Clin Mol Pathol 48:M241–M250

Bostad M, Kausberg M, Weyergang A, Olsen CE, Berg K, Hogset A, Selbo PK (2014) Light-triggered, efficient cytosolic release of IM7-saporin targeting the putative cancer stem cell marker CD44 by photochemical internalization. Mol Pharm 11:2764–2776

Bourguignon LY, Zhu H, Chu A, Iida N, Zhang L, Hung MC (1997) Interaction between the adhesion receptor, CD44, and the oncogene product, p185HER2, promotes human ovarian tumor cell activation. J Biol Chem 272:27913–27918

Bourguignon LY, Wong G, Earle C, Chen L (2012) Hyaluronan-CD44v3 interaction with Oct4-Sox2-Nanog promotes miR-302 expression leading to self-renewal, clonal formation, and cisplatin resistance in cancer stem cells from head and neck squamous cell carcinoma. J Biol Chem 287:32800–32824

Chanmee T, Ontong P, Kimata K, Itano N (2015) Key roles of hyaluronan and its CD44 receptor in the stemness and survival of cancer stem cells. Front Oncol 5:180

Chen P, Huang H, Wu J, Lu R, Wu Y, Jiang X, Yuan Q, Chen Y (2015) Bone marrow stromal cells protect acute myeloid leukemia cells from anti-CD44 therapy partly through regulating PI3K/Akt-p27(Kip1) axis. Mol Carcinog 54:1678–1685

Chen L, Liu Z, Jin R, Yang X, Bai Y, Liu S, Chen X (2018a) Stepwise co-delivery of an enzyme and prodrug based on a multi-responsive nanoplatform for accurate tumor therapy. J Mater Chem B 6:6262–6268

Chen X, Nagai Y, Zhu Z, Ruan H, Peehl DM, Greene MI, Zhang H (2018b) A spliced form of CD44 expresses the unique glycan that is recognized by the prostate cancer specific antibody F77. Oncotarget 9:3631–3640

Chen Y, Wang H, Zuo Y, Li N, Ding M, Li C (2018c) A novel monoclonal antibody KMP1 has potential antitumor activity of bladder cancer by blocking CD44 in vivo and in vitro. Cancer Med 7:2064–2077

Cheng HW, Chiang CS, Ho HY, Chou SH, Lai YH, Shyu WC, Chen SY (2021) Dextran-modified Quercetin-Cu(II)/hyaluronic acid nanomedicine with natural poly(ADP-ribose) polymerase inhibitor and dual targeting for programmed synthetic lethal therapy in triple- negative breast cancer. J Control Release 329:136–147

Cherr GN, Yudin AI, Overstreet JW (2001) The dual functions of GPI-anchored PH-20: hyaluronidase and intracellular signaling. Matrix Biol 20:515–525

Cho JH, Lee SC, Ha NR, Lee SJ, Yoon MY (2015) A novel peptide-based recognition probe for the sensitive detection of CD44 on breast cancer stem cells. Mol Cell Probes 29:492–499

Chong Y, Huang J, Xu X, Yu C, Ning X, Fan S, Zhang Z (2020) Hyaluronic acid-modified au-ag alloy nanoparticles for radiation/nanozyme/ag(+) multimodal synergistically enhanced cancer therapy. Bioconjug Chem 31:1756–1765

Danhier F, Feron O, Preat V (2010) To exploit the tumor microenvironment: passive and active tumor targeting of nanocarriers for anti-cancer drug delivery. J Control Release 148:135–146

D'arena G, Calapai G, Deaglio S (2014) Anti-CD44 mAb for the treatment of B-cell chronic lymphocytic leukemia and other hematological malignancies: evaluation of WO2013063498. Expert Opin Ther Pat 24:821–828

Debele TA, Wu PC, Wei YF, Chuang JY, Chang KY, Tsai JH, Su WP (2021) Transferrin modified GSH sensitive hyaluronic acid derivative micelle to deliver HSP90 inhibitors to enhance the therapeutic efficacy of brain cancers. Cancers (Basel) 13:2375

Deen AJ, Rilla K, Oikari S, Karna R, Bart G, Hayrinen J, Bathina AR, Ropponen A, Makkonen K, Tammi RH, Tammi MI (2014) Rab10-mediated endocytosis of the hyaluronan synthase HAS3 regulates hyaluronan synthesis and cell adhesion to collagen. J Biol Chem 289:8375–8389

Ding L, Jiang Y, Zhang J, Klok HA, Zhong Z (2018) pH-sensitive coiled-coil peptide-cross-linked hyaluronic acid nanogels: synthesis and targeted intracellular protein delivery to CD44 positive cancer cells. Biomacromolecules 19:555–562

Du YR, Chen Y, Gao Y, Niu XL, Li YJ, Deng WM (2013) Effects and mechanisms of anti-CD44 monoclonal antibody A3D8 on proliferation and apoptosis of sphere-forming cells with stemness from human ovarian cancer. Int J Gynecol Cancer 23:1367–1375

Du T, Jia X, Dong X, Ru X, Li L, Wang Y, Liu J, Feng G, Wen T (2020a) Cosmc disruption-mediated aberrant O-glycosylation suppresses breast cancer cell growth via impairment of CD44. Cancer Manag Res 12:511–522

Du Y, Wang S, Zhang T, He D, Tu J, Shen Y (2020b) Enhanced cytotoxicity of a redox-sensitive hyaluronic acid-based nanomedicine toward different oncocytes via various internalization mechanisms. Drug Deliv 27:128–136

Edelman R, Assaraf YG, Levitzky I, Shahar T, Livney YD (2017) Hyaluronic acid-serum albumin conjugate-based nanoparticles for targeted cancer therapy. Oncotarget 8:24337–24353

Fang J, Nakamura H, Maeda H (2011) The EPR effect: unique features of tumor blood vessels for drug delivery, factors involved, and limitations and augmentation of the effect. Adv Drug Deliv Rev 63:136–151

Fedorchenko O, Stiefelhagen M, Peer-Zada AA, Barthel R, Mayer P, Eckei L, Breuer A, Crispatzu G, Rosen N, Landwehr T, Lilienthal N, Mollmann M, Montesinos-Rongen M, Heukamp L, Durig J, Hallek M, Fingerle-Rowson G, Herling M (2013) CD44 regulates the apoptotic response and promotes disease development in chronic lymphocytic leukemia. Blood 121:4126–4136

Fernando J, Malfettone A, Cepeda EB, Vilarrasa-Blasi R, Bertran E, Raimondi G, Fabra A, Alvarez-Barrientos A, Fernandez-Salguero P, Fernandez-Rodriguez CM, Giannelli G, Sancho P, Fabregat I (2015) A mesenchymal-like phenotype and expression of CD44 predict lack of apoptotic response to sorafenib in liver tumor cells. Int J Cancer 136:E161–E172

Finlayson M (2015) Modulation of CD44 activity by A6-peptide. Front Immunol 6:135

Fitzgerald KA, Bowie AG, Skeffington BS, O'neill LA (2000) Ras, protein kinase C zeta, and I kappa B kinases 1 and 2 are downstream effectors of CD44 during the activation of NF-kappa B by hyaluronic acid fragments in T-24 carcinoma cells. J Immunol 164:2053–2063

Flannery CR, Little CB, Hughes CE, Caterson B (1998) Expression and activity of articular cartilage hyaluronidases. Biochem Biophys Res Commun 251:824–829

Flores-Diaz M, Alape-Giron A, Persson B, Pollesello P, Moos M, Von Eichel-Streiber C, Thelestam M, Florin I (1997) Cellular UDP-glucose deficiency caused by a single point mutation in the UDP-glucose pyrophosphorylase gene. J Biol Chem 272:23784–23791

Fraser JR, Laurent TC, Laurent UB (1997) Hyaluronan: its nature, distribution, functions and turnover. J Intern Med 242:27–33

Fujita Y, Kitagawa M, Nakamura S, Azuma K, Ishii G, Higashi M, Kishi H, Hiwasa T, Koda K, Nakajima N, Harigaya K (2002) CD44 signaling through focal adhesion kinase and its anti-apoptotic effect. FEBS Lett 528:101–108

Gadhoum Z, Delaunay J, Maquarre E, Durand L, Lancereaux V, Qi J, Robert-Lezenes J, Chomienne C, Smadja-Joffe F (2004a) The effect of anti-CD44 monoclonal antibodies on differentiation and proliferation of human acute myeloid leukemia cells. Leuk Lymphoma 45: 1501–1510

Gadhoum Z, Leibovitch MP, Qi J, Dumenil D, Durand L, Leibovitch S, Smadja-Joffe F (2004b) CD44: a new means to inhibit acute myeloid leukemia cell proliferation via p27Kip1. Blood 103:1059–1068

Galandrini R, De Maria R, Piccoli M, Frati L, Santoni A (1994) CD44 triggering enhances human NK cell cytotoxic functions. J Immunol 153:4399–4407

Gao Y, Foster R, Yang X, Feng Y, Shen JK, Mankin HJ, Hornicek FJ, Amiji MM, Duan Z (2015) Up-regulation of CD44 in the development of metastasis, recurrence and drug resistance of ovarian cancer. Oncotarget 6:9313–9326

Gao S, Wang J, Tian R, Wang G, Zhang L, Li Y, Li L, Ma Q, Zhu L (2017) Construction and evaluation of a targeted hyaluronic acid nanoparticle/photosensitizer complex for cancer photodynamic therapy. ACS Appl Mater Interfaces 9:32509–32519

Gao S, Islam R, Fang J (2021) Tumor environment-responsive hyaluronan conjugated zinc protoporphyrin for targeted anticancer photodynamic therapy. J Pers Med 11:136

Gatti V, Fierro C, Compagnone M, Giangrazi F, Markert EK, Bongiorno-Borbone L, Melino G, Peschiaroli A (2018) DeltaNp63 regulates the expression of hyaluronic acid- related genes in breast cancer cells. Oncogenesis 7:65

Ghatak S, Misra S, Toole BP (2005) Hyaluronan constitutively regulates ErbB2 phosphorylation and signaling complex formation in carcinoma cells. J Biol Chem 280:8875–8883

Golan M, Feinshtein V, Polyak D, Scomparin A, Satchi-Fainaro R, David A (2016) Inhibition of gene expression and cancer cell migration by CD44v3/6-targeted Polyion complexes. Bioconjug Chem 27:947–960
Gong H, Chao Y, Xiang J, Han X, Song G, Feng L, Liu J, Yang G, Chen Q, Liu Z (2016) Hyaluronidase to enhance nanoparticle-based photodynamic tumor therapy. Nano Lett 16: 2512–2521
Gu W, An J, Meng H, Yu N, Zhong Y, Meng F, Xu Y, Cornelissen J, Zhong Z (2019) CD44-specific A6 short peptide boosts targetability and anticancer efficacy of polymersomal epirubicin to orthotopic human multiple myeloma. Adv Mater 31:e1904742
Gu W, Liu T, Fan D, Zhang J, Xia Y, Meng F, Xu Y, Cornelissen J, Liu Z, Zhong Z (2021) A6 peptide-tagged, ultra-small and reduction-sensitive polymersomal vincristine sulfate as a smart and specific treatment for CD44+ acute myeloid leukemia. J Control Release 329:706–716
Guo J, Cheng H, Zhao S, Yu L (2006) GG: a domain involved in phage LTF apparatus and implicated in human MEB and non-syndromic hearing loss diseases. FEBS Lett 580:581–584
Haist C, Schulte E, Bartels N, Bister A, Poschinski Z, Ibach TC, Geipel K, Wiek C, Wagenmann M, Monzel C, Scheckenbach K, Hanenberg H (2021) CD44v6-targeted CAR T-cells specifically eliminate CD44 isoform 6 expressing head/neck squamous cell carcinoma cells. Oral Oncol 116:105259
Han S, Guo J, Liu Y, Zhang Z, He Q, Li P, Zhang M, Sun H, Li R, Li Y, Zeng W, Liu J, Lian L, Gao Y, Shen L (2015) Knock out CD44 in reprogrammed liver cancer cell C3A increases CSCs stemness and promotes differentiation. Oncotarget 6:44452–44465
Han HS, Choi KY, Lee H, Lee M, An JY, Shin S, Kwon S, Lee DS, Park JH (2016) Gold-nanoclustered hyaluronan nano-assemblies for photothermally maneuvered photodynamic tumor ablation. ACS Nano 10:10858–10868
Hascall VC, Wang A, Tammi M, Oikari S, Tammi R, Passi A, Vigetti D, Hanson RW, Hart GW (2014) The dynamic metabolism of hyaluronan regulates the cytosolic concentration of UDP-GlcNAc. Matrix Biol 35:14–17
Hayward SL, Wilson CL, Kidambi S (2016) Hyaluronic acid-conjugated liposome nanoparticles for targeted delivery to CD44 overexpressing glioblastoma cells. Oncotarget 7:34158–34171
He QY, Liu XH, Li Q, Studholme DJ, Li XW, Liang SP (2006) G8: a novel domain associated with polycystic kidney disease and non-syndromic hearing loss. Bioinformatics 22:2189–2191
Heldin P, Delatorre M, Ytterberg D, Bergh J (1996) Differential synthesis and binding of hyaluronan by human breast cancer cell lines. Oncol Rep 3:1011–1016
Heldin P, Basu K, Kozlova I, Porsch H (2014) HAS2 and CD44 in breast tumorigenesis. Adv Cancer Res 123:211–229
Heldin P, Lin CY, Kolliopoulos C, Chen YH, Skandalis SS (2019) Regulation of hyaluronan biosynthesis and clinical impact of excessive hyaluronan production. Matrix Biol 78-79:100–117
Hibino S, Shibuya M, Hoffman MP, Engbring JA, Hossain R, Mochizuki M, Kudoh S, Nomizu M, Kleinman HK (2005) Laminin alpha5 chain metastasis- and angiogenesis- inhibiting peptide blocks fibroblast growth factor 2 activity by binding to the heparan sulfate chains of CD44. Cancer Res 65:10494–10501
Ishii S, Ford R, Thomas P, Nachman A, Steele G Jr, Jessup JM (1993) CD44 participates in the adhesion of human colorectal carcinoma cells to laminin and type IV collagen. Surg Oncol 2: 255–264
Ishizuka Y, Yokota A, Nishimura M, Saito Y, Nakaseko C (2008) Ligation of CD44 leads to killing activity in human peripheral mononuclear cells via MAP kinase and tyrosine kinases. Hematology 13:230–235
Itano N, Kimata K (2002) Mammalian hyaluronan synthases. IUBMB Life 54:195–199
Itano N, Sawai T, Yoshida M, Lenas P, Yamada Y, Imagawa M, Shinomura T, Hamaguchi M, Yoshida Y, Ohnuki Y, Miyauchi S, Spicer AP, McDonald JA, Kimata K (1999) Three isoforms of mammalian hyaluronan synthases have distinct enzymatic properties. J Biol Chem 274: 25085–25092

Jafari Malek S, Khoshchehreh R, Goodarzi N, Khoshayand MR, Amini M, Atyabi F, Esfandyari-Manesh M, Tehrani S, Mohammad Jafari R, Maghazei MS, Alvandifar F, Ebrahimi M, Dinarvand R (2014) cis-Dichlorodiamminoplatinum (II) glyconanoparticles by drug-induced ionic gelation technique targeted to prostate cancer: preparation, optimization and in vitro characterization. Colloids Surf B Biointerfaces 122:350–358

Jauw YWS, Huisman MC, Nayak TK, Vugts DJ, Christen R, Naegelen VM, Ruettinger D, Heil F, Lammertsma AA, Verheul HMW, Hoekstra OS, Van Dongen G, Menke-Van Der Houven Van Oordt CW (2018) Assessment of target-mediated uptake with immuno-PET: analysis of a phase I clinical trial with an anti-CD44 antibody. EJNMMI Res 8:6

Kajita M, Itoh Y, Chiba T, Mori H, Okada A, Kinoh H, Seiki M (2001) Membrane-type 1 matrix metalloproteinase cleaves CD44 and promotes cell migration. J Cell Biol 153:893–904

Karousou E, Kamiryo M, Skandalis SS, Ruusala A, Asteriou T, Passi A, Yamashita H, Hellman U, Heldin CH, Heldin P (2010) The activity of hyaluronan synthase 2 is regulated by dimerization and ubiquitination. J Biol Chem 285:23647–23654

Karousou E, Misra S, Ghatak S, Dobra K, Gotte M, Vigetti D, Passi A, Karamanos NK, Skandalis SS (2017) Roles and targeting of the HAS/hyaluronan/CD44 molecular system in cancer. Matrix Biol 59:3–22

Khan F, Gurung S, Gunassekaran GR, Vadevoo SMP, Chi L, Permpoon U, Haque ME, Lee YK, Lee SW, Kim S, Lee B (2021) Identification of novel CD44v6-binding peptides that block CD44v6 and deliver a pro-apoptotic peptide to tumors to inhibit tumor growth and metastasis in mice. Theranostics 11:1326–1344

Koyama H, Hibi T, Isogai Z, Yoneda M, Fujimori M, Amano J, Kawakubo M, Kannagi R, Kimata K, Taniguchi S, Itano N (2007) Hyperproduction of hyaluronan in neu-induced mammary tumor accelerates angiogenesis through stromal cell recruitment: possible involvement of versican/PG-M. Am J Pathol 170:1086–1099

Kubens BS, Zanker KS (1998) Differences in the migration capacity of primary human colon carcinoma cells (SW480) and their lymph node metastatic derivatives (SW620). Cancer Lett 131:55–64

Lakshman M, Subramaniam V, Rubenthiran U, Jothy S (2004) CD44 promotes resistance to apoptosis in human colon cancer cells. Exp Mol Pathol 77:18–25

Larsen M, Artym VV, Green JA, Yamada KM (2006) The matrix reorganized: extracellular matrix remodeling and integrin signaling. Curr Opin Cell Biol 18:463–471

Lee SY, Kim KA, Kim CH, Kim YJ, Lee JH, Kim HR (2017) CD44-shRNA recombinant adenovirus inhibits cell proliferation, invasion, and migration, and promotes apoptosis in HCT116 colon cancer cells. Int J Oncol 50:329–336

Lee IC, Wu YC, Hung WS (2018) Hyaluronic acid-based multilayer films regulate hypoxic multicellular aggregation of pancreatic cancer cells with distinct cancer stem-cell-like properties. ACS Appl Mater Interfaces 10:38769–38779

Li L, Heldin CH, Heldin P (2006) Inhibition of platelet-derived growth factor-BB-induced receptor activation and fibroblast migration by hyaluronan activation of CD44. J Biol Chem 281:26512–26519

Li J, Yang J, Yuan J, Li Y, Wang RC, Wang SY, Hao HL (2016) Effect of anti-CD44 monoclonal antibody A3D8 on expression of AP-1 in HL-60 cells. Zhongguo Shi Yan Xue Ye Xue Za Zhi 24:1360–1364

Li W, Jia H, Wang J, Guan H, Li Y, Zhang D, Tang Y, Wang TD, Lu S (2017a) A CD44-specific peptide, RP-1, exhibits capacities of assisting diagnosis and predicting prognosis of gastric cancer. Oncotarget 8:30063–30076

Li W, Ma H, Zhang J, Zhu L, Wang C, Yang Y (2017b) Unraveling the roles of CD44/CD24 and ALDH1 as cancer stem cell markers in tumorigenesis and metastasis. Sci Rep 7:13856

Liang Y, Peng J, Li N, Yu-Wai-Man C, Wang Q, Xu Y, Wang H, Tagalakis AD, Du Z (2019) Smart nanoparticles assembled by endogenous molecules for siRNA delivery and cancer therapy via CD44 and EGFR dual-targeting. Nanomedicine 15:208–217

Liang Y, Wang Y, Wang L, Liang Z, Li D, Xu X, Chen Y, Yang X, Zhang H, Niu H (2021) Self-crosslinkable chitosan-hyaluronic acid dialdehyde nanoparticles for CD44-targeted siRNA delivery to treat bladder cancer. Bioact Mater 6:433–446

Lin C, Yang X, Li H, Zou Y, Mohammad IS, Rong H, Rao Y, Song J, Leung SSY, Hu H (2021) Self-assembled nanomedicine combining a berberine derivative and doxorubicin for enhanced antitumor and antimetastatic efficacy via mitochondrial pathways. Nanoscale 13:6605–6623

Liu T, Yan Z, Liu Y, Choy E, Hornicek FJ, Mankin H, Duan Z (2018) CRISPR-Cas9-mediated silencing of CD44 in human highly metastatic osteosarcoma cells. Cell Physiol Biochem 46: 1218–1230

Liu J, Sun Y, Liu X, Yang Y, Widjaya AS, Long Z, Jiang Y (2020) Efficiency of different treatment regimens combining anti-tumor and anti-inflammatory liposomes for metastatic breast cancer. AAPS PharmSciTech 21:259

Lobo S, Pereira C, Oliveira C, Almeida GM (2020) Skipping exon-v6 from CD44v6-containing isoforms influences chemotherapy response and self-renewal capacity of gastric cancer cells. Cancers (Basel) 12:2378

Lusche DF, Wessels DJ, Reis RJ, Forrest CC, Thumann AR, Soll DR (2021) New monoclonal antibodies that recognize an unglycosylated, conserved, extracellular region of CD44 in vitro and in vivo, and can block tumorigenesis. PLoS One 16:e0250175

Mahira S, Kommineni N, Husain GM, Khan W (2019) Cabazitaxel and silibinin co-encapsulated cationic liposomes for CD44 targeted delivery: A new insight into nanomedicine based combinational chemotherapy for prostate cancer. Biomed Pharmacother 110:803–817

Maiolino S, Russo A, Pagliara V, Conte C, Ungaro F, Russo G, Quaglia F (2015) Biodegradable nanoparticles sequentially decorated with polyethyleneimine and hyaluronan for the targeted delivery of docetaxel to airway cancer cells. J Nanobiotechnology 13:29

Maisel D, Birzele F, Voss E, Nopora A, Bader S, Friess T, Goller B, Laifenfeld D, Weigand S, Runza V (2016) Targeting tumor cells with anti-CD44 antibody triggers macrophage-mediated immune modulatory effects in a cancer xenograft model. PLoS One 11:e0159716

Marengo A, Forciniti S, Dando I, Dalla Pozza E, Stella B, Tsapis N, Yagoubi N, Fanelli G, Fattal E, Heeschen C, Palmieri M, Arpicco S (2019) Pancreatic cancer stem cell proliferation is strongly inhibited by diethyldithiocarbamate-copper complex loaded into hyaluronic acid decorated liposomes. Biochim Biophys Acta Gen Subj 1863:61–72

Matzke-Ogi A, Jannasch K, Shatirishvili M, Fuchs B, Chiblak S, Morton J, Tawk B, Lindner T, Sansom O, Alves F, Warth A, Schwager C, Mier W, Kleeff J, Ponta H, Abdollahi A, Orian-Rousseau V (2016) Inhibition of tumor growth and metastasis in pancreatic cancer models by interference with CD44v6 signaling. Gastroenterology 150:513–525 e10

McClements L, Yakkundi A, Papaspyropoulos A, Harrison H, Ablett MP, Jithesh PV, McKeen HD, Bennett R, Donley C, Kissenpfennig A, Mcintosh S, McCarthy HO, O'neill E, Clarke RB, Robson T (2013) Targeting treatment-resistant breast cancer stem cells with FKBPL and its peptide derivative, AD-01, via the CD44 pathway. Clin Cancer Res 19:3881–3893

Mehic M, De Sa VK, Hebestreit S, Heldin CH, Heldin P (2017) The deubiquitinating enzymes USP4 and USP17 target hyaluronan synthase 2 and differentially affect its function. Oncogenesis 6:e348

Menke-Van Der Houven Van Oordt CW, Gomez-Roca C, Van Herpen C, Coveler AL, Mahalingam D, Verheul HM, Van Der Graaf WT, Christen R, Ruttinger D, Weigand S, Cannarile MA, Heil F, Brewster M, Walz AC, Nayak TK, Guarin E, Meresse V, Le Tourneau C (2016) First-in-human phase I clinical trial of RG7356, an anti- CD44 humanized antibody, in patients with advanced, CD44-expressing solid tumors. Oncotarget 7:80046–80058

Miletti-Gonzalez KE, Chen S, Muthukumaran N, Saglimbeni GN, Wu X, Yang J, Apolito K, Shih WJ, Hait WN, Rodriguez-Rodriguez L (2005) The CD44 receptor interacts with P-glycoprotein to promote cell migration and invasion in cancer. Cancer Res 65:6660–6667

Miletti-Gonzalez KE, Murphy K, Kumaran MN, Ravindranath AK, Wernyj RP, Kaur S, Miles GD, Lim E, Chan R, Chekmareva M, Heller DS, Foran D, Chen W, Reiss M, Bandera EV, Scotto K, Rodriguez-Rodriguez L (2012) Identification of function for CD44 intracytoplasmic domain

(CD44-ICD): modulation of matrix metalloproteinase 9 (MMP-9) transcription via novel promoter response element. J Biol Chem 287:18995–19007

Misra S, Ghatak S, Toole BP (2005) Regulation of MDR1 expression and drug resistance by a positive feedback loop involving hyaluronan, phosphoinositide 3-kinase, and ErbB2. J Biol Chem 280:20310–20315

Misra S, Hascall VC, De Giovanni C, Markwald RR, Ghatak S (2009) Delivery of CD44 shRNA/nanoparticles within cancer cells: perturbation of hyaluronan/CD44v6 interactions and reduction in adenoma growth in Apc Min/+ MICE. J Biol Chem 284:12432–12446

Misra S, Heldin P, Hascall VC, Karamanos NK, Skandalis SS, Markwald RR, Ghatak S (2011) Hyaluronan-CD44 interactions as potential targets for cancer therapy. FEBS J 278:1429–1443

Mizrahy S, Raz SR, Hasgaard M, Liu H, Soffer-Tsur N, Cohen K, Dvash R, Landsman-Milo D, Bremer MG, Moghimi SM, Peer D (2011) Hyaluronan-coated nanoparticles: the influence of the molecular weight on CD44-hyaluronan interactions and on the immune response. J Control Release 156:231–238

Mizrahy S, Goldsmith M, Leviatan-Ben-Arye S, Kisin-Finfer E, Redy O, Srinivasan S, Shabat D, Godin B, Peer D (2014) Tumor targeting profiling of hyaluronan-coated lipid based-nanoparticles. Nanoscale 6:3742–3752

Montanari E, Di Meo C, Oates A, Coviello T, Matricardi P (2018) Pursuing intracellular pathogens with hyaluronan. From a 'Pro-Infection' polymer to a biomaterial for 'Trojan Horse' Systems. Molecules 23:939

Morath I, Hartmann TN, Orian-Rousseau V (2016) CD44: more than a mere stem cell marker. Int J Biochem Cell Biol 81:166–173

Mori H, Tomari T, Koshikawa N, Kajita M, Itoh Y, Sato H, Tojo H, Yana I, Seiki M (2002) CD44 directs membrane-type 1 matrix metalloproteinase to lamellipodia by associating with its hemopexin-like domain. EMBO J 21:3949–3959

Navarro-Marchal SA, Grinan-Lison C, Entrena JM, Ruiz-Alcala G, Tristan-Manzano M, Martin F, Perez-Victoria I, Peula-Garcia JM, Marchal JA (2021) Anti-CD44-conjugated olive oil liquid nanocapsules for targeting pancreatic cancer stem cells. Biomacromolecules 22:1374–1388

Ndinguri MW, Zheleznyak A, Lauer JL, Anderson CJ, Fields GB (2012) Application of collagen-model triple-helical peptide-amphiphiles for CD44-targeted drug delivery systems. J Drug Deliv 2012:592602

Ni J, Cozzi PJ, Hao JL, Beretov J, Chang L, Duan W, Shigdar S, Delprado WJ, Graham PH, Bucci J, Kearsley JH, Li Y (2014) CD44 variant 6 is associated with prostate cancer metastasis and chemo-/radioresistance. Prostate 74:602–617

Nurwidya F, Takahashi F, Kato M, Baskoro H, Hidayat M, Wirawan A, Takahashi K (2017) CD44 silencing decreases the expression of stem cell-related factors induced by transforming growth factor beta1 and tumor necrosis factor alpha in lung cancer: preliminary findings. Bosn J Basic Med Sci 17:228–234

Ogoshi T, Ishii S, Mizoi T, Harada N, Sato W, Saito K, Ogawa H, Inabe K, Shiiba K, Matsuno S (1998) CD44H participates in the intrahepatic growth of murine colon 26 adenocarcinoma cells. Jpn J Cancer Res 89:1160–1168

Ota T, Matsui T, Kohno H, Maeda M, Tanino M, Odashima S (1995) CD44 participates in tumor cell adhesion to endothelial cells in the experimental metastatic process in B16BL6 melanoma cells. Anticancer Res 15:1215–1219

Pandey MS, Harris EN, Weigel JA, Weigel PH (2008) The cytoplasmic domain of the hyaluronan receptor for endocytosis (HARE) contains multiple endocytic motifs targeting coated pit-mediated internalization. J Biol Chem 283:21453–21461

Park HY, Lee KJ, Lee SJ, Yoon MY (2012a) Screening of peptides bound to breast cancer stem cell specific surface marker CD44 by phage display. Mol Biotechnol 51:212–220

Park YS, Huh JW, Lee JH, Kim HR (2012b) shRNA against CD44 inhibits cell proliferation, invasion and migration, and promotes apoptosis of colon carcinoma cells. Oncol Rep 27:339–346

Patel S, Shah K, Mirza S, Shah K, Rawal R (2016) Circulating tumor stem like cells in oral squamous cell carcinoma: An unresolved paradox. Oral Oncol 62:139–146

Peck D, Isacke CM (1998) Hyaluronan-dependent cell migration can be blocked by a CD44 cytoplasmic domain peptide containing a phosphoserine at position 325. J Cell Sci 111 (Pt 11):1595–1601

Peng Y, Prater AR, Deutscher SL (2017) Targeting aggressive prostate cancer-associated CD44v6 using phage display selected peptides. Oncotarget 8:86747–86768

Piotrowicz RS, Damaj BB, Hachicha M, Incardona F, Howell SB, Finlayson M (2011) A6 peptide activates CD44 adhesive activity, induces FAK and MEK phosphorylation, and inhibits the migration and metastasis of CD44-expressing cells. Mol Cancer Ther 10:2072–2082

Porsch H, Bernert B, Mehic M, Theocharis AD, Heldin CH, Heldin P (2013) Efficient TGFbeta-induced epithelial-mesenchymal transition depends on hyaluronan synthase HAS2. Oncogene 32:4355–4365

Qian H, Xia L, Ling P, Waxman S, Jing Y (2012) CD44 ligation with A3D8 antibody induces apoptosis in acute myeloid leukemia cells through binding to CD44s and clustering lipid rafts. Cancer Biol Ther 13:1276–1283

Reinke LM, Xu Y, Cheng C (2012) Snail represses the splicing regulator epithelial splicing regulatory protein 1 to promote epithelial-mesenchymal transition. J Biol Chem 287:36435–36442

Rilla K, Oikari S, Jokela TA, Hyttinen JM, Karna R, Tammi RH, Tammi MI (2013) Hyaluronan synthase 1 (HAS1) requires higher cellular UDP-GlcNAc concentration than HAS2 and HAS3. J Biol Chem 288:5973–5983

Rodriguez MM, Fiore E, Bayo J, Atorrasagasti C, Garcia M, Onorato A, Dominguez L, Malvicini M, Mazzolini G (2018) 4Mu decreases CD47 expression on hepatic cancer stem cells and primes a potent antitumor T cell response induced by interleukin-12. Mol Ther 26: 2738–2750

Sadvakassova G, Dobocan MC, Congote LF (2009) Osteopontin and the C-terminal peptide of thrombospondin-4 compete for CD44 binding and have opposite effects on CD133+ cell colony formation. BMC Res Notes 2:215

Seki K, Yamaguchi A, Goi T, Nakagawara G, Matsukawa S, Urano T, Furukawa K (1997) Inhibition of liver metastasis formation by anti-CD44 variant exon 9 monoclonal antibody. Int J Oncol 11:1257–1261

Shabani Ravari N, Goodarzi N, Alvandifar F, Amini M, Souri E, Khoshayand MR, Hadavand Mirzaie Z, Atyabi F, Dinarvand R (2016) Fabrication and biological evaluation of chitosan coated hyaluronic acid-docetaxel conjugate nanoparticles in CD44(+) cancer cells. Daru 24:21

Shah KN, Ditto AJ, Crowder DC, Overmeyer JH, Tavana H, Maltese WA, Yun YH (2017) Receptor-mediated attachment and uptake of hyaluronan conjugates by breast cancer cells. Mol Pharm 14:3968–3977

Shao J, Fan W, Ma B, Wu Y (2016) Breast cancer stem cells expressing different stem cell markers exhibit distinct biological characteristics. Mol Med Rep 14:4991–4998

Shuttleworth TL, Wilson MD, Wicklow BA, Wilkins JA, Triggs-Raine BL (2002) Characterization of the murine hyaluronidase gene region reveals complex organization and cotranscription of Hyal1 with downstream genes, Fus2 and Hyal3. J Biol Chem 277:23008–23018

Skandalis SS, Kozlova I, Engstrom U, Hellman U, Heldin P (2010) Proteomic identification of CD44 interacting proteins. IUBMB Life 62:833–840

Skandalis SS, Karalis TT, Chatzopoulos A, Karamanos NK (2019) Hyaluronan-CD44 axis orchestrates cancer stem cell functions. Cell Signal 63:109377

Skandalis SS, Karalis T, Heldin P (2020) Intracellular hyaluronan: importance for cellular functions. Semin Cancer Biol 62:20–30

Solis MA, Chen YH, Wong TY, Bittencourt VZ, Lin YC, Huang LL (2012) Hyaluronan regulates cell behavior: a potential niche matrix for stem cells. Biochem Res Int 2012:346972

Soltes L, Mendichi R, Kogan G, Schiller J, Stankovska M, Arnhold J (2006) Degradative action of reactive oxygen species on hyaluronan. Biomacromolecules 7:659–668

Song G, Liao X, Zhou L, Wu L, Feng Y, Han ZC (2004) HI44a, an anti-CD44 monoclonal antibody, induces differentiation and apoptosis of human acute myeloid leukemia cells. Leuk Res 28:1089–1096

Stamenkovic I, Yu Q (2009) Shedding light on proteolytic cleavage of CD44: the responsible sheddase and functional significance of shedding. J Invest Dermatol 129:1321–1324

Su J, Wu S, Wu H, Li L, Guo T (2016) CD44 is functionally crucial for driving lung cancer stem cells metastasis through Wnt/beta-catenin-FoxM1-twist signaling. Mol Carcinog 55:1962–1973

Surace C, Arpicco S, Dufay-Wojcicki A, Marsaud V, Bouclier C, Clay D, Cattel L, Renoir JM, Fattal E (2009) Lipoplexes targeting the CD44 hyaluronic acid receptor for efficient transfection of breast cancer cells. Mol Pharm 6:1062–1073

Takahashi Y, Li L, Kamiryo M, Asteriou T, Moustakas A, Yamashita H, Heldin P (2005) Hyaluronan fragments induce endothelial cell differentiation in a CD44- and CXCL1/GRO1-dependent manner. J Biol Chem 280:24195–24204

Takei J, Kaneko MK, Ohishi T, Hosono H, Nakamura T, Yanaka M, Sano M, Asano T, Sayama Y, Kawada M, Harada H, Kato Y (2020) A defucosylated antiCD44 monoclonal antibody 5mG2af exerts antitumor effects in mouse xenograft models of oral squamous cell carcinoma. Oncol Rep 44:1949–1960

Tan PH, Santos EB, Rossbach HC, Sandmaier BM (1993) Enhancement of natural killer activity by an antibody to CD44. J Immunol 150:812–820

Thankamony SP, Knudson W (2006) Acylation of CD44 and its association with lipid rafts are required for receptor and hyaluronan endocytosis. J Biol Chem 281:34601–34609

Theocharis AD, Skandalis SS, Gialeli C, Karamanos NK (2016) Extracellular matrix structure. Adv Drug Deliv Rev 97:4–27

Tirella A, Kloc-Muniak K, Good L, Ridden J, Ashford M, PURI, S. & Tirelli, N. (2019) CD44 targeted delivery of siRNA by using HA-decorated nanotechnologies for KRAS silencing in cancer treatment. Int J Pharm 561:114–123

Torronen K, Nikunen K, Karna R, Tammi M, Tammi R, Rilla K (2014) Tissue distribution and subcellular localization of hyaluronan synthase isoenzymes. Histochem Cell Biol 141:17–31

Tremmel M, Matzke A, Albrecht I, Laib AM, Olaku V, Ballmer-Hofer K, Christofori G, Heroult M, Augustin HG, Ponta H, Orian-Rousseau V (2009) A CD44v6 peptide reveals a role of CD44 in VEGFR-2 signaling and angiogenesis. Blood 114:5236–5244

Ugarte-Berzal E, Bailon E, Amigo-Jimenez I, Albar JP, Garcia-Marco JA, Garcia-Pardo A (2014) A novel CD44-binding peptide from the pro-matrix metalloproteinase-9 hemopexin domain impairs adhesion and migration of chronic lymphocytic leukemia (CLL) cells. J Biol Chem 289:15340–15349

Vahidian F, Safarzadeh E, Mohammadi A, Najjary S, Mansoori B, Majidi J, Babaloo Z, Aghanejad A, Shadbad MA, Mokhtarzadeh A, Baradaran B (2020) siRNA-mediated silencing of CD44 delivered by jet Pei enhanced doxorubicin chemo sensitivity and altered miRNA expression in human breast cancer cell line (MDA-MB468). Mol Biol Rep 47:9541–9551

Valentine A, O'rourke M, Yakkundi A, Worthington J, Hookham M, Bicknell R, McCarthy HO, McClelland K, McCallum L, Dyer H, McKeen H, Waugh DJ, Roberts J, McGregor J, Cotton G, James I, Harrison T, Hirst DG, Robson T (2011) FKBPL and peptide derivatives: novel biological agents that inhibit angiogenesis by a CD44- dependent mechanism. Clin Cancer Res 17:1044–1056

Van Hal NL, Van Dongen GA, Rood-Knippels EM, Van Der Valk P, Snow GB, Brakenhoff RH (1996) Monoclonal antibody U36, a suitable candidate for clinical immunotherapy of squamous-cell carcinoma, recognizes a CD44 isoform. Int J Cancer 68:520–527

Varghese OP, Sun W, Hilborn J, Ossipov DA (2009) In situ cross-linkable high molecular weight hyaluronan-bisphosphonate conjugate for localized delivery and cell-specific targeting: a hydrogel linked prodrug approach. J Am Chem Soc 131:8781–8783

Vey N, Delaunay J, Martinelli G, Fiedler W, Raffoux E, Prebet T, Gomez-Roca C, Papayannidis C, Kebenko M, Paschka P, Christen R, Guarin E, Broske AM, Baehner M, Brewster M, Walz AC, Michielin F, Runza V, Meresse V, Recher C (2016) Phase I clinical study of RG7356, an anti-CD44 humanized antibody, in patients with acute myeloid leukemia. Oncotarget 7:32532–32542

Vigetti D, Clerici M, Deleonibus S, Karousou E, Viola M, Moretto P, Heldin P, Hascall VC, De Luca G, Passi A (2011) Hyaluronan synthesis is inhibited by adenosine monophosphate-activated protein kinase through the regulation of HAS2 activity in human aortic smooth muscle cells. J Biol Chem 286:7917–7924

Vigetti D, Deleonibus S, Moretto P, Karousou E, Viola M, Bartolini B, Hascall VC, Tammi M, De Luca G, Passi A (2012) Role of UDP-N-acetylglucosamine (GlcNAc) and O-GlcNAcylation of hyaluronan synthase 2 in the control of chondroitin sulfate and hyaluronan synthesis. J Biol Chem 287:35544–35555

Vugts DJ, Heuveling DA, Stigter-Van Walsum M, Weigand S, Bergstrom M, Van Dongen GA, Nayak TK (2014) Preclinical evaluation of 89Zr-labeled anti-CD44 monoclonal antibody RG7356 in mice and cynomolgus monkeys: prelude to phase 1 clinical studies. MAbs 6:567–575

Vyas D, Lopez-Hisijos N, Gandhi S, El-Dakdouki M, Basson MD, Walsh MF, Huang X, Vyas AK, Chaturvedi LS (2015) Doxorubicin-hyaluronan conjugated super-paramagnetic iron oxide nanoparticles (DOX-HA-SPION) enhanced cytoplasmic uptake of doxorubicin and modulated apoptosis, IL-6 release and NF-kappaB activity in human MDA-MB-231 breast cancer cells. J Nanosci Nanotechnol 15:6413–6422

Wang SJ, Wreesmann VB, Bourguignon LY (2007) Association of CD44 V3-containing isoforms with tumor cell growth, migration, matrix metalloproteinase expression, and lymph node metastasis in head and neck cancer. Head Neck 29:550–558

Wang A, Midura RJ, Vasanji A, Wang AJ, Hascall VC (2014) Hyperglycemia diverts dividing osteoblastic precursor cells to an adipogenic pathway and induces synthesis of a hyaluronan matrix that is adhesive for monocytes. J Biol Chem 289:11410–11420

Wang H, Zhu Z, Zhang G, Lin F, Liu Y, Zhang Y, Feng J, Chen W, Meng Q, Chen L (2019) AS1411 aptamer/hyaluronic acid-bifunctionalized microemulsion co-loading shikonin and docetaxel for enhanced antiglioma therapy. J Pharm Sci 108:3684–3694

Wang Z, Li Y, Wang Y, Wu D, Lau AHY, Zhao P, Zou C, Dai Y, Chan FL (2020) Targeting prostate cancer stem-like cells by an immunotherapeutic platform based on immunogenic peptide-sensitized dendritic cells-cytokine-induced killer cells. Stem Cell Res Ther 11:123

Wang Y, Gao D, Liu Y, Guo X, Chen S, Zeng L, Ma J, Zhang X, Tian Z, Yang Z (2021) Immunogenic-cell-killing and immunosuppression-inhibiting nanomedicine. Bioact Mater 6: 1513–1527

Wei W, Hu H, Tan H, Chow LW, Yip AY, Loo WT (2012) Relationship of CD44+CD24−/low breast cancer stem cells and axillary lymph node metastasis. J Transl Med 10(Suppl 1):S6

Weigand S, Herting F, Maisel D, Nopora A, Voss E, Schaab C, Klammer M, Tebbe A (2012) Global quantitative phosphoproteome analysis of human tumor xenografts treated with a CD44 antagonist. Cancer Res 72:4329–4339

Weigel PH (2015) Hyaluronan synthase: the mechanism of initiation at the reducing end and a pendulum model for polysaccharide translocation to the cell exterior. Int J Cell Biol 2015: 367579

Wiranowska M, Ladd S, Moscinski LC, Hill B, Haller E, Mikecz K, Plaas A (2010) Modulation of hyaluronan production by CD44 positive glioma cells. Int J Cancer 127:532–542

Wu J, Deng C, Meng F, Zhang J, Sun H, Zhong Z (2017) Hyaluronic acid coated PLGA nanoparticulate docetaxel effectively targets and suppresses orthotopic human lung cancer. J Control Release 259:76–82

Xiao Z, Wan J, Nur AA, Dou P, Mankin H, Liu T, Ouyang Z (2018) Targeting CD44 by CRISPR-Cas9 in multi-drug resistant osteosarcoma cells. Cell Physiol Biochem 51:1879–1893

Xu H, Tian Y, Yuan X, Wu H, Liu Q, Pestell RG, Wu K (2015) The role of CD44 in epithelial-mesenchymal transition and cancer development. Onco Targets Ther 8:3783–3792

Xu J, Zhang Y, Xu J, Wang M, Liu G, Wang J, Zhao X, Qi Y, Shi J, Cheng K, Li Y, Qi S, Nie G (2019) Reversing tumor stemness via orally targeted nanoparticles achieves efficient colon cancer treatment. Biomaterials 216:119247

Xu K, Yao H, Fan D, Zhou L, Wei S (2021) Hyaluronic acid thiol modified injectable hydrogel: synthesis, characterization, drug release, cellular drug uptake and anticancer activity. Carbohydr Polym 254:117286

Yaghobi Z, Movassaghpour A, Talebi M, Abdoli Shadbad M, Hajiasgharzadeh K, Pourvahdani S, Baradaran B (2021) The role of CD44 in cancer chemoresistance: A concise review. Eur J Pharmacol 903:174147

Yakkundi A, McCallum L, O'kane A, Dyer H, Worthington J, McKeen HD, McClements L, Elliott C, McCarthy HO, Hirst DG, Robson T (2013) The anti-migratory effects of FKBPL and its peptide derivative, AD-01: regulation of CD44 and the cytoskeletal pathway. PLoS One 8:e55075

Yamaguchi Y, Yamamoto H, Tobisawa Y, Irie F (2019) TMEM2: A missing link in hyaluronan catabolism identified? Matrix Biol 78-79:139–146

Yamamoto H, Tobisawa Y, Inubushi T, Irie F, Ohyama C, Yamaguchi Y (2017) A mammalian homolog of the zebrafish transmembrane protein 2 (TMEM2) is the long-sought-after cell-surface hyaluronidase. J Biol Chem 292:7304–7313

Yan Y, Li Z, Kong X, Jia Z, Zuo X, Gagea M, Huang S, Wei D, Xie K (2016) KLF4-mediated suppression of CD44 Signaling negatively impacts pancreatic cancer stemness and metastasis. Cancer Res 76:2419–2431

Yang X, Sarvestani SK, Moeinzadeh S, He X, Jabbari E (2013) Effect of CD44 binding peptide conjugated to an engineered inert matrix on maintenance of breast cancer stem cells and tumorsphere formation. PLoS One 8:e59147

Yang Y, Zhao X, Li X, Yan Z, Liu Z, Li Y (2017) Effects of anti-CD44 monoclonal antibody IM7 carried with chitosan polylactic acid-coated nano-particles on the treatment of ovarian cancer. Oncol Lett 13:99–104

Yao H, Sun L, Li J, Zhou X, Li R, Shao R, Zhang Y, Li L (2020) A novel therapeutic siRNA nanoparticle designed for dual-targeting CD44 and Gli1 of gastric cancer stem cells. Int J Nanomedicine 15:7013–7034

Yasothamani V, Karthikeyan L, Shyamsivappan S, Haldorai Y, Seetha D, Vivek R (2021) Synergistic effect of photothermally targeted NIR-responsive nanomedicine-induced immunogenic cell death for effective triple negative breast cancer therapy. Biomacromolecules 22:2472–2490

Yasuda M, Tanaka Y, Fujii K, Yasumoto K (2001) CD44 stimulation down-regulates Fas expression and Fas-mediated apoptosis of lung cancer cells. Int Immunol 13:1309–1319

Yoo J, Rejinold NS, Lee D, Noh I, Koh WG, Jon S, Kim YC (2020) CD44-mediated methotrexate delivery by hyaluronan-coated nanoparticles composed of a branched cell-penetrating peptide. ACS Biomater Sci Eng 6:494–504

Yoshida H, Okada Y (2019) Role of HYBID (hyaluronan binding protein involved in hyaluronan depolymerization), alias KIAA1199/CEMIP, in hyaluronan degradation in normal and photoaged skin. Int J Mol Sci 20:5804

Yoshida H, Nagaoka A, Kusaka-Kikushima A, Tobiishi M, Kawabata K, Sayo T, Sakai S, Sugiyama Y, Enomoto H, Okada Y, Inoue S (2013a) KIAA1199, a deafness gene of unknown function, is a new hyaluronan binding protein involved in hyaluronan depolymerization. Proc Natl Acad Sci U S A 110:5612–5617

Yoshida H, Nagaoka A, Nakamura S, Sugiyama Y, Okada Y, Inoue S (2013b) Murine homologue of the human KIAA1199 is implicated in hyaluronan binding and depolymerization. FEBS Open Bio 3:352–356

Yu Q, Stamenkovic I (1999) Localization of matrix metalloproteinase 9 to the cell surface provides a mechanism for CD44-mediated tumor invasion. Genes Dev 13:35–48

Zada AA, Singh SM, Reddy VA, Elsasser A, Meisel A, Haferlach T, Tenen DG, Hiddemann W, Behre G (2003) Downregulation of c-Jun expression and cell cycle regulatory molecules in acute myeloid leukemia cells upon CD44 ligation. Oncogene 22:2296–2308

Zhang S, Wu CC, Fecteau JF, Cui B, Chen L, Zhang L, Wu R, Rassenti L, Lao F, Weigand S, Kipps TJ (2013) Targeting chronic lymphocytic leukemia cells with a humanized monoclonal antibody specific for CD44. Proc Natl Acad Sci U S A 110:6127–6132

Zhang D, Jia H, Wang Y, Li WM, Hou YC, Yin SW, Wang TD, He SX, Lu SY (2015) A CD44 specific peptide developed by phage display for targeting gastric cancer. Biotechnol Lett 37: 2311–2320

Zhang D, Jia H, Li W, Hou Y, Lu S, He S (2016) Screening and identification of a phage display derived peptide that specifically binds to the CD44 protein region encoded by variable exons. J Biomol Screen 21:44–53

Zhao C, Thompson BJ, Chen K, Gao F, Blouw B, Marella M, Zimmerman S, Kimbler T, Garrovillo S, Bahn J, Huang L, Huang Z, Shepard HM, Rosengren S, Thanos CD, Maneval DC (2019) The growth of a xenograft breast cancer tumor model with engineered hyaluronan-accumulating stroma is dependent on hyaluronan and independent of CD44. Oncotarget 10: 6561–6576

Zhao Y, Wang K, Zheng Y, Zeng X, Lim YC, Liu T (2020) Co-delivery of salinomycin and curcumin for cancer stem cell treatment by inhibition of cell proliferation, cell cycle arrest, and epithelial-mesenchymal transition. Front Chem 8:601649

Zhong Y, Meng F, Deng C, Mao X, Zhong Z (2017) Targeted inhibition of human hematological cancers in vivo by doxorubicin encapsulated in smart lipoic acid-crosslinked hyaluronic acid nanoparticles. Drug Deliv 24:1482–1490

Zhong W, Pang L, Feng H, Dong H, Wang S, Cong H, Shen Y, Bing Y (2020) Recent advantage of hyaluronic acid for anti-cancer application: a review of "3S" transition approach. Carbohydr Polym 238:116204

Zhou L, Sheng D, Deng Q, Wang D, Liu S (2018) Development of a novel method for rapid cloning of shRNA vectors, which successfully knocked down CD44 in mesenchymal triple-negative breast cancer cells. Cancer Commun (Lond) 38:57

Zhu C, Zhang H, Li W, Luo L, Guo X, Wang Z, Kong F, Li Q, Yang J, Du Y, You J (2018) Suppress orthotopic colon cancer and its metastasis through exact targeting and highly selective drug release by a smart nanomicelle. Biomaterials 161:144–153

Zoller M (2011) CD44: can a cancer-initiating cell profit from an abundantly expressed molecule? Nat Rev Cancer 11:254–267

Zoller M (2015) CD44, hyaluronan, the hematopoietic stem cell, and leukemia-initiating cells. Front Immunol 6:235

Zoltan-Jones A, Huang L, Ghatak S, Toole BP (2003) Elevated hyaluronan production induces mesenchymal and transformed properties in epithelial cells. J Biol Chem 278:45801–45810

Zou L, Song X, Yi T, Li S, Deng H, Chen X, Li Z, Bai Y, Zhong Q, Wei Y, Zhao X (2013) Administration of PLGA nanoparticles carrying shRNA against focal adhesion kinase and CD44 results in enhanced antitumor effects against ovarian cancer. Cancer Gene Ther 20: 242–250

GPSR Compliance
The European Union's (EU) General Product Safety Regulation (GPSR) is a set of rules that requires consumer products to be safe and our obligations to ensure this.

If you have any concerns about our products, you can contact us on

ProductSafety@springernature.com

In case Publisher is established outside the EU, the EU authorized representative is:

Springer Nature Customer Service Center GmbH
Europaplatz 3
69115 Heidelberg, Germany

www.ingramcontent.com/pod-product-compliance
Ingram Content Group UK Ltd.
Pitfield, Milton Keynes, MK11 3LW, UK
UKHW022217051026
15624UKWH00001B/200
9783030997076